# Advances in Near-surface Seismology and Ground-penetrating Radar

Geophysical Developments Series No. 15

Edited by

Richard D. Miller

John H. Bradford

Klaus Holliger

Rebecca B. Latimer, managing editor

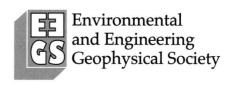

Environmental
and Engineering
Geophysical Society

SOCIETY OF EXPLORATION GEOPHYSICISTS
*The international society of applied geophysics*
Tulsa, Oklahoma, U.S.A.

ISBN 978-0-931830-41-9 (Series)
ISBN 978-1-56080-224-2 (Volume)

Copyright 2010
Society of Exploration Geophysicists
P. O. Box 702740
Tulsa, OK U.S.A. 74170-2740

American Geophysical Union
2000 Florida Avenue N. W.
Washington, D. C., U.S.A. 20009-1277

Environmental and Engineering Geophysical Society
1720 South Bellaire, Suite 110
Denver, Colorado, U.S.A. 80222-4303

Published 2010
Printed in the United States of America


Library of Congress Cataloging-in-Publication Data

Advances in near-surface seismology and ground-penetrating radar / edited by Richard D. Miller,
John H. Bradford, Klaus Holliger ; Rebecca B. Latimer, managing editor.
     p. cm. -- (Geophysical developments series ; no. 15)
  Includes bibliographical references and index.
  ISBN 978-1-56080-224-2 (volume : alk. paper) -- ISBN 978-0-931830-41-9 (series : alk. paper)
  1. Seismic traveltime inversion. 2. Surface waves. 3. Seismic waves--Speed. 4. Ground
penetrating radar.  I. Miller, Richard D. II. Bradford, John H. (John Holloway), 1967- III.
Holliger, Klaus. IV. Latimer, Rebecca B.
  QE539.2.S43A38 2010
  550.28'4--dc22
                              2010050291

This volume is dedicated to the memory of Roger Adams Young (1943–2009). Roger was an outstanding professor and colleague who committed his career to teaching, mentoring, research, and service. His selflessness, integrity, and kindness made him one of the most valued members of our near-surface community.

# Contents

## Section 3: Integrative Approaches

# Section 4: Case Studies

# About the Editors

**Richard D. Miller** has been at Kansas Geological Survey, a research and service division of the University of Kansas, since 1983 and is now a senior scientist and manager of the survey's exploration services section. He also holds a courtesy appointment as associate professor of geology at the University of Kansas. He received a B.A. in physics from Benedictine College in Atchison, Kansas, in 1980, an M.S. in physics with emphasis in geophysics from the University of Kansas in 1983, and a Ph.D. in geophysics from the University of Leoben, Austria, in 2007.

Miller's scientific interests are in shallow high-resolution seismic methods applied to environmental, engineering, energy, groundwater, transportation, and mining problems. From 2005 through 2009, he was a member of SEG's editorial board for THE LEADING EDGE, serving as chairman in 2009. Since 2004, Miller has been guest editor for 12 special sections of *TLE* with a near-surface focus. He has received four research achievement awards, including SEG's Distinguished Achievement Award and Near-surface Geophysics Section Hal Mooney Award, and has written 86 refereed articles.

**John H. Bradford** joined the academic faculty in the Department of Geosciences at Boise State University (BSU) in Idaho in 2005 and is now an associate professor. He was a research professor at the Center for Geophysical Investigation of the Shallow Subsurface (CGISS) at BSU from 2001 to 2005 and was director of CGISS from 2006 to 2009. He received B.S. degrees in physics and engineering physics from the University of Kansas in 1994. He entered Rice University in Houston, Texas, in 1994, where he held a US EPA STAR graduate fellowship from 1995 through 1997 and received a Ph.D. in geophysics in 1999. From 1995 through 1999, Bradford was a research scientist at the Houston Advanced Research Center, working on topics ranging from utility detection with ground-penetrating radar (GPR) to spectral decomposition for seismic exploration. He was a research scientist at the University of Wyoming from 1999 through 2001.

Bradford has worked in methodology development for near-surface seismic and GPR applications with emphasis on imaging, attenuation, offset-dependent reflectivity, and 3D field methods and analysis. He has had articles published on a diverse array of topics, including hydrocarbon detection, hydrogeophysics, and glaciology, and has organized several workshops and technical sessions at national and international meetings. He served as second vice president of SEG in 2010 and associate editor of GEOPHYSICS from 2005 through 2008 and is associate editor of *Near Surface Geophysics.*

**Klaus Holliger** has been at the University of Lausanne, Switzerland, since 2005, where he holds a chaired professorship, recently finished a term as vice dean of research, and is now director of the geophysics institute. He received an M.Sc. in 1987 and a Ph.D. in 1990 in geophysics, both from ETH Zurich in Switzerland, and a postgraduate degree in economics in 2000 from the University of London. After extended postdoctoral studies at Rice University in Houston, Texas, he joined ETH's newly founded applied and environmental geophysics group in 1994. He also has worked for shorter periods of time at the U. S. Geological Survey, Imperial College, and the University of Cambridge.

Holliger has been a member of SEG since 1995, was secretary of the Near-surface Geophysics Section from 2003 through 2005, and is now president-elect of the section. He was an associate editor of GEOPHYSICS from 2004 through 2009 and is now editor in chief of the *Journal of Applied Geophysics.* He has broad scientific interests and has worked in a variety of fundamental and applied research domains. He has been a coauthor of more than 100 peer-refereed publications, 18 of which were published in GEOPHYSICS. Holliger's current main research interest is the emerging and inherently interdisciplinary field of hydrogeophysics. He coorganized SEG's first workshop on that topic in 2006 and recently served as guest editor for a hydrogeophysics special issue of *Near Surface Geophysics.*

# Acknowledgments

We wish to thank all those contributors who have helped to make this a worthwhile and successful endeavor, specifically the authors and reviewers for their time and conscientious work. Serving as editors of a volume in the SEG Geophysical Developments Series has been an honor and a unique challenge. We believe this book will provide a useful overview of current methodologies and practices in near-surface seismology and ground-penetrating radar. This collection of manuscripts grew from a core group of papers presented at a postconvention workshop, "Advances in Near-surface Seismology and Ground-penetrating Radar," held during the 2009 SEG Annual Meeting in Houston, Texas. It is our hope that this volume will be widely read and will promote further growth and progress in this rapidly evolving and expanding discipline.

We would especially like to thank the publications department of SEG. Under the leadership of publications director Ted Bakamjian, staff members Merrily Sanzalone, Jennifer Cobb, and Rowena Mills led us through each step in the publication process, from invitation of authors to approval of final proofs, all using a new Web-based system. Their tireless efforts, relentless attention to details and schedules, and willingness to keep this book project at the top of their "to-do list" allowed this volume to go from workshop to bindery in less than one year. The authors and volume editors greatly benefited from the professionalism and talents of Merrily, Jennifer, and Rowena. We greatly appreciate the editorial work of Rowena Mills and her team of extremely talented copy editors, consisting of Jennifer Baltz, Anne H. Thomas, Kathryne Pile, Paulette Henderson, and Marilyn Perlberg. We are also grateful to Frances Plants Whitehurst for her assistance with proofreading.

We appreciate the confidence that chairman Yonghe Sun and the SEG Publications Committee placed in our commitment to use SEG's resources wisely and to complete this work in an expeditious and technically sound manner. For a book project like this to be successful, it takes the trust and vision of a seasoned and confident managing editor. We greatly appreciate managing editor Rebecca Latimer for her insight and for the latitude she allowed us.

Thanks to the SEG Executive Committee members during the terms of office of Presidents Larry Lines and Steve Hill, who provided the opportunity to propose this project and the flexibility in the formal SEG book publication process that was necessary to meet our aggressive timelines. This book is unique in its publication schedule and in its objective to bring innovative and fresh workshop topics to the membership quickly and in a refereed format. Larry Lines, in a September 2009 letter to the publications and research committee chairmen and the SEG editor, encouraged this type of workshop-based refereed publication, setting a goal to publish within two years of the workshop. This is the first SEG book to accomplish both these goals.

The papers published in this book underwent the same rigorous peer review as papers published in GEOPHYSICS. At least three experts were selected by one of the technical editors to review each manuscript. The rigorous nature of the review process and the insightful critiques provided by the referees mean each chapter conforms to the high-quality standards that readers have come to expect from refereed SEG publications. We graciously and sincerely thank all reviewers for the hours of volunteer work they provided.

## Reviewers in alphabetical order

Ajo-Franklin, Jonathan B.

Arcone, Steven A.

Bano, Maksim

Baron, Ludovic

Buness, Hermann A.

Calderón-Macías, Carlos

Carr, Bradley J.

Cho, Gye-Chun

Dafflon, Baptiste

Dal Moro, Giancarlo

Deparis, Jacques

Doetsch, Joseph A.

Gaines, David

Ghose, Ranajit

Giroux, Bernard

Gloaguen, Erwan

Hagin, Paul N.

Haines, Seth S.

Hanafy, Sherif M.

Haney, Matthew M.

Harris, Brett D.

Harris, James B.

Heincke, Bjoern

Henstock, Timothy J.

Hinz, Emily A.

Hollender, Fabrice

Hu, Wenyi

Irving, James

Ivanov, Julian M.

Johnson, Timothy C.

Jongmans, Denis

Kalinski, Michael E.

Kanlı, Ali I.

Knoll, Michael D.

Lambot, Sebastien

Landa, Evgeny

Lane, John W.

Lin, Chih-Ping

Linde, Niklas

Luke, Barbara

Luo, Yinhe

Marcotte, Denis

Markiewicz, Richard D.

Michaels, Paul

Mikesell, Thomas D.

Mueller, Tobias M.

Murray, Shannon

Musmann, Patrick

Neducza, Boriszlav

Nielsen, Lars

Palmer, Derecke

Park, Choon-Byong

Parolai, Stefano

Peterie, Shelby L.

Pugin, André J. M.

Ribeiro Cruz, Joao Carlos

Routh, Partha S.

Sassen, Douglas S.

Sauck, William A.

Schmitt, Douglas R.

Scholer, Marie

Sloan, Steven D.

Slob, Evert

Socco, L. Valentina

Stephenson, William J.

Stovas, Alexey

Streich, Rita

Tangirala, Seshunarayana

Tsoflias, Georgios P.

Tutuncu, Azra Nur

Versteeg, Roelof J.

Whiteley, Robert J.

Williams, Robert A.

Woelz, Susanne

Xia, Jianghai

Xu, Yixian

Yordkayhun, Sawasdee

Zhang, Jie

The above list should be complete, but if we inadvertently missed someone, please accept our apology.

— Richard D. Miller
John H. Bradford
Klaus Holliger
October 2010

# Introduction

Near-surface seismology and ground-penetrating radar (GPR) have enjoyed success and increasing popularity among a wide range of geophysicists, engineers, and hydrologists since their emergence in the latter half of the twentieth century. With the common ground shared by near-surface seismology and GPR, their significant upside potential, and rapid developments in the methods, a book bringing together the most current trends in research and applications of both is fitting and timely. Conceptually, near-surface seismology and GPR are remarkably similar, and they share a range of attributes and compatibilities that provides opportunities to integrate processing and interpretation workflows, which makes them a perfect pair to share pages in a book.

As pointed out by Don Steeples in his foreword to the 2005 book *Near-surface Geophysics* (SEG Investigations in Geophysics Series No. 13, edited by Dwain K. Butler), the first significant refereed collection of papers on near-surface geophysics was published in a 1988 special issue of GEOPHYSICS, followed two years later by the three-volume compilation *Geotechnical and Environmental Geophysics* (SEG Investigations in Geophysics Series No. 5, edited by Stanley H. Ward). Only a few papers published prior to 1975 provided a glimpse of the potential that near-surface seismic characterization possessed (Evison, 1952; Pakiser and Warrick, 1956; Mooney, 1973). Those authors were true pioneers who masterfully demonstrated that potential with a range of well-orchestrated and curiosity-driven research projects.

Although topics related to near-surface seismology had appeared occasionally in the refereed literature prior to 1980, GPR was a virtual unknown at that time. Of the 71 papers in the 1990 book edited by Ward, there was one GPR paper. In contrast, in Butler's 2005 book, four of the 18 papers in the "Applications and Case Studies" section involved GPR, with a major chapter in the "Concepts and Fundamentals" section dedicated solely to GPR. In this context, it is also noteworthy that of the 31 chapters in Butler's book, eight specifically focused on near-surface seismology.

In the last three decades, near-surface geophysics has steadily built up momentum, and in the last decade, it has seen enormous advancements in technologies, applications, and acceptance. If it is fair to use SEG's THE LEADING EDGE (*TLE*) as a gross measuring stick of trends in professional interest, in the seven years between 1996 and 2003, there were three special sections with a near-surface theme, whereas in the seven years since 2003, there have been

seven. This increase in near-surface topics in the last half decade or so might well be an indicator of what is in store for the geoscience community in the coming decade. In the 2002 SEG Annual Report, then SEG president Walter Lynn discussed the likely diversification of the society's members in the years to come, stating that "exploration geophysics is not just a tool for the petroleum and mining industry...." Recent growth and diversification of SEG are nowhere more evident than at the 2010 annual meeting in Denver, Colorado, at which about 10% of the oral sessions were proposed by or affiliated with SEG's Near-surface Geophysics Section.

With growth in numbers and professional emphasis have come sections, focus groups, and even professional societies specifically promoting near-surface geophysics. The emergence of near-surface geophysics groups, beginning in the late 1990s and extending into the early twenty-first century, has fueled a diversity of opportunities for professional collaborations. A range of workshops and shared publications has been the fruit of collaborative efforts. The near-surface community continues to extend and develop methods and approaches necessary to satisfy increasing demands in some of the socioeconomically pertinent disciplines such as civil and environmental engineering and hydrology. This book represents the first formal cooperative effort undertaken by the near-surface communities of the Society of Exploration Geophysicists, the American Geophysical Union, and the Environmental and Engineering Geophysical Society.

At the 2009 SEG annual meeting in Houston, Texas, representatives from three of the major near-surface groups organized an after-conference workshop titled "Advances in Near-surface Seismology and Ground-penetrating Radar." This workshop was designed to capture both new and innovative methodologies being developed and implemented by leading researchers from around the world. It was also a goal of the workshop to highlight studies that show the applicability of integrating various seismic and GPR methods to enhance near-surface characterizations. Technologies used in the application of near-surface seismology and GPR have benefited from new processing tools, increased computer speeds, and an expanded variety of applications. Many shallow-seismic projects now incorporate analysis results from different parts of the seismic wavefield, allowing for greater redundancy and confidence in interpretations without increased acquisition costs. More information is being extracted from GPR data by adapting and using the wide range of analysis techniques developed for seismic data in concert with new tools specific to high-frequency electromagnetic wave analysis.

Leading investigators were invited to present research at the workshop and submit papers for consideration to be published in this book. To diversify the book as well as to capture many of the most current and significant research developments in near-surface seismic and GPR, more than 60 authors were invited to submit manuscripts for inclusion in this book. From those 60, the cream of the crop appears in the 29 chapters of this book. The book is divided into four principal areas: "Reviews," "Methodology," "Integrative Approaches," and "Case Studies." History will be the judge in determining which of these manuscripts will become landmark works cited as classics for many years to come.

# Reviews

Establishing a vision for future developments requires a thorough understanding of the evolutionary path that a technique or method has taken to reach its current state. The review papers in the first section of this book provide that kind of framework and set the stage for papers in later sections that describe innovative and creative advancements in the use of seismic or GPR. As a good starting point in the review of past developments, the **Linde and Doetsch** paper provides an excellent example of joint inversion of GPR and seismic data and demonstrates improvements in characterization potential using this multimethod approach.

Without a doubt, exploitation of surface waves has been one of the fastest-growing areas in near-surface seismology in the last decade. **Xia and Miller** present a review of the estimation of near-surface shear-wave velocities and quality factors through the inversion of high-frequency Rayleigh waves, a technique now commonly referred to as multichannel analysis of surface waves (MASW). Several real-world examples demonstrate the applicability of inverting high-frequency Rayleigh waves as part of routine MASW applications. This chapter is rounded off by an algorithm for assessing the quality and reliability of MASW inversion results based on the trade-off between model resolution and covariance.

Complementing the Xia and Miller paper is a second surface-wave review paper, by **Xu et al.,** which describes some of the significant and creative developments in China in the last decade. The theory is well developed, and the examples are equally compelling.

With another look back, **Socco et al.** discuss optimal acquisition strategies for obtaining multipurpose seismic data sets and assess the potential improvements that can be achieved by a constrained or joint inversion of various types of seismic data. These concepts are illustrated in several real-world cases extracted from recent projects. With the surge in the use of surface waves in near-surface seismology, it is not surprising that three of the four review papers touch on that part the wavefield.

# Methodology

The past 10 years have seen an explosion in methodology development for near-surface seismic and GPR applications, and new developments continue to accelerate. In contrast with the early days of near-surface geophysics, relative maturity of hardware design and field methodologies has allowed researchers to focus increasingly on data processing and analysis algorithms that enable extraction of detailed quantitative information. It is interesting to note that although many analysis methodologies in the past were borrowed from the oil industry, near-surface researchers are now at the forefront of imaging and inversion, developing tools to solve problems unique to the shallow subsurface.

The papers in the methodology section capture many of these new developments. For example, **Irving et al.** present an effective method for extracting geostatistical structure based on 2D autocorrelation of reflection images.

An innovative approach is taken by **van der Kruk et al.** as they move closer to true-amplitude migration of GPR data and describe an approach that accounts for both the vector nature of electromagnetic wave propagation and the strong directionality of GPR antenna radiation patterns.

The paper by **Gloaguen et al.** develops a multiscale conditional stochastic simulation approach based on the wavelet transform. The proposed method is tested on synthetic crosshole GPR and is applied to a corresponding field data set. Results indicate that the method is capable of reproducing the larger-scale structural grain imaged by the geophysical data and to stochastically "fill in" the smaller-scale texture based on complementary information and/or constraints.

Looking toward a wide range of applications, **Ghose** presents a data-driven method that allows estimations of the in situ horizontal stress in the subsurface and monitoring of its temporal evolution based on fixed-array seismic shear-wave measurements. The corresponding model is validated on data from extensive laboratory experiments, which indicate that predictions based on the shear-wave seismic data are remarkably accurate.

The work by **Baron and Holliger** represents a significant contribution in our quest for a better understanding of the rock physics of unconsolidated sediments. In their paper, they attempt to apply Biot poroelasticity theory to shallow sediments, exploring the possibility of estimating the permeability of saturated surficial alluvial sediments based on the poroelastic interpretation of the velocity dispersion and frequency-dependent attenuation of such broadband sonic-log data.

Using an unconventional approach to extract more information from GPR data, **Haney et al.** analyze the dispersive characteristics of guided GPR waves and interpret the transition from a stream channel to a peat layer along the acquisition line. They find that guided waves capture shallow structure near a stream channel that is not imaged accurately in the reflection profile, thus demonstrating the utility of guided GPR waves for providing information on shallow structure that cannot be obtained from GPR reflection profiling.

Looking at the improvements in image accuracy when all components of the surface wave are correctly identified and included in the analysis, **Calderón-Macías and Luke** analyze the sensitivity of Rayleigh-phase velocity inversion in some specific surficial scenarios and explore the potential of adding higher modes to the analysis of fundamental mode data.

In a continuing effort to identify and classify subsurface anomalies, **Sloan et al.** explore the potential of three seismic methods for the detection of voids in the subsurface: (1) Diffracted body waves are used to identify and locate man-made tunnels in multiple geologic settings, (2) variations in shear-wave reflection velocities are shown to correlate to changes in stress over known void locations, and (3) backscattered surface waves are shown to correlate with a known void location. For all methods, field data correlate well with synthetic data.

Vertical resolution continues to be a research focus. **Deparis and Garambois** develop a dispersive amplitude-and-phase-versus-offset (DAPVA) approach for GPR reflection data to quantitatively analyze reflections from thin beds. Tests on synthetic and real data indicate that this approach carries significant potential for constraining the petrophysical properties of thin beds in general and the filling of fractures in particular.

Introducing a controversial approach to first-arrival analysis, **Palmer** proposes to use seismic attributes as a means to reduce the nonuniqueness of refraction methods. His work incorporates seismic attributes with the generalized reciprocal method and refraction convolution. This paper undoubtedly will stimulate discussion and a look to the future of refraction analysis.

An innovative application of the GPR method detailed in the paper by **Slob and Lambot** demonstrates how frequency-domain analysis of the surface reflection, recorded from an off-ground GPR system, can improve estimates of surface soil permittivity and electric conductivity in contrast to the time-domain method.

In a unique look at shear polarized surface waves, **Michaels and Gottumukkula** present a theory for viscoelastic Love waves by relating viscosity to permeability. The authors explore the method's potential for constraining the permeability of surficial soil and rock layers, which provides several recommendations with regard to the optimal design of corresponding seismic surveys.

# Integrative Approaches

Integrating data from different geophysical methods as a means of reducing nonuniqueness has long been recognized as an important step in the development of geophysical tools. However, methods for effectively implementing this concept have remained elusive, with problems such as differences in volume scaling, errors related to petrophysical assumptions among methods, and computational limitations proving difficult to solve. Researchers have recently made strides in data integration, and the papers in this section provide some exciting examples of what is being done.

In the study of a fault-controlled hydrothermal reservoir, **Musmann and Buness** show how high-resolution seismic reflection can be integrated with conventional industry-scale images to significantly improve the understanding of fault geometry in the near surface.

To assess the threat of landslides, both onshore and within a fjord, **Polom et al.** apply a high-resolution multichannel SH-wave seismic-reflection land survey complemented by a dense network of high-resolution single-channel marine seismic profiles over the deltaic sediments in a fjord to characterize in situ soil conditions. SH-wave seismic reflection provides a nearly direct proxy for in situ soil stiffness, a key geotechnical parameter.

Moving closer to full integration of seismic and GPR, **Bradford** uses 3D GPR and seismic-reflection data to image a shallow aquifer. He demonstrates that through an integrative interpretation approach that accounts for the complementary character of these data, he can provide a reliable 3D image of the major hydrostratigraphic units.

Looking to improve the selection of initial models used for first-arrival inversion, **Ivanov et al.** study the problem of nonuniqueness in refraction traveltime inversion and describe a method which uses the surface-wave-derived shear-wave velocity model to constrain the P-wave refraction problem.

# Case Studies

A critical step in the maturation of a new methodology is its demonstration through careful field study. The papers in this section are just such examples of well-illustrated case studies highlighting the value of an approach. The need to characterize areas with the potential for ground amplification makes the work of **Hunter et al.** significant to zoning and building codes in earthquake-prone areas. The authors first describe downhole and surface methods for measuring shear-wave velocities and then show how those measurements were used to assess seismic hazards in two Canadian cities.

A follow-up study by **Ogunsuyi and Schmitt** demonstrates how additional information can be extracted from conventionally processed data and how important it is to match stacked events on common-midpoint sections with associated reflections on shot gathers at times less than 50 ms. Also significant to many near-surface seismologists is the demonstrated need for an extremely accurate velocity function for effective prestack depth migration and how difficult it is to obtain those accurate velocities. New and previously undetectable features were interpreted on the reprocessed sections presented in this paper.

Stochastic modeling conditioned by crosshole GPR and lithologic data from boreholes allowed **Nielsen et al.** to estimate the fine-scale lithologic heterogeneity of rock from the Chalk Group. The results indicate that with this conditioning, the pursued stochastic simulation approach is capable of modeling the distribution of the pertinent lithologies and produces realistic models of Chalk Group heterogeneity.

The paper by **Renalier et al.** demonstrates the potential for characterizing and monitoring unstable clay slopes based on various active and passive shear-wave measurement techniques. The shear-wave velocity of clayey materials is very sensitive to mechanical disturbances associated with landslide movements and shows a pronounced negative correlation with GPS deformation measurements in such areas.

Time-lapse monitoring of soil moisture using surface seismic methods is a viable approach but has seen little application thus far. **Gaines et al.** use a series of P-wave refraction profiles to monitor variations in a perched water body lying within 4 m of the surface.

In areas with a low GPR velocity underlain by a higher-velocity material, a critically refracted GPR phase can lead to errors in depth estimates when the critical distance for refractions is less than the fixed transmitter-receiver offset. **Hermance et al.** describe a simple composite move-out correction that can correct the problem. They illustrate their approach for a stratified glacial drift site in southeastern New England.

Use of the S-transform on GPR data is not unique, but the study described by **Elwaseif et al.** is a compelling application of the approach. The authors demonstrate the potential of GPR data to locate water-filled fractures down to one-quarter the dominant wavelength and to delineate possible localized transport passages for moisture between fractures via capillary effects.

With the SPAC method first described by Aki's 1957 paper, **Stephenson and Odum** provide a modern look at an application of this ambient noise-analysis method at a small basin scale in the Salt Lake City, Utah, area. With the relationship between shear-wave velocity and amplification, the demonstrated compatibility of the SPAC method with more traditional borehole methods makes it a viable alternative for developing 1D $V_S$ functions.

Coincident MASW and H/V studies allow **Kanlı** to improve the accuracy of the shear-wave velocity function from the near-surface interval down to bedrock. This case study describes an inversion routine that uses a genetic algorithm. Although each component of this study has been described and discussed by other authors, the focus on integration is a clear trend in near-surface seismology and GPR.

It is clear from these summaries that the breadth and depth of this collection of papers is exceptional, touching on a full gamut of current research areas. Near-surface seismologists are making significant progress at unraveling the full wavefield and exploiting all aspects that provide insights into subsurface properties. Likewise, GPR researchers are finding new and innovative ways of using electromagnetic waves to measure electrical properties and relating those to hydrologic and geologic properties. A clear underpinning of many papers in this book is the incorporation of advanced modeling and inversions methods as tools for more accurate and complete characterizations of the subsurface.

Our intent is that the papers in this book are sufficiently forward looking that the volume will serve as a reference for researchers in the next decade and a valuable supplement for graduate or advanced undergraduate courses in near-surface seismology, GPR, or general near-surface geophysics. The credit for making this book project possible and successful goes to the individual contributors. Finally, it should be noted that the origin of this book project (a workshop-based refereed publication) is unique for SEG publications. We intend it to be the first in a series of books highlighting the broad spectrum of techniques, tools, and applications that comprises near-surface geophysics.

# References

Aki, K., 1957, Space and time spectra of stationary stochastic waves, with special reference to microtremors: Bulletin of the Earthquake Research Institute, **35**, 415–457

Evison, F. F., 1952, The inadequacy of the standard seismic techniques for shallow surveying: Geophysics, **17**, 867–875.

Mooney, H. M., 1973, Handbook of engineering geophysics: Bison Instruments Inc.

Pakiser, L. C., and R. E. Warrick, 1956, A preliminary evaluation of the shallow reflection seismograph: Geophysics, **21**, 388–405.

# Section 1

# Reviews

Chapter 1

# Joint Inversion of Crosshole GPR and Seismic Traveltime Data

Niklas Linde[1] and Joseph A. Doetsch[2]

## Abstract

Joint inversion of crosshole ground-penetrating radar and seismic data can improve model resolution and fidelity of the resultant individual models. Model coupling obtained by minimizing or penalizing some measure of structural dissimilarity between models appears to be the most versatile approach because only weak assumptions about petrophysical relationships are required. Nevertheless, experimental results and petrophysical arguments suggest that when porosity variations are weak in saturated unconsolidated environments, then radar wave speed is approximately linearly related to seismic wave speed. Under such circumstances, model coupling also can be achieved by incorporating cross-covariances in the model regularization. In two case studies, structural similarity is imposed by penalizing models for which the model cross-gradients are nonzero. A first case study demonstrates improvements in model resolution by comparing the resulting models with borehole information, whereas a second case study uses point-spread functions. Although radar seismic wave-speed crossplots are very similar for the two case studies, the models plot in different portions of the graph, suggesting variances in porosity. Both examples display a close, quasi-linear relationship between radar seismic wave speed in unconsolidated environments that is described rather well by the corresponding lower Hashin-Shtrikman (HS) bounds. Combining crossplots of the joint inversion models with HS bounds can constrain porosity and pore structure better than individual inversion results can.

## Introduction

Joint inversions of geophysical data can

1) improve model resolution and fidelity of individual models
2) provide consistent geophysical models for interpretation, classification, and petrophysical inference
3) make it easier to identify modeling and geometric errors by comparing the models obtained by individual and joint inversions
4) allow hypotheses testing concerning geologic structure, processes, and petrophysical relationships

Numerous methodologies to jointly invert disparate but colocated geophysical data at different scales and for different applications have been developed and tested in the last decades (e.g., Vozoff and Jupp, 1975; Lines et al., 1988; Tryggvason et al., 2002; Gallardo and Meju, 2003; Musil et al., 2003; Monteiro Santos et al., 2006). Many critical choices for developing joint and individual inversion algorithms are similar. These choices relate to model parameterization, model regularization, model and data norm, type of forward models and equation solvers, and stochastic versus deterministic frameworks. Difficulties related to weighting different data sets (e.g., Lines et al., 1988) are not so different from difficulties that arise when inverting single geophysical data types (e.g., Should one assume absolute or relative errors or a mixture of the two? How are actual errors estimated?). Data weighting for joint inversion needs to consider not only data and modeling errors but

[1]*University of Lausanne, Institute of Geophysics, Lausanne, Switzerland. E-mail: niklas.linde@unil.ch.*
[2]*ETH Zurich, Institute of Geophysics, Zurich, Switzerland. E-mail: doetsch@aug.ig.erdw.ethz.ch.*

also sensitivity with respect to the model parameters of interest and data redundancy that arise when many data points provide very similar information. The fundamental difference between joint and individual inversion is the need to couple the models at the inversion stage.

There are basically four approaches for doing this. The first approach is *structural*, in which it is assumed that models share one or several boundaries or that some measure of model structure is similar over given model domains (e.g., Haber and Oldenburg, 1997; Gallardo and Meju, 2003). In the second approach, models are linked explicitly with known or unknown (i.e., to be determined during the inversion) petrophysical relationships to create *as many inversion models as there are data sets*; one example is the joint inversion of P- and S-wave traveltimes, in which joint inversion is achieved by damping the models against a predefined $V_P/V_S$ ratio (e.g., Tryggvason et al., 2002). For a third approach, the joint inverse problem is formulated in terms of *one inversion parameter type* only that is considered of primary importance, whereas the other data sources provide proxy data related to this primary parameter type through petrophysical relationships. A typical hydrogeophysics example is using measurements of the hydrologic state in boreholes (salinity, pressure, water content) in response to hydrologic testing along with crosshole geophysical data (also sensitive to these state variables) to invert directly for the permeability structure (e.g., Kowalsky et al., 2005). In the final approach, model parameters correspond to *properties that are indirectly related* to the geophysical data at hand (i.e., partial differential equations that describe the physical system are not available). Examples include inverting for the spatially distributed electrical formation factor and surface conductivity using radar traveltimes and attenuation data and inverting for geochemical composition using diverse geophysical data (e.g., Chen et al., 2004).

Structural approaches provide robust solutions for a wide range of application types in deterministic joint inversion. We focus on joint inversion based on the cross-gradient constraints introduced by Gallardo and Meju (2003, 2004). Several other interesting structural approaches are offered in the literature. For example, Hyndman and Harris (1996) present a traveltime inversion scheme for inverting 2D zonal models using crosshole seismic traveltime data. Their technique could be extended easily to joint inversion of radar and seismic traveltime and attenuation data by assuming that all these data sets are sensitive to the same uniform zones and zonal boundaries. Paasche and Tronicke (2007) and Paasche et al. (2008)

discuss an iterative sequential approach to invert crosshole radar traveltime and attenuation data. Their technique combines gradient-based deterministic inversion with a cluster algorithm that is used after each iteration step to classify the models in terms of a number of zones. This zonal model becomes the starting model for the next iteration step.

The examples presented here focus on joint inversion of crosshole radar and seismic traveltimes. It would be rather straightforward to modify the algorithm to accommodate radar attenuation data (Holliger et al., 2001), Fresnel-volume inversion (Vasco et al., 1995), full-waveform inversion (e.g., Pratt, 1999; Ernst et al., 2007; Belina et al., 2009), or joint inversion of surface-based seismic refraction (e.g., Lanz et al., 1998) and ground-penetrataing radar (GPR) reflection data (e.g., Bradford et al., 2009).

In this chapter, the joint-inversion methodology is introduced. Then two case studies are presented with a discussion about cross-property relations of seismic and radar wave speeds.

## Method

Joint inversion based on structural coupling using cross-gradient constraints, introduced by Gallardo and Meju (2003, 2004), has been adapted and applied to a wide range of data types (Gallardo and Meju, 2003, 2004, 2007; Gallardo et al., 2005; Linde et al., 2006; Linde et al., 2008; Tryggvason and Linde, 2006; Gallardo, 2007; Doetsch et al., 2009; Fregoso and Gallardo, 2009; Hu et al., 2009). The normalized cross-gradient function $t'_{qr}(x, y, z)$ of two models $\mathbf{m}_q$ and $\mathbf{m}_r$ at location $x, y, z$ is (Linde et al., 2008)

$$\mathbf{t}'_{qr}(x, y, z) = \frac{\nabla \mathbf{m}_q(x, y, z) \times \nabla \mathbf{m}_r(x, y, z)}{|m_q(x, y, z)| \cdot |m_r(x, y, z)|}, \quad (1)$$

where $\nabla \mathbf{m}_q(x, y, z)$ and $\nabla \mathbf{m}_r(x, y, z)$ are the gradients of models $\mathbf{m}_q$ and $\mathbf{m}_r$ at location $x, y, z$. The original definition of the cross-gradient function $\mathbf{t}_{qr}(x, y, z)$ by Gallardo and Meju (2003) does not include the normalization term, which facilitates comparison of results from different applications and different joint-inversion implementations. Constraints based on the cross-gradient function allow one of the models to change at a given position without requiring the other to change, and it focuses on the direction of the change rather than the magnitude.

The cross-gradient function typically is discretized using forward (e.g., Gallardo and Meju, 2003) or central differences (e.g., Linde et al., 2008). The discretized cross-gradient function based on central differences for the $y$-component $t_{qr}^y(i, j, k)$ for models $\mathbf{m}_q$ and $\mathbf{m}_r$ with a uniform discretization is

$$
\begin{aligned}
t_{qr}^y(i, j, k) = {} & \frac{1}{4\Delta x \Delta z}[m_q(i, j, k+1) - m_q(i, j, k-1)] \\
& \times [m_r(i+1, j, k) - m_r(i-1, j, k)] \\
& - \frac{1}{4\Delta x \Delta z}[m_q(i+1, j, k) - m_q(i-1, j, k)] \\
& \times [m_r(i, j, k+1) - m_r(i, j, k-1)],
\end{aligned} \tag{2}
$$

where $\Delta x$, and $\Delta z$ are the discretizations in the $x$- and $z$-directions and where indices $i$, $j$, and $k$ indicate the corresponding indices of the model cells. Gallardo and Meju (2004) provide a formulation for nonuniform cell spacings with rectangular cells.

The cross-gradient function can be defined for the total model (Gallardo and Meju, 2003) or for the model update with respect to a reference model (Tryggvason and Linde, 2006). The latter definition is useful when including seismic data because there might be strong vertical trends in seismic wave speed that dominate any effects from small-scale variations in lithology.

The cross-gradient function is nonlinear, such that it is necessary to linearize it when performing deterministic inversions. This means an iterative approach is needed, even when solving linear forward problems (e.g., when raypaths are assumed to be straight). Cross-gradient constraints add further nonlinearity to already nonlinear problems. This makes it even more important than for individual inversions to ensure a slow convergence to create final models with the fewest artifacts possible. The visual aspects of the joint-inversion models are not very different when obtained using five to 20 iterations to achieve the target data misfit, but smaller details appear in the scatter plots of the two models, and the resulting magnitude of the cross-gradient function is smaller when using many iterations. Thus, satisfactory results can be obtained using the same number of iterations as for the individual inversions, but the results are improved slightly when using more iterations, which is not a constraint for computationally benign crosshole traveltime tomography applications.

The nonlinearity of the cross-gradient function makes the choice of treating the cross-gradients as hard (Gallardo and Meju, 2003, 2004) or soft (Tryggvason and Linde, 2006) constraints a matter of convenience, with no significant influence on the resulting models. Linde et al. (2008)

suggest that the most important factor to minimize the cross-gradient constraints effectively is to ensure slow convergence (i.e., small model updates) during the inversion. Hu et al. (2009) solve a joint-inversion problem in an iterative sequential manner by applying the cross-gradient constraints with respect to one fixed model and one updated model. This approach decreases computation time and the nonlinearity at each iteration step, but no comparison has been made with results obtained by simultaneous model updates. These authors also improve the convergence by using a Gauss-Newton method (i.e., second-order Taylor expansion of the objective function compared to first-order Taylor expansions in previous work).

When performing joint inversion of geophysical data with cross-gradient constraints, the objective function $\Phi$ is

$$
\Phi = \Phi_d + \Phi_m + \Phi_{CG}, \tag{3}
$$

where $\Phi_d$ is a data-misfit term, $\Phi_m$ is a model-structure term, and $\Phi_{CG}$ is a structural-dissimilarity term as defined by the cross-gradient function. The value $\Phi_d$ is given by

$$
\Phi_d = \sum_{q=1}^{Q} [\mathbf{C}_{d,q}^{-0.5}(\mathbf{d}_q - \mathbf{F}_q(\mathbf{m}_q))]_p - \Phi_d^*, \tag{4}
$$

where $Q$ is the number of data types, $\mathbf{C}_{d,q}$ is the data-error covariance matrix for model $q$ (typically assumed to be a diagonal matrix), $\mathbf{d}_q$ are the observed data for data type $q$, $\mathbf{F}_q(\mathbf{m}_q)$ is the forward response of model $q$, and $\Phi_d^*$ refers to the predefined target data misfit. The forward model usually needs a finer discretization than that used for the inversion, which makes it necessary to interpolate $\mathbf{m}_q$ onto a finer grid to solve the forward problem accurately. A key problem for any inversion strategy is to obtain an accurate representation of $\mathbf{C}_{d,q}$ and to make a good choice of $\Phi_d^*$. This problem is not specific to joint inversion, so it is not discussed here.

The norm $p$ in equation 4 is typically two, which assumes that a Gaussian distribution with zero mean characterizes the data noise. To decrease the sensitivity to outliers or fat tails in the data-error distribution, it is useful to work with approximations of the $l_p$-norms for the case when $p = 1$ by using iteratively reweighted least squares (IRLS) (e.g., Farquharson, 2008). The $l_p$-norm is given by

$$
\|\mathbf{r}\|_p^p = \sum_{n=1}^{N} |r_n|^p, \tag{5}
$$

where the entries in $r_n$ denote data residuals. In practice, one uses a normal least-squares inversion but with a matrix that reweights $\mathbf{C}_{d,q}^{-0.5}$ by multiplying it with a diagonal matrix $\mathbf{R}_{d,q}$ with elements

$$R_{nn} = \sqrt{[p((r_n)^2 + \gamma^2)^{p/2-1}]}, \qquad (6)$$

where $r_n$ is the corresponding data residual at the previous iteration. To approximate an $l_1$-norm, it is common to use $p = 1$ and $\gamma = 0.1$ (Farquharson, 2008). This reweighting yields convergence characteristics similar to those of quadratic functions yet is almost as robust to outliers as $l_1$-norm inversions.

The value $\Phi_m$ is given by

$$\Phi_m = \sum_{q=1}^{Q} \varepsilon_q [\mathbf{C}_{m,q}^{-0.5}(\mathbf{m}_q - \mathbf{m}_q^{\text{ref}})]_p, \qquad (7)$$

where $\varepsilon_q$ acts as a trade-off parameter between data fit and model roughness for model $q$ (the value of $\varepsilon_q$ takes the same value $\varepsilon$ for all data sources and is progressively lowered at each iteration by, e.g., 10% to 50% until the target data misfit $\Phi_d^*$ is reached). The corresponding model covariance matrix is $\mathbf{C}_{m,q}$, and $\mathbf{m}_q^{\text{ref}}$ is the reference model for data type $q$.

The variance of $\mathbf{C}_{m,q}$ is often not known precisely and thus can be used to tune the individual inversions such that they reach the target data misfit at the same value of $\varepsilon_q$ for each data type. We will see later that this is important to avoid too many tuning parameters when performing the joint inversion. In practice, $\mathbf{C}_{m,q}$ often is replaced by damping and smoothness constraints (Maurer et al., 1998). Damping is unsuitable for joint inversion with cross-gradient constraints because these regularization operators have no spatial support. Instead, isotropic (Gallardo and Meju, 2003) and anisotropic (Linde et al., 2006) smoothness constraints have been used.

We have found that stochastic regularization operators (see Appendix A) as introduced by Linde et al. (2006) generally outperform smoothness constraints in terms of convergence, stability, and resultant models that better correspond to complementary ground-truth measurements (Linde et al., 2008). These operators typically are based on an exponential covariance function with integral scales that specify the spatial correlation in each direction. The stochastic regularization operator can be based on geostatistical analysis of geophysical logging data (Linde et al., 2006a; Linde et al., 2008) or on estimated resolution properties of the inverse problem (Doetsch et al., 2009). Stochastic regularization operators define a physical length scale that can be related to field conditions, such that fine tuning is not needed when changing the model

discretization from an initially coarse to a more finely discretized inversion grid. If one knows that two model properties have a strong linear correlation, it is possible to make a joint inversion for these properties by including additional smoothness constraints that operate between model parameters at the same location, as suggested by Gallardo and Meju (2004).

Inversion results generally improve with the quality of $\mathbf{m}_q^{\text{ref}}$. In settings in which the largest variability occurs in the vertical direction (i.e., groundwater table, sediment/bedrock interface, sedimentary layers), it is beneficial to use average zero-offset profiles to define a 1D $\mathbf{m}_q^{\text{ref}}$. To resolve sharper features, it can help to work with IRLS-mimicking $l_1$-norms (defined in an analogous manner as for the data misfit in equations 4 through 6) instead of the traditional $l_2$-norm. Other approaches based on iterative reweighting (e.g., Zhdanov, 2009) may be valuable in traveltime tomography (Ajo-Franklin et al., 2007).

The last cross-gradient component $\Phi_{CG}$ in the objective function ensures coupling between the models. It is given by

$$\Phi_{CG} = \sum_{q=1}^{Q} \sum_{r>q}^{R} \lambda_{qr} \left[ \mathbf{t}_{qr}'(\mathbf{m}_q, \mathbf{m}_r) \right]_p, \qquad (8)$$

where $\lambda_{qr}$ is a constant weight given to the cross-gradient constraints between model types $q$ and $r$, and where $\mathbf{t}_{qr}'(\mathbf{m}_q, \mathbf{m}_r)$ is a vector that consists of the estimated normalized cross-gradient function (see equation 1) in all directions and at all locations where structural similarity is imposed. Linde et al. (2008) explore sensitivities related to the choice of $\lambda_{qr}$. They determine that it can be chosen on the basis of trial inversions in which $\lambda_{qr}$ varies over several orders of magnitude with one or two values of $\lambda_{qr}$ for each order of magnitude. The value chosen is the one for which the mean value of $\mathbf{t}_{qr}'(\mathbf{m}_q, \mathbf{m}_r)$ is the smallest when $\Phi_d^*$ is reached. When jointly inverting three data sets, Doetsch et al. (2009) assign $\lambda = \lambda_{12} = \lambda_{13} = \lambda_{23}$. The number of constraints in equation 8 becomes impractical in three dimensions when jointly inverting more than three data sets. Gallardo (2007) presents an alternative formulation by introducing a reference gradient defined as the strongest model gradient at each location in space.

An iterative solution of the joint-inversion problem is needed because (1) the forward responses typically vary nonlinearly with the model (e.g., rays bend in heterogeneous media); (2) the cross-gradient function is nonlinear, involving the product of two model gradients; and (3) IRLS and other compact-regularization operators use iterative reweighting.

The estimated forward response $\mathbf{d}_q^{l+1}$ of model $\mathbf{m}_q^{l+1}$ at iteration $l+1$ is given by

$$\mathbf{d}_q^{l+1} = \mathbf{F}_q(\mathbf{m}_q^l) + \mathbf{J}_q^l \Delta\mathbf{m}_q^{l+1}, \qquad (9)$$

where $\mathbf{J}_q^l$ is the Jacobian evaluated for model $\mathbf{m}_q^l$, $\mathbf{F}_q(\mathbf{m}_q^l)$ is the forward response of this model, and $\Delta\mathbf{m}_q^l$ is a proposed model update. In traveltime tomography, in which the slowness structure is represented by cells of constant slowness, the elements of the Jacobian are the ray length within each cell.

Here is how we obtain $\Delta\mathbf{m}_q^{l+1}$. Linearization of the cross-gradient function for $\mathbf{t}_y'^{l+1}$ (see equations 1 and 2) is

$$\mathbf{t}_y'^{l+1} \cong \mathbf{t}_y'^l + \mathbf{B}_y^l \left( \begin{array}{c} \Delta\mathbf{m}_1^{l+1} \\ \Delta\mathbf{m}_2^{l+1} \end{array} \right), \qquad (10)$$

where $\mathbf{B}_y^l$ is the Jacobian of the normalized cross-gradient function in the $y$-direction (equation 2) with respect to the model parameters. Extensions of the joint-inversion framework to three or more methods are straightforward (Gallardo, 2007; Doetsch et al., 2009), but we focus on two methods for simplicity. At each iteration, we solve the following system of equations in a least-squares sense:

$$\begin{bmatrix} \mathbf{R}_d^l(\mathbf{C}_d)^{-0.5}\mathbf{J}^l \\ \varepsilon^l \mathbf{R}_m^l \mathbf{C}_m^{-0.5} \\ \lambda \mathbf{B}_x^l \\ \lambda \mathbf{B}_y^l \\ \lambda \mathbf{B}_z^l \end{bmatrix} [\Delta\mathbf{m}^{l+1}] = \begin{bmatrix} \mathbf{R}_d^l(\mathbf{C}_d)^{-0.5}(\mathbf{d} - \mathbf{F}(\mathbf{m}^l)) \\ \varepsilon^l \mathbf{R}_m^l \mathbf{C}_m^{-0.5}(\mathbf{m}^{ref} - \mathbf{m}^l) \\ -\lambda \mathbf{t}_x'^l \\ -\lambda \mathbf{t}_y'^l \\ -\lambda \mathbf{t}_z'^l \end{bmatrix}, \qquad (11)$$

where

$$\Delta\mathbf{m}^{l+1} = \begin{bmatrix} \Delta\mathbf{m}_1^{l+1} \\ \Delta\mathbf{m}_2^{l+1} \end{bmatrix}, \mathbf{m}^l = \begin{bmatrix} \mathbf{m}_1^l \\ \mathbf{m}_2^l \end{bmatrix}, \mathbf{m}^{ref} = \begin{bmatrix} \mathbf{m}_1^{ref} \\ \mathbf{m}_2^{ref} \end{bmatrix},$$

$$\mathbf{d} = \begin{bmatrix} \mathbf{d}_1 \\ \mathbf{d}_2 \end{bmatrix}, \mathbf{F}(\mathbf{m}^l) - \begin{bmatrix} \mathbf{F}_1(\mathbf{m}_1^l) \\ \mathbf{F}_2(\mathbf{m}_2^l) \end{bmatrix}, \mathbf{J}^l = \begin{bmatrix} \mathbf{J}_1^l \\ \mathbf{J}_2^l \end{bmatrix},$$

$$\mathbf{C}_d^{-0.5} = \begin{bmatrix} w_1\mathbf{C}_{d,1}^{-0.5} & 0 \\ 0 & w_2\mathbf{C}_{d,2}^{-0.5} \end{bmatrix},$$

$$\mathbf{C}_m^{-0.5} = \begin{bmatrix} w_1\mathbf{C}_{m,1}^{-0.5} & 0 \\ 0 & w_2\mathbf{C}_{m,2}^{-0.5} \end{bmatrix},$$

and

$$\mathbf{R}_d^l = \begin{bmatrix} \mathbf{R}_{d,1}^l & 0 \\ 0 & \mathbf{R}_{d,2}^l \end{bmatrix}, \mathbf{R}_m^l = \begin{bmatrix} \mathbf{R}_{m,1}^l & 0 \\ 0 & \mathbf{R}_{m,2}^l \end{bmatrix},$$

and where $w_1$ and $w_2$ are the weights given to each data and corresponding model type in the inversion.

Recall that the weight given to the cross-gradient constraints $\lambda$ is constant during the inversion and that the variances of $\mathbf{C}_{m,1}$ and $\mathbf{C}_{m,2}$ are determined from the individual inversions such that the same normalized data misfit is obtained for the same values of $\varepsilon_1^l$ and $\varepsilon_2^l$. It is very important that the final models obtained by the individual and joint inversions have comparable data fit (within a few percent) to make it possible to assess the possible benefits of joint inversions. This objective explains the need for the weights $w_1$ and $w_2$. To ensure that a similar importance is given to each model, we make the first inversions with weights $w_1$ and $w_2$ that are inversely proportional to the number of data of each data type. The weight $w_2$ then is adjusted manually typically $\pm 30\%$ to ensure that the final models have a similar target data misfit (Doetsch et al., 2009).

The resulting system of equations is stored as a sparse matrix and is solved with the conjugate-gradient method LSQR at each iteration (Paige and Saunders, 1982), which has the advantage that the original condition number of equation 8 is preserved. A preconditioner is applied that ensures that the $l_2$-norm of each column on the left side of equation 11 is unity, which avoids unnecessary ill conditioning (Paige and Saunders, 1982).

# Results

## South Oyster case study

We now discuss the 2D joint inversion of radar and seismic data acquired between wells S14 and M3 at the South Oyster Focus Area, Virginia (Hubbard et al., 2001; Linde et al., 2008). These data sets originally were acquired to construct a permeability field to evaluate the role of heterogeneities in controlling the field-scale transport of bacteria injected for remediation. The geology comprises rather coarse and high-porosity marine shoreface deposits. Radar data were acquired using a PulseEKKO 100 system with 100-MHz nominal-frequency antennae and a transmitter-receiver spacing of 0.125 m in each borehole. Seismic data were acquired using a Geometrics Strataview seismic system, a Lawrence Berkeley National Laboratory piezoelectric source, and an ITI string of hydrophone sensors, with a 0.125-m source and sensor spacing. The source pulse had a center frequency of 4 kHz with a bandwidth of approximately 1 to 7 kHz. From these data sets, 3248 radar and 2530 seismic traveltimes were extracted.

We used a cell discretization of 0.125 m × 0.125 m for our forward modeling and 0.25 m × 0.25 m for the inversion. All tomographic inversions were stopped once the target data misfits of 0.5 ns (radar) and 20 μs (seismic) were reached. Stochastic regularization based on an exponential model (see Appendix A; Deutsch and Journel, 1998) with vertical and horizontal integral scales of 0.28 and 1.4 m were used (Hubbard et al., 2001). The traveltimes and Jacobians were calculated in the high-frequency limit (Podvin and Lecomte, 1991; Tryggvason and Bergman, 2006) using the PStomo_eq program (Tryggvason et al., 2002).

The individually inverted radar (Figure 1a) and seismic (Figure 1c) tomograms display predominantly layered structures with small velocity variations and overall low velocities, diagnostic of high-porosity unconsolidated sediments. The joint-inversion tomogram models (Figure 1b and 1d) display slightly more distinct boundaries between facies, but the overall structure is similar to the individually inverted results. Comparison of the cross-gradient function for the individually and jointly inverted data (Figure 1e and 1f) demonstrate that the joint inversion decreases the cross-gradient function by more than two orders of magnitude. Differences between the individual- and joint-inversion results are represented best by scatter plots of the seismic and radar wave speeds (Figure 1g and 1h). Note the much higher scatter of the individual inversion wave speeds (Figure 1g) vis-à-vis the joint-inversion values (Figure 1h).

To determine if the joint-inversion models provide a better representation of subsurface architecture than the individual inversion ones, Linde et al. (2008) compare the models in the vicinity of the right borehole with

**Figure 1.** Radar wave-speed models from the South Oyster site: (a) individual inversion with stochastic regularization; (b) joint inversion with stochastic regularization; (c, d) corresponding seismic wave-speed inversion results; (e, f) cross-gradient functions for the models; (g, h) scatter plots for the models. Depths are given in meters below sea level. Radar wave speed is $v_r$; $\alpha$ is seismic wave speed. Modified from Figure 3 of Linde et al., 2008.

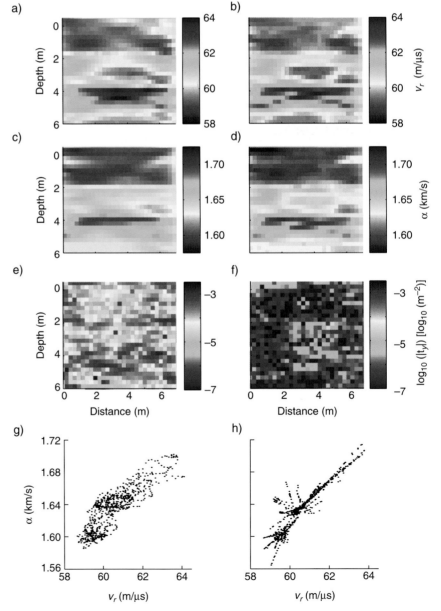

hydraulic-conductivity estimates based on flowmeter measurements and a pumping test (Figure 2a). Trends of the colocated radar (Figure 2b) and seismic (Figure 2c) wave speeds are very similar to the hydraulic-conductivity pattern. Correlation coefficients between log hydraulic conductivity and radar wave speed are 0.72 and 0.78 for the individual and joint inversions. Corresponding values are 0.60 and 0.69 for the seismic wave speed.

This case study demonstrates that joint inversion of crosshole radar and seismic traveltime data somewhat improves resolution compared to individual inversion, yielding an improved hydrogeophysical characterization of the investigation site.

## Thur River case study

Our second case study involves 3D joint inversion of radar and seismic traveltime data acquired in the vicinity of the Thur River, northern Switzerland (Doetsch et al., 2009). These data sets were acquired to delineate the main hydrostratigraphic subunits of a gravel aquifer. The resulting models will be used in an ongoing high-resolution hydrogeophysical study aimed at improving our understanding of groundwater/river-water interactions in Alpine valleys. The geology is composed of coarse, gravelly river deposits with a rather wide grain-size distribution that includes small fractions of fines.

Crosshole radar data at a 0.4-ns sampling rate were acquired using a RAMAC 250-MHz system, which had a center frequency at the site of about 100 MHz with energy in the 50–170-MHz frequency range. A sparker

source was used to generate seismic waves with a center frequency of about 1 kHz, and a Geometrics GEODE system and hydrophone streamer recorded the seismic data at a sampling rate of 21 μs. Borehole deviations were measured with a deviation probe using a three-axis fluxgate magnetometer for bearing and a three-axis accelerometer for inclination.

We inverted the crosshole radar and seismic data acquired between four boreholes located at the corners of a 5-m × 5-m square, approximately 10 m from the Thur River. These data were acquired across all six planes between the four boreholes over the 6-m-thick depth interval that constituted the saturated part of the aquifer. Seismic data were recorded using source-and-receiver spacings of 0.25 m, whereas the radar data were collected with source-and-receiver spacings of 0.5 and 0.1 m, respectively. To ensure symmetric radar coverage, the source and receiver antennae were interchanged and the experiments repeated for each plane. A total of 2661 seismic and 5584 radar traveltimes could be picked reliably (radar traveltimes affected by refractions at the groundwater table were discarded). Examples of the raw data are given in Figure 3.

A cell discretization of 0.0625 m × 0.0625 m was used for the forward modeling and 0.25 m × 0.25 m for the inversion. Target data misfits corresponding to a relative error of 1% for the radar and seismic traveltimes were estimated from reciprocal measurements. All tomographic inversions were stopped once the target misfits were reached. The stochastic regularization was based on an exponential model (see Appendix A; Deutsch and Journel, 1998) with vertical and horizontal integral scales of 0.75 and 1.5 m. This choice

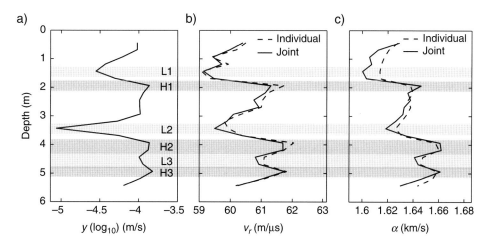

**Figure 2.** (a) Hydraulic-conductivity data from borehole M3 at the South Oyster site (located on the right side of the tomogram in Figure 1). (b) Tomographic radar wave-speed models located two model cells from M3. (c) Tomographic seismic wave-speed models located two model cells from M3. The dashed and solid lines in (b) and (c) represent models from the individual- and joint-inversion models with stochastic regularization. The shaded zones L1–L3 and H1–H3 are locations at which hydraulic conductivities have local minima and maxima. After Figure 6 of Linde et al., 2008.

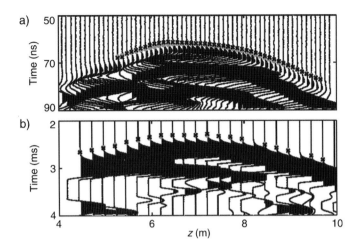

**Figure 3.** Typical raw (a) seismic and (b) radar source gathers for a source depth of 6.75 m. Dots beneath crosses represent calculated forward responses of the final models obtained by joint inversion (see Figure 4e and 4f); black crosses represent the picked first arrivals. (a) Although the seismic data were clipped, first arrivals could be picked reliably. (b) Picked first arrivals in the radar data do not include refracted waves through the unsaturated high-wave-speed layer above 4 m; for the displayed source gather, this means neglecting data collected above 5-m depth.

of weak anisotropy was made to qualitatively honor the subsurface layering seen in the borehole cores without imposing excessive lateral constraints. The integral scales were chosen pragmatically to be comparable to the resolving capabilities of the geophysical data but smaller than the borehole spacing.

The main advantage of performing ray-based 3D inversion of traveltime data at the site compared with a series of six 2D inversions of the data acquired along each tomographic plane is that the regions close to the four boreholes are resolved better and the corresponding models are consistent internally at the borehole locations. The additional constraints offered by the 3D inversion in the near-borehole region also improve the models between the boreholes. Any isolated anomalies located away from the tomographic plane will not be resolved in the 2D or 3D inversions. Rather, the models obtained from the 3D inversion in regions between the planes should be viewed as interpolations between the models along the planes using the stochastic regularization operator.

Individual seismic (Figure 4a) and radar (Figure 4b) inversions resolve a centrally located high-velocity zone embedded in a background of lower velocities. A very similar model has been obtained from inverting crosshole geoelectric data; the high-velocity zone shows up as a region of low resistivity (Doetsch et al., 2009; Doetsch

et al., 2010). The corresponding joint-inversion models (Figure 4e and 4f) are visually very similar to the individual inversion models, but the corresponding cross-gradient functions (Figure 4g) are two to three orders of magnitude smaller than for the individual-inversion models (Figure 4c). In the joint-inversion models, the seismic data have a rather strong influence on the resulting radar wave-speed model in the upper and lower portions of the inversion domain. This is the result of poor GPR ray coverage in these regions because many data were discarded as a result of refractions at the water table at the top and the highly attenuating clay at the bottom. Although the scatter plot for the individual inversion models (Figure 4d) shows a strong correspondence between the seismic and radar wave speeds, just as for the South Oyster case study, the scatter plot for the joint-inversion models (Figure 4h) is defined by much narrower and better-defined correlations.

A useful approach for quantifying improvements in resolution is the point-spread function (PSF), which we calculate following the approach outlined by Alumbaugh and Newman (2000). A PSF can be interpreted as the spatial averaging filter that relates the true underlying model to the resulting inversion model at a specific location for a linearized solution around the final model. Normalization is important for the joint-inversion case, in which the calculated PSFs are normalized with respect to the mean values of the radar and seismic slownesses.

Figure 5 displays normalized PSF volumes at a central location ($x = 2.5$ m, $y = 2.5$ m, $z = 6$ m). These volumes correspond to isosurfaces for which the PSF is 33% of the largest PSF value as suggested by Alumbaugh and Newman (2000). Individual inversions have similar PSFs for the radar (Figure 5a) and seismic (Figure 5b) inversion models. It is clear that the vertical resolution is much higher than the horizontal resolution. When performing joint inversion, the seismic model at this location is influenced by the seismic (Figure 5c) and radar (Figure 5d) properties in the surroundings. Figure 5e and 5f shows the corresponding regions that influence the radar model at this location. The joint inversion may markedly improve the resolution, and the estimated model parameters (e.g., radar or seismic wave speed) at a given point depend on the seismic and radar wave-speed fields in the vicinity of this point. The relative resolution improvements by joint inversion are very similar for other choices of isosurfaces (e.g., 15%). Similar results are presented by Linde et al. (2008) for the South Oyster case study.

Publications on joint inversions based on the cross-gradient function have used least-squares formulations for data and model misfits. Robust inversions based on $l_1$-norms are appealing for applications in which the data are

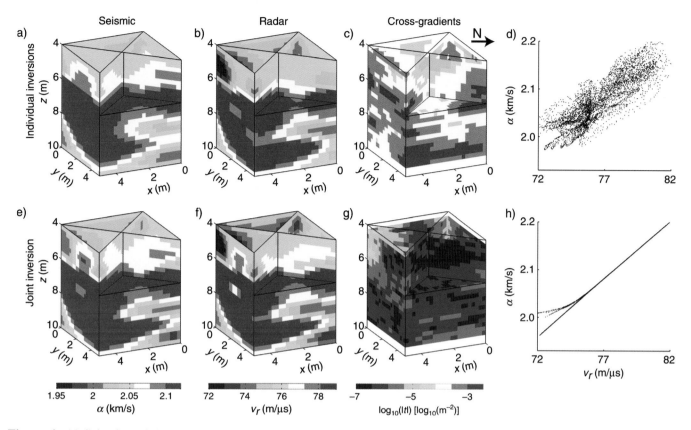

**Figure 4.** (a) Seismic and (b) radar wave-speed models determined from individual inversions of the Thur River site data. (e, f) Corresponding models determined from joint inversion. (c, g) The cross-gradient functions for these models. (d, h) Scatter plots for these models.

noisy or the geology is dominated by a few distinct boundaries, but the computational effort associated with linear programming is often prohibitive. We have investigated the perturbed Ekblom $l_p$-norm for the model norm using IRLS with a common choice of $p = 1.0$ and $\gamma = 0.1$ (see equation 6). Figure 6 displays vertical profiles of the seismic and radar wave speeds obtained from the individual and joint inversions using IRLS and least-squares formulations at $x = 5$ and $y = 2.5$ m (see Figure 4). The largest differences appear between the joint and individual inversions, with the joint-inversion models showing somewhat more variability. The differences in radar wave speeds between the individual and joint inversions in the upper part of the model are from low radar ray coverage (data affected by refractions at the water table were discarded). There are relatively small differences between the joint-inversion results obtained using the IRLS and least-squares model norms; overall, the IRLS inversion results are more variable and less smooth. These results illustrate that joint inversion, at least for the example considered here, has greater impact on the final inversion results than those related to the model norm used in the inversion.

## Seismic and radar wave-speed cross-property relations

The Hashin-Shtrikman (HS) bounds (Hashin and Shtrikman, 1962, 1963) offer an appealing framework for investigating possible relationships between seismic and radar wave speeds. Without restrictive assumptions about pore-space geometry, the HS bounds provide the tightest range of the property values that a mixture of a two-phase media can take with known volume fractions $\phi$ and $1 - \phi$ and properties of each phase. The lower bounds for seismic and radar wave speed correspond to the case in which spherical inclusions (representing the grains) are embedded in a matrix of water and the spheres are not in contact with each other. The upper bounds correspond to the case in which unconnected spherical inclusions of water are embedded in a solid matrix. Pride et al. (2004) argue that the lower HS bound is generally closer to reality in sedimentary settings. Absence of a percolation threshold in most porous media, which indicates that the pore space is connected to very low porosities (Sen et al., 1981), supports this argument.

**Figure 5.** Normalized point-spread functions (PSFs) for the individual (a) seismic and (b) radar inversion models at location $x = 2.5$ m, $y = 2.5$ m, $z = 6$ m for the Thur River site (see Figure 3), where the volume is the region in which the values of the PSFs are at least 33% of the values at the model cell of interest. Normalized PSFs for the seismic model obtained by joint inversion have a smaller spatial support (c) but are also influenced by the radar model (d) over a similar region. Corresponding PSFs for a radar cell show the influence of the (e) seismic and (f) radar models.

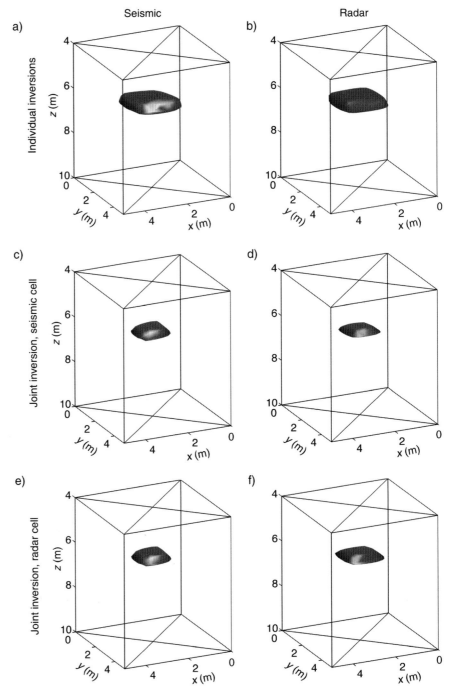

The lower $k_{\mathrm{HSL}}$ and upper $k_{\mathrm{HSU}}$ HS bounds for bulk modulus in water-saturated media are given by (Hashin and Shtrikman, 1963)

$$k_{\mathrm{HSL}} = k_w + \frac{1 - \phi}{\dfrac{1}{k_s - k_w} + \dfrac{3\phi}{3k_w + 4\mu_w}}, \quad (12)$$

$$k_{\mathrm{HSU}} = k_s + \frac{\phi}{\dfrac{1}{k_w - k_s} + \dfrac{3(1 - \phi)}{3k_s + 4\mu_s}}, \quad (13)$$

where $\phi$ is porosity, $k_s$ and $\mu_s$ are the bulk and shear moduli of the solid, and $k_w$ and $\mu_w$ are the corresponding values for the water phase. The lower and upper bounds for shear modulus are (Hashin and Shtrikman, 1963)

$$\mu_{\mathrm{HSL}} = \mu_w + \frac{1 - \phi}{\dfrac{1}{\mu_s - \mu_w} + \dfrac{6}{5\mu_w}\dfrac{(k_w + 2\mu_w)\phi}{(3k_w + 4\mu_w)}}, \quad (14)$$

$$\mu_{\mathrm{HSU}} = \mu_s + \frac{\phi}{\dfrac{1}{\mu_f - \mu_s} + \dfrac{6}{5\mu_s}\dfrac{(k_s + 2\mu_s)(1 - \phi)}{(3k_s + 4\mu_s)}}. \quad (15)$$

The lower and upper bounds for P-wave speed then are given by ($\mu_w = 0$)

$$\alpha_{HSL} = \sqrt{\frac{k_{HSL}}{\rho}}, \qquad (16)$$

$$\alpha_{HSU} = \sqrt{\frac{1}{\rho}\left(k_{HSU} + \frac{4}{3}\mu_{HSU}\right)}. \qquad (17)$$

The lower and upper bounds for dielectric permittivity in saturated media are (e.g., Hashin and Shtrikman, 1962; Brovelli and Cassiani, 2010)

$$\kappa_{HSL} = \kappa_s + \frac{\phi}{\dfrac{1}{\kappa_w - \kappa_s} + \dfrac{1-\phi}{3\kappa_s}}, \qquad (18)$$

$$\kappa_{HSU} = \kappa_w + \frac{1-\phi}{\dfrac{1}{\kappa_s - \kappa_w} + \dfrac{\phi}{3\kappa_w}}, \qquad (19)$$

where $\kappa_s$ and $\kappa_w$ are the dielectric permittivities of the solid and water phases. It is then possible to determine the lower and upper bounds for radar wave speed using

$$v_{HSL} = \frac{c}{\sqrt{\kappa_{HSU}}}, \qquad (20)$$

$$v_{HSU} = \frac{c}{\sqrt{\kappa_{HSL}}}, \qquad (21)$$

where $c = 3 \times 10^8$ m/s is the vacuum speed of light.

Figure 7a and 7b displays the HS bounds for the radar and seismic wave speeds, respectively, for the case of varying $\phi$ with $k_s = 38$ MPa, $k_w = 2.09$ GPa, $\mu_s = 41.5$ GPa, $\mu_w = 0$, $\kappa_s = 6.5$, $\kappa_w = 84$, $\rho_s = 2605$ kg·m$^{-3}$, and $\rho_w = 1000$ kg·m$^{-3}$. These values are representative values of $\alpha$-quartz (Schön, 1996) and water at 10°C (Eisenberg and Kauzmann, 1969; Fine and Millero, 1973). The corresponding relationships between the radar and seismic wave speeds as a function of porosity are shown in Figure 7c with the South Oyster and Thur River scatter plots that result from the joint inversions. The scatter plots lie along or very close to the lower HS bounds. It is well known that tomograms underestimate the variability of the real physical fields (e.g., Day-Lewis and Lane, 2004). Because the estimated radar and seismic slownesses are based on the same inversion processes, they are affected approximately equally by this limitation. Consequently, we expect the cross-property center points and slopes (as revealed by the scatter plots) to be more

a)

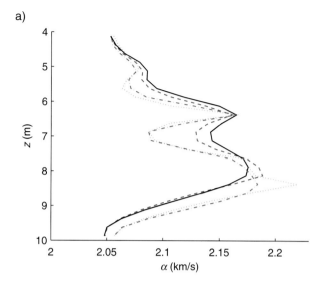

b)

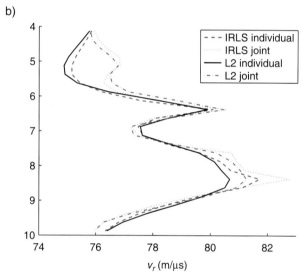

**Figure 6.** (a) Seismic and (b) radar wave-speed models at location $x = 5.0$ m and $y = 2.5$ m (see Figure 3) obtained by individual and joint IRLS and least-squares ($l_2$-norm) inversions.

robust descriptions of the system than the tomograms themselves. For the two case studies, we conclude that the pore space is well connected at both locations and that the South Oyster site has significantly higher porosities than the Thur River site.

An example of the averaging that takes place during inversion is demonstrated in Figure 7d, in which two types of estimates of porosity variations are shown. One is based on neutron-neutron (NN) logs recorded in a borehole at the center of the Thur River inversion domain ($x = 2.5$ m, $y = 2.5$ m), and one is based on the individual and joint inversion wave-speed models at the same location. The NN-to-porosity transform was obtained following Barrash and Clemo (2002), in which the lowest and

**Figure 7.** For the parameters described in the text. (a) Hashin-Shtrikman upper (HSU) and lower (HSL) bounds for seismic wave speed as a function of porosity. (b) HSU and HSL bounds for radar wave speed. (c) Radar-seismic wave-speed relationships for the HSU and HSL bounds as a function of porosity with scatter plots from the South Oyster case study. (d) Porosity log derived from NN log data and porosity estimates obtained from radar and seismic wave speed using the HSL bounds for the seismic and radar wave speeds at the Thur River site.

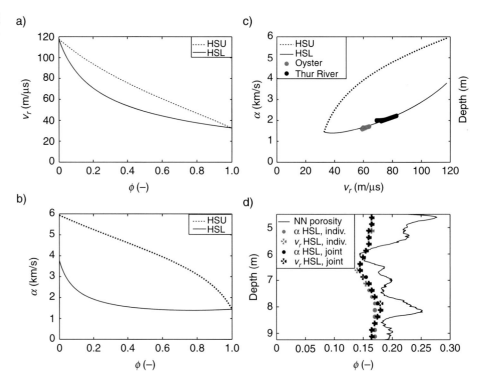

highest NN counts out of 18 borehole logs at the Thur River site are assigned to the highest (0.50) and lowest (0.12) expected end-member porosities for this type of sedimentary setting. Although the resulting absolute porosities obtained from this type of transform might be biased and the variability overestimated, the relative variations with depth are expected to be well resolved. A site-specific NN-to-porosity transform obtained by measuring the porosities on retrieved cores would have helped to improve the absolute porosity values, but no undisturbed cores could be retrieved at our site.

The colocated radar and seismic wave speeds were transformed to porosity via the lower HS bounds using the same parameters as assumed in constructing Figure 7a and 7b. Note that the choice of $\kappa_s = 6.5$ is treated as a fitting parameter to ensure consistent porosity estimates from the seismic and radar joint-inversion models. The wave-speed models provide plausible and fairly tight lower bounds of porosity, but only the main trends of porosity as defined by the NN logs are resolved. A much better correspondence between the overall NN-derived porosity values and those obtained from the radar wave-speed model is obtained using the volume-averaging approach of Pride (1994).

One way to improve the models might be to include the porosity estimates defined by the NN logs into the reference model (e.g., Yeh et al., 2002) or to perform full-waveform inversion. Correlation of the radar and seismic wave speeds with the NN-determined porosities at this location is improved slightly by the joint inversion (the correlation coefficient is increased with 10% to 15% over the individual inversions), and the consistency between the two estimates obtained from joint inversion makes it easier to interpret the results. The scatter plots appearing on the lower HS bounds indicate that the effective porosity is rather similar to the total porosity estimated from the NN logs.

## Discussion

The structural approach to joint inversion using cross-gradient constraints (Gallardo and Meju, 2003, 2004) is a maturing inversion technique that might provide internally consistent geophysical models with improved resolution compared with those obtained from individual inversions. Results presented here and elsewhere indicate that joint inversion using cross-gradient constraints may improve zonation of lithologic subunits (Gallardo and Meju, 2003, 2004, 2007; Doetsch et al., 2009), ratios of physical properties (Tryggvason and Linde, 2006), petrophysical inferences (Linde et al., 2006), and field-scale correlations with hydrologic properties (Linde et al., 2008). These results suggest that joint inversion based on cross-gradient constraints might one day become a standard tool in diverse multimethod geophysical applications. Nevertheless, several questions merit further attention.

How do we justify the assumption of structural similarity for a given field application? Some knowledge about the field site is very important. An example of when the joint-inversion approach is invalid is in a heterogeneous geologic medium with strong gradients in state variables (e.g., salinity), such as in a coastal setting or at a contaminated site, as discussed in Linde et al. (2006). Access to geophysical logging data makes it possible to investigate structural similarity in the vertical direction at a few positions. Note that structural similarity is imposed at the resolution of the resulting models, not at the typically higher resolution of the logging data. We always recommend performing individual and joint inversion of given data sets. If the joint inversion fits the data to the same level as the individual inversions, if the scatter plots of the resulting models display the same main trends, and if the joint-inversion models appear more distinct but no fundamentally new structure is added, then joint inversion with cross-gradient constraints might be a valid approach. If not, it might be possible to impose structural similarity in parts of the model or to decrease the weight given to this constraint. In some cases, it might be possible to reformulate the inverse problem using, for example, time-lapse data to better constrain the properties that are expected to vary. There are many conditions in which the assumption of structural similarity of model parameters is invalid and a careful analysis is needed for each new application.

What is the best discretization of the cross-gradient function? Instead of discretizing using the neighboring model cells, it might be better to define a discretization on the same scale as for the model regularization. This might further stabilize the joint-inverse problem and decrease the sensitivity with respect to model discretization.

How do we determine optimal weights of the components associated with each data set in the objective function, and what are the associated trade-offs? Our approach consists of first giving equal weights to each data set and then reweighting until all data types can be fitted to the same error level as for the individual inversions. There are many alternative approaches in which one could consider the resolution properties, spatial coverage, and so forth. It would probably be quite instructive to perform a detailed analysis of the trade-offs associated with different data types and the different components of the objective function.

How do we transfer the joint-inversion results to geologic models, and how do we make robust petrophysical inferences? Gallardo and Meju (2003) suggest a manual lithologic classification guided by the scatter plots of the different models. Doetsch et al. (2009) develop a formalized classification scheme and perform a zonal inversion for effective petrophysical properties of each zone. A key step in all interpretations is to know what field-scale petrophysical relationships should be used to transform the models into geologic or hydrogeologic properties. It is helpful to use petrophysical relationships that share similar parameterizations and assumptions about the pore structure for all model types, as done here for the HS lower bounds, using a weighted average of the HS bounds (Brovelli and Cassiani, 2010) or volume averaging (Linde et al., 2006). Even if the joint inversion improves resolution, similar resolution-dependent petrophysical relationships, as for individual inversions, remain. A possibility is to focus on the slope of the cross-property relations that might be resolved better than the variability of each model. It also would be useful to extend the method of Day-Lewis et al. (2005) to this type of problem and thereby improve the determination of field-scale petrophysical relationships. One common assumption in petrophysical inference is that some properties such as dielectric permittivity of the solid phase are constant. Relaxing this assumption through Monte Carlo simulations, as suggested by Linde et al. (2006), might help us to better understand different possible explanations, such as the different slopes in the low wave-speed region of Figure 4h.

## Conclusions

Joint inversion of crosshole radar and seismic traveltimes based on cross-gradient constraints using least-squares or $l_1$-norm mimicking measures offers a reliable and robust methodology for improving model resolution in saturated unconsolidated media. With better-resolved models, the confidence in subsequent geophysical-petrophysical analyses is increased. This inversion approach is also expected to apply to consolidated sedimentary environments because porosity is the controlling factor for radar and seismic wave speed, and the two inversion properties are expected to have similar responses to changes in pore structure. For consolidated media, it might be useful to define a 1D reference model and to solve for the model update. No applications of joint inversion of radar and seismic data have been reported under multiphase conditions. Such applications hold considerable promise in the vadose zone and in petroleum-exploration applications, but a detailed assessment of the validity of the cross-gradient function under such conditions remains to be investigated. Similar arguments apply to surface-based data.

It is relatively straightforward to extend our joint-inversion scheme to include improved forward-modeling algorithms based on full-waveform or Fresnel-zone modeling approaches. Significant improvements in hydrogeophysical characterization usually are obtained by including information about the electrical-conductivity distribution obtained when incorporating radar attenuation, full-wave-form modeling of radar data, or geoelectrical data in the joint inversion.

# Acknowledgments

We thank Ari Tryggvason for providing the forward-traveltime modeling codes used in our inversions. The data used in the first case study were provided by John Peterson and Susan Hubbard. We also thank our collaborators within the RECORD project and Ludovic Baron for acquiring the borehole-deviation and neutron-neutron logging data used in the second case study. We thank Alan Green for commenting on a draft of this manuscript. Detailed comments from coeditor John Bradford and two anonymous reviewers improved the manuscript. Funding for this work was provided partly by the Swiss National Science Foundation (SNF) and ETH's Competence Center for Environment and Sustainability (CCES).

# Appendix A

# Stochastic Regularization Operators

The model covariance matrix $C_m$ is a symmetrical Toeplitz matrix if the correlation function is stationary and the grid discretization is uniform in each direction (e.g., Dietrich and Newsam, 1997). Linde et al. (2006) use circulant embedding and the diagonalization theorem for circulant matrices to compute the stochastic regularization operator $C_m^{-0.5}$. Their method is computationally efficient because operations are performed on a vector instead of on a matrix.

Following Dietrich and Newsam (1997), $C_m$ of a stationary process $Y(x)$ with correlation function $C(x)$ sampled on a uniform 1D mesh $\Omega = \{x_0, \ldots, x_m\}$ has values $R_{qr} = r(|x_0 - x_k|)$. A model covariance matrix $C_m$ of size $m \times m$ can be embedded circulantly into a symmetric circulant matrix $S$ of size $2M \times 2M$ by assigning the following entries to the first column $s$ of $S$:

$$s_k = r_k, \quad k = 0, \ldots, m,$$

$$s_{2M-k} = r_k, \quad k = 1, \ldots, m-1, \tag{A-1}$$

where if $M > m$, then the values $s_{m+1}, \ldots, s_{2M-m}$ are chosen arbitrary or conveniently. The next column of $S$ can be obtained by shifting the first column circularly, such that the last element becomes first and all other

elements are shifted forward by one, and so on. Being circulant, $S$ can be decomposed by using the diagonalization theorem of circulant matrices:

$$S = \frac{1}{2M} F \Lambda F^H, \tag{A-2}$$

where $F$ is the fast Fourier transform (FFT) matrix of size $2M$ with entries $F_{pq} = \exp(2\pi i q r / 2M)$, $F^H$ is the conjugate transpose of $F$, and $\Lambda$ is a diagonal matrix whose diagonal entries form the vector $\tilde{s} = Fs$ (e.g., Golub and van Loan, 1996). The matrix $S$ is nonnegative definite if all entries of $\tilde{s}$ are nonnegative. These results are extendable to two and three dimensions (Ranguelova, 2002).

The matrix $S^{-1/2}$ for the 1D case is also circulant, and its first column can be obtained as $F^H \tilde{s}^{-1/2}$; the values corresponding to the first column of $C_m^{-0.5}$ can be retrieved from entries 1 to $m$ (see equation A-1). All other columns of $C_m^{-0.5}$ can be calculated by shifting the first column circularly. To decrease memory requirements, only elements of $C_m^{-0.5}$ larger than 1% of the maximum value of $C_m^{-0.5}$ are stored. In three dimensions, the only difference is that $s$ and $\tilde{s}$ are expressed as 3D arrays and that 3D FFT is applied.

To ensure that $S$ is nonnegative definite in three dimensions when using an exponential correlation function, it is necessary to choose $M$ to be at least seven integral scales in each direction and $s_{m+1}, \ldots, s_{2M-m}$ to be the corresponding values of $r(l)$. The exponential covariance model used to calculate the entries $R_{qr}$ is for a stationary 3D domain defined as

$$C(l) = c e^{-l}, \tag{A-3}$$

where $c$ is the variance, e is the natural logarithm, and $l$ is defined as

$$l = \sqrt{\left(\frac{h_x}{I_x}\right)^2 + \left(\frac{h_y}{I_y}\right)^2 + \left(\frac{h_z}{I_z}\right)^2}, \tag{A-4}$$

where $h_x$, $h_y$, and $h_z$ (all in meters) are the lags (i.e., the distances between a given pair of model parameters) in the $x$-, $y$-, and $z$-directions, respectively, and $I_x$, $I_y$, and $I_z$ (all in meters) are the integral scales specified in the text for the different examples (i.e., the distance at which the correlation between model parameters is $1/e$) in the $x$-, $y$-, and $z$-directions, respectively.

# References

Ajo-Franklin, J. B., B. J. Minsley, and T. M. Daley, 2007, Applying compactness constraints to differential traveltime tomography: Geophysics, **72**, no. 4, R67–R75.

Alumbaugh, D. L., and G. A. Newman, 2000, Image appraisal for 2-D and 3-D electromagnetic inversion: Geophysics, **65**, 1455–1467.

Barrash, W., and T. Clemo, 2002, Hierarchical geostatistics and multifacies systems: Boise Hydrogeophysical Research Site, Boise, Idaho: Water Resources Research, **38**, 1196.

Belina, F. A., J. R. Ernst, and K. Holliger, 2009, Inversion of crosshole seismic data in heterogeneous environments: Comparison of waveform and ray-based approaches: Journal of Applied Geophysics, **68**, 85–94.

Bradford, J. H., W. P. Clement, and W. Barrash, 2009, Estimating porosity with ground-penetrating radar reflection tomography: A controlled 3-D experiment at the Boise Hydrogeophysical Research Site: Water Resources Research, **45**, W00D26.

Brovelli, A., and G. Cassiani, 2010, A combination of the Hashin-Shtrikman bounds aimed at modelling electrical conductivity and permittivity of variably saturated porous media: Geophysical Journal International, **180**, 225–237.

Chen, J. S., S. Hubbard, Y. Rubin, Y. Murray, C. Roden, and E. Majer, 2004, Geochemical characterization using geophysical data and Markov chain Monte Carlo methods: A case study at the South Oyster bacterial transport site in Virginia: Water Resources Research, **40**, W12412.

Day-Lewis, F. D., and J. W. Lane Jr., 2004, Assessing the resolution-dependent utility of tomograms for geostatistics: Geophysical Research Letters, **31**, L07503.

Day-Lewis, F. D., K. Singha, and A. M. Binley, 2005, Applying petrophysical models to radar travel time and electrical resistivity tomograms: Resolution-dependent limitations: Journal of Geophysical Research, **110**, B08206.

Deutsch, C. V., and A. G. Journel, 1998, GSLIB: Geostatistical Software Library and User's Guide: Oxford University Press.

Dietrich, C. R., and G. N. Newsam, 1997, Fast and exact simulation of stationary Gaussian processes through circulant embedding of the covariance matrix: SIAM Journal of Scientific Computing, **18**, 1088–1107.

Doetsch, J., I. Coscia, S. Greenhalgh, N. Linde, A. G. Green, and T. Günther, 2010, The borehole-fluid effect in electrical resistivity imaging: Geophysics, **75**, no. 4, F107–F114. doi:10.1190/1.3467824.

Doetsch, J., N. Linde, I. Coscia, S. Greenhalgh, and A. G. Green, 2009, Joint inversion improves zonation for aquifer characterization: European Geosciences Union Geophysical Research Abstracts, **11**, EGU2009-2303.

Eisenberg, D., and W. Kauzmann, 1969, The structure and properties of water: Oxford University Press.

Ernst, J. R., A. G. Green, H. Maurer, K. Holliger, 2007, Application of a new 2D time-domain full-waveform inversion scheme to crosshole radar data: Geophysics, **72**, no. 5, J53–J64.

Farquharson, C. G., 2008, Constructing piecewise-constant models in multidimensional minimum-structure inversions: Geophysics, **73**, no. 1, K1–K9.

Fine, R. A., and F. J. Millero, 1973, Compressibility of water as a function of temperature and pressure: Journal of Chemical Physics, **59**, 5529–5536.

Fregoso, E., and L. A. Gallardo, 2009, Cross-gradients joint 3D inversion with applications to gravity and magnetic data: Geophysics, **74**, no. 4, L31–L42.

Gallardo, L. A., 2007, Multiple cross-gradient joint inversion for geospectral imaging: Geophysical Research Letters, **34**, L19301.

Gallardo, L. A., and M. A. Meju, 2003, Characterization of heterogeneous near-surface materials by joint 2D inversion of DC resistivity and seismic data: Geophysical Research Letters, **30**, 1658.

———, 2004, Joint two-dimensional DC resistivity and seismic travel time inversion with cross-gradient constraints: Journal of Geophysical Research — Solid Earth, **109**, B03311.

———, 2007, Joint two-dimensional cross-gradient imaging of magnetotelluric and seismic traveltime data for structural and lithological classification: Geophysical Journal International, **169**, 1261–1272.

Gallardo, L. A., M. A. Meju, and M. A. Perez-Flores, 2005, A quadratic programming approach for joint image reconstruction: Mathematical and geophysical examples: Inverse Problems, **21**, 435–452.

Golub, G. H., and C. F. van Loan, 1996, Matrix computations: Johns Hopkins University Press.

Haber, E., and D. Oldenburg, 1997, Joint inversion: A structural approach: Inverse Problems, **13**, 63–77.

Hashin, Z., and S. Shtrikman, 1962, A variational approach to the theory of the effective magnetic permeability of multiphase materials: Journal of Applied Physics, **33**, 3125–3131.

———, 1963, A variational approach to the theory of the elastic behavior of multiphase materials: Journal of the Mechanics and Physics of Solids, **11**, 127–140.

Holliger, K., M. Musil, and H. R. Maurer, 2001, Ray-based amplitude tomography for crosshole georadar data: A numerical assessment: Journal of Applied Geophysics, **47**, 285–298.

Hu, W., A. Abubakar, and T. M. Habashy, 2009, Joint electromagnetic and seismic inversion using structural constraints: Geophysics, **74**, no. 6, R199–R109.

Hubbard, S., J. Chen, J. Peterson, E. Majer, K. Williams, D. Swift, B. Mailliox, and Y. Rubin, 2001, Hydrogeological characterization of the DOE Bacterial Transport Site in Oyster, Virginia, using geophysical data: Water Resources Research, **37**, 2431–2456.

Hyndman, D. W., and J. M. Harris, 1996, Traveltime inversion for the geometry of aquifer lithologies: Geophysics, **61**, 1728–1737.

Lanz, E., H. Maurer, and A. G. Green, 1998, Refraction tomography over a buried waste disposal site: Geophysics, **63**, 1414–1433.

Linde, N., A. Binley, A. Tryggvason, L. B. Pedersen, and A. Revil, 2006, Improved hydrogeophysical characterization using joint inversion of cross-hole electrical resistance and ground-penetrating radar traveltime data: Water Resources Research, **42**, W12404.

Linde, N., A. Tryggvason, J. E. Peterson, and S. S. Hubbard, 2008, Joint inversion of crosshole radar and seismic traveltimes acquired at the South Oyster Bacterial Transport Site: Geophysics, **73**, no. 4, G29–G37.

Lines, L. R., A. K. Schultz, and S. Treitel, 1988, Cooperative inversion of geophysical data: Geophysics, **53**, 8–20.

Kowalsky, M. B., S. Finsterle, J. Peterson, S. Hubbard, Y. Rubin, E. Majer, A. Ward, and G. Gee, 2005, Estimation of field-scale soil hydraulic parameters and dielectric parameters through joint inversion of GPR and hydrological data: Water Resources Research, **41**, W11425.

Maurer, H., K. Holliger, and D. E. Boerner, 1998, Stochastic regularization: Smoothness or similarity?: Geophysical Research Letters, **25**, 2889–2892.

Monteiro Santos, F. A., S. A. Sultan, P. Represesas, and A. L. El Sorady, 2006, Joint inversion of gravity and geoelectrical data for groundwater and structural investigation: Application to the northwestern part of Sinai, Egypt: Geophysical Journal International, **165**, 705–718.

Musil, M., H. R. Maurer, and A. G. Green, 2003, Discrete tomography and joint inversion for loosely connected or unconnected physical properties: Application to crosshole seismic and georadar data sets: Geophysical Journal International, **153**, 389–402.

Paasche, H., and J. Tronicke, 2007, Cooperative inversion of 2D geophysical data sets: A zonal approach based on fuzzy c-means cluster analysis: Geophysics, **72**, no. 3, A35–A39.

Paasche, H., A. Wendrich, J. Tronicke, and C. Trela, 2008, Detecting voids in masonry by cooperatively inverting P-wave and georadar traveltimes: Journal of Geophysics and Engineering, **5**, 256–267.

Paige, C. C., and M. A. Saunders, 1982, LSQR: An algorithm for sparse linear equations and sparse least squares: Transactions of Mathematical Software, **8**, 43–71.

Podvin, P., and I. Lecomte, 1991, Finite difference computation of traveltimes in very contrasted velocity models: A massively parallel approach and its associated tools: Geophysical Journal International, **105**, 271–284.

Pratt, R. G., 1999, Seismic waveform inversion in the frequency domain, Part 1: Theory and verification in a physical scale model: Geophysics, **64**, 888–901.

Pride, S., 1994, Governing equations for the coupled electromagnetics and acoustics of porous media: Physical Review B, **50**, 15678–15696.

Pride, S. R., J. G. Berryman, and J. M. Harris, 2004, Seismic attenuation due to wave-induced flow: Journal of Geophysical Research, **109**, B01201.

Ranguelova, E. B., 2002, Segmentation of textured images on three-dimensional lattices: Ph.D. dissertation, University of Dublin.

Schön, J. H., 1996, Physical properties of rocks: Fundamentals and principles of petrophysics: Elsevier Science Publishing Company, Inc.

Sen, P. N., C. Scala, and M. H. Cohen, 1981, A self-similar model for sedimentary rocks with application to the dielectric constant of fused glass beads: Geophysics, **46**, 781–795.

Tryggvason, T., and B. Bergman, 2006, A traveltime reciprocity inaccuracy in the Podvin & Lecomte *time3d* finite difference algorithm: Geophysical Journal International, **165**, no. 2, 432–435.

Tryggvason, A., and N. Linde, 2006, Local earthquake (LE) tomography with joint inversion for P- and S-wave velocities using structural constraints: Geophysical Research Letters, **33**, L07303.

Tryggvason, A., S. T. Rögnvaldsson, and Ó. G. Flóvenz, 2002, Three-dimensional imaging of the P- and S-wave velocity structure and earthquake locations beneath southwest Iceland: Geophysical Journal International, **151**, 848–866.

Vasco, D. W., J. E. Peterson, and E. L. Majer, 1995, Beyond ray tomography — Wavepaths and Fresnel volumes: Geophysics, **60**, 1790–1804.

Vozoff, K., and D. L. P. Jupp, 1975, Joint inversion of geophysical data: Geophysical Journal of the Royal Astronomical Society, **42**, 977–991.

Yeh, T. C. J., S. Liu, R. J. Glass, K. Baker, J. R. Brainard, D. Alumbaugh, and D. LaBrecque, 2002, A geostatistically based inverse model for electrical resistivity surveys and its applications to vadose zone hydrology: Water Resources Research, **38**, 1278.

Zhdanov, M. S., 2009, New advances in regularized inversion of gravity and electromagnetic data: Geophysical Prospecting, **57**, 463–478.

Chapter 2

# Estimation of Near-surface Shear-wave Velocity and Quality Factor by Inversion of High-frequency Rayleigh Waves

Jianghai Xia[1] and Richard D. Miller[1]

## Abstract

Near-surface shear-wave (S-wave) velocities and quality factors are key parameters for a wide range of geotechnical, environmental, and hydrocarbon-exploration research and applications. High-frequency Rayleigh-wave data acquired with a multichannel recording system have been used to determine near-surface S-wave velocities since the early 1980s. Multichannel analysis of surface waves — MASW — is a noninvasive, nondestructive, and cost-effective acoustic approach to estimating near-surface S-wave velocity. Inversion of high-frequency surface waves has been achieved by the geophysics research group at Kansas Geological Survey during the past 15 years, using surface-wave inversion algorithms of both a layered-earth model (commonly used in the MASW method) and a continuously layered-earth model (Gibson half-space). Comparison of the MASW results with direct borehole measurements reveals that the differences between the two are approximately 15% or less and have a random distribution. Studies show that simultaneous inversion of higher modes and the fundamental mode increases model resolution and investigation depth. Another important seismic property — quality factor ($Q$) — can be estimated with the MASW method by inverting attenuation coefficients of Rayleigh waves. A practical algorithm uses the trade-off between model resolution and covariance to assess an inverted model. Real-world examples demonstrate the applicability of inverting high-frequency Rayleigh waves as part of routine MASW applications.

## Introduction

Elastic properties of near-surface materials, and the effects of those properties on seismic-wave propagation, are of fundamental interest in groundwater, engineering, and environmental studies, and in petroleum exploration. Shear (S)-wave velocity is a key parameter in construction engineering. As an example, Imai and Tonouchi (1982) studied compressional (P)- and S-wave velocities in an embankment, and also in alluvial, diluvial, and tertiary layers, and showed that S-wave velocities in such deposits are related to the N-value (blow count) in the standard penetration test (SPT) (Clayton, 1993; Clayton et al., 1995; Dikmen, 2009), an index value of formation hardness in soil mechanics and foundation engineering. Shear-wave velocity is also an important parameter for evaluating the dynamic behavior of soil in the shallow subsurface and for defining large-scale geologic features such as earthquake zonation (Yilmaz et al., 2009). For example, both the Uniform Building Code (UBC) and Eurocode 8 (EC8) use $V_S30$, the average S-wave velocity for the top 30 m of soil, to classify sites according to soil type for earthquake-engineering design purposes (Sabetta and Bommer, 2002; Sêco e Pinto, 2002; Dobry et al., 2000). In petroleum exploration, a near-surface layer acts as a filter that smears and alters images of deep reflection events. To eliminate such effects, accurate near-surface velocity information is critical. However, determination of these near-surface velocities, generally using S-wave reflection/refraction surveying, is sometimes a troublesome task (e.g., Xia et al., 2002c).

[1]Kansas Geological Survey, University of Kansas, Lawrence, Kansas, U.S.A.

An alternative tool for determining S-wave velocities of near-surface layers relies on Rayleigh-wave techniques. Rayleigh waves are the result of interfering P- and $S_V$-waves. Rayleigh waves are a particular kind of surface wave that travels along a "free" surface, such as the earth-air or the earth-water interface [the latter usually is called a Scholte wave (Scholte, 1947)] and usually are characterized by relatively low velocity, low frequency, high amplitude, and most importantly, by dispersion (Sheriff, 2002). For the case of a solid homogeneous half-space, a Rayleigh wave is not dispersive, travels at a velocity of approximately $0.9194V$ when Poisson's ratio is equal to 0.25 and where $V$ is the S-wave velocity of the half-space (Sheriff and Geldart, 1983), and penetrates to a depth roughly equal to one wavelength. Particle motion of the fundamental mode of Rayleigh waves in the homogeneous medium, moving from left to right, is elliptical in a counterclockwise (retrograde) direction along the free surface. As depth increases to about 20% of a wavelength, the particle motion becomes prograded and still is elliptical throughout its depth of penetration. Particle motion is constrained to a vertical plane consistent with the direction of wave propagation.

Rayleigh waves become dispersive in the case of one layer over a solid homogeneous half-space when their wavelengths are in the range of one to 30 times the thickness of the layer (Stokoe et al., 1994). For a given mode, longer wavelengths penetrate to greater depths, generally exhibit greater phase velocities, and are more sensitive to the elastic properties of deeper layers (Babuska and Cara, 1991). Conversely, shorter wavelengths are sensitive to the physical properties of surface layers. Therefore, a particular mode of surface wave will possess a unique phase velocity for each unique wavelength, thereby leading to the dispersive nature of Rayleigh waves in layered media.

Shear-wave velocities can be derived from inverting the dispersive phase velocity of the surface (Rayleigh and/or Love) wave (e.g., Dorman and Ewing, 1962). Near-surface S-wave velocity can be determined by inverting high-frequency Rayleigh waves. Several seismic methods use dispersion of Rayleigh waves to determine S-wave velocities of near-surface materials. Stokoe and Nazarian (1983) and Nazarian et al. (1983) present a surface-wave method — spectral analysis of surface waves (SASW) — that analyzes the dispersion curve of Rayleigh waves by using dual-trace approaches to produce near-surface S-wave velocity profiles. Matthews et al. (1996) summarize the SASW method and the continuous-surface wave (CSW) method (Abbiss, 1981; Tokimatsu et al., 1991) with detailed diagrams.

For the last 15 years, the Kansas Geological Survey (KGS) at the University of Kansas has worked to develop a method called multichannel analysis of surface waves (MASW), which can be traced back to early work by Song et al. (1989). The method includes acquisition of high-frequency broadband Rayleigh waves, extraction of Rayleigh-wave dispersion curves from Rayleigh-wave energy, and inversion of these dispersion curves to obtain near-surface S-wave velocity profiles. Principal advantages of the MASW method compared with dual-trace approaches are (1) ease in recognizing and interpreting surface waves (distinguishing fundamental from higher modes), (2) effectiveness in eliminating body waves, and (3) accuracy in picking phase velocities and estimating S-wave velocities by inversion.

Systematic verification of inverted S-wave velocities resulting from Rayleigh-wave inversion was undertaken at various sites in North America in the late 1990s. Shear-wave velocity profiles derived from those MASW method tests compared favorably with direct borehole measurements at sites in Kansas (Xia et al., 1999), Vancouver, British Columbia, Canada (Xia et al., 2002a), and Wyoming (Xia et al., 2002c). Effects of changing the total number of recording channels, sampling interval, source offset, and receiver spacing on the inverted S-wave velocity were studied at a test site in Kansas. On the average, the difference between MASW-calculated $V_S$ and borehole-measured $V_S$ in eight wells along the Fraser River in Vancouver, British Columbia, Canada, was less than 15%. One of the eight wells was a blind-test well with a calculated overall difference between MASW and borehole measurements of less than 9%. No systematic differences were observed in derived $V_S$ values from any of the eight test sites.

The MASW method has been given increasingly more attention by the near-surface geophysical community in the last decade, with studies that have focused on field-data acquisition, accuracy of dispersion-energy images, forward modeling, and applications to a variety of near-surface geologic and geophysical problems (e.g., Miller et al., 1999; Xia et al., 2002a; Xia et al., 2002c; Beaty et al., 2002; Beaty and Schmitt, 2003; Tian et al., 2003a; Tian et al., 2003b; Xia et al., 2004; Ivanov et al., 2006a; Ivanov et al., 2006b; Xia et al., 2006a; Xu, et al., 2006; Chen et al., 2006; Xu et al., 2007 and 2009; Luo et al., 2008; Luo et al., 2009a; Luo et al., 2009b; Luo et al., 2009c; Yilmaz et al., 2009). The method routinely has proved to be nondestructive, noninvasive, low-cost, and relatively accurate. It has become a primary tool for determining S-wave velocities that are to be used in applications relating to near-surface geology, the environment, and engineering.

Herein, the discussion focuses on inversion results and observations of the MASW method obtained by researchers at the KGS. We briefly describe inversion algorithms that invert the surface waves of a layered-earth model (commonly used in the MASW method) and a non-layered-earth model (Gibson half-space). The discussion also includes advantages of inverting multimodes into S-wave velocities and the feasibility of estimating quality factors through inverting Rayleigh-wave attenuation coefficients. We conclude the paper by introducing a practical approach that uses the trade-off between model resolution and covariance to assess an inverted model.

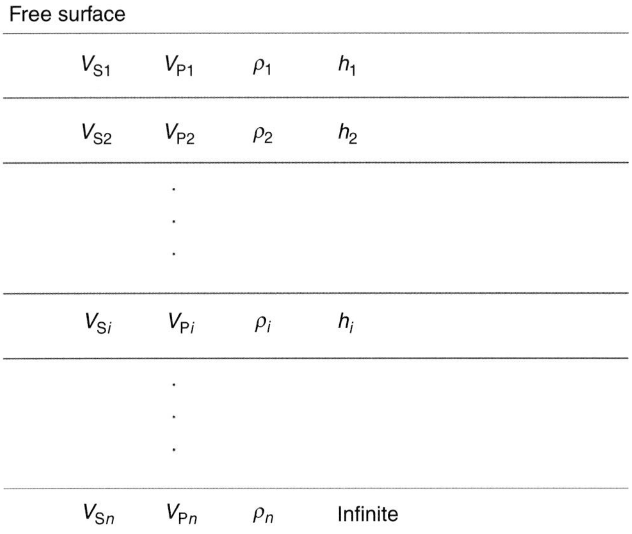

# Inversion of Rayleigh Waves Using a Layered-earth Model

Dispersion is the main characteristic of Rayleigh waves that makes them distinctive and a key component of the near-surface wavefield. Being dispersive means that their velocities change as a function of frequency. A second important characteristic of Rayleigh waves is that their velocities are affected primarily by S-wave velocities. A systematic research program using MASW as a tool for accurately determining near-surface S-wave velocities was launched in the early 1990s at the KGS. Specific treatments in determining S-wave velocities may be given for the subsurface with a high-velocity layer (Calderón-Macías and Luke, 2007) or a low-velocity layer (Lu et al., 2007; Liang et al., 2008).

Rayleigh-wave phase velocity of a model that is composed of $n-1$ solid homogenous layers over a homogenous half-space (Figure 1; we call this model an $n$-layer model hereafter) is a function of frequency and four earth properties: $V_P$, S-wave velocity, density, and thickness of the layers. Analysis of the Jacobian matrix **J** provides a measure of dispersion-curve sensitivity to earth properties (Xia et al., 1999). The sensitivity **S** of one earth property can be calculated by $S = Jdx$, where **dx** is a vector of changes in one earth property. Xia et al.'s (1999) analysis concluded that a 25% error in estimated $V_P$ or rock density results in a difference of less than 10% between the modeled and actual dispersion curves. Because in the real world it is relatively easy to obtain density information that has accuracy greater than 25% (Carmichael, 1989), densities are assumed to be known in our inversion procedure. It also is reasonable to suggest that relative variations in P-wave velocities can be estimated within 25% of actual P-wave velocities, and therefore P-wave velocities are assumed to be known. Inverting Rayleigh-wave phase velocity for layer thickness is more feasible than for $V_P$ or

**Figure 1.** A layered-earth model with shear-wave velocity ($V_S$), compressional-wave velocity ($V_P$), density ($\rho$), and thickness (**h**).

density because the sensitivity indicator is greater for thickness variation than for P-wave velocity or density.

However, because the subsurface always can be subdivided into a reasonable number of layers, each possessing an approximately constant $V_S$, thickness also can be eliminated as a variable in our inversion procedure. Thus, only S-wave velocities are left as unknowns in our inversion procedure. With these assumptions, the inversion algorithm contains only $n$ unknowns ($n$ layers) instead of $4n-1$ unknowns. The fewer unknowns there are in an inversion procedure, the more efficient and stable the process is, and also the more reliable the solution is. It should be pointed out that an abrupt change in $V_P$ and density may increase their sensitivity (Foti and Strobbia, 2002). Special treatment may be needed in such a case.

There are two groups of algorithms to solve inverse problems: search methods and gradient methods. Search methods, such as the simplex algorithm or the genetic algorithm, also can be applied to surface-wave inversion. The main advantage of search methods is that they possess a higher possibility of finding the global minimum than gradient methods do, but they also require much longer computer time than do gradient methods. Herein, we focus our discussion on using a gradient method (the Levenberg-Marquardt or L-M method; Marquardt, 1963) to solve surface-wave inversion problems.

Analysis of the Jacobian matrix (Xia et al., 1999) suggests that S-wave velocities fundamentally and most significantly control changes in Rayleigh-wave phase velocities for a layered-earth model. Therefore, S-wave velocities can be inverted adequately from Rayleigh-wave phase velocities. Xia et al. (1999) (1) start with linearization, then (2) define an objective function and a weighting matrix, (3) give a solution by minimizing the objective function using the L-M method and the singular-value-decomposition (SVD) technique (Golub and Reinsch, 1970), and finally (4) discuss formulas used to determine initial values.

For a layered-earth model with $n$ layers, Rayleigh-wave dispersion curves can be calculated by Knopoff's method (Schwab and Knopoff, 1972). [Note that in the following equations, an uppercase bold letter represents a matrix, a lowercase bold letter represents a vector, and a nonbold italic letter represents a scalar.]

Rayleigh-wave phase velocity, $c_{Rj}$, is determined by a characteristic equation $F$, in its nonlinear, implicit form (Xia et al., 1999):

$$F(f_j, c_{Rj}, \mathbf{v}_S, \mathbf{v}_P, \boldsymbol{\rho}, \mathbf{h}) = 0 \quad (j = 1, 2, \ldots, m), \quad (1)$$

where $f_j$ is the frequency in Hz, $c_{Rj}$ is the Rayleigh-wave phase velocity at frequency $f_j$, $\mathbf{v}_S = (V_{S1}, V_{S2}, \ldots, V_{Sn})^T$ is the S-wave velocity vector with $V_{Si}$ being the S-wave velocity of the $i$th layer, $n$ is the the number of layers, $\mathbf{v}_P = (V_{P1}, V_{P2}, \ldots, V_{Pn})^T$ is the P-wave velocity vector, with $V_{Pi}$ being the P-wave velocity of the $i$th layer; $\boldsymbol{\rho} = (\rho_1, \rho_2, \ldots, \rho_n)^T$ is the density vector, with $\rho_i$ being the density of the $i$th layer; $\mathbf{h} = (h_1, h_2, \ldots, h_{n-1})^T$ is the thickness vector, with $h_i$ being the thickness of the $i$th layer; and the superscript $^T$ represents the transpose operation of a vector or a matrix. Given a set of model parameters ($\mathbf{v}_S$, $\mathbf{v}_P$, $\boldsymbol{\rho}$, and $\mathbf{h}$) and a specific frequency ($f_j$), the roots of equation 1 are the phase velocities. If a dispersion curve consists of $m$ data points, a set of $m$ equations in the form of equation 1 can be used to solve for the phase velocities at frequencies $f_j$ ($j = 1, 2, \ldots, m$) using the bisection method (Press et al., 1992).

S-wave velocities (earth-model parameters) can be represented as the elements of a vector $\mathbf{x}$ of length $n$, $\mathbf{x} = (V_{S1}, V_{S2}, V_{S3}, \ldots, V_{Sn})^T$. Similarly, the measurements (data) of Rayleigh-wave phase velocities at $m$ different frequencies can be represented as the elements of a vector $\mathbf{b}$ of length $m$, $\mathbf{b} = (b_1, b_2, b_3, \ldots, b_m)^T$. Because the Rayleigh-wave phase-velocity vector $\mathbf{c}_R = (c_{R1}, c_{R2}, \ldots, c_{Rm})^T$ (equation 1) is a nonlinear function of frequency and earth properties, equation 1 must be linearized by Taylor-series expansion to employ the matrix theory:

$$\mathbf{J}\Delta\mathbf{x} = \Delta\mathbf{b}, \quad (2)$$

where $\Delta\mathbf{b}[= \mathbf{b} - \mathbf{c}_R(\mathbf{x}_0)]$ is the difference between measured data and model response to the initial estimation; $\mathbf{c}_R(\mathbf{x}_0)$ is the model response to the initial S-wave velocity

estimates, $\mathbf{x}_0$; $\Delta\mathbf{x}$ is a modification of the initial estimation; and $\mathbf{J}$ is the Jacobian matrix with $m$ rows and $n$ columns ($m > n$). The elements of the Jacobian matrix are the first-order partial derivatives of $\mathbf{c}_R$ with respect to S-wave velocities.

Because the number of data points contained in the dispersion curve is generally much larger than the number of layers used to define the subsurface, equation 2 usually is solved by optimization techniques. The objective function is

$$\Phi = \|\mathbf{J}\Delta\mathbf{x} - \Delta\mathbf{b}\|_2 \mathbf{W} \|\mathbf{J}\Delta\mathbf{x} - \Delta\mathbf{b}\|_2 + \lambda\|\Delta\mathbf{x}\|_2^2, \quad (3)$$

where $\|\ \|_2$ is the $l_2$-norm length of a vector, $\lambda$ ($>0$) is the damping factor, and $\mathbf{W}$ is a weighting matrix. Each element of the weighting matrix $\mathbf{W}$ can be determined on the basis of phase-velocity changes within the element neighborhood, because a rapid change in phase velocity indicates a high sensitivity at a particular frequency. Weighting matrix $\mathbf{W}$ also can be determined using the error in phase velocity. Because the weighting matrix $\mathbf{W}$ (equation 3) is both diagonal and positive, it can be written that $\mathbf{W} = \mathbf{L}^T\mathbf{L}$, where $\mathbf{L}$ is also a diagonal matrix.

Marquardt (1963) pointed out that the damping factor ($\lambda$) controls the direction of $\Delta\mathbf{x}$ and the speed of convergence. The damping factor also acts as a constraint on the model space (Tarantola, 1987). By adjusting the damping factor, we can improve processing speed and guarantee the stable convergence of the inversion. In practice, we need to try several different values of $\lambda$ to find a proper damping factor. Employing the SVD technique to minimize the objective function (equation 3) allows us to change the damping factor ($\lambda$) without recalculating the inverse matrix of $(\mathbf{A}^T\mathbf{A} + \lambda\mathbf{I})$, where $\mathbf{A} = \mathbf{L}\mathbf{J}$. The solution is

$$\Delta\mathbf{x} = \mathbf{V}(\Lambda^2 + \lambda\mathbf{I})^{-1}\Lambda\mathbf{U}^T\mathbf{d}, \quad (4)$$

where matrix $\mathbf{A}$ is decomposed as $\mathbf{A} = \mathbf{U}\Lambda\mathbf{V}^T$; $\mathbf{U}$ and $\mathbf{V}$ are a semiorthogonal and an orthogonal matrix, respectively; $\Lambda$ is a diagonal matrix that holds singular values $\Lambda_1 \geq \Lambda_2 \geq \cdots \Lambda_i \geq \Lambda_{i+1} \geq \cdots \geq \Lambda_{n-1} \geq \Lambda_n \geq 0$; $\mathbf{d} = \mathbf{L}\Delta\mathbf{b}$; and $\mathbf{I}$ is the identity matrix.

A solution to this constrained (weighted) least-squares problem (equation 3) with minimum modification to model parameters is sought so that the convergence procedure will remain stable for each iteration. This does not necessarily mean that the final model will be closer to the initial model than with other optimization techniques, such as the Newton method. The modification determined at each iteration (equation 4) is added to the previous model, thereby producing an updated model ($\mathbf{x}^{k+1} = \mathbf{x}^k + \Delta\mathbf{x}$).

The iteration process will stop when $\Phi$ reaches a given threshold value or the number of iterations meets the given maximum iteration number.

Positivity constraints on S-wave velocities have been imposed implicitly in calculating phase velocities, so it is not necessary to add additional positivity constraints on S-wave velocities in the inversion.

A good initial model is important if one is to obtain an acceptable model at the end of inversion, especially for the phase-velocity inversion that is strongly nonlinear and ill-posed (normally its condition number is in the magnitude of $10^6$). If no other information is available regarding S-wave velocity, selection of a consistent initial model becomes necessary. Because Rayleigh waves with different wavelengths have different maximum penetration depths, it is reasonable to use Rayleigh-wave phase velocities selectively at specific frequencies to define initial estimates of depth-dependent S-wave velocities. In our iterative procedure, initial S-wave velocities are determined using the following formulas:

$$V_{S1} = c_R(\text{high})/\beta, \text{ (for the first layer)},$$
$$V_{Sn} = c_R(\text{low})/\beta, \text{ (for the half-space), and}$$
$$V_{Sk} = c_R(f_k)/\beta, \ (k = 2, 3, \ldots, n-1), \quad (5)$$

where $\beta$ is a constant ranging from 0.874 to 0.955 for Poisson's ratio ranges from 0.0 to 0.5 (Stokoe et al., 1994). On the basis of our modeling studies, $\beta$ is chosen as 0.88. Asymptotic approximations of Rayleigh-wave phase velocities in the higher frequency range $c_R(\text{high})$ and the lower frequency range $c_R(\text{low})$ are defined when the measured dispersion curve clearly shows asymptotes on both ends. If the asymptotes do not appear on the dispersion curve, the highest and lowest phase velocities are chosen as $c_R(\text{high})$ and $c_R(\text{low})$, respectively. Our modeling results suggest that Rayleigh-wave velocity $c_R(f_k)$ with a wavelength ($W_L$) can be used in equation 5 to find the initial values for the S-wave velocity of a layer at a depth of $0.63W_L$ for layers between the first layer and the half-space. The initial values of S-wave velocities determined in this way may not be optimal, but they are good enough to start the inversion algorithm and they usually converge to models that are very close to the true model in our numerical tests and real-world examples.

A real-world example in which S-wave velocities were to be determined for the upper 7 m of the earth in Wyoming during the fall of 1998 demonstrates the approach and its benefits over traditional refraction analysis. A shallow SH-wave refraction survey failed to define S-wave velocities because wave-type conversion (S-P-S) occurred along

a slightly dipping interface and prompted the undertaking of a surface-wave survey (Xia et al., 2002c). Surface-wave data were collected using 48 8-Hz vertical-component geophones on a 0.9-m interval, with a nearest source-to-receiver offset of 1.8 m and a 6.3-kg hammer seismic source vertically impacting a metal plate. Data were acquired (Figure 2a) off both ends of the same west-east line along which the shallow SH-wave refraction data had been acquired previously, for comparison of inverted S-wave velocities with traditional traveltime analysis. An image of dispersion energy (Figure 2b) from Figure 2a

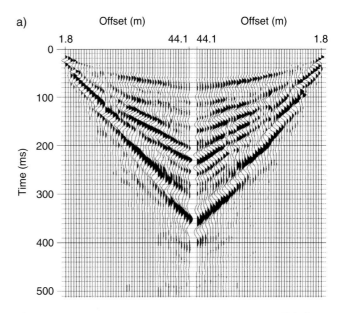

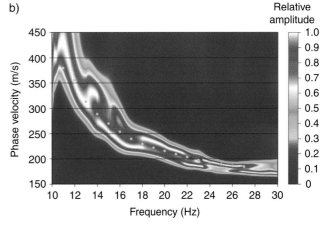

**Figure 2.** (a) Forty-eight-channel surface-wave data (Xia et al., 2002c) acquired with a source off both ends of a west-east line in Wyoming using 8-Hz vertical-component geophones deployed at 0.9-m intervals and a nearest-source offset of 1.8 m. The source was a 6.3-kg hammer vertically impacting a metal plate. (b) A dispersion image in the *f-v* domain from the raw data shown in the left panel of (a), with superimposed cyan dots representing picked phase velocities.

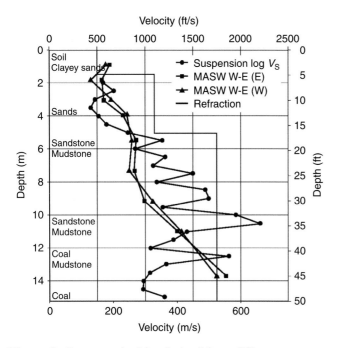

**Figure 3.** S-wave velocities derived from different approaches. S-wave velocities labeled as MASW W-E (E) and MASW W-E (W) represent the inverted results from surface-wave data with the source at the east or west end of the survey line, respectively (Figure 2a). The suspension log results are labeled as Suspension log $V_S$. The average relative difference between two velocity models (or phase velocities) is defined as $(\sum_{i=1}^{k} |V_{1i} - V_{2i}|/V_{1i})/k \times 100\%$, where $k$ is the dimension of models/data. The average relative difference between the dispersion curve calculated from the suspension log and the dispersion curve from the inverted MASW model is 12%, with phase velocities from 11 Hz to 27 Hz. The three-layer velocity model derived from the SH refraction survey also is presented by a solid line, labeled "Refraction" (Xia et al., 2002c). The meter lines serve as lithologic boundaries between units.

was generated using the high-resolution linear Radon transform (Luo et al., 2008). Using this kind of high-resolution approach to image dispersion energy makes phase-velocity picking easy.

A 10-layer model was used to invert dispersion curves in which layer thickness increased gradually from 0.9 m to 2.7 m. With the data-acquisition geometry, the shortest wavelength of Rayleigh waves that could be recorded and that is most sensitive to the top layer was 0.9 m, so the thinnest layer (0.9 m) was chosen as the geophone interval. The density of each layer was chosen to be 2.0 g/cm³, and P-wave velocities of each layer were estimated using first arrivals (Figure 2a). Inverted S-wave velocities for these two data sets (Figure 2a) differ an average of 8% (Figure 3), and the largest difference is 52 m/s with the half-space, which likely results from lateral heterogeneity. To confirm the inverted S-wave velocity derived from the MASW method, a borehole was drilled and suspension-logged near the profile (Figure 3). In the range of 0 to 6 m below the ground surface, the average difference between S-wave velocities estimated from the MASW method and those measured from suspension logging is less than 15% (Xia et al., 2002c).

# Inversion of Rayleigh Waves Using a Continuous-earth Model

A continuously layered model (a compressible Gibson half-space) is the shear modulus that varies linearly with depth in an inhomogeneous elastic half-space and is used in near-surface geophysics. The relationship between S-wave velocities and phase velocities of this model is defined analytically by a closed-form function. The algebraic form of an analytical dispersion law of Rayleigh-type waves in a compressible Gibson half-space (Vardoulakis and Vrettos, 1988) makes our inversion processing extremely simple and fast (Xia et al., 2006b). This is useful in practice for situations in which Rayleigh-wave energy is recorded only in a limited frequency range or at certain frequencies, such as with data acquired at human-made structures such as dams and levees.

A Gibson half-space (Gibson, 1967) is defined as an inhomogeneous elastic half-space $z \geq 0$, with a constant density $\rho$ and Poisson's ratio $\nu$, and with a dynamic shear modulus $G$ that increases linearly with depth (Figure 4). The shear-modulus variation in the Gibson half-space is given by

$$G = G_0(1 + mz), \quad\quad\quad (6)$$

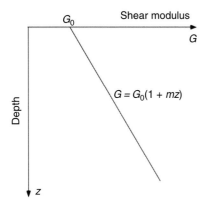

**Figure 4.** A Gibson half-space (see Gibson, 1967).

where $G_0 > 0$ is the shear modulus at the free surface, and $m$ ($\geq 0$) is a measure of inhomogeneity and possesses the dimension of inverse length. The limiting value $m = 0$ corresponds to the homogeneous elastic half-space, for which Rayleigh waves do not exhibit dispersion.

Vardoulakis and Verttos (1988) derived an approximate algebraic form of the dispersion law for the fundamental mode of the Rayleigh wave in the Gibson half-space (equation 6):

$$C = c_R / V_{S0} \cong \frac{1}{\Omega_v} + \sqrt{\frac{1}{\Omega_v^2} + \frac{1}{0.35(3.6 - v)}},$$

$$(0.25 \leq v \leq 0.5), \qquad (7)$$

where $c_R$ is the Rayleigh-wave phase velocity, $V_{S0} = \sqrt{G_0/\rho}$ is the S-wave velocity at the free surface,

$$\Omega_v = \frac{0.56(3.6 - v)}{1.5 + v} \times \frac{2\pi f}{mV_{S0}}, \qquad (8)$$

$f$ is frequency in Hz, and $C$ is defined as dimensionless velocity. The relative error induced by this approximation (equation 7) to true phase velocities is 1–3% (Vardoulakis and Vrettos, 1988).

On the basis of equation 6, S-wave velocity at depth $z$ can be written as

$$V_S(z) = \sqrt{G/\rho} = \sqrt{G_0(1 + mz)/\rho} = V_{S0}\sqrt{(1 + mz)}. \quad (9)$$

If the Gibson half-space density $\rho$ can be determined by another means, the shear-modulus, Young's modulus, and the bulk modulus at depth $z$ can be estimated by S-wave velocities and Poisson's ratio at that depth.

Equation 7 shows that in the Gibson half-space, phase velocities of the fundamental mode of Rayleigh waves $c_R$ are a function of the S-wave velocity at surface $V_{S0}$, the measure of inhomogeneity $m$, and Poisson's ratio $v$. Xia et al. (2006b) gave the partial derivatives with respect to these parameters, so the accurate Jacobian matrix for the same inversion algorithm discussed in the previous section can be calculated. A constraint on Poisson's ratio ($0.25 \leq v \leq 0.5$) is critical to ensure convergence of the inversion. A minimum of three data points is required in the inversion. The inversion algorithm (Xia et al., 1999) discussed in the previous section was used to invert Rayleigh waves to estimate three parameters that determine the S-wave velocities.

The same real-world data (Figure 2a) used for the layered-earth model were employed again to demonstrate the use of the continuously layered-earth model. Xia et al. (2006b) inverted the Rayleigh-wave dispersion curve (Figure 2b) with the algorithm discussed in this section. The initial model parameters were the S-wave velocity at surface $V_{S0} = 100$ m/s, the measure of inhomogeneity $m = 2.0$ 1/m, and Poisson's ratio $v = 0.320$. Poisson's ratio was constrained in the range of 0.27 to 0.37. Iterations stopped and the final inverted model ($V_{S0} = 49$ m/s, $m = 6.018$ 1/m, and $v = 0.272$) was obtained when the root-mean-square (rms) error between measurements and model responses was less than 40 m/s. The rms error between measured data and modeled data was reduced from the initial error of 130 m/s to 12 m/s (Figure 5a). From equation 9, S-wave velocities (Figure 5b) can be calculated using the final estimates of S-wave velocity at the surface $V_{S0}$ and the measure of inhomogeneity $m$. For comparison, the inverted results that are based on the layered model (Xia et al., 2002c) are shown in Figure 5b (squares).

Xia et al. (2006b) also estimated S-wave velocities by inverting only three data points ($f = 12$, 15, and 20 Hz) and keeping all other conditions and parameters the same as in the previous case. The final inverted model ($V_{S0} = 67$ m/s, $m = 3.112$ 1/m, and $v = 0.274$) was obtained (Figure 5b). Compared with the suspension-log results (Figure 5b), the Gibson half-space model ("Non-layered") does a good job from the subsurface down to 12 m. The Gibson half-space model does not appear to be a good model for a subsurface deeper than 12 m, however, which might be the result of oversimplification of the assumed earth model.

## Simultaneous Inversion of Multimodes

A series of Rayleigh waves sampled at different frequencies can possess the same phase velocity. Different-frequency Rayleigh waves with the same phase velocity are known as modes. The lowest velocity for any given frequency is called the fundamental-mode velocity (or the first mode). The next velocity higher than the fundamental-mode phase velocity is called the second-mode velocity (or the first higher mode), and so on. All phase velocities that are higher than the fundamental-mode velocities are referred to as higher modes. Theoretically, all modes always exist; the questions are how to record them and how to use them if they are recorded.

Higher modes are independent from the fundamental-mode phase velocities in the sense that higher modes cannot be recovered directly from the fundamental mode. They appear in the data under a specific set of frequency

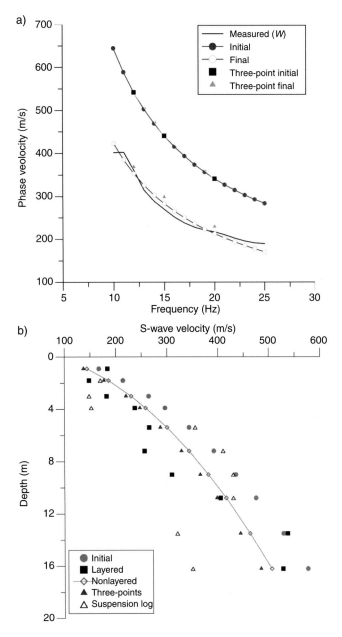

**Figure 5.** The inverse results from Figure 2a, using a continuously layered-earth model. (a) The average dispersion curve was calculated from two shot gathers (Figure 2a) and compared with model dispersion curves. Calculated phase velocities resulting from the initial model using equation 7 are labeled "Initial." (b) S-wave velocity profiles from a layered model, a continuously layered-earth model inverted from the whole data set shown in Figure 2b (labeled "Nonlayered"), from three phase-velocity data points at 12 Hz, 15 Hz, and 20 Hz (labeled "Three-points"), and results of borehole measurements whose depths are the same as the S-wave velocity model. Initial S-wave velocities using equation 9 are labeled "Initial" (Xia et al., 2006b). For the first eight layers, the average relative difference (the definition in the caption of Figure 3) between the results from "Layered" and "Three-points" is 14%.

conditions (Aki and Richards, 1980). Stokoe et al. (1994) have reported that higher modes have been associated with the presence of a velocity reversal (a lower-S-wave-velocity layer between two higher-S-wave-velocity layers). It also has been reported that higher-mode surface waves, when trapped in a layer, are much more sensitive to the fine structure of the S-wave velocity field within that layer than the fundamental mode is (Kovach, 1965). Many authors have shown that very often the measured dispersion curve can follow higher modes also at low frequencies (Socco and Strobbia, 2004; Cercato, 2009) when a significant velocity contrast is present. Reliable observation and separation of higher modes is possible and generally optimal with multichannel recording (Luo et al., 2009b).

Most surface-wave researchers are aware that accuracy of the inverted S-wave velocity can be improved significantly by incorporating higher-mode data when available (Xia et al., 2000; Beaty et al., 2002; Xia et al., 2003; Luo et al., 2007; Song et al., 2007; Liang et al., 2008). With numerical modeling, Xia et al. (2000) and Xia et al. (2003) identified two significant higher-mode properties through analysis of the Jacobian matrix when populated with high-frequency Rayleigh-wave data. First, higher-mode Rayleigh waves penetrate deeper than the fundamental mode does for the same wavelength of fundamental- and higher-mode Rayleigh-wave energy. Second, higher-mode data also can increase resolution of inverted S-wave velocities. In addition, modeling results demonstrate that P-wave velocities affect higher modes much less than they affect the fundamental mode, which provides additional support for suggesting that inversion of higher modes produces more accurate S-wave velocities. Studies (Xia et al., 2008) on the data-resolution matrix (Minster et al., 1974) of the inversion system of surface waves provided insight into the intrinsic characteristic that higher modes normally are predicted much more easily than the fundamental mode is because of restrictions on the data kernel for the inversion system.

Another reason to use higher modes is the fact that in some situations higher modes extract a higher percentage of the energy than the fundamental mode does in the relatively higher frequency range. This means that the fundamental-mode energy sometimes is not available in the higher-frequency range, thereby leaving higher modes as the only choice. We demonstrate this characteristic by using real data (Figure 6) in which the fundamental-mode energy represents a much smaller portion of the total energy than higher-mode energies do for frequencies greater than 20 Hz. Higher modes must be included in the analysis for this case to obtain an accurate S-wave velocity profile. The previously discussed algorithm (Xia et al., 1999) was used to invert these multimode surface-wave data.

High-frequency surface-wave data (Figure 6a) acquired in San Jose, California, were used to determine S-wave velocities of near-surface materials to a depth of 10 m below the ground surface. Higher modes are obvious on the dispersion-curve image (Figure 6b) in the frequency-velocity (*f-v*) domain. The second mode is apparent from 20 to 50 Hz and the third mode starts at 35 Hz. Three data sets were generated and inverted for comparison. The first set comprised only fundamental-mode surface-wave data (Figure 6b). The second data set comprised fundamental-mode data with deliberately introduced noise in the frequency range from 13 to 19 Hz. The noise was designed to simulate the case in which fundamental-mode data are contaminated with higher modes and/or body waves. In our experience, the shape of the second noise-enhanced data set is seen commonly in real dispersion curves. The standard deviation between these two data sets is only 16 m/s. The third data set included the second set (noise-enhanced data) and also second-mode surface-wave data. A fourteen-layer model with each layer 1 m thick was chosen to test these three data sets.

All rms errors between the measured-dispersion curves and calculated-dispersion curves from each of these three S-wave velocity models (Figure 7a) are less than 5 m/s. Because the fundamental-mode data (the first data set) were extracted accurately, the inverted S-wave velocities (solid squares in Figure 7a) are geologically reasonable. The inverted S-wave velocities from the fundamental mode increase smoothly from shallow to deep layers. However, the appearance of smoothness disappears with inversion of data set two (noise added). The S-wave velocity model (diamonds with a solid line in Figure 7a) changes irrationally in the 3- to 7-m depth range. This instability is caused by forcing the response of the inverted model to fit the signal and noise.

In the real world, it is common to provide an error threshold (determined by errors in phase velocities and used to stop the iterations) that inadvertently could force an inverted model into an unreasonable space. We have experienced this situation several times when processing surface-wave data. Better results are obtained when higher-mode surface-wave data (the third data set) are inverted simultaneously with the fundamental-mode data. The S-wave velocity model with the abrupt variation (diamonds with a solid line in Figure 7a) was rejected because of the higher rms error in calculated second-mode data. Inverted S-wave velocities (triangles in Figure 7a) that include the second modes were similar to those obtained from data set one (squares in Figure 7a). The inversion

process is more stable when higher-mode data are included in the inversion of surface-wave data. This stability indeed improves the resolution of inversion results.

So what if no higher modes are available? We must choose between error and resolution of the inverted

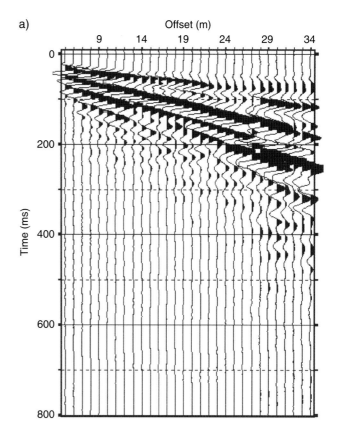

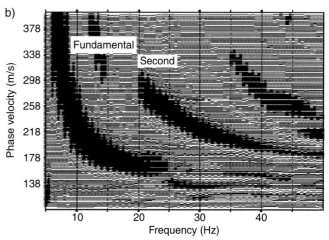

**Figure 6.** An example from San Jose, California (Xia et al., 2003). (a) Raw surface-wave data acquired using 30 4.5-Hz vertical-component geophones with a 1-m receiver interval and the nearest source offset of 4 m. The source was a 6.3-kg hammer vertically impacting a metal plate. (b) Dispersion image is in the *f-v* domain.

model. Making a deliberate trade-off between error and resolution of a model to obtain stable results is a wise strategy (Backus and Gilbert, 1970). Errors in the inverted S-wave velocity model can be reduced by reducing the resolution

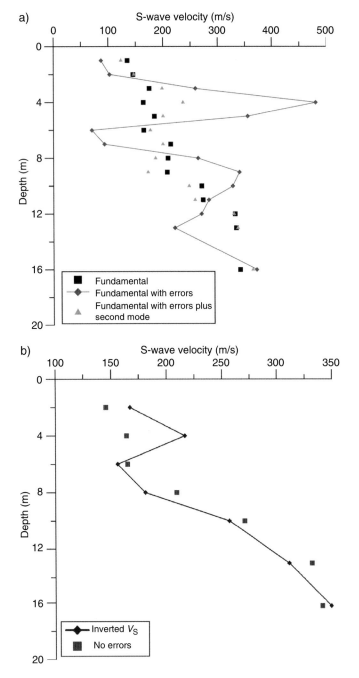

a)

b)

**Figure 7.** (a) Inverted S-wave velocity profiles from three data sets. The average relative difference (the definition is in the caption of Figure 3) between the results from "fundamental" and "fundamental with error" is 49% and the average relative difference between the results from "fundamental" and "fundamental with error plus second mode" is only 9%. (b) Inverted S-wave velocity from the "fundamental with error" data set using a 2-m layer thickness.

of the model (increasing thickness of layers). For the San Jose example, Xia et al. (2003) inverted data set two (the fundamental mode with error) — again using a seven-layer model but with each layer being 2 m thick, so the model possesses only half the resolution of the previous model (a 1-m-thick model in Figure 7a). Data set two underwent the same inversion procedure used for the previous San Jose example. Clearly, the inverted S-wave velocity model with reduced resolution (a 2-m-thick model, diamonds with a solid line in Figure 7b) was smoother and geologically more feasible than the inverted mode depicted by diamonds with a solid line (Figure 7a).

## Estimation of the Near-surface Quality Factor

The seismic quality factor $Q$ as a function of depth, which is directly related to the material damping ratio $D$ ($=0.5Q^{-1}$) (Rix et al., 2000), is of fundamental interest in geotechnical engineering, groundwater, and environmental studies, as well as in oil exploration and earthquake seismology. A desire to understand the attenuative properties of the earth is based on the observation that seismic-wave amplitudes are affected by the elastic moduli of the material.

Laboratory experiments (Johnston et al., 1979) show that $Q$ may be independent of frequency over a broad bandwidth ($10^{-2}$–$10^7$ Hz), especially for some dry rocks. The value of $Q^{-1}$ in liquids, however, is proportional to frequency, so in some highly porous and permeable rocks $Q^{-1}$ may contain a frequency-dependent component. This component may be negligible at seismic frequency, even in unconsolidated marine sediments (Johnston et al., 1979). Mitchell (1975) investigated $Q$ structure of the upper crust in North America by inverting Rayleigh-wave attenuation coefficients in a layered-earth model. In his work, $Q$ was independent of frequency. Although some authors suggest that near-surface $Q$ values may be frequency dependent (Jeng et al., 1999), we will follow the laboratory results (Johnston et al., 1979) and Mitchell's work (1975) and consider $Q$ to be independent of frequency. $Q$ as a function of depth can be determined on the basis of amplitude attenuation of Rayleigh-wave energy. Use of high-frequency Rayleigh waves is essential in estimating the quality factors of near-surface materials. The relationship between Rayleigh-wave attenuation coefficients and the quality factors for P- and S-waves of a layered model were given by Anderson et al. (1965) as

$$\alpha_R(f) = \frac{\pi f}{c_R^2(f)}\left[\sum_{i=1}^{n} P_i(f)Q_{Pi}^{-1} + \sum_{i=1}^{n} S_i(f)Q_{Si}^{-1}\right], \quad (10)$$

where $P_i(f) = V_{Pi}(\partial c_R(f)/\partial V_{Pi})$, $S_i(f) = V_{Si}(\partial c_R(f)/\partial V_{Si})$, and $\alpha_R(f)$ are Rayleigh-wave attenuation coefficients in units of 1/length, and $f$ is frequency in Hz. $Q_{Pi}$ and $Q_{Si}$ are the quality factors for P- and S-waves of the $i$th layer, respectively; $V_{Pi}$ and $V_{Si}$ are the P-wave velocities and S-wave velocities of the $i$th layer, respectively; $c_R(f)$ is Rayleigh-wave phase velocity; and $n$ is the number of layers in the earth model. Xia et al. (2002b) adopted Kudo and Shima's work (1970) to calculate the attenuation coefficients. The attenuation coefficient is defined by

$$A_R(x + dx) = A_R(x)e^{-\alpha dx}, \qquad (11)$$

where $A_R$ is Rayleigh-wave amplitude, $\alpha$ is a Rayleigh-wave attenuation coefficient, and $x$ and $dx$ are a source-geophone offset and a geophone interval, respectively. After the Fourier transform with respect to time, we obtain

$$\alpha_R(f) = -\frac{\ln\left[\left|\dfrac{W(x + dx, f)}{W(x, f)}\right|\sqrt{\dfrac{x + dx}{x}}\right]}{dx}, \qquad (12)$$

where $\alpha_R(f)$ is the Rayleigh-wave attenuation coefficient as a function of frequency $f$, $W$ is the amplitude of a specific frequency, and $\sqrt{(x + dx)/x}$ is a scaling factor used in calculating the attenuation coefficient.

Sensitivity of Rayleigh-wave attenuations associated with $Q_P$ is directly proportional to the product of P-wave velocity and the derivative of Rayleigh-wave phase velocity with respect to P-wave velocity, not just to the derivative itself (Equation 10). That is why we may able to invert Rayleigh-wave attenuation coefficients to estimate $Q_P$ in a certain range of $V_S/V_P$. Modeling results (Xia et al., 2002b) suggest that it is feasible to solve for P-wave quality factor $Q_P$ and S-wave quality factor $Q_S$ in a layered-earth model by inverting Rayleigh-wave attenuation coefficients when $V_S/V_P$ is at or above 0.45. Only $Q_S$ can be estimated from Rayleigh-wave attenuation coefficients when $V_S/V_P$ is less than 0.45. Sensitivity analysis showed that errors in inverted quality factors could reach 1 to 1.5 times the error in attenuation coefficients. Compared with the inversion system for Rayleigh waves (Xia et al., 1999) (a 10% error in surface-wave phase velocity will result in 6% error in S-wave velocity), the inversion system for $Q$ (Xia et al., 2002b) possesses less stability. Hence, accurate calculation of Rayleigh-wave attenuation coefficients is critical. That being said, it is important to note that the inversion system for $Q$ is more stable than is the AVO (amplitude variation with

offset) analysis that has been a mainstay in the oil industry for the last 30 years. A 10% error in determining true incident angles in AVO analysis could result in a 40% error in reflection coefficients (Jin et al., 2000).

Because equation 10 is a linear system, the same method used in Xia et al. (1999) can be employed directly to solve for $Q_P$ and/or $Q_S$ from Rayleigh-wave attenuation coefficients. In many cases, only a single iteration is necessary to obtain quality factors. Menke (1984) discussed an algorithm to solve a linear inversion problem with positivity constraints:

$$\mathbf{Ax} = \mathbf{b}, \, (x_i > 0), \qquad (13)$$

where $\mathbf{A}$ is a data kernel matrix determined by equation 10, $\mathbf{x}$ is a reciprocal of quality factors (a model vector $1/Q$) with $x_i$ as the $i$th component, and $\mathbf{b}$ is attenuation coefficients (a data vector $\alpha_R(f)$). Equation 13 provides accurate estimates of $Q_P$ and $Q_S$ if attenuation coefficients contain no error, as in a synthetic example shown in Xia et al. (2002b). If attenuation coefficients possess errors, solutions of equation 13 may not exist. In such a case, a damping factor $\lambda$ must be introduced (Xia et al., 2002b):

$$(\mathbf{A} + \lambda \mathbf{I})\mathbf{x} = \mathbf{b}, \, (x_i > 0), \qquad (14)$$

where $\mathbf{I}$ is the identity matrix. In practice, $\lambda$ is set to be a small value (say, $10^{-7}$) at the beginning of the inversion. On the basis of the inverted results of $Q_P$ and/or $Q_S$, $\lambda$ will be increased systematically until a smooth solution is obtained.

A successful example of this approach is with data acquired in the Arizona desert (Figure 8a). S-wave velocities (Figure 8b) of a 10-layer model were calculated using the MASW method. P-wave velocities of the model were determined by the first arrivals of the data (Figure 8a). Data denoted as "Measured" (Figure 8c) were calculated from raw data using equation 12. Because an average value of the ratio $V_S/V_P$ for the model is approximately 0.4, only $Q_S$ can be inverted confidently from attenuation coefficients. Under the assumption that $Q_P$ was equal to twice $Q_S$ (Figures 10 and 11 in Toksöz et al., 1979), we inverted attenuation coefficients to obtain $Q_S$ (Figure 8d). Data denoted as "Final" (Figure 8c) were calculated on the basis of an inverted quality-factor model (Figure 8d). Values of $Q_S$ were in the range of 7 to 25. Confidence in this approach can be drawn from the observation that modeled Rayleigh-wave attenuation coefficients (labeled "Final" in Figure 8c) agree well with the measured coefficients.

**Figure 8.** An example from the Arizona desert (Xia et al., 2002a). (a) Sixty-channel surface-wave data acquired using 4.5-Hz vertical geophones deployed at 1.2-m intervals and with a nearest source offset of 4.8 m. The seismic source was an accelerated weight drop designed and built by the KGS. (b) Inverted S-wave velocities of a 10-layer model using the MASW method and known P-wave velocities. (c) Measured and modeled Rayleigh-wave attenuation coefficients. Data points identified as "Measured" were calculated from raw data and those labeled "Final" were calculated on the basis of the inverted quality-factor model (d).

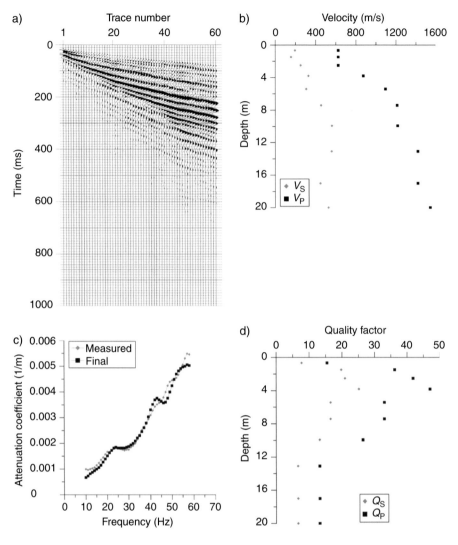

## Appraisal of Inverted S-wave Velocities

Appraisal of the accuracy of S-wave velocity models derived from equation 4 is critical in geologic interpretation of these models, because a general inverse matrix normally is a function of the damping factor. For the regularized least-squares inversion, Zhdanov and Tolstaya (2006) introduced a method for model appraisal and resolution analysis based on evaluating the spatial distribution of the upper bounds of the model variations. Other documented techniques for assessing a geophysical inverse model derived from a linear system are based on calculating model resolution (Wiggins, 1972) and covariance matrices (Tarantola, 1987; Menke, 1984). The model resolution and covariance matrices of the regularized solutions are controlled by the regularization parameter. Because of the complexity of the objective functions of geophysical inverse problems in

general, the regularization parameter can vary dramatically in the vicinity of a solution.

Appraisal of inverse models in terms of accuracy and resolution is essential for a meaningful interpretation of these models. Because of uncertainties associated with the damping factor, extra conditions usually are required to determine a proper regularization parameter for assessing inverse models. Xia et al. (2010b) propose using the trade-off between model resolution and model covariance matrices in the vicinity of a regularized solution to find the proper regularization parameter. They present a practical way to determine the proper regularization parameter, and they demonstrate it by inverting real surface-wave data. The covariance matrix can be used to calculate error bars of the inverse model.

The model-resolution matrix (Wiggins, 1972) for an overdetermined system ($m > n$, $m$ denotes the number of data and $n$ the number of unknowns) with the regularized least-squares method is

$$\mathbf{m}^{\text{est}} = \mathbf{HGm}^{\text{true}} = [\mathbf{G}^{\text{T}}\mathbf{G} + \lambda\mathbf{I}]^{-1}\mathbf{G}^{\text{T}}\mathbf{Gm}^{\text{true}}$$

$$= \mathbf{Rm}^{\text{true}}, \tag{15}$$

where $\mathbf{G}$ is an $m \times n$ matrix and stands for the Jacobian matrix of an inversion system that embodies $\mathbf{m}^{\text{true}}$ and includes the experimental geometry, $\mathbf{H}$ is an $n \times m$ matrix and a generalized inverse of $\mathbf{G}$, and $\mathbf{m}^{\text{est}}$ and $\mathbf{m}^{\text{true}}$ are an estimated earth-model and the true earth-model vectors, respectively. The model-resolution matrix $\mathbf{R} = [\mathbf{G}^{\text{T}}\mathbf{G} + \lambda\mathbf{I}]^{-1}\mathbf{G}^{\text{T}}\mathbf{G}$ is an $n \times n$ matrix.

For a linear system, if data are assumed to be uncorrelated, the unit covariance matrix of an inverted model is $\mathbf{C} = \mathbf{HH}^{\text{T}}$ (Menke, 1984; Tarantola, 1987). Therefore, the unit covariance matrix of a regularized least-squares solution is

$$\mathbf{C} = \mathbf{HH}^{\text{T}} = [\mathbf{G}^{\text{T}}\mathbf{G} + \lambda\mathbf{I}]^{-1}\mathbf{G}^{\text{T}}([\mathbf{G}^{\text{T}}\mathbf{G} + \lambda\mathbf{I}]^{-1}\mathbf{G}^{\text{T}})^{\text{T}}. \tag{16}$$

With the singular-value decomposition $\mathbf{G} = \mathbf{U\Lambda V}^{\text{T}}$, where $\mathbf{U}$ and $\mathbf{V}$ are semiorthogonal and orthogonal matrices, respectively, $\mathbf{\Lambda}$ is a diagonal matrix that holds singular values, and $\mathbf{G}^{\text{T}}\mathbf{G} + \lambda\mathbf{I} = \mathbf{V}(\mathbf{\Lambda}^2 + \lambda\mathbf{I})\mathbf{V}^{\text{T}}$, we can rewrite the model-resolution matrix (equation 15) and the unit-covariance matrix (equation 16) in the following forms:

$$\mathbf{R} = \mathbf{V}\left\{\frac{\mathbf{\Lambda}^2}{\mathbf{\Lambda}^2 + \lambda\mathbf{I}}\right\}\mathbf{V}^{\text{T}}, \tag{17}$$

and

$$\mathbf{C} = \mathbf{V}\left\{\frac{\mathbf{\Lambda}^2}{(\mathbf{\Lambda}^2 + \lambda\mathbf{I})^2}\right\}\mathbf{V}^{\text{T}}, \tag{18}$$

respectively.

Two extreme cases are interesting to us and also provide key insights. When $\lambda$ approaches zero, $\mathbf{R}$ moves toward the identity matrix, thereby indicating a solution that possesses the highest resolution possible. Diagonal elements of $\mathbf{C}$, however, move toward infinity if there are any singular values near zero, thereby indicating a solution that possesses the highest error possible. When $\lambda$ approaches infinity, diagonal elements of $\mathbf{R}$ move toward zero, which indicates a solution that possesses the lowest resolution possible. Diagonal elements of $\mathbf{C}$, however, move toward zero, which indicates a solution that possesses the lowest error possible. These two extreme cases imply

that a choice of $\lambda$ exists that can be selected and that produces a solution with acceptable resolution and variance.

Estimating the model resolution and covariance with equations 17 and 18 in the vicinity of a solution is possible after a proper regularization parameter has been determined. This is possible because an objective function of a small-dimensional problem in the vicinity of a local or global minimum often can be approximated as a linear function. A regularized least-squares solution generally results in an erratic unit-covariance matrix, as a result of uncertainty regarding the regularization parameter in the vicinity of a solution. Therefore, we seek a solution with a constraint on the regularization parameter, which provides the trade-off between model resolution and covariance (Backus and Gilbert, 1967, 1968). The trade-off can be achieved equivalently by cutting off small singular values, as Lay and Wallace (1995) pointed out.

The trade-off model between the model resolution and variance in the vicinity of a regularized solution can be found by selecting the $k$th singular value $\Lambda_k$ as a specific regularization parameter $\lambda$. To determine $\Lambda_k$, singular values from large to small are plotted to form a singular-value distribution curve; $\Lambda_k$ is the first singular value that approaches zero, and the rest of the singular values $\Lambda_j$ ($j > k$) are positioned along an almost horizontal line in the plot. In practice, with inversion of high-frequency surface-wave data, we found that $\Lambda_k$ normally is less than or approximately equal to 0.01. The following real-world example shows the ease with which $\Lambda_k$ can be determined using the singular-value distribution curve. If the diagonal elements of the unit-covariance matrices are larger than we expected, we can choose a larger regularization parameter (at the cost of reducing the model's resolution).

After an inverse solution is found by the regularized least-squares inversion, we can calculate the trade-off model's resolution and unit-covariance matrices using equations 17 and 18 with the regularization parameter $\Lambda_k$ ($=\lambda$). In practice, it is possible then to estimate the standard deviation of an inverse model with the formula, if we assume that the errors in data are uncorrelated:

$$\Delta m_i = \Delta d \sqrt{\sum_{j=1}^{n} \gamma_j v_{ij}^2}, \tag{19}$$

where $\Delta m_i$ is the $i$th element of the standard deviation $\Delta m$ of an inverse model; $\Delta d$ is the data standard deviation, which could be replaced by a threshold of terminating iterations, $\gamma_j = \Lambda_j^2[\Lambda_j^2 + \Lambda_k]^{-2}$; $\Lambda_j$ is the $j$th singular value; and $v_{ij}$ is the element of matrix $\mathbf{V}$ (an $n \times n$ matrix

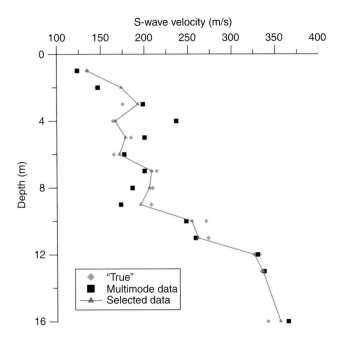

**Figure 9.** Inversion results from the San Jose, California, example. Model "True" was obtained by inverting "error-free" data. "Multimode data" were the inverted model from 28 multimode data points (Xia et al., 2003). "Selected data" were the inverted model from 16 data points selected based on data-resolution functions (Xia et al., 2008).

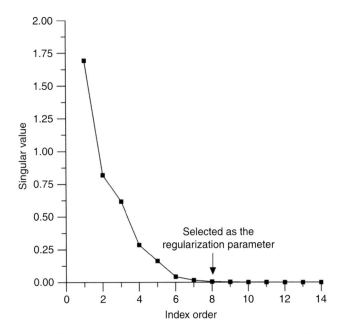

**Figure 10.** A singular value plot from the San Jose, California, data set (Xia et al., 2003) with the eighth singular value of 0.00559 used as the regularization parameter and the smallest value of 0.00028 (Xia et al., 2010b).

after singular-value decomposition of a data kernel) at the $i$th row and the $j$th column.

Data acquired in San Jose, California (Figure 6a) possessed the well-defined fundamental and higher modes in the frequency-velocity domain needed to provide convincing results (Figure 6b) (Xia et al., 2003). A 14-layer model with layer thicknesses of 1 m was used to invert the Rayleigh-wave data. Xia et al. (2008) calculated the data-resolution matrix (Menke, 1984; $\mathbf{N} = \mathbf{G}[\mathbf{G}^{\mathrm{T}}\mathbf{G} + \lambda\mathbf{I}]^{-1}\mathbf{G}^{\mathrm{T}} = \mathbf{G}[\mathbf{G}^{\mathrm{T}}\mathbf{G}]^{-1}\mathbf{G}^{\mathrm{T}}$ with $\lambda = 0$) of the multilayer model, with 35 data points at different frequencies as high as the second mode, to show the predictability of each datum and each mode. They selected 16 data points that were based on the data-resolution matrix and that possessed diagonal values of 0.45 and higher, to perform the inversion. Sixteen data points are almost the minimum number necessary to solve a problem with 14 unknowns. The initial model and initial legalization parameter ($\lambda_0 = 1$) was the same used in Xia et al. (2003). Inverted results from the selected data (16 data points) and inverted results from multimode data (28 data points, with the fundamental-mode data as high as 23 Hz; Xia et al., 2003) are shown in Figure 9, with the best inverted model that was obtained from "error-free" data (we labeled this model as "True" in Figure 9). Overall, inversion results from the 16 selected data points (Xia et al., 2003, Xia et al., 2008) are closer to the true model than are results obtained from multimode data.

The inversion of the 28-data-point set converged to the multimode model denoted by squares in Figure 9, with $\lambda = 10^{-1}$ (Xia et al., 2003). The inversion of 16 selected data points converged to the model denoted by triangles in Figure 9, with $\lambda = 10^{-3}$. The inverted model from 16 selected data points possesses the perfect model resolution when no damping is applied, but the model has extremely high variance (as high as $10^5$, which could result in errors in an inverse mode as high as more than 300 times the errors in the data). As discussed in the previous section, the singular-values distribution is plotted (Figure 10) and the best regularization parameter $\Lambda_8 = 0.00559$ is determined.

The trade-off model with the regularization parameter possesses a much smaller variance and reasonable model resolution (Table 1). Model variance can be reduced dramatically at the cost of reducing model resolution. The Euclidean length ($l^2$-norm) between the two points, determined by the trade-off model (Figure 11) and the model from the selected data (triangles in Figure 9), is 57 m/s. Using equation 19 and assuming the data are uncorrelated, we can estimate a standard deviation of the trade-off model using the threshold of terminating iterations (2.07 m/s) as the standard deviation of the data. The maximum standard

**Table 1.** Model resolution and variance (diagonal elements of **R** and **C**) of the trade-off model from the San Jose data set (Xia et al., 2010b).

| Layer no. | 1 | 2 | 3 | 4 | 5 | 6 | 7 | 8 | 9 | 10 | 11 | 12 | 13 | 14 |
|---|---|---|---|---|---|---|---|---|---|---|---|---|---|---|
| Variance | 4.79 | 9.71 | 2.98 | 3.62 | 10.20 | 2.88 | 5.94 | 6.20 | 6.80 | 4.12 | 8.82 | 7.15 | 7.85 | 3.07 |
| Resolution | 0.16 | 0.33 | 0.44 | 0.35 | 0.45 | 0.52 | 0.32 | 0.41 | 0.37 | 0.21 | 0.27 | 0.14 | 0.11 | 0.98 |

deviation of the trade-off model is therefore $\pm 6.64$ m/s and is associated with layer 5 (Figure 11).

## Future Studies

Mode identification on dispersion curves intended for inversion is a challenge both in theory and in practice, if no supporting information is available on S-wave velocity. Mode identification is difficult for cases of missing modes as well as for cases of "mode kissing" (we thank Robert Stewart for suggesting this term), which is the result of a very small difference in phase velocities between two modes at a certain frequency (see Figures 8 and 10 in Xia et al., 2006a). With the current resolution potential of dispersive images, it is very difficult to determine exactly which mode is at a frequency. Mode kissing causes mode misidentification (Zhang and Chan, 2003), thereby resulting in inverted S-wave velocities that are higher than true velocities.

Seismic numerical modeling that treats the free-surface condition as an explicit acoustic/elastic boundary (Xu et al., 2007) provides a tool for evaluating and studying seismic characteristics of near-surface features and builds a basis for full-wavefield or Rayleigh-wave inversion in the time-offset domain. Obviously, mode identification is not a problem with full-wavefield inversion. However, for full-wavefield inversion, we face a huge system with several thousand unknowns if we have a few or no constraints on an earth model. The most challenging problems with full-waveform inversion are how to determine a model update efficiently and how to maintain stability of the inversion effectively.

Mapping specific near-surface features or anomalies such as faults and voids using the MASW method remains a challenge to the surface-wave research community. Current work is limited to simple models. Surface-wave diffraction (Xia et al., 2007a) and scattering (Blonk and Herman, 1994) have potential applications to several problems troubling the near-surface geophysics community. A few real-world examples (Xia et al., 2007a; Putnam

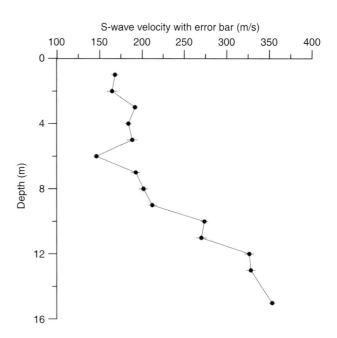

**Figure 11.** A trade-off model from the San Jose, California, data set that is obtained using a regularization parameter of 0.00559 in the last iteration of the model labeled "Selected data" (Figure 9). Error bars were computed using equation 19, which were calculated on the basis of an assumption of uncorrelated errors in data. Resolution and variance of the inverted model are shown in Table 1 (Xia et al., 2010b).

et al., 2008) show that it is possible to detect anomalous bodies at shallow depths using surface-wave diffraction. Physical modeling of surface-wave responses in various geologic settings needs to be done to better understand and verify phenomena observed in real and numerical modeling data and to establish the limitations of the MASW method.

Studies on limitations of the MASW method in mapping near-surface features have been reported in the literature. For example, Luo et al. (2009c) demonstrated, using synthetic and real-world examples, that a dipping interface with a slope smaller than 15° can be mapped successfully by the separated fundamental mode, using a high-resolution linear Radon transform. Successful separation of surface-wave modes (Luo et al., 2009b) makes it possible to perform high-resolution S-wave profiling with the MASW method. Small-scale geologic features can be detected and in some cases delineated, using the mode-separation technique (Luo et al., 2009b).

Group velocities may possess higher resolution in the $f$-$v$ domain than do the phase velocities, especially at low frequencies. Group velocities of high-frequency Rayleigh waves can be determined using significantly fewer traces (Luo et al., 2010a) than are needed to determine conventional phase velocities. Fewer traces mean less horizontal smearing at shallow depth, thereby increasing horizontal resolution of S-wave profiling. Mode identification, however, may be more problematic with group-velocity data than with phase-velocity data. Slant stacking (Xia et al., 2007b) has the potential to generate a dispersive-energy image from data acquired using an arbitrary acquisition geometry (for example, a fan configuration), and that lays the foundation for true 3D surface-wave analysis using dispersion curves. Research results on 3D surface-wave techniques such as tomography have been presented in the refereed literature (e.g., Long and Kocaoglu, 2001; Chevrot and Zhao, 2007; Masterlark et al., 2010).

A very limited number of publications on high-frequency Love waves exist for near-surface geophysics applications (Steeples, 2005). This may result from the difficulty in acquiring broadband SH-wave data. Advantages of extracting SH-wave data from Love-wave analysis are obvious and include sharper images of dispersive energy, simpler dispersion curves, and, therefore, more stable inversions (Xia et al., 2009; Xia et al., 2010a; Luo et al., 2010b). Numerical-modeling results also suggest that Love waves are more sensitive to changes in S-wave velocities than are Rayleigh waves at most frequencies (0–100 Hz) (Zeng et al., 2007). Eslick et al. (2008) concluded that a minimum thickness of 1 m of low-velocity material in the near-surface layer is necessary to generate and record usable Love-wave data in a frequency range that is appropriate for near-surface investigations (5–50 Hz). Joint inversion of Rayleigh and Love waves could increase the reliability of S-wave velocity estimates in some geologic settings (Misiek et al., 1997), if objective functions can be determined properly. In certain settings, compared with Rayleigh waves, Love waves may produce poorer resolution in deeper depths but better resolution for the first few meters (e.g., Calderon-Macias and Simmons, 2008). Separate analysis of high-frequency Rayleigh waves and Love waves at the same site could provide a valuable tool for studying near-surface anisotropy.

# Conclusions

Inverting high-frequency Rayleigh-wave dispersion data by the L-M method and the SVD technique provides reliable near-surface S-wave velocities because the inversion system is normally small (usually less than 30 unknowns). Inversion systems of a layered model or the Gibson half-space for S-wave velocities are numerically stable. Accuracy can be increased by simultaneously inverting the fundamental mode with higher modes. An inversion system using multimodes can produce an S-wave velocity model with even higher resolution. It is feasible to determine near-surface quality factors $Q_S$ and/or $Q_P$, depending on a $V_S/V_P$ ratio, from Rayleigh-wave attenuation coefficients with a constraint on Poisson's ratio. Assessing regulated inverse results is possible using a proper regularization parameter determined by the trade-off between model resolution and model covariance matrices.

# Acknowledgments

The authors appreciate Matt Haney, Barbara Luke, Carlos Calderón-Macías, and an anonymous reviewer for their detailed comments, constructive suggestions, and numerous questions, which improved the paper. The authors thank Marla Adkins-Heljeson of Kansas Geological Survey for editing the manuscript.

# References

Abbiss, C. P., 1981, Shear wave measurements of the elasticity of the ground: Géotechnique, **31**, no. 1, 91–104.

Aki, K., and P. G. Richards, 1980, Quantitative seismology: W. H. Freeman and Company.

Anderson, D. L., A. Ben-Menahem, and C. B. Archambeau, 1965, Attenuation of seismic energy in upper mantle: Journal of Geophysical Research, **70**, 1441–1448.

Babuska, V., and M. Cara, 1991, Seismic anisotropy in the earth: Kluwer Academic Publishers.

Backus, G. E., and J. F. Gilbert, 1967, Numerical applications of a formalism for geophysical inverse problems: Geophysical Journal of the Royal Astronomical Society, **13**, 247–276.

———, 1968, The resolving power of gross earth data: Geophysical Journal of the Royal Astronomical Society, **16**, 169–205.

———, 1970, Uniqueness in the inversion of gross earth data: Philosophical Transactions, Royal Society of London, Series A, **266**, 123–192.

Beaty, K. S., and D. R. Schmitt, 2003, Repeatability of multimode Rayleigh-wave dispersion studies: Geophysics, **68**, 782–790.

Beaty, K. S., D. R. Schmitt, and M. Sacchi, 2002, Simulated annealing inversion of multimode Rayleigh-wave dispersion curves for geological structure: Geophysical Journal International, **151**, 622–631.

Blonk, B., and G. C. Herman, 1994, Inverse scattering of surface waves: A new look at surface consistency: Geophysics, **59**, 963–972.

Calderón-Macías, C., and B. Luke, 2007, Addressing nonuniqueness in inversion of Rayleigh-wave data for shallow profiles containing stiff layers: Geophysics, **72**, no. 1, U1–U10.

Calderón-Macías, C., and J. Simmons, 2008, Constrained surface wave inversion from 9-component seismic reflection data: 78th Annual International Meeting, SEG, Expanded abstracts, 1063–1067.

Carmichael, R. S., 1989, Practical handbook of physical properties of rocks and minerals: CRC Press.

Cercato, M., 2009, Addressing non-uniqueness in linearized multichannel surface wave inversion: Geophysical Prospecting, **57**, 27–47.

Chen, C., J. Liu, J. Xia, and Z. Li, 2006, Integrated geophysical techniques in detecting hidden dangers in river embankments: Journal of Environmental and Engineering Geophysics, **11**, no. 2, 83–94.

Chevrot, S., and L. Zhao, 2007, Multiscale finite-frequency Rayleigh wave tomography of the Kaapvaal craton: Geophysical Journal International, **169**, 201–215.

Clayton, C. R. I., 1993, The standard penetration test (SPT): Methods and use: Construction Industry Research and Information Association: Funder Report CP/7.

Clayton, C. R. I., M. C. Matthews, and N. E. Simons, 1995, Site investigation: Blackwell Science.

Dikmen, U., 2009, Statistical correlations of shear wave velocity and penetration resistance for soils: Journal of Geophysics and Engineering, **6**, 61–72.

Dobry, R., R. D. Borcherdt, C. B. Crouse, I. M. Idriss, W. B. Joyner, G. R., Martin, M. S. Power, E. E. Rinne, and R. B. Seed, 2000, New site coefficients and site classification system used in recent building seismic code provisions: Earthquake Spectra, **16**, 41–67.

Dorman, J., and M. Ewing, 1962, Numerical inversion of seismic surface wave dispersion data and crust-mantle structure in the New York–Pennsylvania area: Journal of Geophysical Research, **67**, 5227–5241.

Eslick, R., G. Tsoflias, and D. W. Steeples, 2008, Field investigation of Love waves in near-surface seismology: Geophysics, **73**, no. 3, G1–G6.

Foti, S., and C. Strobbia, 2002, Some notes on model parameters for surface wave data inversion: Symposium on the Application of Geophysics to Engineering and Environmental Problems (SAGEEP), CD-ROM.

Gibson, R. E., 1967, Some results concerning displacements and stresses in a non-homogeneous elastic half-space: Géotechnique, **17**, no. 1, 58–67.

Golub, G. H., and C. Reinsch, 1970, Singular value decomposition and least-squares solution: Numerische Mathematik, **14**, no. 5, 403–420.

Imai, T., and K. Tonouchi, 1982, Correlation of N-value with S-wave velocity: Proceedings of the Second European Symposium on Penetration Testing, 67–72.

Ivanov, J, R. D. Miller, P. Lacombe, C. D. Johnson, and J. W. Lane, Jr., 2006a, Delineating a shallow fault zone and dipping bedrock strata using multichannel analysis of surface waves with a land streamer: Geophysics, **71**, no. 5, A39–A42.

Ivanov, J., R. D. Miller, J. Xia, D. W. Steeples, and C. B. Park, 2006b, Joint analysis of refractions with surface waves: An inverse solution to the refraction-traveltime problem: Geophysics, **71**, no. 6, R131–R138.

Jeng, Y., J. Tsai, and S. Chen, 1999, An improved method of determining near-surface $Q$: Geophysics, **64**, 1608–1617.

Jin, S., G. Cambois, and C. Vuilermoz, 2000, Shear-wave velocity and density estimation from PS-wave AVO analysis: Application to an OBS dataset from the North Sea: Geophysics, **65**, 1446–1454.

Johnston, D. H., M. N. Toksöz, and A. Timur, 1979, Attenuation of seismic waves in dry and saturated rocks, II: Mechanisms: Geophysics, **44**, 691–711.

Kovach, R. L., 1965, Seismic surface waves: Some observations and recent developments: *in* L. H. Ahrens, F. Press, S. K. Runcorn, and H. C. Urey, eds., Physics and chemistry of the earth, **6**, Pergamon Press, 251–314.

Kudo, K., and E. Shima, 1970, Attenuation of shear wave in soil: Bulletin of the Earthquake Research Institute, **48**, 145–158.

Lay, T., and T. C. Wallace, 1995, Modern global seismology: International Geophysics Series, **58**, Academic Press.

Liang, Q., C. Chen, C. Zeng, Y. Luo, and Y. Xu, 2008, Inversion stability analysis of multimode Rayleigh wave dispersion curves using low-velocity-layer models: Near Surface Geophysics, **6**, no. 3, 157–165.

Long, L. T., and A. H. Kocaoglu, 2001, Surface-wave group-velocity tomography for shallow structures: Journal of Environmental and Engineering Geophysics, **6**, no. 2, 71–81.

Lu, L., C. Wang, and B. Zhang, 2007, Inversion of multimode Rayleigh waves in the presence of a low velocity layer: Numerical and laboratory study: Geophysical Journal International, **168**, 1235–1246.

Luo, Y., J. Xia, J. Liu, Q. Liu, and S. Xu, 2007, Joint inversion of high-frequency surface waves with fundamental and higher modes: Journal of Applied Geophysics, **62**, no. 4, 375–384.

Luo, Y., J. Xia, J. Liu, Y. Xu, and Q. Liu, 2009a, Research on the middle-of-receiver-spread assumption of the MASW method: Soil Dynamics and Earthquake Engineering, **29**, no. 1, 71–79.

Luo, Y., J. Xia, R. D. Miller, Y. Xu, J. Liu, and Q. Liu, 2008, Rayleigh-wave dispersive energy imaging by high-resolution linear Radon transform: Pure and Applied Geophysics, **165**, no. 5, 903–922.

———, 2009b, Rayleigh-wave mode separation by high-resolution linear Radon transform: Geophysical Journal International, **179**, no. 1, 254–264.

Luo, Y., J. Xia, Y. Xu, and C. Zeng, 2010a, Group velocity dispersion analysis of high-frequency Rayleigh waves for near-surface applications: Proceedings of the 4th International Conference on Environmental and Engineering Geophysics (ICEEG), Science Press USA Inc., 189–194.

Luo, Y., J. Xia, Y. Xu, C. Zeng, and J. Liu, 2010b, Finite-difference modeling and dispersion analysis of high-frequency Love waves for near-surface applications: Pure and Applied Geophysics, accessed 8 August 2010, doi: 10.1007/s00024-010-0144-7.

Luo, Y., J. Xia, Y. Xu, C. Zeng, R. D. Miller, and Q. Liu, 2009c, Dipping interface mapping using mode-separated Rayleigh waves: Pure and Applied Geophysics, **166**, no. 3, 353–374.

Marquardt, D. W., 1963, An algorithm for least squares estimation of nonlinear parameters: Journal of the Society for Industrial and Applied Mathematics, **2**, 431–441.

Masterlark, T., M. Haney, H. Dickinson, T. Fournier, and C. Searcy, 2010, Rheologic and structural controls on the deformation of Okmok volcano, Alaska: FEMs, InSAR, and ambient noise tomography: Journal of Geophysical Research, **115**, B02409, accessed 8 August 2010, doi: 10.1029/2009JB006324.

Matthews, M. C., V. S. Hope, and C. R. I. Clayton, 1996, The use of surface waves in the determination of ground stiffness profiles: Proceedings of the Institution of Civil Engineers — Geotechnical Engineering, **119**, 84–95.

Menke, W., 1984, Geophysical data analysis: Discrete inversion theory: Academic Press.

Miller, R. D., J. Xia, C. B. Park, and J. Ivanov, 1999, Multichannel analysis of surface waves to map bedrock: The Leading Edge, **18**, 1392–1396.

Minster, J. B., T. J. Jordan, P. Molnar, and E. Haines, 1974, Numerical modeling of instantaneous plate tectonics: Geophysical Journal of the Royal Astronomical Society, **36**, 541–576.

Misiek, R., A. Liebig, A. Gyulai, T. Ormos, M. Dobroka, and L. Dresen, 1997, A joint inversion algorithm to process geoelectric and surface wave seismic data, part II: Application: Geophysical Prospecting, **45**, 65–85.

Mitchell, B. J., 1975, Regional Rayleigh wave attenuation in North America: Journal of Geophysical Research, **80**, 4904–4916.

Nazarian, S., K. H. Stokoe II, and W. R. Hudson, 1983, Use of spectral analysis of surface waves method for determination of moduli and thicknesses of pavement systems: Transportation Research Record No. 930, 38–45.

Press, W. H., S. A. Teukosky, W. T. Vetterling, and B. P. Flannery, 1992, Numerical recipes in C, 2nd ed.: Press Syndicate of the University of Cambridge, New York.

Putnam, N., A. Nasseri-Moghaddam, O. Kovin, and N. Anderson, 2008, Preliminary analysis using surface wave methods to detect shallow manmade tunnels: Symposium on the Application of Geophysics to Environmental and Engineering Problems (SAGEEP), 679–688.

Rix, G. J., C. D. Lai, and A. W. Spang Jr., 2000, In situ measurement of damping ratio using surface waves: Journal of Geotechnical and Geoenvironmental Engineering, **126**, no. 5, 472–480.

Sabetta, F., and J. Bommer, 2002, Modification of the spectral shapes and subsoil conditions in Eurocode 8: 12th European Conference on Earthquake Engineering, Paper 518.

Scholte, J. C., 1947, The range of existence of Rayleigh and Stoneley waves: Monthly Notices of the Royal Astronomical Society, Geophysical Supplement 5, 120–126.

Schwab, F. A., and L. Knopoff, 1972, Fast surface wave and free mode computations; in B. A. Bolt, ed., Methods in computational physics: Academic Press, 87–180.

Sêco e Pinto, P. S., 2002, Eurocode 8 — Design provisions for geotechnical structures: Special Lecture, 3rd Croatian Soil Mechanics and Geotechnical Engineering Conference CD-ROM.

Sheriff, R. E., 2002, Encyclopedic dictionary of applied geophysics, 4th ed.: SEG Geophysical References Series No. 13.

Sheriff, R. E., and L. P. Geldart, 1983, Exploration seismology, volume 1: History, theory, and data acquisition: Cambridge University Press.

Socco, L. V., and C. Strobbia, 2004, Surface wave methods for near-surface characterisation: A tutorial: Near Surface Geophysics, **2**, 165–185.

Song, X., H. Gu, J. Liu, and X. Zhang, 2007, Estimation of shallow subsurface shear-wave velocity by inverting fundamental and higher-mode Rayleigh waves: Soil Dynamics and Earthquake Engineering, **27**, no. 7, 599–607.

Song, Y. Y., J. P. Castagna, R. A. Black, and R. W. Knapp, 1989, Sensitivity of near-surface shear-wave velocity determination from Rayleigh and Love waves: 59th Annual International Meeting, SEG, Expanded Abstracts, 509–512.

Steeples, D. W., 2005, Near-surface geophysics: 75 years of progress: Supplement to The Leading Edge, **24**, no. 1, S82–S85.

Stokoe, K. H. II, and S. Nazarian, 1983, Effectiveness of ground improvement from spectral analysis of surface waves: Proceeding of the Eighth European Conference on Soil Mechanics and Foundation Engineering, **1**, 91–95.

Stokoe, K. H. II, S. G. Wright, J. A. Bay, and J. M. Roësset, 1994, Characterization of geotechnical sites by SASW method, *in* R. W. Woods, ed., Geophysical characterization of sites: ISSMFE Technical Committee No. 10, Oxford & IBH Publishing, 15–25.

Tarantola, A., 1987, Inverse problem theory: Elsevier Science.

Tian, G., D. W. Steeples, J. Xia, R. D. Miller, K. T. Spikes, and M. D. Ralston, 2003a, Multichannel analysis of surface wave method with the autojuggie: Soil Dynamics and Earthquake Engineering, **23**, no. 3, 243–247.

Tian, G., D. W. Steeples, J. Xia, and K. T. Spikes, 2003b, Useful resorting in surface wave method with the Autojuggie: Geophysics, **68**, 1906–1908.

Toksöz, M. N., D. H. Johnston, and A. Timur, 1979, Attenuation of seismic waves in dry and saturated rocks: I. Laboratory measurements: Geophysics, **44**, 681–690.

Tokimatsu, K., S. Kuwayama, S. Tamura, and Y. Miyadera, 1991, Vs determination from steady state Rayleigh wave method: Soils and Foundations, **31**, no. 2, 153–163.

Vardoulakis, I., and C. Vrettos, 1988, Dispersion law of Rayleigh-type waves in a compressible Gibson half-space: International Journal for Numerical and Analytical Methods in Geomechanics, **12**, 639–655.

Wiggins, R. A., 1972, The general linear inverse problem: Implication of surface waves and free oscillations for Earth structure: Reviews of Geophysics and Space Physics, **10**, 251–285.

Xia, J., R. Cakir, R. D. Miller, C. Zeng, and Y. Luo, 2009, Estimation of near-surface shear-wave velocity by inversion of Love waves: 79th Annual International Meeting, SEG, Expanded Abstracts, 1390–1395.

Xia, J., C. Chen, P. H. Li, and M. J. Lewis, 2004, Delineation of a collapse feature in a noisy environment using a multichannel surface wave technique: Géotechnique, **54**, no. 1, 17–27.

Xia, J., R. D. Miller, R. Cakir, Y. Luo, Y. Xu, and C. Zeng, 2010a, Revisiting SH-wave data with Love-wave analysis: Symposium on the Application of Geophysics to Environmental and Engineering Problems (SAGEEP), 569–580.

Xia, J., R. D. Miller, and C. B. Park, 1999, Estimation of near-surface shear-wave velocity by inversion of Rayleigh wave: Geophysics, **64**, 691–700.

———, 2000, Advantage of calculating shear-wave velocity from surface waves with higher modes: 70th Annual International Meeting, SEG, Expanded Abstracts, 1295–1298.

Xia, J., R. D. Miller, C. B. Park, J. A. Hunter, J. B. Harris, and J. Ivanov, 2002a, Comparing shear-wave velocity profiles from multichannel analysis of surface wave with borehole measurements: Soil Dynamics and Earthquake Engineering, **22**, no. 3, 181–190.

Xia, J., R. D. Miller, C. B. Park, and G. Tian, 2002b, Determining *Q* of near-surface materials from Rayleigh waves: Journal of Applied Geophysics, **51**, nos. 2–4, 121–129.

———, 2003, Inversion of high frequency surface waves with fundamental and higher modes: Journal of Applied Geophysics, **52**, no. 1, 45–57.

Xia, J., R. D. Miller, C. B. Park, E. Wightman, and R. Nigbor, 2002c, A pitfall in shallow shear-wave refraction surveying: Journal of Applied Geophysics, **51**, no. 1, 1–9.

Xia, J., R. D. Miller, and Y. Xu, 2008, Data-resolution matrix and model-resolution matrix for Rayleigh-wave inversion using a damped least-square method: Pure and Applied Geophysics, **165**, no. 7, 1227–1248.

Xia, J., J. E. Nyquist, Y. Xu, M. J. S. Roth, and R. D. Miller, 2007a, Feasibility of detecting near-surface feature with Rayleigh-wave diffraction: Journal of Applied Geophysics, **62**, no. 3, 244–253.

Xia, J., Y. Xu, C. Chen, R. D. Kaufmann, and Y. Luo, 2006a, Simple equations guide high-frequency surface-wave investigation techniques: Soil Dynamics and Earthquake Engineering, **26**, no. 5, 395–403.

Xia, J., Y. Xu, and R. D. Miller, 2007b, Generating image of dispersive energy by frequency decomposition and slant stacking: Pure and Applied Geophysics, **164**, no. 5, 941–956.

Xia, J., Y. Xu, R. D. Miller, and C. Chen, 2006b, Estimation of elastic moduli in a compressible Gibson half-space by inverting Rayleigh wave phase velocity: Surveys in Geophysics, **27**, no. 1, 1–17.

Xia, J., Y. Xu, R. D. Miller, and C. Zeng, 2010b, A trade-off solution between model resolution and covariance in surface-wave inversion: Pure and Applied Geophysics, accessed 8 August 2010, doi: 10.1007/s00024-010-0107-z.

Xu, Y., J. Xia, and R. D. Miller, 2006, Quantitative estimation of minimum offset for multichannel surface-wave survey with actively exciting source: Journal of Applied Geophysics, **59,** no. 2, 117–125.

———, 2007, Numerical investigation of implementation of air-earth boundary by acoustic-elastic boundary approach: Geophysics, **72,** no. 5, SM147–SM153.

————, 2009, Approximation to cutoffs of higher modes of Rayleigh waves for a layered earth model: Pure and Applied Geophysics, **166**, no. 3, 339–351.

Yilmaz, Ö., M. Eser, and M. Berilgen, 2009, Applications of engineering seismology for site characterization: Journal of Earth Science, **20,** no. 3, 546–554.

Zeng, C., J. Xia, Q. Liang, and C. Chen, 2007, Comparative analysis on sensitivities of Love and Rayleigh waves: 77th Annual International Meeting, SEG, Expanded Abstracts, 1138–1141.

Zhang, S. X., and L. S. Chan, 2003, Possible effects of misidentified mode number on Rayleigh wave inversion: Journal of Applied Geophysics, **53**, 17–29.

Zhdanov, M. S., and, E. Tolstaya, 2006, A novel approach to the model appraisal and resolution analysis of regularized geophysical inversion: Geophysics, **71**, no. 6, R79–R90.

Chapter 3

# Investigation and Use of Surface-wave Characteristics for Near-surface Applications

Yixian Xu[1,2], Yinhe Luo[1], Qing Liang[1], Liming Wang[1], Xianhai Song[1,3], Jiangping Liu[1], Chao Chen[1], and Hanming Gu[1]

## Abstract

High-frequency surface-wave methods can provide reliable near-surface shear-wave (S-wave) velocity, which is a key parameter in many shallow-engineering applications, groundwater and environmental studies, and petroleum exploration. Recent research and key accomplishments at the China University of Geosciences at Wuhan into near-field effects on surface-wave analysis provide not only insight into minimum-source geophone offsets required for generating high-quality surface-wave images but also provide a better understanding of the propagation characteristics of seismic wavefields through near-surface materials. New numerical modeling and dispersion-analysis algorithms are key tools used routinely in those studies. The modeling results illustrate very different energy-partitioning characteristics for Rayleigh and Love waves. Using a high-resolution linear Radon transform produces dispersion images with much better resolution and therefore represents a tool for more accurate separation and determination of phase velocities for different modes. Mode separation results in wavefield components that individually possess great potential for increasing horizontal resolution of S-wave velocity-field determinations. Amplitude corrections can significantly improve the accuracy of phase-velocity estimates from mixed-modal wavefields. Results from two simple models demonstrate how dramatic topographic changes can distort wavefields. This finding was the catalyst for suggesting that a topographic correction should be considered for surface-wave data acquired on a rugged ground surface. Phase-velocity inversion is an ill-posed problem. Rayleigh-wave sensitivity analysis reveals the difficulty in estimating S-wave velocities for a model with a low-velocity layer. Constraints in the model space are therefore necessary. Approximating cutoffs could help build a better initial model and provide critical information about the subsurface when higher modes are present.

## Introduction

From a view outside the United States, there are two research groups in the United States that have made great contributions in developing surface-wave techniques for the near-surface community in the past 25 years. A research group at the University of Texas, Austin, developed spectral analysis of surface waves (SASW) and was the first to use surface waves in civil engineering (e.g., Nazarian, 1984; Nazarian and Stokoe, 1986; Sheu et al., 1988; Stokoe et al., 1994). The research group at Kansas Geological Survey at the University of Kansas (KGS/KU) developed another method, multichannel analysis of surface waves (MASW), which estimates shear-wave velocity based on surface-wave arrivals from a single wave train at multiple receiver locations, thereby improving the resolution and accuracy of surface-wave applications (e.g., Song et al., 1989; Miller et al., 1999; Park et al., 1999; Xia et al., 1999).

Essentially, the MASW method changes the concept of using surface-wave energy from a single mode to multimodes, which has made quantitative analysis and accurate

---

[1]*Institute of Geophysics and Geomatics, China University of Geosciences, Wuhan, Hubei, China.*
[2]*State Key Laboratory of Geological Processes and Mineral Resources, China University of Geosciences, Wuhan, Hubei, China.*
[3]*Changjiang River Scientific Research Institute, Wuhan, Hubei, China.*

inversion possible. The MASW method allows the use of advanced technologies in digital signal processing for multichannel records (Dal Moro et al., 2003; Xia et al., 2007b) and therefore takes advantage of time-space information of seismic wavefields (Forbriger, 2003; Zhang and Chan, 2003; Zhang et al., 2004; Xia et al., 2006; Luo et al., 2008). In addition, the method can be more flexible in discriminating and integrating fundamental and higher modes of Rayleigh waves (Xia et al., 2003). These are demonstrated by numerous applications of MASW found in the geosciences and engineering literature in the last decade.

Surface-wave methods first were applied to exploration programs for route expansion of railways in China using GR-810 automatic exploration machines made by the Japanese VIC company in the mid-1980s (e.g., Huang et al., 1991; Guan et al., 1993). As civil-engineering needs increased rapidly after 1993, SASW was used widely in China to estimate surface-wave (S-wave) velocity of near-surface materials. MASW has become the more popular tool in engineering applications since the beginning of the century. This paper aims to provide a scope of contributions to research and applications of surface waves made by the near-surface geophysical community in China, especially research conducted in the last six years at China University of Geosciences (CUG), Wuhan. As a result of the close cooperation between CUG and KU since 2004, this paper will complement those written by researchers at KU.

## Tools for Analysis of Characteristics of Surface-wave Propagation

### Simulation of surface-wave propagation (P-SV and SH systems)

Despite the reflectivity method (e.g., Kennett, 2001) often used in seismology and the finite-element method commonly used in civil-engineering communities (e.g.,

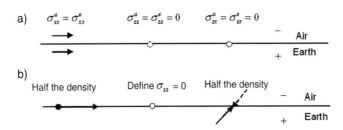

**Figure 1.** Cartoon illustrating (a) air/earth boundary conditions and (b) computing implementations of air/earth boundary conditions.

Gucunski and Woods, 1992), the staggered-grid finite-difference method in the time domain (FDTD) is most popular in exploration seismology because of its flexible implementation and balance between efficiency and accuracy (e.g., Virieux, 1984, 1986; Levander, 1988; Graves, 1996; Virieux and Operto, 2009).

Because modeling of Rayleigh-wave propagation is affected seriously by implementation of the air/earth boundary (Mittet, 2002; Bohlen and Saenger, 2006), one needs to pay special attention to this aspect. Although many schemes for implementing the air/earth boundary have existed for some time (e.g., vacuum formalism or heterogeneous approximation [e.g., Zahradník and Priolo, 1995], the stress image method [SIM] [Levander, 1988; Robertsson, 1996], and the transversely isotropic approach [Mittet, 2002]), the acoustic/elastic interface approach (AEA) seems more effective at characterizing the dispersive Rayleigh waves without losing body-wave accuracy (Xu et al., 2007).

The elastodynamic equation of motion in an isotropic media on staggered grids can be written using H-formulation (Kristek et al., 2002):

$$
\begin{cases}
\dot{v}_x = \rho_x^{-1}[D_x^+ \sigma_{xx} + D_z^- \sigma_{zx}], & (1) \\
\dot{v}_z = \rho_z^{-1}[D_z^+ \sigma_{zz} + D_x^- \sigma_{zx}], & (2) \\
\dot{\sigma}_{xx} = (\lambda + 2\mu)D_x^- v_x + \lambda D_z^- v_z, & (3) \\
\dot{\sigma}_{zz} = (\lambda + 2\mu)D_z^- v_z + \lambda D_x^- v_x, & (4) \\
\dot{\sigma}_{zx} = \mu_{zx}(D_x^+ v_x + D_z^+ v_z), & (5)
\end{cases}
$$

where $D_i^+$ and $D_i^-$ represent forward and backward spatial-differencing operators, respectively, in the $i$-direction. Other material properties are represented by conventional symbol nomenclature. By our not thinking of the free surface as "free," at least in principle, the air/earth boundary can be viewed as an acoustic/elastic interface. This allows energy transmission to continue across the interface. This requirement can be described as (Figure 1a)

$$
\sigma_{zz}^a = \sigma_{zz}^e = 0, \quad \sigma_{xz}^a = \sigma_{xz}^e = 0, \quad \text{and} \quad \sigma_{xx}^a = \sigma_{xx}^e.
$$

Superscripts $a$ and $e$ represent the air and the earth, respectively. These air/earth boundary conditions can be implemented simply by keeping $2\mu$ unchanged and reducing the density on the air/earth interface by half, as illustrated in Figure 1b. Because harmonic averaging can be applied to the Lamé constant lambda (Moczo et al., 2002), one can choose lambda on the interface to be approximately the same as its value in air (acoustic media). The scheme was named the acoustic/elastic interface approach (AEA).

Updating the wavefield components at time $t + \Delta t$ relative to time $t$ at the air/earth interface can be written therefore as

$$\dot{v}_x = 2\rho_x^{-1}[D_x^+ \sigma_{xx} + D_z^- \sigma_{zx}]. \tag{6}$$

$$\dot{v}_z = 2\rho_z^{-1}[D_z^+ \sigma_{zz} + D_x^- \sigma_{zx}]. \tag{7}$$

$$\sigma_{xx} = 2\mu \int_t^{t+\Delta t} D_x^- v_x \, dt. \tag{8}$$

$$\sigma_{zz} = 2\mu \int_t^{t+\Delta t} D_z^- v_z \, dt. \tag{9}$$

$$\sigma_{zx} = \mu_{zx} \int_t^{t+\Delta t} (D_x^+ v_z + D_z^+ v_x) \, dt. \tag{10}$$

The advantages of AEA over other schemes are discussed and illustrated using numerical comparisons in Xu et al. (2007). This implementation differs from the SIM by the requirement of continuous normal horizontal stress that is incorporated explicitly into the AEA method. SIM implementations implicitly assume that the normal horizontal stress vanishes in the air layer. Even though AEA is a minor modification to Mittet's method (Mittet, 2002), key advantages include that it is physically reasonable and easily programmable and it effectively reduces numerical dispersion that is important for modeling wave propagation. The AEA method also was used to implement free-surface conditions for Love-wave modeling. Examples of P-SV and SH modeling will be shown in the following sections.

The technical details of the FD method for seismic wave propagation, e.g., the grid, the equation system, non-reflection boundary conditions, numerical dispersion, stability, and so forth, are all important for accurate modeling. Several publications address those aspects (e.g., Graves, 1996, and references therein). A comprehensive summary written by Moczo et al. (2007) can be used as a manual for experienced researchers.

## Dispersion images and analysis

Although some researchers have made efforts to separate Rayleigh modes for use in the SASW method, e.g., a multifilter technique developed by Karray and Lefebvre (2009), mode information of Rayleigh-wave data is very difficult to identify from the wavefield alone. That is because the modes are strongly dependent on earth-layer composition and structure (including geometric, Poisson's ratio, and S-wave contrasts between layers, and so forth), source-frequency spectrum, and depth. Those aspects all

relate to spatial sampling configuration, e.g., a spread length and a receiver interval, and so forth (Zhang et al., 2004; Xia et al., 2006; Luo et al., 2009a). It is possible that this drawback of the SASW method could be overcome by incorporating energy terms into the period equation. In spite of difficulty in distinguishing modes, the SASW method remains attractive because of its ease in implementation.

Generating a high-resolution image of dispersive energy in the frequency-velocity (*f-v*) domain is a principal task in the MASW method. Four algorithms are available to calculate high-frequency dispersive energy images: the *f-k* transformation (e.g., Yilmaz, 1987), the tau-p transform (McMechan and Yedlin, 1981), the phase-shift method (Park et al., 1998), and the frequency-decomposition and slant-stacking approach (Xia et al., 2007b). Dal Moro et al. (2003) evaluated the effectiveness of the first three of the four above-mentioned algorithm types and concluded that the phase-shift approach was the best in terms of performance sensitivity relative to data processing and the number of traces. The fourth algorithm, developed by Xia et al. (2007b), consists of two steps: (1) stretching data into pseudo-vibroseis data or frequency-swept data and (2) slant-stacking frequency-swept data. The main contribution of this fourth algorithm is its ability to produce dispersion images from seismic data acquired in an arbitrary acquisition geometry.

Recently, Luo et al. (2008) pointed out that all four schemes are derivatives of the standard discretized linear Radon transform (LRT). It is well known that standard LRT suffers from the loss of resolution and spatial aliasing as a consequence of incomplete information, including limited aperture and discretization (Trad et al., 2003). Hence, none of the previously stated algorithms can produce true high-resolution dispersion images of surface waves.

### High-resolution LRT and mode separation

The forward LRT can be written in a matrix form as

$$\mathbf{d} = \mathbf{L}\mathbf{m} \tag{11}$$

and the adjoint transformation as

$$\mathbf{m}_{\text{adj}} = \mathbf{L}^{\text{T}}\mathbf{d}, \tag{12}$$

where $\mathbf{L} = e^{i2\pi f p x}$ is the forward LRT operator; $\mathbf{d}$ and $\mathbf{m}$ represent a shot gather and a Radon panel, respectively; and $\mathbf{m}_{\text{adj}}$ denotes a low-resolution Radon panel using the transpose or adjoint operator $\mathbf{L}^{\text{T}}$. It is clear that $\mathbf{L}^{\text{T}}$ does not define the inverse of $\mathbf{L}$ because $\mathbf{L}$ is not a unitary matrix.

Sparsity constraint should be considered when finding a model $\mathbf{m}$ that best fits the data while minimizing the number of model-space parameters necessary to represent the data in the Radon domain. The model $\mathbf{m}$ can be found by solving the following equation (Trad et al., 2002, 2003):

$$(\mathbf{W}_m^{-T}\mathbf{L}^T\mathbf{W}_d^T\mathbf{W}_d\mathbf{L}\mathbf{W}_m^{-1} + \lambda\mathbf{I})\mathbf{W}_m\mathbf{m} = \mathbf{W}_m^{-T}\mathbf{L}^T\mathbf{W}_d^T\mathbf{W}_d\mathbf{d},$$
$$(13)$$

where $\mathbf{I}$ denotes the identity matrix; $\mathbf{W}_d$ is a matrix of data weights; $\mathbf{W}_m$ is a matrix of model weights (model weights play an extremely important role in the design of high-resolution Radon operators, affecting both resolution and smoothness); $\mathbf{W}_d^{-1}$ and $\mathbf{W}_m^{-1}$ are the inverses of $\mathbf{W}_d$ and $\mathbf{W}_m$, respectively; $\mathbf{W}_d^{-T}$ and $\mathbf{W}_m^{-T}$ are the transpose matrixes of $\mathbf{W}_d^{-1}$ and $\mathbf{W}_m^{-1}$, respectively; and finally, $\lambda$ is the regularization parameter which maintains balance between data misfit and model constraints.

Formula 13 can be solved efficiently using the conjugate-gradient (CG) algorithm. Details of that algorithm's strategy can be found in several publications (e.g., Sacchi and Porsani, 1999; Trad et al., 2002, 2003). On the $i$th iteration of the CG algorithm, the weighting and preconditioning matrixes $\mathbf{W}_d$ and $\mathbf{W}_m$ are determined by using diagonal matrices $\mathrm{diag}(\mathbf{W}_d)_i = |\mathbf{r}_i|^{-1/2}$ and $\mathrm{diag}(\mathbf{W}_m)_i = |\mathbf{m}_i|^{1/2}$, where $\mathbf{m}_i$ is the Radon model at the $i$th iteration of the CG algorithm which is set initially as the identity vector, and $\mathbf{r}_i$ is the standard deviation of residual $\mathbf{r} = \mathbf{L}\mathbf{W}_m^{-1}\mathbf{W}_m\mathbf{m} - \mathbf{d}$ (Herrmann et al., 2000; Trad et al., 2002, 2003; Ji, 2006). Rayleigh waves are generally dominant on shot gathers, and high-resolution LRT is a robust method of imaging dispersive energy.

Generating dispersive energy images using high-resolution LRT is straightforward. First the shot gather is transformed from the time domain to the frequency domain. Then the high-resolution LRT is applied to each frequency slice, with the associated Radon panel transformed from the frequency-slowness domain to the $f$-$v$ domain using a linear interpolation operation. Rayleigh-wave mode separation by high-resolution LRT requires three additional steps: (1) select modes, (2) transform the different modes back to the $t$-$x$ domain, and (3) apply the high-resolution LRT again to transform shot gathers to the $f$-$v$ domain (Luo et al., 2008). Examples of dispersion images generated by high-resolution LRT are shown in the following sections.

## Amplitude effects in dispersion analysis

Some seismologists pay considerable attention to how to correctly estimate structure phase velocity from surface waves (e.g., Wielandt, 1993; Friederich et al., 2000; Bodin and Maupin, 2008). Generally speaking, the structure phase velocity for surface waves at a given point in a medium is the wavefront velocity of a plane and local single-mode wave in a laterally homogeneous area around that point, whereas dynamic phase velocity is defined as the local phase velocity of a wavefield (complex wave trains) (Wielandt, 1993). Wielandt (1993) demonstrates that the horizontal coordinates of single-mode seismic surface waves in a laterally homogeneous half-space obey the Helmholtz equation, and the Helmholtz equation remains valid in acoustic media with nonuniform compressibility and density.

The difference between the structure phase velocity and dynamic phase velocity is based on how well the eikonal equation approximates the acoustic equation. In fact, as Friederich et al. (2000) point out, surface waves in a laterally heterogeneous half-space do not obey the Helmholtz equation even when we consider only one mode at a time. However, a local mode of surface waves can be restored in the Helmholtz equation if mode conversions and the directivity of scattered waves from a point heterogeneity are neglected. That unexpected energy, which generally is thought to be small, can distort the analysis of dispersion curves dramatically (Friederich et al., 2000; Bodin and Maupin, 2008). It is therefore very important to impose some statistical and/or a priori constraints to prevent biases and discontinuities in the calculated dispersion curve.

The structure phase velocity in the 1D case (horizontal coordinate) is determined uniquely by three sample points of a wavefield (Wielandt, 1993). Active-source applications of the MASW method generally involve small inline and dense sampling arrays. The recorded wavefield (shot gather) will contain the direct, reflected, refracted, and diffracted waves. Therefore, the phase velocity estimated from MASW dispersion analysis is not the structure phase velocity because the phase velocity determined from dispersive analysis of a shot gather (entire wave train) is in fact the dynamic phase velocity. The key question that needs to be resolved is whether the amplitude effects described by the structure phase velocity exist in the widely used MASW method (Xu et al., 2008).

The wavefield can be described for any monotone plane wave propagating in a homogeneous acoustic medium by the Helmholtz equation:

$$(\nabla^2 + k^2)F = 0, \qquad (14)$$

where $\ln F(x) = a(x) + i\phi(x)$ represents 1D wavefield propagation along the $x$-direction; $a(x)$ and $\phi(x)$ are amplitude and phase, respectively; and $k$ is the structure wavenumber, which relates to continuous wavefields as

$$k^2 = -F^{-1}\nabla^2 F. \tag{15}$$

On the other hand, the dynamic wavenumber vector $\mathbf{w}$ is obtained by calculating the negative spatial gradient of the wavefield's phase

$$\mathbf{w} = -\nabla(\text{Im} \ln F) = -\nabla\phi. \tag{16}$$

It is obvious that $w = |\mathbf{w}|$ is generally different from structure wavenumber $k$, but each can be converted into the other when the distribution of amplitudes around the observation point is known (Friederich et al., 2000):

$$k^2 = w^2 - (\nabla a)^2 - \nabla^2 a. \tag{17}$$

Then the structure phase velocity $c$ is calculated by the modified classical equation with an amplitude-correction term $\phi_a$ as

$$c = \frac{2\pi}{T\sqrt{\|\nabla\phi\|^2 - \phi_a}}, \quad \phi_a = (\nabla a)^2 + \nabla^2 a, \tag{18}$$

where $T$ is the period of the Rayleigh wave.

Two models (M1 and M2 in Table 1) were used to demonstrate the theory. M1 is a step model with layer velocities increasing incrementally with depth, and M2 is a model modification of M1, making the third layer a low-velocity layer. The shot gathers have a 1-m receiver interval and were generated using the AEA scheme with a staggered-grid time-domain finite-difference algorithm

(Xu et al., 2007). To avoid near-field effects (Xu et al., 2006), the near-source traces are excluded from dispersion analysis (Figures 2b and 3b). The dispersion curves were calculated using the Xu et al. (2008) method.

First the fundamental-mode Rayleigh wave is extracted from the shot gather using mode separation (Luo et al., 2009b). The phase velocities (Figures 2 and 3) are calculated directly on shot gathers from M1 and M2 (Figures 2b and 3b) and include amplitude correction (Figures 2c and 3c). The phase velocities of the single fundamental mode (Figures 2d and 3d) and the single fundamental mode with amplitude correction (Figure 2e and 3e) are calculated from the separated single fundamental energy.

To avoid numerical errors, frequency points are excluded from the calculation when their amplitude values are less than 10% of the maximum amplitude. Averaged deviations from the true model for M1 in Figures 2 and 3 are (b) 26.91, (c) 25.82, (d) 22.18, and (e) 21.54 m/s, respectively, and for M2, they are (b) 84.64, (c) 75.23, (d) 14.70, and (e) 14.71 m/s, respectively. It is easy to see (Figure 3b and c) that phase velocities estimated directly from the total wavefield for traces 10 through 13 and 35 through 41 possess severe distortion. Those phenomena result from strong local interference between existing and new joined modes, e.g., the leaky mode waves for M2.

Those numerical results have shown that the phase velocity estimated from the whole wavefield can be improved greatly by amplitude correction. Those results also suggest that amplitude corrections might not need to be done exclusively from phase velocities estimated by using a single fundamental mode.

**Table 1.** Models and parameters used in the text.

| M1 | Velocity step increasing model | $V_P$ = 800, 1200, 1400, 1600, 1800, 2000, 2200, 2400 m/s; $V_S$ = 200, 400, 500, 650, 800, 950, 1100, 1250 m/s; $\rho$ = 2000 kg/m³; thickness = 2 m for every layer except for half-space. | | | |
|---|---|---|---|---|---|
| M2 | LVL model | $V_P$ = 800, 1200, 800, 1600, 1800, 2000, 2200, 2400 m/s; $V_S$ = 200, 400, 200, 650, 800, 950, 1100, 1250 m/s; $\rho$ = 2000 kg/m³; thickness = 2 m for every layer except for half-space. | | | |
| M3 | Two-layer model | $V_P$ = 800, 1200 m/s; $V_S$ = 200, 400 m/s; $\rho$ = 2000 kg/m³; thickness of the top layer = 10 m. | | | |
| M4 | Half-space model with different Poisson's ratio | $V_P$ = 1000 m/s; $\rho$ = 1800 kg/m³. | | | |
| | | **Poisson's ratio** | **Spatial step (m)** | **Time step (µs)** | **Receiver interval (m)** |
| | | 0.25 | 0.53 | 100 | 1.06 |
| | | 0.4 | 0.384 | 100 | 0.768 |
| | | 0.48 | 0.186 | 50 | 0.372 |
| M5 | Two-layer model | $V_P$ = 800, 1200 m/s; Poisson's ratio 0.25, 0.4, 0.48, respectively, for every layer; $\rho$ = 2000 kg/m³; thickness of the top layer = 10 m. | | | |

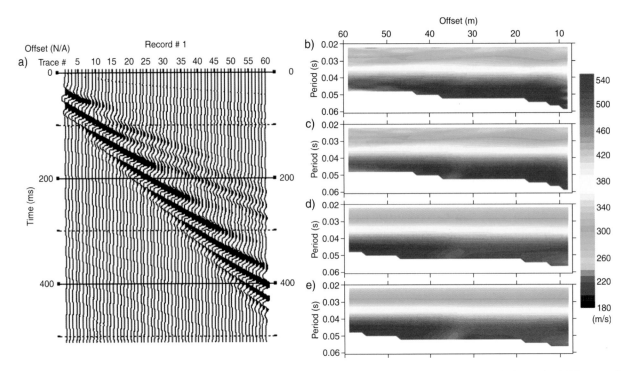

**Figure 2.** (a) Synthetic shot gather of M1 (Table 1) using the AEA algorithm (Xu et al., 2007). (b through e) Phase-velocity profiles estimated from the shot gather with different methods: (b) estimated directly from the shot gather, (c) estimated directly from the shot gather with amplitude corrected, (d) estimated from the separated single fundamental mode, and (e) estimated from the separated single fundamental mode with amplitude corrected.

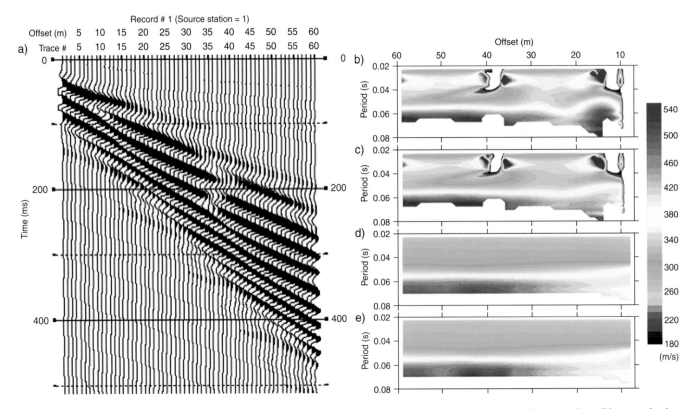

**Figure 3.** (a) Synthetic shot gather of M2 (Table 1) using the AEA algorithm (Xu et al., 2007). (b through e) Phase-velocity profiles estimated from the shot gather with different methods: (b) estimated directly from the shot gather, (c) estimated directly from the shot gather with amplitude corrected, (d) estimated from the separated single fundamental mode, and (e) estimated from the separated single fundamental mode with amplitude corrected.

# Surface-wave Excitement and Acquisition

## Partition of surface-wave energy

It is enlightening to compare Rayleigh- and Love-wave propagation characteristics using the same layered half-space model. For model M3 (Table 1), the shot gathers (Figures 4a and 5a) are based on P-SV and SH systems, which were simulated using the AEA developed by Xu et al. (2007). The most evident difference in dispersion images of Rayleigh and Love waves (Figures 4b and 5b, respectively) is energy partition of fundamental and higher modes: The majority of energy is in the fundamental mode of Rayleigh waves, whereas the energy difference is reduced notably between the fundamental and higher modes for Love waves. Those partitioning characteristics might result in different constructive interference mechanisms for Rayleigh and Love waves.

In addition, this phenomenon provides a potential explanation for how the standard SASW method, which uses only dispersive Rayleigh-wave energy, generally gives S-wave velocities that are consistent with conventional testing methods. Because most near-surface settings are defined by a soft soil layer underlain by rock, the experimental dispersion curve extracted from a wavefield image

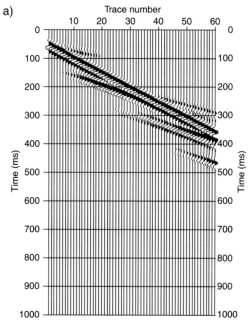

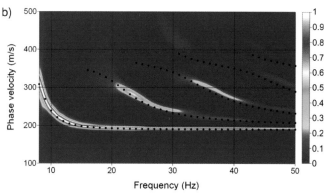

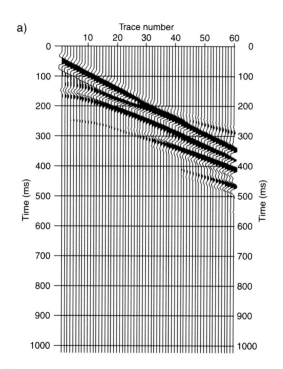

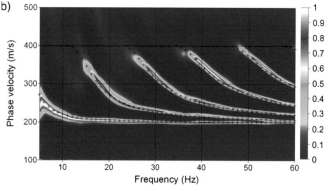

**Figure 4.** (a) Synthetic shot gather of M3 (Table 1) using the AEA algorithm (Xu et al., 2007). (b) Image of Rayleigh-wave dispersive energy in the *f-v* domain generated by high-resolution LRT (Luo et al., 2008). Black dots indicate analytical results calculated using the Knopoff method (Schwab and Knopoff, 1972).

**Figure 5.** (a) Synthetic shot gather of M3 (Table 1) using the AEA algorithm (Xu et al., 2007). (b) Image of Love-wave dispersive energy in the *f-v* domain generated by high-resolution LRT (Luo et al., 2008). Plus symbols represent analytical results calculated using the Knopoff method (Schwab and Knopoff, 1972).

generally will match the fundamental Rayleigh waves in spite of the fact that higher modes can interfere with the fundamental Rayleigh mode (Tokimatsu et al., 1992).

## Near-field effects

At short source-to-spread distances, dispersion images calculated using either MASW or SASW will be distorted. This situation often is referred to as near-field effects. It results from various physical effects, including but not limited to nonlinear deformation (e.g., plastic deformation in the case of soil evaluation), strong body-wave interference (Sanchez-Salinero, 1987), a leaking mode from the PL wave and equivalent counterpart (Oliver and Major, 1960; Roth and Holliger, 2000), and nonplane-wave propagation (Xu et al., 2006).

Zhang et al. (2004) demonstrate that the maximum velocity difference within an energy band is inversely proportional to frequency. That suggests that greater errors will be observed in the extracted phase velocities at lower frequencies and therefore inversely proportional to the source-receiver distance, implying that better resolution will be obtained at greater source-receiver offsets. Because of interference among frequency, wavenumber, and phase velocity, the optimal acquisition parameters can be determined only when one of them is fixed within the assumed plane-wave propagation pattern. That leads to the suggestion that the nearest usable offset cannot be resolved by harmonic wavefield analysis.

Based on ray theory, the smallest offset that can be estimated when receivers are laid on the surface of a layered half-space model can be written as (Xu et al., 2006)

$$x_o = 2h/\sqrt{\delta^2 - 1}, \quad \delta = \alpha/\beta, \tag{19}$$

where $h$, $\alpha$, and $\beta$ are the thickness, P-wave velocity, and S-wave velocity of the top layer, respectively. When the offset is far more than $x_o$, the normal modes of dispersive Rayleigh waves, including the fundamental and higher modes, are well defined. When the offset is less than $x_o$, however, only the fundamental mode and/or a leaking mode can be observed. Therefore, as a rule of thumb for Rayleigh waves, only dispersion data extracted from far-offset traces can be used as input for inversion in which forward computation is based on the period equation.

To further justify our conclusion previously derived from ray theory, it is necessary to assume an earth model that consists of a single surface layer overlying the half-space. The recorded upgoing waves at the free surface relate to the incident wave as (Kennett, 2001)

$$\begin{pmatrix} P_u \\ S_u \end{pmatrix} = \mathbf{R}_D \begin{pmatrix} P_D \\ S_D \end{pmatrix} \mathrm{diag}\{e^{-2\omega|q_\alpha|h}, e^{-2\omega|q_\beta|h}\},$$

$$\mathbf{R}_D = \begin{pmatrix} R_D^{PP} & R_D^{PS} \\ R_D^{SP} & R_D^{SS} \end{pmatrix}, \tag{20}$$

where subscripts $U$ and $D$ denote upgoing and downgoing waves, respectively; $h$ is the thickness of the surface layer; $q_\alpha = (\alpha^{-2} - p^2)^{1/2}$, $q_\beta = (\beta^{-2} - p^2)^{1/2}$; $\alpha$ and $\beta$ are P- and S-wave velocities, respectively; $p$ is ray parameter or horizontal slowness; and $R_D$ is a matrix consisting of reflection coefficients that include those of converted waves.

It should be noted that the reflection-coefficient matrix is phase related. For a single layer underlain by a half-space model, some conclusions can be drawn from inspection of the reflection-coefficient matrix (Kennett, 2001):

1) $c < \beta_1 < \beta_2 < \alpha_1 < \alpha_2$, where $c$ is phase velocity of the Rayleigh wave; $\alpha$ and $\beta$ are the P-, and S-wave velocities, respectively; and subscripts $_1$ and $_2$ denote the top layer and the half-space, respectively. P- and S-waves are all evanescent, with only the fundamental mode existing across the model space; *evanescent* as used here means the wave energy exponentially decreases from the free surface. That case corresponds to the leaking mode.

2) $\beta_1 < c < \beta_2 < \alpha_1 < \alpha_2$, where the S-wave is propagating and the P-wave is evanescent. That situation corresponds to case I discussed in Xu et al. (2006). Note that at high frequencies, the higher modes of Rayleigh waves are very similar to Love waves except that there is a velocity-dependent phase shift.

3) $\beta_1 < \alpha_1 < c < \beta_2 < \alpha_2$, where the P-wave is propagating and the S-wave is evanescent. The dispersive behavior is controlled by P- and S-waves and their coupling. That situation corresponds to case II in Xu et al. (2006).

It is interesting to characterize near-field effects in seismic wavefields quantitatively by physical and numerical analysis. Bodet et al. (2009) investigate the near-field effects by laser-Doppler physical experiments on both a homogeneous and a two-layer model. They find that phase velocities are underestimated and do not depend on the propagation medium, whether homogeneous or normally dispersive. The bias becomes significant ($\geq 5\%$) once the measured wavelength exceeds 50% of the spread length. The bias is the result of aliasing expected, based on sampling theory. However, yet to be explained is why that

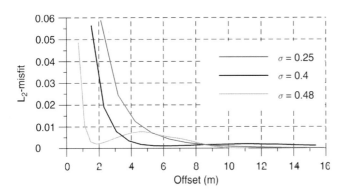

**Figure 6.** Misfits of two neighboring traces versus offset for M4 (Table 1).

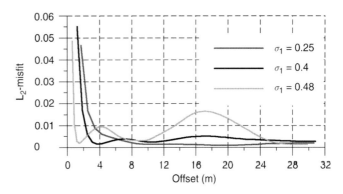

**Figure 7.** Misfits of two neighboring traces versus offset for M5 (Table 1).

can occur as a result of spatial variation in the wavefields and how that relates to elastic properties.

We examined the spatial variation of synthesized wavefields using the criterion of L2-misfit between two neighboring traces after the normal-moveout (NMO) correction. To avoid complexity, comparisons first were based on an elastic homogeneous half-space model with different Poisson's ratios (M4 in Table 1). The seismic source was a vertical point force with a peak frequency of 25 Hz (Figure 6). The main conclusion that can be drawn from this examination is that the offset distances affected by near-field effects are a function inversely proportional to Poisson's ratio. That is consistent with results derived from ray theory in a layered half-space model (Xu et al., 2006).

Comparisons also have been made using M5 (Table 1), in which the top layer possesses different Poisson's ratios with the P-wave velocity fixed. The L2-misfit versus offset (Figure 7) showed similar characteristics to those of M4. At certain offset distances near to the source (Figures 6 and 7), L2-misfits for all modeled Poisson's ratios show very large biases that are the effects of non-plane-wave propagation. There is evidence that the L2-misfit for M5 possesses two local peaks (Figure 7), whereas the L2-misfit of M4 has only one local peak for the case of a Poisson's ratio of 0.48 (Figure 6). The results are consistent with previous analyses, implying that the first peak is the result of a leaking mode and the second is caused by higher modes of Rayleigh waves and/or reflected body-wave energy. The larger the Poisson's ratio, the stronger the disturbed peak.

## Topographic effects

To investigate topographic effects on Rayleigh-wave propagation, Wang (2009) develops 2D and 3D modeling codes using the AEA scheme and incorporating topography

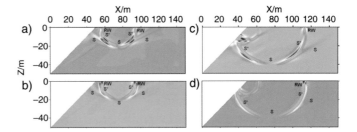

**Figure 8.** Snapshots of wavefields at (a and b) 0.15 s and (c and d) 0.25 s generated by a vertical point source at the ground surface with a slope-step topography modeled by the AEA algorithm (Xu et al., 2007). Parts (a) and (c) represent horizontal particle velocity and parts (b) and (d) vertical particle velocity.

using Robertsson's method (Robertsson, 1996). Pure topography effects will be discussed for a model consisting of an isotropic homogeneous half-space with a slope-step surface. The P- and S-wave velocities are 1000 m/s and 200 m/s, respectively. At 0.15 s (Figure 8a and 8b) and 0.25 s (Figure 8c and 8d), both the horizontal and vertical components possess reflected Rayleigh-wave energy and converted body waves that were generated as the Rayleigh waves passed the corner. That observation can be confirmed further by studying the seismograms (Figures 9 and 10).

To model small topographic undulation of the ground surface, a negative topography formed by a half circle with a 3-m radius was configured on the surface of M3 (Figure 11). Some scattering and converging energies can be observed on the shot gather (Figure 12). There is some variability between the extracted dispersion image and theoretical values of the first three modes (Figure 13). The positive topography also affects estimation of the dispersion curve to a smaller degree than the negative one at the same spatial dimension (Wang, 2009). Optimizing the criterion for accurate topographic effects is a current area of study at CUG.

**Figure 9.** (a and b) Horizontal and (c and d) vertical particle velocities recorded at the left and right sides with offset of 15 m (surface positions at 60 m and 90 m in the model of Figure 8). RW denotes Rayleigh waves.

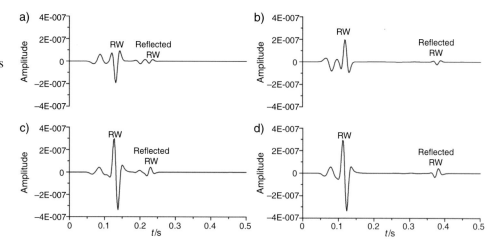

**Figure 10.** (a and b) Horizontal and (c and d) vertical particle velocities recorded at the left and right sides with offset of 40 m (surface position at 35 m and 115 m in the model of Figure 8). RW and BW denote Rayleigh and body waves, respectively.

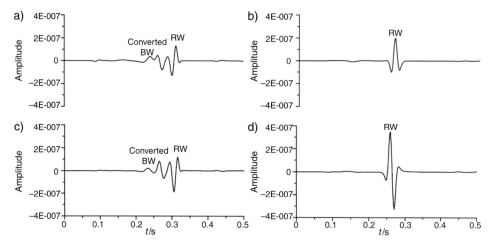

## Solvable Models by Inverting Dispersion Curves

### Elementary aspects of inverting dispersion curves

The mathematical term *well-posed problem* originally came from a definition given by Hadamard (1902). Today, one defines a well-posed problem as a uniquely solvable problem described by mathematical models of physical phenomena such that the solution depends in a continuous way on the data. Because dispersion analysis gives rise to an ill-posed inverse problem, a generalized definition of solvable models by dispersion analysis is necessary.

Here we propose completely solvable models based on dispersion analysis as those with model parameters that possess a sensitivity of 5% or more within a frequency range more than three times the resolution frequency as determined by the particular technique used in the dispersion

analysis. The misfit between the inverted and true model parameters also must be controllable. The proposed definition is based on the common knowledge that parameters of models that possess low sensitivity (e.g., less than 5%) means that changing a parameter's value has very little effect on measured data. Hence, we cannot uniquely invert incompletely measured and noisy data regardless of whether the value of the parameter is acceptable. Furthermore, partly solvable models can be thought of as approximate subsets of the inverted model parameters and as uniquely solvable.

Numerous methods have been proposed to invert fundamental or multimode dispersion data. Those methods generally can be divided into linear and nonlinear strategies. There are completely nonlinear methods, e.g., the genetic-algorithm (GA) method (Yamanaka and Ishida, 1996; Feng et al., 2005; Nagai et al., 2005; Pezeshk and Zarrabi, 2005; Dal Moro and Pipan, 2007), the simulated-annealing algorithm (Beaty et al., 2002), the neighborhood algorithm (Wathelet et al., 2004), pattern-search algorithms (Song et al., 2008; Song et al., 2009), and functional evolutionary algorithms (Liang et al., 2008).

Final solutions for those methods do not allow unique evaluation of the accuracy or uncertainty of the resultant model. That shortcoming can be resolved by the Monte Carlo ensemble analysis (O'Neill, 2004; Dal Moro and Pipan, 2007). However, resultant models from the Monte Carlo method can be dramatically different in terms of different parameterization and controlling of evolutionary process as well as time-consuming calculation. Therefore, that kind of inversion method was excluded in our discussion. We now move our attention to traditional methods in which the formulation is mathematically unique and every operator and its influence can be interpretable and controllable for geophysics problems (e.g., Xia et al., 1999; Song et al., 2007).

Solving an inverse problem requires the finding of a model that could reproduce the observed data within a certain degree of accuracy and at the same time satisfy one specified set of constraints. To achieve that, an objective function is constructed and minimized. A generic form of an objective function therefore usually is comprised of two terms: data misfit and model constraints. The first term is used to quantify how well the observed data are reproduced by the solution, and the second term contains imposed constraints in the model space that are based on a priori information and/or smoothness of the expected recoverable model. A commonly used objective function in geophysics is given as the following (Oldenburg and Li, 2005):

$$\Phi(\mathbf{m}) = \Phi_d + \beta\Phi_m, \qquad (21)$$

where $\Phi_d$ is the data-misfit term, $\Phi_m$ is the regularization term that determines constraints on the recovered model, and $\beta$ is known as the (Tikhonov) regularization parameter that balances $\Phi_d$ and $\Phi_m$. If $\beta$ is too small, the data might be overfitted and produce a model with excessive structure caused by noise in data (Constable et al., 1987) and an inherent error resulting from the approximation to a nonlinear forward problem. That situation can be very severe for the inversion of dispersion curves when using only the fundamental mode. On the other hand, if heavy weighting is put on the regularization term, i.e., if $\beta$ is very big, a model constrained by poor structure cannot reproduce the observed data at the required degree.

Several methods, e.g., L-curve criterion (Hansen, 1992, 1998), generalized cross-validation (GCV) (Wahba, 1990; Golub and von Matt, 1997), and so forth, have been suggested to find the best regularization parameter to produce a

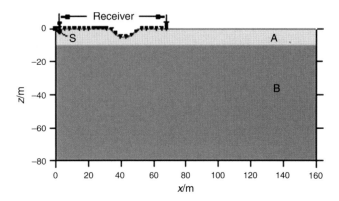

**Figure 11.** Topography model of a lower half circle with a radius of 3 m. Model parameters are the same as for M3 (Table 1). S indicates the source position.

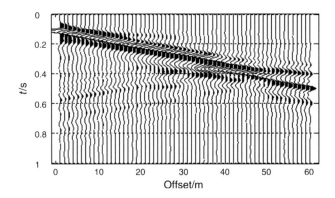

**Figure 12.** Synthetic shot gather for the model shown in Figure 11.

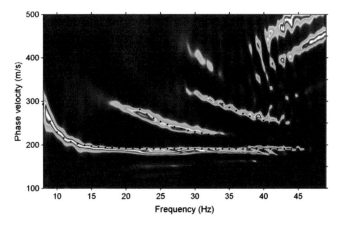

**Figure 13.** Image of Rayleigh-wave dispersive energy in the *f-v* domain generated by high-resolution LRT (Luo et al., 2008) for the lower half circle (radius is 3 m) topography model. Black dots denote analytical results calculated using the Knopoff method (Schwab and Knopoff, 1972).

final acceptable model. However, selection of a simple and time-saving scheme for establishing the best regularization parameter in iterative processes has not been determined. The principles and practical implementations of inversion were well documented by Oldenburg and Li (2005).

Reconstructed subsurface S-wave velocity profiles estimated by the inverting of dispersion curves are essentially one dimensional in structure. Many researchers have attempted to use multimode dispersion data as input to inversion routines as a means to improve resolution and increase depth of the recovered model (e.g., Xia et al., 2003; Luo et al., 2007; Song et al., 2007). Liang et al. (2008) provide insights into how the Rayleigh-wave modes are sensitive to characteristics of strata structure. That study provides guidelines that help us to understand which 1D models are solvable by inverting the multimode dispersion data in accordance with the previously proposed definition.

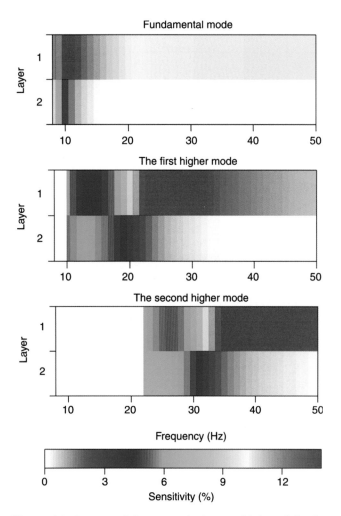

**Figure 14.** Images of S-wave velocity sensitivity of the first three Rayleigh-wave modes for M3 (Table 1).

## Sensitivity of Rayleigh-wave modes to the structure of strata

For simplicity, we first discuss the results of M3 (Table 1) using the method described by Liang et al. (2008). Note in Figure 14 that the high-sensitivity frequency bands of the first layer using the first and second higher modes are above 18 Hz and 29 Hz, respectively. Meanwhile, the high-sensitivity frequency bands for the half-space of the fundamental mode are below 8 Hz and for the first and second higher modes are 10–16 Hz and 23–30 Hz, respectively. For the phase velocities within those high-sensitivity frequency bands, S-wave velocities of the first layer and the underlying half-space are solvable using one Rayleigh-wave mode or a combination of the first three. When we compare the sensitivity-frequency image (Figure 14) and the energy image (Figure 4), the energy distribution appears to be well matched with the sensitivity variation through the first three modes. These are significant examples of how higher modes can increase the resolution and distinguishable depth of an inverted model for high-frequency Rayleigh-wave exploration.

The S-wave velocity of a shallow low-velocity layer (LVL) under a caprock can be resolved well. However, the S-wave velocity of a layer beneath the LVL might be resolved poorly (O'Neill, 2004). It is therefore very important to understand how the sensitivity variation versus frequency through the different modes changes in the presence of an LVL. We look to the results of Liang et al. (2008) in discussing the LVL effects on resolvable strata structure.

For an LVL existing within a six-layer model (Figure 15a), the theoretical dispersion curves of the first four Rayleigh-wave modes possess irregular variations (Figure 15b). That phenomenon is consistent with numerous field experiments in which strong evidence supports the existence of one or more LVL or horizontal low-velocity inhomogeneity. Because most of the energy is trapped within the LVL, the top surface layer is resolved poorly by the fundamental Rayleigh-wave mode alone (the top of Figure 16b), and the half-space can be resolved by using the first and/or second higher modes.

It is interesting to note that the layers between the LVL and the half-space might not be resolvable in spite of the fact that all four modes are used during the inversion process (Figure 16b), whereas the LVL is completely resolvable using only one of the four modes. This example shows that selectivity of frequencies and modes in surface-wave propagation characteristics is not only valuable from a theoretical perspective but is also very important for evaluating resolvable models.

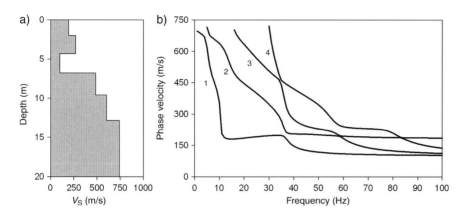

**Figure 15.** (a) Low-velocity-layer model and (b) its dispersion curves.

Although defining resolvable models from inverting dispersion curves should be limited to a plane-layered structure, it is useful to evaluate the limitations of that approach for real-world applications. A stepped-plane-layered model can approximate a model with small horizontal undulation in an interface (Kuo and Nafe, 1962). Luo et al. (2009c) present synthetic models containing dipping layers with slopes of 5°, 10°, 15°, 20°, or 30° and a real-world example to assess the accuracy of reconstructed phase-velocity sections using the current dispersion-analysis technique (Luo et al., 2009c).

Results of the analysis of those slope models demonstrate two things: (1) If a dipping angle is smaller than 15°, phase-velocity sections generated by a pair of consecutive traces after mode separation possess less than 10% relative error when compared with analytical results. As a rule of thumb, therefore, inverted S-wave velocities generally will possess more than 90% accuracy (Xia et al., 1999; Xia et al., 2002a). (2) If the dip of a layer is larger than 15°, Rayleigh-wave modes interfere strongly with one another in the *f-v* domain, and therefore, high-accuracy dispersion curves are nearly impossible to calculate with only a pair of consecutive traces.

## Roles of cutoffs

Cutoff is the long-period termination of a surface-wave higher mode and therefore can be defined by a discrete point in the phase-velocity frequency domain. For a layered-earth model, phase velocities of higher-mode Rayleigh waves approach the S-wave velocity of the underlying half-space at the cutoffs (e.g., Lay and Wallace, 1995). It is therefore reasonable to think intuitively of a cutoff as the point at which energy begins to transform into S-wave radiation in the bounding half-space (Oliver and Ewing, 1957). From numerical investigation of the first three Rayleigh modes for a layered crust-mantle model, Mooney and Bolt (1966) conclude that the main factor controlling cut-

offs is the ratio of S-wave velocities in the surface layer to the lower half-space.

Explicit formulations used to calculate cutoffs can be developed only for the simplest single layered-earth model (Newlands, 1952; Kuo and Nafe, 1962). It is therefore difficult to determine analytically the characteristic relation between cutoffs and strata parameters. Xu et al. (2009) developed an approximation method for estimating the cutoffs of a multilayered earth model. The method effectively approximates multiple layers using a single layer model whose velocity is defined as the harmonic averaging of the multiple layer velocities, with density being the arithmetic average of the multiple layer densities.

The accuracy of the proposed approximation method was investigated numerically. The numerical results are in good agreement with cutoffs that are determined by using modeling and the approximation method. The use of Rayleigh-wave cutoffs can possess a great deal of potential for S-wave velocity estimations, thereby providing a useful tool for constructing initial models for inversion of multimode data.

The advantages of increasing exploration depths and improving model resolution using higher modes (Xia et al., 2003) have encouraged the near-surface community to develop and now occasionally to use joint inversion of multimode Rayleigh waves (e.g., Xia et al., 2006; Dal Moro and Pipan, 2007). There are two main reasons for using cutoffs in the practical inversion of higher modes: (1) Higher-mode surface waves are dominant at sites where large stiffness contrasts or velocity reversals exist. A high-resolution linear Radon transform method (Luo et al., 2008) has proved to be feasible for extracting high-quality dispersion curves from higher modes, making the use of cutoffs possible. (2) Because the dispersion images of surface waves inherently possess low resolution at low frequencies, it is often difficult to pick the dispersion curve at lower frequencies from fundamental-mode data. In those cases, cutoffs therefore play a critical role in defining the depth to the half-space or an initial model.

## Outlook

Numerical surface-wave modeling is a fundamental tool in understanding surface-wave characteristics at different near-surface settings as well as a base for true 2D inversion. Although our recent research is limited to elastic media, we have started to study surface-wave behaviors in a viscoelastic medium and Biot's two-phase medium, and both bring us closer to the real world. Wavefield inversion relies on efficient and accurate forward-modeling algorithms. Reducing spurious reflections from artificial boundaries, especially for Rayleigh-wave modeling, is still

**Figure 16.** Sensitivity-frequency images of the first four Rayleigh-wave modes for each layer of the model shown in Figure 15a.

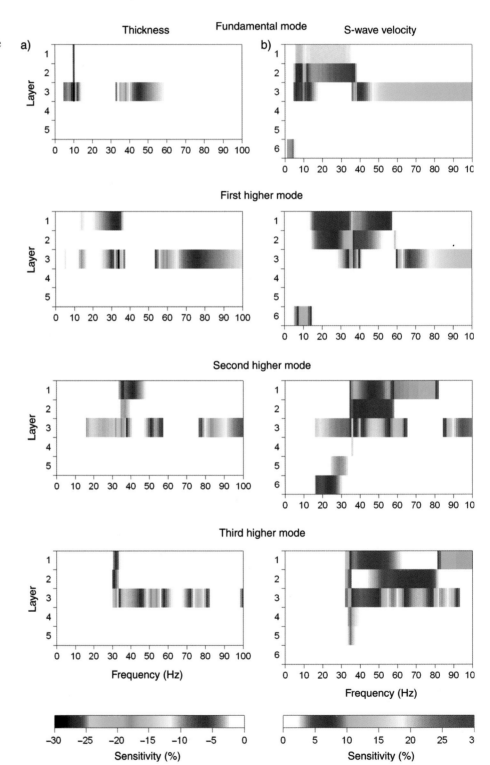

a challenge for the near-surface community. Using accurate modeling techniques for arbitrary subsurface materials and structures, which is necessary to study surface-wave reflection, refraction, diffraction (Xia et al., 2007a), attenuation, and so forth, can further improve our ability to perform surface-wave exploration and investigation.

Some new research directions and paths ahead along current research lines also have been illuminated recently to our near-surface seismic community. For example, arbitrary acquisition configurations and data-processing algorithms are now feasible for active- and passive-source approaches, which undoubtedly will lead to 3D surface-wave techniques. Passive-source surface-wave techniques such as the spatial autocorrelation technique (SPAC) (Aki, 1957; Okada, 2003) and its derivative — ambient noise-based Rayleigh-wave empirical Green's function method (e.g., Asten et al., 2005; Campillo, 2006; Larose et al., 2006; Halliday et al., 2008) — can improve the imaging depth and spatial coverage of current active methods and applications. Those passive approaches will continue to be very useful tools for defining the S-wave velocity function in areas inaccessible to MASW or SASW, e.g., lakes, larger buildings, and so forth.

## Conclusions

Accurate phase velocities are a key product of surface-wave techniques. Estimation of phase velocities from surface waves requires careful design of data-acquisition geometry and high-resolution data-processing techniques. Discussion of near-offset effects provides insights into minimum-source geophone offsets required for generating high-quality surface-wave images.

Amplitude effects and their use and effects in dispersive analysis are new to the near-surface community. We numerically demonstrated the necessity of amplitude corrections for multimode data. Topographic effects from two simple models demonstrated how topographic changes could significantly affect distribution patterns of surface waves. The correction criterion for topography is being studied at China University of Geosciences. Dispersion images generated by the high-resolution LRT are the basis for accurate determinations of phase velocities, especially in the case of higher modes and mode separation.

Modeling results related to partitioning of surface-wave energy suggest that the fundamental mode contains most of the energy when a soft layer overlies a hard rock. Processing surface-wave data using mode-separation techniques possesses great potential for increasing horizontal resolution of calculated S-wave velocity fields. Approximating and

incorporating cutoffs represent a useful tool for constructing the good initial model necessary for the inversion of multimode data. Sensitivity analysis of surface waves indicates that layers below a low-velocity layer possess low sensitivity, making detection of those sublow-velocity layers difficult when using inversion analysis.

Integration of different geophysical methods with surface-wave investigations provides untold potential for improving the investigative potential of near-surface geophysics. For example, because most low-resistivity subsurface materials also possess low-velocity characteristics, the trend is for electrical current to pass through a low-resistivity zone, whereas Rayleigh waves are inclined to propagate in a low-velocity zone. Hence, it is suggested to use the MASW method with a multielectrode electrical-profiling survey to enhance confidence in determining a low-velocity layer.

## Acknowledgments

This work is supported by the National Nature Science Foundation of China (No. 40974079 and No. 40904031). The authors thank Richard Miller, editor of this volume, for his invitation to contribute to this book and his kind instructions for writing the paper. Thanks also go to Jianghai Xia, who discussed main aspects of the paper with us by e-mail and telephone and carefully reviewed the manuscript. An anonymous reviewer is gratefully acknowledged for his constructive suggestions.

## References

Aki, K., 1957, Space and time spectra of stationary stochastic waves, with special reference to microtremors: Bulletin of the Earthquake Research Institute, **35**, 415–456.

Asten, M. W., W. R. Stephenson, and P. N. Davenport, 2005, Shear-wave velocity profile for Holocene sediments measured from microtremor array studies, SCPT, and seismic refraction: Journal of Environmental and Engineering Geophysics, **10**, 235–242.

Beaty, K. S., D. R. Schmitt, and M. Sacchi, 2002, Simulated annealing inversion of multimode Rayleigh-wave dispersion curves for geological structure: Geophysical Journal International, **151**, 622–631.

Bodet, L., O. Abraham, and D. Clorennec, 2009, Near-offset effects on Rayleigh-wave dispersion measurements: Physical modeling: Journal of Applied Geophysics, **68**, 95–103.

Bodin, T., and V. Maupin, 2008, Resolution potential of surface wave phase velocity measurements at small arrays: Geophysical Journal International, **172,** 698–706.

Bohlen, T., and E. H. Saenger, 2006, Accuracy of heterogeneous staggered-grid finite-difference modeling of Rayleigh waves: Geophysics, **71,** no. 4, T109–T115.

Campillo, M., 2006, Phase and correlation in "random" seismic fields and the reconstruction of the Green's function: Pure and Applied Geophysics, **163,** 475–502.

Constable, S. C., R. L. Parker, and C. G. Constable, 1987, Occam's inversion: A practical algorithm for generating smooth models from EM sounding data: Geophysics, **52,** 289–300.

Dal Moro, G., and M. Pipan, 2007, Joint inversion of surface wave dispersion curves and reflection travel times via multi-objective evolutionary algorithms: Journal of Applied Geophysics, **61,** 56–81.

Dal Moro, G., M. Pipan, E. Forte, and I. Finetti, 2003, Determination of Rayleigh wave dispersion curves for near surface applications in unconsolidated sediments: 73rd Annual International Meeting, SEG, Expanded Abstracts, 1247–1250.

Feng, S., T. Sugiyama, and H. Yamanaka, 2005, Effectiveness of multi-mode surface wave inversion in shallow engineering site investigations: Exploration Geophysics, **36,** 26–33.

Forbriger, T., 2003, Inversion of shallow-seismic wavefields: I. Wavefield transformation: Geophysical Journal International, **153,** 719–734.

Friederich, W., S. Hunzinger, and E. Wielandt, 2000, A note on the interpretation of seismic surface waves over three-dimensional structures: Geophysical Journal International, **143,** 335–339.

Golub, G. H., and U. von Matt, 1997, Generalized cross-validation for large scale problems: Journal of Computational and Graphic Statistics, **6,** 1–34.

Graves, R. W., 1996, Simulating seismic wave propagation in 3D elastic media using staggered-grid finite differences: Bulletin of the Seismological Society of America, **86,** 1091–1106.

Guan, X., J. Huang, and H. Zhou, 1993, Probe the interpretation theory of steady-state Rayleigh-wave method in engineering exploration: Journal of Geophysics (Chinese edition with English abstract), **36,** 96–105.

Gucunski, N., and R. D. Woods, 1992, Numerical simulation of the SASW test: Soil Dynamics and Earthquake Engineering, **11,** 213–227.

Hadamard, J., 1902, Sur les problèmes aux dérivées partielles et leur signification physique: Princeton University Bulletin, **13,** 49–52.

Halliday, D. F., A. Curtis, and E. Kragh, 2008, Seismic surface waves in a suburban environment — Active and passive interferometric methods: The Leading Edge, **27,** 210–218.

Hansen, P. C., 1992, Analysis of discrete ill-posed problems by means of the L-curve: Society of Industrial and Applied Mathematics Review, **34,** 561–580.

——, 1998, Rank-deficient and discrete ill-posed problems: Numerical aspects of linear inversion: Society of Industrial and Applied Mathematics.

Herrmann, P., T. Mojesky, and P. Hugonnet, 2000, Dealiased high-resolution Radon transforms: 70th International Annual Meeting, SEG, Extended Abstracts, 1953–1956.

Huang, J., H. Zhou, and X. Guan, 1991, Theoretical study of Rayleigh-wave technique in engineering geology: Geophysical and Geochemical Exploration (Chinese edition with English abstract), **15,** 268–277.

Ji, J., 2006, CGG method for robust inversion and its application to velocity-stack inversion: Geophysics, **71,** no. 4, R59–R67.

Karray, M., and G. Lefebvre, 2009, Techniques for mode separation in Rayleigh wave testing: Soil Dynamics and Earthquake Engineering, **29,** 607–619.

Kennett, B. L. N., 2001, The seismic wavefield, v. 1: Introduction and theoretical development: Cambridge University Press.

Kristek, J., P. Moczo, and R. J. Archuleta, 2002, Efficient methods to simulate planar free surface in the 3D 4th-order staggered-grid finite-difference schemes: Studia Geophysica et Geodaetica, **46,** 355–381.

Kuo, J. T., and J. E. Nafe, 1962, Period equation for Rayleigh waves in a layer overlying a half space with a sinusoidal interface: Bulletin of the Seismological Society of America, **52,** 807–822.

Larose, E., L. Margerin, A. Derode, B. van Tiggelen, M. Campillo, N. Shapiro, A. Paul, L. Stehly, and M. Tanter, 2006, Correlation of random wavefields: An interdisciplinary review: Geophysics, **71,** no. 4, SI11–SI21.

Lay, T., and T. C. Wallace, 1995, Modern global seismology: Academic Press.

Levander, A. R., 1988, Fourth-order finite-difference P-SV seismograms: Geophysics, **53,** 1425–1436.

Liang, Q., C. Chen, C. Zeng, Y. Luo, and Y. Xu, 2008, Inversion stability analysis of multimode Rayleigh wave dispersion curves using low-velocity-layer models: Near Surface Geophysics, **6,** 157–165.

Luo, Y., J. Xia, J. Liu, Q. Liu, and S. Xu, 2007, Joint inversion of high-frequency surface waves with fundamental and higher modes: Journal of Applied Geophysics, **62,** 375–384.

Luo, Y., J. Xia, R. Miller, Y. Xu, J. Liu, and Q. Liu, 2008, Rayleigh-wave dispersive energy imaging by high-resolution linear Radon transform: Pure and Applied Geophysics, **165,** 903–922.

Luo, Y., J. Xia, J. Liu, Y. Xu, and Q. Liu, 2009a, Research on the MASW middle-of-the-spread-results assumption: Soil Dynamics and Earthquake Engineering, **29**, 71–79.

Luo, Y., J. Xia, R. Miller, Y. Xu, J. Liu, and Q. Liu, 2009b, Rayleigh-wave mode separation by high-resolution linear Radon transform: Geophysical Journal International, **179**, 254–264.

Luo, Y., J. Xia, Y. Xu, C. Zeng, R. Miller, and Q. Liu, 2009c, Dipping-interface mapping using mode-separated Rayleigh waves: Pure and Applied Geophysics, **166**, 353–374.

McMechan, G. A., and M. J. Yedlin, 1981, Analysis of dispersive waves by wave field transformation: Geophysics, **46**, 869–874.

Miller, R., J. Xia, C. B. Park, and J. Ivanov, 1999, Multichannel analysis of surface waves to map bedrock: The Leading Edge, **18**, 1392–1396.

Mittet, R., 2002, Free-surface boundary conditions for elastic staggered-grid modeling schemes: Geophysics, **67**, 1616–1623.

Moczo, P., J. Kristek, V. Vavryčuk, R. J. Archuleta, and L. Halada, 2002, 3D heterogeneous staggered-grid finite-difference modeling of seismic motion with volume harmonic and arithmetic averaging of elastic moduli and densities: Bulletin of the Seismological Society of America, **92**, 3042–3066.

Moczo, P., J. O. A. Robertsson, and L. Eisner, 2007, The finite-difference time-domain method for modelling of seismic wave propagation, *in* R. S. Wu, V. Maupin, and R. Dmowska, eds., Advances in wave propagation in heterogeneous earth: Elsevier–Academic Press Advances in Geophysics Series No. 48, 421–516.

Mooney, H. M., and B. A. Bolt, 1966, Dispersive characteristics of the first three Rayleigh modes for a single surface layer: Bulletin of the Seismological Society of America, **56**, 43–67.

Nagai, K., A. O'Neill, Y. Sanada, and Y. Ashida, 2005, Genetic algorithm inversion of Rayleigh wave dispersion from CMPCC gathers over a shallow fault model: Journal of Environmental and Engineering Geophysics, **10**, 275–286.

Nazarian, S., 1984, In-situ determination of elastic moduli of soil deposits and pavement systems by spectral-analysis-of-surface-waves method: Ph.D. dissertation, University of Texas, Austin.

Nazarian, S., and K. H. Stokoe II, 1986, Use of surface waves in pavement evaluation: Transportation research record 1070, TRB: National Research Council, 132–144.

Newlands, M., 1952, The disturbance due to a line source in a semi-infinite elastic medium with a single surface layer: Philosophical Transactions of the Royal Society of London, Series A, **245**, 213–308.

Oldenburg, D. W., and Y. Li, 2005, Inversion for applied geophysics: A tutorial, *in* D. K. Bulter, ed., Near-surface

geophysics, Chapter 5: SEG Investigations in Geophysics Series No. 13, 89–150.

Oliver, J., and M. Ewing, 1957, Higher modes of continental Rayleigh waves: Bulletin of the Seismological Society of America, **47**, 187–204.

Oliver, J., and M. Major, 1960, Leaking modes and the PL phase: Bulletin of the Seismological Society of America, **50**, 165–180.

Okada, H., 2003, The microtremor survey method (K. Suto, trans.): SEG Geophysical Monograph Series No. 12.

O'Neill, A., 2004, Shear velocity model appraisal in shallow surface wave inversion, *in* A. Viana, da Fonseca, and P. W. Mayne, eds., Proceeding of the ISC-2 on Geotechnical and Geophysical Site Characterization: Millpress, 539–546.

Park, C. B., R. Miller, and J. Xia, 1998, Imaging dispersion curves of surface waves on multi-channel record: 68th Annual International Meeting, SEG, Expanded Abstracts, 1377–1380.

———, 1999, Multi-channel analysis of surface waves: Geophysics, **64**, 800–808.

Pezeshk, S., and M. Zarrabi, 2005, A new inversion procedure for spectral analysis of surface waves using a genetic algorithm: Bulletin of the Seismological Society of America, **95**, 1801–1808.

Robertsson, J. O. A., 1996, A numerical free-surface condition for elastic/viscoelastic finite-difference modeling in the presence of topography: Geophysics, **61**, 1921–1934.

Roth, M., and K. Holliger, 2000, The non-geometric PS wave in high-resolution seismic data: Observations and modeling: Geophysical Journal International, **140**, F5–F11.

Sanchez-Salinero, I., 1987, Analytical investigation of seismic methods used for engineering application: Ph.D. dissertation, University of Texas, Austin.

Sacchi, M., and M. Porsani, 1999, Fast high resolution parabolic RT: 69th Annual International Meeting, SEG, Expanded Abstracts, 1477–1480.

Schwab, F. A., and L. Knopoff, 1972, Fast surface wave and free mode computations, *in* B. A. Bolt, ed., Methods in computational physics: Academic Press, 87–180.

Sheu, J. C., K. H. Stokoe II, and J. M. Roesset, 1988, Effect of reflected waves in SASW testing of pavements, Transportation Research Record 1196, TRB: National Research Council, 51–61.

Song, X., H. Gu, J. Liu, and X. Zhang, 2007, Estimation of shallow subsurface shear-wave velocity by inverting fundamental and higher-mode Rayleigh waves: Soil Dynamics and Earthquake Engineering, **27**, 599–607.

Song, X., H. Gu, X. Zhang, and J. Liu, 2008, Pattern search algorithms for nonlinear inversion of high-frequency

Rayleigh-wave dispersion curves: Computers & Geosciences, **34**, 611–624.

Song, X., D. Li, H. Gu, Y. Liao, and D. Ren, 2009, Insights into performance of pattern search algorithms for high-frequency surface wave analysis: Computers & Geosciences, **35**, 1603–1619.

Song, Y. Y., J. P. Castagna, R. A. Black, and R. W. Knapp, 1989, Sensitivity of near-surface shear-wave velocity determination from Rayleigh and Love waves: 59th Annual International Meeting, SEG, Expanded Abstracts, 509–512.

Stokoe, K. H. II, S. G. Wright, J. A. Bay, and J. M. Roësset, 1994, Characterization of geotechnical sites by SASW method, *in* R. D. Woods, ed., Geophysical characterization of sites: ISSMFE Technical Committee No. 10, Oxford & IBH Publishing, 15–25.

Trad, D., T. Ulrych, and M. Sacchi, 2002, Accurate interpolation with high-resolution time-variant Radon transforms: Geophysics, **67**, 644–656.

———, 2003, Latest views of the sparse Radon transform: Geophysics, **68**, 386–399.

Tokimatsu, K., S. Tamura, and H. Kojima, 1992, Effects of modes on Rayleigh wave dispersion characteristics: Journal of Geotechnical and Engineering, ASCE, **118**, 1529–1543.

Virieux, J., 1984, SH wave propagation in heterogeneous media: Velocity stress finite-difference method: Geophysics, **49**, 1933–1957.

———, 1986, P-SV wave propagation in heterogeneous media: Velocity-stress finite-difference method: Geophysics, **51**, 889–901.

Virieux, J., and S. Operto, 2009, An overview of full-waveform inversion in exploration geophysics: Geophysics, **74**, no. 6, WCC1–WCC26.

Wahba, G., 1990, Spline models for observational data: Society of Industrial and Applied Mathematics.

Wang, L., 2009, Technique study of staggered-grid FD schemes in 3D Rayleigh-wave modeling with surface topography: M.S. thesis, China University of Geosciences, Wuhan.

Wathelet, M., D. Jongmans, and M. Ohrnberger, 2004, Surface-wave inversion using a direct search algorithm and its application to ambient vibration measurements: Near Surface Geophysics, **2**, 211–221.

Wielandt, E., 1993, Propagation and structural interpretation of non-plane waves: Geophysical Journal International, **113**, 45–53.

Xia, J., R. Miller, and C. B. Park, 1999, Estimation of near-surface shear-wave velocity by inversion of Rayleigh wave: Geophysics, **64**, 691–700.

Xia, J., R. Miller, C. B. Park, J. A. Hunter, J. B. Harris, and J. Ivanov, 2002a, Comparing shear-wave velocity profiles from multichannel analysis of surface wave with borehole measurements: Soil Dynamics and Earthquake Engineering, **22**, 181–190.

Xia, J., R. Miller, C. B. Park, and G. Tian, 2003, Inversion of high frequency surface waves with fundamental and higher modes: Journal of Applied Geophysics, **52**, 45–57.

Xia, J., J. E. Nyquist, Y. Xu, M. J. S. Roth, and R. Miller, 2007a, Feasibility of detecting near-surface feature with Rayleigh-wave diffraction: Journal of Applied Geophysics, **62**, 244–253.

Xia, J., Y. Xu, C. Chen, R. D. Kaufmann, and Y. Luo, 2006, Simple equations guide high-frequency surface-wave investigation techniques: Soil Dynamics and Earthquake Engineering, **26**, 395–403.

Xia, J., Y. Xu, and R. Miller, 2007b, Generating image of dispersive energy by frequency decomposition and slant stacking: Pure and Applied Geophysics, **164**, 941–956.

Xu, Y., J. Xia, Y. Luo, and R. Miller, 2008, Amplitude effects in phase velocity estimation from shot gather, *in* Y. Xu and J. Xia, eds., Near-surface geophysics and human activities, Proceedings of the 2008 International Conference on Environmental and Engineering Geophysics (ICEEG): Science Press USA Inc., 75–80.

Xu, Y., J. Xia, and R. Miller, 2006, Quantitative estimation of minimum offset for multichannel surface-wave survey with actively exciting source: Journal of Applied Geophysics, **59**, 117–125.

———, 2007, Numerical investigation of implementation of air-earth boundary by acoustic-elastic boundary approach: Geophysics, **72**, no. 5, SM147–SM153.

———, 2009, Approximation to cutoffs of higher modes of Rayleigh waves for a layered earth model: Pure and Applied Geophysics, **166**, 339–351.

Yamanaka, H., and H. Ishida, 1996, Application of genetic algorithm to an inversion of surface wave dispersion data: Bulletin of the Seismological Society of America, **86**, 436–444.

Yilmaz, Ö., 1987, Seismic data processing: SEG Investigations in Geophysics Series No. 2.

Zahradník, J., and E. Priolo, 1995, Heterogeneous formulations of elastodynamic equations and finite-difference schemes: Geophysical Journal International, **120**, 663–676.

Zhang, S. X., and L. S. Chan, 2003, Possible effects of misidentified mode number on Rayleigh wave inversion: Journal of Applied Geophysics, **53**, 17–29.

Zhang, S. X., L. S. Chan, and J. Xia, 2004, The selection of field acquisition parameters for dispersion images from multichannel surface wave data: Pure and Applied Geophysics, **161**, 185–201.

Chapter 4

# Advances in Surface-wave and Body-wave Integration

Laura Valentina Socco[1], Daniele Boiero[1], Sebastiano Foti[2], Claudio Piatti[1]

## Abstract

Seismic methods are the primary characterization tools for several engineering and near-surface problems. Noninvasive and invasive methods based on the propagation of either body or surface waves are used widely. Often, more than one method is applied at the same site. In spite of possible synergies that exist between different methods, the data are often processed and interpreted independently. The integration of different data sets could provide more reliable final models and comprehensive site characterization. Acquisition can be optimized to obtain a multipurpose data set. Additional improvements might be obtained by a constrained or joint inversion of different seismic data. These can be demonstrated with real-world examples.

## Introduction

Seismic characterization plays an important role in engineering because it provides parameters relative to the mechanical behavior of geomaterials at small strains for large subsoil volumes in undisturbed conditions. This information is of paramount importance, particularly for dynamic studies related to seismic hazard (Semblat et al., 2005; Yilmaz et al., 2006) and vibration propagation (Comina and Foti, 2007). Moreover, seismic surveys supply a fast and cost-effective way to estimate the geometry of geologic formations at engineering scale (Socco et al., 2008).

Body-wave techniques, mainly based on seismic refraction, are used widely to estimate depth and morphology of the bedrock and to define geologic boundaries and property distribution in sedimentary environments. High-resolution seismic reflection is sometimes applied to delineate formation boundaries and bedrock topography (O'Donnell et al., 2001), even though costly acquisition limits the use of this survey technique in engineering projects. In the last decade, surface-wave methods have gained great popularity because of the possibility of efficiently estimating S-wave velocity, which represents a key parameter for geotechnical and engineering characterization. Besides surface-based methods, invasive tests, such as crosshole or downhole tests, often are applied to retrieve local seismic properties. In several applications, different geophysical surveys are performed for a comprehensive and reliable site characterization.

The survey layouts for body and surface-wave acquisition are similar, allowing a significant optimization of the field work (Foti et al., 2003; Ivanov et al., 2006; Yilmaz et al., 2006). Indeed, if acquisition parameters are designed properly, surface-wave data can be extracted from P-wave refraction data and vice-versa. Joint processing and interpretation of the two data sets can produce a significant improvement of the accuracy and reliability of the results, mitigating the effect of interpretation ambiguities, which are different for each of the two methods (Foti et al., 2003; Ivanov et al., 2006). For example, hidden layers related to low-velocity layers or produced by strong velocity contrasts in seismic refraction can be detected through surface-wave analysis. On the other hand, the solution nonuniqueness that affects surface-wave inversion can be mitigated by introducing information from P-wave refraction.

When P-wave seismic reflection data are available at a site, the ground roll (surface waves) can be processed and inverted to supply S-wave velocity along the seismic line.

[1]Politecnico di Torino — DITAG, Torino, Italy.
[2]Politecnico di Torino — DISTR, Torino, Italy.

This strategy also takes advantage of the higher resolution of surface waves at shallow depth with respect to seismic reflection. The velocity retrieved by surface-wave data might help in the interpretation of the uppermost portion of the seismic section and can be used for static corrections (Ernst, 2007). The overabundance of data in seismic-reflection surveys could be exploited to extract several dispersion curves along the seismic line to get a pseudo-2D model (Socco et al., 2009).

Another important issue for data integration involves borehole seismic methods. When downhole or crosshole tests are available, they are seldom integrated during the inversion process, but rather used as a benchmark for non-invasive methods. Indeed, in many fields of application, there is the general feeling that borehole seismic data are more reliable than surface-based data. In fact, the former are often assumed as the ground truth, even though they are affected by several sources of uncertainties whose effect is difficult to quantify.

In only a few cases, the stratigraphy obtained from borehole logs is used explicitly to drive the parameterization of the initial model for seismic data inversion. Even simple information (such as the water-table level) could be used as a constraint to improve the consistency of inversion results in seismic refraction and surface-wave analysis.

In this context, data integration can provide more reliable information by joint or constrained inversion procedures and also an overall increase of knowledge through the estimation of additional soil parameters, which could not be inferred using a single technique.

Here, we show the advantages of the integration of body- and surface-wave techniques at different levels. We first describe the different investigation techniques focusing on specific potentials and limitations. Then we discuss synergies in acquisition and interpretation of the different methods, and we show how different data can be integrated. We present several real-world examples from recent research projects of the applied geophysics research team of Politecnico di Torino.

# Methods

In this section, we discuss the main features of body- and surface-wave methods focusing on their intrinsic limitations and on the synergies that might arise from their integration.

## Body waves — Seismic refraction

Seismic refraction has been adopted widely for near-surface characterization since the 1920s for its simplicity in acquisition and processing. For near-surface characterization, acquisition usually is performed with multichannel spreads of vertical geophones and hammer sources that allow P-wave first arrivals to be detected. By far, S-wave refraction is used less than P-wave refraction because of the difficulties and limitations in data acquisition. Indeed, the amount of energy that can be transmitted by horizontally polarized sources is lower than that generated by vertical-impact sources used for P-wave surveys. Hence, the length of acquisition spreads used for S-wave seismic refraction is limited by the signal-to-noise that rapidly decreases with offset, making an accurate first-break picking difficult. This limitation affects the investigation depth.

Even though seismic refraction data acquired on layered media often are interpreted with very simple approaches based on the linearization of traveltime graphs (Ewing et al., 1939), more sophisticated inversion approaches have been developed over the last decades to derive the velocity distribution in complex environments: wavefront refraction methods (Thornburg, 1930; Rockwell, 1967; Aldridge and Oldenburg, 1992), model-based or tomographic methods (Zhu et al., 1992; Stefani, 1995; Lanz et al., 1998; Zhang and Toksöz, 1998) and reciprocal methods (Hawkins, 1961; Palmer, 1981).

Using various refraction inversion algorithms, it is possible to generate quite different velocity models that can be considered equally acceptable because all of them are consistent with the traveltime data to sufficient accuracy (Palmer 2010a). Moreover, irrespective of the inversion approach, there are limitations in seismic refraction techniques because the arrival times are related to unknown travel paths and there are many stratigraphic conditions that cannot be resolved by first-arrival analysis. Low velocity layers embedded in stiffer layers or gradual velocity increases over interfaces with strong velocity contrast produce "hidden layers" and significant errors in the final velocity profile, even if very sophisticated ray-tracing algorithms are used as forward modeling to invert the traveltimes. These problems are well known but often disregarded by practitioners who tend to consider seismic refraction results as a deterministic truth. On the contrary, several authors have evidenced that strong solution non-uniqueness affects seismic refraction results if no independent information is introduced to constrain traveltime interpretation. Foti et al. (2003) show that in the case of a gravel deposit (with gradually increasing velocity) that overlies stiff bedrock, traveltime interpretation produced an error of 150% in the estimation of the bedrock depth, which was known from a nearby borehole (Figure 1). Foti et al. (2003) also show that a surface-wave survey, performed using the same acquisition layout used for the seismic refraction survey, provided a good estimation of the velocity profile.

Ivanov et al. (2005a) and Ivanov et al. (2005b) use synthetic data for simple three-layer models to evidence that very different stratigraphic conditions could give exactly the same P-wave traveltimes in presence of a hidden layer (Figure 2). They also show that, even if a correct parameterization is adopted for the inverse problem (i.e., the number of layers in the reference model is the same as the true one) and the data are uncertainty free, the misfit function of the inverse problem does not present a unique minimum. Rather, it has a "valley shape," in which a broad band of solutions are equally possible. This problem can be overcome only by introducing some a priori information in the inversion process.

Palmer (2010b) shows that final velocity models obtained by tomographic inversion usually are very similar to the starting model. If important features, such as low-velocity zones, are not included in the starting model, it is unlikely that they will be reconstructed in the final refraction tomogram.

As far as 2D and 3D inversion approaches are concerned, if nonuniqueness is not adequately addressed, the results constituted by a single refraction tomogram, which fits the traveltime data to sufficient accuracy, do not guarantee that the result is either correct or even the most probable (Palmer 2010b). Enhancements in the reduction of lateral ambiguities have been obtained through the integration of amplitudes and traveltimes with the refraction convolution section (RCS), which is generated by the convolution of forward and reverse shot records (Palmer 2001).

## Body waves — Seismic reflection

Seismic reflection is the most widely adopted method for deep seismic exploration and the primary investigation technique in hydrocarbon prospecting. In engineering and near-surface applications, seismic reflection is applied to map boundaries between different geologic formations and to delineate the bedrock topography (Woolery, 2002). High-resolution seismic-reflection surveys can be used to characterize the shape of complex targets and discontinuities related to sinkholes and shallow faults (Lambrecht and Miller, 2006). Such applications to shallow targets require very short receiver spacing, dense shot coverage, and high-frequency sources.

The lateral resolution for seismic-reflection surveys can be quantified through the Fresnel radius. For different materials, the resolution rapidly decreases with depth (Lai et al., 2000) (Figure 3). On the other hand, strong heterogeneities typical in shallow layers can make the data very noisy.

Seismic reflection is performed mainly using P-waves because of the faster and simpler acquisition than for S-waves. However, the latter provide higher resolution, supplying a very detailed picture of the subsoil stratigraphy at shallow depth (Deidda et al., 2006).

Near-surface complexity often imposes careful processing to improve the resolution and the signal-to-noise ratio (S/N). Migration must be applied to locate dipping interfaces properly and to mitigate diffraction effects. All the processing steps benefit from accurate knowledge of the velocity distribution, which is difficult to obtain through velocity analyses only. For this reason, first arrivals of acquired signals can be interpreted with tomographic approaches to supply the velocity of the shallower layers along the seismic line. Ground-roll inversion also might be used to infer shallow S-wave velocity to be used for SH statics (Mari, 1984).

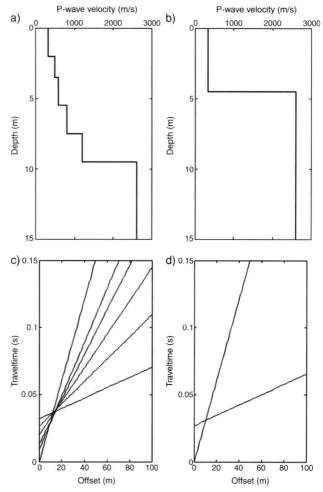

**Figure 1.** Traveltimes for equivalent velocity profiles: (a) multilayer profile derived from surface-wave inversion results; (b) two-layer profile from seismic refraction; (c) traveltimes for model (a); (d) traveltimes for model (b). From Foti et al., 2003. Used by permission of EAGE.

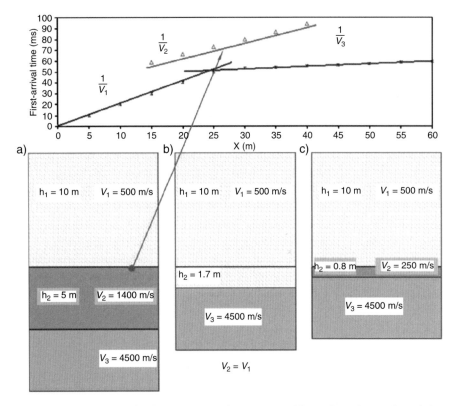

a)  b)  c)

**Figure 2.** Example of refraction nonuniqueness problem: three layered models generate the same first arrivals. Layers 1 and 3 have the same velocities and thickness, whereas layer 2 varies. (a) Layer 2 is a high-velocity layer. (b) Layer 2 has the same velocity as layer 1. (c) Layer 2 is a low-velocity layer. From Ivanov et al., 2006, based on an example from Burger, 1992. Used by permission.

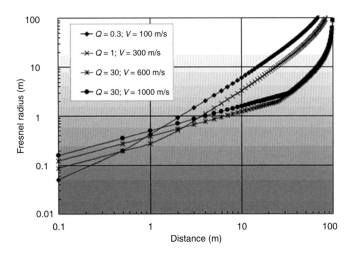

**Figure 3.** Resolution versus travel path for an amplitude attenuation of 20 dB corresponding to a signal-amplitude reduction of 100 times with respect to the source. Values of $Q$ and $V$ are representative for uncompacted soil, compacted soil, shale, and competent rock. After Lai et al., 2000. Used by permission of AGI.

## Surface waves

Surface-wave measurements have gained great popularity in the last decade. They allow the estimation of S-wave velocity profiles in layered media and are based on inversion of the dispersion of surface waves (mainly Rayleigh waves in land surveys, even if a few applications based on Love waves are reported in the literature). The experimental dispersion curve can be estimated using a variety of acquisition and processing strategies (SASW, MASW, microtremors). The inverse problem can be solved using deterministic or stochastic inversion approaches to provide a layered S-wave velocity model of the subsoil to a depth that depends on the maximum retrieved wavelength. The surface-wave inverse problem is ill-posed, mix-determined, and strongly nonlinear; hence, it suffers from severe solution nonuniqueness.

Different models might fit the experimental dispersion curve with sufficient accuracy and the identification of the solution within a set of possible models might not be feasible if a priori information is not available to constrain the inversion. For complex velocity profiles (e.g., stiff inclusions, low-velocity layers, or strong velocity contrasts) the sensitivity of the dispersion curve to model parameters might increase the uncertainty. Calderón-Macías and Luke (2007) show through sensitivity analysis that stiff inclusions might be poorly resolved if a priori information is not included (Figure 4). Socco and Boiero (2008) use a statistical test to select acceptable S-wave velocity profiles in a Monte Carlo inversion algorithm. They show that inverting the same data set with different initial models using a local search method might lead to very different results. They also show how the nonuniqueness of inversion results increases if the data are noisy or have high experimental uncertainty (Figure 5). In this context, the use of a priori information might significantly increase the reliability of the final velocity model. A priori information can be derived from boreholes or other investigation techniques and can be used to define a consistent initial model or to constrain the inversion process (Comina et al., 2002; Hayashi et al., 2005; Socco et al., 2009).

## Synergies

In spite of the described drawbacks and interpretation ambiguities, seismic body-wave and surface-wave methods have strong potentials in site characterization and present noticeable synergies that suggest integration of the methods. First, we deal with the synergies between seismic refraction and surface-wave analysis. Next we analyze the role of seismic reflection and borehole measurements.

The first synergy regards the data acquisition that is performed with analogous survey layouts for seismic refraction and surface waves, even if spatial samplings of the two methods obey different criteria. For seismic refraction, the spread must be long enough to record first arrivals from the deepest refractor of interest. For surface-wave acquisition, spread length is related to the maximum investigated wavelength. In both cases, we can state as a rule of thumb that investigation depth increases with length of the receiver spread. Receiver spacing for seismic refraction must be dense enough to sample different traveltime-graph segments properly, and for surface waves, spacing is related to the minimum recorded wavelength (Socco and Strobbia, 2004). In both cases, receiver spacing usually is defined as densely as possible, given the required spread length and the available number of channels of the recording equipment. The optimum acquisition layout for seismic refraction is often suitable for multichannel acquisition of surface waves and vice versa.

The two techniques impose opposite requirements on sampling in time. For seismic refraction, we need a high sampling rate to perform the first-break picking accurately, whereas the sampling frequency for surface waves might be lower because the highest frequency of interest is usually quite low (below 100 Hz in the majority of cases). The acquisition window for seismic refraction can be limited to the first arrival at the farthest offset receiver, whereas for surface waves, the whole wavetrain must be recorded. Hence, for a contemporary acquisition of seismic refraction and surface-wave data, long acquisition windows and high sampling rates must be adopted. This strategy originates large data sets. This is not a major problem with modern acquisition equipment, but separate shots can be performed,

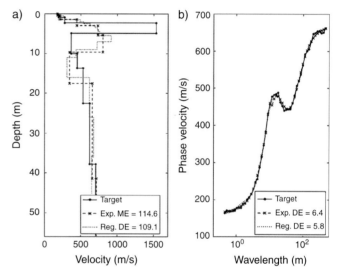

**Figure 4.** (a) Inverted shear-wave velocity profiles using linearized inversion starting from an exponential-layer (Exp) and a regular-layer (Reg) geometry; MEs represent model errors with respect to the true one (Target). (b) Corresponding Rayleigh-wave phase-velocity curves and misfits (DEs) with respect to the true one. From Calderón-Macías and Luke, 2007.

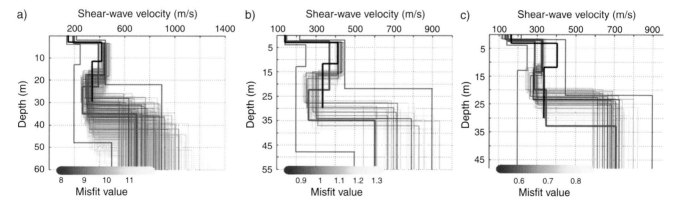

**Figure 5.** Some examples of Monte Carlo inversion results: the ensemble of acceptable models is compared with the results of a least-squares inversion of the same data (magenta) and of a downhole test (black) performed in a site nearby; the boundaries of model parameter space also are depicted (red lines). Examples with (a) fair data quality, (b) good data quality, and (c) poor data quality. From Socco and Boiero, 2008. Used by permission of EAGE.

adopting different time sampling for seismic refraction and surface waves for data-storage convenience.

Concerning the recording equipment, low-frequency geophones, which are required in surface-wave acquisition, might be adopted also for seismic refraction data. Impulsive sources, traditionally used in seismic refraction, are suitable for surface-wave acquisition. The acquisition setup for seismic refraction requires at least the acquisition of two opposite end-off shots and it is also suitable for surface-wave data. If further shots are acquired along the receiver spread, they can be used to assess the presence of lateral variations. Both methods benefit from stacking of different shots for each source position to improve the S/N.

Hence, it is convenient to design seismic acquisition to acquire simultaneously seismic refraction and surface-wave data, collecting a double-purpose data set.

Apart from acquisition synergies, integration of the two methods can be profitable in inversion. The two techniques present different sensitivities to model parameters, and hence their integration might mitigate the problem of solution nonuniqueness. In particular, surface waves also can provide reliable information in the case of velocity decreases and other velocity profiles that present hidden layers for seismic refraction. On the other hand, seismic refraction can contribute to constraint of the depth of deeper interfaces, which often are poorly resolved by surface-wave inversion. In addition, the information on water-table position, which typically is obtained from P-wave refraction in soft sediments, can be used to improve model parameterization for surface-wave inversion. Indeed, the latter usually is performed assuming a priori the Poisson's ratio values for the linear elastic model, and soil saturation affects the undrained Poisson's ratio dramatically. In this respect, Foti and Strobbia (2002) show on synthetic and experimental data that large errors can be induced by the wrong choice of a priori assumptions of the Poisson's ratio.

Finally, the combined use of P-wave refraction and surface-wave analysis provides information on both P- and S-wave velocity models, leading to a comprehensive site characterization.

Considering seismic reflection data, several examples presented in literature have shown that surface-wave dispersion curves can be extracted conveniently from seismic reflection records (Park et al., 2005; Neducza, 2007). Because acquisition of reflection data usually is not optimized for surface waves, we first must assess whether the data fulfill the quality requirements for surface-wave analysis. The data must have a good S/N in a wide frequency band and sufficient spectral resolution to allow for modal separation (Socco and Strobbia, 2004). Choice of the source type, sensor frequency, and sampling in time

and space might not be adequate in traditional seismic reflection data. Seismic reflection sources are often powerful enough to supply very high S/N surface-wave signals at far offsets. The time sampling and trace length are usually adequate and should contain only the traces in which the surface waves are not truncated. In deep investigations, the offset is usually sufficient to guarantee high wavenumber resolution, and hence good modal separation, although spatial sampling is often quite coarse. However, in very high-resolution surveys, receiver spacing is always small enough to retrieve the short wavelengths of the dispersion curve, but the offset, defined according to the optimum offset for seismic reflection, might be too short for acquiring long wavelengths.

Because low-frequency surface waves are considered as coherent noise in seismic-reflection surveys, often they are filtered using high-frequency sensors, sensor groups, source arrays, and/or low-cut filters in acquisition. The ideal recording for surface-wave analysis would be single, low-frequency sensors (no sensor groups) without any filter. If these requirements are not fulfilled, the data should be evaluated carefully before using them for surface-wave analysis. One of the most attractive aspects of surface waves in seismic reflection records is the large amount of available data that represent a resource, but which also requires an automatic processing approach to handle the full data set efficiently (Socco et al., 2009).

## Data Integration

The general concept of data integration is applied widely in geophysics. Data integration often refers to advanced data representation tools that allow different retrieved parameters and measured data to be displayed together in 2D or 3D plots to highlight common features and consistent spatial distributions.

Different levels of data integration might be conceived in geophysical applications, depending on the strategies for the solution of the inverse problem and the use of the final results:

1) independent inversions and qualitative or quantitative comparison of the results
2) independent inversions, but the results of the inversion of one data set are used for the model parameterization and the definition of constraints for the other data set
3) mutually and/or spatially constrained inversion
4) joint inversion

The different levels are discussed in the following.

## Data comparison

The first level often is applied in near-surface applications, when different survey methods are adopted at the same site. The results of the inversion of each data set are compared visually or numerically to one another during interpretation, thus exploiting complementary information and validating common features. This approach does not produce any improvement in the reliability of the results because the inversions are performed independently. Possible pitfalls related to intrinsic ambiguities of the different methods can be evidenced through the comparison but not mitigated.

Two real-world examples of a posteriori comparison of different results are reported. The first refers to a site, located near the leaning tower of Pisa (Italy) and characterized within remediation studies for the tower (Foti, 2003). The subsoil of this site has been very well characterized and is constituted mainly of clayey formations with some slightly stiffer layers. The surface-wave dispersion curve is obtained by merging active data, acquired with a linear array of 24 geophones with 4.5-Hz frequency (2.5-m geophone spacing and two end-off shots performed with an hammer source) and passive data, acquired with a circular array (diameter 40 m) of eight 2-Hz vertical geophones evenly spaced along the circumference (15.3-m spacing). The time sampling is 2 ms for both data sets and the time windows are 4 s and 688 s for active and passive data, respectively. The active data are processed by picking the maxima in the *f-k* domain and transforming them into phase velocity–frequency data points through the relation $V = 2\pi f/k$ (Gabriels et al., 1987) using the surface-wave-analysis-tool (SWAT) code, developed in the MATLAB environment at Politecnico di Torino (Strobbia, 2003). The passive data are processed using the frequency-domain beam-former technique (FDBF), implemented in a MATLAB code by Zywicki (1999). Because several acquisitions were repeated for the same receiver configuration, it is possible to estimate the experimental uncertainty for each frequency value.

The two experimental dispersion curves cover different frequency bands ranging from 4 to 33 Hz, for active data, and from 2.5 to 5.5 Hz for passive data. The two branches of dispersion curve are merged in a unique dispersion curve with a wide frequency band (Figure 6a). The obtained dispersion curve is inverted with a Monte Carlo algorithm (Socco and Boiero, 2008) that provides a set of final equivalent profiles. The equivalent profiles are selected by a statistical test that accounts for both the data uncertainty and the chosen parameterization. The results of the Monte Carlo inversion (Figure 6b) are compared with the results of a crosshole test, showing good agreement. In particular, both results identify a velocity inversion between 5 and 10 m. The velocity of the deeper layer is slightly overestimated by the surface-wave results with respect to the crosshole test.

The second real-world example refers to the Rojo Piano site (Italy). This site is located near L'Aquila and has been characterized within the activities for the selection of new construction sites in the aftermath of the 2009 earthquake. The site is located on the lacustrine deposits of the Aquilian Basin, which are composed mainly of fines and sands (Monaco et al., 2010). Limestone bedrock (more than 50-m depth) is expected at the site. In addition, in this case, both active and passive surface-wave data are acquired and merged. Active data are acquired with a linear array of 48 vertical geophones (4.5 Hz) with 1.5-m spacing, 0.5-ms sampling, and a 2-s acquisition window. Passive data are acquired with a circular array (diameter 50 m) of 12 2-Hz

a)

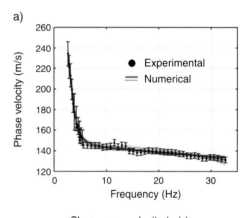

b)

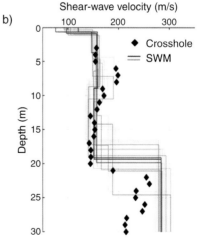

**Figure 6.** Monte Carlo inversion of active and passive surface-wave data at Pisa site: (a) Experimental and numerical dispersion curves. (b) Shear-wave velocity profiles from Monte Carlo analysis compared to crosshole test result. Used by permission of C. Comina.

geophones, evenly spaced along the circumference (12.9-m spacing) with 8-ms sampling interval, and a 524-s acquisition window.

The data are processed with the same approach adopted for the Pisa site to extract the dispersion curves. The two dispersion curves obtained cover different frequency bands, ranging from 8 to 52 Hz for active data, and from 3 to 11 Hz for passive data. The two branches of dispersion curve are merged in a unique dispersion curve with a wide frequency band (Figure 7a). The results of the Monte Carlo inversion are reported in Figure 7b.

For this site, both the results of a downhole test and of a seismic dilatometer test (SDMT) are available for comparison (Monaco et al., 2010) (Figure 7b). The seismic dilatometer is similar to a downhole test but the receivers are pushed in the ground within a steel rod rather then inserted in a preformed borehole. The SDMT data are available only up to 20 m of depth, while downhole testing and surface-wave analysis provided S-wave velocity down to 30 m.

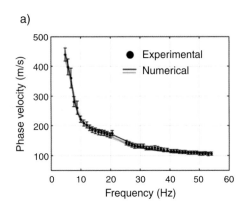

a)

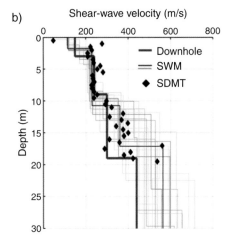

b)

**Figure 7.** Monte Carlo inversion of active and passive surface-wave data at Rojo Piano site: (a) Experimental and numerical dispersion curves. (b) Shear-wave velocity profiles from Monte Carlo analysis compared to crosshole and SDMT test results. Used by permission of C. Comina.

The investigation depth of surface-wave data is obtained through integration of active and passive data. Except for the first 10 m, in which the results of the three methods are in very good agreement, the results of the two invasive surveys show different velocities (the SDTM provides velocity about 30% higher than the downhole test). Surface-wave results in the deeper part of the model show lower resolution with respect to borehole results.

## A priori information for model parameterization

The second level of integration is applied sometimes when borehole logs are available at the site and are used as a priori information to define a consistent initial model for inversion or to determine one or more model parameters (for instance, the depth of a layer interface) to be kept fixed during inversion. Wathelet et al. (2004) show, on both synthetic and field data, that introduction of a priori information concerning the bedrock depth in stochastic inversion of surface-wave dispersion curves produces a significant improvement in the final results, also confirmed by sensitivity analysis. Hayashi et al. (2005) use S-wave velocity models from surface-wave analysis and empirical laws obtained in the lab to define a density model used later as the initial model for inversion of gravity measurements. Concerning the integration of seismic refraction and surface-wave data, Ivanov et al. (2006) demonstrate on synthetic and field data that the inversion of seismic traveltimes can benefit from a priori information derived by surface-wave dispersion inversion.

One example of integration of high-resolution refraction data into near-surface seismic reflection data processing and interpretation is represented by the work by Miller et al. (1998). They use turning-ray tomography to model near-surface velocities from seismic reflection profiles recorded in the Hueco Basin of west Texas and southern New Mexico. The first-arrival data provided velocity information for the shallow subsurface, where commonly imaging is poor in shallow reflection data because of low fold and noise, allowing for reprocessing of field data with better results.

## Mutually and spatially constrained inversion

The third level of integration consists of the simultaneous inversion of different data sets in which the model parameters for different methods are mutually or spatially constrained. This approach, introduced by Auken and Christiansen (2004) for resistivity data, also has been applied successfully to seismic surface-wave data (Wisén and Christiansen, 2005; Socco et al., 2009; Boiero and Socco, 2010; Socco et al., 2010). The spatial constraints can be applied to neighboring models, imposing similarity of interface depth and seismic velocity of the same layer. The strength of the constraints is set according to expected lateral variability or following strategies based on normalized misfit values at the last iteration (Boiero and Socco, 2010). Experiments of integration between surface-wave dispersion curves and resistivity data with mutual constraints imposed only on the depth of the layer interfaces have been performed successfully by Wisén and Christiansen (2005).

Socco et al. (2009) show that further improvement can be obtained with the introduction of additional mutual constraints from borehole S-wave soundings. Seismic refraction traveltimes and surface-wave data also can benefit from a mutually constrained approach. This inversion strategy is very promising for further development in data integration because the constraints between different parameters also can be tuned according to physical laws and to other possible a priori knowledge about the site.

The constrained inversion scheme minimizes the objective function $L$ (Menke, 1989):

$$L = \left( \frac{1}{N+M+A} [(\mathbf{d}_{\text{obs}} - g(\mathbf{m}))^{\text{T}} \mathbf{C}_{\text{obs}}^{-1} (\mathbf{d}_{\text{obs}} - g(\mathbf{m}))] \right.$$
$$\left. + [(-\mathbf{Rm})^{\text{T}} \mathbf{C}_{Rp}^{-1} (-\mathbf{Rm})] \right)^{\frac{1}{2}}, \quad (1)$$

where the model parameters $\mathbf{m}$ are the spatially distributed 1D $V_S$ profiles considered in the inversion process. The $\mathbf{m}$ are linked to the observed data set $\mathbf{d}_{\text{obs}}$ (ensemble of experimental dispersion curves) with the associated observational covariance matrix $\mathbf{C}_{\text{obs}}$, $N$ is the number of data points, $M$ is the total number of the model parameters, $A$ is the number of constraints, while $\mathbf{R}$ is the lateral regularization matrix for thicknesses and velocities. The effectiveness of the $\mathbf{R}$ matrix depends on the strength of the constraints described in the covariance matrices $\mathbf{C}_{Rp}$ (Auken and Christiansen, 2004; Tarantola, 2005).

For surface-wave analysis, the forward response $g(\mathbf{m})$ is, for instance, given by the Haskell (1953) and Thomson (1950) approach as modified by Dunkin (1965) for a stack of homogeneous elastic isotropic layers, assuming the experimental curves are fundamental modes for the pseudo-Rayleigh waves. All the dispersion curves along the line are inverted simultaneously, minimizing the common objective function of equation 1, and the number of output models is equal to the number of 1D soundings located along the seismic line. The constraints and the data are both part of the inversion. Information from one model will spread to neighboring models through the lateral constraints; the result is a smoothly varying pseudo-2D model. The output models represent a balance between the constraints, the physics, and the experimental data. Model parameters with little influence on the data will be controlled by the constraints. The lateral and a priori constraints are scaled according to the model separation so they are weakened with increased separation.

Using a deterministic local-search approach, the model solution updated at the *nth* iteration can be expressed as:

$$\mathbf{m}_{n+1} = \mathbf{m}_n + ([\mathbf{G}^{\text{T}} \mathbf{C}_{\text{obs}}^{-1} \mathbf{G} + \mathbf{R}_{\mathbf{p}}^{\text{T}} \mathbf{C}_{Rp}^{-1} \mathbf{R}_{\mathbf{p}} + \lambda \mathbf{I}]^{-1}$$
$$\times [\mathbf{G}^{\text{T}} \mathbf{C}_{\text{obs}}^{-1} (\mathbf{d}_{\text{obs}} - g(\mathbf{m}_n)) + \mathbf{R}_{\mathbf{p}}^{\text{T}} \mathbf{C}_{Rp}^{-1} (-\mathbf{R}_{\mathbf{p}} \mathbf{m}_n)]), \quad (2)$$

where the Jacobian $\mathbf{G}$ represents the sensitivity matrix. Considering nonlinearity of the problem, the iterative procedure is stabilised by the Marquart damping parameter $\lambda$ (Marquart, 1963).

We present three real-world examples of data integration obtained with laterally and mutually constrained inversion.

The first example refers to a data set acquired in a populated Alpine valley for a seismic response study (Socco et al., 2009). Site geology is characterized by shallow fluvial sediments with an expected thickness of 10 to 50 m, overlying lacustrine sediments. The bedrock depth is expected to be more than 100 m in the central part of the valley. High-resolution reflection surveys were performed along two lines across the valley, with the main task of identifying the bedrock position. The survey also contained P-wave and S-wave downhole tests.

Significant ground roll is present in the seismic records and we have analyzed it to retrieve information about the $V_S$ distribution in the overburden. Each profile is about 800 m long and has 240 active channels with 10-Hz vertical geophones, 2-m geophone spacing, 6-m shot spacing, 1-ms sampling rate, 2-s recording time, antialias filter, and no low-cut filters. The seismic lines are characterized by varied topography and therefore have been subdivided into data sets with a constant slope. The data contain a variable S/N, which is caused by human activities. Because of the dense spatial sampling and long offsets, the data allow for a good spectral resolution and recovery of dispersion curves over a wide frequency band. The presented example is for a subsection of one of the two seismic lines.

The overabundance of records in the seismic reflection acquisition allowed a multifold processing approach to be used also for surface waves (Socco et al., 2009; Boiero and Socco, 2010). Several windows were considered along the survey line and all the shots with active channels in the selected windows were used. The different shots were stacked in the *f-k* domain to improve the S/N of retrieved dispersion curves, and single-fold shots were used to retrieve the experimental uncertainty of each dispersion-curve data point. In particular, four dispersion curves were retrieved in 100 m of seismic line. A preliminary Monte Carlo inversion was performed to select model parameterization (Figure 5). The Monte Carlo inversion suggested the presence of a velocity decrease (Figure 5), which is in good agreement with the geology of the site. A velocity decrease is a very challenging and interesting target for surface-wave analysis. The fundamental mode can come very close to higher modes so a very accurate zero search is needed for the solution of the forward problem to avoid misidentifying different modes.

The laterally constrained inversion of the four dispersion curves confirmed the presence of a low-velocity layer embedded in stiffer layers for three of the inverted curves. Further improvement of the laterally constrained inversion solution was obtained by introducing independent a priori information from a downhole test in the vicinity of seismic line. We introduced the downhole S-wave velocity profile as a fixed velocity model that influences the surface-wave models through lateral constraints. Because the a priori information is located at a distance from the seismic line, its influence is reduced because the lateral constraints are scaled with distance. The comparison in Figure 8 shows that the velocity inversion, confirmed by the downhole test, also becomes evident in the velocity profile located at 24 m,

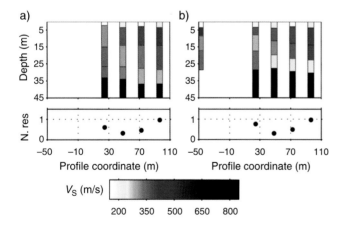

**Figure 8.** Laterally constrained inversion of dispersion curves extracted by seismic reflection data at Torre Pellice (Italy); the first three profiles refer to the same data of Figure 5. (a) Surface waves only; (b) surface-wave inversion is constrained by the results of a downhole test performed nearby. After Socco et al., 2006.

making the whole model more internally consistent. This indicates that the result obtained for the profile at 24 m without introducing a priori information falls in a local minimum because of the high uncertainties of the dispersion curve.

The second example refers to data acquired over a wide alluvial fan located in La Salle (in the Aosta Valley, Italy). The maximum thickness of the Quaternary deposits is 200 m and the fan is mainly composed of alluvial deposits (sands and gravels), polygenic slivers, pebbles, and blocks. Boreholes show the typical chaotic sequences of gravelly soils in alpine alluvial fans, with no marked layering down to a depth of about 50 m.

The survey performed at the La Salle site is described in detail in Socco et al. (2008). Two crossing seismic reflection lines about 1 km long were acquired using the same acquisition parameters used at the Torre Pellice site. Two P- and S-wave downhole tests were performed down to a depth of 50 m and five passive surface-wave acquisitions were acquired using circular arrays of 24 2-Hz vertical geophones with variable diameter ranging from 45 to 70 m. The seismic reflection data were processed to obtain the seismic cross section down to bedrock depth. In addition, P-wave first arrivals were inverted with a tomographic algorithm to obtain the P-wave velocity distribution and the surface-wave dispersion curves extracted from the same data set were inverted to retrieve S-wave information.

Surface waves present in seismic reflection records were processed with the same approach used for the data from Torre Pellice. Because of the very good quality of the surface-wave data, it was possible to reduce the number of stacked shots and retrieve a dense series of dispersion curves along the two seismic lines (58 and 38 curves along line L1 and line L2, respectively). Based on the result of a preliminary Monte Carlo inversion, we chose a multilayer model with a fixed geometry and vertical constraints on the velocities as the initial model for the laterally constrained inversion; hence, we inverted only for $V_S$.

Results obtained from the different surveys and analyses were integrated to supply a full and detailed seismic characterization of the site. Figure 9 shows the P-wave tomograpic imaging superimposed onto the seismic reflection sections for the two lines. The $V_P$ profiles retrieved through downhole tests also are depicted using the same color scale adopted for the tomographic results. The investigation depth of the downhole tests was limited to 50 m because of the borehole length. Downhole tests validate the velocity values obtained by P-wave tomography in the shallower part of the model; in turn, the P-wave tomography confirms the first reflection is related to the water table ($V_P$ about

1500 m/s). As far as the deeper reflection is concerned, the P-wave tomography had only a sufficient penetration depth on a portion of line 2, which confirms the reflection is related to contact with the bedrock ($V_P$ about 4400 m/s). A very detailed $V_S$ model was obtained combining the downhole tests, the passive surface-wave data, and the surface waves extracted from seismic reflection records. The seismic reflection sections are compared to all $V_S$ results in Figure 9.

Analysis of the surface waves extracted from seismic reflection records supplied the S-wave velocity of the first 60 m along the sections. Obtained values are in good agreement with both downhole test results and passive data inversion. Because of the merging of active and passive surface-wave data, the investigation depth was extended to more than 100 m, allowing shear-wave velocities to be estimated down to the bedrock. The bedrock depth from surface-wave analysis agrees well with the deeper reflection in the seismic reflection sections.

This case history shows that it is possible to obtain complementary information from different seismic surveys. The result is a comprehensive characterization of a site and the same set of experimental data can be analyzed profitably with different techniques to maximize the amount of retrieved information.

The third example refers to a geophysical campaign conducted on an alpine site in Linthal (Switzerland) in the canton of Glarus, in collaboration with Joseph Fourier of the University of Grenoble. Details about the application can be found in Socco et al. (2010). The aim of the geophysical investigation was to characterize the volume of the Sandalp rock avalanche deposit. Information was available about the topography before the rock falls that occurred in the valley. The deposit material is strongly heterogeneous with an irregular topography. Several P-wave tomography and resistivity tomography were acquired over the deposit.

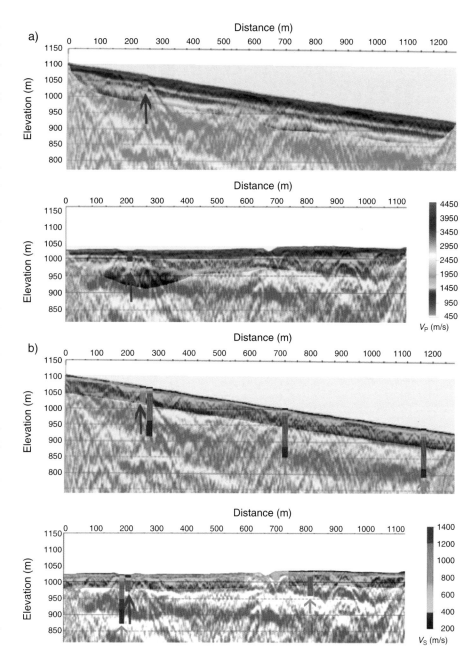

**Figure 9.** (a) $V_P$ model obtained through P-wave tomography and downhole tests (red arrows), compared with seismic reflection results (top graph, line 1; bottom graph, line 2). (b) $V_S$ model obtained through surface-wave active and passive data (cyan arrows), laterally constrained inversion of surface waves extracted from seismic reflection records and down hole tests (red arrows), compared with seismic reflection results (top graph, line 1; bottom graph, line 2). After Socco et al., 2008. Used by permission of EAGE.

The acquisition was performed with a typical seismic refraction scheme using three 24-channel Geode (Geometrix) seismographs (for a total of 72 channels), 4.5-Hz vertical geophones with 2-m geophone spacing, a weight-drop source with 12-m source spacing, a time sampling of 0.250 ms, and an acquisition window of 2 s.

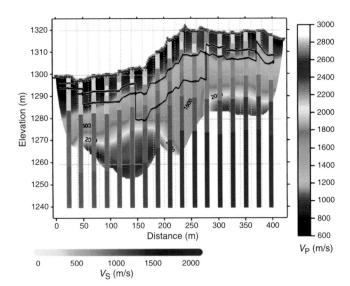

**Figure 10.** The Sandalp rock avalanche deposit: the S-wave velocity model from surface wave laterally constrained inversion superimposed to the P-wave velocity model from the P-wave traveltime tomography. The two analyses have been conducted on the same data set. Reprinted from Boiero and Socco, 2010. Used by permission of EAGE.

Surface waves present in the data were processed with the same scheme described for the Torre Pellice and La Salle data sets. We obtained 17 dispersion curves along the seismic line. The data quality is rather critical, and the frequency band of the dispersion curve ranges from 8 to 50 Hz, with strong variations along the seismic line. The data were inverted through a laterally constrained inversion algorithm. The initial model was selected through a preliminary analysis of some experimental dispersion curves (velocity versus wavelength) and transformed into a layered model by the operator, according to a minimum parameterization criterion. A single initial model was adopted for the whole data set.

A comparison between the $V_S$ profiles obtained by the laterally constrained inversion of surface waves and the $V_P$ section obtained by the tomography is shown in Figure 10. Black lines represent the boundary of the site elevation before the rock avalanche with an accuracy that ranges from 2 to 10 m. Results of the two seismic surveys allowed the deposit to be delineated. The S-wave velocity retrieved from surface-wave analysis shows the strong heterogeneities of the deposit materials, although the P-wave tomography points out that the deposit materials present a lower velocity than host materials. In terms of lateral variation identification, the surface waves in the zone between 90 and 140 m show a stiff uppermost layer overlying a softer one. The P-wave tomography shows an increase in

the average velocity of the deposit in the same portion of the seismic line. Very similar patterns are also evident between 240 and 300 m.

The investigation depth obtained with surface-wave data is greater than the one obtained with P-wave refraction tomography, given the same acquisition array. In the shallower portion of the model, where both S- and P-wave velocities have been estimated, the S-wave velocities vary between 200 and 1600 m/s. These values are reasonable, considering the P-wave velocity range (600 to 2500 m/s). In the deepest part of the model, where only S-wave velocity is retrieved, the S-wave velocity is higher (1800 to 2000 m/s) and can be associated with the top of the bedrock.

The presented examples show that spatial and mutual integration of different data might be used profitably to improve knowledge about the velocity model of a site.

## Joint inversion

The highest level of integration is reached by joint inversion of different data sets. Contrary to mutually constrained inversion that produces separate models for different data sets, joint inversion provides a unique model that must honor all available experimental data. Joint inversion can be performed using both local or global search methods and a unique objective function that accounts for all the data. The different nature of the data requires the adoption of an appropriate form of normalization to provide an adequate weighting of different contributions to the misfit function. The coupling of the models can be specified in several ways, e.g., accounting only for the geometry, or introducing physical laws to link model parameters. The joint inversion is, in general, a very promising approach because it exploits different sensitivities to model parameters and allows model parameters which are poorly resolved by one method to be better constrained by the other and vice versa.

Hering et al. (1995) and Comina et al. (2002) apply a joint inversion algorithm to synthetic and field data in which surface-wave data and vertical electric-sounding data are coupled only on layer geometry. In spite of the very different nature of model parameters (seismic velocities and soil resistivities), significant improvements are obtained with respect to individual inversions. Dal Moro (2008) uses a biobjective evolutionary algorithm to perform a joint inversion of refraction traveltimes and surface-wave dispersion curves. The coupling of the two seismic velocities is obtained using a range of Poisson's ratio both for the generation of initial models and in the crossover and mutation of the genetic algorithm. He shows that intrinsic ambiguities related to hidden layers can be resolved with this approach. The joint approach of Dal Moro (2008) also is used for joint inversion of surface-wave dispersion curves and reflection traveltimes (Dal Moro and Pipan, 2007). Picozzi and Albarello (2007) combine genetic and linearized algorithms for a two-step joint inversion of Rayleigh wave dispersion and H/V spectral-ratio curves.

An important category of algorithms is represented by joint inversion with cross-gradient constraints, which exploit structural resemblance between two property images. Gallardo and Meju (2003, 2004), Linde et al. (2006), Tryggvason and Linde (2006), and Gallardo and Meju (2007) use this approach to jointly invert paired combinations of geoelectromagnetic and traveltime data sets. Other applications of the cross-gradient approach include integrating ground-penetrating-radar traveltime data, cross-hole electrical-resistivity data, and seismic traveltime data for better determination of lithologic boundaries in hydrogeologic studies (Linde et al., 2006; Linde et al., 2008).

Schuler (2008) presents a method that inverts dispersive Love and Rayleigh waves together with P- and S-wave first arrivals to obtain medium properties of a layered earth model with increasing velocity with depth. Synthetic data sets were inverted jointly, using both a Pareto and relative objective approach. The obtained results demonstrate that joint inversion of Rayleigh, Love, and refracted waves return more accurate and better-constrained results than individual inversions. In particular, shallow layer information is constrained better by the surface-wave method, whereas deeper layer properties are reconstructed by using first-arrival inversion.

Here we present a joint inversion algorithm of P-wave refraction and surface-wave data. The joint-inversion scheme is very similar to the one used for the laterally constrained inversion (equation 2), but here model **m** refers to a single model that includes the thicknesses, densities, and S- and P-wave velocities of a 1D linear elastic model.

The dispersion curve and the P-wave traveltimes are inverted simultaneously, by minimizing a common objective function that includes both $V_P$ and $V_S$:

$$\hat{L} = \left( \frac{1}{\hat{N}} [(\hat{\mathbf{d}}_{obs} - \hat{\mathbf{g}}(\hat{\mathbf{m}}))^T \hat{\mathbf{C}}_{obs}^{-1} (\hat{\mathbf{d}}_{obs} - \hat{\mathbf{g}}(\hat{\mathbf{m}}))] \right)^{\frac{1}{2}}. \quad (3)$$

The model update at the $n$th iteration is given by:

$$\hat{\mathbf{m}}_{n+1} = \hat{\mathbf{m}}_n + ([\hat{\mathbf{G}}^T \hat{\mathbf{C}}_{obs}^{-1} \hat{\mathbf{G}} + \lambda \mathbf{I}]^{-1}$$
$$\times [\hat{\mathbf{G}}^T \hat{\mathbf{C}}_{obs}^{-1} (\hat{\mathbf{d}}_{obs} - \hat{\mathbf{g}}(\hat{\mathbf{m}}_n))]). \quad (4)$$

In equations 3 and 4, the 1D model $\hat{\mathbf{m}}$ is linked to the experimental data set $\hat{\mathbf{d}}_{obs}$, which includes both P-wave traveltimes and surface-wave phase velocities, while $\hat{\mathbf{C}}_{obs}$ is the observational covariance matrix associated to $\hat{\mathbf{d}}_{obs}$. The Jacobian **G** represents the sensitivity matrix, $N$ is the number of data points (sum of dispersion curve and P-refraction traveltimes data points), and $\hat{\mathbf{g}}(\hat{\mathbf{m}})$ includes the forward responses for both surface waves and P-refraction traveltimes.

The forward response for surface waves is the same as the one used in the laterally constrained inversion. The forward modeling for P-wave refraction traveltimes has been implemented with a 1D ray tracing that computes P-wave traveltimes for any velocity distribution, also including possible velocity decreases. The $V_P$ values enter both in the P-wave traveltime forward model and in the dispersion-curve calculation, while $V_S$ values affect only the latter.

The initial model is set on the basis of a preliminary analysis of the experimental data. The P-wave traveltimes are processed with the intercept time method to get a preliminary P-wave velocity model. The S-wave velocity of each layer is obtained from the experimental dispersion curve. The dispersion curve is analyzed in an alternative domain obtained by multiplying the phase velocity by 1.1 and computing the wavelength $\lambda/2$ for each data point. The wavelength $\lambda/2$ is considered as a pseudodepth and the velocity is assigned to the corresponding layer.

After retrieving this preliminary model, the number of layers is increased by splitting each layer in two, to account for possible hidden layers and the water table. The position of the interfaces is the same for $V_P$ and $V_S$ profiles but the layer properties might change in one profile and not in the other. For instance, in soft sediments, the P-wave velocity changes significantly in saturated media with respect to unsaturated media, hence, the water table represents a strong variation of $V_P$, but it has a negligible effect on $V_S$. Considering that the dispersion curve is poorly sensitive to density, the latter is fixed a priori to reasonable values.

Tests on synthetic data (Piatti and Socco, 2010) show that the proposed algorithm exploits the different sensitivities of the two data sets to model parameters and allows hidden layers and velocity decreases also to be resolved properly in terms of P-wave velocity.

In the following, we present an application of the proposed algorithm to a field case. The data set was collected in Tarcento (Friuli, Italy) within a research project aimed at building the near-surface velocity model of the sedimentary basin to be used in a seismic response study. The stratigraphy of the site is known from a borehole drilled nearby. It is composed of a gravel layer down to 17 m, overlying a layer of clay and silt with a thickness of about 6 m, and a stiff flysch layer down to the investigation depth. The borehole log is used as a benchmark for the joint-inversion results. The water table is expected at a depth between 11 and 14 m from the ground surface. The seismic data have been acquired using a 48-channel spread of 4.5-Hz vertical geophones, with 2-m geophone spacing. The source was a 5-kg hammer and two shot points at the beginning and end of the spread with 2-m minimum offset were acquired.

For data storage convenience, separate shots were performed adopting 0.5-ms sampling and a 2-s acquisition window for surface waves, and 0.125-ms sampling and a 200-ms window for seismic refraction. Several shots were stacked in the time domain during acquisition for seismic refraction data. For surface waves, the shot repetitions were recorded separately and stacked in the *f-k* domain during processing. After assessing the absence of lateral variations by analyzing the two reciprocal shots, the data of the two shots were averaged.

Surface-wave data were processed in the *f-k* domain to extract the experimental dispersion curve by picking of spectral maxima on the stacked spectra. The P-wave first arrivals were picked manually for reciprocal shots and averaged. The surface-wave dispersion curve and P-wave traveltimes are shown in Figure 11c and 11d, respectively. Initial models of $V_P$ and $V_S$ are shown in Figure 11a and 11b, along with the obtained final profiles of both joint and individual inversions. Results are compared to individual inversions performed with the same inversion algorithm, the same

forward modeling, and the same initial models. The results also are compared with borehole information and show better agreement of joint-inversion results with independent geologic information compared to individual inversions.

## Constrained joint inversion

If other a priori information about physical properties of materials is available at the site, a further constraint can be added in the joint-inversion scheme. For instance, if the water-table level is known with good accuracy, the Poisson's ratio of different materials can be constrained to reasonable values. This introduces additional information in the inversion process and allows different model parameters to be coupled from a physical point of view.

The objective function can be expressed as

$$\hat{L} = \left( \frac{1}{\hat{N} + \hat{M} + \hat{A}} [(\hat{\mathbf{d}}_{\mathbf{obs}} - \hat{\mathbf{g}}(\hat{\mathbf{m}}))^T \hat{\mathbf{C}}_{\mathbf{obs}}^{-1} (\hat{\mathbf{d}}_{\mathbf{obs}} - \hat{\mathbf{g}}(\hat{\mathbf{m}}))] \right.$$
$$\left. + [(\nu_{\mathrm{prior}} - \nu(\hat{\mathbf{m}}))^T \mathbf{C}_{\mathbf{Pp}}^{-1} (\nu_{\mathrm{prior}} - \nu(\hat{\mathbf{m}}))] \right)^{\frac{1}{2}}, \quad (5)$$

where $\hat{N}$ is the number of data points (sum of dispersion curve and P-refraction traveltime data), $\hat{M}$ is the number of the model parameters, and $\hat{A}$ is the number of constraints.

Similarly to the laterally constrained case, the model solution updated at the *n*th iteration can be expressed as indicated in equation 6, where suitable constraints terms

**Figure 11.** The Tarcento site (Italy): initial and final models for individual and joint inversions: (a) $V_S$ profiles, (b) $V_P$ profiles, (c) fitting on surface-wave dispersion curves, (d) fitting on P-wave traveltimes.

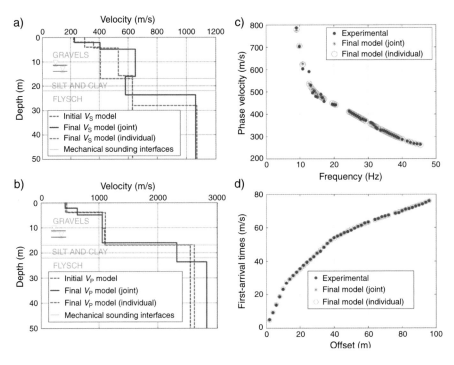

are added with respect to equation 4:

$$\hat{\mathbf{m}}_{n+1} = \hat{\mathbf{m}}_n + ([\hat{\mathbf{G}}^T \hat{\mathbf{C}}_{obs}^{-1} \hat{\mathbf{G}} + \mathbf{P}_p^T \mathbf{C}_{Pp}^{-1} \mathbf{P}_p + \lambda \mathbf{I}]^{-1}$$

$$\times [\hat{\mathbf{G}}^T \hat{\mathbf{C}}_{obs}^{-1} (\hat{\mathbf{d}}_{obs} - \hat{\mathbf{g}}(\hat{\mathbf{m}}_n))$$

$$+ \mathbf{P}_p^T \mathbf{C}_{Pp}^{-1} (\nu_{prior} - \nu(\hat{\mathbf{m}}_n))]). \quad (6)$$

As in the joint approach, here the $\hat{\mathbf{m}}$ model refers to a single spatial model that includes thicknesses, densities, and the S- and P-wave velocities of a 1D linear elastic model. The additional terms in equation 6 with respect to equation 4 contain the constraints on the a priori information on a petrophysical parameter (in this case, the Poisson's ratio $\nu$) relative to the model $\hat{\mathbf{m}}$. The constraint terms are formally similar as the ones used in the laterally constrained approach (equation 2), but they are related to the physical relation between $V_P$ and $V_S$ through the Poisson's ratio instead of the spatial constraints on velocities and thicknesses of the model set **m**.

The algorithm uses the constraints on Poisson's ratio values to converge towards a consistent model in terms of $V_S$ and $V_P$ relationship that complies with the data through the relation

$$\nu_n = \frac{(V_P/V_S)^2 - 2}{2 \cdot (V_P/V_S)^2 - 2}. \quad (7)$$

The constraint strength can be tuned acting on the covariance matrix $\mathbf{C}_{Pp}$ of equation 6. If $\mathbf{C}_{Pp}^{-1}$ is set to zero, no a priori information on the Poisson's ratio is considered in the inversion process, i.e., $V_P$ and $V_S$ values are varied in the inversion process according to experimental data only, as for joint inversion. Tests on synthetic data (Piatti and Socco, 2010) show that the introduction of further constraints for Poisson's ratio improves the results of the joint inversion.

The advantage of using the indicated objective functions (equations 1, 3, and 5), which are based on the same framework, is the possibility of adding suitable constraints between model parameters of single models, between model parameters of different models and with respect to any type of a priori information or physical law that can be traduced in terms of useful information. Furthermore, the introduction of covariance matrices in the formulations allows for tuning of the constraints, giving the possibility of strengthening or relaxing the interdependence between the constrained parameters on the basis of their reliability.

A limitation for the presented joint approach is related to its 1D nature: At this stage, the joint surface wave and P-refraction traveltime inversion algorithm can be applied only in the case of stratified earth because both P-wave refraction and surface-wave forward models are one dimensional.

# Conclusions

Even though the improvements that might be obtained by data integration (particularly for levels 3 and 4) are well recognized in literature, these approaches are not applied extensively. Review of the existing literature on surface-wave analysis in different fields of application shows that only in a very limited number of cases are other data integrated in the inversion process, even though they are often available.

We showed that a strong synergy exists in surface-based seismic data acquisition. A unique data set can provide different information if the acquisition is designed properly to accomplish the requirements imposed by different methods. High-resolution seismic reflection and seismic refraction tomography data provides high-coverage data sets that can be processed profitably to extract surface-wave dispersion curves along the seismic line.

If several surface-wave soundings are available along a line or over an area, they can be integrated with spatially constrained inversion to get an internally consistent S-wave velocity model even in the case of complex systems and smooth lateral heterogeneities. Reliability of the final model can be improved by the introduction of further constraints from borehole data and invasive seismic tests. A posteriori integration of the retrieved S-wave velocity models with the results of seismic reflection processing and/or P-wave tomography leads to a comprehensive characterization of the site that also allows Poisson's ratio and other properties of the materials to be retrieved.

A posteriori comparison and constrained inversion represent different ways to integrate different data during interpretation. With these approaches, different models are retrieved from different data sets and similarities between models can be imposed by constraints. If a higher level of integration is preferred, a joint-inversion scheme should be adopted. With joint inversion, a unique model that complies with different data sets is obtained.

We have proposed a joint inversion of P-wave refraction traveltimes and surface-wave dispersion curves. The joint-inversion approach solves for a unique P- and S-wave velocity model. The P-wave velocities used in both seismic refraction and surface-wave forward modeling produce a physical coupling of the two problems. Because P-wave velocity value can vary according to the water-table level (which does not affect S-wave velocity) the parameterization of the reference model should be selected carefully.

Initial model selection is based on the apparent P-wave velocity profile retrieved by the intercept time method. This initial model is characterized by the correct estimation of the velocity of some layers and provides a good fitting with experimental traveltimes. Therefore, it is placed in the "valley shape minimum" of the solution space for seismic refraction. This model is modified splitting the layers in two and selecting S-wave velocity on the basis of the surface-wave dispersion curve. This approach accounts for the two data sets from the beginning of the inversion process.

The improvement obtained by joint inversion with respect to individual inversions is related to the different sensitivities of the two data sets with respect to model parameters. Model parameters that are poorly resolved or not resolved by one method are constrained properly by the other.

A remarkable feature of the proposed approach derives from the implicit improvement of surface-wave inversion. Indeed, in most approaches, values of $V_P$ (or Poisson's ratio) are set a priori together with soil density, relying on the reduced sensitivity of the Rayleigh-wave dispersion curve to these parameters. This is an attempt to reduce the number of unknowns in the ill-posed inversion problem. Nevertheless, the difference between dry and fully saturated conditions causes an abrupt change in the values of apparent Poisson's ratio in soils. In dry conditions, values of the Poisson's ratio are related to the solid skeleton (0.2 to 0.3), although in fully saturated conditions (accounting for the theory of wave propagation in porous media) the Poisson's ratio is associated to undrained conditions (no relative motion between solid skeleton and interstitial fluid with Poisson's ratio close to 0.5). If this aspect is not properly recognized in the choice of the fixed Poisson's ratio values, it can lead to large errors in the estimated S-wave velocity profile.

In our proposed joint inversion, the P-wave velocity model in the inversion of surface-wave data is constrained by P-wave refraction data and the presence of a water table is taken into account implicitly and automatically. Moreover, a further improvement might be obtained if the Poisson's ratio value is constrained on the basis of the a priori knowledge of the water-table depth.

The data integration, performed at different levels, is a powerful and promising approach for shallow seismic data interpretation. In particular, when data are integrated in the inversion, the final results are more consistent and reliable than the ones obtained with single inversions. Even though the use of constrained or joint inversion is still limited to a minority of cases, the advantages of this procedure suggest adopting it whenever different data are available at a site.

Future developments should focus on the physical coupling of different parameters and data in inversion processes, also including field potential, resistivity, and electromagnetic data to achieve a comprehensive characterization of sites.

## Acknowledgments

We would like to thank all the authors who gave their permission to include their results in this chapter, especially J. Ivanov, R. D. Miller, J. Xia, D. Steeples, C. B. Park, C. Calderón-Macías, B. Luke, C. G. Lai, A. Godio, G. J. Rix, L. Sambuelli, C. Comina, and R. Wisén. We also would like to thank our colleagues, students, and the technical staff who contributed to the data acquisition, in particular D. Jongmans, D. Hantz, S. Stocco, C. Calzoni, E. Bena, and G. Bianchi. Data for the La Salle and Torre Pellice sites were collected within the EU SISMOVALP Interreg IIIB Alpine Space Programme, project for seismic hazard and alpine-valley response analysis. Data for the Rojo Piano site were collected within the characterization program aimed at reconstruction coordinated by DPC (Department of Civil Protection, Italy). Data for the Tarcento site were acquired within the PRIN-2007 Project on prediction of earthquake seismic motion in proximity of seismic sources, financed by the Italian Ministry of Research. Particular thanks are due to K. Holliger, Ö. Yilmaz, and two anonymous reviewers who provided very useful and constructive remarks to improve this contribution.

## References

Aldridge, D. F., and D. W. Oldenburg, 1992, Refractor imaging using an automated wavefront reconstruction method: Geophysics, **57**, 378–385.

Auken, E., and A. V. Christiansen, 2004, Layered and laterally constrained 2D inversion of resistivity data: Geophysics, **69**, 752–761.

Boiero, D., and L. V. Socco, 2010, Retrieving lateral variations from surface wave dispersion curves: Geophysical Prospecting, **58**, doi: 10.1111/j.1365.2478.2010.00877.x.

Burger, H. R., 1992, Exploration geophysics of the shallow subsurface: Prentice Hall, Inc.

Calderón-Macías, C., and B. Luke, 2007, Improved parameterization to invert Rayleigh-wave data for shallow profiles containing stiff inclusions: Geophysics, **72**, no. 1, U1–U10.

Comina, C., and S. Foti, 2007, Surface wave tests for vibration mitigation studies: Journal Of Geotechnical And Geoenvironmental Engineering, **133**, 1320–1324.

Comina, C., S. Foti, L. Sambuelli, L. V. Socco, and C. Strobbia, 2002, Joint inversion of VES and surface wave data: 2002 Symposium on the Application of Geophysics to Environmental and Engineering Problems, Proceedings (CD-ROM).

Dal Moro, G., 2008, $V_S$ and $V_P$ vertical profiling via joint inversion of Rayleigh waves and refraction traveltimes by means of bi-objective evolutionary algorithm: Journal of Applied Geophysics, **66**, 15–24.

Dal Moro, G., and M. Pipan, 2007, Joint inversion of surface wave dispersion curves and reflection traveltimes via multi-objective evolutionary algorithms: Journal of Applied Geophysics, **61**, 56–81.

Deidda, G. P., G. Ranieri, G. Uras, P. Cosentino, and R. Martorana, 2006, Geophysical investigations in the Flumendosa River Delta, Sardinia, Italy — Seismic reflection imaging: Geophysics, **71**, no. 4, B121–B128.

Dunkin, J., 1965, Computation of modal solutions in layered, elastic media at high frequencies: Bulletin of the Seismological Society of America, **55**, 335–358.

Ernst, F., 2007, Long-wavelength statics estimation from guided waves: 69th EAGE Conference and Exhibition, Expanded Abstracts, E033.

Ewing, M., G. P. Woollard, and A. C. Vine, 1939, Geophysical investigations in the emerged and submerged Atlantic coastal plain. Part 3: Barnegat Bay, New Jersey section: GSA Bulletin, **50**, 257–296.

Foti, S., 2003, Small strain stiffness and damping ratio of Pisa clay from surface wave tests: Géotéchnique, **53**, 455–461.

Foti, S., L. Sambuelli, L. V. Socco, and C. Strobbia, 2003, Experiments of joint acquisition of seismic refraction and surface wave data: Near-surface Geophysics, **1**, 119–129.

Foti, S., and C. Strobbia, 2002, Some notes on model parameters for surface wave data inversion: 2002 Symposium on the Application of Geophysics to Environmental and Engineering Problems, Proceedings, CD-ROM.

Gabriels, P., R. Snieder, and G. Nolet, 1987, In situ measurements of shear-wave velocity in sediments with higher-mode Rayleigh waves: Geophysical Prospecting, **35**, 187–196.

Gallardo, L. A., and M. A. Meju, 2003, Characterization of heterogeneous near-surface materials by joint 2D inversion of DC resistivity and seismic data: Geophysical Research Letters, **30**, 1658.

———, 2004, Joint two-dimensional DC resistivity and seismic traveltime inversion with cross gradient constraints: Journal of Geophysical Research, **109**, 1–11.

———, 2007, Joint two-dimensional cross-gradient imaging of magnetotelluric and seismic traveltime data for structural and lithological classification: Geophysical Journal International, **169**, 1261–1272.

Haskell, N., 1953, The dispersion of surface waves on multilayered media: Bulletin of the Seismological Society of America, **43**, 17-34.

Hawkins, L. V., 1961, The reciprocal method of routine shallow seismic refraction investigations: Geophysics, **26**, 806–819.

Hayashi, K., T. Matsuoka, and H. Hatakeyama, 2005, Joint analysis of a surface-wave method and micro-gravity survey: Journal of Environmental and Engineering Geophysics, **10**, 175–184.

Hering, A., R. Misiek, A. Gyulai, T. Ormos, M. Dobroka, and L. Dresen, 1995, A joint inversion algorithm to process geoelectric and surface wave seismic data. Part I: Basic ideas: Geophysical Prospecting, **43**, 135–156.

Ivanov, J., R. D. Miller, J. Xia, and D. Steeples, 2005a, The inverse problem of refraction traveltimes, part II: Quantifying refraction nonuniqueness using a three-layer model: Pure and Applied Geophysics, **162**, 461–477.

Ivanov, J., R. D. Miller, J. Xia, D. Steeples, and C. B. Park, 2005b, The inverse problem of refraction traveltimes, part I: Types of geophysical nonuniqueness through minimization: Pure and Applied Geophysics, **162**, 447–459.

———, 2006, Joint analysis of refractions with surface waves: An inverse solution to the refraction-traveltime problem: Geophysics, **71**, no. 6, R131–R138.

Lai, C. G., S. Foti, A. Godio, G. J. Rix., L. Sambuelli, and L. V. Socco, 2000, Caratterizzazione geotecnica dei terreni mediante l'uso di tecniche geofisiche: Rivista Italiana di Geotecnica, Pàtron editore, XXXIV Supplemento al No. 3, 99–118 (in Italian).

Lambrecht, J. L., and R. D. Miller, 2006, Catastrophic sinkhole formation in Kansas: A case study: The Leading Edge, 324.

Lanz, E., H. Maurer, and A. G. Green, 1998, Refraction tomography over a buried waste disposal site: Geophysics, **63**, 1414–1433.

Linde, N., A. Binley, A. Tryggvason, L. B. Pedersen, and A. Revil, 2006, Improved hydrogeophysical characterization using joint inversion of crosshole electrical resistance and ground-penetrating radar traveltime data: Water Resources Research, **42**, W12404.

Linde, N., A. Tryggvason, J. E. Peterson, and S. S. Hubbard, 2008, Joint inversion of crosshole radar and seismic traveltimes acquired at the South Oyster Bacterial Transport Site: Geophysics, **73**, no. 4, G29–G37.

Mari, J. L., 1984, Estimation of static corrections for shear-wave profiling using the dispersion properties of Love waves: Geophysics, **49**, 1169–1179.

Marquart, D., 1963, An algorithm for least squares estimation of nonlinear parameters: SIAM Journal of Applied Mathematics, **11**, 431–441.

Menke, W., 1989, Geophysical data analysis: Discrete inverse theory: Academic Press Inc.

Miller, K. C., S. H. Harder, D. C. Adams, and T. O'Donnell Jr., 1998, Integrating high-resolution refraction data into near-surface seismic reflection data processing and interpretation: Geophysics, **63**, 1339–1347.

Monaco, P., G. Totani, G. Barla, A. Cavallaro, A. Costanzo, A. D'Onofrio, L. Evangelista, S. Foti, S. Grasso, G. Lanzo, C. Madiai, M. Maraschini, S. Marchetti, M. Maugeri, A. Pagliaroli, O. Pallara, A. Penna, F. Santucci de Magistris, A. Saccenti, G. Scasserra, F. Silvestri, A. L. Simonelli, G. Simoni, P. Tommasi, G. Vannucchi, and L. Verrucci, 2010, Geotechnical aspects of 2009 l'Aquila earthquake: Earthquake Geotechnical Engineering Satellite Conference: XVIIth International Conference on Soil Mechanics and Geotechnical Engineering, http://www.reluis.it/images/stories/Monaco_et_al_2010.pdf, accessed 21 July 2010.

Neducza, B., 2007, Stacking of surface waves: Geophysics, **72**, no. 2, V51–V58.

O'Donnell, T. M. Jr., K. C. Miller, and J. C. Witcher, 2001, A seismic and gravity study of the McGregor geothermal system, southern New Mexico: Geophysics, **66**, 1002–1014.

Palmer, D., 1981, An introduction to the generalized reciprocal method of seismic refraction interpretation: Geophysics, **46**, 1508–1518.

——, 2001, A new direction for shallow refraction seismology: integrating amplitudes and traveltimes with the refraction convolution section: Geophysical Prospecting, **49**, 657–673.

——, 2010a, Non-uniqueness with refraction inversion — A syncline model study: Geophysical Prospecting, **58**, 203–218, doi: 10.1111/j.1365-2478.2009.00818.x.

——, 2010b, Non-uniqueness with refraction inversion — The Mount Bulga shear zone: Geophysical Prospecting, doi: 10.1111/j.1365-2478.2009.00855.x.

Park, C. B., R. Miller, J. Xia, J. Ivanov, G. V. Sonnichsen, J. A. Hunter, R. L. Good, R. A. Burns, and H. Christian, 2005, Underwater MASW to evaluate stiffness of water-bottom sediments: The Leading Edge, 724–728.

Piatti, C., and L. V. Socco, 2010, Joint inversion of P-refraction traveltimes and surface waves dispersion curves: EAGE, 72nd Technical Conference and Exhibition, Extended Abstracts.

Picozzi, M., and D. Albarello, 2007, Combining genetic and linearized algorithms for a two-step joint inversion of Rayleigh wave dispersion and H/V spectral ratio curves: Geophysical Journal International, **169**, 189–200.

Rockwell, D. W., 1967, A general wavefront method, in A. W. Musgrave, ed., Seismic refraction prospecting: SEG, 363–415.

Schuler, J., 2008, Joint inversion of surface waves and refracted P- and S-waves: M.S. thesis, Eidgenössische Technische Hochschule Zurich.

Semblat, J. F., M. Kham, E. Parara, P. Y. Bard, K. Pitilakis, K. Makra, and D. Raptakis, 2005, Seismic wave amplification: Basin geometry vs. soil layering: Soil Dynamics and Earthquake Engineering, **25**, 529–538.

Socco, L. V., and D. Boiero, 2008, Improved Monte Carlo inversion of surface wave data: Geophysical Prospecting, **56**, 357–371.

Socco, L. V., D. Boiero, C. Comina, S. Foti, and R. Wisén, 2008, Seismic characterisation of an alpine site: Near-surface Geophysics, **6**, 253–265.

Socco, L. V., D. Boiero, S. Foti, and R. Wisén, 2009, Laterally constrained inversion of ground roll from seismic reflection records: Geophysics, **74**, no. 6, G35–G45.

Socco, L. V., D. Boiero, R. Wisén, and S. Foti, 2006, Laterally constrained inversion of ground roll of seismic reflection records: 76th Annual International Meeting, SEG, Expanded Abstracts, **25**, 1441.

Socco, L. V., D. Jongmans, D. Boiero, S. Stocco, M. Maraschini, K. Tokeshi, and D. Hantz, 2010, Geophysical investigation of the Sandalp rock avalanche deposits: Journal of Applied Geophysics, **70**, 277–291.

Socco, L. V., and C. Strobbia, 2004, Surface wave methods for near-surface characterisation: A tutorial: Near-surface Geophysics, **2**, 165–185.

Stefani, J. P., 1995, Turning-ray tomography, Geophysics, **60**, 1917–1929.

Strobbia, C., 2003, Surface wave methods: Acquisition, processing and inversion: Ph.D. dissertation, Politecnico di Torino.

Tarantola, A., 2005, Inverse problem theory and methods for model parameter estimation: Society for Industrial and Applied Mathematics.

Thomson, W. T, 1950, Transmission of elastic waves through a stratified solid medium: Journal of Applied Physics, **21**, 89.

Thornburg, H. R., 1930, Wavefront diagrams in seismic interpretation: AAPG Bulletin, **14**, 18–200.

Tryggvason, A., and N. Linde, 2006, Local earthquake (LE) tomography with joint inversion for P- and S-wave velocities using structural constraints: Geophysical Research Letters, **33**, L07303.

Wathelet, M., D. Jongmans, and M. Ohrnberger, 2004, Surface-wave inversion using a direct search algorithm and its application to ambient vibration measurements: Near-surface Geophysics, **2**, 211–221.

Wisén, R., and Christiansen, A. V., 2005, Laterally and mutually constrained inversion of surface wave seismic data and resistivity data: Journal of Environmental and Engineering Geophysics, **10**, 251–262.

Woolery, E. W., 2002, SH-wave seismic reflection images of anomalous foundation conditions at the Mississinewa Dam, Indiana: Journal of Environmental and Engineering Geophysics, **7**, 161–168.

Yilmaz, Ö., M. Eser, and M. Berilgen, 2006, A case study of seismic zonation in municipal areas: The Leading Edge, **25**, 319–330.

Zhang, J., and M. N. Toksöz, 1998, Nonlinear refraction traveltime tomography: Geophysics, **63**, 1726–1737.

Zhu, X., D. P. Sixta, and B. G. Andstman, 1992, Tomostatics: Turning-ray tomography + static corrections: The Leading Edge, **11**, 15–23.

Zywicki, D. J., 1999, Advanced signal processing methods applied to engineering analysis of seismic surface waves: Ph.D. dissertation, Georgia Institute of Technology.

# Section 2

# Methodology

Chapter 5

# Inversion for the Stochastic Structure of Subsurface Velocity Heterogeneity from Surface-based Geophysical Reflection Images

James Irving[1], Marie Scholer[2], and Klaus Holliger[2]

## Abstract

Much previous seismic and ground-penetrating radar (GPR) research has focused on investigating, theoretically and empirically, the relationship between the statistical characteristics of subsurface velocity heterogeneity and those of the associated surface-based reflection image. However, an effective and robust method for solving the corresponding inverse problem has not been presented. Assuming that waves are weakly scattered in the subsurface, a relatively simple relationship can be derived between the 2D autocorrelation of a geophysical reflection image and that of the underlying velocity field. A Monte Carlo inversion strategy based on this relationship can then be used to generate sets of parameters describing the autocorrelation of velocity that are consistent with recorded reflection data. Results of applying that strategy to realistic synthetic seismic and GPR data indicate that the inverse solution is inherently nonunique in that many combinations of the vertical and horizontal correlation lengths that describe the velocity heterogeneity can yield reflection images with the same 2D autocorrelation structure. However, the ratio of each of those combinations is approximately the same and corresponds to the aspect ratio of the velocity heterogeneity, which suggests that the aspect ratio is a quantity that can be recovered reliably from geophysical-reflection-survey data.

## Introduction

For the past two decades, estimation of the stochastic structure of subsurface velocity heterogeneity from back-scattered wavefields recorded at the earth's surface has been of significant interest in many areas of geophysics. Specifically, geophysicists have questioned whether the second-order statistical properties of seismic and/or high-frequency electromagnetic reflection data can be used to gain information about the correlation structure of the subsurface velocity field through which the waves have traveled. In that context, the lateral correlation properties are of particular interest because they cannot be obtained from the analysis of single borehole data. In crustal seismic studies, such information might yield important clues regarding the nature of the crystalline crust, which could be used to unravel the tectonic history and/or petrology of the probed region in addition to providing estimates of physical properties for use in a variety of geodynamical problems. In exploration seismology, information on the lateral correlation structure of velocity could provide critical constraints for the geostatistical characterization of petroleum reservoirs and/or mineral deposits between sparsely spaced wells. Finally, in ground-penetrating-radar (GPR) research, the velocity of high-frequency electromagnetic waves in the subsurface depends largely on soil water content, which is related to key hydrologic variables above and

---

[1]*Formerly at University of Lausanne, Institute of Geophysics, Lausanne, Switzerland; presently at University of Guelph, School of Engineering, Guelph, Ontario, Canada.*
[2]*University of Lausanne, Institute of Geophysics, Lausanne, Switzerland.*

below the water table. Consequently, information regarding the correlation structure of GPR velocity might be useful for constraining stochastic models of groundwater flow and contaminant transport.

Numerous studies investigate the relationship between the second-order spatial statistics of subsurface velocity heterogeneity and those of the corresponding seismic or GPR reflection data (Gibson and Levander, 1990; Gibson, 1991; Holliger et al., 1992; Holliger et al., 1994; Hurich, 1996; Pullammanappallil et al., 1997; Line et al., 1998; Rea and Knight, 1998; Bean et al., 1999; Hurich and Kocurko, 2000; Knight et al., 2004; Oldenborger et al., 2004; Poppeliers and Levander, 2004; Dafflon et al., 2006; Carpentier and Roy-Chowdhury, 2007; Knight et al., 2007; Poppeliers, 2007; Irving et al., 2009). In most of those studies, the aim is to understand the link between the lateral correlation properties of the velocity field and reflection data, again because information about velocity heterogeneity in the horizontal direction cannot be obtained from single boreholes and because the vertical correlation properties of a reflected wavefield tend to contain little information about anything other than the source wavelet (e.g., Yilmaz, 1987; Ulrych, 1999). Although seismic and GPR reflection data have been considered separately in all of those works, it is important to note that the results and conclusions obtained are mutually transferable because of the strong mathematical analogies that exist between seismic and electromagnetic wave propagation (e.g., Carcione and Cavallini, 1995).

The current consensus in the seismic community appears to be that although a relationship clearly exists between the horizontal correlation structure of seismic data and that of the underlying velocity distribution for weakly scattering media, the nature of the relationship is largely unclear. Initial theoretical work by Gibson (1991), Holliger et al. (1992, 1994), and Pullammanappallil et al. (1997) attempts to show that the lateral statistics of those two fields should be equivalent for small-magnitude velocity fluctuations. That was done using a primary-reflectivity-section (PRS) model for a seismic section, which considers the imaged/migrated section as the convolution product of the subsurface reflectivity field and source wavelet. However, empirical studies by Hurich (1996) and by Hurich and Kocurko (2000) show that although a strong degree of correlation does exist between the lateral statistics of velocity and the resulting seismic-reflection image for weakly scattering media, those statistics are certainly not equivalent. In fact, Bean et al. (1999) and Carpentier and Roy-Chowdhury (2007) point out the fundamental dependence of the lateral correlation structure of a seismic image on bandwidth and on the vertical-derivative operator that acts

to create reflection coefficients from a velocity field. Both of these elements are key features of the PRS model, and thus it can be concluded that they have not been considered properly in the previous theoretical work.

In the GPR literature, similar research efforts lead to essentially identical results and conclusions. Rea and Knight (1998) and Dafflon et al. (2006) see good agreement between the lateral geostatistics of a cliff-face photograph and those of a GPR image collected along the top of the cliff, which leads them to conclude that the lateral statistics of a GPR image and the underlying water-content distribution are likely equivalent. In both cases, the rather weak assumption is made that the grayscale tones in the photograph are related to sediment grain size and thus to water content. Knight et al. (2007) also observe similarities between the horizontal correlation statistics of GPR data and neutron-probe measurements. However, Knight et al. (2004) and Oldenborger et al. (2004) notice that the lateral correlation structure of a GPR image is affected significantly by its vertical resolution, which is controlled by the GPR antenna frequency. Again, this indicates that the assumption of equivalence between the lateral statistics of water content and GPR reflection data is not generally valid and that the physics of the underlying wave-propagation phenomena must be accounted for more appropriately. Thus, what is needed in the seismic and GPR domains is a better conceptual and theoretical understanding of the link between the spatial correlation statistics of a velocity field and the corresponding geophysical reflection image, which can account for the observations noted above and can be used to design an effective strategy to solve the inverse problem.

We address the above concern and present a consistent methodological framework for estimating the essential parameters describing the 2D autocorrelation of subsurface velocity fluctuations when provided with the 2D autocorrelation of a seismic or GPR reflection image. Our work builds on recent research by Irving et al. (2009), who address the problem of estimating the lateral correlation statistics of water content for the GPR case through development of a numerical model linking the 2D autocorrelations, which then is used in a Bayesian Markov-chain Monte Carlo (MCMC) inversion strategy.

We begin by developing a similar numerical model for the general case of seismic or GPR data, which links the 2D autocorrelation of a reflection image with that of the underlying wave-velocity distribution. Next we introduce a Monte Carlo inversion methodology based on this model that we consider not only simpler conceptually than the MCMC approach of Irving et al. (2009) but also more effective for determining the parameters of the velocity autocorrelation model. Finally, we show the application of our inversion methodology to realistic, zero-offset, finite-difference-modeled GPR data and to finite-difference-modeled and stacked seismic data.

## Model Derivation

To relate the second-order statistics of a seismic or GPR reflection section with those of the underlying velocity distribution, we begin with the common assumption that the subsurface velocity field can be written as the sum of a slowly varying deterministic or background component $v_0(x, z)$ and a stochastic component exhibiting zero-mean fluctuations $\Delta v(x, z)$:

$$v(x, z) = v_0(x, z) + \Delta v(x, z). \tag{1}$$

It is the stochastic component, or velocity-perturbation field, that gives rise to the reflections recorded at the surface and whose second-order statistics we wish to estimate. For both seismic and GPR data, the subsurface reflectivity field $r(x, z)$ can be approximated by taking the vertical derivative of the velocity-perturbation field (Pullammanappallil et al., 1997; Dafflon et al., 2006; Poppeliers, 2007):

$$r(x, z) \approx \frac{\partial}{\partial z} \Delta v(x, z). \tag{2}$$

If we assume that (1) the single scattering of incident wave energy prevails (an assumption inherent to the majority of seismic and GPR processing, imaging, and interpretation strategies), (2) dispersion in the reflection data is minimal or has been corrected so that a constant wavelet shape can be assumed, and (3) mode conversions in the seismic case are absent or have been removed by processing, then a recorded zero-offset reflection wavefield $p(x, t)$, after proper processing and migration, can be expressed approximately as a simple vertical convolution of the reflectivity field with the input seismic or GPR wavelet $w(t)$. That yields

$$\begin{aligned}p(x, t) &= w(t) \ * \ r(x, t) \\ &= w(t) \ * \ \left[\frac{\partial t}{\partial z}\frac{\partial}{\partial t}\Delta v(x, t)\right],\end{aligned} \tag{3}$$

where the asterisk denotes the convolution operator and $t$ is traveltime. Here, reflectivity is mapped to time using an estimate of $v_0(x, z)$, which can be obtained from common-midpoint (CMP) analysis or tomography. For a con-

stant value of $v_0$, the depth-derivative term in equation 3 is given by

$$\frac{\partial t}{\partial z} = \frac{2}{v_0}. \tag{4}$$

Equation 3 is the PRS model for a geophysical reflection section. It considers the section as a collection of 1D, vertical-incidence, primaries-only recordings and is regarded widely as the ideal reflection image. Consequently, it is the goal that processing and migration hope to achieve (e.g., Claerbout, 1985; Gibson, 1991). When the underlying assumptions are satisfied, the PRS model has the remarkable ability to capture the key features of realistic reflection data, especially their statistical characteristics. Indeed, as we will see in the examples presented later, finite-difference-modeled GPR and seismic data from weakly scattering heterogeneous structures are well predicted by this formulation.

In previous work on the topic of estimating the correlation statistics of subsurface velocity variability from reflection-survey data, the PRS model in equation 3 has been used in seismic and GPR studies to justify theoretically a rather simple approach to addressing that problem, which has been to assume that the lateral correlation statistics of a reflection image are the same as those of the underlying velocity field (Gibson and Levander, 1990; Gibson, 1991; Holliger et al., 1992; Holliger et al., 1994; Pullammanappallil et al., 1997; Rea and Knight, 1998; Dafflon et al., 2006). In particular, previous work has justified that because the wavelet convolution and derivative in equation 3 operate only along the time/depth coordinate, those operations have no effect on the lateral correlation structure of the reflection image, and thus the average horizontal statistics of the two fields should be equivalent.

However, recent work by Carpentier and Roy-Chowdhury (2007) has demonstrated clearly that the process of differentiating in time or depth (and indeed, any filtering operation along that dimension such as convolution with the source wavelet) has a profound effect on the lateral correlation structure. As mentioned previously, empirical studies support those conclusions, showing a lack of equivalence between the lateral statistics of velocity and the corresponding reflection image (Hurich, 1996; Hurich and Kocurko, 2000) and a strong dependence of the lateral image-correlation structure on the spectral content of the source wavelet (Bean et al., 1999; Knight et al., 2004; Oldenborger et al., 2004). In other words, recent work and empirical results demonstrate that the vertical convolution and differentiation operations, which are key features of the PRS model, cannot be separated from the horizontal correlation structure of a seismic or GPR image. Thus, proper incorporation of those features into a conceptual model is absolutely critical for development of a reliable inversion strategy to estimate the correlation structure of velocity from that of the image.

In our work, we also make the assumption that a properly processed and migrated geophysical reflection image can be expressed as the convolution product of the subsurface reflectivity field with the source wavelet. Unlike in previous research, however, we express the PRS model in the depth domain, which allows us to formulate a convenient relationship between the 2D autocorrelations of velocity and the image that accounts properly for the observations noted above. This is given by

$$p(x, z) = w(z) * r(x, z)$$
$$= w(z) * \frac{\partial}{\partial z} \Delta v(x, z). \tag{5}$$

Inherent to the depth-domain formulation in equation 5 is the assumption that the background velocity function $v_0(x, z)$ does not change significantly over the subsurface region that is being analyzed so that the wavelet in depth $w(z)$ remains approximately constant. That implies that $w(z)$ is not stretched or compressed because of significant velocity changes, which we think is justified in many cases. It is important to note that if this assumption is violated severely, regions between which significant changes in wavelet shape occur always could be considered separately in an analysis.

As mentioned, the PRS model, expressed in time or depth, does a good job of capturing the overall behavior of realistic reflection-survey data when the underlying assumptions are satisfied approximately. We have found that to be particularly the case for the correlation statistics of the data. However, accounting for the horizontal resolution limits of a migrated reflection image is one addition to this formulation that we have found necessary for its general applicability under a wide range of scenarios.

It is well known that horizontal resolution in an unmigrated reflection section is limited by the Fresnel zone, which describes the area on a subsurface reflector that contributes to the recorded data (e.g., Berkhout, 1984). The radius of the Fresnel zone increases with depth and wavelength, and hence, unmigrated reflection sections have a horizontal resolution that worsens with increasing depth and decreasing frequency. The imaging/migration process acts to collapse the size of the Fresnel zone to a uniform theoretical value on the order of the dominant wavelength and thus to improve the lateral resolution everywhere in a

reflection image (Berkhout, 1984; Stolt and Benson, 1986). However, this lateral-resolution limit is not predicted by equation 5 because all mathematical operations occur along the vertical dimension. Thus, to capture the correlation statistics of realistic reflection data accurately under a wide range of conditions, it is necessary to modify the PRS formulation. We do that by adding a horizontal-resolution filter $h(x)$ to equation 5:

$$p(x, z) = w(z) * \frac{\partial}{\partial z} \Delta v(x, z) * h(x). \tag{6}$$

Based on methodological considerations (e.g., Chen and Schuster, 1999) and extensive empirical testing involving comparison of finite-difference-modeled seismic and GPR data with those predicted by the PRS model, we have found that a simple Gaussian low-pass filter, whose width is determined by the dominant signal wavelength, is an effective choice for $h(x)$. The filter is of the form

$$h(x) = \exp\left(-\frac{x^2}{2c^2}\right), \tag{7}$$

where $c$ determines the filter width and is set so that the distance between the two points where the Gaussian reaches 1% of its maximum amplitude is equal to the dominant wavelength. We fully acknowledge that other horizontal filter operators, possibly taking into account more details of the reflection experiment such as migration aperture or, in the GPR case, the transmitter radiation pattern, could be considered instead of equation 7 for greater accuracy. Nevertheless, we have found that $h(x)$, as expressed above, is effective for the purposes set out here. A primary reason for choosing the Gaussian operator is its combination of simplicity and demonstrated effectiveness.

Noting that the derivative operator in equation 6 can be treated as a filter whose position in the equation can be shifted to act on the wavelet, we also can express the modified PRS model as

$$p(x, z) = \Delta v(x, z) * f(x, z), \tag{8}$$

where

$$f(x, z) = \frac{\partial}{\partial z} w(z) * h(x). \tag{9}$$

Here, we simply have lumped together all items acting on the velocity-perturbation field (that is, the vertical derivative, input wavelet, and horizontal-resolution filter) into a single 2D filter operator $f(x, z)$, which yields a simple con-

a)

$p(x,z)$     $\Delta v(x,z)$     $f(x,z)$

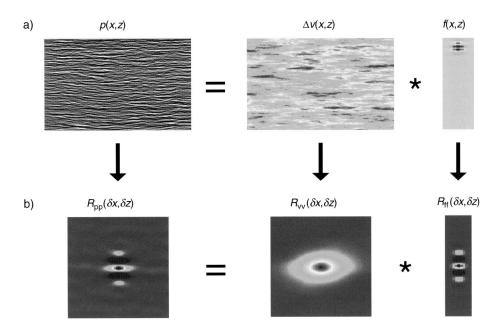

**Figure 1.** Schematic illustration of (a) our modified PRS model, which links a GPR or seismic-reflection image to the underlying velocity-perturbation field, and (b) the corresponding relationship between the 2D autocorrelations.

b)

$R_{pp}(\delta x, \delta z)$     $R_{vv}(\delta x, \delta z)$     $R_{ff}(\delta x, \delta z)$

volutional relationship between $\Delta v(x, z)$ and $p(x, z)$. Now taking the 2D Fourier transform of equation 8 and calculating the power spectrum of both sides, we have

$$|P(k_x, k_z)|^2 = |\Delta V(k_x, k_z)|^2 \, |F(k_x, k_z)|^2, \qquad (10)$$

where $k_x$ and $k_z$ are the horizontal and vertical wavenumbers, respectively. Taking the inverse Fourier transform and making use of the Wiener-Khintchine theorem linking the power spectrum and autocorrelation functions (e.g., Couch, 2001), we arrive at the final result:

$$R_{pp}(\delta x, \ \delta z) = R_{vv}(\delta x, \ \delta z) \ * \ R_{ff}(\delta x, \ \delta z). \qquad (11)$$

Equation 11 states that the 2D spatial autocorrelation of the PRS image $R_{pp}(\delta x, \delta z)$, where $\delta x$ and $\delta z$ refer to the horizontal and vertical lags, respectively, is related to the 2D spatial autocorrelation of the velocity-perturbation field $R_{vv}(\delta x, \delta z)$ through 2D convolution with the filter autocorrelation $R_{ff}(\delta x, \delta z)$. This result, albeit extremely simple, has profound implications in the sense that it provides us with an effective link between the second-order spatial statistics of a seismic or GPR velocity field and those of the corresponding processed and migrated reflection image. Most important, equation 11 represents the basis for the solution of the corresponding inverse problem, which we discuss in the following section. Figure 1 shows pictorially the meaning of and relationship between equations 8 and 11.

## Inversion Methodology

Now we will describe how equation 11 can be used to develop an inversion strategy to estimate a small number of parameters describing the 2D autocorrelation of subsurface velocity when provided with the 2D autocorrelation of a seismic or GPR reflection image. We assume knowledge of the general parametric form of the velocity autocorrelation model (e.g., spherical, exponential, von Kármán) so that only a small number of parameters is required for its description. Although we consider this to be reasonable because such information likely could be obtained from the analysis of borehole data or prior knowledge about the geologic environment (e.g., Kelkar and Perez, 2002), it is important to note that the inversion strategy outlined below is flexible and easily could allow for consideration of multiple parametric velocity autocorrelation models if required.

Notice that in equation 11, to relate $R_{vv}(\delta x, \delta z)$ and $R_{pp}(\delta x, \delta z)$ and thus to invert the 2D autocorrelation of a seismic or GPR reflection image for a set of parameters describing $R_{vv}(\delta x, \delta z)$, we require knowledge of $R_{ff}(\delta x, \delta z)$, the autocorrelation of the filter operator in equation 9. That can be obtained from the autocorrelation of the seismic or GPR source wavelet or equivalently its power spectrum because the horizontal-resolution filter in equation 9 is known. In Irving et al. (2009), such information regarding the source wavelet was obtained through spectral analysis of the direct ground wave traveling between radar antennae at a far-horizontal offset, where it could be isolated from the direct arrival traveling through the air. Although

Irving et al. (2009) prove that methodology to be quite effective for the GPR case, we propose a simpler approach to estimate the autocorrelation of the source wavelet $R_{\mathrm{ww}}(\delta z)$. This involves approximating it as the vertical autocorrelation of the recorded reflection image, which we denote by $R_{\mathrm{dd}}(0, \delta z)$. Here, subscript $_\mathrm{d}$ refers to the observed data and not to the data predicted by the PRS formulation, which is denoted by subscript $_\mathrm{p}$.

Our approach to estimating $R_{\mathrm{ww}}(\delta z)$ is based on the assumption that the vertical correlation properties of a seismic or GPR reflection image will be dominated by the source wavelet. This is supported by increasing evidence that regardless of the geologic environment, the power spectra of sonic logs uniformly obey a flicker-noise-type scaling behavior, as characterized by a decay of the power spectra proportional to $k_z^{-\beta}$, with $\beta$ ranging approximately from 0.5 to 1.5 (e.g., Kelkar and Perez, 2002; Holliger and Goff, 2003). The corresponding vertical autocorrelations differ from a Dirac impulse only in terms of their first zero crossing (Ulrych, 1999). This then implies that the vertical autocorrelation of the PRS resulting from convolution of the corresponding reflectivity series with the source wavelet essentially corresponds to the autocorrelation of the wavelet. Indeed, that assumption forms the basis for standard stochastic seismic-deconvolution algorithms, which have been proved through years of successful application in the petroleum industry (Yilmaz, 1987) despite the fact that the underlying assumption of whiteness of the reflectivity spectrum is clearly incorrect. We shall see later that the approximation $R_{\mathrm{ww}}(\delta z) \approx R_{\mathrm{dd}}(0, \delta z)$ is valid and robust, and hence it is effective for the correlation parameter inversion of both seismic and GPR data. In that context, it is also important to note that the more conventional explicit approaches for estimation of the source-wavelet characteristics of seismic and GPR data are notoriously difficult and prone to error (e.g., Belina et al., 2009).

Because the problem of estimating model parameters describing $R_{\mathrm{vv}}(\delta x, \delta z)$ given $R_{\mathrm{dd}}(\delta x, \delta z)$ is inherently nonlinear, is dimensionally small, and has the potential for multiple acceptable solutions, we consider a stochastic inversion approach. Stochastic inverse methods naturally account for the existence of multiple solutions, are inherently flexible with regard to how parameter sets are judged as acceptable, and avoid the use of gradient measures on a global objective function that are not suitable in the presence of multiple local minima. To address the similar problem of estimating the autocorrelation-model parameters describing subsurface water-content heterogeneity from reflection GPR data, Irving et al. (2009) use a Bayesian MCMC stochastic inversion methodology. Bayesian methods have the major benefit of embracing formal statistical theory, thus allowing for the determination of valid posterior statistics for the model parameters involved. However, in practice, such posterior uncertainty estimates rely strongly on our having an error-free forward model (or knowing the full impact of the model errors on the observed data) and having detailed knowledge of the statistical distribution of the data-measurement errors. In our case, and indeed in the case of most geophysical inverse problems, those conditions are not satisfied. That is, the forward model in equation 11 relating the velocity and reflection-image autocorrelations is only approximate and thus contains structural errors that are difficult to quantify, and we have little knowledge of the nature of the errors in the observed autocorrelation data.

For those reasons, we present a comparatively simple Monte Carlo inversion approach here. Although MCMC methods can offer significant computational benefits over standard Monte Carlo approaches because of the construction of the Markov chain, the inverse problem in our case contains very few parameters, so computational expense is not a critical issue. The Monte Carlo inversion strategy that we use to invert equation 11 consists of the following steps:

1) Define prior probability distributions or ranges for each model parameter to be estimated, which describe the 2D autocorrelation of velocity $R_{\mathrm{vv}}(\delta x, \delta z)$.

2) Define criteria describing an acceptable fit to the observed lateral autocorrelation of the processed and migrated reflection image $R_{\mathrm{dd}}(\delta x, 0)$. We have found that fitting only in the horizontal direction is necessary because the vertical correlation structure of the image is controlled completely by the source wavelet, as described above. In other words, if the fit to $R_{\mathrm{dd}}(\delta x, 0)$ is adequate, then we should have an acceptable fit to the whole 2D reflection-image autocorrelation.

3) Draw a proposed set of values for the parameters describing the 2D autocorrelation of velocity from the prior distributions defined in step 1 and calculate the corresponding $R_{\mathrm{vv}}(\delta x, \delta z)$ using the assumed parametric autocorrelation model (e.g., equation 12).

4) Calculate the predicted reflection-image autocorrelation $R_{\mathrm{pp}}(\delta x, \delta z)$ using the $R_{\mathrm{vv}}(\delta x, \delta z)$ obtained in step 3 and equation 11. To determine $R_{\mathrm{ff}}(\delta x, \delta z)$ in equation 11, we convolve $R_{\mathrm{dd}}(0, \delta z)$ with the autocorrelation of a finite-difference vertical-derivative operator and that of the known horizontal-resolution filter $h(x)$. As mentioned previously, we have found that $R_{\mathrm{dd}}(0, \delta z)$ gives a good estimate of the autocorrelation of the source wavelet $R_{\mathrm{ww}}(\delta z)$.

5) Compare the predicted and observed lateral autocorrelations $R_{pp}(\delta x, 0)$ and $R_{dd}(\delta x, 0)$. If the prediction fits the criteria defined in step 2, which generally means it lies within bounds prescribed around $R_{dd}(\delta x, 0)$, then the proposed set of velocity-autocorrelation-model parameters is considered to be possible given the reflection-image autocorrelation, and it is accepted. Otherwise, the proposed set of model parameters is rejected.

6) Return to step 3 and repeat the above procedure until the desired number of accepted realizations for the velocity-autocorrelation-model parameters has been obtained.

The above inversion procedure is summarized conceptually by the flowchart shown in Figure 2. Specific details regarding its use will be given in the next two sections, where we show the application of the methodology to realistic GPR and seismic data simulated using finite-difference methods over known velocity fields. For both the GPR and seismic cases, the inversion is performed using two sets of prior information, one in which the vertical correlation structure of velocity has been well constrained and the other in which it has not.

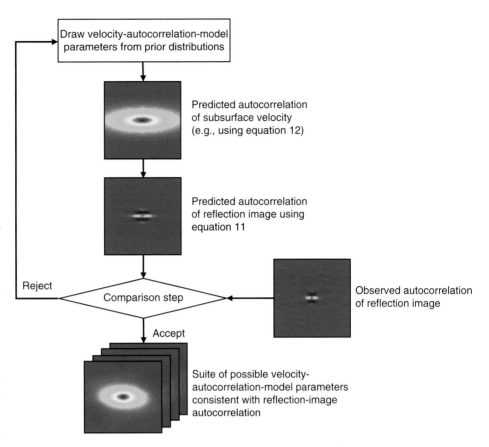

**Figure 2.** Flowchart illustrating our Monte Carlo approach for estimating the model parameters that describe the autocorrelation of the velocity-perturbation field given the autocorrelation of a GPR or seismic-reflection image.

## GPR Data Example

First we will show the application of the Monte Carlo inversion methodology described above to a realistic synthetic GPR data example. Figure 3a shows the subsurface velocity field that we consider for this case. The field is 50 m long and 15 m deep and is discretized on a grid that has uniform cell dimensions of $5 \times 5$ cm. It was generated based on an anisotropic exponential autocorrelation model whose form is given by

$$R(\delta x, \delta z) = \exp\left[-\sqrt{\left(\frac{\delta x}{a_x}\right)^2 + \left(\frac{\delta z}{a_z}\right)^2}\right], \qquad (12)$$

where model parameters $a_x$ and $a_z$ are the horizontal and vertical correlation lengths, respectively. We assume with equation 12 that the axis of anisotropy for the velocity heterogeneity is aligned with the $x$- and $z$-coordinates. To obtain the stochastic realization shown, we used a spectral-simulation technique that involved assigning a uniformly random phase spectrum to the amplitude spectrum corresponding to equation 12 and then taking the inverse Fourier transform (e.g., Goff et al., 1994). The parameters input into the spectral simulation were $a_x = 4$ m and $a_z = 0.4$ m.

The mean and standard deviation of the velocity field shown in Figure 3a are 0.08 and 0.002 m/ns, respectively. That is, the standard deviation of the velocity heterogeneity was set to 2.5% of the mean value. The minimum and maximum velocity values are 0.072 and 0.09 m/ns, which correspond to water-content values of approximately 0.19 and 0.31, respectively. That range for water content is representative of a saturated-zone scenario. However, our analysis methodology could be applied equally well to vadose-zone environments and has been tested successfully in that regard (Irving et al., 2009).

Figure 3b shows the 2D autocorrelation corresponding to Figure 3a, which was calculated using an algorithm based on the fast Fourier transform (e.g., Carpentier and Roy-Chowdhury, 2007). As specified, the autocorrelation follows an exponential form and was found to have best-fitting parameter values of $a_x = 0.34$ m and $a_z = 3.49$ m. Those values are slightly different from the input values noted above because of the inherent boxcar filtering effect related to limited extent of the modeling domain (Western and Blöschl, 1999). The best-fitting parameter values were obtained using a procedure that sought to minimize the least-squares misfit between the calculated autocorrelation and that predicted by equation 12 (e.g., Hurich and Kocurko, 2000; Carpentier and Roy-Chowdhury, 2007).

To simulate a GPR survey over the velocity field in Figure 3a, we used the MATLAB finite-difference-time-domain (FDTD) solution of Maxwell's equations in 2D Cartesian coordinates of Irving and Knight (2006). Values for the relative dielectric permittivity $\varepsilon_r$ were obtained for the numerical modeling using

$$\varepsilon_r = \frac{c^2}{v^2}, \tag{13}$$

where $c$ is the speed of light in free space. Because most surficial earth materials are nonmagnetic, the magnetic permeability was set equal to its free-space value everywhere in the simulation grid. For the electrical conductivity, we prescribed a uniform value of 1 mS/m. That is reasonable for the sand- and gravel-type materials amenable to GPR surveys because the effects of conductivity variations with water content predicted by Archie's law (Archie, 1942) would be minimal for the velocity variations shown. A thin layer that has $\varepsilon_r = 1$ was included at the top of the velocity field in Figure 3a to represent the air-earth interface, and the modeling domain was surrounded on all sides by perfectly matched-layer (PML) absorbing boundaries. Zero-offset GPR traces were simulated along the earth's surface at lateral increments of 0.2 m. The source-wavelet function that was input into the FDTD grid (shown in Figure 4a) had a center frequency of approximately 50 MHz and a bandwidth of two to three octaves. The modeling was performed using a time step of 0.25 ns, which was small enough to satisfy the Courant stability criterion. The grid spacing of 5 cm provided more than 10 points per minimum wavelength and thus allowed us to avoid numerical dispersion in the resulting data.

**Figure 3.** (a) Synthetic GPR velocity field generated using an exponential autocorrelation model. (b) Corresponding 2D autocorrelation. (c) Migrated and gained 50-MHz, zero-offset GPR reflection image, obtained by finite-difference modeling on (a). (d) Corresponding 2D autocorrelation. (e) PRS reflection image obtained from (a) using equation 8 and the same source wavelet used in (c). (f) Corresponding 2D autocorrelation.

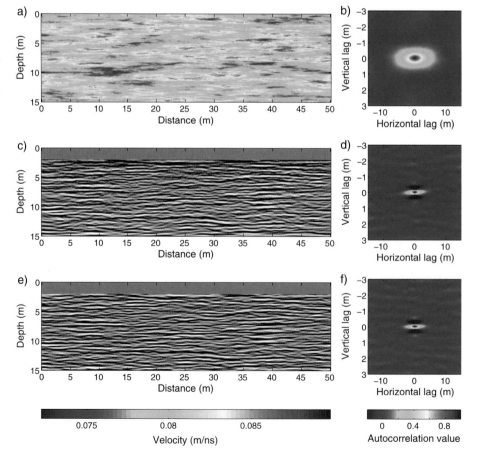

After the FDTD simulations were complete, the synthetic GPR data were gained in time using a smoothly varying function inversely proportional to the average trace envelope. That type of gain preserves the relative amplitudes of reflections along a trace but compensates for the effects of geometric energy spreading and attenuation. Note, however, that we have found the correlation statistics of GPR and seismic-reflection images to be relatively insensitive to the type of gain used. Next, the GPR data were depth-migrated in the frequency-wavenumber (*f-k*) domain using the constant-velocity algorithm of Stolt (1978) and the mean subsurface velocity value of 0.08 m/ns, which was assumed to be known. Figure 3c shows the resulting GPR image. The uppermost part of the image has been muted to exclude arrivals traveling directly between antennae that would bias our correlation analysis. The muted zone, although included in Figure 3c for easier comparison of the velocity field and reflection image, was not considered in the calculation of the image's 2D autocorrelation.

For comparison, Figure 3e shows the GPR section computed from the velocity field in Figure 3a using the modified PRS formulation in equation 8. To create that section, we first approximated the subsurface reflectivity field by taking the vertical finite-difference derivative of velocity, and then we convolved the result with the wavelet shown in Figure 4a and $h(x)$ in equation 7. Notice that the PRS does an excellent job of predicting the behavior of the more realistic FDTD-modeled data, which include the effects of 2D wave propagation and multiple scattering. The two images look very similar with regard to the location of major reflecting interfaces and, more important for our work, in terms of their statistical characteristics.

The latter point becomes more evident in Figure 3d and 3f, which shows the calculated 2D autocorrelations of the finite-difference and PRS images, respectively. The autocorrelations are almost identical, and therefore we can conclude that the PRS model can capture enough of the physics of the GPR experiment to represent properly the second-order statistics of a realistic GPR reflection image. Also notice the distinct difference between the autocorrelations in Figure 3d and 3f and that of the underlying velocity model in Figure 3b. In going from velocity to the corresponding reflection image, we have a significant change in the 2D autocorrelation in the vertical and horizontal directions, mostly as a result of differentiation and convolution with the source wavelet, which operate only along the vertical dimension. As mentioned, those vertical operations cannot be separated with regard to their effect on the lateral stochastic structure.

In Figures 5 through 8 below, we show the results of applying our Monte Carlo inversion strategy to estimate the exponential model parameters describing the velocity autocorrelation in Figure 3b, when provided with the finite-difference GPR image autocorrelation in Figure 3d. This is done for two sets of prior information, which correspond to our having detailed versus limited knowledge about the vertical correlation structure of the velocity heterogeneity. Table 1 summarizes the results of all of the inversions we performed.

As mentioned, for our inversion procedure, we require knowledge of the filter autocorrelation $R_{\text{ff}}(\delta x, \delta z)$, which requires knowledge of the autocorrelation or spectral con-

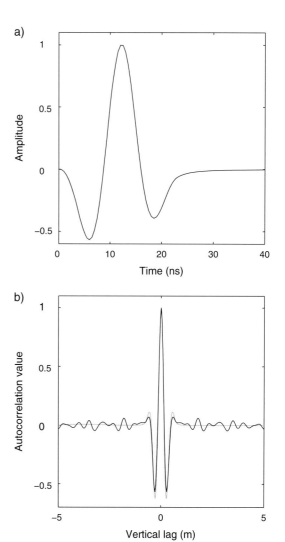

**Figure 4.** (a) Source wavelet used for the GPR finite-difference modeling. (b) Autocorrelation of the source wavelet converted to depth (gray) and vertical autocorrelation of the finite-difference-based GPR image (black).

**Table 1.** Summary of the Monte Carlo inversion results for the GPR and seismic data examples presented in Figures 3 through 15. The mean and standard deviation for each parameter were calculated based on 2000 output realizations.

| Data | Figures | Parameter | True value | Prior range | Mean | Standard deviation |
|------|---------|-----------|-----------|-------------|------|--------------------|
| GPR | 5, 6 | $a_z$ | 0.34 m | [0.25, 0.4] m | 0.33 m | 0.43 m |
| | | $a_x$ | 3.49 m | [0.1, 20] m | 3.35 m | 0.39 m |
| | | $a_x/a_z$ | 10.3 | | 10.2 | 0.4 |
| | 7, 8 | $a_z$ | 0.34 m | [0.1, 1] m | 0.68 m | 0.23 m |
| | | $a_x$ | 3.49 m | [0.1, 20] m | 6.50 m | 2.14 m |
| | | $a_x/a_z$ | 10.3 | | 9.7 | 0.7 |
| Seismic | 12, 13 | $a_z$ | 34 m | [30, 40] m | 35.2 m | 2.9 m |
| | | $a_x$ | 242 m | [10, 1000] m | 244.9 m | 22.5 m |
| | | $a_x/a_z$ | 7.1 | | 7.0 | 0.4 |
| | 14, 15 | $a_z$ | 34 m | [10, 500] m | 104.2 m | 37.2 m |
| | | $a_x$ | 242 m | [10, 1000] m | 649.6 m | 222.3 m |
| | | $a_x/a_z$ | 7.1 | | 6.4 | 1.1 |

tent of the source wavelet. Figure 4b plots the autocorrelation of the wavelet shown in Figure 4a along with the vertical autocorrelation of the finite-difference GPR image in Figure 3c. Notice in Figure 4b that our approximation of $R_{ww}(\delta z) \approx R_{dd}(0, \delta z)$ is indeed remarkably accurate and robust.

Also required for our inversion procedure is some measure of what represents an acceptable fit of the predicted lateral autocorrelation data $R_{pp}(x, 0)$ to the observed lateral autocorrelation of the reflection image $R_{dd}(x, 0)$. For all examples presented in this study, this was specified by prescribing bounds around $R_{dd}(x, 0)$ within which acceptable $R_{pp}(x, 0)$ curves must lie. We have found that the best results are obtained when such bounds are specified along both the horizontal and vertical axes. In other words, a predicted $R_{pp}(x, 0)$ curve is considered acceptable if at each point on the curve, it lies vertically or horizontally within a defined distance from the observed $R_{dd}(x, 0)$ curve. For our GPR inversions, lying within 0.4 m from the observed curve along the horizontal-lag axis or within 0.055 units from the observed curve along the autocorrelation-value axis meant that a predicted curve could be accepted. Choosing such bounding values is subjective, but the general idea is that they should best reflect our opinion about the maximum distance that acceptable autocorrelation curves can stray from the observed curve. Although not without problems, that choice is far less difficult than the one that

must be made in the MCMC inversion approach proposed by Irving et al. (2009), in which the generally unknown statistical distribution of the errors in the GPR image autocorrelation data must be specified.

We first performed the Monte Carlo inversion of the autocorrelation data in Figure 3d under the assumption that we had reasonably good prior knowledge of the vertical correlation length of the subsurface velocity heterogeneity $a_z$. In practice, that knowledge might come from analysis of complementary borehole data or, if they are unavailable, possibly from a promising new strategy to recover such information from reflection traces through spiking deconvolution, numerical integration, and subsequent statistical analysis (Poppeliers and Levander, 2004; Poppeliers, 2007). Conversely, we assume to have limited information regarding the lateral correlation length of the velocity heterogeneity $a_x$. For the prior distribution for $a_z$ in the inversion procedure, we used a uniform probability density function that has lower and upper bounds of 0.25 and 0.4 m, respectively, which rather narrowly contains the true value of 0.34 m. For $a_x$, we also used a uniform prior distribution, but we provided much broader bounds, between 0.1 m and 20 m, with the true value being equal to 3.49 m. We ran the inversion algorithm until 2000 realizations for $a_x$ and $a_z$ were accepted. On a 3.16-GHz computer with 3.23 GB of RAM, that procedure took approximately 10 hours.

Figures 5 and 6 show the results. In Figure 5a, the observed lateral GPR image autocorrelation is plotted in black, whereas all 2000 predicted autocorrelations that were accepted are plotted in gray. Notice that all of the accepted curves match the observed curve closely because of the fitting constraints imposed along both axes described above. Figure 5b, on the other hand, shows the observed and predicted vertical seismic data autocorrelations. We see that despite our not having imposed any fitting constraints in the vertical direction in the inversion procedure, all accepted sets of model parameters allow us also to match $R_{dd}(0, z)$ reasonably well. Again, that is because the vertical correlation structure of a seismic or GPR reflection image is controlled largely by the source wavelet. Finally, Figure 5c and 5d shows the horizontal and vertical autocorrelations of velocity corresponding to all 2000 accepted sets of model parameters, respectively, along with the corresponding "true" curves for the velocity field in Figure 3a. Although the accepted parameter sets predict a close match to the observed autocorrelations of the GPR reflection image, notice that they represent a slightly broader spread about the true velocity-field autocorrelations. Nevertheless, the range of the predicted curve behavior is quite limited in Figure 5c and 5d, which suggests that the inversion has reduced our uncertainty significantly regarding $a_x$ and $a_z$ compared with the input prior ranges for those parameters.

In Figure 6, we show the marginal histograms for $a_z$, $a_x$ and the aspect ratio $a_x/a_z$, which were computed from the 2000 parameter sets obtained in the inversion procedure. Figure 6a demonstrates that the inversion of the GPR image autocorrelation data does not allow for any further refinement in our knowledge of $a_z$. The histogram in Figure 6a is rather uniform in appearance, with values near the upper end of the prescribed range being slightly preferred. Conversely, in Figure 6b, we see that the inversion has done an excellent job of significantly narrowing our uncertainty regarding $a_x$, which was prescribed

initially to lie between 0.1 and 20 m. Here, the output histogram is narrow and peaked, with the mean and standard deviation being equal to 3.35 m and 0.39 m, respectively. These agree very well with the true value of 3.49 m and confirms that when $a_z$ is relatively well known and a restricted prior range is provided for this parameter, we can recover $a_x$ successfully from the second-order statistics of a GPR reflection image. Finally, Figure 6c shows that the aspect ratio of the velocity heterogeneity is well recovered by the inversion procedure (as could be expected given Figure 6a and 6b). The histogram is also narrow and peaked, with a mean and standard deviation of 10.2 and 0.4, respectively, compared with the true value of 10.3.

We now consider the inversion of the GPR image autocorrelation data in Figure 3d assuming that we have limited prior knowledge regarding the vertical correlation structure

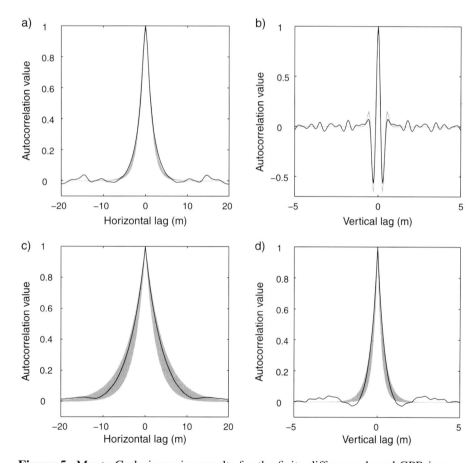

**Figure 5.** Monte Carlo inversion results for the finite-difference-based GPR image autocorrelation shown in Figure 3d, using a narrow prior range for the vertical correlation length. (a and b) Horizontal and vertical autocorrelations of the GPR image (black) and the predicted horizontal and vertical image autocorrelations corresponding to all the accepted parameter sets, obtained using equation 11 (gray). (c and d) Horizontal and vertical autocorrelations of the true velocity model in Figure 3a (black) and those corresponding to all accepted parameter sets (gray).

of velocity. All inversion parameters were kept the same as before except the prior uniform range for $a_z$, which was changed to have significantly broader bounds of 0.1 to 1 m. Figures 7 and 8 show the inversion results, which are summarized in Table 1.

In Figure 7a and 7b, we see again that all accepted sets of exponential model parameters provide a close fit to the horizontal and vertical GPR image autocorrelations, despite the fitting having been evaluated only in the horizontal direction. In Figure 7c and 7d, however, notice that

the accepted parameter sets represent a wide range of horizontal and vertical autocorrelation behavior compared with the case in which $a_z$ was constrained well. That is, the spread of the velocity autocorrelation curves corresponding to the accepted parameter sets is large, which suggests that without prior knowledge of $a_z$, there are many $a_x$-$a_z$ combinations describing the velocity heterogeneity that will yield almost identical reflection-image autocorrelations. That means the inverse problem for estimating $a_x$ and $a_z$ is inherently nonunique given only the image autocorrelation data, which we consider to have tremendous importance for future efforts in this domain of research.

The output histograms provide further insight into the nature of the nonuniqueness. In the marginal histogram for $a_z$ in Figure 8a, we see that the inversion essentially provides no useful information about the vertical correlation length of velocity. In fact, larger values in the considered interval can be seen to be strongly preferred, despite the true value being only 0.34 m. Similarly, in the marginal histogram for $a_x$ in Figure 8b, we see that the inversion provides completely unreliable results. Although the histogram clearly is peaked, the calculated mean and standard deviation are 6.50 m and 2.14 m, respectively, which are not at all helpful considering that the true value is 3.49 m.

In Figure 8c, however, we observe how all the accepted parameter sets in this nonunique inverse problem are related. Here, we plot the contoured joint histogram for $a_x$ and $a_z$, which allows us to identify any correlation between the accepted model parameters. The figure shows that all accepted $a_x$-$a_z$ combinations fall along an approximately linear trend, which demonstrates that they all have approximately the same aspect ratio. In Figure 8d, we indeed see that despite the seemingly inconsistent results in Figure 8a and 8b, the histogram for $a_x/a_z$ is narrow and peaked, with a mean and standard deviation of 9.74 m and 0.7 m, respectively. Those values agree very well with the true aspect ratio of the velocity heterogeneity of 10.3, which demonstrates that the aspect ratio still can be well recovered when we are provided with limited prior information regarding $a_x$ and $a_z$. Given the increasing evidence for the seemingly ubiquitous and remarkably uniform nature of scale invariance in the distribution of petrophysical parameters (Kelkar and Perez, 2002; Holliger and Goff, 2003), it is arguable that the aspect ratio is the most important and valuable parameter with regard to geostatistical characterization of the probed subsurface region.

The above results show that stochastic velocity fields that have different lateral and vertical correlation lengths but the same aspect ratio will generate reflection data that appear visually different but have approximately the same

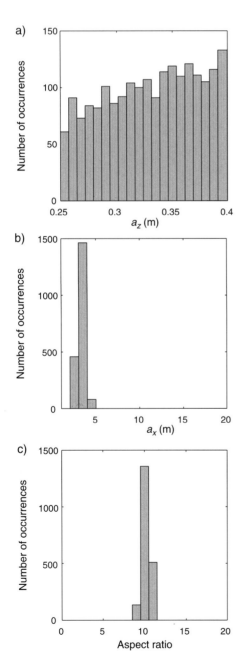

**Figure 6.** Parameter histograms for the inversion results shown in Figure 5.

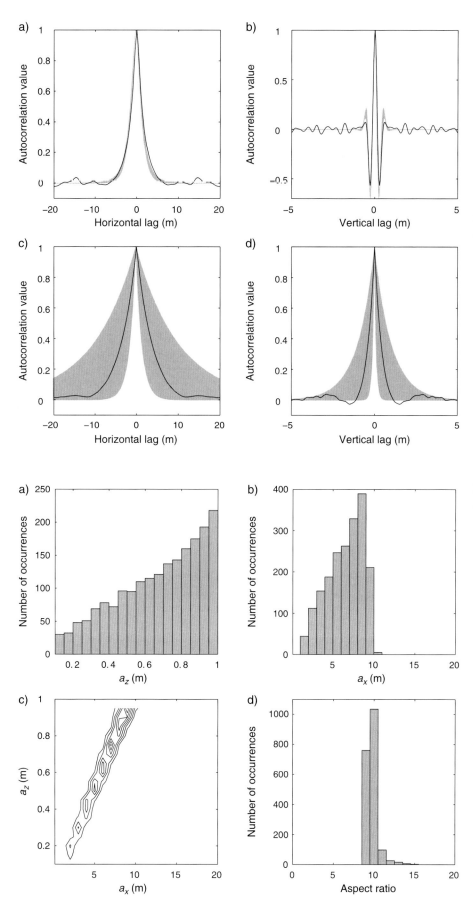

**Figure 7.** Monte Carlo inversion results for the finite-difference-based GPR image autocorrelation in Figure 3d, using a broad prior range for the vertical correlation length. (a and b) Horizontal and vertical autocorrelations of the GPR image (black) and the predicted horizontal and vertical image autocorrelations corresponding to all the accepted parameter sets, obtained using equation 11 (gray). (c and d) Horizontal and vertical autocorrelations of the true velocity model in Figure 3a (black) and those corresponding to all accepted parameter sets (gray).

**Figure 8.** Parameter histograms for the inversion results shown in Figure 7.

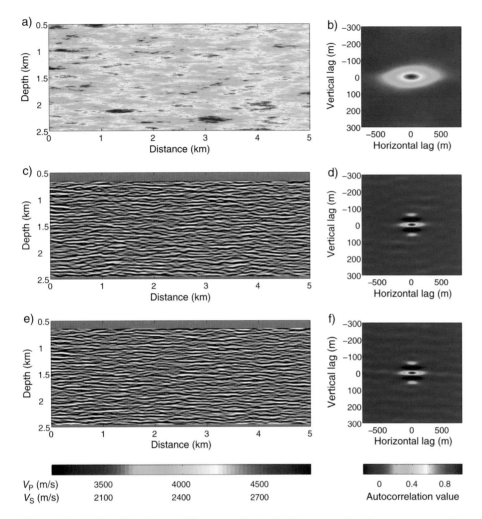

$V_P$ (m/s)   3500   4000   4500
$V_S$ (m/s)   2100   2400   2700

0   0.4   0.8
Autocorrelation value

**Figure 9.** (a) Combined P- and S-wave velocity field generated using an exponential autocorrelation model. (b) Corresponding 2D autocorrelation. (c) Stacked, migrated, and gained 25-Hz, zero-offset seismic-reflection image obtained by viscoelastic finite-difference modeling on (a). (d) Corresponding 2D autocorrelation. (e) PRS reflection image obtained from (a) using equation 8 and the same source wavelet used in (c). (f) Corresponding 2D autocorrelation.

the same rate (i.e., the aspect ratio is held constant), the above two effects work against each other, resulting in no net effect on the lateral correlation behavior. In other words, the average length of horizontal reflectors increases, but at the same time, we have less merging of horizontally discontinuous reflection events through convolution with the source wavelet.

## Seismic Data Example

In an identical manner to the GPR example discussed above, we will now show the application of our autocorrelation-parameter inversion methodology to a realistic seismic data example, which provides a significantly more challenging test case. Figure 9a shows the combined P- and S-wave velocity field considered. The field is 5 km long and 2 km deep and is discretized on a grid that has uniform cell dimensions of 5 × 5 m. As in the GPR case, an exponential autocorrelation model was used to generate the velocity field through a spectral-simulation approach, with input values of $a_x = 250$ m and $a_z = 35$ m. The P-wave distribution in Figure 9a has a mean of 4000 m/s and a 5% standard deviation of 200 m/s. The minimum and maximum P-wave velocities are 3158 and 4827 m/s, respectively. The S-wave velocities were set to be equal to 0.6 times their P-wave counterparts, which is a realistic approximation for consolidated sediments and crystalline rocks (e.g., Schön, 1998). Figure 9b shows the calculated 2D autocorrelation of the velocity field in Figure 9a, which was found to have best-fitting parameter values of $a_x = 242$ m and $a_z = 34$ m, resulting in an aspect ratio of 7.1.

To generate seismic-reflection data corresponding to the P- and S-wave velocity distribution in Figure 9a, we used the numerical modeling code of Robertsson et al. (1994), which uses an explicit staggered-grid finite-difference scheme to solve the viscoelastic wave equations. The velocity field was specified to lie between depths of 0.5 and 2.5 km and was sandwiched between two 0.5-km-

horizontal correlation structure. Although this requires further research, we feel that the phenomenon can be explained heuristically as follows: If we increase $a_x$ but keep $a_z$ fixed, the average length of reflection events in the reflection image clearly will increase, and thus we will have an increase in the image's apparent lateral correlation length. Conversely, if we increase $a_z$ but keep $a_x$ fixed, the average vertical distance between reflectors will increase, and consequently there will tend to be fewer occurrences of reflectors that are in fact horizontally discontinuous but effectively "line up" (i.e., are seen as continuous) when convolved with the source wavelet. That causes a decrease in the apparent correlation length of the reflection image. If $a_x$ and $a_z$ are both increased at thick layers that have a constant velocity of 4000 m/s. Density was assumed to be uniformly constant at 2800 kg/m³. Quality factors for P- and S-waves also were set to a constant value of 1000 throughout the modeling region. Absorbing boundaries were placed along the sides and bottom of the domain, and a free-surface boundary condition was prescribed along the top (Robertsson et al., 1994). Seismic modeling then was carried out using a source wavelet with a dominant frequency of 25 Hz and a bandwidth of two to three octaves, which is shown in Figure 10a. Shots were simulated every 20 m along the earth's surface, and receivers were placed every 10 m out to 1000 m on either side of the shot location. For each shot, the vertical component of velocity was recorded at

each of the receiver locations. Again, the spatial discretization interval of 5 m and the time step of 0.4 ms were enough to contain numerical dispersion in the generated data and to satisfy the Courant stability criterion, respectively.

After the seismic modeling was complete, each simulated shot gather was subjected to *f-k* filtering to attenuate steeply dipping S-wave arrivals (e.g., Line et al., 1998) before the data were muted, sorted, NMO-corrected, and stacked to form the zero-offset P-wave seismic-reflection section. Figure 11a and 11b shows an example of one of the simulated shot gathers before and after the *f-k* filtering and muting, respectively. The filtering is clearly effective at removing much of the scattered S-wave noise in the original gather, which appears as steeply dipping coherent energy. NMO correction was performed using the average background velocity of 4000 m/s, which was assumed to be known. After stacking, the seismic section was gained

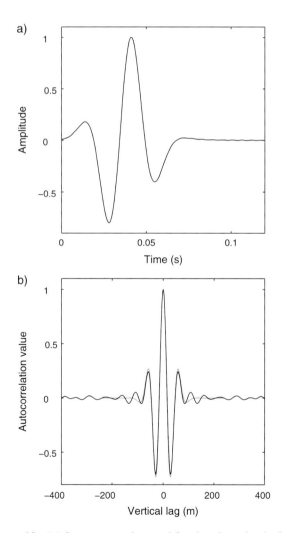

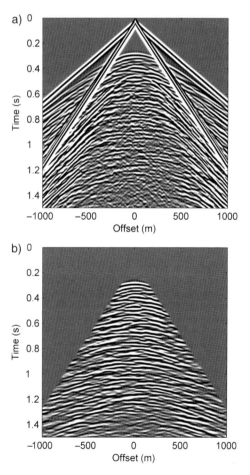

**Figure 10.** (a) Source wavelet used for the viscoelastic finite-difference seismic modeling. (b) Autocorrelation of the source wavelet converted to depth (gray) and vertical autocorrelation of the finite-difference-based seismic image (black).

**Figure 11.** (a) Raw shot gather from the finite-difference seismic modeling. (b) The same shot gather after *f-k* filtering and muting, which then was NMO-corrected and stacked to produce the zero-offset section in Figure 9c.

in time using the same smooth-scaling approach as in our GPR example, and it was depth-migrated using the *f-k* algorithm of Stolt (1978). Figure 9c shows the resulting P-wave reflection image. For that figure, data between depths of 0.5 and 0.65 km and between depths of 2.45 and 2.5 km were muted to suppress horizontal reflections coming from the upper and lower bounding interfaces of the random medium. Again, the muted regions are included in Figure 9c for easier comparison with Figure 9a, but they were not considered in the calculation of the corresponding autocorrelation.

For comparison, Figure 9e shows the PRS modeled seismic section, which again was obtained by convolving the numerical vertical derivative of the P-wave velocity field with the source wavelet in Figure 10a. Notice how the PRS once more provides a close approximation to the finite-difference-modeled reflection image, especially with regard to stochastic structure. This is seen most clearly in the corresponding 2D autocorrelations shown in Figure 9d and 9f, and it occurs despite the significantly greater complications involved in obtaining the finite-difference image here than in our GPR example. For instance, here we consider stronger velocity heterogeneity characterized by a standard deviation of 5% as compared with the 2.5%

used in the GPR example, which means that the detrimental effects of multiple scattering (not predicted by the PRS formulation) will be greater. Unlike in the quasi-acoustic GPR case, the finite-difference-modeled seismic data also contain the realistic effects of S-wave contamination, which is not included in the PRS formulation and therefore was removed by filtering. Finally, the seismic-reflection image in Figure 9c was obtained through NMO correction and stacking, whereas the GPR image in Figure 3c was obtained by simply modeling the zero-offset section. Stacking has the potential to alter the lateral correlation statistics compared with the PRS model (Gibson, 1991; Line et al., 1998) and thereby affect our inversion procedure. However, Figure 9c through 9f illustrates that these effects are minimal for the considered example.

In an identical manner to the GPR case, Figures 12 through 15 show the results of inverting the seismic autocorrelation data in Figure 9d for the exponential model parameters describing the velocity-field autocorrelation in Figure 9b. Again, we performed the Monte Carlo inversion considering two sets of prior information, one in which knowledge about $a_z$ was assumed to be available and the other in which it was not. The prior parameter ranges used in each inversion, along with the true values and mean

**Figure 12.** Monte Carlo inversion results for the finite-difference-based seismic image autocorrelation in Figure 9d. A narrow prior range was used for the vertical correlation length. (a and b) Horizontal and vertical autocorrelations of the seismic image (black) and the predicted horizontal and vertical image autocorrelations corresponding to all accepted parameter sets, obtained using equation 11 (gray). (c and d) Horizontal and vertical autocorrelations of the true velocity model in Figure 9a (black) and those corresponding to all accepted parameter sets (gray).

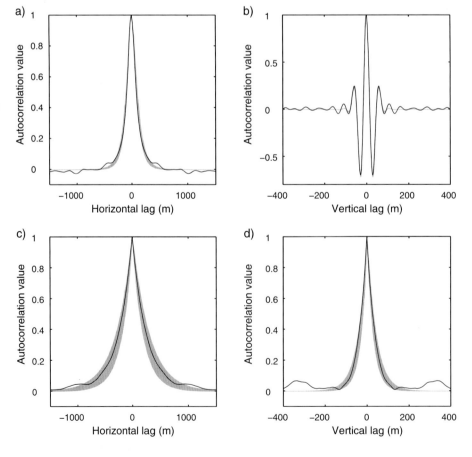

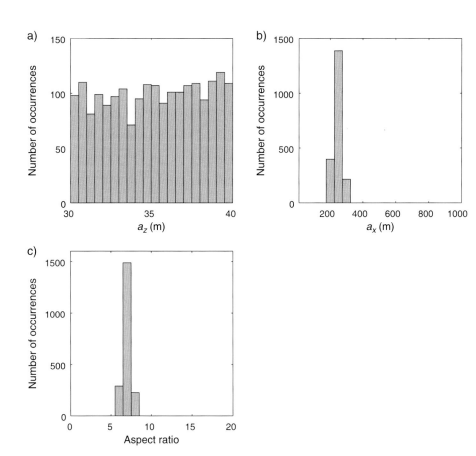

**Figure 13.** Parameter histograms for the inversion results shown in Figure 12.

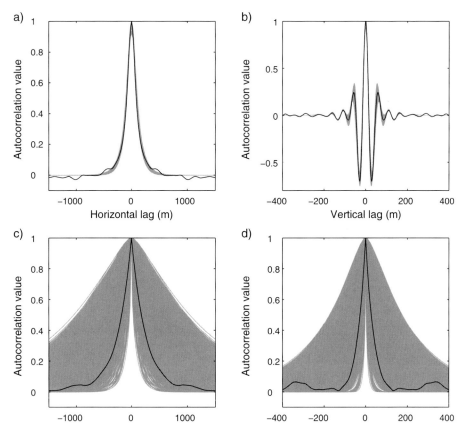

**Figure 14.** Monte Carlo inversion results for the finite-difference-based seismic image autocorrelation in Figure 9d. A broad prior range was used for the vertical correlation length. (a and b) Horizontal and vertical autocorrelations of the seismic image (black) and the predicted horizontal and vertical image autocorrelations corresponding to all accepted parameter sets, obtained using equation 11 (gray). (c and d) Horizontal and vertical autocorrelations of the true velocity model in Figure 9a (black) and those corresponding to all accepted parameter sets (gray).

**Figure 15.** Parameter histograms for the inversion results shown in Figure 14.

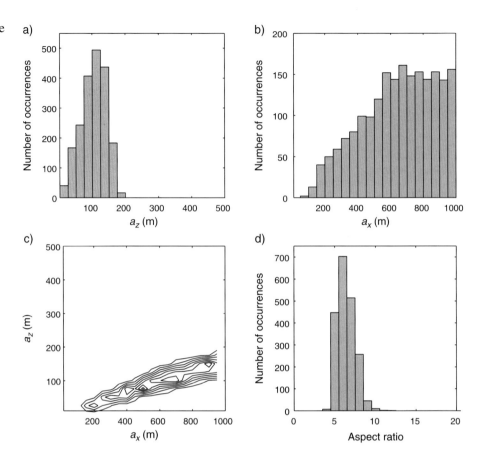

and standard deviation of the output realizations, are summarized in Table 1. We again used the approximation $R_{ww}(\delta z) \approx R_{dd}(0, \delta z)$ in the inversions to calculate $R_{ff}(\delta x, \delta z)$, which Figure 10b shows to be reasonable. Also required for the seismic inversions were new criteria describing an acceptable fit to the observed horizontal autocorrelation of the seismic image, which we defined as lying within 25 m from the observed curve along the horizontal-lag axis or within 0.03 units from the observed curve along the autocorrelation-value axis.

Figures 12 and 13 show the inversion results obtained for the case in which uniform prior ranges of $a_z = [30, 40]$ m and $a_x = [10, 1000]$ m were used, whereas Figures 14 and 15 show the results for which the prior range for $a_z$ was extended to [10, 500] m. We see that the results are largely identical to those obtained in our GPR example. In the case in which $a_z$ was constrained narrowly about the true value, the inversion provides little further information about $a_z$ but significantly reduces our uncertainty regarding $a_x$. In that case, the accepted realizations have a mean $a_z$ value of 35.2 m with standard deviation of 2.9 m (true value 34 m) and a mean $a_x$ value of 244.9 m with standard deviation of 22.5 m (true value 242 m). Thus we have a reliable recovery of both $a_z$ and $a_x$, and hence, the aspect ratio of the heterogeneity.

On the other hand, in the case in which $a_x$ and $a_z$ are prescribed broad uniform prior distributions, we clearly see again the nonuniqueness of the inverse problem in the sense that a wide variety of velocity autocorrelation models can fit the observed seismic-image autocorrelation. In that case, the accepted realizations have a mean $a_z$ value of 104.2 m with standard deviation of 37.2 m and a mean $a_x$ value of 649.6 m with standard deviation of 222.3 m. However, as in the GPR case, all of the acceptable $a_x$-$a_z$ combinations for velocity have approximately the same correct aspect ratio (mean 6.4 m with a standard deviation of 1.1 m compared with the true value of 7.1 m), which again indicates that $a_x/a_z$ is a quantity that can be recovered reliably in any case.

## Conclusions

We have presented a generic method for relating the spatial statistics of GPR and seismic-reflection images to the stochastic structure of the underlying subsurface velocity distributions. Unlike previous related efforts, our formulation correctly predicts the often significant changes in the lateral autocorrelation behavior of GPR and seismic images caused by vertical filtering operations on the velocity field and thus is well suited for an inversion strategy to estimate the stochastic parameters that describe the correlation structure of velocity heterogeneity from the second-order statistics of the image. Key assumptions for our methodological framework are that (1) weak scattering prevails, (2) dispersion and mode conversions are negligible or can be removed through processing, and (3) the observed GPR or seismic-reflection data can be imaged/migrated adequately.

As opposed to the case of quasi-acoustic GPR data acquired in the commonly practiced zero-offset mode, seismic-reflection data are acquired in the form of multiple-offset shot gathers, which must be cleaned from scattered S-waves, sorted into CMPs, corrected for NMO, and stacked prior to being migrated. Despite those rather fundamental differences with regard to the nature of the data and their conditioning, we found our method to be equally accurate and robust for GPR and seismic-reflection data when inverting the second-order spatial statistics of the images for the statistics of the underlying velocity structure.

Our results indicate that when the vertical correlation length of the underlying velocity heterogeneity is well constrained, for example from borehole log information, it is possible to estimate the horizontal correlation length reliably from the autocorrelation of GPR and seismic-reflection images. Without such prior knowledge, our inversion procedure still can resolve reliably the aspect ratio of the velocity heterogeneity but not the horizontal and vertical correlation lengths independently. Given increasing evidence that the spatial distributions of most if not all pertinent petrophysical properties are scale invariant and that the corresponding scaling laws are remarkably uniform, the structural aspect ratio arguably represents the most valuable and most important unknown geostatistical parameter to be retrieved from GPR and seismic-reflection images.

We also have found that the vertical autocorrelation of GPR or seismic images is not at all sensitive to the stochastic structure of the underlying velocity field, but rather only to the second-order statistics of the seismic source wavelet. Indeed, that result is expected, based on the seemingly ubiquitous flicker-noise-type scaling behavior of sonic logs, but it is critically important for our work because it provides an effective means of estimating the autocorrelation of the source wavelet, which is required to use our developed model in the corresponding inversion.

Finally, it is important to note that we also have investigated the impact of the presence of band-limited ambient noise on the effectiveness of our inversion procedure. Our results indicate that small to moderate amounts of ambient noise of as much as 10% of the root-mean-square (rms) amplitude of the data do not have any noticeable effects, although large amounts of noise of as much as 20% of the rms amplitude tend to systematically shorten our estimates of the lateral correlation length in the single-digit percentage range.

# Acknowledgments

This work was supported by a grant from the Swiss National Science Foundation.

# References

Archie, G., 1942, Electrical resistivity as an aid in core analysis interpretation: Transactions of the American Institute of Mining Engineers, **146**, 54–62.

Bean, C., D. Marsan, and F. Martini, 1999, Statistical measures of crustal heterogeneity from reflection seismic data: The role of seismic bandwidth: Geophysical Research Letters, **26**, 3241–3244.

Belina, F., J. Irving, K. Holliger, and J. Ernst, 2009, Evaluation of the viability and robustness of an iterative deconvolution approach for estimating the source wavelet during waveform inversion of crosshole ground-penetrating radar data: 79th Annual International Meeting, SEG, Expanded Abstracts, 1370–1374.

Berkhout, A., 1984, Seismic resolution: A quantitative analysis of resolving power of acoustical echo techniques: Geophysical Press.

Carcione, J., and F. Cavallini, 1995, On the acoustic-electromagnetic analogy: Wave Motion, **21**, 149–162.

Carpentier, S., and K. Roy-Chowdhury, 2007, Underestimation of scale lengths in stochastic fields and their seismic response: A quantification exercise: Geophysical Journal International, **169**, 547–562.

Chen, J., and G. Schuster, 1999, Resolution limits of migrated images: Geophysics, **64**, 1046–1053.

Claerbout, J., 1985, Imaging the earth's interior: Blackwell.

Couch, L., 2001. Digital and analog communications systems: Prentice Hall.

Dafflon, B., J. Tronicke, and K. Holliger, 2006, Inferring the lateral subsurface correlation structure from georadar data: Methodological background and experimental evidence, *in* P. Renard, H. Demougeot-Renard, and R. Froidevaux, eds., Geostatistics for environmental applications: Springer, 467–478.

Gibson, B., 1991, Analysis of lateral coherency in wide-angle seismic images of heterogeneous targets: Journal of Geophysical Research, **96**, 10261–10273.

Gibson, B., and A. Levander, 1990, Apparent layering in common-midpoint stacked images of two-dimensionally heterogeneous targets: Geophysics, **55**, 1466–1477.

Goff, J., K. Holliger, and A. Levander, 1994, Modal fields: A new method for characterization of random seismic velocity heterogeneity: Geophysical Research Letters, **21**, 493–496.

Holliger, K., R. Carbonell, and A. Levander, 1992, Sensitivity of the lateral correlation function in deep seismic reflection data: Geophysical Research Letters, **19**, 2263–2266.

Holliger, K., and J. Goff, 2003, A generalized model for the 1/f-scaling nature of seismic velocity fluctuations, *in* J. Goff, and K. Holliger, eds., Heterogeneity in the crust and upper mantle: Nature, scaling, and seismic properties: Kluwer Academic/Plenum Scientific Publishers, 131–154.

Holliger, K., A. Levander, R. Carbonell, and R. Hobbs, 1994, Some attributes of wavefields scattered from Ivrea-type lower crust: Tectonophysics, **232**, 267–279.

Hurich, C., 1996, Statistical description of seismic reflection wavefields: A step towards quantitative interpretation of deep seismic reflection profiles: Geophysical Journal International, **125**, 719–728.

Hurich, C., and A. Kocurko, 2000, Statistical approaches to interpretation of seismic reflection data: Tectonophysics, **329**, 251–267.

Irving, J., K. Holliger, and R. Knight, 2009, Estimation of the lateral correlation structure of subsurface water content from surface-based ground-penetrating radar reflection images: Water Resources Research, **45**, W09407, doi:09410.01029/02008WR007646.

Irving, J., and R. Knight, 2006, Numerical modeling of ground-penetrating radar in 2D using MATLAB: Computers and Geosciences, **32**, 1247–1258.

Kelkar, M., and G. Perez, 2002, Applied geostatistics for reservoir characterization: SPE.

Knight, R., J. Irving, P. Tercier, G. Freeman, C. Murray, and M. Rockhold, 2007, A comparison of the use of radar images and neutron probe data to determine the horizontal correlation length of water content: AGU Geophysical Monograph Series 171, 31–44.

Knight, R., P. Tercier, and J. Irving, 2004, The effect of vertical measurement resolution on the correlation structure of a ground penetrating radar reflection image: Geophysical Research Letters, **31,** 21607.

Line, C., R. Hobbs, and D. Snyder, 1998, Estimates of upper-crustal heterogeneity in the Baltic Shield from seismic scattering and borehole logs: Tectonophysics, **286**, 171–183.

Oldenborger, G., M. Knoll, and W. Barrash, 2004, Effects of signal processing and antenna frequency on the geostatistical structure of ground-penetrating radar data: Journal of Environmental and Engineering Geophysics, **9**, 201–212.

Poppeliers, C., 2007, Estimating vertical stochastic scale parameters from seismic reflection data: Deconvolution with non-white reflectivity: Geophysical Journal International, **168**, 769–778.

Poppeliers, C., and A. Levander, 2004, Estimation of vertical stochastic scale parameters in the earth's crystalline crust from seismic reflection data: Geophysical Research Letters, **31**, L13607.

Pullammanappallil, S., A. Levander, and S. Larkin, 1997, Estimation of crustal stochastic parameters from seismic exploration data: Journal of Geophysical Research, **102**, 15269–15286.

Rea, J., and R. Knight, 1998, Geostatistical analysis of ground-penetrating radar data: A means of describing spatial variation in the subsurface: Water Resources Research, **34,** 329–340.

Robertsson, J., J. Blanch, and W. Symes, 1994, Viscoelastic finite difference modeling: Geophysics, **59,** 1444–1456.

Schön, J., 1998, Physical properties of rocks: Fundamentals and principles of petrophysics: Pergamon Press.

Stolt, R., 1978, Migration by Fourier transform: Geophysics, **43**, 23–48.

Stolt, R., and A. Benson, 1986, Seismic migration: Geophysical Press.

Ulrych, T. J., 1999, The whiteness hypothesis: Reflectivity, inversion, chaos, and Enders: Geophysics, **64**, 1512–1523.

Western, A., and G. Blöschl, 1999, On the spatial scaling of soil moisture: Journal of Hydrology, **217**, 203–224.

Yilmaz, Ö., 1987, Seismic data processing: SEG Investigations in Geophysics Series No. 2.

Chapter 6

# Toward True-amplitude Vector Migration of GPR Data Using Exact Radiation Patterns

J. van der Kruk[1], R. Streich[2], and M. Grasmueck[3]

## Abstract

The amplitudes and phases of raw ground-penetrating-radar (GPR) data depend on the antenna radiation patterns, the vector nature of electromagnetic (EM) wave propagation, and the EM properties of the subsurface. Migrated GPR data should accurately represent the subsurface EM property contrasts alone. To achieve this, migration algorithms must explicitly account for the radiation patterns and vector wave propagation. A specific vector-migration algorithm models and corrects for exact-field radiation patterns, which include far-, intermediate-, and near-field contributions and propagation effects. When applied to GPR data containing dipping planar reflections, the algorithm produces images largely invariant to the relative orientations of the antennae and reflectors, indicating that most radiation-pattern effects are corrected for. In contrast, strongly orientation-dependent amplitudes and phases in scalar Gazdag and far-field vector images show that these algorithms do not adequately account for radiation-pattern effects. For polarization-dependent features (e.g., most underground utilities), the exact-field vector-migration algorithm produces images with orientation-dependent amplitude variations in qualitative agreement with theoretical expectations, suggesting that the algorithm may serve as a starting point for reconstructing the scattering properties of the targets. In contrast, the scalar Gazdag and far-field algorithms yield distinctly false amplitude variations.

## Introduction

Correctly imaged (migrated) 3D ground-penetrating-radar (GPR) data sets have the potential to provide reliable, high-resolution information about the shallow subsurface. Ideally, a GPR imaging algorithm should supply quantitative amplitude information that is related directly to the contrasts in subsurface physical properties. Although several approximately true amplitude-migration techniques exist for seismic data (Gray, 1997), there are few analogous approaches for quantitative imaging of GPR data. Instead, many GPR data sets are migrated using scalar algorithms explicitly written for migrating seismic data (Johansson and Mast, 1994; Mast and Johansson, 1994; Grandjean and Gourry, 1996; Grasmueck, 1996; Grasmueck et al., 2005; Bradford, 2008; Sassen and Everett, 2009). Such algorithms yield image amplitudes that depend on the antenna radiation patterns and therefore on the relative orientations of the antennae and target reflectors.

To determine quantitatively correct GPR images, it is necessary to account for the antenna radiation patterns and the vector nature of electromagnetic (EM) wavefield propagation. Early attempts at quantitative GPR imaging include the schemes of Molyneux and Witten (1993) and Witten et al. (1994), which address these issues for 2D objects embedded in a homogeneous medium. Other attempts to derive 3D migration algorithms that include the effects of the antenna radiation patterns and vector wavefields are based on far-field approximations of the

[1]*Formerly ETH Zurich; currently Research Center Juelich, Agrosphere, Germany.*
[2]*Formerly ETH Zurich; currently GFZ German Research Centre for Geosciences, Potsdam, Germany.*
[3]*University of Miami, Rosenstiel School of Marine and Atmospheric Science (RSMAS), Miami, Florida, U.S.A.*

radiation patterns. For example, Moran et al. (2000) present a Kirchhoff migration technique that empirically incorporates the far-field radiation patterns derived by Engheta et al. (1982), and Wang and Oristaglio (2000) use the same patterns in an algorithm based on the generalized Radon transform. Diffraction tomographic approaches, in which the approximations of Green's functions are expressed in the horizontal wavenumber-frequency domain, have been introduced by Hansen and Johansen (2000) and by Meincke (2001).

In contrast to these single-component methods, other techniques are based on multicomponent data recorded with different antenna configurations. In principle, combining different components of GPR data should improve signal-to-noise ratios and yield additional information on subsurface physical properties. GPR data recorded using different configurations of copolarized (parallel) antennae are investigated by Lehmann et al. (2000), and combinations of co- and cross-polarized data are analyzed by van Gestel and Stoffa (2000), Seol et al. (2001), van der Kruk et al. (2003a, 2003b), and Streich et al. (2006). Although cross-polarized data generally have much lower amplitudes than copolarized data, they may contain significant information (Guy et al., 1999; Streich et al., 2006; Orlando and Slob, 2009).

A migration that uses co- and cross-polarized data is presented by van Gestel and Stoffa (2000), who combine far-field radiation patterns and Alford rotation to extract the azimuths of the targets and then migrate their data over limited azimuth ranges. Expanding, van der Kruk et al. (2003a, 2003b) present a multicomponent imaging algorithm that explicitly corrects for far-field radiation patterns by migrating co- and cross-polarized components jointly; information from copolarized and cross-polarized data are merged in the images.

Far-field imaging approaches have been used widely, primarily because of their computational efficiency. For a fast, efficient imaging algorithm, it is indispensable to have closed-form expressions of the EM field. The far-field Green's functions for the commonly assumed model of horizontal dipole antennae on the surface of a homogeneous half-space can be computed readily from analytic expressions (Annan, 1973; Engheta et al., 1982; Smith, 1984). However, many environmental and engineering targets of interest are located within a few wavelengths of the antennae at distances where far-field conditions are not yet satisfied. In fact, far-field conditions, in the sense that the exact Green's functions are approximated adequately by far-field ones, are practically unreached within the distance range commonly penetrated by GPR (Streich et al., 2007). At a distance of seven wavelengths ($7\lambda$), the far

field still differs significantly from the exact field, especially near the critical angle (Streich and van der Kruk, 2007a), which results in erroneous phases and amplitudes in far-field migrated images. It is therefore important to know the precise propagation behavior of the exact EM field in this region. Unfortunately, exact-field expressions for the EM field in a homogeneous half-space do not exist in closed form and are time-consuming to compute.

To overcome this trade-off, a fast, accurate, and practical method for computing exact-field radiation patterns has been introduced by Streich and van der Kruk (2007a); the exact-field radiation patterns include the near-, intermediate-, and far-field contributions of infinitesimal dipoles located on a homogeneous half-space. Combining this novel method with the imaging algorithm described by van der Kruk et al. (2003a) results in an accurate multicomponent vector-migration scheme that produces 3D images nearly devoid of radiation-pattern effects.

This vector-migration scheme uses a phase-shift migration approach, which is well suited for efficiently incorporating exact-field radiation patterns and in principle allows us to consider the full EM wave equation. In contrast, Kirchhoff algorithms (e.g., Schneider, 1978; Bleistein, 1987; Docherty, 1991) are numerically less convenient and conceptually less suited for exact-field GPR migration because they are based inherently on high-frequency (i.e., far-field) approximations. Reverse-time-migration algorithms (Baysal et al., 1983; Symes, 2007) would facilitate incorporating exact-field radiation patterns; such schemes likely will become computationally feasible.

In this chapter, we give an overview of our developments that were aimed at devising an amplitude-preserving 3D migration for GPR data. First, we discuss the forward models of the far-field and exact-field expressions for a dipole present on the interface between two homogeneous half-spaces. These expressions are used for deriving single- and multicomponent forward wavefield extrapolators that describe the measured electric field resulting from an arbitrary point scatterer. These forward operators include the source-and-receiver radiation patterns and polarization of the EM field. To achieve amplitude-preserving imaging, we then derive a frequency-dependent inverse wavefield extrapolator that describes the reverse process (i.e., back propagation). All back-propagated frequency components are summed using the imaging principle of Claerbout (1971). The returned final image should be representative of physical-property contrasts. Characteristics of the far- and exact-field imaging schemes are compared to those of the standard scalar synthetic-aperture-radar (SAR) (Curlander and McDonough, 1991) and Gazdag (1978) imaging algorithms. We then demonstrate the performance of the exact-field imaging schemes on synthetic data sets for models containing dipping planar reflectors and dielectric pipes. Finally, we present imaging results for two field data sets.

# Forward Model for a GPR Antenna: Electric Fields of a Horizontal Electric Dipole

The configuration for the forward-source problem consists of an unbounded inhomogeneous medium $D$ with known EM properties. In this medium, sources $\hat{J}_\alpha$ are present that occupy the bounded domain $D^e$, a subdomain of $D$. The incident electric-field values $\hat{E}_k^i$ at any point $\mathbf{x}^r$ ∈ D are obtained in the space-frequency $\mathbf{x}$-$\omega$ domain as (de Hoop, 1995)

$$\hat{E}_k^i(\mathbf{x}^r) = \int_{x \in D^e} \hat{G}_{kr}(\mathbf{x}^r|\mathbf{x}, \omega)\hat{J}_k(\mathbf{x}, \omega)dV. \quad (1)$$

Here, $\mathbf{x} = (x_1, x_2, x_3)$, Latin subscripts can take the values {1,2,3}, and Einstein's summation convention applies for repeated subscripts. Equation 1 shows that the EM field from a known electric source $\hat{J}_k$ in a known medium can be calculated in principle in all space once the fields radiated by appropriate point sources (the Green's tensor function $\hat{G}_{kr}$) have been calculated. In the following sections, we compare the exact- and far-field expressions for the EM field for a homogeneous space and a homogeneous half-space.

## Dipole source in a homogeneous space

The exact-field expression for the Green's function in a homogeneous space is given by (de Hoop, 1995)

$$\hat{G}_{kr}(\mathbf{x}, \omega) = \eta^{-1}[\partial_k\partial_r - \gamma^2\delta_{kr}]\frac{\exp(-\gamma|\mathbf{x}|)}{4\pi|\mathbf{x}|}, \quad (2)$$

where $\delta_{kr} = 1$ for $k = r$ and $\delta_{kr} = 0$ otherwise. In addition, $\gamma = \sqrt{\eta s_0}$ is the propagation factor in the $\mathbf{x}-\omega$ domain. The medium parameters $\eta$ and $\zeta$ are given by $\eta = \sigma + j\omega\varepsilon$ and $s_0 = j\omega\mu_0$, the angular frequency $\omega = 2\pi f$, $\sigma$ is the electric conductivity, $\varepsilon$ is the dielectric permittivity, and $\mu_0$ is the vacuum magnetic permeability. Because of the spatial derivatives $\partial_k\partial_r$, particular direction patterns arise that depend on the orientation of the source and the direction of observation. The closed-form representation obtained by evaluating the spatial derivatives (e.g., de Hoop, 1995) enables a thorough analysis of the radiation characteristics

of the near, intermediate, and far fields, which have amplitudes proportional to $|\mathbf{x}|^{-3}$, $|\mathbf{x}|^{-2}$, and $|\mathbf{x}|^{-1}$, respectively.

## Dipole source on the interface between two homogeneous half-spaces

Unfortunately, closed-form analytic solutions of Maxwell's equations for a horizontal electric dipole present on the interface between two homogeneous half-spaces do not exist in the $\mathbf{x}$-$\omega$ domain (Annan, 1973; Valle et al., 2001; Radzevicius et al., 2003). The exact analytic expressions for the corresponding horizontal wavenumber-frequency ($k_1$-$k_2$-$x_3$-$\omega$) domain Green's functions $\widetilde{G}_{m\alpha}$ are given by (van der Kruk, 2001)

$$\begin{pmatrix} \widetilde{G}_{11} & \widetilde{G}_{12} \\ \widetilde{G}_{21} & \widetilde{G}_{22} \\ \widetilde{G}_{31} & \widetilde{G}_{32} \end{pmatrix} = -\zeta \begin{pmatrix} k_1^2\widetilde{V} + \widetilde{U} & k_1 k_2\widetilde{V} \\ k_1 k_2\widetilde{V} & k_2^2\widetilde{V} + \widetilde{U} \\ -jk_1\Gamma_0\widetilde{V} & -jk_2\Gamma_0\widetilde{V} \end{pmatrix}, \quad (3)$$

where

$$\widetilde{U} = \frac{\exp(-\Gamma_1 x_3)}{\Gamma_0 + \Gamma_1},$$

$$\widetilde{V} = \frac{\exp(-\Gamma_1 x_3)}{\gamma_1^2\Gamma_0 + \gamma_0^2\Gamma_1},$$

$$\Gamma_i = \sqrt{\gamma_i^2 + k_1^2 + k_2^2}, \quad (4)$$

and where $\Gamma$ is the propagation factor in the $k_1$-$k_2$-$x_3$-$\omega$ domain and $i = \{0,1\}$ denotes air and the subsurface, respectively. Note that the hat (caret) indicates the $\mathbf{x}$-$\omega$ domain and the tilde indicates the $k_1$-$k_2$-$x_3$-$\omega$ domain. Because we assume horizontal sources $J_\alpha$ ($\alpha = \{1,2\}$), we need to consider only two columns of the full Green's function matrix. Exact-field solutions in the $\mathbf{x}$-$\omega$ domain can be obtained by numerically evaluating 2D spatial inverse Fourier transform integrals of these expressions. This transformation is complicated because of the presence of branch-point singularities along the two circles:

$$k_1^2 + k_2^2 = -\gamma_i^2. \quad (5)$$

To account for these critical regions in the numerical integral evaluation, it is necessary to sample at very dense wavenumber intervals in the vicinity of these circles. We can do this with adaptive integration, which automatically reduces the wavenumber sampling until a predefined relative accuracy is reached (Piessens et al., 1983). However, adaptive routines require that the Fourier integrals be evaluated separately at every subsurface point, so they are prohibitively slow for 3D imaging algorithms that require Green's-function values on a large 3D spatial grid.

Far-field asymptotic solutions in the $\mathbf{x}$-$\omega$ domain that are proportional to radius $1/R$ can be derived by inverse spatial Fourier transformation of equation 3 using the stationary phase or saddle-point method of integration (Engheta et al., 1982; Smith, 1984).

## Fast approximate evaluation of exact-field half-space expressions

A fast alternative for calculating the Green's functions at a certain depth level for all horizontal positions simultaneously is using an inverse fast Fourier transform (FFT). The main drawback of using inverse FFT is the requirement of equidistant sampling of the input functions $\widetilde{G}_{m\alpha}$. If we use the data-defined wavenumber grid spacings

$$dk_\alpha = dk_\alpha^{\text{data}} = \frac{2\pi}{N_\alpha dx_\alpha},  \quad (6)$$

where $N_\alpha$ are the numbers of traces and $dx_\alpha$ are the spatial sample intervals in the $x_1$ and $x_2$ directions, the critical regions near the branch points in the $k_1$-$k_2$-$x_3$-$\omega$ domain are likely to be sampled inadequately. This results in highly inaccurate $\mathbf{x}$-$\omega$-domain values of $\hat{G}_{m\alpha}$. We have tested sampling $\widetilde{G}_{m\alpha}$ using

$$dk_\alpha = \frac{dk_\alpha^{\text{data}}}{p},  \quad (7)$$

where the oversampling factor $p$ varies as powers of two up to 32. For the GPR survey geometries we considered, $k_1$-$k_2$-$x_3$-$\omega$-domain oversampling at rates of 8–16 relative to $dk_\alpha^{\text{data}}$ enormously improved the accuracy of inverse FFT-derived $\hat{G}_{m\alpha}$ (Streich and van der Kruk, 2007a).

For computing large 3D Green's function volumes, we also need to optimize the calculations of the $k_1$-$k_2$-$x_3$-$\omega$-domain expressions $\widetilde{G}_{m\alpha}$. For the homogeneous half-space

model, explicit calculation of $\widetilde{G}_{m\alpha}$ is necessary only at an initial subsurface depth level $x_3^{(0)}$. We then apply the vertical wavenumber phase shift (Slob et al., 2003)

$$\widetilde{G}_{m\alpha}(k_1, k_2, \omega, x_3^{(n)}) = \widetilde{G}_{m\alpha}(k_1, k_2, \omega, x_3^{(0)})$$
$$\times \exp[-\Gamma(x_3^{(n)} - x_3^{(0)})]  \quad (8)$$

to derive the $\widetilde{G}_{m\alpha}$ values at the $n$th depth level $x_3^{(n)}$, where $x_3^{(n)} > x_3^{(0)}$ and the propagation parameter $\Gamma_1$ is given by equation 4. Using these approaches, we can rapidly calculate practically exact $\mathbf{x}$-$\omega$-domain Green's functions at a given depth for a range of horizontal positions.

## Comparison between far-field and exact-field radiation patterns in a homogeneous half-space

In the following, we demonstrate the amplitude behavior of the electric field in spherical coordinates $(R, \varphi, \theta)$. The coordinate system and model configuration are displayed in Figure 1a. The analysis is carried out in two planes: the E-plane ($\varphi = 0$) and the H-plane ($\varphi = \pi/2$). The E-plane is parallel to and the H-plane is perpendicular to the direction of the current-source dipole. The electric-current source is positioned at the origin and is oriented in the $x_1$-direction ($\varphi = 0$). A dielectric medium is present in the lower half-space (i.e., for $0 < \theta < \pi/2$); in the upper half-space ($\pi/2 < \theta < \pi$), air is present. The different contributions to the electric field are shown in Figure 1b. The far-field expressions describe only the body wave in the air and the subsurface and are zero at the interface, whereas the exact-field expressions also describe the head wave in the ground and the inhomogeneous wave in the air.

In Figure 2, we compare, for a subsurface relative permittivity $\varepsilon_r = 4$, a frequency $f$ of 500 MHz and a radial distance of 1 m from the source (corresponding to $\approx 3.3$ wavelengths in the subsurface), the far-field and exact-field spherical electric-field components $E_\varphi$ and $E_\theta$.

**Figure 1.** (a) Two planes of investigation: E and H. (b) Wavefronts generated by a dipole source on the surface of a half-space earth with ① body wave in air, ② body wave in ground, ③ inhomogeneous wave in air, and ④ head wave in ground. The critical angle is given by $\theta_c = \sin^{-1}(1/\sqrt{\varepsilon_r})$.

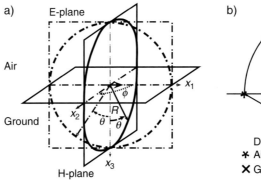

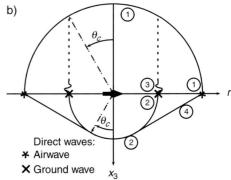

Figure 2a and 2b shows results for a source present on the interface. The far-field contribution attains a maximum in the H-plane and a minimum in the E-plane at the critical angle $\theta_c = \sin^{-1}(1/\sqrt{\varepsilon_r})$. However, the maximum and minimum responses for the exact-field radiation pattern in the H- and E-planes appear at angles larger than the critical angle and are less pronounced than those computed for far-field models. A possible explanation is the large contribution of the intermediate field near the critical angle $\theta_c$, which describes the head wave. Similar observations have been made in comparisons of far-field expressions to exact-field FDTD modeling results (Holliger and Bergmann, 1998).

In Figure 2c and 2d, the far field and exact field are compared for a source elevated 0.1 $\lambda$ above the interface. In this case, a much larger portion of the field is radiated into the air (Smith, 1984; Slob and Fokkema, 2003a, 2003b; van der Kruk, 2004). The far-field amplitudes differ strongly from the exact-field ones, not only near the critical angle but also in the vertical below the source

(i.e., for $\theta = 0$). These examples clearly demonstrate that (1) for accurate imaging, it is important to consider the true height of the antennae above the ground surface and (2) exact-field radiation patterns and their far-field approximations exhibit distinctly different directivities, such that using far-field expressions within GPR migration schemes likely will be insufficient for removing radiation-pattern effects from GPR data.

## Forward Model: The Response of an Isotropic Point Scatterer

We investigate the scattering of EM fields by a contrasting domain of bounded extent present in an unbounded homogeneous medium $D$ that has conductivity $\sigma$ and permittivity $\varepsilon$ (Figure 3). The term $D^s$ is the bounded domain occupied by the scatterer with conductivity $\sigma^s(\mathbf{x})$ and permittivity $\varepsilon^s(\mathbf{x})$. The contrast-source volume density $J_k^s$ of the electric current, also denoted as a scatter source, is

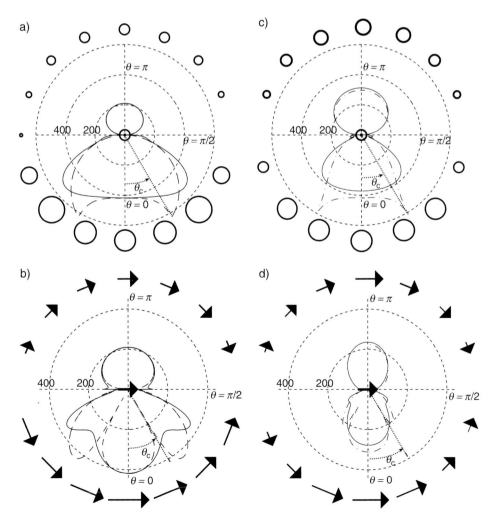

**Figure 2.** Comparison between the far-field and exact-field spherical electric-field components at a radius of 1 m and $f = 500$ MHz. Solid lines are exact field; broken lines are far field. (a) $E_\phi$ in the H-plane and (b) $E_\theta$ in the E-plane for antennae present on the interface. (c) $E_\phi$ in the H-plane and (d) $E_\theta$ in the E-plane for antennae elevated 0.1 $\lambda$ above the interface. The electric-field orientation is (a, c) perpendicular (indicated by circles) and (b, d) parallel (indicated by arrows) to the plane displayed. The circle radii and arrow lengths indicate the relative electric-field amplitudes.

given by

$$\hat{J}_k^s = \chi^\eta \hat{E}_k, \qquad (9)$$

where the medium contrast function is

$$\chi^\eta = \hat{\eta}^s - \hat{\eta} = (\sigma^s - \sigma) + j\omega(\varepsilon^s - \varepsilon). \qquad (10)$$

Once the total electric field $\hat{E}_k$ (equation 9) inside the scattering volume is known, the scattered field values in all space can be obtained by substituting the relevant values for the scatter source $\hat{J}_k^s$ into

$$\hat{E}_k^s = \int_{x \in D^s} \hat{G}_{kr}(\mathbf{x}^r|\mathbf{x}, \omega)\hat{J}_r^s(\mathbf{x}, \omega)dV, \qquad (11)$$

which is similar to equation 1. Using the first-order Born approximation (Born and Wolf, 1965), we assume that $\hat{J}_k^s$ can be approximated using the incident instead of the total field:

$$\hat{J}_k^s = \chi^\eta \hat{E}_k^i. \qquad (12)$$

Substituting equations 1 and 12 into equation 11 yields (van der Kruk, 2001)

$$\hat{E}_\alpha^s = \int_{V(\mathbf{x}^c)} \hat{G}_{\alpha l}(\mathbf{x}^r|\mathbf{x}^c, \omega)\chi^\eta(\mathbf{x}^c)\hat{G}_{l\beta}(\mathbf{x}^c|\mathbf{x}^s, \omega)\hat{J}_\beta(\mathbf{x}^s, \omega)dV,$$

$$(13)$$

where subscripts $\alpha, \beta = \{1,2\}$ imply that we consider only horizontal sources and receivers, $\mathbf{x}^c$ is a point within the scattering volume $V(\mathbf{x}^c)$, and $\mathbf{x}^s$ and $\mathbf{x}^r$ are the source and receiver locations, respectively.

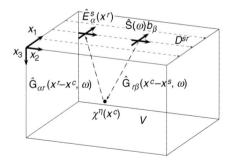

**Figure 3.** The configuration of the four possible source-receiver setups. Solid arrows indicate antenna orientations, $\hat{S}(\omega)b_\beta$ indicates the dipole source wavelet and orientation, and $\hat{E}_\alpha^s(\omega)$ indicates the dipole receiver. Dashed arrows indicate the EM wave-propagation paths from source to scatterer and from point scatterer $\chi^\eta(x^c)$ to receiver described by $\hat{G}_{r\beta}(x^c - x^s, \omega)$ and $\hat{G}_{\alpha r}(x^r - x^c, \omega)$, respectively.

The Green's function $\hat{G}_{l\beta}(\mathbf{x}^c|\mathbf{x}^s, \omega)$ describes the propagation of the vectorial electric field emitted by a point source $\hat{J}_\beta$ (electric-current density) at position $\mathbf{x}^s$ to a scatterer with physical-property contrast $\chi$ at position $\mathbf{x}^c$ (Figure 3). This scatterer can be considered as a secondary source, such that the propagation from $\mathbf{x}^c$ toward the receiver at $\mathbf{x}^r$ is described by $\hat{G}_{\alpha l}(\mathbf{x}^r|\mathbf{x}^c, \omega)$. Both Green's functions include the vectorial radiation characteristics of the dipole antennae. The point source can be described by

$$\hat{J}_\beta(\mathbf{x}^s, \omega) = \hat{S}(\omega)b_\beta, \qquad (14)$$

where $\hat{S}$ is the source wavelet and where $b_1$ and $b_2$ identify source-antenna orientations along the $x_1$- and $x_2$-directions (see Figure 3). We assume that the source and receiver are located at the same height.

## Forward and Inverse Wavefield Extrapolators

To simplify the inverse formulation we discuss later, here we define the source-receiver midpoints $\mathbf{x}^m = (\mathbf{x}^s + \mathbf{x}^r)/2$ and half-offsets $\mathbf{x}^h = (\mathbf{x}^r - \mathbf{x}^s)/2$. Then the scattered electric field $\hat{E}_\alpha^s$ from a source oriented in the $x_\beta$-direction and a receiver oriented in the $x_\alpha$-direction can be written as

$$\hat{E}_\alpha^s(\mathbf{x}^s, \mathbf{x}^r, \omega) = \hat{E}_{\alpha\beta}^s(\mathbf{x}^m, \mathbf{x}^h, \omega)b_\beta, \qquad (15)$$

where

$$\hat{E}_{\alpha\beta}(\mathbf{x}^m, \mathbf{x}^h, \omega) = \hat{S}(\omega)\int_{V(\mathbf{x}^c)} \hat{D}_{\alpha\beta}(\mathbf{x}^m, \mathbf{x}^h|\mathbf{x}^c, \omega)\chi(\mathbf{x}^c)dV$$

$$(16)$$

and the forward wavefield extrapolator $\hat{D}_{\alpha\beta}$ is defined as the inner product of the Green's functions describing downward propagation from the source to the scatterer and upward propagation from the scatterer to the receiver:

$$\hat{D}_{\alpha\beta}(\mathbf{x}^r, \mathbf{x}^s, \omega) = \hat{G}_{\alpha l}(\mathbf{x}^r|\mathbf{x}^c, \omega)\hat{G}_{l\beta}(\mathbf{x}^c|\mathbf{x}^s, \omega). \qquad (17)$$

For zero-offset source-receiver configurations ($\mathbf{x}^s = \mathbf{x}^r = \mathbf{x}^m$; $\mathbf{x}^h = 0$), the forward wavefield extrapolator can be written in the $\mathbf{x}$-$\omega$ domain as the product of an amplitude and a phase-shift term (van der Kruk et al., 2003a):

$$\hat{D}_{\alpha\beta}(\mathbf{x}, \omega) = \hat{A}_{\alpha\beta}(\mathbf{x}, \omega)\exp(-2jkR), \qquad (18)$$

where $k = \omega\sqrt{\mu_0(\varepsilon - j\sigma/\omega)}$. For homogeneous media, the amplitude term $\hat{A}_{\alpha\beta}$ can be determined analytically.

Corresponding $k_1$-$k_2$-$x_3$-$\omega$-domain operators are obtained by the horizontal spatial Fourier transform and generally can be represented as the products of amplitude and wavenumber phase-shift terms:

$$\widetilde{D}_{\alpha\beta}(k_1, k_2, x_3, \omega) = \widetilde{d}_{\alpha\beta}(k_1, k_2, x_3, \omega)$$
$$\times \exp(-jk_3|x_3|), \qquad (19)$$

where

$$k_3 = \sqrt{4k^2 - k_1^2 - k_2^2} \qquad \text{for } k_1^2 + k_2^2 \leq 4k^2,$$

$$k_3 = -j\sqrt{k_1^2 + k_2^2 - 4k^2} \quad \text{for } k_1^2 + k_2^2 \geq 4k^2. \qquad (20)$$

For homogenous media, analytic far-field expressions for $\widetilde{D}_{\alpha\beta}$ can be obtained using the method of stationary phase (Bleistein, 1984).

The inclusion of amplitude terms constitutes an important difference between our vector-migration algorithms and conventional scalar algorithms; the latter use wavefield extrapolators that account only for the phase shift in the $\mathbf{x}$-$\omega$ or $k_1$-$k_2$-$x_3$-$\omega$ domain. For a given forward wavefield extrapolator $D_{\alpha\beta}$, a corresponding inverse wavefield extrapolator that describes reverse propagation is obtained by taking, in some sense, the reciprocal of $D_{\alpha\beta}$. In the following sections, we discuss different inverse scalar and vector-wavefield extrapolators.

## Scalar inverse wavefield extrapolators

The conventional SAR and Gazdag scalar inverse wavefield extrapolators do not consider the vector characteristics of GPR propagation and do not include amplitude information in the domains in which they are defined. SAR imaging originally was developed for remote sensing (Curlander and McDonough, 1991). The SAR inverse wavefield extrapolator is defined in the $\mathbf{x}$-$\omega$ domain by the complex conjugate of the phase shift in equation 18. Its $k_1$-$k_2$-$x_3$-$\omega$-domain equivalent can be determined using the method of stationary phase. The SAR inverse extrapolators in the $\mathbf{x}$-$\omega$ and $k_1$-$k_2$-$x_3$-$\omega$ domains are

$$\hat{H}^{\mathrm{SAR}} = \exp(2jkR),$$

$$\widetilde{H}^{\mathrm{SAR}} = \frac{4\pi jk|x_3|}{(k_3^*)^2} \exp(jk_3^*|x_3|). \qquad (21)$$

Analysis of SAR images of a point diffractor embedded in a homogeneous space shows that for a real-valued contrast, the SAR operator returns a real-valued but negative image (van der Kruk et al., 2003a). Thus, a modified

operator is introduced by multiplying equation 21 by $-1$ to obtain a positive, real-valued image. In addition, the source-receiver offset can be incorporated by separating the phase shifts for the source and receiver (with correction factors $j$ included) as (van der Kruk et al., 2003b)

$$\hat{H}^{m\mathrm{SAR}} - j\exp(jkR^s) \times j\exp(jkR^r), \qquad (22)$$

where

$$R^s = |\mathbf{x}^m - \mathbf{x}^h - \mathbf{x}^d|,$$
$$R^r = |\mathbf{x}^m + \mathbf{x}^h - \mathbf{x}^d|.$$

The inverse wavefield extrapolator that forms the basis for Gazdag phase-shift migration (see equation 45 of Gazdag [1978]) is defined in the $k_1$-$k_2$-$x_3$-$\omega$ domain by the complex conjugate of the phase term in equation 19. Its equivalent in the $\mathbf{x}$-$\omega$ domain is determined by the method of stationary phase. The Gazdag inverse extrapolators in the $\mathbf{x}$-$\omega$ and $k_1$-$k_2$-$x_3$-$\omega$ domains are (van der Kruk et al., 2003b)

$$\hat{H}^{\mathrm{Gzd}} = \frac{-jk|x_3|}{\pi R^2}\exp(j2kR),$$

$$\widetilde{H}^{\mathrm{Gzd}} = \exp(jk_3^*|x_3|). \qquad (23)$$

Analysis of Gazdag images of a point diffractor shows that for a real-valued contrast, the Gazdag operator does not correct for all geometric propagation effects; it returns an image that depends on the depth $|x_3|$ of the scatterer. Moreover, the amplitudes are imaginary valued (van der Kruk et al., 2003a). A modified Gazdag operator (mGzd) thus is introduced by scaling with $|x_3|$ and multiplication by $-j$ to obtain real-valued images corrected for geometric spreading. Similar to the SAR operator, the offset between the source and receiver can be incorporated, and it renders (van der Kruk et al., 2003b)

$$\hat{H}^{m\mathrm{Gzd}}(x^m, x^h, x^d) = j^{3/2}\sqrt{\frac{k|x_3^d|}{\pi}}\frac{1}{R^s}\exp(jkR^s)j^{3/2}$$

$$\times \sqrt{\frac{k|x_3^d|}{\pi}}\frac{1}{R^r}\exp(jkR^r). \qquad (24)$$

## Multicomponent inverse vector-wavefield extrapolators

Ideally, inverse wavefield extrapolation should compensate for all effects of forward propagation described by equation 16. This description of the scattered electric field as a convolution of the forward wavefield extrapolator with the contrast function is converted into a multiplication upon transformation to the $k_1$-$k_2$-$x_3$-$\omega$ domain. Accordingly, we can, in principle, readily obtain an inverse operator in the $k_1$-$k_2$-$x_3$-$\omega$ domain by transforming the forward wavefield extrapolator given in equation 17 to the $k_1$-$k_2$-$x_3$-$\omega$ domain and then taking its inverse. To ensure stability, the inverse wavefield extrapolator is determined only in the propagating wave region; evanescent waves are ignored.

For analytic far-field forward wavefield extrapolators $\widetilde{D}_{\alpha\beta}$ obtained from equation 18, corresponding inverse operators $\widetilde{H}_{\alpha\beta}$ formed for an individual source-receiver configuration are unbounded and thus are not usable for migration (van der Kruk, 2001). To obtain a suitable approximate inverse of the forward wavefield, we combine the wavefield extrapolators for the four source-receiver configurations as given in equation 18 and introduce a multicomponent forward wavefield extrapolator as (van der Kruk et al., 2003a)

$$\hat{\mathbf{D}} = \begin{bmatrix} \hat{D}_{11} & \hat{D}_{12} \\ \hat{D}_{21} & \hat{D}_{22} \end{bmatrix}. \tag{25}$$

A corresponding matrix expression $\widetilde{\mathbf{D}}$ can be obtained in the $k_1$-$k_2$-$x_3$-$\omega$ domain. It has been shown analytically for homogeneous-space far-field expressions that the multicomponent inverse wavefield extrapolator $\widetilde{\mathbf{H}}$ obtained by inverting the matrix $\widetilde{\mathbf{D}}$ is bounded and thus suitable for GPR migration (van der Kruk et al., 2003a). This multicomponent inverse wavefield extrapolator takes into account the vectorial radiation characteristics of the dipole antennae and the propagating wavefield.

For a homogeneous half-space, inverse wavefield extrapolators can be derived only numerically, such that a rigorous proof of numerical stability or instability of inverse wavefield extrapolators for a single source-receiver configuration has not yet been accomplished. Nevertheless, it can be expected that for far-field half-space operators, the sharp E-plane minima present in the far-field radiation patterns (see Figure 2b) will cause such single-component far-field inverse operators to be numerically unstable. In contrast, for the exact-field expressions, in which sharp amplitude minima are absent, the multicomponent approach and the formation of inverse wavefield extrapolators for individual source-receiver configurations should result in stable images. The exact-field Green's functions are feasible by using the above-described techniques for fast computation of exact-field Green's functions.

## Imaging scheme

To obtain an image of the scattering domain, we apply the imaging principle of Claerbout (1971), which states

that the data at zero traveltime of the inverse-extrapolated recordings relate to a band-limited version of the physical-property contrasts of the medium of investigation. Consequently, we can sum the results for all (positive and negative) frequency components to obtain the imaged contrast in the space domain, and we can write the multicomponent (MC) imaged contrast at a specific depth as

$$\left\langle \begin{bmatrix} \chi_{11}^{MC} & \chi_{12}^{MC} \\ \chi_{21}^{MC} & \chi_{22}^{MC} \end{bmatrix} \right\rangle = \frac{1}{2\pi} \int \frac{d\omega}{\hat{S}(\omega)} \int\limits_{(x_1^m,x_2^m) \in \mathrm{D}^{sr}} \begin{bmatrix} \hat{H}_{11}^{MC} & \hat{H}_{12}^{MC} \\ \hat{H}_{21}^{MC} & \hat{H}_{22}^{MC} \end{bmatrix}$$
$$\times (\mathbf{x} - \mathbf{x}^m, \omega) \begin{bmatrix} \hat{E}_{11} & \hat{E}_{12} \\ \hat{E}_{21} & \hat{E}_{22} \end{bmatrix} (\mathbf{x}^m, \omega) dA. \tag{26}$$

For the single-component (SC) inverse wavefield extrapolators and the SAR and Gazdag operators, the imaged contrast at a specific depth can be written as

$$\chi_{\alpha\beta} = \frac{1}{2\pi} \int \frac{d\omega}{\hat{S}(\omega)} \int\limits_{(x_1^m,x_2^m) \in \mathrm{D}^{sr}} \hat{H}_{\alpha\beta}(\mathbf{x} - \mathbf{x}^m, \omega)$$
$$\times \hat{E}_{\alpha\beta}(\mathbf{x}^m, \omega) dA. \tag{27}$$

In both cases, we write the imaged contrast as a convolution of the inverse wavefield extrapolator with the measured electric field in the space-frequency domain. In practice, these calculations are carried out more efficiently in the frequency-wavenumber domain.

Figure 4 describes the exact-field migration process. First, we calculate $\mathbf{x}$-$\omega$-domain Green's functions for a given depth level (steps 1–3). The antenna offset can be accounted for by simple space-domain shifting of the Green's functions if the offset is a multiple of the spatial sampling (step 4). Next, the two-way forward wavefield extrapolator $\hat{\mathbf{D}}$ is formed by multiplying the appropriate source and receiver Green's functions. This forward operator now contains the exact radiation patterns of the source and receiver antennae and all propagation effects of the vectorial electric field between the antennae and the assumed reflection point. Each element of $\hat{\mathbf{D}}$ is then transformed to the $k_1$-$k_2$-$x_3$-$\omega$ domain (step 6) to allow the inverse wavefield extrapolator to be computed (step 7). This can be done by inverting $\widetilde{\mathbf{D}}$ (Streich and van der Kruk, 2007a). Alternatively, with the ability of including exact-field radiation patterns in the 3D imaging, it is also possible to use a single-component inverse wavefield extrapolator to image single-component data.

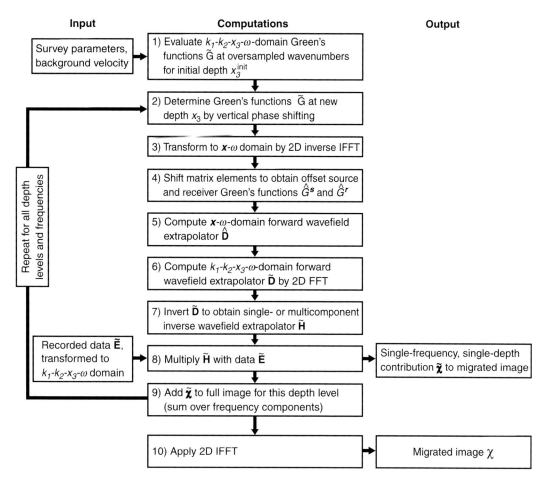

**Figure 4.** Process flowchart showing the computation of exact-field vector-migrated images. Bold letters are scalar quantities in single-component imaging and $2 \times 2$ matrices in multicomponent imaging.

A variety of tests demonstrates the stability of exact-field single-component operators (Streich et al., 2007). Applying the inverse wavefield extrapolator to the GPR data (step 8) provides the contribution of a single-frequency component to the image at a given depth level. The contributions of all frequency components are summed (step 9), and the process is repeated for the entire depth range of interest. Finally, the images are transformed back from $k_1$-$k_2$-$x_3$-$\omega$ to the space domain (step 10).

## Spatial-resolution Functions

To investigate the performance of the scalar, far-, and exact-field 3D inverse wavefield extrapolators, we analyze imaging results of a point scatterer for one frequency component. Such images are referred to as *spatial-resolution functions*. For a single diffraction point at position $\mathbf{x}^d$ with a unit-amplitude real-valued contrast, the ideal spatial-resolution function would be

$$\chi^\eta(\mathbf{x}) = \delta(\mathbf{x} - \mathbf{x}^d). \qquad (28)$$

In practice, the spatial-resolution function is band limited but should be real valued. In Figure 5, we show spatial-resolution functions for a point scatterer embedded in a homogeneous half-space at a depth of 1 m. Synthetic electric-field data for this configuration were calculated by numerically evaluating the exact-field integral expressions using highly accurate adaptive integration; various inverse wavefield extrapolators then were applied to the data.

In Figure 5a and 5b, the normalized real and imaginary parts of the modified SAR and Gazdag images are shown for $f = 500$ MHz. For both images, we obtain a noncircularly symmetrical resolution function and a nonzero imaginary part. The modified SAR image exhibits more oscillations in the tails of the resolution function than the Gazdag image does. The central peak of the Gazdag image is slightly wider than that of the SAR image. In Figure 5c, the normalized real and imaginary parts of the far-field multicomponent image are displayed. This image is more circularly symmetrical, and its imaginary part is smaller than those of the scalar images. Nevertheless, the oscillations,

**Figure 5.** Normalized real and imaginary parts of spatial-resolution functions in two homogeneous half-spaces. The inverse wavefield extrapolators were calculated using the scalar (a) modified SAR and (b) modified Gazdag operator and using the multicomponent imaging algorithm with the (c) far-field and (d) exact-field Green's functions.

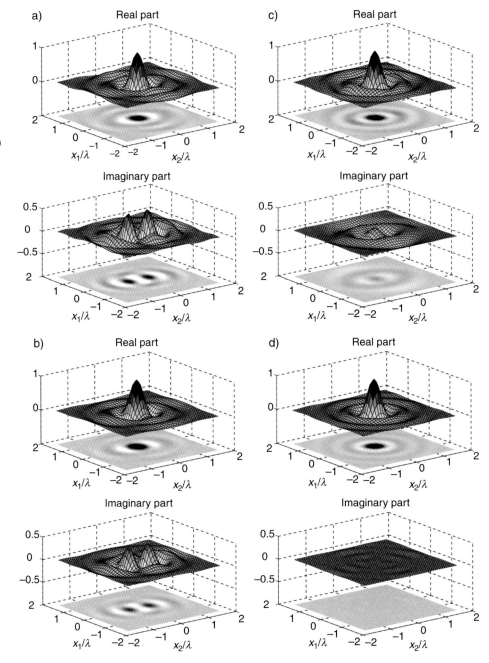

slight deviations from circular symmetry, and nonzero imaginary part of the far-field resolution function demonstrate that the far-field expressions do not completely correct for the propagation effects. This is expected because of the differences between the exact-field and far-field radiation patterns shown in Figure 2. In Figure 5d, the normalized real and imaginary parts of the exact-field multicomponent image are shown. This resolution function is positive, real valued with a near-zero imaginary part, and circularly symmetrical; it accurately represents the properties of the point scatterer.

## Tests on Synthetic Data

Synthetic GPR data were generated using the 3D FDTD algorithm of Lampe et al. (2003). This versatile FDTD modeling tool is second order in space and time and uses a standard staggering scheme and generalized perfectly matched layer (PML) boundary conditions. Two models were used, each approximately $7 \times 7 \times 5$ m, one containing a dipping planar layer boundary and the other containing a pipe embedded in a homogeneous subsurface. Sources emitting a wavelet represented by the derivative of a Gaussian function with a center frequency of 70 MHz were placed on a 0.1- × 0.1-m surface grid. The source-receiver offset was 1 m. Co- and cross-polarized data were generated for dipole antennae oriented at 0° and 90° relative to the trend of the plane reflector and the pipe. In all synthetic GPR data, we muted the direct waves, which are not accounted for in our imaging schemes. We then migrated the data using different imaging algorithms.

## Dipping plane

The first model (Figure 6a) contains an 18° dipping boundary with relative permittivities $\varepsilon_r = 7$ above and $\varepsilon_r = 10$ below and a constant conductivity of 0.2 mS/m. For the small antenna offset, the angles of incidence are small, such that the amplitudes of reflections from these planes should be practically independent of the polarization of the incident wavefield. Accordingly, eliminating the antenna radiation patterns and vector-wavefield effects with accurate vector migration should result in practically equal image amplitudes and phases independent of antenna orientation. This is best tested by comparing the images $\chi_{11}$ and $\chi_{22}$ because the amplitudes differ most strongly between the raw data acquired with these particular orthogonal antenna orientations. For the central survey area of this model, the average peak-to-peak reflection amplitude of the nonmigrated data for the 0°-oriented antennae is 39% higher than that for the 90°-oriented antennae.

We migrated the synthetic 3D GPR data using the scalar Gazdag algorithm and the far-field multicomponent (van der Kruk et al., 2003a), far-field single component, exact-field multicomponent (Streich and van der Kruk, 2007a), and exact-field single component (Streich et al., 2007) vector-migration schemes. From each algorithm, we obtained the $\chi_{11}$ and $\chi_{22}$ images (Figure 7). In applying the exact-field algorithms, oversampling was used 16 times to compute the $\mathbf{x}$-$\omega$-domain Green's functions (i.e., we evaluated $k_1$-$k_2$-$x_3$-$\omega$-domain Green's functions and carried out FFTs on $2048 \times 2048$ matrices). On a laptop computer with a 2.2-GHz CPU, the computation took 4.2 minutes for Gazdag imaging, 6.1 minutes for far-field single-component imaging, 7.1 minutes for far-field multicomponent imaging, 1.45 hours for exact-field single-component imaging, and 2.42 hours for exact-field multicomponent imaging.

For the scalar Gazdag images, the average peak-to-peak reflection amplitude in the $\chi_{11}^{\text{Gzd}}$ image is 41% higher than that in the $\chi_{22}^{\text{Gzd}}$ image (Figure 7a through 7c), slightly higher than the 39% amplitude difference of the nonmigrated images. Qualitatively similar results can be expected from the scalar SAR algorithm, as indicated by the spatial-resolution functions (Figure 5). The far-field multi- and

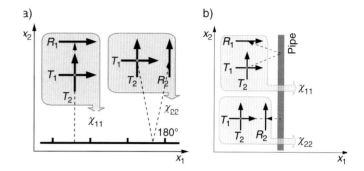

**Figure 6.** Geometry of synthetic surveys for (a) a reflector trending parallel to the $x_1$-axis (the horizontal line with tick marks indicates the trend and dip of the reflector) and (b) a pipe oriented parallel to the $x_2$-axis. Within each set of antennae (underlain shaded in gray), $T_\alpha$ and $R_\beta$ denote the transmitting and receiving antennae used to obtain the $E_{\alpha\beta}$ components. The values $\chi_{11}$ and $\chi_{22}$ are the images associated with the different antenna sets. Dashed lines indicate typical wave-propagation paths.

single-component images are practically identical for each antenna orientation (compare Figure 7d through 7g with 7e through 7h). Nevertheless, the average peak-to-peak reflection amplitudes of the $\chi_{11}^{\text{far,MC}}$ and $\chi_{11}^{\text{far,SC}}$ images differ from those of the $\chi_{22}^{\text{far,MC}}$ and $\chi_{22}^{\text{far,SC}}$ images by 30% (Figure 7f and 7i). In addition, $\chi_{22}^{\text{far,MC}}$ and $\chi_{22}^{\text{far,SC}}$ are phase-shifted relative to $\chi_{11}^{\text{far,MC}}$ and $\chi_{11}^{\text{far,SC}}$, which makes the amplitude differences appear even higher than for the scalar phase-shift images. Clearly, the Gazdag and far-field images are unsatisfactorily dependent on antenna orientation (Streich et al., 2007).

In contrast, the peak-to-peak reflection amplitudes of all exact-field multi- and single-component images are very close to each other (Figure 7j through 7o), with average differences of about 2.6%. These small residual differences likely are caused by forward-modeling artifacts and slight reflection-coefficient variations for the different antenna orientations (rotating the 1-m-offset broadside antennae 90° around a fixed midpoint yields maximum amplitude variations of about 1% because of changes in the reflection coefficients that result from the small antenna offset). These results confirm that the exact-field multi- and single-component vector-migration schemes equivalently correct for the radiation patterns. For polarization-independent scatterers, only one copolarized antenna pair is required for imaging the subsurface accurately.

## Metallic and dielectric pipes

For objects such as pipes, the data amplitudes are influenced not only by the radiation patterns but also by

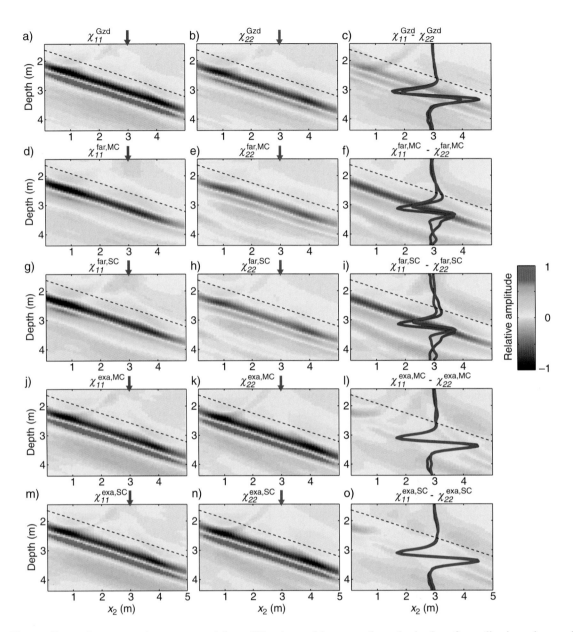

**Figure 7.** The $x_2$-directed cross sections extracted from 3D migrated images of synthetic data for a dipping plane using (a, b) scalar modified Gazdag, (d, e) far-field multicomponent, (g, h) far-field single-component, (j, k) exact-field multicomponent, and (m, n) exact-field single-component imaging schemes. Subscripts $_{11}$ and $_{22}$ correspond to copolarized antennae oriented in the $x_1$- and $x_2$- directions (Figure 6). The angle between the reflector trend and the copolarized antennae is $0°$ for $\chi_{11}$ and $90°$ for $\chi_{22}$ (see Figure 6a). The right column (c, f, i, l, o) shows the respective differences $\chi_{11} - \chi_{22}$, overlain by traces extracted from $\chi_{11}$ (blue) and $\chi_{22}$ (red) at $x_2 = 3$ m.

the target size and shape and by polarization effects that depend on the polarization and incidence angle of the EM field relative to the reflecting object. Polarization effects comprise preferential scattering (i.e., variations of the scattered wave amplitudes depending on the polarization of the incident field relative to the reflecting object) and depolarization (i.e., changes of the relative amplitudes of the scattered field components compared to the incident-field components). Scattering amplitudes for reflection at a pipe

are thus functions of the relative antenna-to-pipe orientation, pipe and background properties, scattering angle, and pipe thickness with respect to GPR wavelength.

Scattering strengths of pipes as functions of the material properties for a fixed antenna-to-pipe orientation have been analyzed by Zeng and McMechan (1997), and polarization effects as functions of the material properties and antenna-to-pipe orientations have been investigated by, e.g., Roberts and Daniels (1996) and Radzevicius and

Daniels (2000). Their results show that for metallic pipes and dielectric pipes with permittivities larger than that of the surrounding medium, the direction of preferential scattering usually is aligned with pipe orientation. They also show that for dielectric pipes with permittivities lower than that of the surrounding medium, the direction of preferential scattering depends on pipe diameter (with respect to the dominant GPR wavelength) and typically is perpendicular to pipe orientation for small pipe-diameter-to-wavelength ratios.

We simulated dielectric pipes with a diameter of 0.2 m and relative permittivities of $\varepsilon_r = 4$ and 12 and a metallic pipe with a conductivity of $10^7$ S/m, embedded in a homogeneous medium with $\varepsilon_r = 8$. The pipes were oriented parallel to the $x_2$-axis (Figure 6b), and their top was located at a depth of 1.8 m. A constant conductivity of $\sigma = 0.2$ mS/m was used for the background medium and the dielectric pipes.

After using the vector migration scheme to eliminate nearly all radiation-pattern effects from the data, the amplitudes and phases of the resulting images should be influenced only by the physical characteristics of the pipes. Amplitude variations with the orientations of the antennae to the pipe direction then should be solely from scattering effects.

In Figures 8 and 9, we display results of the modified Gazdag and single-component exact-field migrations; the various images $\chi_{11}$ and $\chi_{22}$ were obtained from the $E_{11}$ and $E_{22}$ configurations, respectively (see Figure 6b). The Gazdag images (left two columns of Figure 8 and left column of Figure 9) are somewhat less focused than the vector-migrated images. Differences in amplitude and phase between the $\chi_{11}$ and $\chi_{22}$ Gazdag images of each pipe are a result of polarization and radiation-pattern effects and thus are difficult to interpret in terms of pipe properties.

For the low-permittivity pipe in Figure 8a through 8d and Figure 9a and 9b, the two Gazdag images show different signatures (i.e., shapes of the imaged waveforms) and a significant phase difference, resulting in an apparent time shift between $\chi_{11}^{\text{Gzd}}$ and $\chi_{22}^{\text{Gzd}}$ (Figure 9a). In contrast, for the exact-field vector images of the same pipe, the waveforms and phases are nearly equal, whereas the reflection amplitudes are approximately 22% higher in $\chi_{11}^{\text{exa}}$ than in $\chi_{22}^{\text{exa}}$ (Figure 8c and 8d and Figure 9b). This exact-field imaging result agrees qualitatively with results of Roberts and Daniels (1996) and Radzevicius and Daniels (2000), who also find, for comparable models of low-permittivity dielectric pipes, that scattering amplitudes are largest for antennae oriented perpendicular to the pipes.

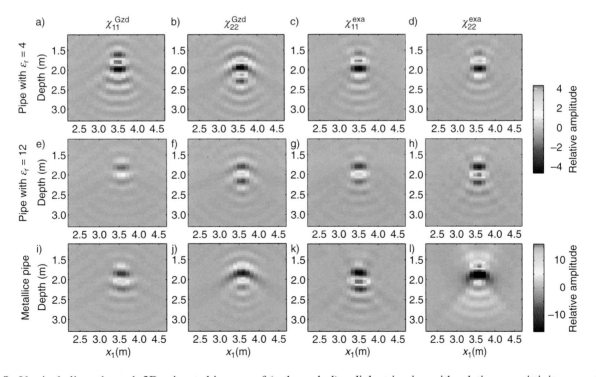

**Figure 8.** Vertical slices through 3D migrated images of (a through d) a dielectric pipe with relative permittivity $\varepsilon_r = 4$, (e through h) a pipe with $\varepsilon_r = 12$, and (i through l) a metal pipe with conductivity $\sigma = 10^7$ S/m. All pipes are embedded in a homogeneous medium with $\varepsilon_r = 8$ and $\sigma = 0.2$ mS/m (see Figure 6b). Shown are (a, e, i) Gazdag images $\chi_{11}^{\text{gzd}}$ of the $E_{11}$ component, (b, f, j) Gazdag images $\chi_{22}^{\text{gzd}}$ of the $E_{22}$ component, (c, g, k) vector images $\chi_{11}^{\text{exa}}$ of the $E_{11}$ component, and (d, h, l) vector images $\chi_{22}^{\text{exa}}$ of the $E_{22}$ component. Note the different color scales for the images of the dielectric and metallic pipes.

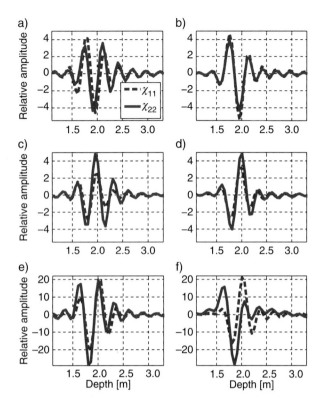

**Figure 9.** Traces extracted from the images shown in Figure 8 at $x = 3.5$ m for (a, b) the low-permittivity dielectric pipe with $\varepsilon_r = 4$, (c, d) the high-permittivity dielectric pipe with $\varepsilon_r = 2$, and (e, f) the metal pipe. Gazdag images are displayed in the left column and exact-field vector images in the right column. Dotted blue lines are $\chi_{11}$; solid red lines are $\chi_{12}$.

For the images of the high-permittivity dielectric pipe (Figure 8e through 8h and Figure 9c and 9d), the Gazdag images again have different signatures. By contrast, the two vector images have very similar waveforms and phases, and the amplitudes of $\chi_{22}^{\text{exa}}$ (antennae parallel to the pipe) are approximately 52% higher than those of $\chi_{11}^{\text{exa}}$ (antennae perpendicular to the pipe), which is also consistent with analytic results (Radzevicius and Daniels, 2000). Note that the images of the low- and high-permittivity pipes have opposite polarities (as is the case for specular reflections from planar layer boundaries).

The images of the metallic pipes have, as expected, much higher amplitudes than those of the dielectric pipes (Figure 8i through 8l and Figure 9e and 9f). Despite the use of a weak scattering approximation in the vector migration, the images are well focused. As expected, the amplitudes are higher in $\chi_{22}^{\text{exa}}$ than in $\chi_{11}^{\text{exa}}$. An approximately 90° phase shift is present between the two exact-field vector-migrated images, which does not appear in the two Gazdag images. Further analysis is required to explain this phase difference. One possible explanation could be the presence

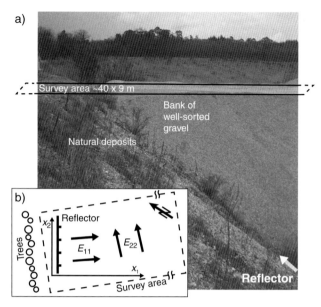

**Figure 10.** (a) Photograph showing the Weiach survey site (Switzerland). Measurements were made on the planar surface of an artificial gravel bank. The arrow identifies the near-planar dipping boundary between this gravel unit and the underlying natural gravel deposits. (b) Top view, showing the relative orientations of the antennae, survey area (dashed box), coordinate system adopted for displaying the results, and trees near the northwest edge of the survey area.

of creeping waves that travel around the pipe and might contribute differently to the $E_{11}$ and $E_{22}$ components.

## Application to Field Data

Two experimental data sets, each comprised of data for two orthogonal copolarized antenna configurations, were acquired at field sites that contained a dipping plane and pipes, respectively. Both data sets were subjected to basic processing, including time-zero alignment, regularization of the data for both antenna configurations onto the same grid, and application of a time-variant gain function that preserved relative amplitudes between different traces and between the two data volumes but enhanced the signal at later times within each trace.

### Dipping plane

Data containing a dipping-plane reflection were acquired at a gravel pit near Weiach, northern Switzerland (Streich et al., 2007). In the process of quarrying at this site, an artificial bank of well-sorted gravel was deposited temporarily on the dipping surface of less homogeneous natural gravel deposits (Figure 10a). Preliminary tests

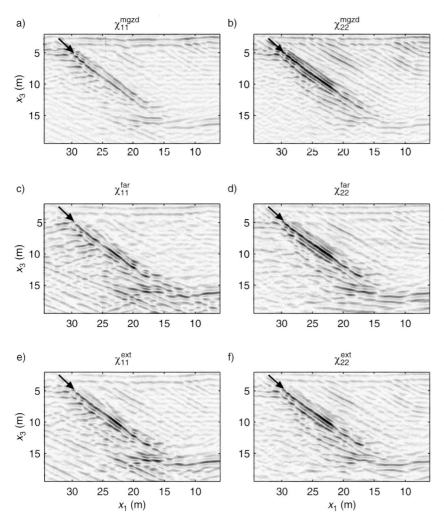

**Figure 11.** Vertical slices extracted at $x_2 = 11.25$ m from the (a, b) modified scalar Gazdag, (c, d) far-field single-component vector, and (e, f) exact-field single-component vector images obtained from the 3D field data. Subscripts $_{11}$ and $_{22}$ correspond to copolarized antennae oriented approximately in the $x_1$- and $x_2$-directions in Figure 10b. Arrows identify the principal dipping reflector on which the amplitude analysis of Figure 12 was performed. Reflections below the black lines in (a), (c), and (e) originate from trees present near the survey area (van der Kruk and Slob, 2004).

demonstrated that the interface between these two gravel units, which dipped at approximately 35° and was nearly planar, produced a strong GPR reflection.

The 3D GPR data were collected in a 40- × 9-m area that extended across the surface of the artificial bank and adjacent natural deposits (Figure 10b). Ramac GPR antennae with a nominal center frequency of 100 MHz were used, such that the reflecting interface was present at distances of 5–10 $\lambda$, where the far field still differs significantly from the exact field (Streich and van der Kruk, 2007a). The trace spacing was about 12.5 × 25 cm in the inline and crossline directions, which is significantly smaller than the antenna length ($\sim$1 m), suggesting that ideally, the antenna dimension should be considered in exact-field imaging. However, this would require precise, site-specific determination of antenna characteristics (Streich and van der Kruk, 2007b) and might be pursued as a future extension of the exact-field imaging algorithms.

After preprocessing, we obtained two data volumes, each containing 321 × 200 traces on a 0.125- × 0.125-m

grid. A velocity of 0.13 m/ns (i.e., $\varepsilon_r \sim 5.3$) determined from common-midpoint data recorded parallel to the trend of the reflecting interface was used for migrating the data. The processed data were imaged using the modified scalar Gazdag scheme and the far- and exact-field single-component vector-migration schemes (Streich et al., 2007). The computation took 1.79 hours for a scalar image, 2.40 hours for a far-field image, and 6.91 hours for an exact-field image. The cost of scalar and far-field imaging is determined primarily by the size of the data set and is thus substantially higher than for imaging the synthetic data. In contrast, the cost of exact-field imaging depends primarily on the required sampling of the Green's functions. As for the synthetic data, FFT matrix sizes of 2048 × 2048 points were used, resulting in computation times comparable to those required for imaging the small synthetic data sets.

Vertical sections of the images are displayed in Figure 11. The reflection from the interface between the artificial bank and the naturally deposited gravels (marked by arrows in Figure 11) is the most prominent event in all

**Figure 12.** Results of analyzing the amplitudes of the principal reflection in the (a, b) modified scalar Gazdag, (c, d) far-field single-component vector, and (e, f) exact-field single-component vector images of the Weiach field data. The left column shows amplitude ratios $\chi_{22}/\chi_{11}$ across the dipping reflector. Bold numbers indicate the average amplitude ratios. The right column shows corresponding scatter plots of the reflection amplitudes. Ideally, the ratios in the left column should uniformly equal one, and the points in the right column should lie along the diagonals.

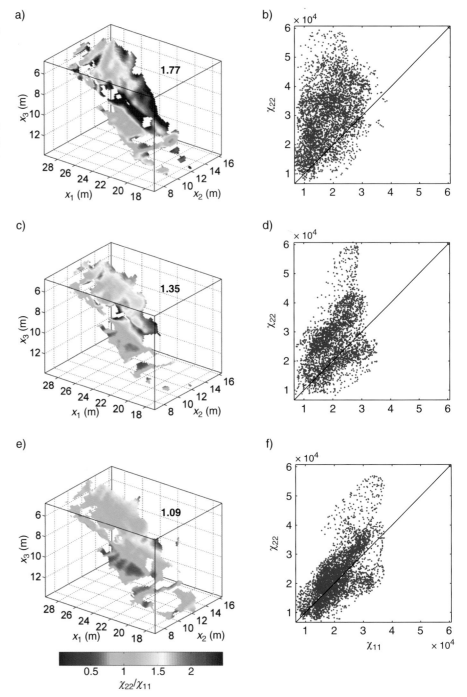

images. The amplitudes of all subsurface reflections differ significantly between the modified scalar Gazdag images $\chi_{11}^{mGzd}$ and $\chi_{22}^{mGzd}$ and between the far-field vector images $\chi_{11}^{far}$ and $\chi_{22}^{far}$, but they are similar in the exact-field vector images $\chi_{11}^{exa}$ and $\chi_{22}^{exa}$. To assess quantitatively the performance of the different migration schemes, we compare the amplitudes of the principal interface reflection observed in the various $\chi_{11}$ and $\chi_{22}$ images. We pick peak-to-peak amplitudes using the strongest negative troughs (colored brown and marked by arrows in Figure 11) and the positive peaks below. The amplitudes determined from the $\chi_{11}$ and $\chi_{22}$ images are compared for each of the migration schemes in Figure 12b, 12d, and 12f, and the distributions of amplitude ratios $\chi_{22}/\chi_{11}$ are shown superimposed on the picked reflection surface in Figure 12a, 12c, and 12e. For each pair of migrated images, we show only picked reflection amplitudes that are well above background values in both images.

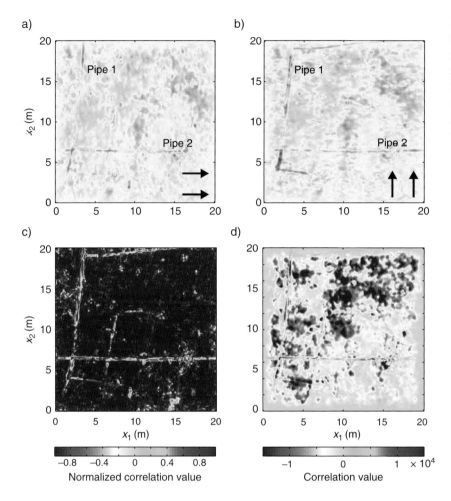

**Figure 13.** The (a) $\chi_{11}$ and (b) $\chi_{22}$ horizontal slices extracted from exact-field vector images of the Miami field data at a depth of $x_3 = 0.78$ m. Arrows in the lower right corners indicate antenna orientations. (c) Normalized and (d) nonnormalized zero-lag crosscorrelation between images $\chi_{11}$ and $\chi_{22}$ over a depth window of 0.45 m centered at $x_3 = 0.78$ m.

For the modified Gazdag migrated images, the amplitude analysis yields an average $\chi_{22}/\chi_{11}$ amplitude ratio of 1.77 with a relatively high standard deviation of 0.74 (Figure 12a and 12b); for the far-field vector images, the average amplitude ratio is 1.35 and the standard deviation is 0.46 (Figure 12c and 12d). In marked contrast, for the exact-field vector-migrated images, the amplitude ratio is reduced to 1.09, which is close to the ideal value of one, and the standard deviation is only 0.30 (Figure 12e and 12f). These results indicate that exact-field single-component vector migration has eliminated most radiation-pattern and vector-wave-propagation effects from the field data.

## Polarization effects of pipes

High-density data that contained pipe reflections and thus allowed us to investigate polarization effects were acquired across a 20- × 20-m area in Miami, Florida, U.S.A. High data quality was achieved by using a centimeter-precise, laser-positioned GPR system (Grasmueck and Viggiano, 2007). From previous surveys (Grasmueck et al., 2004), several pipes and other objects distinguished

by polarization-dependent characteristics were known to be present at the study site. In the new survey, we used a 250-MHz shielded Ramac GPR antenna and a trace spacing of 6 cm in the inline and crossline directions. On the same day, we repeated the 6-cm spaced 3D GPR survey with orthogonal polarization, rotating the antenna orientation by 90°. We then imaged the preprocessed data using the exact-field vector-migration algorithm. The depth sampling of the images was 3 cm.

Figure 13a and 13b shows horizontal slices from the resulting images. Several elongated structures are visible in these slices, of which those annotated pipe 1 and pipe 2 are most prominent. A zero-lag crosscorrelation of the two images in the $z$-direction highlights some remarkable features (Figure 13c and 13d). First, the overall correlation level is exceptionally good, with normalized correlation values (Figure 13c) near unity throughout most of the surveyed volume (the same is true for dipping, near-planar geologic structures present in deeper parts of the volume). In contrast, the two pipes exhibit negative correlation values along large parts of their lengths. The nonnormalized correlation (Figure 13d) shows that for pipe 1, the signal is strongest in the regions with strongly negative correlation and weaker in regions with positive correlation. There may be interference between the pipe structure and an adjacent planar reflector and differences in coupling or slight velocity variations in the overburden between the two surveys. For pipe 2, the signal is similarly strong in regions with positive and negative correlations.

Clearly, the negative correlation (i.e., opposite polarity of the pipe images) does not correspond to any of our synthetic models or theoretical expectations. It later was verified by excavations that pipe 1 is actually a set of three 1-cm-diameter plastic irrigation lines spaced 2.5 cm apart, and pipe 2 is a 5-cm diameter PVC pipe. Consequently, our simple synthetic models cannot be expected to represent the true structures faithfully, and further investigation and modeling studies are required. These should include pipes that have a smaller diameter-to-wavelength ratio than those we have modeled, hollow or water-filled pipes, interference from multiple thin pipes, and the possibility of modeling anisotropic targets. In addition, amplitude effects should be analyzed in detail. (Although we did observe somewhat higher amplitudes for both structures in the image $\chi_{22}$, attempting to interpret amplitude effects at this stage would be premature.)

## Conclusions

To obtain GPR images in which the reflection amplitudes accurately represent the physical-property contrasts of the causative structures, 3D imaging that fully corrects for the antenna radiation patterns and EM wave propagation is required. To facilitate such imaging, we have developed a rapid method for computing practically exact Green's functions that involves inverse FFT oversampled representations of closed-form horizontal wavenumber-frequency-domain Green's functions, taking advantage of vertical-wavenumber phase shifting. Comparison of radiation patterns determined by our method with exact radiation patterns computed via adaptive integration proves that the new approach is accurate for target depths typical of most GPR investigations.

We have formulated an exact-field, single-component vector-migration scheme that can be applied readily to most standard GPR data acquired using a single pair of copolarized antennae. Our tests on synthetic and experimental GPR data containing dipping reflectors, which are polarization-independent scatterers, demonstrate that the exact-field multi- and single-component vector-migration schemes equivalently account for radiation-pattern and vector wave-propagation effects. The images of our data

produced by the two schemes are nearly identical to each other and are almost independent of antenna-to-reflector orientations. In contrast, the amplitudes and phases differ markedly in the corresponding Gazdag and far-field images.

We also have delineated various scattering effects for different dielectric and metallic pipes in vector-migrated images that are practically devoid of radiation-pattern effects. For synthetic pipes and different antenna-to-pipe orientations, exact-field vector migration of the data yielded images with scatterer amplitudes in qualitative agreement with theoretical expectations. Vector images of dipping planar reflectors have amplitudes nearly independent of the antenna-to-plane orientation. In contrast, in our images of elongated targets, we observe considerable amplitude variations with the relative antenna-to-pipe orientation. For metallic pipes, phases depend strongly on the antenna-to-pipe orientation. The Gazdag images did not delineate these effects. Because Gazdag migration does not account for radiation-pattern effects and returns images that still depend on the radiation patterns, it is unsuitable for investigating polarization effects.

In a field data set acquired at a site containing several objects with polarization-dependent scattering properties, we identified objects with different polarization properties. However, strikingly negative crosscorrelations between signals from a pipe and a set of three pipes in our real-data images cannot be explained by our simplistic synthetic model studies; more complex models are required to interpret such phenomena. Future work will aim at obtaining a more quantitative understanding of our observations to infer scatterer properties more reliably from vector-migrated images.

## Acknowledgments

This work was funded by grants from ETH Zurich and the Swiss National Science Foundation. We thank Tim Henstock and two anonymous reviewers for their comments. The University of Miami acknowledges the support of this research by the National Science Foundation (grants 0323213 and 0440322).

## References

Annan, A. P., 1973, Radio interferometry depth sounding: Part 1 — Theoretical discussion: Geophysics, **38**, 557–580.

Baysal, E., D. D. Kosloff, and J. W. C. Sherwood, 1983, Reverse time migration: Geophysics, **48**, 1514–1524.

Bleistein, N., 1984, Mathematical methods for wave phenomena: Academic Press, Inc.

———, 1987, On the imaging of reflectors in the earth: Geophysics, **52**, 931–942.

Born, M., and E. Wolf, 1965, Principles of optics: Pergamon Press.

Bradford, J. H., 2008, Measuring water content heterogeneity using multifold GPR with reflection tomography: Vadose Zone Journal, **7**, 184–193.

Claerbout, J., 1971, Toward a unified theory of reflector mapping: Geophysics, **36**, 467–481.

Curlander, J. C., and R. N. McDonough, 1991, Synthetic aperture radar, systems and signal processing: Wiley Interscience.

de Hoop, A. T., 1995, Handbook of radiation and scattering of waves: Academic Press Inc.

Docherty, P., 1991, A brief comparison of some Kirchhoff integral formulas for migration and inversion: Geophysics, **56**, 1164–1169.

Engheta, N., C. H. Papas, and C. Elachi, 1982, Radiation patterns of interfacial dipole antennas: Radio Science, **17**, 1557–1566.

Gazdag, J., 1978, Wave equation migration with the phase-shift method: Geophysics, **43**, 1342–1351.

Grandjean, G., and J. C. Gourry, GPR data processing for 3D fracture mapping in a marble quarry (Thassos, Greece): Journal of Applied Geophysics, **36**, 19–30.

Grasmueck, M., 1996, 3D Ground penetrating radar applied to facture imaging in gneiss: Geophysics, **61**, 1050–1064.

Grasmueck, M., and D. A. Viggiano, 2007, Integration of ground-penetrating radar and laser positioning sensors for real-time 3D data fusion: IEEE Transactions on Geoscience and Remote Sensing, **45**, 130–137.

Grasmueck, M., R. Weger, and H. Horstmeyer, 2004, Three-dimensional ground-penetrating radar imaging of sedimentary structures, fractures, and archeological features at submeter resolution: Geology, **32**, 933–936.

———, 2005, Full-resolution 3D GPR imaging: Geophysics, **70**, no. 1, K12–K19.

Gray, S., 1997, True-amplitude seismic migration: A comparison of three approaches: Geophysics, **62**, 929–936.

Guy, E. D., J. J. Daniels, and S. J. Radzevicius, 1999, Demonstration of using crossed dipole GPR antennae for site characterization: Geophysical Research Letters, **26**, 3421–3424.

Hansen, T., and P. Johansen, 2000, Inversion scheme for ground penetrating radar that takes into account the planar air-soil interface: IEEE Transactions on Geoscience and Remote Sensing, **38**, 496–506.

Holliger, K., and T. Bergmann, 1998, Accurate and efficient modeling of ground-penetrating radar antenna radiation: Geophysical Research Letters, **25**, 3883–3886.

Johansson, E., and J. Mast, 1994, Three-dimensional ground penetrating radar imaging using synthetic aperture time-domain focusing: Proceedings of the SPIE, **2275**, 205–214.

Lampe, B., K. Holliger, and A. G. Green, 2003, A finite-difference time-domain simulation tool for ground-penetrating radar antennas: Geophysics, **68**, 971–987.

Lehmann, F., D. Boerner, K. Holliger, and A. G. Green, 2000, Multicomponent georadar data: Some implications for data acquisition and processing: Geophysics, **65**, 1542–1552.

Mast, J. E., and E. M. Johansson, 1994, Three-dimensional ground penetrating radar imaging using multi-frequency diffraction tomography: Proceedings of the SPIE, **2275**, 196–203.

Meincke, P., 2001, Linear GPR inversion for lossy soil and a planar air-soil interface: IEEE Transactions on Geoscience and Remote Sensing, **39**, 2713–2721.

Molyneux, J. E., and A. Witten, 1993, Diffraction tomographic imaging in a monostatic measurement geometry: IEEE Transactions on Geoscience and Remote Sensing, **31**, 507–511.

Moran, M., R. Greenfield, S. Arcone, and A. Delaney, 2000, Multidimensional GPR array processing using Kirchhoff migration: Journal of Applied Geophysics, **43**, 281–295.

Orlando, L., and E. C. Slob, 2009, Using multicomponent GPR to monitor cracks in a historical building: Journal of Applied Geophysics **67**, 327–334.

Piessens, R., E. deDoncker Kapenga, C. W. Uberhuber, and D. K. Kahaner, 1983, Quadpack: A subroutine package for automatic integration: Springer-Verlag Berlin.

Radzevicius, S. J., C. C. Chen, L. Peters Jr., and J. J. Daniels, 2003, Near-field dipole radiation dynamics through FDTD modeling: Journal of Applied Geophysics, **52**, 75–91.

Radzevicius, S. J., and J. J. Daniels, 2000, Ground penetrating radar polarization and scattering from cylinders: Journal of Applied Geophysics, **45**, 111–125.

Roberts, R., and J. J. Daniels, 1996, Analysis of GPR polarization phenomena: American Journal of Environmental and Engineering Geophysics, **1**, 139–157.

Sassen, D. S., and M. E. Everett, 2009, 3D polarimetric GPR coherency attributes and full-waveform inversion of transmission data for characterizing fractured rock: Geophysics, **74**, no. 3, J23–J34.

Schneider, W. A., 1978, Integral formulation for migration in two and three dimensions: Geophysics, **43**, 49–76.

Seol, S. J., J. H. Kim, Y. Song, and C. H. Chung, 2001, Finding the strike direction of fractures using GPR: Geophysical Prospecting, **49**, 300–308.

Slob, E. C., R. F. Bloemenkamp, and A. G. Yarovoy, 2003, Efficient computation of the wavefield in a two media configuration emitted by a GPR system from incident field measurements in air: Proceedings of the 2nd

International Workshop on Advanced Ground Penetrating Radar, 60–65.

Slob, E. C., and J. T. Fokkema, 2002a, Interfacial dipoles and radiated energy: Subsurface Sensing Technologies and Applications, **3**, 347–367.

———, 2002b, Coupling effects of two electric dipoles on an interface: Radio Science, **37**, 1073, doi:10.1029/2001RS002529.

Smith, G. S., 1984, Directive properties of antennas for transmission into a material half-space: IEEE Transactions on Antennas and Propagation, **32**, 232–246.

Streich, R., and J. van der Kruk, 2007a, Accurate imaging of multicomponent GPR data based on exact radiation patterns: IEEE Transactions on Geoscience and Remote Sensing, **45**, 93–103.

———, 2007b, Characterizing a GPR antenna system by near-field electric field measurements: Geophysics, **72**, no. 5, A51–A55.

Streich, R., J. van der Kruk, and A. G. Green, 2006, Three-dimensional multicomponent georadar imaging of sedimentary structures: Near Surface Geophysics, **4**, 39–48.

———, 2007, Vector-migration of standard copolarized 3D GPR data: Geophysics, **72**, no. 5, J65–J75.

Symes, W. W., 2007, Reverse time migration with optimal checkpointing: Geophysics, **72**, no. 5, SM213–SM221.

Valle, S., L. Zanzi, M. Sgheiz, G. Lenzi, and J. Friborg, 2001, Ground penetrating radar antennas: Theoretical and experimental directivity functions: IEEE Transactions on Geoscience and Remote Sensing, **39**, 749–758.

van der Kruk, J., 2001, Three-dimensional imaging of multi-component ground penetrating radar: Ph.D. dissertation, Delft University of Technology.

———, 2004, Three-dimensional GPR imaging in the horizontal wavenumber domain for different heights of source and receiver antennas: Near Surface Geophysics, **2**, 25–32.

van der Kruk, J., and E. C. Slob, 2004, Reduction of reflections from above surface objects in GPR data: Journal of Applied Geophyscis, **55**, 271–278.

van der Kruk, J., C. P. A. Wapenaar, J. T. Fokkema, and P. M. van den Berg, 2003a, Three-dimensional imaging of multicomponent ground-penetrating radar data: Geophysics, **68**, 1241–1254.

———, 2003b, Improved three-dimensional image reconstruction technique for multicomponent ground penetrating radar data: Subsurface Sensing Technologies and Applications, **4**, 61–99.

van Gestel, J. P., and P. L. Stoffa, 2000, Migration using multi-configuration GPR data: Proceedings of the 8th International Conference on Ground Penetrating Radar, 300–308.

Wang, T., and M. L. Oristaglio, 2000, GPR imaging using the generalized Radon transform: Geophysics, **65**, 1553–1559.

Witten, A. J., J. E. Molyneux, and J. E. Nyquist, 1994, Ground penetrating radar tomography: Algorithms and case studies: IEEE Transactions on Geoscience and Remote Sensing, **32**, 461–467.

Zeng, X., and G. A. McMechan, 1997, GPR characterization of buried tanks and pipes: Geophysics, **62**, 797–806.

Chapter 7

# Multiple-scale-porosity Wavelet Simulation Using GPR Tomography and Hydrogeophysical Analogs

Erwan Gloaguen[1], Bernard Giroux[1], Denis Marcotte[2], Camille Dubreuil-Boisclair[1], and Patrick Tremblay-Simard[1]

## Abstract

A novel approach can be used to simulate porosity fields constrained by borehole-radar tomography images. The cornerstone of the method is statistical analysis of the approximation wavelet coefficients of a petrophysical analog scenario. The method is tested with a 2D synthetic porosity field generated from a digital picture of a real sand deposit. The porosity field is translated into electrical properties and a crosshole tomography synthetic survey is built using a finite-difference modeling algorithm. Hereafter, this synthetic survey is considered as the measured one. In parallel, an analog deposit is created based on geologic knowledge of the area under study. The analog porosity field is converted into electrical property fields using the same equation. A synthetic ground-penetrating-radar (GPR) tomography also is computed from the latter. Then, wavelet decomposition of both measured and analog tomography and porosity analog fields is calculated. Based on the assumption that geophysical data carry essentially large-scale information about the geology, statistical analysis of the approximation wavelet coefficients of each variable is carried out. From measured tomographic approximation coefficients and cross statistics evaluated on the analogs, approximation of the real porosity field is inferred. Finally, based on the statistical relationships between wavelet coefficients across the different scales, all porosity wavelet-detail coefficients are simulated using a standard geostatistical cosimulation algorithm. The wavelet coefficients then are back-transformed in the porosity space. The final simulated porosity fields contain the large wavelengths of the measured radar tomographic images and the texture of the analog porosity field.

## Introduction

Groundwater is one of the major water sources that sustain life on earth. During the last century, human activities (industrial, domestic or agricultural) have led to an enormous pressure on groundwater systems (United Nations Environment Programme, 2003; Vereecken et al., 2005). Even in countries where water is abundant, human activity greatly influences groundwater quality near the water-catchment areas, usually located close to end users. Hence, there is a need to characterize the subsurface aquifer to optimize its sustainable use. In Canada, most groundwater studies are based on the construction of 3D numerical hydrogeologic models to study and validate groundwater flow and transport. However, conventional groundwater assessment is based on soil samples and well tests (slug test or pumping test). The scarcity and limited volume of soil supported by these data do not permit modeling the hydrogeologic parameters at the appropriate scale (de Marsily, 2002) and at acceptable costs.

Many studies show the potential of geophysical methods to complement conventional hydrogeologic surveys both for characterizing the geologic settings at a watershed scale and the hydrogeologic parameters, such as porosity and hydraulic conductivity, at a local scale (Rubin and

[1]*Institut National de la Recherche Scientifique, Québec City, Québec, Canada.*
[2]*École Polytechnique de Montréal, Département CGM, Montréal, Québec, Canada.*

Hubbard, 2005). Inference of hydrogeologic parameters from geophysical data at the local scale is one of the most important research topics in environmental geophysics. In conventional hydrogeophysical studies, in situ measured geophysical data are converted into hydrogeologic parameters using empirical relations (Butler, 2005). Usually these relations are obtained by fitting parameters of a polynomial on measured hydrogeologic parameters or inferred from the literature. However, use of such hydrogeophysical relations generally does not permit the representation of spatial variability of the lithology, allow integration of multiple geophysical data, or permit taking into account the smoothing effect inherent to geophysical data (Hyndman et al., 2000).

As tools for measuring both in situ geophysical and hydrogeologic data improve, geostatistics becomes more and more attractive to generate hydrogeologic scenarios using site-calibrated hydrogeophysical relations. For example, Cassiani et al. (1998) use cokriging to estimate 2D hydraulic conductivity fields from P-wave velocity tomograms. Gloaguen et al. (2004) and Tronicke and Holliger (2005) use conditional geostatistical simulations to generate porosity models from ground-penetrating-radar (GPR) tomography and borehole logs. Hyndman et al. (2000) use sequential Gaussian cosimulation of crosshole seismic tomograms and hydraulic conductivity measurements at boreholes to obtain 3D realizations of hydraulic conductivity and also estimate the field-scale petrophysical transform using dependent hydrological data. In addition, Chen et al. (2001), use Bayesian theory to generate porosity fields using collocated hydraulic conductivity and geophysical data. However, these techniques imply a constant relationship between geophysical and hydrogeologic properties across different scales. In addition, in situ measured hydrogeologic data do not represent the relation of spatial connectivity that influences the groundwater flow and the amount of measured data is usually too small to compute reliable statistics.

Researchers from the petroleum industry (Doyen, 1988) and hydrology (Macfarlane et al., 1994) propose the use of analog scenarios to infer the field-scale relations between hydrogeologic and geophysical parameters (Day-Lewis et al., 2005; Kowalsky et al., 2005; Linde et al., 2005; Moysey et al., 2005). An analog scenario is thought to represent the hydrogeologic knowledge of the ground under study. This knowledge could come from in situ measurements (core samples), from previous studies in similar geologic settings, or from geologic knowledge. The geologic analog can be converted into a geophysical model using a priori or site-calibrated petrophysical relations. Geophysical forward modeling is carried out using the physical analog scenario followed by the geo-

physical inversion. This allows estimating the statistical relations between the geologic model analog and the associated inverted geophysical model. In particular, it permits highlighting the smoothing effect of inverse modeling (Day-Lewis et al., 2005). In addition, the relation between geologic and inverse geophysical models can be complex, especially in the presence of noise and lateral variability.

Consequently, it can be stated that the geophysical model is an indirect low-resolution representation that only carries information about large-scale features of the geology. The problem of downscaling a fuzzy indirect-parameter image to a high-resolution image of the parameter of interest can be handled by using statistical imagery algorithms (Mallat, 1989). In particular, the relation between patterns at different scales can be addressed through discrete wavelet transform (DWT; Portilla and Simoncelli, 2000; Romberg and Baraniuk, 2001). The DWT coefficients fully characterize an image in terms of scale dependencies, features, and statistics (Mallat, 1989). The DWT decomposes an image into wavelet coefficients that are regrouped in directional subbands related to a given scale (or level) of an image and it allows multiscale analysis and geostatistical modeling (Flandrin, 1992). Statistical dependencies of wavelet coefficients at different scales (namely, high-order dependencies of wavelet coefficients) were studied for statistical image modeling with applications to texture analysis and texture synthesis (Crouse et al., 1998; Choi and Baraniuk, 2001).

Limitations of these techniques include their inability to allow direct conditioning to data, the fact that only correlation between scales is taken into account, and that likelihood modeling is a computer-intensive iterative process. Using a different wavelet-based approach, Tran et al. (2002) show that unconditional Haar wavelet simulations give results equivalent to the classical sequential Gaussian simulations in the case of multi-Gaussian data.

Here we propose a method to generate conditional simulation of hydrogeologic parameters, based on statistical analysis of DWT coefficients of geologic and geophysical analogs. First, we show how DWT of the geologic and corresponding geophysical analogs captures the high-order dependencies between features at different scales from the geologic and geophysical data. In addition, high-order dependencies of the geologic features are explored through covariance and cross-covariance modeling of DWT coefficients. The covariance captures the spatial relation between directional wavelet subbands representing a given resolution and the cross covariance captures the frequency relation between the directional wavelet subbands at different scales. Subsequently, joint simulation of wavelet coefficients is presented. Finally, the proposed method is tested on a synthetic example.

# Overview of the 2D Wavelet Transform

This section gives an overview of the 2D wavelet transform in the context of the proposed simulation algorithm. A complete review can be found in Mallat (1989) and Daubechies (1992). By construction, the wavelet transform provides both spatial and frequency-domain information about regularly sampled data such as an image. The wavelet coefficients represent a measure of similarity in the frequency content between an image and a chosen wavelet function. These coefficients are computed as a convolution of the image and the scaled wavelet function, which can be interpreted as a dilated bandpass filter because of its band-passlike spectrum. The scale is inversely proportional to the frequency. Consequently, low frequencies correspond to high scales and dilated wavelet functions. By wavelet analysis at high scales, we extract global information (approximations) from an image, whereas at low scales, we extract fine information (details) from an image. Images are usually band-limited, which is equivalent to having finite energy, and therefore, we need to use only a constrained interval of scales. The DWT is computed by passing the image successively through a high-pass and a low-pass filter. For each decomposition level, high-pass filter $h_d$ forming the wavelet function produces details **D**. The complementary low-pass filter $l_d$ representing the scaling function produces approximation **A**. This computational algorithm is called subband coding. In image processing, the details **D** consist of three directional subbands corresponding to horizontal H, vertical V, and diagonal D filters (Figure 1).

The filtering process alters the resolution, and the scale is changed by upsampling and downsampling by two. This is described by the equations

$$\mathbf{D_1}[n] = \sum_{k=-\infty}^{\infty} h_d[k]\mathbf{x}[2n - k]$$

$$\mathbf{A_1}[n] = \sum_{k=-\infty}^{\infty} l_d[k]\mathbf{x}[2n - k], \tag{1}$$

where $n$ and $k$ denote discrete time coefficients and **x** stands for the given image.

Half-band filters form orthonormal bases, and therefore make reconstruction straightforward. The synthesis

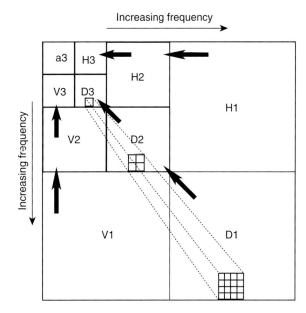

**Figure 1.** Illustration of three DWT scales. All subgrids represent the same area. Arrows point from the parent (scale $j - 1$) toward the child (scale $j$). Hj, Vj, and Dj are the directional wavelet detail coefficients at scale $j$ and a3 is the approximation at scale $J = 3$.

consists of upsampling by two and filtering

$$\mathbf{x}[n] = \sum_{k=-\infty}^{\infty} (\mathbf{D_1}[k]h_r[2k - n] + \mathbf{A_1}[k]l_r[2k - n]). \tag{2}$$

Downsampling by $p$ is defined as taking every $p$th sample, starting with sample one. Upsampling is the process of lengthening a signal component by inserting zeros between samples.

Reconstruction filters (inverse wavelet transform, or IDWT) $l_r$ and $h_r$ are identical to the decomposition filters $l_d$ and $h_d$, respectively, except the reverse time course. These filters produce exact image reconstruction from DWT coefficients, provided that the signal is of finite energy and that the wavelet satisfies the admissibility condition. Both of these conditions are satisfied with natural images and usual wavelets. Figure 1 shows the 2D decomposition tree of an image $\mathbf{x}[n]$ for $J = 3$ scales. The box labeled a3 corresponds to the approximation coefficients at scale 3. Boxes H3, V3, and D3 correspond to detail coefficients in horizontal, vertical, and diagonal directions, respectively. The link between the different scales is represented by arrows pointing toward the scale $j$ from the generating scale $j - 1$.

In theory, DWT should decompose an image with uncorrelated wavelet coefficients representing the most important features of an image, but Crouse et al. (1998)

demonstrate there are still considerable dependencies between scales, as observed from the characteristics of wavelet-coefficient distribution. This proves that some scale dependency still remains in the wavelet-coefficient space. The dependencies between scales are classified in two groups: intrascale and interscale, as discussed in the next sections.

## Intrascale dependencies

By definition, at a given scale, the three directional details are collocated and represent the directional component of an image at a given frequency (for example, H3, V3, D3, and a3 in Figure 1). The cross covariance between subbands belonging to the same scale characterizes the spatial link between directional features of an image at this scale.

## Interscale dependencies

Figure 1 also represents the link between the smallest structures of an image and the largest. The cross covariance between two subbands belonging to two different scales characterizes the spectrospatial information shared between the two subbands (for example, H3 and V1 in Figure 1). Wavelet coefficients at different scales are nested. Consequently, starting from the coarsest level to a point in the finest level is similar to zooming without conflict between scales. This property is particularly important for scaling purposes (Mallat, 1989).

## Some practical aspects of the wavelet coefficients

In this section, we enumerate the two most important properties of DWT in the context of geostatistical simulation. First, DWT provides a set of directional and frequency-dependent coefficients that can be represented on a tree structure (Kumar and Foufoula-Georgiou, 1997). This property is important as DWT decomposes the image in sets of oriented coefficients at different scales allowing characterization of possible geometric anisotropies at different scales. Note that each subband represents the same area as the original image $\mathbf{x}(s, t)$.

Second, DWT allows decomposing an image into its multiple-frequency features at their appropriate spatial resolution (illustrated in Figure 1). This property allows consistent modeling of the covariance relation between features at different scales and is particularly interesting for downscaling purposes. The small grids on the diagonal

coefficients represent the same area, but the resolution of each child is half the resolution of its parent. Each of the $J - 1$ scales has one-quarter the resolution of the generating J scale because the algorithm resamples the approximation of the scale J by two in both $(s, t)$ directions.

Intrascale and interscale cross variances of this particular spatial decomposition implicitly convey information relative to high-order statistics (Portilla and Simoncelli, 2000) because each point at the coarsest scale is related to each of its generating points belonging to the finer scales. Considering both interscales and intrascales, the number of spatial statistics to be computed is of the order of $3J^2$, where J is the number of scales considered and there are three subbands.

## Inverse discrete wavelet transform (IDWT)

Covariance modeling is done in the wavelet-coefficient space. The IDWT is the back transformation from the wavelet space to the original space. It can be computed by inverting the orthogonal matrix that transforms the original image into coefficients in the wavelet basis. In practice, IDWT is computed starting from the coarsest scale and combining the approximation and the details into the approximation at the next finer scale, and so on. Where DWT involves filtering and downsampling, IDWT consists of upsampling and filtering.

## The geologic analog as scale-dependency model

Guardiano and Srivastava (1992) introduce the idea in geostatistics of modeling the scale dependency of an attribute from a geologic analog. The geologic analog describes geometric facies patterns believed to be present in the subsurface and reflects a prior geologic/structural concept.

The need for a geologic analog in the proposed DWT simulation comes from three sources.

First, the scale dependency might not be present in the finite set of measured points. Second, DWT works only on full and regular grids. And third, in many cases, the specialist has a clear understanding of global features present in the attribute under interest. For example, it can be a geologic model inferred from indirect geophysical measurements or from geologic knowledge.

The geologic analog represents the different textures present in the studied ground. The analog should have global statistics equivalent to the field under study, i.e., equivalent histogram but not necessarily the same spatial distribution, because spatial variability is supported by the geophysical data. To study the statistical relationship between geology and geophysical images, the geologic analog is translated in terms of geophysical properties using known petrophysical relationships or in situ calibrated relations typically obtained from borehole logs and cores. In the case presented here, full-waveform GPR forward modeling is applied on the geophysical analog. Then picked traveltimes and amplitude are inverted to produce the velocity and attenuation analogs. Both geologic and geophysical analogs are wavelet transformed.

Conditional simulation of wavelet coefficients will allow generation of a geologic image that should have the same wavelet-domain characteristics as the geologic analog but is constrained by the measured geophysical data at the appropriate scale.

To summarize, DWT allows computation of the directional spectrospatial characteristics of a geologic analog. The DWT (and IDWT) is a 1:1 transformation (or bijunctive) without loss (equations 1 and 2). Conditional simulation of the wavelet coefficients will allow generating sets of wavelet coefficients that can be used to obtain an image with the same wavelet-domain characteristics as the geologic analog.

# Two-dimensional Conditional Simulation with DWT

Here we present the two main steps of the method. The first step consists of estimating the large wavelengths (wavelet approximation coefficients) of the geology constrained by large wavelengths of the measured geophysical data. The second step allows generation of different high-frequency scales of the geology, based on geostatistical analysis of the wavelet coefficient of the geologic analog.

## Simulation of approximation coefficients

In geoscience, it is important that modeling algorithms honor the measured data, which are called conditioning data in geostatistics. The conditioning data, here the approximation coefficients of the measured geophysical data, are measured on a regular grid and influence only the local trend of the variable to be simulated. Using the idea that geophysical measurements give only information on the large-scale features of the ground, in this step, we focus only on the coarsest level of the wavelet approximation coefficients. Our chosen algorithm is based on the nonparametric kernel-estimation method. The kernel is calculated by interpolating the scatter plot of the geologic and geophysical approximation coefficients using a Gaussian window. The kernel permits us to determine the prob-

ability of occurrence of the geologic approximation coefficients knowing the geophysical ones, i.e., also called likelihood. Once the kernel between geophysical and geologic-analog approximation coefficients is calculated, the measured geophysical wavelet approximation coefficients are transformed into geologic wavelet approximation coefficients using the Bayesian multivariate estimation method (Chilès and Delfiner, 1999).

## Simulation of geologic detail coefficients

After approximation coefficients of the geology are estimated, high frequencies (namely the detail coefficients) have to be simulated. Under the hypothesis that geophysical data measure only the low frequencies of the hydrogeologic patterns, geophysical data cannot be used to infer the high-frequency patterns of the geology. In the proposed method, statistical relations between geologic detail DWT coefficients across scales are inferred from the detail coefficients of the geologic analog. To cosimulate the detail coefficients conditioned to approximation coefficients estimated as described in the previous section, the auto and cross covariance of the geologic detail coefficients are modeled on the geologic detail coefficients of the analog. The following sections describe in detail the chosen simulation algorithm in the context of wavelet simulation.

### Sequential Gaussian cosimulation

The conditional simulation algorithm of wavelet coefficients is based on the well-known sequential Gaussian cosimulation (SGCS) (Rubinstein, 1981; Ripley, 1987; Verly, 1994). In the proposed method, the approximation and the detail coefficients (equation 1) are treated as regionalized covariables (Journel, 1989; Chilès and Delfiner, 1999). The cosimulation algorithm must reproduce the statistical distribution of the intrascale or interscale subbands cross variograms and any known wavelet-coefficient data. In the following paragraph, SGCS is presented in a 2D wavelet-coefficient framework. The detailed theory of SGCS can be found in Verly (1994).

For mathematical simplicity, we present only the estimation of wavelet coefficients belonging to the same scale $J = 1$ at location $x0$. There are four covariables to estimate: $\mathbf{z}^{a1}$, $\mathbf{z}^{H1}$, $\mathbf{z}^{V1}$, and $\mathbf{z}^{D1}$. Assuming the variables have joint multinormal distribution and considering the previously simulated vector (of size $4\,nd$), $\mathbf{z} = [\mathbf{z}_{nd}^{a1}, \mathbf{z}_{nd}^{H1}, \mathbf{z}_{nd}^{D1}, \mathbf{z}_{nd}^{V1}]^{\mathrm{T}}$, the conditional distribution of $\mathbf{z}_{x0}^{a1}, \mathbf{z}_{x0}^{H1}, \mathbf{z}_{x0}^{D1}, \mathbf{z}_{x0}^{V1}$ at $x0$ is defined by its conditional mean $\mathbf{m}(\mathbf{z}_{x0|nd}^{a1,H1,D1,V1}) = \mathbf{C}_{x0,nd}\,\mathbf{C}_{nd,nd}^{-1}\mathbf{z}$ and its conditional covariance matrix (of size $4 \times 4$)

$$\mathbf{C}_{x0|nd}^{a1,H1,D1,V1} = \mathbf{C}_{x0,x0} - \mathbf{C}_{x0,nd}\mathbf{C}_{nd,nd}^{-1}\mathbf{C}_{nd,x0}, \qquad (3)$$

where $\mathbf{C}_{x0,x0}$ is the $4 \times 4$ matrix of covariance of the coefficients, $\mathbf{C}_{x0,nd} = \mathbf{C}_{nd,x0}^{\mathrm{T}}$ is the $(4 \times 4\ nd)$ covariance matrix between coefficients to estimate and the already measured coefficients, and $\mathbf{C}_{nd,nd}$ is the $(nd \times nd)$ covariance matrix between already measured coefficients. A realization of the conditional random vector is given by (Rubinstein, 1981):

$$\mathbf{z}_{x0|nd}^{a1,H1,D1,V1} = \mathbf{C}_{x0,nd}\mathbf{C}_{nd,nd}^{-1}\mathbf{z} + \mathbf{L}_{x0|nd}\boldsymbol{\varepsilon}, \qquad (4)$$

where $\boldsymbol{\varepsilon}$ is a vector of four independent standard Gaussian values and $\mathbf{L}_{x0|nd}$ is the lower triangular matrix of the Cholesky decomposition of $\mathbf{C}_{x0|nd}^{a1,H1,D1,V1}$. If only a subset $nds$ of the $nd$ conditioning data is considered based on the neighborhood of $x0$, equation 4 becomes:

$$\mathbf{z}_{x0|nds}^{a1,H1,D1,V1} \approx \mathbf{C}_{x0,nds}\mathbf{C}_{nds,nds}^{-1}\mathbf{z} + \mathbf{L}_{x0|nds}\boldsymbol{\varepsilon}, \qquad (5)$$

where $\mathbf{z}_{x0|nds}^{a1,H1,D1,V1}$ is a $4 \times 1$ vector with the simulated values of $\mathbf{z}_{x0}^{a1}$, $\mathbf{z}_{x0}^{H1}$, $\mathbf{z}_{x0}^{D1}$, and $\mathbf{z}_{x0}^{V1}$ at $x0$ that respect auto and cross covariances and the conditioning data.

## Intrascale and interscale covariance

Complete modeling of the auto and cross covariances of the DWT coefficient might be a drawback of the method because the linear model of coregionalization has to be positive definite (Chiles and Delfiner, 1999). Under the hypothesis of intrinsic correlation, all auto and cross covariances are proportional (Chiles and Delfiner, 1999). As the scales have different support, two cases have to be considered: intrascale, involving wavelet coefficients belonging to the same scale, and interscale, involving coefficients belonging to two scales.

### Intrascale models

The intrascale coefficients are collocated and have the same support size, hence, the correlation coefficient $r$ is straightforward to compute. For example, the correlation between $\mathbf{z}^{H1}$ and $\mathbf{z}^{V1}$ with zero mean is given by

$$r_{\mathbf{z}^{H1},\mathbf{z}^{V1}} = \frac{\mathrm{cov}(\mathbf{z}^{H1}, \mathbf{z}^{V1})}{\sqrt{\mathrm{var}(\mathbf{z}^{H1})\mathrm{var}(\mathbf{z}^{V1})}}, \qquad (6)$$

where cov and var stand for sample covariance and variance, respectively.

### Interscale models

In the interscale case, the variables belong to two scales with different spatial sampling. Similar to the intrascale model, the cross covariance between interscale coefficients is computed using the correlation coefficient, but spatial sampling has to be taken into account. Boxes in the diagonal coefficient scales in Figure 1 represent the spatial relation between wavelet coefficients belonging to different scales. Each box represents the same area, but for example, the resolution of the box in D1 is four times the resolution of the box in D2. To compute the correlation using equation 6, values inside the highest-resolution grid are averaged at the undersampled grid resolution. In the intrinsic model, the correlation coefficient is unchanged by this averaging (Chilès and Delfiner, 1999).

## Case Study

The synthetic case study consists of a picture showing a section of a shallow sedimentary deposit that has been converted into porosities (Figure 2) ranging between 10% and 50% (Figure 3). The porosity field in Figure 2 was converted into dielectric permittivity $\boldsymbol{\varepsilon}$ and electrical conductivity $\boldsymbol{\sigma}$ fields, using the Maxwell-Garnett model (Sihvola, 2000) and the macroscopic model of Pride (1994), respectively. The implementation of Giroux and Chouteau (2008), which allows the computation of $\boldsymbol{\varepsilon}$ and $\boldsymbol{\sigma}$ for unsaturated sand-clay mixtures, was used for that purpose. Here, a clay fraction is not considered and the medium is fully saturated with fresh water (1000 mg/l total dissolved solids at 15°C, yielding a water conductivity of 128 mS/m). A synthetic cross-borehole GPR survey then is simulated using the electrical property fields. Virtual transmitters and receivers are located every 0.25 m along the boreholes leading to 1296 transmitter-receiver pairs.

Using permittivity and conductivity fields and assuming magnetic permeability of free space, a synthetic radargram is computed using a finite-difference time-domain (FDTD) implementation of Maxwell's equations in cylindrical coordinates. A Gaussian noise with standard deviation of 3% of the maximum amplitude was added to the raw traces. Traveltimes were picked manually and the slowness covariance was modeled using an anisotropic exponential variogram with ranges of 6 m along $-10°$ perpendicular to the wells and 2 m in the orthogonal direction. Variance of the velocity is $2 \times 10^{-2}\,(\text{ns/m})^2$. Then the velocity field is inverted using a stochastic tomography algorithm (Gloaguen et al., 2004). This algorithm is implemented in bh_tomo (Giroux et al., 2007) and allows us to estimate the slowness covariance parameters based on measured traveltimes and permits us to perform iterative cokriging and cosimulation of the slowness constrained by the picked traveltimes and any available slowness data. The velocity tomography that corresponds to Figure 2 after two curved-ray iterations is shown in Figure 4. In addition, the peak-to-peak amplitudes, reduced to pseudotraveltimes (Giroux et al., 2007), were picked manually and attenuation tomography was computed for the same data set using a modeled spherical covariance with variance of $2 \times 10^{-4}\,\text{mV}^2$ and the same ranges as the slowness (Figure 5). These data will be considered further as measured data. As can be seen in Figures 4 and 5, the inverted data are much smoother than the original porosity model (Figure 2). This is mainly because of the data coverage and the smoothing effect of the tomographic process. Correlations between the slowness and attenuation tomograms with the porosity field (Figure 2) are 0.4 and −0.3, respectively.

The analog comes from another picture of a section of the same sedimentary deposit. The gray intensity of the picture was converted into porosity. The porosity field shows a different basic statistical distribution and different dip angles for the layers compared to the first model

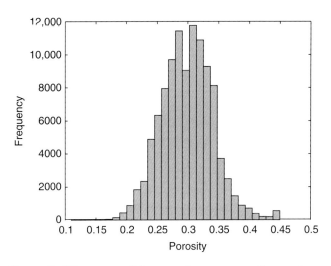

**Figure 3.** Histogram of the porosity corresponding to the model shown in Figure 2.

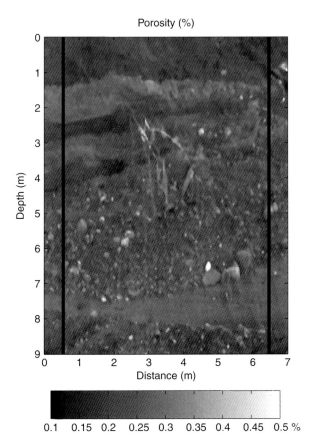

**Figure 2.** Synthetic porosity field. The black lines represent the vitual well locations where the synthetic GPR tomography was calculated.

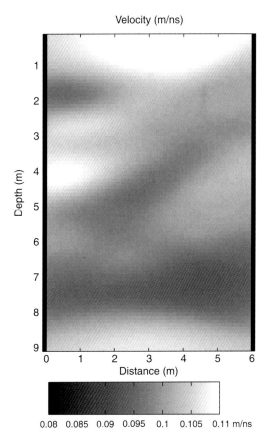

**Figure 4.** Tomogram of the velocity corresponding to the model shown in Figure 2.

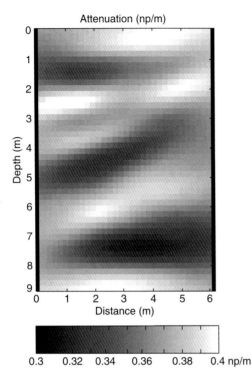

Attenuation (np/m)

Figure 5. Tomogram of the attenuation corresponding to the model shown in Figure 2.

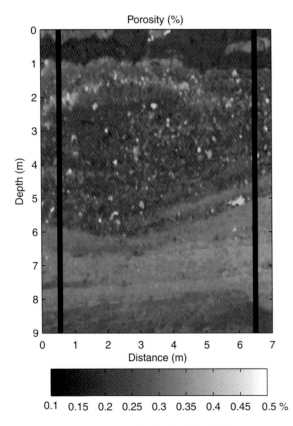

Porosity (%)

Figure 6. Analog porosity field. The black lines represent the virtual well locations where the synthetic GPR tomography was calculated.

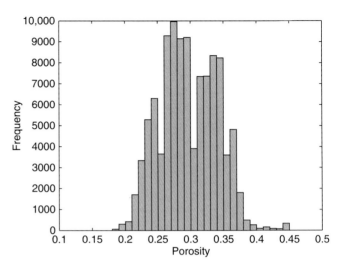

Figure 7. Histogram of the porosity analog shown in Figure 6.

(Figures 6 and 7). This analog porosity field is converted into permittivity and conductivity fields using the same model generator as for the first model. Figure 8a and 8b shows the real part of the relative permittivity and conductivity fields corresponding to the porosity analog, respectively. The survey design is identical to that of the previous model. Full-waveform forward modeling of the radar traces is computed using the two physical models (Figure 8a and 8b). A Gaussian noise with standard deviation of 3% of the maximum amplitude was added to the raw traces.

Using manually picked traveltimes and the fitted covariance model (exponential model with ranges of 5 m perpendicular to the boreholes and 2.5 m along the boreholes and variance of $4 \times 10^{-2}$ [ns/m]$^2$) stochastic tomography was computed to yield velocity and attenuation tomograms. Figure 9 shows the results of stochastic velocity tomography after two curved ray iterations. Figure 10 shows the stochastic attenuation tomogram obtained with the reduced peak-to-peak amplitudes and the associated curved rays calculated in the velocity tomography and the modeled covariance (exponential model with ranges of 6 m perpendicular to boreholes and 2 m parallel to the boreholes and variance of $4 \times 10^{-4}$ mV$^2$). Smoothing effect of the tomographic process appears to be less than in Figures 4 and 5 corresponding to the measured data. The main reasons are that the thin layers are smaller compared to the wavelength and that intervals between layers play the role of waveguides that increase the smoothing parameter of the covariance. The correlation coefficients between the porosity analog (Figure 6) and the corresponding permittivity and conductivity are 0.6 and −0.5, respectively.

Two-dimensional DWT is applied on both measured and analog geophysical tomograms (Figures 4, 5, 9, and 10) and on the image of the geologic analog (Figure 6). One of the most interesting things that appeared during the data analysis is that nonlinear relationships observed

a)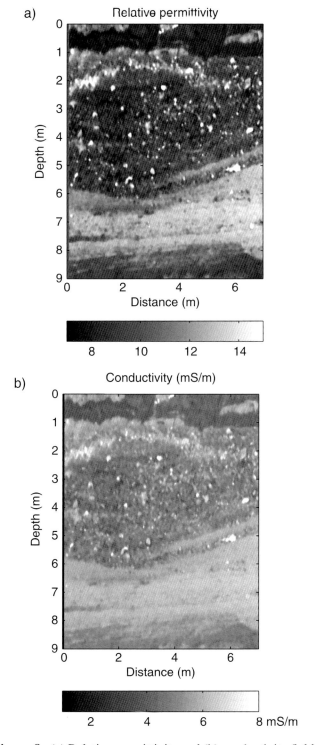

b)

**Figure 8.** (a) Relative permittivity and (b) conductivity fields generated from the analog porosity field (Figure 6).

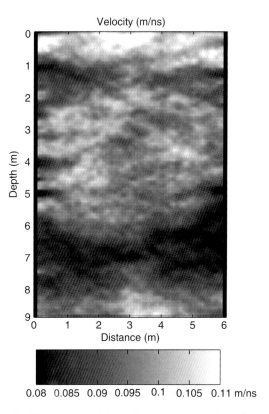

**Figure 9.** Tomogram of the velocity computed on the analog (Figure 6).

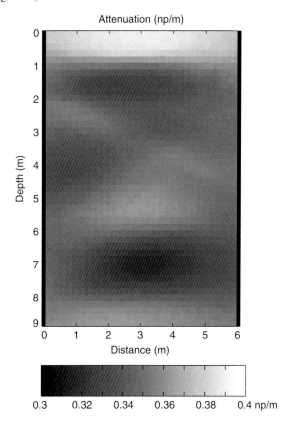

**Figure 10.** Tomogram of the attenuation computed on the analog (Figure 6).

**Table 1.** Correlation between geologic and geophysical approximation coefficients of the analog at different scales.

| Correlation between approximation coefficients at scale | Porosity/velocity | Porosity/attenuation | Velocity/attenuation |
|---|---|---|---|
| Scale 1 | 0.6 | −0.5 | −0.6 |
| Scale 2 | 0.65 | −0.55 | −0.6 |
| Scale 3 | 0.7 | −0.6 | −0.7 |
| Scale 4 | 0.8 | −0.65 | −0.7 |
| Scale 5 | 0.8 | −0.6 | −0.7 |

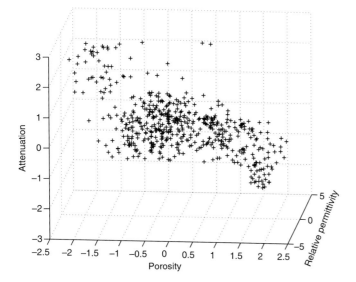

**Figure 11.** Scatter plot of the approximation coefficient of porosity, relative permittivity, and conductivity analog (Figure 6), at scale 3.

between geophysical and geologic analogs in the original space tend to become more linear in the coarsest wavelet space. This occurs principally because of the physics of the phenomenon, that is, the geophysical models are coarse-scale representations of the geologic model. The optimal level of decomposition is determined when the correlation coefficient between approximation coefficients of geologic and geophysical analogs stop increasing. The correlations between geologic and geophysical analog approximation coefficients at scale one to six are shown in Table 1. In this particular case, the correlation between geologic and attenuation approximations stops increasing at the fourth scale. Hence, the fourth scale of decomposition was chosen as the appropriate level of decomposition. For illustrative purposes, Figure 11 shows the

scatter plot of the analog porosity, permittivity, and conductivity at scale 4.

At the fourth wavelet-decomposition-level space, the correlations between the approximation coefficients of the porosity and the relative permittivity and conductivity of the analog fields reach 0.8 and −0.65, respectively. Because all the data have been processed, the reconstruction algorithm can be applied. Figure 12 shows the general flowchart of the proposed method. First, geostatistical structural analysis was applied on the approximation coefficients of the three analog models (one geologic and two geophysical) to define the geostatistical parameters (variograms and cross variograms) between porosity and geophysical analog approximation coefficients. This relation makes it possible to infer porosity approximation coefficients using the measured geophysical approximation coefficients. The optimal fitted model was an exponential model with ranges of 7 m perpendicular to the boreholes and 2 m along the boreholes. Assuming intrinsic relation between variables (Chilès and Delfiner, 1999), the variogram and cross variograms of the measured geophysical data and the porosity model to be simulated (Figure 2) are inferred. Second, cokriging of the porosity approximation coefficients are calculated using the linear model of coregionalization and the full sets of measured geophysical wavelet-approximation coefficients. Finally, starting from cokriged-porosity wavelet-approximation coefficients, the porosity wavelet details are simulated one scale at a time (equation 5), starting from the coarsest (corresponding to the approximation) to the finest scale.

Once all coefficients have been calculated, the inverse wavelet transform is applied to the simulated wavelet coefficients. Figure 13 shows three statistically equivalent realizations having the same set of conditioning data, i.e., the measured geophysical tomograms and the inferred statistical relation (least common multiple [LCM] model and

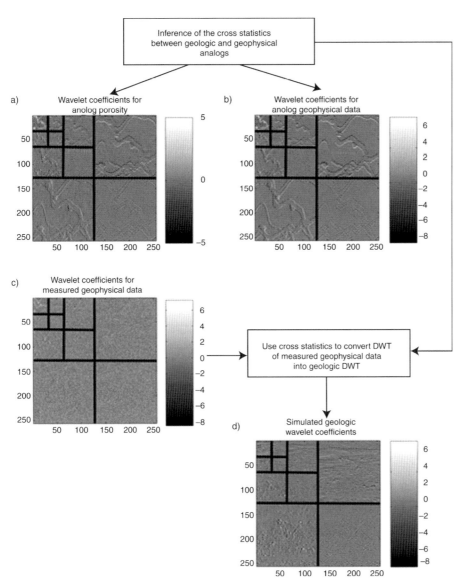

**Figure 12.** Flowchart of the proposed method. First, measured geophysical data and the analogs are decomposed using DWT. The statistical analysis is computed between geophysical and geologic analogs in the DWT space. These statistics are used to translate the measured geophysical wavelet coefficients into geologic wavelet coefficients.

interscale correlations) inferred from the geologic and geophysical analogs. Except for the pixel shape of the simulated porosity fields, all simulations are in good agreement with large-scale features of the porosity field shown in Figure 2 and, as expected, only the short-scale features show some variability. The pixel shape of the simulated fields occurs because of the choice of coarse simulation grid (128 × 256) and the choice of the wavelet itself (Haar wavelet). Smoother simulation would have been obtained using smoother wavelet and finer grid. The correlation coefficients between each of the hundred porosity simulations with the target porosity field shown in Figure 2 are between 0.6 and 0.72. Other simulation techniques like sequential Gaussian simulation, cosimulation, kernel-based, and Bayesian cosimulation were calculated on the same data set by Dubreuil-Boisclair et al. (2008). They show that kernel and sequential Gaussian simulation

have an average correlation of 0.43 and 0.56, respectively, whereas cosimulation and Bayesian cosimuation have an average correlation of 0.63 with the targeted porosity field. In addition, the average of the hundred simulations (E-type) was calculated (Figure 14). The correlation between the E-type (Figure 14) and the porosity model (Figure 2) is 0.68 indicating a good estimation of the porosity field.

## Discussion

In the presented example, it is assumed that the texture of the geology can be reproduced in the geologic analog, contrary to the large features in the case of geophysical data. This is the main weakness of the method. However, acquisition of logging and cone-penetrating test data (Rubin

Porosity (%)

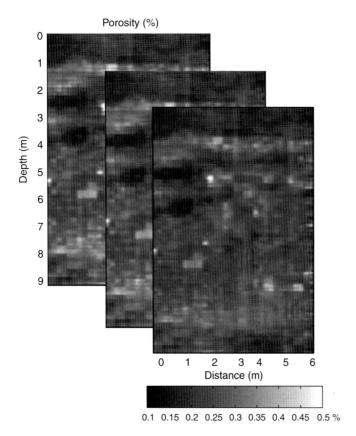

Figure 13. Three realizations of the porosity chosen randomly among 100 realizations.

Porosity (%)

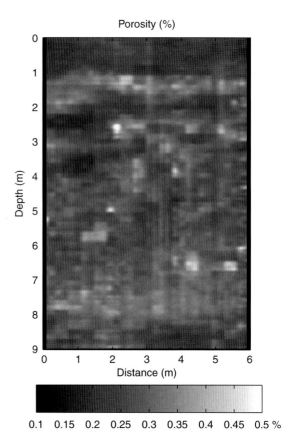

Figure 14. E-type or mean of the 100 realizations.

and Hubbard, 2005) provides very accurate estimates of the physical and geologic parameters. This allows populating the analog models with realistic petrophysical data that respect the measured basic statistics (mean, variance). In addition, the texture of quaternary deposits usually is well understood by geologists. Hence, even if there is a risk in not choosing the appropriate analog, its construction permits us to integrate geologic knowledge into the estimation process. Similar to multiple-point simulation (Strebelle, 2002), the inference of geologic parameters is not only data driven but also explicitly includes the conceptual model provided by geologists.

Another drawback of the method is that it assumes a constant resolution within the tomograms. Day-Lewis and Lane (2004) show this is not the case, because of the difference of ray coverage throughout the tomogram. This topic is subject to ongoing work: it is handled within the approach by using multiple realizations of the stochastic tomography instead of cokriged ones (Gloaguen et al., 2005). Slowness and attenuation variance of each pixel of the tomograms is included in a Bayesian manner to weight the posterior estimates of the geologic approximation coefficients (Gloaguen et al., 2010).

## Conclusion

Realizations of porosity fields generated using the proposed method reproduce the major large-scale heterogeneities observed in the geophysical models, the conditioning data and the wavelet-coefficient spatial and statistical dependencies. In addition, the method still works if the analog shows different spatial anisotropy than in the analog because the spatial anisotropy is carried out by the geophysical data. Hence, the use of a priori rotation matrices is not required (for example, as in the multiple-point technique). Choice of the wavelet-decomposition scale is supported by the physics of the phenomenon when the correlation between the geologic and geophysical analog is maximized. This approach allows fusion of the geologic and geophysical data at the appropriate scale.

The methodology can be extended easily to more than two covariate sets of geophysical data. The only change from the proposed example is the cokriging step of the approximation coefficient, which becomes $n$-variable cokriging instead of three-variable cokriging. The method can be extended to three-dimensions by using any 3D wavelet-decomposition package.

However, the major limitation of the method is that the geologic analog has to contain the same wavelet statistics as the field to be simulated. Hence, the difficulty is to find an appropriate geologic analog.

# References

Butler, D. K., 2005, ed., Near-surface geophysics: SEG Investigations in Geophysics Series No. 13.

Cassiani, G., G. Böhm, A. Vesnaver, and R. Nicolich, 1998, A geostatistical framework for incorporating seismic tomography auxiliary data into hydraulic conductivity estimation: Journal of Hydrology, **206**, 58–74.

Chen, J., S. Hubbard, and Y. Rubin, 2001, Estimating hydraulic conductivity at the South Oyster site from geophysical tomographic data using Bayesian techniques based on the normal linear regression model: Water Resources Research, **37**, 1603–1613.

Chilès and Delfiner, 1999, Modeling spatial uncertainty: Wiley Interscience.

Choi, H., and R. Baraniuk, 2001, Multiscale image segmentation using wavelet domain hidden Markov models: IEEE Transactions on Image Processing, **10**, 1309–1321.

Crouse, M. S., R. D. Nowak, and R. G. Baraniuk, 1998, Wavelet-based statistical signal processing using hidden Markov models: IEEE Transaction on Signal Processing, **46**, 886–902.

Daubechies, I., 1992, Ten lectures on wavelets: CBMS-NSF regional conference series in applied mathematics: Society for Industrial and Applies Mathematics.

Day-Lewis, F. D., and Lane, J. W. Jr., 2004, Assessing the resolution-dependent utility of tomograms for geostatistics: Geophysical Research Letters, **31**, L07503

Day-Lewis, F.D., K. Singha, and A. M. Binley, 2005, Applying petrophysical models to radar travel time and electrical resistivity tomograms: Resolution-dependent limitations: Journal of Geophysical Research, **110**, B08206.

De Marsily, G., F. Delay, J. Gonçavès, P. Renard, V. Teles, and S. Violette, 2005, Dealing with spatial heterogeneity: Hydrogeology Journal, **13**, 161–183.

Doyen, P. M., 1988, Porosity from seismic data — A geostatistical approach: Geophysics, **53**, 1263–1275.

Dubreuil-Boisclair, C., E. Gloaguen, D. Marcotte, B. Giroux, and M. Chouteau, 2008, Comparison between classical geostatistical simulation algorithms and Bayesian simulation for the simulation of porosity fields based on borehole geological data and GPR tomography: Presented at the Conference of the Association of Hydrogeophysicists.

Flandrin, P., 1992, Wavelet analysis and synthesis of fractional Brownian motion: IEEE Transactions on Information Theory, **35**, 197–199.

Giroux, B., E. Gloaguen, and M. Chouteau, 2007, Bh_tomo — A Matlab borehole georadar 2D tomography package: Computers and Geosciences, **33**, 126–137

Giroux, B., and M. Chouteau, 2008, A hydrogeophysical synthetic model generator: Computers and Geosciences, **34**, 1080–1092.

Gloaguen, E., G. Giroux, C. Durbeuil, D. Marcotte, and P. Simard, 2010, Simulation of porosity field using wavelet Bayesian inversion of crosswell GPR and log data: Presented at GPR 2010, 16th International Conference on Ground Penetrating Radar.

Gloaguen, E., B. Giroux, D. Marcotte, and M. Chouteau, 2005, Borehole radar velocity inversion using cokriging and cosimulation : Journal of Applied Geophysics, **57**, 242–259.

Gloaguen, E., D. Marcotte, and M. Chouteau, 2004, Borehole radar velocity imaging using slowness covariance estimation and cokriging: GPR 2004: Proceedings of the Tenth International Conference on Ground Penetrating Radar, 75–78.

Guardiano, F., and R. M. Srivastava, 1992, Multivariate geostatistics: Beyond bivariate moments, *in* A. Soares, ed., Geostatistics Troia 92: Kluwar Academic Publishers, 133–144.

Hyndman, D. W., J. M. Harris, and S. M. Gorelick, 2000, Inferring the relationship between seismic slowness and hydraulic conductivity in heterogeneous aquifers: Water Resources Research, **36**, 2121–2132.

Journel, A. G., 1989, Imaging of spatial uncertainty: A non-Gaussian approach, *in* B. E. Buxton, ed., Proceedings of the conference on geostatistical, sensitivity, and uncertainty methods for ground-water flow and radionuclide transport modeling: Batelle Press, 585–599.

Kowalsky, M. B., S. Finsterle, J. Peterson, S. Hubbard, Y. Rubin, E. Majer, A. Ward, and G. Gee, 2005, Estimation of field-scale soil hydraulic and dielectric parameters through joint inversion of GPR and hydrological data: Water Resource Research, **41**, W11425.

Kumar, P., and E. Foufoula-Georgiou, 1997, Wavelet analysis for geophysical applications: Reviews of Geophysics, **35**, 385–412.

Linde, N., K. Singha, and A. Binley, 2005, Applying petrophysical models to radar traveltime and electrical resistivity tomograms: Resolution-dependent limitations: Journal of Geophysical Research, **110**, B08206.

Macfarlane, P. A., J. H. Doveton, H. R. Feldman, J. J. Butler, Jr., J. M. Combes, and D. R. Collins, 1994, Aquifer/aquitard units of the Dakota aquifer system in Kansas: Methods of delineation and sedimentary architecture effects on ground-water flow and flow properties: Journal of Sedimentary Research, **B64**, 464–480.

Mallat, S., 1989, Multifrequency channel decompositions of images and wavelet models: IEEE Transaction in Acoustic Speech and Signal Processing, **37**, 2091–2110.

Moysey, S., K. Singha, and R. Knight, 2005, A framework for inferring field-scale rock physics relationships through numerical simulation: Geophysical Research Letters, **32**, L08304.

Portilla, J., and E. P. Simoncelli, 2000, A parametric texture model based on joint statistics of complex wavelet coefficients: International Journal of Computer Vision, **40**, 182–192.

Pride, S., 1994, Governing equations for the coupled electromagnetics and acoustic of porous media: Physical Review B, **50**, 15678–15696.

Romberg, J. K., and R. G. Baraniuk, 2001, Bayesian tree-structured image modeling using wavelet-domain hidden Markov models: IEEE Transactions on Image Processing, **10**, 1056–1068.

Ripley, B. D., 1987, Stochastic simulation: John Wiley & Sons, Inc.

Rubin, Y., and S. Hubbard, 2005, Hydrogeophysics: Springer Water Science and Technology Library, No. 50.

Rubinstein, R. Y., 1981, Simulation and the Monte-Carlo method (Wiley series in probability and statistics): John Wiley & Sons, Inc.

Sihvola, A., 2000, Mixing rules with complex dielectric coefficients: Subsurface Sensing Technologies and Applications, **1**, 393–415.

Strebelle, S., 2002, Conditional simulation of complex geological structures using multiple-point statistics: Mathematical Geology, **34**, 1–21.

Tran, T., U. A. Mueller, and L. M. Bloom, 2002, Multiscale conditional simulation of two-dimensional random processes using Haar wavelets: Quantifying risk and error: Proceedings of the Geostatistical Association of Australia Symposium, 56–78.

Tronicke, J., and K. Holliger, 2005, Quantitative integration of hydrogeophysical data: Conditional geostatistical simulation for characterizing heterogeneous alluvial aquifers: Geophysics, **70**, no. 3, H1–H10.

United Nations Environment Programme, 2003, Groundwater and its susceptibility to degradation: A global assessment of the problem and options for management: http://www.unep.org/dewa/water/groundwater/ groundwater_pdfs.asp, accessed 30 June 2010.

Vereecken, H., A. Binley, G. Cassiani, A. Révil, and K. Titov, 2005, Applied hydrogeophysics (NATO Science Series IV: Earth and Environmental Sciences), Springer.

Verly, G. W., 1994, Sequential Gaussian cosimulation: A simulation method integrating several types of information, *in* A. Soares, ed., Geostatistics Troia 92: Kluwer Academic Publishers, 85–94.

Chapter 8

# Estimating In Situ Horizontal Stress in Soil Using Time-lapse $V_S$ Measurements

Ranajit Ghose[1]

## Abstract

The magnitude and temporal changes of in situ horizontal stress at shallow depths in the subsoil are crucial information in geotechnical engineering. Although various methods of monitoring in situ horizontal stress have shown some success, such monitoring remains extremely challenging, especially for sands and stiff clays, and large uncertainties are usual. Laboratory experiments are performed that involve realistic values of stress and porosity, combined with seismic-array data acquisition, to monitor changes in shear-wave velocity ($V_S$) induced solely by changes in horizontal stress. Seismic-array data have been instrumental in distinguishing the small velocity changes associated with horizontal stress changes. Stress-porosity empirical models and micromechanical models have predicted quite accurately the observed trend of variation in $V_S$ as a function of horizontal stress. This trend is unique for a given combination of vertical stress, porosity, and soil type. Therefore, by monitoring the temporal change of $V_S$ by means of a seismic receiver array fixed at a given depth range and then by using the velocity–horizontal stress trend predicted by the model, one can estimate the temporal change and magnitude of in situ effective horizontal stress. A data-driven inversion approach has been tested on laboratory-experiment data for which the effective horizontal stress is known. The results demonstrate the possibility of estimating in situ effective horizontal stress at a given depth in subsoil, with an uncertainty of less than 15–20%, even when the porosity, vertical stress, and field factor are unknown. This approach shows potential for use on real field data.

## Introduction

Monitoring the in situ state of stress in soil is important in a wide variety of geotechnical problems, such as for (1) designing tunnels and other underground structures and maintaining their safety (e.g., Schmertmann, 1985; Andrea and Pietro, 2008), (2) constructing and maintaining retaining walls, diaphragm walls, and embankments (e.g., Ng and Lei, 2003; Stamatopoulos et al., 2005), (3) dealing with slope stability problems, especially in unsaturated expansive soils (i.e., soils showing a marked increase or decrease in volume as water content is increased or decreased, respectively, e.g., Nelson and Miller, 1992; Duncan and Wright, 2005), (4) counteracting natural hazards, such as landslides (e.g., Löfroth, 2008), (5) evaluating the risk of structural instability that results from a change in moisture content or pore pressure (e.g., Sheng et al., 2003), and (6) dealing with mine stability problems (e.g., Mark and Mucho, 1994; Zhal et al., 2000; Esterhuizen et al., 2008).

The in situ stress state generally is described by the orientations and the magnitudes of the three principal stresses. Numerous past investigations addressing the issue of such stress-state monitoring have achieved varying degrees of success. Although a substantial database has been developed over the years, it still is not possible to predict satisfactorily the in situ state of stress in most natural soil deposits, because they have undergone a complex history of loading and unloading that is difficult to evaluate precisely.

Geostatic vertical stress $\sigma_v$ can be estimated with a relatively small error from a profile of the overburden stress with depth ($\sigma_v = \int_0^H \rho g\, dh$, where $\rho$ is bulk density, $h$ is vertical distance from the ground surface, $H$ is the

[1]*Department of Geotechnology, Delft University of Technology, Delft, Netherlands. E-mail: r.ghose@tudelft.nl.*

depth from the surface ($h = 0$) to the point in subsurface where $\sigma_v$ is being estimated, and $g$ is the acceleration resulting from gravity). The in situ horizontal stress $\sigma_h$, however, is very dependent on the geologic history of the soil and is extremely difficult to estimate or monitor. Ladd et al. (1977), Schmertmann (1985), Jamiolkowski and Lo Presti (1994), Fahey (1998), and others have discussed many of the common field and laboratory methods for in situ estimation of $\sigma_h$. These approaches can be classified broadly under three categories: earth pressure cells, hydraulic fracturing, and self-boring devices. The first two categories have serious problems because of difficulty in calibrating the interaction of soil and the measuring device inserted, or because of difficulty in interpretation, or both.

Of the in situ methods, the self-boring pressuremeter (SBP) tests appear to be the most accurate direct method for determining $\sigma_h$. In SBP, a device is inserted into the ground with a very small amount of disturbance to the soil. The pressure required to produce a given change in the radius of an expanded cavity in the soil is measured (expansion pressuremeter) or load cells are used to sense the pressure directly (load-cell pressuremeter). In spite of the apparent success of the SBP with measurement of stress in soft clays, there is no consensus about the reliability or accuracy of the SBP for this purpose in stiff clays and sands (e.g., Fahey, 1998). To overcome the problem of $\sigma_h$ estimation in cohesionless sandy soils, the technique of in situ freezing has been used (e.g., Yoshimi et al., 1978; Hatanaka and Suzuki, 1995). Such an approach of freezing the sample is relatively expensive and demands utmost care.

Excepting in active tectonic areas and unstable zones, the difference in magnitude between the two horizontal stress components is not considered to be large, and an average value is used. If the stress state in a soil mass is below the Mohr-Coulomb failure envelope (or critical state line), the soil is in an elastic equilibrium. The soil mass then is said to be at rest (or at $K_0$ state), and the effective horizontal stress $\sigma_h'$ ($\sigma_h' = \sigma_h - \sigma_p$, where $\sigma_h$ = total average horizontal stress and $\sigma_p$ = pore pressure) is given by

$$\sigma_h' = K_0 \sigma_v', \tag{1}$$

where $K_0$ is the coefficient of earth pressure at rest and $\sigma_v'$ is the effective vertical stress ($\sigma_v' = \sigma_v - \sigma_p$, where $\sigma_v$ = total vertical stress). Consideration of elastic equilibrium (assuming confined one-dimensional compression with zero lateral strain) within a soil mass gives

$$K_0 = \frac{\nu}{1 - \nu}, \tag{2}$$

where $\nu$ is Poisson's ratio (e.g., Whitlow, 1995). The value of $K_0$ depends on the angle of friction $\phi'$, in addition to the loading/unloading history. For normally consolidated soils, the following relationship was proposed (Jaky, 1944), which seems to correlate well with observed values (e.g., Bishop, 1958; Brooker and Ireland, 1965):

$$K_0 = 1 - \sin \phi_c', \tag{3}$$

where $\phi_c'$ is the critical angle of friction. $K_0$ increases to approximately 1.0 for lightly overconsolidated soils, and to even higher values with an increasing overconsolidation ratio $R_0$:

$$K_0 = (1 - \sin \phi_c')\sqrt{R_0}. \tag{4}$$

The overconsolidation ratio is the ratio of the highest stress experienced by a soil mass to the current stress. A soil that currently is experiencing the highest stress is said to be normally consolidated ($R_0 = 1$). A soil can be considered to be underconsolidated immediately after a new load is applied but before the excess pore pressure has had time to dissipate. The typical range of values for $K_0$ is 0.45–0.6 for loose sand, 0.3–0.5 for dense sand, 0.5–0.7 for newly consolidated clay, 1.0–4.0 for overconsolidated clay, and 0.7–2.0 for compacted clay (Whitlow, 1995). In active soil, which can crack during shrinking and can generate large confining pressures during swelling, $K_0$ can vary widely; on the basis of field observations of heave and shrinkage, empirical relationships between $K_0$ and $\phi_c'$ have been proposed for different soil conditions (e.g., Lytton et al., 2005). These approaches for estimating $K_0$ generally rely on laboratory tests on soil samples, which often do not represent the in situ soil conditions and also have large uncertainties.

In spite of the difficulty of predicting the magnitude of in situ average values of $\sigma_h'$ or $K_0$ in soil, this parameter remains crucial in geotechnical engineering. In this paper, we explore a new possibility for estimating $\sigma_h'$ in soil using temporal variations of seismic shear-wave velocity ($V_S$). Geophysicists and reservoir engineers have done a large volume of work regarding prediction of stress in hydrocarbon reservoirs (for an update see, e.g., *The Leading Edge*, 2005, and Sayers, 2010). These works focus predominantly on hard rocks and not on near-surface soft soils of geotechnical interest. The stress-strain characteristics are very different between hard-reservoir rocks and soft soils. Seismic velocities are considerably lower in near-surface soils than they are in hard rocks. The research presented here focuses on $\sigma_h'$ estimation in soft soil. The goals are to (1) distinguish the sensitivity of $V_S$ to realistic changes in $\sigma_h'$ through careful laboratory tests on saturated, loose sands ($V_S$ = 200–300 m/s representing the near-surface condition), (2) investigate whether the existing empirical models involving stress and porosity, and micromechanical models, can be applied to discern the effect of $\sigma'_h$ on $V_S$ in soil, and (3) o explore a new concept for monitoring in situ $\sigma'_h$ by temporally monitoring $V_S$ accurately at a given depth. In this vein, the potential of a fixed seismic receiver array to capture small changes in $V_S$ is examined.

In the following sections, first we shall discuss briefly the previously proposed laboratory- and field-experiment relations and the micromechanical models that link $V_S$ to in situ stress in soil. Then we will present the results of new laboratory experiments involving seismic acquisition with a source and a receiver array in the transmission geometry. We will look at empirical stress-porosity models and micromechanical models for $V_S$ in soil in order to understand the observed $\sigma'_h$-dependence of $V_S$. Finally, we will discuss the possibility of capturing small changes in $V_S$ and the practical applicability of the proposed approach.

## Stress in Soil Determined from Seismic Waves: Previously Established Experimental Laws and Micromechanical Models

The stress-sensitivity of seismic waves is well known and has been used widely in hydrocarbon-reservoir exploration and monitoring, as mentioned in the previous section. Various rock-physics models have been developed and used for this purpose, as have geomechanical modeling studies. Time and amplitude data for seismic waves of different polarization and propagation directions have been employed, and phenomena such as borehole breakouts and shear-wave splitting have been used. Such investigations have targeted hard rocks primarily, because of their relevance to hydrocarbon reservoirs. Several studies that looked at unconsolidated marine sediments have used various extensions of the poroelasticity or visco-elasticity theories (e.g., Leurer and Brown, 2006; Leurer and Dvorkin, 2006); but those situations are not entirely comparable with soft, relatively loose, near-surface soil layers on land. In this paper, we concentrate on unconsolidated, near-surface soils on land. We look at several experimentally derived laws and micromechanical models that are relevant in that context.

### Experimentally derived $V_S$-stress-porosity relationships for soil

For many decades, geotechnical engineers have used empirical relations derived from laboratory experiments to estimate small-strain elastic moduli of soil from seismic-wave velocities. Numerous laboratory studies have been conducted that relate shear modulus ($G_0 = \rho V_S^2$, where $\rho$ is bulk density) to mean effective stress $\sigma'_m$, void ratio $e$, or porosity $n$ where $e = n/(1 - n)$, cyclic strain level, cementation, stress history (i.e., overconsolidation ratio), soil age, and other factors. These experimental investigations, as well as theoretical studies, suggest that the functional form of the real part of $G_0$ should be

$$G_0 = f(e)\sigma'^q_m, \qquad (5)$$

where $f(e)$ is a function of void ratio and $q$ typically has a value between 0.33 and 0.67. The function $f(e)$ traditionally has been chosen to fit a particular data set that usually is associated with one soil type, over a limited range of void ratios. Much effort has been devoted in the past to proposing alternative expressions for the function $f(e)$ (e.g., Hardin and Black, 1966; Ishihara, 1982; Stokoe et al., 1985; Saxena et al., 1988; Jamiolkowski et al., 1994).

One of the earliest and most widely used stress-porosity models of $G_0$ is one proposed by Hardin and Richart (1963), Hardin and Black (1968), Drnevich and Richart (1970), Richart et al. (1970), Hardin and Drnevich (1972), Richart (1977), etc., and is given by

$$G_{0-R} = (FF_1)a_1 p_a^{1-q}\sigma'^q_m \frac{(e_g - e)^2}{(1 + e)}, \qquad (6)$$

where $a_1$, $e_g$, and $q$ represent a set of regression coefficients obtained from laboratory tests, $FF_1$ is a field factor, $e$ is the void ratio, $p_a$ is atmospheric pressure ($=100$ kPa), and $\sigma'_m$ is the mean effective stress and is equal to $(\sigma'_v + 2\sigma'_h)/3$, assuming the two effective horizontal stress components to be the same. Hardin and Blandford (1989) proposed similar experimental laws that include also overconsolidation ratio, material constants reflecting soil fabric, and varying contributions between $\sigma'_v$ and $\sigma'_h$; in the past these experimental relationships have been used, among others, to find an anisotropic shear modulus in soil (e.g., Pennington et al., 1997; Wang and Mok, 2008).

To find a more general form for $f(e)$ that might be used as a first approximation for a wide range of soil types, Bryan and Stoll (1988) proposed a functional form such that

$$G_{0-BS} = (FF_2)a_2 p_a^{1-q}\sigma'^q_m \exp(be) \qquad (7)$$

where $a_2$, $b$, and $q$ represent a single set of regression coefficients that are obtained from an empirical test data set and that adequately represent a wide variety of soil types, including sand, silt, and clay, $e$ is the void ratio, $\sigma'_m$ is the mean effective stress, and $G_{0\text{-BS}}$ is the shear modulus estimated using this relation, and $FF_2$ is a field factor that typically takes a value in the range of 1.3 to 2.5 and that allows this laboratory empirical relation to be used on actual field data — including the effect of aging, stress history, cyclic shear-strain amplitude, and sampling disturbance. The model given in equation 6 is not as general as that in equation 7, in the sense that the values of the regression coefficients $a_1$ and $e_g$ in equation 6 depend on grain angularity, soil type, void-ratio range, and the like, whereas in equation 7 one single set of coefficients applies to a wide variety of soil types and all void ratios. Only the functional form $f(e)$ is different between the two models.

Figure 1a shows, following equation 6, a plot of $G_{0\text{-R}}$ as a function of varying porosity $n = e/(1 + e)$ and $K_0$ ($\sigma'_h$ varies, $\sigma'_v$ remains constant). Here, we have taken $\sigma'_v = 150$ kPa. With $FF_1 = 1$ and the global values (for angular grains) for the other regression coefficients $a_1$, $e_g$, and $q$ ($a_1 = 1230$, $e_g = 2.97$, $q = 0.45$) in equation 6, the resulting surface is illustrated in Figure 1a. Next, we have taken $\sigma'_v$, $\sigma'_h$, and $n$ from one data point of the second laboratory experiment to be discussed later in this paper ($\sigma'_v$ and $\sigma'_h$ are shown in Figure 11, and the point is marked as "ref" in Figure 15a). For this data point, the bulk density $\rho$ and shear-wave velocity $V_S$ are known, and hence $G_0$ ($= \rho V_S^2$) can be calculated. Let us call that $G_{0\text{-VS}}$.

Now, for this particular point, we use the global values for $a_1$, $e_g$, and $q$, and scale $FF_1$ such that $G_{0\text{-R}}$ equals $G_{0\text{-VS}}$. Finally, with this scaled $FF_1$, and still using the global values for $a_1$, $e_g$, and $q$, we again let $n$ and $K_0$ vary. The resulting surface is marked as $G_{0\text{-R}} = G_{0\text{-VS}}$ in Figure 1a.

The same is done using equation 7, the results of which are shown in Figure 1b. In this case, the surface marked as $G_{0\text{-BS}} = G_{0\text{-VS}}$ represents the one obtained with a scaled value of $FF_2$ and with global values for the other regression coefficients $a_2$, $b$, and $q$ ($a_2 = 2526$, $b = -1.5$, $q = 0.45$) in equation 7. In Figure 1c, the surfaces $G_{0\text{-R}} = G_{0\text{-VS}}$ and $G_{0\text{-BS}} = G_{0\text{-VS}}$ are plotted together. Note that these two surfaces are remarkably close, suggesting that if the two field factors $FF_1$ and $FF_2$ are calibrated (using field data), the difference between the two models given by equations 6 and 7 is very small. Therefore, to estimate in situ stress, it appears to be less important which of the two models is used. However, Bryan and Stoll's (1988) model has the advantage of being more general and applicable for wide varieties of soil and has therefore been used in this research.

Roesler (1979), on the basis of laboratory tests on homogeneous isotropic sand samples, showed that $V_S$ depends on the normal effective stresses both in the direction of wave propagation and the direction of particle motion:

$$V_S = C\sigma_a^m \sigma_b^n \sigma_c^o, \tag{8}$$

where $\sigma_a$, $\sigma_b$, and $\sigma_c$ are the components of effective stress in the direction of S-wave propagation, S-wave polarization, and orthogonal to $\sigma_a$ and $\sigma_b$, respectively. The term $C$ is a constant. The values of $m$, $n$, and $o$ were found empirically to be 0.149, 0.107, and 0.0, respectively (Roesler, 1979). The fact that $o = 0.0$ implies that the effective stress in the direction orthogonal to $\sigma_a$ and $\sigma_b$ has no influence on $V_S$.

On the basis of the field data sets of Redpath et al. (1982) and Redpath and Lee (1986) of cross-polarized shear waves with near-vertical propagation direction and recorded in boreholes, Lynn (1991) estimated birefringence between north-south and east-west polarized shear waves. This observed shear-wave birefringence was interpreted to result from a combination of factors, such as unequal $\sigma'_h$ in north-south and east-west directions, fabric anisotropy introduced by a depositional agent, and stress-aligned fluid-filled microcracks and pores. To estimate the birefringence and relate it to the stress anisotropy, the ratio of the velocity of two cross-polarized shear waves traveling the same path and equation 8 were used:

$$\frac{V_{S(N\text{-}S)}}{V_{S(E\text{-}W)}} = \frac{C\sigma_a^m \sigma_{N\text{-}S}^n \sigma_c^o}{C\sigma_a^m \sigma_{E\text{-}W}^n \sigma_c^o} = \frac{\sigma_{N\text{-}S}^n}{\sigma_{E\text{-}W}^n}, \tag{9}$$

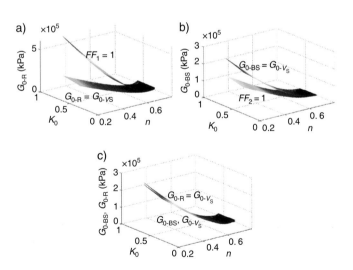

**Figure 1.** Small-strain shear modulus $G_0$, as a function of porosity $n$ and the coefficient of earth pressure at rest ($K_0 = \sigma'_h/\sigma'_v$), estimated using two empirical stress-porosity models presented in equations 6 and 7, is illustrated in (a) and (b), respectively. When the field factors ($FF_1$ and $FF_2$) are calibrated using one experimental data point [where $G_0$ ($= \rho V_S^2$), $n$, $\sigma'_v$, and $\sigma'_h$ are known], the difference between the two models is found to be negligibly small, as is evident in (c).

where $V_{S(N-S)}$ and $V_{S(E-W)}$ are respectively the velocities of north-south and east-west polarized shear waves traveling along the same near-vertical travel path, and $\sigma_{N-S}$ and $\sigma_{E-W}$ are the $\sigma'_h$ values in those two directions (Lynn, 1991). Obviously, an estimate of stress anisotropy using equation 9 will provide only an upper limit because the contributions of fabric anisotropy and stress-aligned cracks and pores are ignored. Separation of the effect of stress from those of fabric and aligned microcracks and pores remains one of the main challenges in this approach.

Assuming crosshole geometry and $C$ in equation 9 to be the same for shear waves of all possible polarizations for a given horizontal propagation direction, when we take the ratio of the velocity of two shear waves of different polarizations (in this case one in the horizontal and the other in the vertical direction), we get

$$\frac{V_{S(h-h)}}{V_{S(h-v)}} = \frac{C\sigma_a^m \sigma_h'^n \sigma_c^o}{C\sigma_a^m \sigma_v'^n \sigma_c^o} = \frac{\sigma_h'^n}{\sigma_v'^n} = K_0^n, \qquad (10)$$

where $V_{S(h-h)}$ and $V_{S(h-v)}$ are the velocities of horizontally propagating shear waves polarized in the horizontal and vertical directions, respectively. Therefore,

$$K_0 = \left(\frac{V_{S(h-h)}}{V_{S(h-v)}}\right)^{\frac{1}{n}}. \qquad (11)$$

This is different from the idea proposed by Fahey (1998), who suggested taking the ratio of $V_{S(h-h)}$ and $V_{S(v-h)}$ ($V_{S(v-h)}$ is the velocity of a vertically propagating shear wave with a horizontal polarization direction). That would imply two propagation directions but the same constant $C$ for both of them, which is not realistic. Jamiolkowski and Manassero (1995) suggested taking two constants instead of one. But in that case, one needs to know both of those constants in addition to $n$, in order to estimate $K_0$. Butcher and Powell (1995a, 1995b) applied the approach suggested by Jamiolkowski and Manassero (1995) to field data.

In comparison with the suggestions of Fahey (1998) and Jamiolkowski and Manassero (1995), equation 11 offers a relatively straightforward means to estimate $K_0$, although generation of both horizontally and vertically polarized shear waves in a borehole is not easy. Sully and Campanella (1995) and Zeng and Ni (1999), among others, tested a similar idea for estimating $K_0$. The issue of combining the effects of stress with those of fabric, aligned microcracks, pores, and fine layering still remains problematic.

## Micromechanical models for soil

Another approach to predicting $V_S$ as a function stress in soil is based on micromechanics in which the granular material is considered to be an assembly of discrete particles (e.g., Deresiewicz, 1973; Petrakis and Dobry, 1987; Chang et al., 1989; Ng and Petrakis, 1996; Santamarina and Cascante, 1996; Liao et al., 2000; Hicher and Chang, 2006). Analysis of particle-to-particle interaction using contact theories such as the Hertz-Mindlin theory explains the inherently nonlinear and nonelastic nature of soils (Mindlin, 1949). The micromechanical representation involves particle orientations, contacts, and forces of interaction among particles. It is possible theoretically to relate the micromechanical characteristics of the particulate medium to the macrocharacteristics of wave propagation (i.e., to wave velocity and attenuation). The physical and geometrical properties of particles and their relative arrangement influence the microbehavior.

Such micromechanical analyses can be performed analytically or numerically to study soils. Regular and isotropic random packings of monosized spheres have been studied extensively, in part because of the relative simplicity of such a study. Interparticle forces depend on the applied stresses and the degree of effective connections among particles — that is, on coordination number $\bar{n}$. Real soils can be analyzed as a combination of porosity, density, and coordination number (e.g., Deresiewicz, 1973; Petrakis and Dobry, 1987; Santamarina and Cascante, 1996).

Petrakis and Dobry (1987), Wang and Nur (1992), and Santamarina and Cascante (1996) have derived expressions for the small-strain shear modulus $G_0$ under isotropic loading and regular packings. Petrakis and Dobry (1987) have used an incremental approach to derive $G_0$ for regular packings subjected to anisotropic loading at constant fabric. In the case of anisotropic loading of a simple cubic array, $G_0$ is governed by the stresses in the direction of wave propagation and in the direction of particle motion (Petrakis and Dobry, 1987) as

$$G_0 = 2G^{\frac{2}{3}}\left(\frac{3(1-v)}{2(2-v)}\right)^{\frac{1}{3}}\left(\frac{\sigma_a \sigma_b}{\sigma_a + \sigma_b}\right)^{\frac{1}{3}}, \qquad (12)$$

where $\sigma_a$ and $\sigma_b$ are stresses in the directions of wave propagation and particle motion, respectively. In the case of a fluid-saturated medium, these can be considered the effective stresses. The term $v$ is Poisson's ratio, and $G$ is the small-strain shear modulus of regular packings under isotropic stress. Such expressions have been proposed also for other packing types (e.g., Santamarina and Cascante, 1996).

Both the empirical and the micromechanical models of $G_0$ and $V_S$ discussed above suggest the potential of using $V_S$ as an indicator of in situ stress — particularly stress in the direction perpendicular to the direction of wave propagation (and hence of $\sigma'_h$), provided $V_S$ can be measured with sufficient accuracy and the effect of stress on $V_S$ can be separated from the effects of other factors. One major advantage of using $V_S$ is that it represents the in situ soil condition, which otherwise is very difficult to capture (and that translates into large uncertainties in the estimated values of stress). With other approaches that attempt to address the in situ stress state of soil, the measurement becomes too cumbersome and expensive.

The main difficulty associated with using seismic shear waves in quantitative monitoring of $\sigma'_h$ or $K_0$ in soil, however, relates to the degree of stress sensitivity of $V_S$. Very often, a change of $\sigma'_h$, which is quite significant for geotechnical considerations, produces a small change in the $V_S$, and it is difficult to resolve this small $V_S$ change reliably, especially given the short travel distance for the seismic waves, the ambient noise, and the limited-frequency bandwidth for seismic waves in soft soil.

Second, taking the velocity ratio of two differently polarized shear waves traveling along the same path to obtain estimates of birefringence and $K_0$ (see equations 9–11) has generally limited accuracy because other factors, such as fabric, aligned microcracks, pores, and thin layers, work in combination with stress and make it difficult to separate the effect of $\sigma'_h$ alone, as discussed earlier.

Third, in soil-physics terms, $V_S$ also depends strongly on in situ porosity ($V_S = (G_0/\rho)^{1/2}$, where both $G_0$ and $\rho$ are functions of porosity). Thus, one should take into consideration the effect of a possible change in porosity in addition to a change in $\sigma'_h$. When the ratio of the velocities of two differently polarized shear waves traveling along the same path is used to estimate $K_0$, it is assumed that the porosity is constant, which is likely not to be true in the case of fabric anisotropy, aligned pores, and the like. In the following sections, we shall discuss results of new research in which attempts have been made to overcome these specific difficulties.

## Fixed-array Shear-wave Seismic: Laboratory Experiments

It is difficult to determine through field measurements whether an estimated value of in situ $\sigma'_h$ is correct, because there is no in situ measurement that can guarantee high accuracy. It is, therefore, useful to test a new methodology first through realistic laboratory experiments in which the magnitude of the applied effective stress is known better. Further, given a challenging target such as stress monitoring, laboratory a priori testing of a specific field-acquisition geometry and set of parameters is advantageous and helps improve the chance of success in the field. We have conducted laboratory experiments that focused on distinguishing the effect of just $\sigma'_h$ on $V_S$ in soil. The goals of our experiments are (1) to test whether a fixed seismic-receiver array can provide sufficient resolution and accuracy to enable us to distinguish small changes in $\sigma'_h$ from the observed $V_S$, which is difficult in a noisy environment and using single source-receiver geometry, and (2) to test the empirical and micromechanical stress-porosity models of $V_S$, with implications for estimating in situ $\sigma'_h$.

We have experimented on water-saturated, loose sands that are representative of near-surface soft soil. To simulate the shallow subsoil condition, representative values of stress and porosity have been used. Further, the size of the laboratory experimental facility has been made large enough to observe the far-field seismic response. The frequency of the shear waves in our experiment is not as high as in many earlier tests conducted in the ultrasonic frequency range (e.g., Prioul et al., 2004). We have used combined P- and S-wave bender elements in our experiments. These are piezoelectric transducers that are coupled directly to the soil sample. Each element can be used as a P-wave source or an S-wave receiver, depending on the external wiring made. Leong et al. (2005), Lee and Santamarina (2005), and Viana de Fonseca et al. (2008), among others, have explored various aspects of bender element installations. Before performing our main experiments, we have carried out preparatory tests to improve the performance.

Figure 2a shows the received wavefield for four conditions of the shear-wave (horizontal) bender receiver coupling. Here, the thickness of the cylindrical sand sample is 235 mm, the source is located at the bottom, and the receiver is at the top of the sample (travel distance is 235 mm), the S-wave source frequency is 6 kHz, and there is no applied external stress. Without touching the source and the receiver, the receiver coupling is varied by changing the water saturation at the location of the bender receiver element. The four receiver-coupling conditions represent when (1) the water is completely overflowing the top surface where the receiver is coupled, (2) the water is barely flowing over the top surface, (3) the water is not overflowing and the water height is at the top surface level, and (4) after an overflow, the water level is lowered artificially so that the full suction pressure is effective for coupling the receiver element to the sand.

The difference in wavefields for the four receiver couplings is clear. There is much P-wave energy in the received data when there is no suction coupling and the S-wave energy is less than the P-wave energy. However, when the horizontal receiver element is coupled very well through suction (see trace iv in Figure 2a), the S-wave signal-to-noise ratio is very good and the P-wave amplitude is considerably smaller. Evidently, the coupling of the bender element to the sample is very important because it affects the amplitude, time picks, and waveform. In our final tests, therefore, we have always used good suction-induced coupling of the receiver elements, and the source element is coupled through axial loading.

Figure 2b illustrates the S-wave data for different orientations of the horizontal S-wave bender receiver elements, for fixed S1, S2, and P bender source motions. Sources S1 and S2 have opposite horizontal source motions, whereas the P source has a vertical motion. The length of the cylindrical sample is 215 mm, and the source frequency is 6 kHz. The shear-wave arrivals are clear because they have much lower velocity than do the P-waves, the frequency content also is much lower, and most importantly, the shear waves being polarized in the plane perpendicular to the direction of wave propagation, they show maximum amplitude at 0° and 180° receiver orientations (i.e., when the source and receiver are both oriented in the same direction). The orientation is exactly opposite between S1 and S2 source motions, as expected. For the P source we see only P-wave arrivals in the horizontal bender component; these appear to be near-receiver converted waves.

Figure 2c shows the wavefield with a fixed S-wave-source and S-wave-receiver pair (both oriented in the same direction, with the location of one just above the other on two sides of the cylindrical sample in the length direction) for different source frequencies in the range of 4 to 60 kHz. The sample length is 490.4 mm, and $\sigma'_v$ and $\sigma'_h$ are constant at 235 kPa and 107 kPa, respectively. Mainly because of the higher stress, $V_S$ is significantly higher here compared with the situations in Figure 2a and 2b. Clearly, the S-wave amplitude is highest at the lowest source frequency, and above 12 Hz, the S-wave amplitude drops gradually. This occurs primarily because of dispersion (exhibiting frequency-dependent attenuation), although a near-field frequency-dependent radiation impedance might also contribute. The results shown in Figure 2b and 2c have been useful in interpreting shear-wave events in our final tests.

One main argument for using seismic S-waves and not P-waves to monitor in situ $\sigma'_h$ is the fact that S-waves are sensitive to in situ stress in the directions of wave propagation and particle motion, whereas P-waves are sensitive to stress change in the direction of wave propagation only. Therefore, if we generate an SH-wave by using a

surface-seismic source and then we measure the velocity of the arriving SH-wave in a nearby borehole with a fixed receiver, as is illustrated schematically in Figure 3b, it might be possible to monitor the change in stress for the

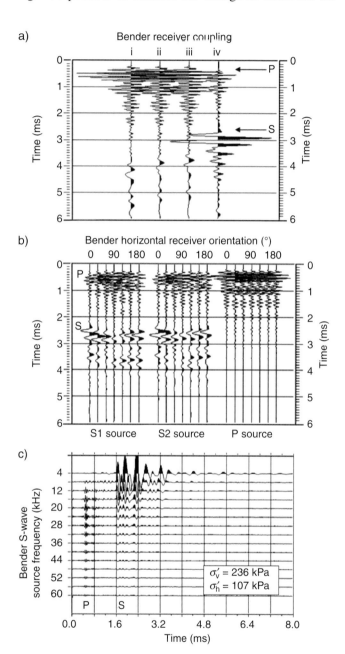

**Figure 2.** Preparatory laboratory experiments (seismic transmission) in water-saturated sand showing (a) the effect of bender receiver-element coupling on the received wavefield — the four traces represent four coupling conditions (see text for more details), (b) rotation of the horizontal receiver for fixed S1, S2, and P source motions, (c) received wavefield with a fixed S-wave source-receiver pair for a range of bender S-wave source frequencies. In these figures, P- and S-wave arrivals can be distinguished from their velocities, polarizations, and frequencies.

case in which the change in $V_S$ that is induced by stress change is captured accurately.

In Figure 3a, the case with a P-wave source is illustrated. Although P- and SH-waves have the same direction of propagation, their polarization directions are different, as indicated in Figure 3a and 3b by the small arrows. Figure 3c and 3d shows the seismograms observed in the laboratory for P- and S-wave sources (the source frequency is 4 kHz for both), respectively, for four $\sigma_h'$ values. Evidently, whereas the P-wave is not sensitive to a change in $\sigma_h'$ in this loose, water-saturated sand, the S-wave clearly is. In this case, the thickness of the sand sample is 490.4 mm, and $\sigma_v'$ is 200 kPa. We also notice here (Figure 3c) that the

bender P source generates, in addition to a P-wave, significant S-wave (SV) energy, whereas with an S source, the P-wave energy in the horizontal receiver is nearly absent (Figure 3d). This is because the source frequency is 4 kHz here, and in that case the generated P-wave from a bender S source, in the case of a good coupling, is almost negligible (see Figure 2c). The arrival of the S-wave is later than that in Figure 2c because whereas the source-bender element is located at the center of the 400-mm-diameter cylindrical sample, the receiver element is located at a 170-mm offset from the center, on the other side of the sample.

## Laboratory seismic-array experiments

Figure 4 shows schematically the basic features of the laboratory experimental setup. The facility consists of a large biaxial pressure chamber that can house a cylindrical sample 494 mm in height and 400 mm in diameter. Although any kind of soil sample can be tested in this facility, we have used water-saturated sands for the present research. The sample is prepared by pluviation of medium-grained sand (mean grain size is 90 μ) in water (with sand particles raining slowly on the water and then, with the help of a rotating sieve, settling by gravitation evenly over the entire surface). Special measures are taken not to induce any extra compaction of the sample and to make the sample as homogeneous as possible.

The top and bottom platens are designed to install an array of source and receiver bender elements directly in

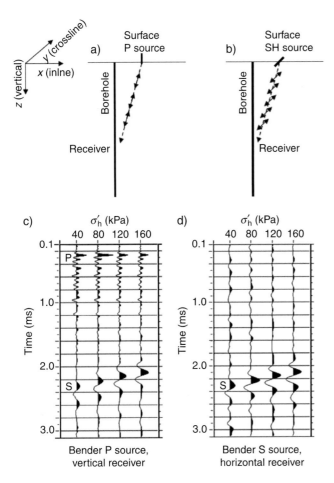

**Figure 3.** Preparatory laboratory experiments (seismic transmission) in water-saturated sand to test the relative sensitivities of P- and S-wave velocities to $\sigma_h'$ changes in soil: (a) and (b) show schematically the field downhole seismic measurements using surface P- and SH-wave sources, respectively. (c) and (d) The laboratory-observed seismograms for a fixed source-receiver pair are shown for four $\sigma_h'$ values, for P- and S-wave sources, respectively. Note in (c) and (d) the clear change in S-wave arrival times as $\sigma_h'$ changes, but the P-wave arrival time remains unaffected.

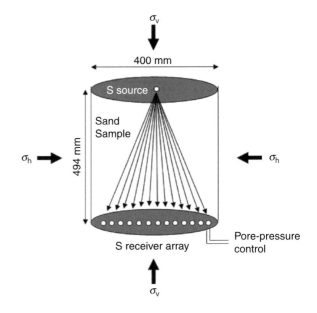

**Figure 4.** Schematic illustration of the seismic-transmission experiments, with a seismic receiver array installed in a large biaxial pressure chamber.

contact with the sample. In Figure 4, only one source position is illustrated, but in reality more sources are located on the source platen. On the receiver platen, 13 receiver elements lie along a line with a 25-mm interval separating each pair of elements (Figure 4). We have acquired seismic-transmission data, although acquisition of reflection data also is possible in this facility. A linear response of the acquisition system for the frequency range of our interest is ensured.

The masses of water and sand are measured, and the volumes and densities are known; hence, the average porosity is known. Any slight amount of water leaving or entering the chamber during loading/unloading also is measured accurately to monitor any possible change in the average porosity. Terzaghi and Peck (1967) and Berry and Reid (1987), among others, give 40–46% as the porosity for loose, well-graded, uniform natural sands. We maintained the porosity in our laboratory experiments in this range. A linear variable differential transformer (LVDT) was placed on top of the cylindrical sample to monitor accurately any change in the sample height during loading/unloading. To use realistic values of stress in our experiments, we have considered a soil profile made of 0–4 m being sand, 4–7 m being clay, 7–11 m being gravel, and below 11 m being sand, with the groundwater table (GWT) located at a depth of 2 m. Given the bulk density of sand above the GWT as $1.70 \, \mathrm{Mg/m^3}$, and the saturated densities of the sand, clay, and gravel below the GWT as $2.05 \, \mathrm{Mg/m^3}$, $1.9 \, \mathrm{Mg/m^3}$, and $2.15 \, \mathrm{Mg/m^3}$, respectively, a range of 80–250 kPa for $\sigma'_v$ corresponds to a depth range of about 7 to 23 m. We have performed our experiments in this range of $\sigma'_v$. The value for $K_0$ ($= \sigma'_h/\sigma'_v$) varies between 0.2 and 0.7. The value for $\sigma'_h$ is chosen accordingly. The pore pressure ($\sigma_p$) is kept constant at 100 kPa (1 bar), so the applied total vertical and horizontal stresses are $\sigma_v = \sigma'_v + \sigma_p$ and $\sigma_h = \sigma'_h + \sigma_p$, respectively.

The values for $\sigma_v$, $\sigma_h$, and $\sigma_p$ are controlled with the help of three separate pumps. The horizontal stress is azimuthally constant. The cylindrical sample has a rubber membrane on the side; this is a good absorbing boundary for the seismic waves and thereby minimizes the interference of side-reflected/refracted seismic events that might interfere with the direct arrivals. Outside the rubber wall there is a perforated cylindrical metal wall through which oil enters the space between the rubber wall and the metal wall and applies the horizontal/radial stress to the sample. The vertical stress is applied by an axial hydraulic piston.

First we start with a value of $\sigma'_v$ that corresponds to a depth of several meters in the subsoil. Keeping this value constant, we change $\sigma'_h$ gradually so that $K_0$ ($= \sigma'_h/\sigma'_v$)

increases, at a fixed interval, from the lower to the upper end of a chosen range. Next, we increase $\sigma'_v$ to represent an increase in depth by several meters, and then we change $\sigma'_h$ gradually again to produce the same set of $K_0$ values as before. These cycles are repeated until a maximum value of $\sigma'_v$ is reached. That completes the uploading cycle. After that, we gradually decrease the $\sigma'_v$, following the steps as before while repeating the same $K_0$ steps each time. At each step of this loading-and-unloading cycle, shear-wave transmission data are acquired using the receiver array. Because of the large size of the sample and the high seismic attenuation in the water-saturated uncompacted sand, it is necessary to run the bender source elements at maximum power and also to perform source stacking.

Next we shall discuss results of two sets of laboratory experiments, and we shall examine the empirical stress-porosity model on the first data set and both the empirical and micromechanical models on the second data set. The two laboratory experiments differ from each other in loading accuracy, sample porosity, bulk density, and source location. Table 1 summarizes the $\sigma'_v$ and $\sigma'_h$ values used in the two experiments.

## Results of laboratory experiment 1

Porosity of the sample is 42%. The source element is located on the source platen just above the location of the eighth bender element of the 13-element receiver array (see Figure 4). The maximum uncertainty in the applied stresses is $\pm 7$ kPa. That accuracy is improved greatly in the second experiment. Figure 5 shows single-receiver data for two values of $\sigma'_h$: 48 kPa and 107 kPa, and for a constant value of $\sigma'_v = 238$ kPa. In geotechnical laboratory tests, such a single pair of source and receiver is used commonly. In the case of a good signal-to-noise ratio, it is not difficult to estimate the difference in arrival times. However, the estimate of the magnitude of velocity will depend on time picking of the phase arrival (see the arrows in Figure 5), which is not trivial; this introduces considerable uncertainty in the estimated velocity and hence in the estimated magnitude of $\sigma'_h$.

Figure 6 shows the shear-wave seismic-array data (source frequency 9 kHz) for two $\sigma'_v$ levels, and for each $\sigma'_v$ level there are three values of $\sigma'_h$. Receiver elements 5 and 6 have poor coupling; these traces have been muted. Because the observed wavelength is 3–4 cm, we have more than 10 wavelengths between source and receiver, and the receivers are in the far field. At normal moveout, the arrival-time difference is rather small com-

**Table 1.** Porosity and stress values used in laboratory experiments 1 and 2.

| | Porosity $n$ | Pore pressure $\sigma_p$ (kPa) | Effective vertical stress $\sigma_v'$ (kPa) | Effective horizontal stress $\sigma_h'$ (kPa) | $K_0$ (= $\sigma_h'/\sigma_v'$) |
|---|---|---|---|---|---|
| Laboratory experiment 1 | 0.420 | 100 | 93 | 25 | 0.27 |
| | | | 85 | 32 | 0.38 |
| | | | 99 | 74 | 0.75 |
| | | | 235 | 48 | 0.20 |
| | | | 236 | 107 | 0.45 |
| | | | 240 | 165 | 0.69 |
| Laboratory experiment 2 | 0.464 | 100 | 100 | 30 | 0.30 |
| | | | 100 | 40 | 0.40 |
| | | | 100 | 50 | 0.50 |
| | | | 100 | 60 | 0.60 |
| | | | 100 | 70 | 0.70 |
| | | | 150 | 45 | 0.30 |
| | | | 150 | 60 | 0.40 |
| | | | 150 | 75 | 0.50 |
| | | | 150 | 90 | 0.60 |
| | | | 150 | 105 | 0.70 |
| | | | 200 | 60 | 0.30 |
| | | | 200 | 80 | 0.40 |
| | | | 200 | 100 | 0.50 |
| | | | 200 | 120 | 0.60 |
| | | | 200 | 140 | 0.70 |

pared with the difference for larger moveouts. Note the clear change in the moveout velocity in the receiver array. It is possible to obtain quite accurate $V_S$ estimates from the moveout velocities.

Figure 7a shows a plot of picked arrival times versus source-receiver distances; the velocities are estimated from the best-fitting line. Figure 7b shows the lag time estimated by crosscorrelation, for a given receiver channel between two $\sigma_h'$ values, plotted against channel number (shown for channels 8–13); the slope of the best-fitting line gives the velocity change with respect to the reference-stress level. The values of $\sigma_{h1}' \ldots \sigma_{h6}'$ are indicated in Figure 7a. The values for $\sigma_{h1}'$ and $\sigma_{h4}'$ are used as references for the two $\sigma_v'$ levels. Note that the estimated changes in $V_S$ agree well between the two approaches (Figure 7a and 7b). Use of a fixed seismic array to distinguish small changes in $V_S$ resulting from a change in $\sigma_h'$ in soil

clearly is advantageous. For a given $\sigma_v'$ (representing a given depth in the field), as the $\sigma_h'$ increases $V_S$ increases markedly, and that increase can be distinguished well by the array seismic data.

When the empirical stress-porosity model shown in equation 7 is applied to the data set of experiment 1, the known values of porosity, $\sigma_v'$, and $\sigma_h'$ together with the global values of $a_2$, $b$, and $q$ ($a_2 = 2526$, $b = -1.5$, and $q = 0.45$), and an $FF_2$ of 1.85, offer a good fit. This is shown in Figure 8. The values 92 kPa and 237 kPa represent the two levels of $\sigma_v'$ used in this experiment. It is possible to draw the following conclusions. (1) The slope of the $V_S$-$\sigma_h'$ plot clearly differs between the two $\sigma_v'$ levels; this difference agrees generally between the laboratory observation and the estimate. The $\sigma_h'$ sensitivity of $V_S$ is greater at lower values of $\sigma_v'$ and hence at shallower depths in the subsoil. (2) For a given combination of porosity, $\sigma_v'$, and

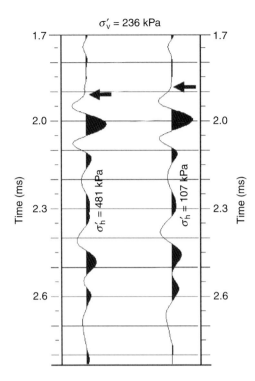

**Figure 5.** Laboratory experiment 1: seismic shear-wave (transmission) data acquired with only one source-receiver pair, for two values of $\sigma_h'$. The value of $\sigma_v'$ is constant (thereby representing a fixed depth in the subsoil). Depending on the time pick (the arrow), the estimated velocity ($V_S$) will change. This will be especially problematic in the case of noisy data.

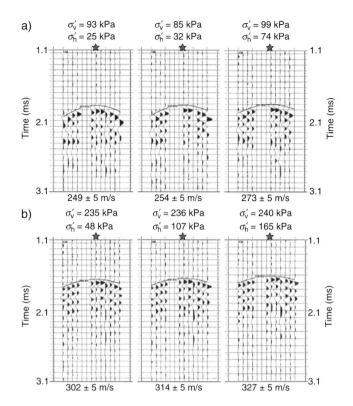

**Figure 6.** Laboratory experiment 1: seismic-array shear-wave (transmission) data for (a) mean $\sigma_v'$ value of 92 kPa (range 85–99 kPa), and (b) mean $\sigma_v'$ value of 237 kPa (range 235–240 kPa); for each $\sigma_v'$ level there are three $\sigma_h'$ values. The star on the top margin shows the source location. The value for $V_S$ that is estimated from the receiver array (estimation procedure illustrated in Figure 7) is marked below each trace gather.

soil type (which means that the empirical coefficients remain constant), the slope of the $V_S$-$\sigma_h'$ curve is unique. This is confirmed both by laboratory data and porosity-stress empirical models. The practical implication of this observation is important, and we will examine it more critically later in this paper.

## Results of laboratory experiment 2

In our second experiment, we have used an improved loading system. The addition of differential pressure controllers and of special O-rings and the reduction of friction have enabled us to implement stress changes with an accuracy of a fraction of a kilopascal. The loading system is shown schematically in Figure 9. The gas pressure exerted on the small surface area of oil in the pressure cylinders (separately for radial and axial loading) is controlled precisely to achieve loading accuracy and stability. Any slight change in the average porosity during loading/unloading is monitored by measuring accurately any water leaving or entering the chamber (see the balance in Fig-

ure 9). The LVDT monitors any slight change in the height of the sample during change of stress.

Figure 10 shows the S-wave seismic-array data for $\sigma_v' = 100$ kPa and $\sigma_h' = 30$ KPa, 40 kPa, 50 KPa, 60 kPa, and 70 kPa. The source element (frequency 6 kHz) is located vertically above the second element of the 13-element receiver array. The upper and lower panels in Figure 10 correspond to the same stress conditions, but before and after a whole set of uploading and downloading cycles. In other words, the $\sigma_v'$ is changed gradually — 100 KPa, 150 kPa, 200 kPa, and then back to 150 KPa and 100 kPa. Each time, for a given $\sigma_v'$ value the different magnitudes of $\sigma_h'$ corresponding to $K_0 = 0.3, 0.4, 0.5, 0.6,$ and 0.7 are applied and array seismic data are acquired. The data shown in Figure 10 correspond to the first series in the uploading cycles and the last one in the downloading cycles. The close similarity between the two data sets shown in Figure 10 indicates the reproducibility and stability of the measurements and the fact that the tests are performed in the elastic domain, with little permanent compaction of the sample. The effect of the change in $\sigma'_h$ is distinct on the array seismic data. We have noted negligible change in the LVDT data, indicating that the length of the sample has not been altered during loading/unloading. We also find little water leaving or entering the system during loading/unloading, which suggests that the average porosity has remained practically unaltered. The change in stress is likely to be accommodated entirely by grain contacts and possible rearrangements (e.g., Molenkamp, 2006).

Figure 11 shows array shear-wave seismic data (with a source frequency of 9 kHz) for three levels of $\sigma'_v$, which represent approximately a 10- to 20-m depth in realistic sand-clay-gravel subsoil conditions, as discussed earlier. Porosity of the sample in this second experiment is

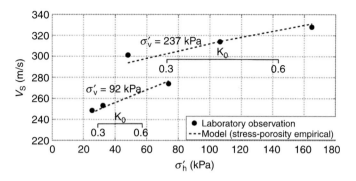

**Figure 8.** Laboratory experiment 1: estimated $V_S$ as a function of $\sigma'_h$, for two levels of $\sigma'_v$ (see Figure 6). The black circles represent laboratory observations. The dashed line shows the prediction using the empirical stress-porosity model given in equation 7 and explained in the text. Note that the trend of the $V_S$-$\sigma'_h$ distribution is clearly different between the two $\sigma'_v$ levels.

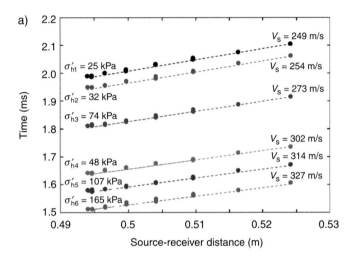

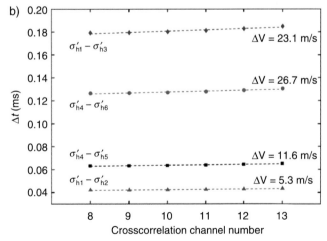

**Figure 7.** Laboratory experiment 1: estimation of $V_S$ using the seismic-array shear-wave (transmission) data shown in Figure 6. (a) Fitting the first-arrival time picks (time picking performed digitally) and (b) fitting the lag times estimated by time-domain crosscorrelation for each channel between two $\sigma'_h$ values, thereby obtaining the $V_S$ change ($\Delta V$) with respect to a reference value.

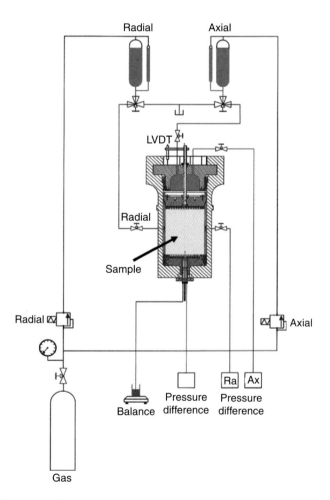

**Figure 9.** Laboratory experiment 2: the improved loading system.

46.4%. We observe wavelengths for a shear wave of about 3 cm and we have more than 15 wavelengths between source and receiver, so the receivers are in the far field. For each $\sigma'_v$ value, five values of $\sigma'_h$ are applied that correspond to $K_0 = 0.3, 0.4, 0.5, 0.6,$ and $0.7$ (Figure 11). The change

in $\sigma'_h$ between two adjacent $K_0$ values is small; with the new loading system this small change is implemented accurately. Note that there is a visible change in the moveout velocity of the first-arrival shear waves. In receivers 8 through 13, we notice interference of the first arrival with

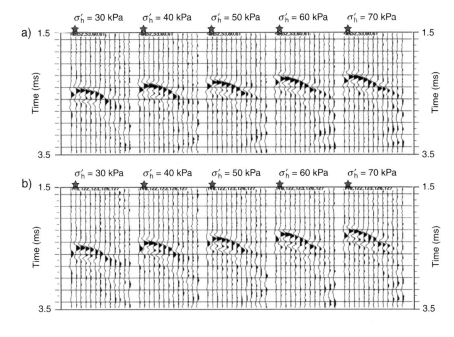

**Figure 10.** Laboratory experiment 2: checking the stability and elastic response of loading and acquisition systems. Array shear-wave seismic-transmission data acquired with one source and a receiver array, for five values of $\sigma'_h$, whereas $\sigma'_v$ is constant at 100 kPa. The star on the top margin shows the source location. (a) Data at the start of a loading cycle and (b) at the very end of an entire series of loading and unloading cycles, when the same $\sigma'_v$ and $\sigma'_h$ values as in (a) are reached again. Note the very similar arrival times and similar waveforms between (a) and (b).

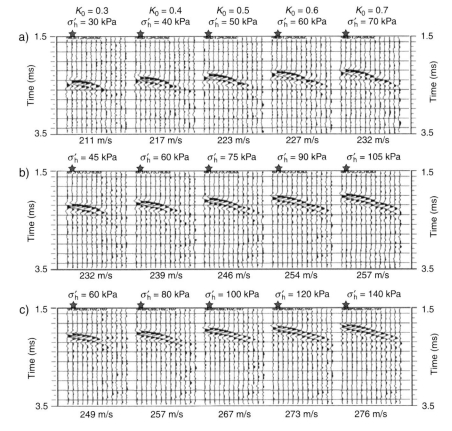

**Figure 11.** Laboratory experiment 2: array seismic shear wave (transmission) data for three levels of $\sigma'_v$ ([a] 100 kPa, [b] 150 kPa, and [c] 200 kPa) and for each $\sigma'_v$, five $\sigma'_h$ values that represent $K_0 (= \sigma'_h/\sigma'_v) = 0.3, 0.4, 0.5, 0.6,$ and $0.7$. The star on the top margin shows the source location. $V_S$ estimated from the receiver-array data (estimation procedure illustrated in Figures 13 and 14) is marked below each trace gather.

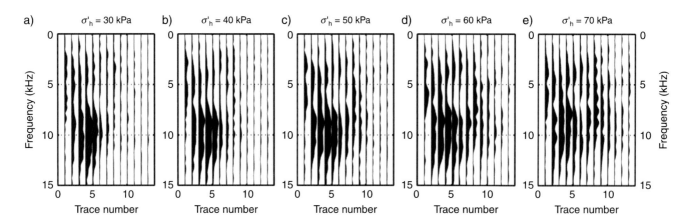

**Figure 12.** Laboratory experiment 2: Amplitude spectrum of each trace in the array seismic data, for five $\sigma'_h$ values and $\sigma'_v = 100$ kPa (see the uppermost panel in Figure 11).

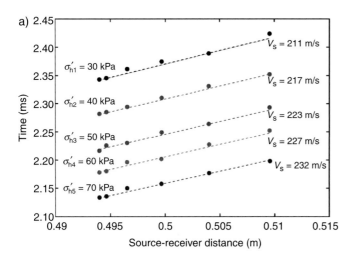

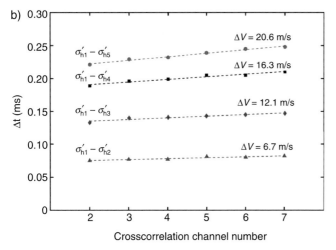

**Figure 13.** Laboratory experiment 2: estimation of $V_S$ using the seismic-array shear-wave (transmission) data shown in Figure 11. (a) Fitting the first-arrival time picks (time picking performed digitally) and (b) fitting the lag times estimated by time-domain crosscorrelation for each channel between two $\sigma'_h$ values, thereby obtaining the $V_S$ change ($\Delta V$) with respect to a reference value.

another phase, particularly at low values of $\sigma'_v$. We have excluded those arrivals during $V_S$ estimation. We also have identified reflections from the side of the pressure chamber (in Figures 10 and 11, see the area at about 3.0–3.3 ms in the five receivers farthest from the source, showing negative moveout velocity); these arrivals have not affected the $V_S$ estimates made on first arrivals.

Figure 12 shows the amplitude spectrum of each trace in the array for five $\sigma'_h$ values and for $\sigma'_v = 100$ kPa (i.e., for the top panel in Figure 11). The source and receivers are the same and untouched, and only $\sigma'_h$ is changed. We notice that as $\sigma'_h$ increases there is more energy at the receivers at larger offsets (in channel numbers beyond 6), which is indicative of the change in grain contacts.

As in experiment 1, $V_S$ is estimated here by best-fitting the first-arrival times over the whole array (Figure 13a), as well as by channel-by-channel crosscorrelation (Figure 13b). These estimates are quite stable and they agree with each other. It is evident that from the seismic data acquired using a fixed-receiver array and high temporal sampling, one can distinguish fairly accurately a small change in $V_S$ associated with a change in $\sigma'_h$ in the soil. Because the loading system is very accurate in this experiment, the data quality is much better than in the first experiment. It is therefore possible to estimate $V_S$ also using the phase difference ($\Delta\phi$) between signals at two receivers:

$$V_S(\omega) = \frac{\Delta x}{\Delta t} = \frac{\Delta x \omega}{\Delta \phi}, \qquad (13)$$

where $\Delta x$ is the difference in distance from the source between the two receivers and $\omega$ is frequency. This is illustrated in Figure 14 for traces 2 and 6 for values of $\sigma'_v = 100$ kPa and $\sigma'_h = 30$ kPa. First the signal is band-pass-filtered (Figure 14a) and time-windowed. The phase difference then is estimated through Fourier transform

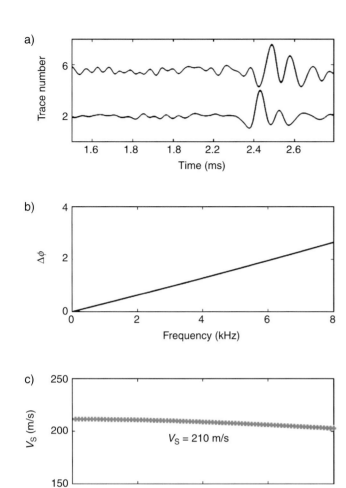

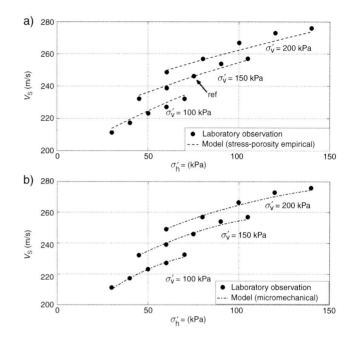

**Figure 15.** Laboratory experiment 2: estimated $V_S$ as a function $\sigma'_h$, for three levels of $\sigma'_v$. The black circles represent laboratory observations. (a) The dashed lines show the prediction using the empirical stress-porosity model given in equation 7. (b) The dash-dotted line shows the prediction using the micromechanical model given in equation 12. These models are explained in the text. Note that the trend of the $V_S$-$\sigma'_h$ distribution clearly differs among three $\sigma'_v$ levels.

**Figure 14.** Laboratory experiment 2: estimation of $V_S$ from the phase difference between two traces. (a) Trace 2 and trace 6 for $\sigma'_v = 100$ kPa and $\sigma'_h = 30$ kPa (see Figure 11), representing different source-receiver distances, (b) phase difference between two traces versus frequency, (c) estimated $V_S$ versus frequency (using equation 13). The value of $V_S$ averaged over the spectral-frequency range of the signal is obtained. This is done for all receivers in the array (taking one of them as the reference), and their average value is determined.

(Figure 14b). Finally, the velocity is obtained using the known $\Delta x$ (Figure 14c). This has been done for all traces, using one of them as a reference, and the estimated $V_S$ values are averaged. We found that these estimates are quite close to those estimated from the best-fit line through the first-arrival picks (Figure 13a) and by time-domain crosscorrelation (Figure 13b). These results collectively show the potential of a fixed-array seismic measurement for providing reliable estimates of changes in $V_S$ as a result of changes in $\sigma'_h$ in the soil. The main requirements are a sufficiently high temporal sampling in combination with fixed-

array-seismic data acquisition. We shall show later in this paper that such accuracy is realizable also in the field.

Figure 15 shows laboratory-observed $V_S$ together with $V_S$ estimated by two approaches: (1) using the empirical stress-porosity model shown in equation 7 and (2) using the micromechanical model explained in equation 12. These are illustrated respectively in Figure 15a and 15b. Unlike the case in experiment 1 (Figure 8), where the field factor ($FF_2$) was chosen to achieve a general fit for all data, in this case we have used one central data point (marked by "ref" in Figure 15a) to estimate the $FF_2$; then this value ($FF_2 = 1.67$) is used for all other points. It appears that a small $\sigma'_v$-dependent tuning of $FF_2$ is necessary for the best fit at all $\sigma'_v$ levels. In actual field conditions, such tuning is not necessary because one can estimate $FF_2$ from time-lapse $V_S$ data. Note the very good match of the $V_S$-$\sigma'_h$ trend between the laboratory observation and the prediction obtained using the stress-porosity empirical model.

Because the data quality is better for this experiment than for the previous one and the stress values are much more reliable here, we also have looked at the micromechanical model for this data set. For the high porosity

(46.4%) of our sand sample, a simple cubic packing is a reasonable assumption (Wang and Mok, 2008). We assume a cubic array of particles subjected to anisotropic loading. The values $\sigma_a$ and $\sigma_b$ in equation 12 can be substituted by $\sigma'_v$ and $\sigma'_h$, respectively. The value of Poisson's ratio ($\nu$) is calculated from the known $V_P$ and $V_S$ in the experimental data: $\nu = [0.5(V_P/V_S)^2 - 1]/[(V_P/V_S)^2 -- 1]$. The values of $V_S$ used are shown in Figures 11 and 13. The value for $V_P$ is constant for a given $\sigma'_v$. From our laboratory data, we have taken $V_P = 1100$ m/s, $1150$ m/s, and $1200$ m/s for $\sigma'_v = 100$ kPa, $150$ kPa, and $200$ kPa, respectively. For saturated soil these $V_P$ values appear to be low, but we find such values consistently in our experiments. To use the micromechanical model, the only unknown is the small-strain shear modulus ($G$) of regular packings under isotropic stress. The small-strain shear modulus has been estimated as the one that fits best the five data points for each $\sigma'_v$ value (Figure 11). This is sufficient for investigating the trend of change in $V_S$ as a function of varying $\sigma'_h$ between laboratory observation and prediction from the model. In Figure 15b, we notice that the $V_S$-$\sigma'_h$ trend exhibits a remarkably good match between the observation and the prediction using the micromechanical model.

As in experiment 1 (Figure 8), in experiment 2 we also see a clear change in the trend of the $V_S$-$\sigma'_h$ curve for different $\sigma'_v$ values (Figure 15). The $\sigma'_h$ sensitivity of $V_S$ is greater at lower values of $\sigma'_v$ and hence at shallower depths. A careful look at the results in Figure 15 suggests that for a given sand and constant values for porosity and $\sigma'_v$, the trend of the $V_S$-$\sigma'_h$ curve is unique. Experimental data and the models show nonlinear changes in $V_S$ as a function of $\sigma'_h$. The curvature (nonlinearity) is higher for the micromechanical model compared with the stress-porosity empirical model (Figure 15). Although on the basis of just this data set it is difficult to compare the performance of these two models, clearly both models can predict the $V_S$-$\sigma'_h$ trend quite well. That predictive ability has been instrumental in the conception of a new methodology for in situ, quantitative monitoring of $\sigma'_h$ with the use of reliable time-lapse estimates of $V_S$. This will be discussed in the next section.

# In Situ Monitoring of $\sigma'_h$ in Soil Using Time-lapse $V_S$ Measurements

The results discussed in the previous section suggest that (1) it is possible to distinguish small changes in $V_S$ that have been caused by changes in $\sigma'_h$ in soil, in the case in which high-quality data are acquired using a fixed seismic array, and (2) both the stress-porosity empirical

models and the micromechanical models, with proper calibration, can predict well the trend or the shape of $V_S$ variation as a function of $\sigma'_h$. On the basis of such results, we explore the possibility of using temporal $V_S$ measurements to monitor $\sigma'_h$ in situ.

From equation 7, the change in the mean effective stress in the subsoil from one time to another can be estimated directly from the measured change in $V_S$:

$$\frac{\sigma'_{m-t_2}}{\sigma'_{m-t_1}} = \left(\frac{V_{S-t_2}}{V_{S-t_1}}\right)^{\frac{2}{q}}, \tag{14}$$

where $V_{S-t_1}$ and $V_{S-t_2}$ are the values of $V_S$ measured respectively at times $t_1$ and $t_2$ by a fixed seismic array, and $\sigma'_{m-t_1}$ and $\sigma'_{m-t_2}$ are the mean effective stresses respectively at those two times. An unchanged overburden and a good repeatability of the seismic-acquisition parameters are assumed. Further, because the receiver is fixed at a given depth, $\rho$, $n$, and $\sigma'_v$ are considered to be practically unchanged between the two measurements. Equation 14 can be expressed in terms of $K_0$ and $\sigma'_h$ as follows:

$$\frac{2K_{0-t_2} + 1}{2K_{0-t_1} + 1} = \left(\frac{V_{S-t_2}}{V_{S-t_1}}\right)^{\frac{2}{q}} = R \tag{15}$$

and

$$\sigma'_{h-t_2} = R\sigma'_{h-t_1} + (R - 1)\frac{\sigma'_v}{2}, \tag{16}$$

where $K_{0-t_1}$ and $K_{0-t_2}$ are the $K_0$ values at times $t_1$ and $t_2$, respectively. Clearly, unlike with $\sigma'_m$ (equation 14), monitoring $\sigma'_h$ or $K_0$ directly from $V_S$ is not straightforward. However, equations 15 and 16 can be used to estimate $K_{0-t_2}$ and $\sigma'_{h-t_2}$ for the case in which some independent estimates of $K_{0-t_1}$ and $\sigma'_{h-t_1}$ are available.

Because the trend or shape of the $V_S$-$\sigma'_h$ curve is unique for a given combination of $\sigma'_v$ (and thus at a given depth), porosity, and soil type, and because this trend can be predicted quite accurately by stress-porosity empirical and micromechanical models, we have attempted to use that trend to extract the value of in situ $\sigma'_h$. If at a given depth the value for $\sigma'_h$ changes with time $t$, then this will cause a temporal change in the value for $V_S$. If this time-varying $V_S$ is monitored correctly, then the $V_S$-$t$ curve will correspond uniquely to a certain $V_S$-$\sigma'_h$ curve. The observed $V_S$-$t$ curve can, in principle, be inverted to derive the corresponding $\sigma'_h$-$t$ curve. To accomplish that, a global search algorithm is developed to look for the $\sigma'_h$ values for all different measurements while the other variables remain constant; the minimum differences between the observation and the model for all time-lapse $V_S$ values and the corresponding $\sigma'_h$ values are searched for simultaneously. This is done for all possible combinations of the other variables, and the global minimum is found. The set of $\sigma'_h$ values corresponding to this global minimum offers the final $\sigma'_h$-$t$ curve.

For the data of laboratory experiment 2 (Figure 11), we have applied this inversion algorithm together with the model given by equation 7. The values for $FF_2$, $n$, $\sigma'_v$, and $\sigma'_h$ are varied while $a_2$, $b$, and $q$ are kept constant (for the reasons explained earlier in this paper in the context of equation 7 and Figure 1), and $V_S$ is estimated. Bulk density $\rho$ is calculated as a function of porosity: $\rho = \rho_g(1 - n) + \rho_w n$, where $n$ is porosity and $\rho_g$ and $\rho_w$ are densities of dry sand and water, respectively. In the first step, for a constant $\sigma'_v$ (thus for a given depth in the field), we search for a $\sigma'_h$ value that minimizes the difference between the observed and estimated $V_S$ for a given measurement point (which corresponds to each time of measurement in the field). We do this for all the measurement points (which correspond to all time-lapse measurements for $V_S$ using a fixed seismic array). An average minimum of all five points is estimated. In the second step, for all possible combinations of $FF_2$, $n$, and $\sigma'_v$, we calculate the average minima, and we search for the global minimum of all these averages. The $\sigma'_h$ values corresponding to this global minimum give the final estimated $\sigma'_h$ values as a function of time.

Figure 16 illustrates the convergence when the inversion scheme is applied to the data set of our laboratory experiment 2. The values for $FF_2$, $n$, $\sigma'_v$, and $\sigma'_h$ all are considered to be unknown. The proposed approach takes advantage of the unique trend of the entire $V_S$-$t$ curve. The set of $\sigma'_h$ values corresponding to the entire $V_S$-$t$ curve is estimated simultaneously. Figure 16a through 16c shows the convergence in terms of porosity $n$, whereas Figure 16d through 16f illustrates the same in terms of field factor $FF_2$. The point marked as "ref" in Figure 15a is used here. The true value of $n$ is 0.464, and the value of $FF_2$ for an exact fit

(between observation and model) at this data point is 1.67. The results are shown for one, three, and five measurement points used in minimization (corresponding to five $\sigma'_h$ values, which on field data will imply five time-lapse measurement points). When more measurement points are used, the estimates clearly get better (Figure 16). The values for $n$ and $FF_2$ corresponding to the minimum difference between the observation and model are quite close to the true ones. The sharpness of the minima will be improved if a finer parameter search is implemented together with a more suitable inversion scheme. That will be the goal of future research.

Figure 17 shows the estimated values (triangles and squares) of $\sigma'_h$ and the known values (circles) for all data points of our laboratory experiment 2 (see Figures 11 and 15). The three colors represent three $\sigma'_v$ values. The triangles represent estimated $\sigma'_h$ when $FF_2$ and $n$ are unknown but $\sigma'_v$ is known (which in the field implies that the density-depth profile down to the depth of the measurement is known). The squares represent the estimated values of $\sigma'_h$ when $FF_2$, $n$, and $\sigma'_v$ all are unknown. The three sample points highlighted by dotted circles are the ones at which the data uncertainty is large (see in Figure 15a the

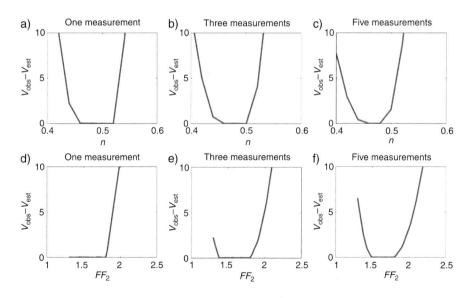

**Figure 16.** Convergence of the parameters $n = e/(1 + e)$ and $FF_2$ during minimization of the difference between observed $G_0$ ($= \rho V_S^2$) and modeled $G_0$ (as a function of $FF_2$, $n$, $\sigma'_v$, and $\sigma'_h$, see equation 7), when $\sigma'_v$ and $\sigma'_h$ are kept constant. (a) through (c) Convergence for porosity $n$; (d) through (f) convergence for field factor $FF_2$. The true value for $n$ is 0.464 and the value for field factor $FF_2$ for an exact fit between observation and model is known to be 1.67. Note that the inverted values are close to the true values: (a) and (d) when only one measurement point (observed $V_S$) is used, (b) and (e) when two time-lapse measurement points are used, and (c) and (f) when five time-lapse measurement points are used. The convergence becomes sharper when more data points are used simultaneously. The values used here are from laboratory experiment 2, for $\sigma'_v = 100$ kPa.

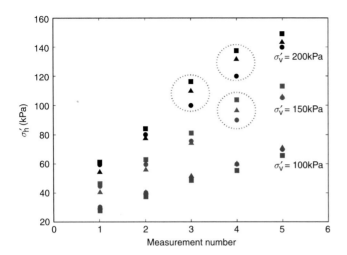

**Figure 17.** Estimating $\sigma'_h$ and its temporal change from the time-lapse $V_S$ measurements. A global minimization algorithm has been employed to invert the entire $V_S$-$t$ curve (in this case, five points simultaneously) to $\sigma'_h$-$t$ distribution, shown here for three $\sigma'_v$ values (see text for details). The data from laboratory experiment 2 (Figures 11 and 15) are used here. Circles indicate $\sigma'_h$ values that are known from laboratory experiment, triangles indicate $\sigma'_h$ values estimated from time-lapse $V_S$ by using global minimization, assuming $FF_2$ and $n$ to be unknown but $\sigma'_v$ to be known, and squares indicate $\sigma'_h$ values estimated from time-lapse $V_S$ by using global minimization when $FF_2$, $n$, and $\sigma'_v$ are all unknown. Three locations marked by dotted circles correspond to data points at which the observed $V_S$ deviates more, compared with other points, from the overall $V_S$-$t$ trend (see Figure 15; the $\sigma'_h$ axis corresponds to the $t$-axis (i.e., measurement number) in this estimation and in field data).

large deviation at these points between observed and modeled $V_S$). Clearly, if the deviation of some data point from the modeled $V_S$-$t$ trend is large, it will translate to a somewhat large error in the estimated value of $\sigma'_h$. That is expected, because this approach is based on the assumption that a good prediction of the trend or the shape of the $V_S$-$t$ curve is possible. The results presented in the previous section (Figures 8 and 15) give credence to that assumption. Further, the systematic difference between the predicted $\sigma'_h$ and the known values also can be explained by the deviation of the modeled trend from the observed data points.

Nevertheless, it is evident in Figure 17 that for all data points, the estimated value of $\sigma'_h$, even when $FF_2$, $n$, and $\sigma'_v$ are unknown, has an error of less than 15% to 20%. When $\sigma'_v$ is known, the error is even smaller. Until now, it has remained impossible to predict $\sigma'_h$ in situ in soil with that level of accuracy. Thus, the results shown in Figure 17 appear to be promising.

## Discussion

To extend this approach successfully to real field applications, the following challenges must be met: (1) we must have access to high-quality time-lapse $V_S$ measurements in shallow subsoil, (2) we must have reliable estimates of small $V_S$ changes that result from changes in $\sigma'_h$, (3) we must be able to assume that $n$, $\sigma'_v$, and other intrinsic parameters remain unchanged while $\sigma'_h$ changes, and (4) we must be able to model sufficiently accurately the temporal variation of $V_S$ that results from changes in $\sigma'_h$ in soil. We shall address these issues briefly here.

The effects of the top soil, weathering layer, weather changes (rain, snow, freezing, thawing), and the like can hinder acquisition of acceptable near-surface time-lapse seismic data from which small meaningful changes in $V_S$ values (as an effect of $\sigma'_h$ changes) can be extracted reliably. The possibility of acquiring high-quality time-lapse $V_S$ information increases if the effective seismic source signature has good repeatability and a constant receiver location and receiver coupling can be achieved. The problem with repeatability of the seismic sources is well known. However, with developments such as controlled and well-monitored vibratory seismic sources, the source signature can be deterministically deconvolved, and shot-to-shot variation of the source wavelet can be minimized (Ghose, 2002; Ghose and Goudswaard, 2004).

After source-signature deconvolution, if the source function still varies slightly from one measurement to another, a seismic receiver array (instead of a single receiver) fixed at a given depth in the subsoil for the entire period of monitoring improves the chance of capturing meaningful small changes in $V_S$, as the laboratory data presented in this paper have demonstrated. To realize these possibilities, we have conducted research and developed an array seismic cone penetrometer (SCPT). This is a digital, seven-level, three-component downhole seismic tool attached to a high-precision cone penetrometer. The cone and the seismic array are pushed into the soft soil. The coupling of the tool with the surrounding soil is excellent and is generally uniform for all seismic sensors. Figure 18 shows data acquired with a prototype array SCPT tool fixed at approximately 2.8 m in depth in a very soft, peaty soil. The receiver interval is 25 cm, and the total length of the seven-receiver vertical array is 1.5 m.

The horizontal-component seismograms are shown in Figure 18a without and with band-pass filtering. On the day of our field experiment, the muddy surface conditions in the field and the stormy, rainy weather were not favorable for good-quality data acquisition; we still had clear S-wave alignment in the seismic-array data acquired in the field.

Only channel 4 was noisy and could not be used in estimating $V_S$. In this case, because the surface seismic source (located at 0.5 m offset from the SCPT location) was a sledgehammer source, the frequency content is low. The frequency content can be increased greatly if vibratory shear-wave sources are used (Ghose et al., 1996; Ghose and Goudswaard, 2004). For this prototype array-SCPT data set (Figure 18a), the time sampling is 5 kHz. For time-lapse surveys, to estimate small changes in $V_S$ one needs to use even higher sampling, which is realizable with such specialized downhole seismic-array tools.

Time-domain crosscorrelation and phase differences estimated in the frequency domain both offer very stable velocity estimates, which are illustrated in Figure 18b and 18c. Deterministic source-signature deconvolution in conjunction with such downhole array-seismic reception fixed at a given depth should allow high-quality time-lapse $V_S$ measurements. Such an acquisition can be carried out at a site where a change in $\sigma'_h$ is expected in the subsoil (e.g., next to a large excavation or in tectonically active areas). Because such a vertical receiver array is kept fixed at a certain depth range for the entire monitoring period, any change in the overburden will be less problematic, and the assumption of constant $\rho$, $n$, and $\sigma'_v$ is likely to be met. The inversion scheme, discussed above, can be adapted for the case in which these assumptions need to be relaxed. Future tests should aim at gaining more understanding on these practical issues. Tests on laboratory experimental data have shown that both porosity-stress empirical models and micromechanical models can capture well the trend of variation of $V_S$ versus $\sigma'_h$ in soil. This is crucial to the success of the proposed methodology.

Sarkar et al. (2003) have discussed seismic anisotropy resulting from nonhydrostatic stress. In this paper, we have considered average effective horizontal stress. If three-component receivers are used, it will be possible to incorporate anisotropy also. However, that may not be a trivial issue because polarization and principal symmetry directions may not necessarily match. The anisotropy of

the overburden is not of major concern because we are using local (representing a very limited depth range) seismic-array data. By using seismic anisotropy detected in local three-component array data, one can address the issue of orientation of anisotropic horizontal stresses.

## Conclusions

Although $\sigma'_h$ or $K_0$ is one of the most important parameters in geotechnical design and safety considerations, in situ estimation of $\sigma'_h$ in the shallow subsoil is very difficult. In situ $\sigma'_h$ is strongly dependent on stress history. The existing approaches for such estimation are either grossly inaccurate or cumbersome and expensive. Laboratory experimental results obtained in this research have distinguished the sensitivity of $V_S$ to variations in $\sigma'_h$, in the case

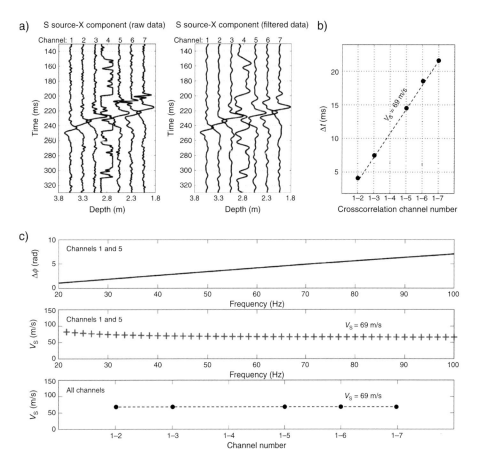

**Figure 18.** Array seismic cone penetration test (SCPT) data acquired in a near-surface peat layer. The possibility of stable and reliable $V_S$ estimation is examined on this seismic-array data set: (a) field seismic data (raw and band-pass-filtered) acquired with a prototype array SCPT tool fixed at a depth range of 2.05–3.55 m, (b) $V_S$ obtained by fitting the lag time of crosscorrelation of channels 2, 3, and 5–7 with respect to channel 1, and (c) $V_S$ estimated from the phase difference ($\Delta\phi$) (see equation 13) and using the array data. The surface S-wave source is a wooden sledgehammer.

of constant $n$, $\sigma'_v$, and soil type. Fixed-array seismic acquisition has been instrumental in capturing reliably the $\sigma'_h$ dependence of $V_S$. Fitting of first-arrival picks, time-domain crosscorrelation, and frequency-domain estimates of phase difference between two traces all provide robust estimates of small $V_S$ changes, for the case in which high-quality seismic-array data are available. The stress-porosity empirical model and the micromechanical model both have been found to be capable of predicting the observed $V_S$-$\sigma'_h$ trend. This trend is unique for a given $\sigma'_v$, $n$, and soil type. That has been the basis for proposing a new concept for monitoring the in situ $\sigma'_h$ in near-surface soil. To estimate the in situ $\sigma'_h$, a data-driven inversion scheme based on minimization of the difference between the observed and the predicted (via model) $V_S$-$t$ distribution has been implemented.

The approach is applied to our laboratory experimental data set for which the value of $\sigma'_h$ is known and $V_S$ is measured. The results suggest that the estimated value of $\sigma'_h$, even when all other parameters (including $n$, $\sigma'_v$, and the field factor) are unknown, generally has an error of less than 15–20%. The error is even smaller when $\sigma'_v$ is known. We also have considered the practicability of this approach in real field situations. With a newly developed downhole seismic-array tool together with seismic sources designed to provide relatively repeatable source wavelets, the proposed approach appears promising. More laboratory and field checks and more suitable inversion schemes should be the targets for future research.

## Acknowledgments

This research was financed by the Dutch Technology Foundation STW, the applied science division of NWO, and the technology program of the Ministry of Economic Affairs (grant DAR.5761). The detailed comments by the reviewers were very useful. The contribution of Karel Heller, Jan Etienne and Andre Hoving in the laboratory experiments is gratefully acknowledged. Thanks are due to Alimzhan Zhubayev for his assistance in the final stage of preparing this manuscript.

## References

Andrea, A., and C. Pietro, 2008, The influence of in situ stress state on tunnel design, in V. K. Kanjila, T. Ramamurthy, P. P. Wahi, and A. C. Gupta, eds.: Proceedings of the World Tunnel Congress 2008, **1**, paper no. 22.

Berry, P. L., and D. Reid, 1987, An introduction to soil mechanics: McGraw-Hill Book Co.

Bishop, A. W., 1958, Test requirements for measuring the coefficient of earth pressure at rest: Proceedings of the Conference on Earth Pressure Problems, **1**, 2–14.

Brooker, E. W., and H. O. Ireland, 1965, Earth pressures at rest related to stress history: Canadian Geotechnical Journal, **2**, 1–15.

Bryan, G. M., and R. D. Stoll, 1988, The dynamic shear modulus of marine sediments: Journal of the Acoustical Society of America, **83**, 2159–2164.

Butcher, A. P., and J. J. M. Powell, 1995a, Practical considerations for field geophysical techniques used to assess ground stiffness: Proceedings of the International Conference on Advances in Site Investigation Practice, 701–714.

———, 1995b, The effects of geological history on the dynamic stiffness in soils: Proceedings of the 11th European Conference on Soil Mechanics and Foundation Engineering, 27–36.

Chang, C. S., S. S. Sundaram, and A. Misra, 1989, Initial moduli of particulated mass with frictional contacts: International Journal of Numerical and Analytical Methods in Geomechanics, **13**, no. 6, 629–644.

Deresiewicz, H., 1973, Bodies in contact with applications to granular media, in G. Hermann, ed., R. D. Mindlin and applied mechanics: Pergamon Press.

Drnevich, V. P., and F. E. Richart Jr., 1970, Dynamic prestraining of dry sand: Journal of the Soil Mechanics Foundation Division, ASCE, **96**, 453–469.

Duncan, J. M., and S. G. Wright, 2005, Soil strength and slope stability: John Wiley & Sons, Inc.

Esterhuizen, G. S., D. R. Dolinar, and A. T. Iannacchione, 2008, Field observations and numerical studies of horizontal stress effects on roof stability in U. S. limestone mines: Journal of the Southern African Institute of Mining and Metallurgy, **108**, no. 6, 345–352.

Fahey, M., 1998, Deformation and in situ stress measurement, in P. K. Robertson and P. W. Mayne, eds., Geotechnical site characterisation: Balkema.

Ghose, R., 2002, High-frequency shear-wave reflections from shallow subsoil layers using a vibrator source: Sweep cross-correlation versus deconvolution with groundforce derivative: 72nd Annual International Meeting, SEG, Expanded Abstracts, 1408–1411.

Ghose, R., J. Brouwer, and V. Nijhof, 1996, A portable S-wave vibrator for high-resolution imaging of the shallow subsurface: Proceedings of the 58th EAGE Conference & Exhibition, Extended Abstracts, M037.

Ghose, R., and J. Goudswaard, 2004, Integrating S-wave seismic-reflection data and cone-penetration-test data using a multiangle multiscale approach: Geophysics, **69**, no. 2, 440–459.

Hardin, B. O., and W. L. Black, 1966, Sand stiffness under various triaxial stresses: Journal of the Soil Mechanics Foundation Division, ASCE, **92**, 27–42.

————, 1968, Vibration modulus of normally consolidated clay: Journal of the Soil Mechanics Foundation Division, ASCE, **94**, 353–399.

Hardin, B. O., and G. Blandford, 1989, Elasticity of particulate materials: Journal of Geotechnical Engineering, ASCE, **115**, no. 6, 788–804.

Hardin, B. O., and V. P. Drnevich, 1972, Shear modulus and damping in soils: design equations and curves: Journal of the Soil Mechanics Foundation Division, ASCE, **98**, 667–692.

Hardin, B. O., and F. E. Richart, 1963, Elastic wave velocities in granular soils: Journal of the Soil Mechanics Foundation Division, ASCE, **89**, no. SM1, 33–65.

Hatanaka, M., and Y. Suzuki, 1995, Two methods for the determination of lateral stress in sand: Soils and Foundations, **35**, no. 2, 77–84.

Hicher, P.-Y., and C. S. Chang, 2006, Anisotropic nonlinear elastic models for particulate materials: Journal of Geotechnical and Geoenvironmental Engineering, **132**, no. 8, 1052–1061.

Ishihara, K., 1982, Evaluation of soil properties for use in earthquake response analysis: Proceedings of the International Symposium on Numerical Models in Geomechanics, 237–259.

Jaky, J., 1944, A nyugalmi nyomás tényezöje [The coefficient of earth pressure at rest]: Magyar Mernok es Epitesz-Egylet Kozlonye [Journal of the Society of Hungarian Architects and Engineers], October, 355–362.

Jamiolkowski, M., and D. C. F. Lo Presti, 1994, Validity of in situ tests related to real behaviour: Proceedings of the 13th International Conference on Soil Mechanics and Foundation Engineering, **5**, 51–55.

Jamiolkowski, M., and M. Manassaro, 1995, The role of in situ testing in geotechnical engineering — Thoughts about the future: Proceedings of the International Conference on Advances in Site Investigation Practice, 929–951.

Jamiolkowski, M., R. Lancellotta, and D. C. F. Lo Presti, 1994, Remarks on the stiffness at small strains of six Italian clays: Proceedings of the 1st International Conference on Prefailure Deformation Characteristics of Geomaterials, 817–836.

Ladd, C. C., R. Foott, K. Ishihara, F. Schlosser, and H. G. Poulos, 1977, Stress-deformation and strength characteristics: Proceedings of the 9th International Conference on Soil Mechanics and Foundation Engineering, **2**, 421–494.

The Leading Edge, 2005, Rocks under stress: TLE Special section, **24**, no. 12, 1213–1286.

Lee, J.-S., and J. C. Santamarina, 2005, Bender elements: Performance and signal interpretation: Journal of Geotechnical and Geoenvironmental Engineering, **131**, no. 9, 1063–1070.

Leong, E. C., S. H. Yeo, and H. Rahardjo, 2005, Measuring shear-wave velocity using bender elements: Geotechnical Testing Journal, **28**, no. 5, GTJ 12196.

Leurer, K. C., and C. Brown, 2008, Acoustics of marine sediments under compaction — Binary grain size model and viscoelastic extension of Biot's theory: Journal of the Acoustical Society of America, **123**, no. 4, 1941–1951.

Leurer, K. C., and J. Dvorkin, 2006, Viscoelasticity of precompacted unconsolidated sand with viscous cement: Geophysics, **71**, no. 2, T31–T40.

Liao, C. C., T. C. Chan, A. S. J. Suiker, and C. S. Chang, 2000, Pressure-dependent elastic moduli of granular assemblies: International Journal of Numerical and Analytical Methods in Geomechanics, **24**, 265–279.

Löfroth, H., 2008, Undrained shear strength in clay slopes — Influence of stress conditions. A model and field test study: Ph.D. thesis, Chalmers University of Technology.

Lynn, H. B., 1991, Field measurement of azimuthal anisotropy: First 60 meters, San Francisco Bay area, CA, and estimation of the horizontal stresses' ratio from $V_{s1}/V_{s2}$: Geophysics, **56**, no. 6, 822–832.

Lytton, R. L., C. P. Aubeny, and R. Bulut, 2005, Design procedure for pavements on expansive soils: Texas Transport Institute Technical Report 0-4518-1.

Mark, C., and T. P. Mucho, 1994, Longwall mine design for control of horizontal stress, *in* Proceedings of the USBM Technology Transfer Seminar, U. S. Bureau of Mines Special Publications 01-94: New Technology for Longwall Ground Control, 53–76.

Mindlin, R. D., 1949, Compliance of elastic bodies in contact: Journal of Applied Mechanics, **16**, 259–268.

Molenkamp, F., 2006, Effective stress formulation of unsaturated soil mechanics, *in* W. Wu and H.-S. Yu, eds., Modern trends in geomechanics, Part 1: Springer Proceedings of Physics, **106**, 61–75.

Nelson, J. D., and D. J. Miller, 1992, Expansive soil — Problems and practice in foundation and pavement engineering: John Wiley & Sons.

Ng, C. W. W., and G. H. Lei, 2003, An explicit analytical solution for calculating horizontal stress changes and displacements around an excavated diaphragm wall panel: Canadian Geotechnical Journal, **40**, 780–792.

Ng, T. T., and E. Petrakis, 1996, Small-strain response of random arrays of spheres using discrete elements method: Journal of Engineering Mechanics, **122**, no. 3, 239–244.

Pennington, D. S., D. F. T. Nash, and M. L. Lings, 1997, Anisotropy of $G_0$ shear stiffness in Gault clay: Géotechnique, **47**, no. 3, 391–398.

Petrakis, E., and R. Dobry, 1987, Micromechanical modeling of granular soil at small strain by arrays of elastic spheres: Rensselaer Polytechnic Institute Dept. of Civil Engineering, Report CE-87-02.

Prioul, R., A. Bakulin, and V. Bakulin, 2004, Nonlinear rock physics model for estimation of 3D stress in anisotropic formations: Theory and laboratory verification: Geophysics, **69**, no. 2, 415–425.

Richart, F. E. Jr., 1977, Field and laboratory measurements of dynamic soil properties, *in* B. Prange, ed., Proceedings of dynamical methods in soil and rock mechanics: Balkema, 3–36.

Richart, F. E. Jr., J. R. Hall, and R. D. Woods, 1969, Vibrations of soil and foundations: Prentice Hall.

Redpath, B. B., and R. C. Lee, 1986, In-situ measurements of shear-wave attenuation at a strong-motion recording site: Report prepared for U.S.G.S. Contract No. 14-08-001-21823: John A. Blume and Associates.

Redpath, B. B., R. B. Edwards, R. J. Hale, F. C. Kintzer, 1982, Development of field techniques to measure damping values for near-surface rocks and soils: Report: John A. Blume and Associates.

Roesler, S. K., 1979, Anisotropic shear modulus due to stress anisotropy: Journal of the Geotechnical Engineering Division, ASCE, **105**, GT7 871–880.

Santamarina, J. C., and G. Cascante, 1996, Stress anisotropy and wave propagation: A micromechanical view: Canadian Geotechnical Journal, **33**, 770–782.

Sarkar, D., A. Bakulin, R. L. Kranz, 2003, Anisotropic inversion of seismic data for stressed media: Theory and a physical modeling study on Berea Sandstone: Geophysics, **68**, no. 2, 690–704.

Saxena, S. K., A. S. Avramidis, and K. R. Reddy, 1988, Dynamic moduli and damping ratios for cemented sands at low strains: Canadian Geotechnical Journal, **25**, 353–368.

Sayers, C. M., 2010, Geophysics under stress: Geomechanical applications of seismic and borehole acoustic waves: SEG and EAGE 2010 Distinguished Instructor Short Course, Distinguished Instructor Series No. 13.

Schmertmann, J. H., 1985, Measure and use of the in situ lateral stress, *in* The practice of foundation engineering, a volume honoring Jorj O. Osterberg: Department of Civil Engineering, Northwestern University, 189–213.

Sheng, Z., D. C. Helm, and J. Li, 2003, Mechanism of earth fissuring caused by groundwater withdrawal: Environmental and Engineering Geosciences, **9**, no. 4, 351–362.

Stamatopoulos, C., P. Petridis, M. Bassanou, and A. Stamatopulos, 2005, Increase in horizontal stress induced by preloading: Ground Improvement, **9**, no. 2, 47–57.

Stokoe, K. H., S. H. Lee, and D. P. Knox, 1985, Shear moduli measurements under true triaxial stresses: Proceedings of Advances in the Art of Testing Soils under Cyclic Conditions: Geotechnical Engineering Division, ASCE, 166–185.

Sully, J. P., and R. G. Campanella, 1995, Evaluation of in-situ anisotropy from crosshole and downhole shear wave velocity measurements: Géotechnique, **45**, no. 2, 267–282.

Terzaghi, K., R. B. Peck, and G. Mesri, 1996, Soil mechanics in engineering practice: John Wiley & Sons, Inc.

Viana de Fonseca, A., C. Ferreira, and M. Fahey, 2008, A frame work interpreting bender element tests, combining time-domain and frequency-domain methods: Geotechnical Testing Journal, **32**, no. 2, GTJ 100974.

Wang, Y. H., and C. M. B. Mok, 2008, Mechanism of small-strain shear-modulus anisotropy in soil: Journal of Geotechnical and Geoenvironmental Engineering, **134**, no. 10, 1516–1530.

Wang, Z., and A. Nur, 1992, Elastic wave velocities in porous media: A theoretical recipe, *in* Z. Wang and A. Nur, eds., Seismic and acoustic velocities in reservoir rocks: SEG Geophysics Reprint Series, No. 10, vol. 2, 1–35.

Whitlow, R., 1995, Basic soil mechanics: Longman Scientific & Technical.

Yoshimi, Y., M. Hatanaka, and H. Ohoka, 1978, Undisturbed sampling of saturated sand by freezing: Soils and Foundations, **25**, no. 3, 59–73.

Zeng, X., and B. Ni, 1999, Stress-induced anisotropic $G_{max}$ of sands and its measurement: Journal of Geotechnical and Geoenvironmental Engineering, **125**, no. 9, 741–749.

Zhal, E. G., J. P. Dunford, A. Schissler, M. K. Larson, P. A. Pierce, and F. M. Jones, 2000, Development of stress measurement techniques in bump-prone coal for safety decisions, *in* S. S. Peng and C. Mark, eds., Proceedings of the 19th International Conference on Ground Control in Mining, 101–111.

Chapter 9

# Analysis of the Velocity Dispersion and Attenuation Behavior of Multifrequency Sonic Logs

Ludovic Baron[1] and Klaus Holliger[1]

## Abstract

Modern slim-hole sonic-logging tools designed for surficial environmental and engineering applications allow for measurements of the phase velocity and the attenuation of P-waves at multiple emitter frequencies over a bandwidth covering five to 10 octaves. One can explore the possibility of estimating the permeability of saturated surficial alluvial deposits based on the poroelastic interpretation of the velocity dispersion and frequency-dependent attenuation of such broadband sonic-log data. Methodological considerations indicate that for saturated, unconsolidated sediments in the fine silt to coarse sand range and typical nominal emitter frequencies ranging from approximately 1 to 30 kHz, the observable P-wave velocity dispersion should be sufficiently pronounced to allow for reliable first-order estimations of the underlying permeability structure based on the theoretical foundation of poroelastic seismic-wave propagation. Theoretical predictions also suggest that the frequency-dependent attenuation behavior should show a distinct peak and detectable variations for the entire range of unconsolidated lithologies. With regard to the P-wave velocity dispersion, results indicate that the classical framework of poroelasticity allows for obtaining first-order estimates of the permeability of unconsolidated clastic sediments with granulometric characteristics ranging between fine silts and coarse sands. The results of attenuation measurements are more difficult to interpret because the inferred attenuation values are systematically higher than the theoretically predicted ones, and the form of their dependence on frequency is variable and is only partially consistent with theoretical expectations.

## Introduction

Permeability arguably is the most important but most elusive hydraulic parameter in native earthen materials, and it commonly can be measured only through dedicated hydrologic laboratory and field experiments (e.g., Butler, 2005). Knowledge of the permeability distribution within an aquifer is a key prerequisite for reliable predictions of fluid flow and contaminant transport. This information is critical for the effective protection, remediation, and sustainable management of increasingly scarce and fragile groundwater resources in densely populated and/or highly industrialized regions.

Geophysical constraints with regard to aquifer structure in general and to the distribution of hydraulic parameters in particular are considered to be especially valuable. The underlying methods are comparatively cheap and noninvasive. In addition, in terms of spatial resolution and coverage, they have the potential to bridge the gap between traditional hydrogeologic methods, such as core analyses and tracer or pumping tests (e.g., Hubbard and Rubin, 2005). Although standard traditional geophysical techniques cannot provide any direct information on the permeability of the integrated medium, more specialized approaches exhibit a more or less direct sensitivity to this important parameter. Along with nuclear magnetic resonance and spectrally induced polarization measurements, the interpretation of seismic data in a so-called poroelastic context arguably represents the most promising avenue to this end (e.g., Holliger, 2008).

The methodological foundations of seismic-wave propagation in saturated porous media generally are credited to Biot (1956a, 1956b). The corresponding theoretical

[1]Institute of Geophysics, University of Lausanne, Lausanne, Switzerland.

153

framework and its subsequent extensions now are commonly referred to as poroelasticity or Biot theory (e.g., Bourbié et al., 1987; Pride, 2005; Carcione, 2007). Biot theory establishes critical links between the seismic and hydraulic properties of saturated porous media and thus holds the promise of deriving permeability estimates from seismic observations. Probably the most tangible results of corresponding research efforts in the poroelastic interpretation of seismic data are permeability estimates obtained from the attenuation of Stoneley waves (e.g., Tang and Cheng, 1996; Tang and Cheng, 2004).

In the given context, Stoneley waves are seismic surface waves that propagate along the fluid-rock interface of a borehole. For uncased, open-hole conditions typical in hydrocarbon exploration, Stoneley-wave logging has reached a high degree of maturity and is capable of providing remarkably reliable and vertically quasicontinuous permeability estimates. We are not aware of any successful applications of this borehole technique to unconsolidated alluvial sediments. This is because, at least in part, corresponding surficial boreholes tend to be cased with plastic tubes, and the region affected by drilling-induced disturbances surrounding the boreholes is relatively wide. Because of their nature as surface waves, the penetration of Stoneley waves is related to frequency and is inherently shallow, and their propagation is likely to be dominated by the effects of the casing and the disturbed parts of the geologic formation in the vicinity of the borehole (Tang and Cheng, 2004).

Given that Biot theory predicts a characteristic frequency dependence of the seismic phase velocity and attenuation of seismic body waves in saturated porous media, this problem potentially can be alleviated by analyzing the velocity dispersion and the frequency-dependent attenuation behavior of broadband, multifrequency shallow sonic logs. The objective of this study is to explore the feasibility of such an approach and its applicability to observed data. For this pilot study, we base our analyses on classical Biot theory, knowing that the inherent assumptions of a stiff, porous frame and structural and compositional homogeneity of the rock matrix, on which this theory is based, are unlikely to be fully valid in unconsolidated sediments (e.g., Mavko et al., 2009). We choose this approach because we want to explore to what extent and/or which parts of classical Biot theory, whose parameterization already is quite extensive and cumbersome, are applicable in this context and because corresponding extensions of Biot theory require additional and generally rather poorly constrained parameters (e.g., Jackson and Richardson, 2007).

First we present the overall methodological framework and analyze under what conditions we ideally could expect

this approach to provide reliable first-order permeability estimates. Then we apply this technique to a pertinent field data set consisting of multifrequency, broadband P-wave sonic-log data acquired in a surficial alluvial aquifer and attempt to predict the velocity dispersion and attenuation as a function of frequency.

## Methodology Background

The equations describing the propagation of P-waves in saturated porous media can be written as (e.g., Schock, 2004)

$$\Delta(He - C\zeta) = \frac{\partial^2}{\partial t^2}(\rho_b e - \rho_f \zeta), \qquad (1)$$

$$\Delta(Ce - M\zeta) = \frac{\partial^2}{\partial t^2}(\rho_f e - m\zeta) - \frac{F\eta}{k_d}\frac{\partial\zeta}{\partial t}, \qquad (2)$$

where $F$ commonly is referred to as Biot's (1956a, 1956b) viscous correction factor, as explained in detail below, and $H$, $C$, and $M$ denote the so-called generalized parameters defined as

$$H = \frac{(K_s - K_b)^2}{D - K_b} + K_b + \frac{4}{3}G, \qquad (3)$$

$$C = \frac{K_s(K_s - K_b)}{D - K_b}, \qquad (4)$$

$$M = \frac{K_s^2}{D - K_b}, \qquad (5)$$

with

$$D = K_s\left(1 + \phi\left(\frac{K_s}{K_f} - 1\right)\right). \qquad (6)$$

In the above equations, $e$ represents the total volume dilatation and $\zeta$ the accumulation or depletion of fluid. The expressions $K_f$, $K_s$ and $K_b$ denote the bulk moduli of the saturating pore fluid, the grains constituting the rock matrix or frame, and the drained frame, respectively. $G$ is the shear modulus of the frame, $\phi$ the porosity, $k_d$ the permeability, $\eta$ the viscosity of the fluid filling the pores, $\rho_f$ the density of the pore fluid, $\rho_s$ the density of dry rock matrix, and

$$\rho_b = \phi\rho_f + (1 - \phi)\rho_s \qquad (7)$$

the bulk density of the saturated porous medium. The so-called added mass $m$ is defined as

$$m = \frac{\gamma\rho_f}{\phi}, \qquad (8)$$

with $\gamma$ denoting the tortuosity defined as (e.g., Lesmes and Friedman, 2005)

$$\gamma = \left(\frac{l}{L}\right)^2, \tag{9}$$

where $l$ corresponds to the effective length of the fluid-flow path between two points in the studied porous medium and $L$ to the length of the corresponding straight line connecting those two points.

Expressions $K_f$, $\rho_f$, and $\eta$ generally are assumed to be known, and $K_s$ and $\rho_s$ are based on the prevailing lithologies. In the given context of borehole logging, $G$ can be estimated from the S-wave log data; $\rho_b$ from the gamma-gamma log data; $\phi$ from the gamma-gamma and/or neutron-neutron log data; and $\gamma$ from the electrical resistivity, water conductivity, and porosity log data (Brown, 1980), or it can be based on specific assumptions regarding the geometry of the pore space (Berryman, 1981).

Finally, Biot's (1956a, 1956b) viscous correction factor is defined as

$$F(\xi) = \frac{1}{4} \frac{\xi T(\xi)}{1 - \frac{2T(\xi)}{i\xi}}, \tag{10}$$

with

$$T(\xi) = \frac{ber'(\xi) + ibei'(\xi)}{ber(\xi) + ibei(\xi)}, \tag{11}$$

and

$$\xi = a\sqrt{\omega\rho_f/\eta}, \tag{12}$$

where *ber* and *bei* denote the real and imaginary parts of the zero-order Kelvin function, respectively; *ber'* and *bei'* are the corresponding derivatives; $\omega$ is the angular frequency; and $a$ is the pore-size parameter, defined as (Hovem and Ingram, 1979)

$$a = \frac{d}{3} \frac{\phi}{1 - \phi}, \tag{13}$$

where $d$ is the average grain diameter. Using the Kozeny-Carman equation, we obtain (Schock, 2004)

$$d = 6\sqrt{\frac{k_d k_0 (1 - \phi)^2}{\phi^3}} \tag{14}$$

and thus

$$\xi = 2\sqrt{\frac{k_d k_0 \omega \rho_f}{\phi\eta}}, \tag{15}$$

where $k_0$ denotes a constant whose magnitude is determined by the geometric characteristics of the pore space. It might vary approximately from two for cylindrical pores to five for a pore space bounded by spherical grains (Hovem and Ingram, 1979; Schock, 2004). Throughout this study, we use an intermediate value of 3.5 for $k_0$. It can be shown that the sensitivity of the seismic-velocity dispersion and attenuation behavior with regard to this parameter is very minor. Biot (1956b) refers to $\xi$ in equations 10 through 12 as the "frequency parameter" and uses the symbol "$\kappa$" to denote it.

To derive the poroelastic P-wave dispersion curve, we solve equation 1 and equation 2 using a harmonic ansatz of the form $e = A_e \exp\{i(\omega t - kx)\}$ and $\zeta = A_\zeta \exp\{i(\omega t - kx)\}$. This then yields the following expression for the complex wavenumber (e.g., Schock, 2004)

$$k^2 = \frac{-a_2 \pm \sqrt{a_2^2 - 4a_1 a_3}}{2a_1}, \tag{16}$$

with

$$a_1 = C^2 - HM, \tag{17}$$

$$a_2 = \left(\frac{\omega^2 \gamma \rho_f}{\beta} - i\omega\frac{F\eta}{k_d}\right)H + M\rho\omega^2 - 2C\rho_f\omega^2, \tag{18}$$

$$a_3 = (\rho_f\omega^2)^2 - \rho\omega^2\left(\frac{\omega^2 \gamma \rho_f}{\beta} - i\omega\frac{F\eta}{k_d}\right). \tag{19}$$

The P-wave velocity $V_P$ and the attenuation $\alpha$ as a function of the frequency $f$ then is given by

$$V_P(f) = \frac{\omega}{\Re(k)} = \frac{2\pi f}{\Re(k)} \quad \text{with } \Re(k) > 0 \tag{20}$$

and

$$\alpha(f) = \Im(k), \tag{21}$$

where $\Re$ and $\Im$ denote the real and imaginary parts. These expressions form the basis for our endeavor of relating the observed velocity dispersion and the frequency-dependent attenuation behavior of P-waves to the permeability of an unconsolidated saturated porous medium.

## Database

A multifrequency slim-hole sonic tool has been used to log an $\sim$45-m-deep borehole penetrating $\sim$30 m of unconsolidated Pleistocene glaciofluvial clastic sediments before penetrating the bedrock consisting of Miocene sandstones. The lithostatic pressure at the base of the alluvial deposits is less than 1 MPa, which implies that relatively little force is acting on the grain contacts. The corresponding implications with regard to poroelastic wave propagation are still a matter of debate. In the ocean-acoustic community, the possibility of excess energy dissipation from relative movements along the grain contacts was considered early on to be important and was accounted for by an empirically motivated extension of classical Biot theory through complex and potentially frequency-dependent frame moduli (e.g., Stoll, 1977). The corresponding methodological framework nowadays generally is referred to as Biot-Stoll theory or poroviscoelasticity (e.g., Jackson and Richardson, 2005; Carcione, 2007). Conversely, in the seismic community, the consensus seems to remain that the permeability of unconsolidated clastic sediments is sufficiently high so that such movements along the grain contacts are likely to be very small, and hence classical Biot theory is likely to remain largely valid (e.g., King, 2005; Mavko et al., 2009).

The considered borehole has a diameter of 10 cm and is equipped with a continuously perforated plastic casing. For the purpose of this study, we consider the material between the water table and the top of the bedrock. Evidence from the retrieved core material indicates that the mineralogical matrix consists primarily of quartz with varying amounts of calcite and generally minor amounts of other mineralogical constituents, such as clay. At the time of our sonic-log measurements, the groundwater table was at a depth of approximately 4 m. Seasonal fluctuations of the depth to the groundwater table are on the order of $\pm 1$ m, and from a practical hydrologic point, complete saturation can be assumed below the groundwater table.

It is important to note that because of the shallow nature of this borehole, saturation might not be complete, and minuscule amounts of gas/air could be trapped in the pore space below the groundwater table. If remnant gas/air is present in the form of bubbles rather than in dissolved form, this could have a strong effect on the seismic attenuation behavior (e.g., Pride, 2005). Given the highly speculative nature of this issue, we consider it beyond the scope of this study.

Sonic logs were acquired using 11 evenly spaced nominal emitter frequencies ranging from 1 to 30 kHz and a vertical sampling interval of 2 cm with four stacks at each measurement point. The record length was 0.4 ms with a temporal sampling rate of 4 μs and a recording instrument with 12-bit analog-to-digital conversion. The sonic tool was centralized relative to the borehole, and the spacings between the dipole-type emitter and the two receivers were 1 m and 1.302 m, respectively. Each nominal frequency is associated with a constant and well-controlled gain level. No processing was applied, and the analyses described here were carried out on the raw sonic-log data.

In addition to these broadband sonic-log data, a comprehensive suite of complementary petrophysical logging data — including S-wave, electrical resistivity, natural-gamma, gamma-gamma, and neutron-neutron log data as well as measurements of the electrical conductivity of the groundwater — were also available for this borehole. The natural-gamma-log data were used to assess the mineralogical clay content. The active gamma-gamma-log data allowed for estimating the bulk density $\rho_b$ as defined in equation 7, which then, along with S-wave log data, allowed for estimating the shear modulus $G$. The gamma-gamma and neutron-neutron logs were inverted jointly for porosity $\phi$ using the neural-network-based approach proposed by Baron and Chapellier (2000). The resistivity-log data and the electrical conductivity of groundwater filling the borehole served to evaluate the formation factor which, according to Archie's law, is given by

$$\tilde{F} = \frac{\sigma}{\sigma_w} = \phi^{-\tilde{m}}, \tag{22}$$

where $\sigma$ is the electrical conductivity of the saturated porous medium under consideration, $\sigma_w$ the electrical conductivity of the pore fluid, and $\tilde{m}$ the cementation exponent of the porous matrix. The corresponding definition of the tortuosity then is given by (e.g., Brown, 1980; Lesmes and Friedman, 2005)

$$\gamma = \tilde{F}\phi = \phi^{-\tilde{m}+1}. \tag{23}$$

The physical basis for this relation is that Archie's law in its original form given by equation 22 assumes that the rock matrix of the porous medium is perfectly resistive and that all electrical-current conduction occurs through the fluid filling the pore space. This implies that the hydraulic and electrical pathways are essentially identical. The path taken by the current under those conditions then depends essentially on the porosity as well as on the interconnection of the individual pores. The latter is characterized by the cementation exponent $\tilde{m}$, which is related to the amount of dead-end pore space in the probed rocks (e.g., Revil and Cathles, 1999). The symbol $\tilde{m}$ is equal to one for the hypothetical case of a medium whose pore space is fully interconnected, and it generally increases with increasing consolidation and cementation to maximum values of three or more for very tight formations. Because of the general scarcity of mineralogical clay in the probed aquifer, as evidenced by the overall low intensity of the natural-gamma log, the assumption of a highly resistive rock matrix is fulfilled for the considered aquifer.

Another common approach for estimating tortuosity is the one proposed by Berryman (1981), which requires relatively strong assumptions regarding the geometry of the pore space. It can be shown that Berryman's approach for perfectly spherical grains is in reasonably close agreement with corresponding tortuosity estimates based on the formation factor for a cementation exponent $\tilde{m}$ of $\sim 1.7$, which is close to the commonly observed values of $\tilde{m}$ for unconsolidated sediments (e.g., Lesmes and Friedman, 2005).

Based on evidence from the core material, we assumed that the rock matrix was dominated by quartz, which then allows for constraining density $\rho_s$ and the bulk modulus of grains $K_s$ (Schoen, 2004). Using the approximation of Geertsma and Smit (1961) for the high-frequency limit of the fast P-wave velocity, which is given by

$$V_{P\infty} = \sqrt{\frac{1}{\rho_s(1-\phi) + \phi\rho_f(1-\gamma^{-1})}}$$

$$\times \left[ K_b + \frac{4}{3}G + \frac{\phi\dfrac{\rho}{\rho_f}\gamma^{-1} + \left(1 - \dfrac{K_b}{K_s}\right)\left(1 - \dfrac{K_b}{K_s} - 2\phi\gamma^{-1}\right)}{\left(1 - \dfrac{K_b}{K_s} - \phi\right)\dfrac{1}{K_s} + \dfrac{\phi}{K_{fl}}} \right],$$

(24)

provides a first-order estimate of the drained bulk modulus $K_b$, which was the last major unconstrained parameter in the poroelastic equations. Equation 24 tends to overestimate the high-frequency limit of the P-wave velocity $V_{P\infty}$ by a few percent (Mavko et al., 2009). The resulting bias toward values that are somewhat too low in our estimates of the drained bulk modulus $K_b$ is not of any practical importance because the inferred values of $K_b$ are used only as first-order constraints in the inversion of the dispersion curves outlined below.

## Interpretation

### Velocity dispersion

To explore the potential range of practical applicability of the poroelastic inversion of P-wave velocity dispersion, we first consider Biot's critical frequency, which defines the limit between the poroelastic low-frequency and high-frequency regimes in terms of phase coupling. More specifically, the critical frequency separates the viscous

and inertial coupling domains between the solid and fluid phases. In the viscous coupling mode, which prevails at lower frequencies, fluid movements through the pore space are characterized by Poiseuille flow, whereas the inertial coupling mode prevails at higher frequencies, when most of the fluid displacement occurs freely with regard to the matrix. It can be shown that the critical frequency is located in the region of the steepest gradient of the dispersion curve (e.g., Pride, 2005) and therefore can serve as an indicator of the frequency range that allows for relating seismic-wave phenomena to the underlying permeability structure. Biot's critical frequency, which depends essentially on the key hydraulic parameters porosity and permeability as well as on the frequency, is given by (Biot, 1956a, 1956b)

$$f_c = \phi\eta / 2\pi\rho_f k_d.$$

(25)

Figure 1 shows typical permeability ranges for common unconsolidated clastic sediments (Schoen, 2004) as a function of the critical frequency for porosities of 10% and 50%. This first-order analysis indicates that the poroelastic velocity dispersion of broadband P-wave sonic-log measurements in the frequency range of 1 kHz through 30 kHz are likely to contain pertinent information on the permeability of lithologies ranging from fine silts to coarse sands. In a subsequent step, we need to evaluate the expected magnitude of this velocity dispersion and thus its practical detectability. To this end, we consider typical porosities, permeabilities, and P-wave velocities for saturated unconsolidated clastic sediments and assume that

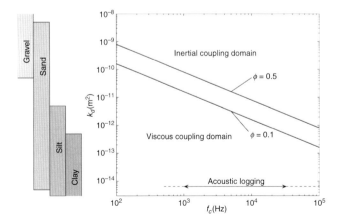

**Figure 1.** Plot showing permeability $k_d$, along with corresponding ranges of values for typical unconsolidated clastic sediments, as a function of Biot's critical frequency $f_c$ for porosities $\phi$ of 10% and 50%. The horizontal arrow denotes the emitter frequency range covered by modern sonic-logging tools.

the S-wave velocity is given as $V_S = 0.15 \, V_P$ (e.g., Schoen, 2004). For the considered frequency range of 1 kHz through 30 kHz, this indicates that P-wave velocity differences related to poroelastic dispersion might reach values in excess of 3%, and hence for high-quality, broadband sonic-log data, they should be clearly detectable. In agreement with Figure 1, the most prominent and hence most readily detectable and interpretable velocity differences are expected to occur at intermediate permeabilities, which are characteristic of unconsolidated silty to sandy deposits.

Figure 2 shows examples of the corresponding full-waveform sonic logs with nominal emitter center frequencies of 1, 15, and 30 kHz. Despite the significant changes in frequency content and thus in the support volumes of the measurements, these logs are remarkably uniform and mutually consistent in overall character and appearance as well as noise level. In this context, it is important to note that the individual source signals are not monochromatic but generally have a spectral bandwidth of approximately two octaves with regard to the corresponding nominal source frequency.

To construct the velocity-dispersion curves from our multifrequency sonic-log data, we determined the difference in arrival time with regard to the two receivers of the sonic tool at the nominal emitter center frequency of each data set. To do that, we first took the crosscorrelation of the signals recorded at the two receivers and then its Fourier transform. The slope of the corresponding phase spectrum $\varphi$ at the considered emitter center frequency $f_0$ then is related to the difference in arrival time as

$$\Delta t(f_0) = \frac{1}{2\pi} \frac{\partial \varphi}{\partial f}\bigg|_{f_0}. \qquad (26)$$

The velocity at the considered frequency is given by dividing the observed arrival-time difference by the constant spacing between the two receivers $V(f_0) = \Delta t(f_0)/\Delta z_{receiver}$. Taking a conservative approach, the average uncertainty of the inferred velocities is estimated to be about $\pm 0.5\%$ to $\pm 0.75\%$. As a consequence, we set the threshold for the difference in P-wave velocity observed at 1 kHz and 30 kHz to be analyzed in terms of poroelastic dispersion to 1.5%, which implies that the corresponding measurement uncertainties of $\pm 0.75\%$ do not overlap. Figure 3a shows the picked P-wave velocities at 1 kHz and 30 kHz, Figure 3b the corresponding percentage discrepancy log $\delta$, and Figure 3c a grayscale plot of the observed P-wave velocity dispersion between 1 kHz and 30 kHz. Also shown in Figure 3a is the S-wave velocity log, which was used to constrain the shear modulus $G$.

Using the procedure outlined previously (equation 26), we constructed the corresponding dispersion relations for each depth level for which the observed velocity discrepancy exceeds the above uncertainty threshold. These velocity-dispersion curves then were fitted with the theoretical dispersion relation using a standard linearized

**Figure 2.** Full-waveform sonic logs acquired at nominal emitter frequencies of (a) 1 kHz, (b) 15 kHz, and (c) 30 kHz.

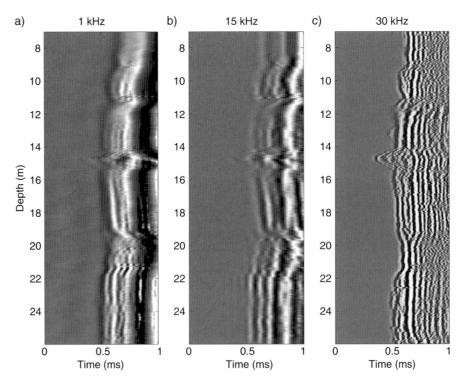

least-squares procedure that jointly inverts for the bulk modulus $K_b$ of the drained frame and the permeability $k_d$. As outlined previously, initial estimates of $K_b$ were obtained through equation 24. It can be shown that the corresponding inverse problem is well posed and that within the expected value ranges, there is essentially no trade-off between the permeability and bulk modulus of the drained frame. A corresponding example is shown in Figure 4,

with the vertical dashed line denoting the corresponding critical frequency as defined by equation 25.

Based on the inferred best-fit poroelastic dispersion curves, we can estimate permeability $k_d$ and its variability for the pertinent depth intervals. The corresponding results are shown in Figure 5, along with the natural-gamma and active nuclear logs and the lithologic sequence reconstructed from core material. We also show a log of the

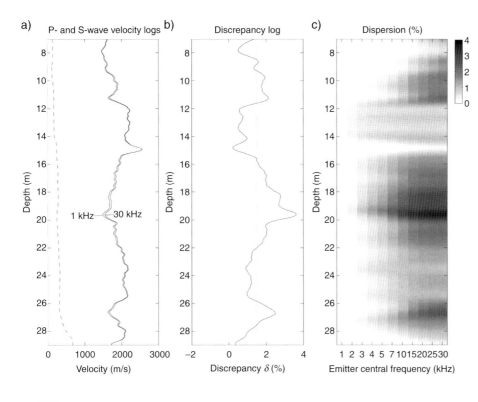

**Figure 3.** (a) Standard S-wave log (dashed line) along with multifrequency P-wave sonic logs (solid lines) acquired and picked at nominal emitter frequencies of 1 kHz and 30 kHz, (b) corresponding percentage discrepancy log $\delta$ of the P-wave velocity observed between 1 kHz and 30 kHz, and (c) P-wave velocity dispersion.

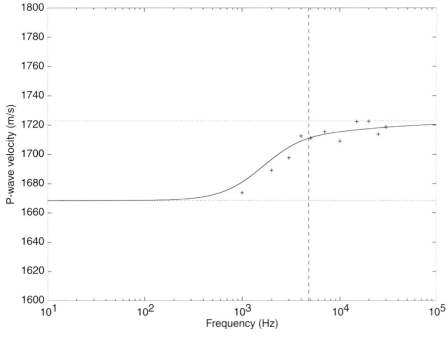

**Figure 4.** Example of inferred P-wave phase-velocity dispersion data (crosses) observed at a depth of 21.0 m and best fitting, in a least-squares sense, the poroelastic model dispersion curve (solid line). The vertical dashed line denotes the corresponding critical frequency $f_c$. Porosity estimated at the considered depth level is 30%. See also Figure 1 and Figure 3.

percentage difference $\delta$ of the estimated P-wave phase velocity at 1 kHz and 30 kHz and a grayscale plot of the velocity dispersion as a function of frequency and depth.

The isolated peak in the natural-gamma log between depths of 14 and 15 m seems to be related to a large gneissic block. Otherwise, the natural radioactivity of the entire lithologic sequence is uniformly moderate to low, shows little variation with depth, and largely is uncorrelated with the lithologic variations. This indicates that the clays listed in the litholog are primarily granulometric rather than mineralogical and thus form a mechanical and hydrologic continuum with fine silts. Conversely, the log denoting the

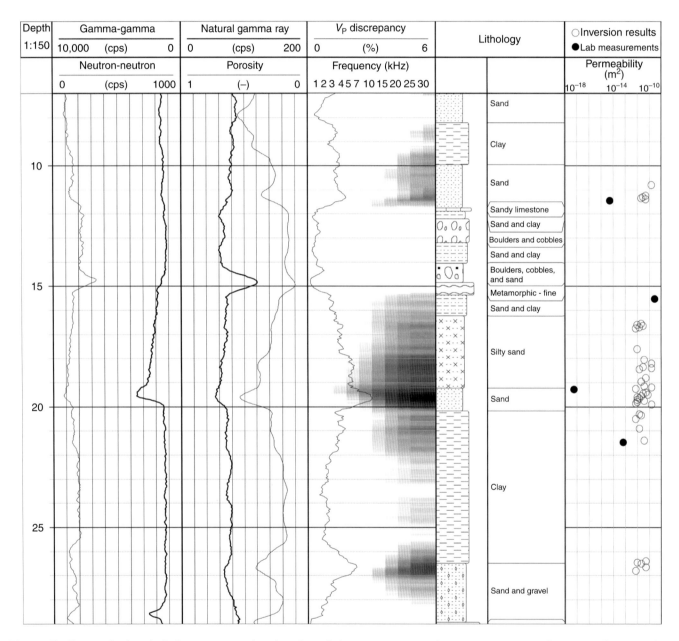

**Figure 5.** Composite borehole log sequence showing, from left neutron-neutron log, gamma-gamma log, natural-gamma log, porosity inferred jointly from neutron-neutron and gamma-gamma logs, percentage difference sonic log $\delta$ for nominal emitter frequencies of 1 kHz and 30 kHz, grayscale display of P-wave phase-velocity dispersion, litholog inferred from cores and cuttings, and permeability estimates inferred from the poroelastic inversion of the observed P-wave phase-velocity dispersion (open circles). Also shown are the available laboratory-based permeability measurements made on core material (filled circles). The core samples retrieved from 11.5-m and 19.5-m depths all had gravelly components, characterized by grain diameters in excess of 3 mm, which were removed. Hence, the samples are biased strongly toward low-permeability values. The units used for the nuclear logs denote counts per second (cps).

percentage difference $\delta$ of the P-wave phase-velocity difference measured at 1 kHz and 30 kHz shows a pronounced variability with depth, a clear relation to the interpreted lithologic sequence, and a strong positive correlation with the porosity log, which is fully consistent with our methodological considerations. As predicted by our feasibility analysis (Figure 1), we obtain meaningful results primarily for intervals dominated by silty to sandy lithologies. In those intervals, the inferred permeability values as well as their spatial variability are consistent with the corresponding range of values cited in the literature (e.g., Schoen, 2004).

Superimposed on our poroelastic permeability estimates, we show four laboratory measurements made on core material retrieved from 11.5 m, 15.5 m, 19.5 m, and 21.5 m (Huot, 1999). Those measurements were made on samples loaded into a Darcy-type permeametry cylinder with a diameter of 5 cm and a length of 10 cm (e.g., Butler, 2005). It is critical to note that only the samples from 15.5-m and 21.5-m depths contained the original core material, whereas the samples from 11.5-m and 19.5-m depths contained coarse granulometric fractions, as characterized by grain diameters larger than 3 mm, which were removed selectively to assure a certain degree of textural homogeneity in the size of the measurement cylinder (Huot, 1999).

Not surprisingly, the permeability estimates for the altered samples are much lower than those inferred from our dispersion analysis. Conversely, the permeability measurements made on the original samples are largely compatible with our estimates. This is consistent with the prediction that classical Biot theory should be largely valid at the high permeabilities and relatively high frequencies considered in this study (Mavko et al., 2009). Those results are also consistent with evidence from ocean acoustics, which indicates that over a frequency range of $\sim 3$ orders of magnitude, classical Biot theory is generally capable of explaining the observed P-wave velocity dispersion in medium- to fine-grained seabed sediments (e.g., Jackson and Richardson, 2007).

It is important to remember that our permeability-estimation procedure requires neither prior calibration nor any empirical corrections or adjustments of any input parameters. Finally, it is essential to note that the texture and hence the hydraulic properties of the samples used for laboratory measurements are likely to bear only a rather vague resemblance to their in situ state because the material was mixed and recomposed at least twice, once on retrieval of the core material and again when being placed into the measurement cylinder. This underscores the potential value of geophysical estimates of permeability in unconsolidated sediments because despite their high cost, corresponding laboratory measurements are likely to provide only order-of-magnitude constraints.

## Attenuation

As opposed to the velocity dispersion of our P-wave sonic-log data, which can be measured only for a limited range of lithologies (Figure 1 and Figure 5), the corresponding attenuation and its dependence on frequency do not seem to suffer from this limitation. To estimate the attenuation from the observed sonic-log data (Figure 2), we first correct all recordings for amplitude losses related to spherical spreading by multiplying each sample with its traveltime. Then we take the Fourier transform of the first-arriving P-wave train and collate the spectra for the various nominal emitter frequencies. At each measurement location along the borehole, the frequency-dependent attenuation $\alpha$ then is evaluated as (e.g., Bourbié et al., 1987; Pride, 2005)

$$\alpha(f) = -\frac{\ln(A_1(f)/A_0(f))}{\Delta z_{\text{receiver}}}, \qquad (27)$$

where $A_0$ and $A_1$ denote the spectral amplitudes recorded at the proximate and distant receivers, respectively. A representative example of this procedure is shown in Figure 6. We tested and verified our spectral estimates of $\alpha$ through corresponding time-domain measurements at the various nominal emitter frequencies based on the maximum first-cycle amplitude as well as the maximum of the P-wave train envelope.

Figure 6d shows the inverse of the quality factor $Q^{-1}$, which is another common measure of seismic energy dissipation and is related to attenuation $\alpha$ as (e.g., Pride, 2005)

$$Q^{-1}(f) = \frac{\alpha V_{\text{P}}(f)}{\pi f}. \qquad (28)$$

The major difference between $\alpha$ and $Q^{-1}$ is that the former quantifies attenuation with regard to a metric distance scale, whereas the latter uses the wavelength as the corresponding reference frame. This also explains why $\alpha$ tends to increase with frequency, whereas $Q^{-1}$ tends to decrease (Figure 6c and 6d). In the following, we work mostly with $Q^{-1}$ because according to the theoretical framework of poroelasticity, this parameter is expected to be characterized by a pronounced relaxation peak at the critical frequency and hence should allow for the corresponding location of our observed data based on the trends of their slopes (e.g., Pride, 2005).

**Figure 6.** Representative example of spectral amplitudes and attenuations inferred from sonic-log data recorded at 22.5-m depth. (a) Amplitude spectra $A_0$ and $A_1$ of the P-wave trains recorded at near and distant receivers, respectively. (b) Corresponding amplitude ratio, (c) attenuation $\alpha$ derived from (b), and (d) corresponding inverse quality factor $Q^{-1}$.

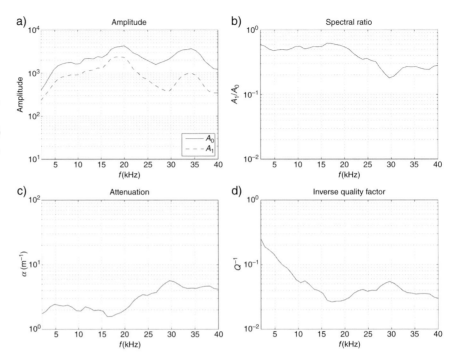

Values of $Q^{-1}$ close to one mark the transition from the wave-propagation regime to a physical environment governed by diffusive phenomena. In this context, it is important to note that $Q^{-1}$ values on the order of 0.1 are observed relatively commonly in near-surface seismic and ocean-acoustic studies (e.g., Malagnini, 1996; Schoen, 2004; Jackson and Richardson, 2007).

Figure 7 shows $Q^{-1}$ for the entire frequency and depth range considered along with the litholog. Ideally, this database now can be inverted for the permeability $k_d$ in a largely analogous manner to the one outlined for the velocity dispersion. The major lithologic units seem to be characterized by more or less distinct and internally consistent patterns of $Q^{-1}$. However, a more in-depth look at these data indicates that the $Q^{-1}$ values are consistently too high to be explained based on classical Biot theory alone, that they do not show the predicted relaxation peak related to the critical frequency, and that their frequency dependence on $Q^{-1}$ is quite variable and largely inconsistent with that predicted by Biot theory. This is illustrated in Figure 8, which shows a comparison of the measured $Q^{-1}$ values with corresponding simulations.

The observed attenuation data represent a comprehensive suite of all data subsampled in the depth domain by a factor of 50 for display purposes as well as the representative individual measurement at 22.5-m depth shown in Figure 6. The corresponding synthetic curves have been evaluated for a range of permeabilities $k_d$ and drained bulk moduli $K_b$ using otherwise constant petrophysical

parameters typical of a water-saturated quartz sand (e.g., Schoen, 2004).

We also have explored the impact of extensive variations of various parameter combinations on the modeled curves of $Q^{-1}$. Figure 9 gives a representative example of this analysis, which illustrates the impact of variations of porosity $\phi$ and shear modulus $G$ while keeping all other parameters invariant. The shapes and positions of the modeled curves do not change significantly, and our basic observation that the measured attenuation values exceed those predicted by Biot theory remains valid. Comparable results were obtained for similarly extensive variations of other parameter combinations, as well as when replacing our quartz matrix by a calcite matrix.

These results are somewhat disturbing because at least in the seismic community, there seems to be some consensus that classical Biot theory should be largely valid for the very high permeabilities and relatively high frequencies considered in this study (e.g., King, 2005; Mavko et al., 2009). In this context, it is important to note that as pointed out by Pride (2005), the available evidence from observed data is very thin because seismic data generally lack the necessary spectral bandwidth. What makes these results quite interesting is that in the past, similar observations — that classical Biot theory was systematically predicting attenuation values that were too low compared to the corresponding observations — were made for crystalline and lithified sedimentary rocks. This led to the development of modifications of the classical theoretical

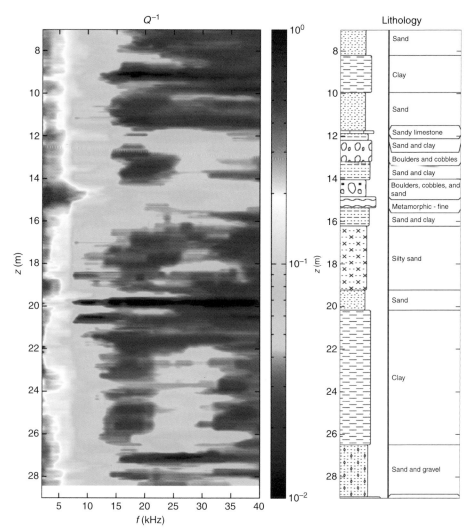

**Figure 7.** Frequency-dependent attenuation behavior as quantified by the inverse quality factor $Q^{-1}$ compared to the lithology observed in the core material.

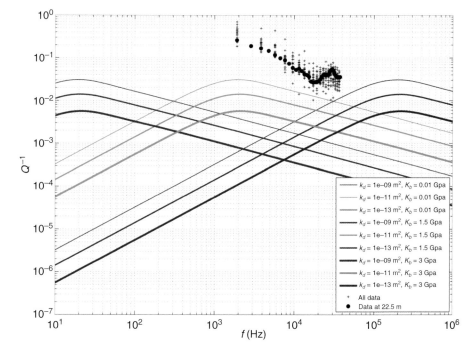

**Figure 8.** Comparison of observed and modeled inverse quality factors $Q^{-1}$ for a range of permeabilities $k_d$ and drained frame bulk moduli $K_b$.

**Figure 9.** Impact of uncertainties in porosity $\phi$ and shear modulus $G$ on the modeled inverse quality factor $Q^{-1}$. The various colored lines lie too close together to be distinguishable.

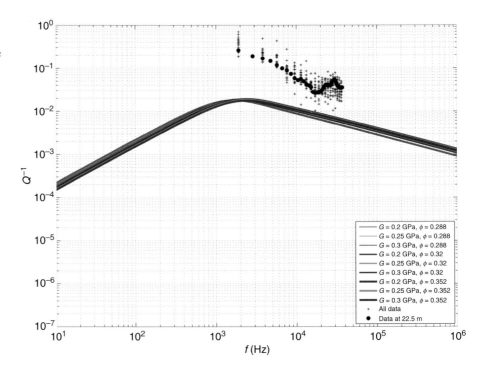

framework, such as the squirt-flow model to account for wave-induced fluid movements in pores with high aspect ratios (Dvorkin et al., 1995) and the double porosity model to account for the effects of structural and/or compositional heterogeneity at the mesoscopic scale (Pride and Berryman, 2003). These methodological extensions now often allow more adequate modeling of observed attenuation in consolidated sediments and crystalline rocks (e.g., Pride, 2005; Mavko et al., 2009). Whereas the classical squirt-flow model is unlikely to account for excess attenuation in unconsolidated granular sediments, accounting for heterogeneity, which definitely is present at our site, could be an interesting avenue to explore.

Ocean-acoustic data generally have very large spectral bandwidths compared to seismic data, and the surficial seabed commonly has lithologic and petrophysical characteristics that are reasonably similar to those of shallow alluvial aquifers.

In agreement with our results, Biot theory systematically underestimates the acoustic attenuation observed in surficial seabed sediments for frequencies above 5 kHz while providing adequate predictions of the velocity dispersion (Jackson and Richardson, 2007). Correspondingly, evidence suggests that velocity-based modeling of ocean-acoustic data seems to be capable of yielding realistic estimates of the elastic and hydraulic seabed properties (e.g., Schock, 2004). Extensions and modifications of classical poroelastic theory have been developed more or less

specifically for oceanic environments (Jackson and Richardson, 2007).

As outlined previously, one of the oldest and arguably still most widely used empirical extensions of classical Biot theory, generally referred to as Biot-Stoll theory, is based on the use of complex-valued frame moduli to account for excess attenuation in unconsolidated sediments (e.g., Stoll, 1977). Current evidence indicates that the choice of sufficiently large imaginary values for the frame moduli to account for excess attenuation at high frequencies leads to physically unreasonable attenuation behavior at lower frequencies (Jackson and Richardson, 2007). This has led to the development of various physically motivated extensions and modifications of classical Biot theory for unconsolidated seabed sediments, such as the consideration of specific micro- and mesoscopic heterogeneities (Chotiros, 2002) and adaptation of the squirt-flow model (Chotiros and Isakson, 2004). These extensions seem to reproduce the observed dispersion and attenuation behavior in medium- to fine-grained seabed sediments over a very broad range of frequencies covering ∼3 orders of magnitude.

Finally, there is the possibility that the mismatch of the observed and predicted attenuation behavior might be related partly to drilling-induced disturbances around the borehole, the presence of plastic casing, or the fact that the formula for estimating the attenuation $\alpha$ (equation 27) is strictly valid only for waves traveling in a homogeneous full-space and not for waves refracted from a strongly heterogeneous medium, as is the case for our observations. What is not immediately obvious is why these potential violations of the fundamental assumption of wave propagation in a homogeneous porous full-space would affect the velocity dispersion seemingly so much less than the attenuation behavior. This and all other fundamental questions related to our observations of the attenuation behavior of broadband sonic-log data in unconsolidated sediments can be addressed only through realistic modeling based on numerical solutions of the poroelastic equations and/or suitable extensions thereof. This is an endeavor which we intend to pursue energetically in the future.

## Conclusions

We have explored the feasibility of obtaining constraints on the permeability structure of saturated alluvial sediments based on a poroelastic approach to the interpretation of the phase-velocity dispersion and attenuation behavior of broadband multifrequency sonic-log data. In agreement with methodological considerations, we found the analysis of the P-wave velocity dispersion to be feasible for field data. The inferred permeability values and their variability are consistent with sparse laboratory measurements and generally are in the expected ranges for the considered lithologies. Importantly, our permeability estimates did not require prior calibration or adjustments of input parameters. In agreement with our methodological considerations, our results demonstrate that the range of permeabilities that can be resolved is limited by the spectral bandwidth of the sonic-log measurements. For typical commercial slim-hole multifrequency sonic tools operating in the frequency range of approximately 1 kHz through 30 kHz, meaningful permeability estimates can be obtained for unconsolidated clastic sediments with granulometric characteristics in the range of fine silt to coarse sand.

Although it would be desirable to cover a more comprehensive range of permeabilities and lithologies, it is important to note that the intermediate permeability range resolved by the method in this study, filling the poorly documented gap between "nonpermeable" clays and "infinitely permeable" gravels, is of particular interest in many practical situations. With that said, it is conceivable to extend this frequency range by approximately one octave at each end of the current frequency band based on signal processing alone. That would offer the perspective of extending the granulometric range amenable to our poroelastic dispersion-analysis approach from coarse clays to fine gravels, based on existing logging hardware.

Conversely, results of the frequency-dependent attenuation behavior of the same data set are seemingly inconsistent with the considered classical theoretical framework of poroelasticity. The observed attenuation values are systematically too high and generally lack a characteristic peak near the critical frequency, and their frequency dependence is variable and inconsistent, at least partly, with Biot theory. Those observations regarding the frequency-dependent attenuation behavior are enigmatic but are consistent with related observations in ocean acoustics. Realistic numerical modeling based on the poroelastic equations and/or suitable extension is most likely to provide deeper insights into the underlying physical processes governing the seismic attenuation in unconsolidated clastic sediments. This also indicates that robust permeability estimates from seismic body-wave attenuation measurements are likely to be a much more distant prospect than those for corresponding phase-velocity dispersion data. This study indicates that the latter approach seems to offer the prospect of reliable geophysically based first-order estimates of the in situ permeability in most surficial environments.

## Acknowledgments

We are very grateful to Paul Hagin for providing an exceptionally thorough and insightful review that greatly helped us to improve the quality of this manuscript. The revision process also benefited substantially from comments and suggestions by German Rubino. This work was supported by a grant from the Swiss National Science Foundation.

## References

Baron, L., and D. Chappelier, 2000, Calibration of environmental nuclear tools based on core sample analysis: Root mean square and neural network approaches: European Journal of Environmental and Engineering Geophysics, **4**, 129–149.

Berryman, J. G., 1981, Elastic wave propagation in fluid-saturated porous media: Journal of the Acoustical Society of America, **69**, 416–424.

Biot, M. A., 1956a, Theory of propagation of elastic waves in a fluid-saturated porous solid: I. Low-frequency range: Journal of the Acoustical Society of America, **28**, 168–178.

———, 1956b, Theory of propagation of elastic waves in a fluid-saturated porous solid: II. Higher frequency range: Journal of the Acoustical Society of America, **28**, 179–191.

Bourbié, T., O. Coussy, and B. Zinszer, 1987. Acoustics of porous media: Editions Technip.

Brown, R. J .S., 1980, Connection between formation factor for electrical resistivity and fluid-solid coupling factor in Biot's equations for acoustic waves in fluid-filled porous media: Geophysics, **45**, 1269–1275.

Butler, J. J. Jr., 2005, Hydrogeological methods for estimation of spatial variations in hydraulic conductivity, *in* Y. Rubin and S. S. Hubbard, eds., Hydrogeophysics: Water Science and Technology Library Series No. 50, Springer, 23–58.

Carcione, J. M., ed., 2007, Wave fields in real media: Wave propagation in anisotropic, anelastic, porous and electromagnetic media: Handbook of geophysical exploration, Elsevier Seismic Exploration Series No. 38.

Chotiros, N. P., 2002, An inversion for Biot parameters in water-saturated sand: Journal of the Acoustical Society of America, **112**, 1853–1868.

Chotiros, N. P., and M. J. Isakson, 2004, A broadband model of sandy ocean sediments: Biot-Stoll with contact squirt flow and shear drag: Journal of the Acoustical Society of America, **116**, 2011–2022.

Dvorkin, J., G. Mavko, and A. Nur, 1995. Squirt flow in fully saturated rocks: Geophysics, **60**, 97–107.

Geertsma, J., and D. C. Smit, 1961, Some aspects of elastic wave propagation in fluid-saturated porous solids: Geophysics, **26**, 169–181.

Holliger, K., 2008, Groundwater geophysics: From structure and porosity towards permeability?, *in* C. J. G. Darnault, ed., Overexploitation and contamination of shared groundwater resources: Springer NATO Science for Peace and Security Series C, Environmental Security, 49–66.

Hovem, J. M., and G. D. Ingram, 1979, Viscous attenuation of sound in saturated sand: Journal of the Acoustical Society of America, **66**, 1807–1812.

Hubbard, S. S., and Rubin, Y., 2005. Introduction to hydrogeophysics, *in* Y. Rubin and S. S Hubbard, eds., Hydrogeophysics: Springer Water Science and Technology Library Series No. 50, 3–21.

Huot, F., 1999, Caractéristiques élastiques des sols: Du comportement pseudo-statique à la propagation des ultrasons: Ph.D. thesis, University of Lausanne.

Jackson, D. R., and M. D. Richardson, 2007, High-frequency seafloor acoustics: Springer.

King, M. S., 2005, Rock-physics developments in seismic exploration: A personal 50-year perspective: Geophysics, **70**, no. 6, 3ND–8ND.

Lesmes, D. P., and S. P. Friedman, 2005, Relationships between the electrical and hydrogeological properties of rocks and soils, *in* Y. Rubin and S. S. Hubbard, eds., Hydrogeophysics: Springer Water Science and Technology Library Series No. 50, 87–128.

Malagnini, L., 1996, Velocity and attenuation structure of very shallow soils: Evidence for a frequency-dependent $Q$: Bulletin of the Seismological Society of America, **86**, 1471–1486.

Mavko, G., T. Mukerji, and J. Dvorkin 2009, The rock physics handbook: Tools for seismic analysis in porous media, 2nd ed.: Cambridge University Press.

Pride, S. R., 2005, Relationships between seismic and hydrological properties, *in* Y. Rubin and S. Hubbard, eds., Hydrogeophysics: Springer Water Science and Technology Library Series, No. 50, 253–290.

Pride, S. R., and J. G. Berryman, 2003, Linear dynamics of double-porosity dual-permeability materials: I. Governing equations and acoustic attenuation: Physical Review E, **68**, 036603.

Revil, A., and L. M. Cathles III, 1999. Permeability of shaly sands: Water Resources Research, **35**, 651–662.

Schock, S. G., 2004, A method for estimating the physical and acoustic properties of the sea bed using chirp sonar data: IEEE Journal of Oceanic Engineering, **29**, 1200–1217.

Schoen, J. H., 2004, Physical properties of rocks: Fundamentals and principles of petrophysics: Elsevier Handbook of geophysical exploration: Seismic Exploration Series No. 18.

Stoll, R. D., 1977, Acoustic waves in ocean sediments: Geophysics, **42**, 715–725.

Tang, X. M., and A. Cheng, 2004, Quantitative borehole acoustic methods: Handbook of geophysical exploration, Elsevier Seismic Exploration Series No. 24.

Tang, X., and C. H. Cheng, 1996, Fast inversion of formation permeability from Stoneley wave logs using a simplified Biot-Rosenbaum model: Geophysics, **61**, 639–645.

Chapter 10

# Permittivity Structure Derived from Group Velocities of Guided GPR Pulses

Matthew M. Haney[1], Kathryn T. Decker[1], and John H. Bradford[1]

## Abstract

On a 2D profile of subsurface permittivity structure derived from guided GPR pulses recorded in the Kuparuk River watershed, Alaska, the transition from a stream channel to a peat layer is interpreted. Although multichannel data are used, guided waves are analyzed using single-channel analysis, which sidesteps assumptions regarding lateral homogeneity within receiver arrays. As a result, 2D structure is obtained along a profile using an inversion procedure. These data were processed in three steps: (1) picking group traveltimes, (2) performing tomography in the lateral direction, and (3) inverting local group-velocity dispersion curves. When the permittivity profile obtained from the guided waves is compared to a GPR reflection profile, it is clear that the guided waves capture shallow structure near a stream channel that is not imaged accurately on the reflection profile. This demonstrates the utility of using guided waves to provide information on shallow structure that cannot be obtained from reflections.

## Introduction

Knowledge of the permittivity structure of the near surface is valuable for understanding hydrology of the vadose zone (Strobbia and Cassiani, 2007) and the hyporheic zone (Brosten et al., 2009b). This is because permittivity is a reliable diagnostic of material type and water content. Moveout analysis, such as reflection tomography (Brosten et al., 2009b), must accompany GPR reflection surveys to determine this structure. In the presence of guided waves, moveout analysis can be difficult because guided waves act as a type of unwanted source-generated noise. In this study, instead of treating the guided waves as noise, we exploit their properties along a single transect. We develop a methodology for processing and interpretation based entirely on the unique characteristics of guided waves in GPR data.

Guided waves are a well-documented phenomenon that occurs over a wide variety of length scales in the earth. Seismologists have benefited greatly from the sensitivity to the structure of the crust and upper mantle of low-frequency guided waves produced by large earthquakes (Dorman and Ewing, 1962; Aki and Richards, 1980). At higher source frequencies, tube waves excited in borehole experiments, for example, sense only rock formations in the immediate vicinity of a well. Regardless of scale, a hallmark of guided-wave propagation is dispersion — the dependence of velocity on frequency. Because, for guided waves, this property arises as a result of lower frequencies penetrating deeper into the earth, the behavior is described as geometric dispersion (Socco and Strobbia, 2004) in contrast to, for instance, anelastic dispersion. For dispersive waves, phase and group velocities are generally unequal: The discrepancy between the velocities leads to an appearance commonly referred to as "shingling" in seismic shot gathers.

Recently, evidence of guided-wave propagation has been found to exist for ground-penetrating-radar (GPR) reflection surveys (Arcone, 1984; Arcone et al., 2003b; van der Kruk et al., 2006; van der Kruk et al., 2007; van der Kruk et al., 2009). Ground-penetrating-radar guided waves are advantageous because they have a high signal-to-noise ratio (S/N) and can travel laterally relatively far

[1]*Boise State University, Department of Geosciences, Center for the Geophysical Investigation of the Shallow Subsurface, Boise, Idaho, U.S.A.*

compared to other wave types such as reflected and refracted waves. In the subsurface, GPR guided waves propagate at speeds dictated in large part by permittivity. Because permittivity values are determined by chemical composition and water content, features of geologic interest such as stratigraphic layers and water-saturated zones are often evident in GPR data.

Several similarities besides the occurrence of guided waves exist between GPR and seismic data. For instance, GPR data can be collected to enhance wide-angle reflection and refraction (WARR) energy by increasing the distance between the transmitting and receiving antennae (Fisher et al., 1992). Furthermore, the governing equations of seismic and electromagnetic wave propagation can be shown to have a one-to-one correspondence (Carcione and Cavallini, 1995).

Two distinctive cases of lossless and lossy waveguides exist for GPR guided waves (Arcone et al., 2003b; van der Kruk et al., 2007). Here, we use the terms *guided waves* exclusively for lossless waveguides and *leaky waves* for lossy waveguides. For a simple layer at the surface, the presence of guided waves implies that the substrate has a relatively lower permittivity than the layer. In this case, total internal reflection occurs at the air-surface interface and between the surface layer and the lower permittivity medium below it when the angle of incidence exceeds the critical angle at both interfaces (Arcone et al., 2003b; van der Kruk et al., 2007). In the case of leaky waves, the substrate has a higher permittivity and can be a strong reflector. A common example of leaky waves is found where ice has formed over water. The higher permittivity of water results in a strong radar reflection; however, some of the energy is transmitted into the lower layer. Ground-penetrating-radar guided waves can exist in a simple high-permittivity surface layer model.

In general, GPR guided waves can exist whenever the permittivity of a layered structure decreases with depth. Because of this requirement, GPR guided waves are not commonly encountered in GPR data; that is, permittivity usually increases with depth (Fisher et al., 1992; Becht et al., 2006; Strobbia and Cassiani, 2007). However, permittivity does decrease with depth in certain geologic environments (for instance, when unfrozen sediments cover a layer of permafrost). Note that the typical "negative velocity gradient" encountered in GPR studies is therefore the opposite of the commonly observed increase of velocity with depth in seismology. For this reason, guided waves historically have been used in seismology. In addition, a prominent type of guided wave in seismology, the Rayleigh wave, can exist regardless of whether the velocity gradient is positive or negative. There is no analogy to the Rayleigh wave in GPR; GPR guided waves are most similar to Love waves in seismology.

Many techniques exist to analyze guided seismic waves. When dense arrays of receivers are available, phase velocity can be measured by transforming time traces into the frequency domain and then plotting the spectral amplitudes from a suite of stacks with different linear delays (Park et al., 1999). Because stacking is applied over a receiver array, it is a multichannel method. In contrast, group velocity can be measured with single-channel methods. For example, frequency-time analysis, or FTAN (Ritzwoller and Levshin, 2002; Abbott et al., 2006; Bussat and Kugler, 2009), decomposes the signal with a cascade of narrowband filters that allows the frequency dependence of the envelope peak over the suite of narrowband signals to be measured. Frequency-time analysis thus leads to construction of group-velocity dispersion curves, which can be inverted for an average layered structure between a source-receiver pair.

Another technique for guided-wave analysis is known as group-velocity tomography (Ritzwoller and Levshin, 2002; Abbott et al., 2006; Gerstoft et al., 2006; Bussat and Kugler, 2009; Masterlark et al., 2010). For a narrowband signal, group velocity determined from FTAN applied to a source-receiver pair can be converted to group traveltime because the source-receiver offset is known. By measuring group traveltime between many source-receiver pairs, a lateral map of local group velocity can be produced through application of tomography.

The distinction between single-channel methods used to measure group-velocity and multichannel methods used to measure phase velocity is an important one. Previous research on guided waves in GPR data emphasizes the analysis applied to single profiles (Arcone, 1984; Arcone et al., 2003b; van der Kruk et al., 2006; van der Kruk et al., 2007; van der Kruk et al., 2009). A single phase-velocity spectrum can be computed for a profile, and a layered model can be found that explains the observed phase-velocity dispersion. Implicit in this approach is the assumption of lateral homogeneity beneath the profile. However, detection of lateral variations of the waveguide within a profile is straightforward using single-channel methods based on group velocity. A 2D subsurface model resulting from this analysis would provide structure within a wavelength of the surface, depths which are inaccessible in GPR reflection profiles. Such information could be useful for statics corrections applied to GPR reflections from deeper interfaces.

In this study, our objective is to develop a methodology that exploits GPR pulses when strong permittivity contrasts act as waveguides. The approach is based on the analysis of GPR data collected by using a roll-along survey approach.

Emphasis is placed on resolving the laterally varying properties of the guided waves, with the result being a 2D model of subsurface permittivity. We apply group-velocity tomography to obtain local group-velocity dispersion curves followed by the inversion of local dispersion curves for defining depth structure. Our application is a natural extension of methods originally developed in seismology. We provide details of this approach in later sections.

## Previous Work on GPR Guided Waves

Dispersive guided and leaky waves are not discussed commonly in GPR reflection surveys. However, they occur for specific conditions that are of great interest. Collin (1960) is responsible for some of the pioneering work on guided-wave phenomena in electromagnetics. Application to geophysical investigations originates in the work of Annan (1973). Several subsequent studies shed light on guided-wave phenomena in GPR data sets and use dispersion to determine characteristics of the subsurface. Most of these studies focus on guided waves in single profiles. Few studies highlight the analysis of guided-wave phenomena in a roll-along profile, as is done by Arcone et al. (2003a) to estimate attenuation rates. Here, we briefly review previous research on GPR guided waves.

Arcone (1984) provides one of the earliest accounts of GPR guided-wave propagation. In that study, the author observed GPR guided waves using 200-MHz antennae with a bandwidth from 100 to 300 MHz. The data were collected with WARR (moveout) profiles over three high-permittivity layers to study the dielectric behavior of the materials. Two profiles were collected over ice that had formed on lakes, and the third was collected over a body of granite divided into two layers by a section of joints filled by water. The dielectric constants and layer thicknesses for each case were well constrained. At all three sites, both transverse-electric (TE) and transverse-magnetic (TM) configurations were attempted; however, only the TE surveys provided useful data, possibly because of a lack of a cutoff frequency for the lowest-order TM mode. This might have resulted in sensitivity to small-scale variations in the layers, effectively breaking up the waveguide. An alternative explanation, discovered in later studies and described below, was that direct airwaves were excited more strongly than guided waves.

Liu and Arcone (2003) numerically model electromagnetic energy in waveguides by the finite-difference time-domain (FDTD) method. The authors also simulate field GPR data from Fort Richardson, Alaska, and Hanover, New Hampshire. The study found that GPR guided waves occurred because of total internal reflection at the site in Alaska, where a sandy material with a relatively low dielectric constant overlaid a sandy silt layer with a higher dielectric constant. At the New Hampshire site, the top layer was sandy, with a silty layer beneath it. The contrast in dielectric constant for the New Hampshire site was low enough that total reflection did not occur at the lower boundary and the waveguide was leaky. It was found that the generation of ground waves mostly occurred for TE mode, although the airwave was generated most successfully for the TM mode for all geologic conditions, supporting the earlier field observations by Arcone (1984).

Arcone et al. (2003b) compare field observations of GPR guided waves to predicted results from FDTD modeling. Using a field site in Anchorage, Alaska, GPR data were collected using 100- and 400-MHz antennae aligned in the TE mode, although actual center frequencies were lower than the manufacturer-designated values. The survey was conducted over an area known to have stratified gravels. A well log showed that the surface layer was relatively thin and composed of wet cobbles, silty sand, and organics. Beneath the surface layer was a thicker layer of moist gravel mixed with sand. Further below was a dry sand layer with a thin, perched water table at its base. A 1200-MHz antenna was used to obtain estimates of dielectric constants by gathering reflection profiles on boulders of material expected to have electrical properties identical to those of the gravel in the subsurface. The theoretical calculations yielded the fundamental TE mode propagating at low frequencies, with higher modes appearing at higher frequencies. Values of permittivity deduced from theoretical predictions of the cutoff frequency were close to those observed in the field data. Arcone et al. (2003b) conclude that dispersive waves found in GPR data sets under the right conditions provide useful information about subsurface electrical properties.

Extending the analysis of GPR guided waves, van der Kruk et al. (2006) test the joint inversion of TE and TM modes of propagation. To obtain layer thickness and permittivity, van der Kruk et al. (2006) computed phase-velocity spectra by transforming data from the time-distance domain into the phase velocity-frequency domain. Then they constructed dispersion curves from the spectra, which they inverted for the desired properties. Both separate and joint TE and TM inversion were performed on dispersion curves extracted from field data taken in New Zealand over braided river-channel sediments and in close proximity to the Alpine Fault zone. The separate TE and TM inversions produced models that fit the data poorly; however, joint TE-TM inversion produced a more consistent model. The authors conclude that joint inversion of TE and TM data provides a wider bandwidth of frequencies and more realistic models, especially for the case in which some degree of heterogeneity exists.

In the case of leaky waves, van der Kruk et al. (2007) show that TE modes propagate with less loss than TM modes, and the authors develop an inversion technique involving the higher-order modes of TE and TM propagation for increased accuracy. The authors collected GPR data in both TE and TM configurations using 700-MHz antennae over a smooth, bare-ice layer at a pond in New Hampshire. Results for the leaky waves demonstrated that the electrical properties of the ice could be determined using the same guided-wave techniques and inversion scheme as used previously for GPR guided waves. Applications of both GPR guided waves and leaky waves are described further by van der Kruk et al. (2009).

## Equipment and Field Procedures

The GPR data analyzed in the present study were acquired with similar techniques and during the same field experiments described in Brosten et al. (2006) and Brosten et al. (2009b). We briefly review the equipment and procedures here. The GPR system was a Pulse EKKO 100-MHz antenna, which in the field had a center frequency set to approximately 70 MHz (Brosten et al., 2006). The need for a center frequency of 70 MHz is reflected in the fact that, as described later, we analyze guided waves over the frequency band from 20 to 120 MHz, which is centered $\pm 50$ MHz on either side of 70 MHz. We specifically chose to analyze GPR data acquired with this antenna because of its relatively low frequency compared to other available data sets. For guided waves, lower frequencies imply deeper resolution. The GPR data were acquired using the TE configuration and had the following acquisition parameters: 0.9-m (3-ft) near offset, 0.6-m (2-ft) source spacing, and 0.3-m (1-ft) receiver spacing. Time sampling was set at 0.8 ns, and the total recording time was 200 ns. Each of the 62 total transmitter locations had a maximum of 48 receivers. Multioffset data were acquired by leaving the transmitter in a fixed position and progressively moving the receiver. Once an entire common-source gather was obtained, the transmitter was moved and the process was repeated. An important detail of the acquisition scheme concerned stream-channel traverses. Boards 25 to 30 cm wide were extended across stream channels to provide an accurate base elevation for the survey.

## Methodology

We build on previous studies of GPR guided waves to produce a method that inverts for a 2D subsurface model of permittivity, instead of an average 1D depth profile. In this and the following five sections, we describe the processing and inversion steps for constructing a 2D model from observations of GPR guided waves. The entire procedure is the same as that described in Abbott et al. (2006) and by Masterlark et al. (2010) for seismic waves except that here we specialize to sources and receivers distributed along a line instead of over an area. The approach is based on the assumption that lateral propagation properties of the guided waves and their velocity dispersion from depth structure are decoupled (Aki and Richards, 1980). This means that the propagation of guided waves at a particular location are determined by a local 1D depth model. The complete procedure consists of three steps:

1) Apply frequency-time analysis and assemble a table of group traveltime as a function of source, receiver, and frequency: $t_u (x_s, x_r, f)$.
2) Perform group-velocity tomography in the lateral direction at each frequency to produce local group-velocity dispersion curves: $u(x, f)$.
3) Invert dispersion curves at each lateral point for a depth model of permittivity: $\epsilon(x, z)$.

Of these three steps, the third is by far the most computationally intensive. For the GPR field data analyzed in a later section, the three steps had compute times of seven minutes, 15 seconds, and three hours, respectively, on a desktop computer with a 2.66-GHz processor. The first step is more demanding than the second because of the need for Fourier transforms in FTAN. The computational demand of the third step originates in its use of an eigenvector and eigenvalue solver.

### Group traveltimes

The construction of a group-traveltime table as a function of source, receiver, and frequency requires the picking of group traveltimes and a set of selection criteria for quality control. Because selection criteria are inherently ad hoc and survey dependent, we delay the discussion of the criteria until we fully introduce the GPR field data. Here, we discuss and demonstrate how group-traveltime measurements are made using FTAN.

To illustrate FTAN, we show the process on two GPR traces from the field data set. The first 39 shot gathers (out of 62 total) from the field data set are plotted in Figure 1. Figure 2 shows more detailed evidence for dispersion and shingling on two shot gathers. Evidence for the existence of GPR guided waves comes from the strong, late-arriving energy beginning at approximately trace number 1300. Prior to trace number 1300, it is likely that GPR leaky

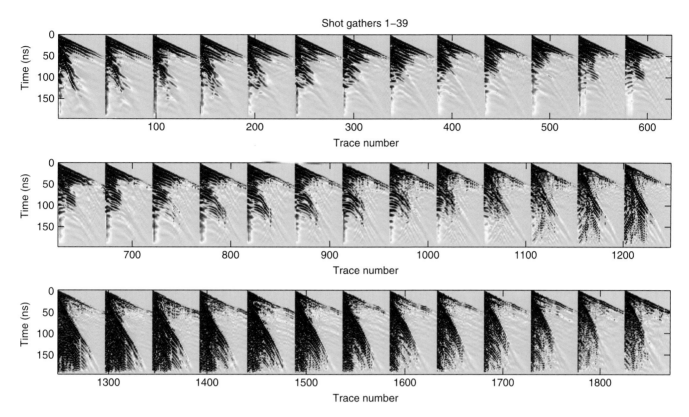

**Figure 1.** The first 39 shot gathers from the GPR survey. GPR guided waves develop by trace number 1300, as seen in the large amplitude energy within the cone defined by the lower guided-wave velocity.

waves exist instead of guided waves. To investigate this, we applied FTAN (Ritzwoller and Levshin, 2002; Abbott et al., 2006) to obtain path-averaged group-velocity dispersion curves for traces nos. 640 and 1555 (Figures 3 and 4). Note that these dispersion curves represent the average velocity structure between the source and receiver. In contrast, we invert *local* dispersion curves that have had propagation effects removed by group-velocity tomography in the third step of the procedure described above.

In Figure 3a, there is a large amplitude arrival with a traveltime of approximately 100 ns. By decomposing the GPR recording with FTAN, we observe the dispersive character of this arrival. Figure 3b shows the output of FTAN, called an energy diagram. The energy diagram is a surface defined by the envelopes of a series of narrowband filtered versions of the signal in Figure 3a. However, instead of plotting the surface as a function of traveltime and frequency, the traveltime axis is transformed to a velocity axis because the source-receiver distance is known. The group-velocity dispersion curve is found by tracing the peak power in the energy diagram across the frequency band. To build a group-traveltime table, the group-velocity dispersion curve can be transformed to group traveltime as

a function of frequency, again because the source-receiver distance is known. For trace number 1555, we find a dispersion curve (Figure 3b) consistent with the existence of GPR guided waves: higher velocities for lower frequencies and vice versa. Therefore, we can conclude that GPR guided waves do exist at trace number 1555.

In contrast, the dispersion curve extracted for trace number 640 has a velocity dispersion characteristic of a leaky wave. Figure 4 shows how FTAN recovers a dispersion curve with lower velocities for lower frequencies. Moreover, the amplitude of the dispersive arrival at approximately 100 ns is considerably smaller than observed for the guided wave in Figure 3. This comparison is valid given the offset for the two traces: 5.5 m (number 640) and 6.4 m (number 1555). The offset for the dispersive arrival in trace number 640 is weaker than in trace number 1555, even though the offset is smaller for trace number 640. As a result, we conclude that the dispersive arrival in trace number 640 is a leaky wave. It should be noted that because of the usual negative velocity gradient encountered in GPR, leaky waves, as in trace number 640, are more common than guided waves (Fisher et al., 1992; Becht et al., 2006; Strobbia and Cassiani, 2007).

a)

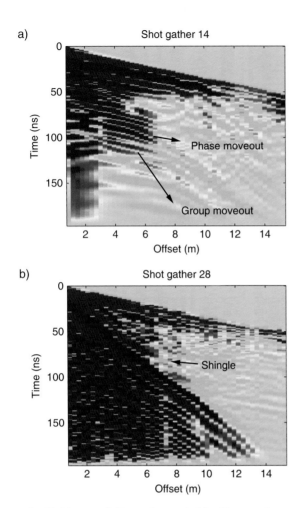

b)

Figure 2. Evidence of dispersion and shingling on shot gathers 14 and 28. (a) Leaky wave; (b) guided wave.

a)

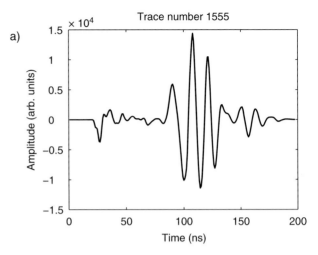

b)

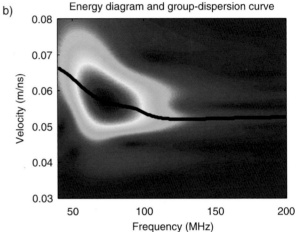

Figure 3. Example of a GPR guided wave at trace number 1555. Shown are (a) the trace and (b) its energy diagram. The energy diagram represents a scan over many possible group velocities at each frequency. The actual group velocity at that frequency is given by the maximum of the energy diagram.

Because we are interested in the inversion of GPR guided waves and not leaky waves, we chose to analyze traces beyond trace number 1300 in Figure 1. A more robust version of the three-step procedure outlined previously, which could accept leaky and guided waves on an equal footing, is an area of active research. This issue arises because of the fundamentally different behavior and sensitivity to subsurface structure for guided versus leaky waves.

## Group-velocity tomography

The group traveltimes measured using FTAN reflect the average group slowness (the inverse of group velocity $u$) between source-receiver pairs. To resolve lateral heterogeneity in group velocity at a particular frequency, a tomographic inversion of the group traveltimes is necessary. By tomography, we mean the estimation of a lateral group-velocity model that is consistent with the set of group-

traveltime measurements between sources and receivers. Note that the tomography is applied independently for each frequency. This problem is similar to the inversion of arrival times in a VSP seismic experiment for a layered velocity structure (van Wijk et al., 2002). However, instead of a single source at the surface and an array of seismometers in a well, GPR surveys typically involve many source locations and a receiver array that shifts location along with the source.

Unlike tomography in higher dimensions, group-velocity tomography for surface waves along an acquisition line does not require ray tracing — the guided waves are assumed to propagate within the vertical plane defined by the acquisition line. After defining the lateral characteristics of a 2D permittivity model, we construct the tomography kernel $\mathbf{K}_s^t$ relating observed group-traveltime perturbations

a)

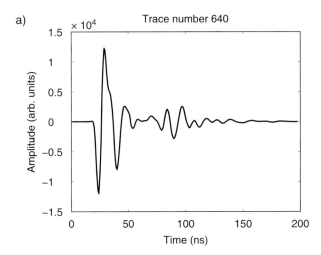

b)

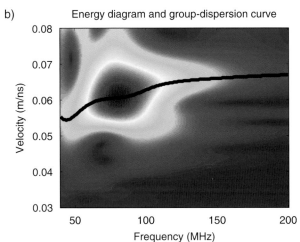

**Figure 4.** Example of a GPR leaky wave at trace number 640. Shown are (a) the trace and (b) its energy diagram.

$\delta t$ to group slowness perturbations $\delta s$ for each frequency under consideration such that

$$\vec{\delta t}_u = \mathbf{K}_s^t\, \vec{\delta s}. \qquad (1)$$

In this equation, $\vec{\delta t}_u$ is the vector of group-traveltime perturbations from an initial model for each source-receiver pair, $\vec{\delta s}$ is the vector of group slowness perturbations for each frequency, and $\mathbf{K}_s^t$ is the matrix kernel relating group slowness (subscript s) to group traveltime (superscript $^t$). This linear matrix relation is inverted with a weighted-damped least-squares scheme described below. Note that in contrast to the inversion of group-velocity dispersion curves, this inverse problem is linear. In Figure 5, we show an example of the tomography kernel for the GPR field data discussed in a later section. The lateral model we use consists of 60 cells each 0.3 m (1 ft) in extent. The tomography kernel $\mathbf{K}_s^t$ in this case is simple — it takes on one of two values depending on whether the path between source and

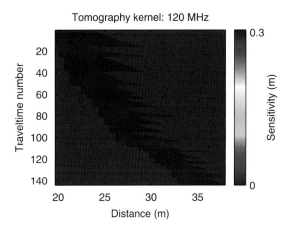

**Figure 5.** Tomography kernel for traveltimes measured at 120 MHz. The kernel takes on only two values, representing whether or not the path between source and receiver includes a cell.

receiver included a cell or not. The structure of $\mathbf{K}_s^t$ shows which source-receiver pairs had group traveltimes that satisfied certain selection criteria, which is why not all source-receiver pairs are evident in Figure 5.

For the inversion of equation 1, data covariance $\mathbf{C}_d$ and model covariance $\mathbf{C}_m$ matrices are chosen, as in Gerstoft et al. (2006). The data covariance matrix is assumed to be diagonal such that

$$\mathbf{C}_d = \sigma_d^2 \mathbf{I}, \qquad (2)$$

where $\mathbf{I}$ is the identity matrix and $\sigma_d$ is the data standard deviation. For simplicity, the standard deviation is assumed to be the same for all data points. The model covariance matrix has the form

$$\mathbf{C}_m(i, j) = \sigma_m^2 \exp(-|x_i - x_j|/\ell), \qquad (3)$$

where $\sigma_m$ is the model standard deviation, $x_i$ and $x_j$ are the lateral locations of the $i$th and $j$th cells, and $\ell$ is a smoothing length.

With the covariance matrices so chosen, the inversion proceeds by forming the augmented system of equations (Snieder and Trampert, 1999; Aster et al., 2004):

$$\begin{bmatrix} \mathbf{C}_d^{-1/2} \\ 0 \end{bmatrix} (\vec{t} - \mathbf{K}_s^t \vec{s}_0) = \begin{bmatrix} \mathbf{C}_d^{-1/2}\mathbf{K}_s^t \\ \mathbf{C}_m^{-1/2} \end{bmatrix} (\vec{s}_1 - \vec{s}_0), \qquad (4)$$

where $\vec{t}$ is the vector of group traveltimes, $\vec{s}_1$ is the inverted group-slowness model, and $\vec{s}_0$ is the initial guess. The augmented matrix-vector relation can be passed to a conjugate gradient solver — for instance, the implementation of sparse least squares, known as LSQR (Paige and Saunders,

1982). Because the relation in equation 1 is linear for this 1D geometry, the augmented matrix-vector relation is inverted once instead of being solved and updated iteratively, as for the case of nonlinear inversion.

## Forward modeling of GPR guided waves

Before discussing the inversion of group-velocity dispersion curves, the forward modeling of the modal properties of GPR guided waves needs to be described because it is not trivial. We have developed a forward-modeling code using a technique first applied in seismology for modeling Rayleigh waves (Lysmer, 1970) and recently extended to the inverse problem (Haney, 2009; Masterlark et al., 2010). The method has been referred to as the "thin-layer method" (TLM) in civil engineering (Kausel, 2005). A type of finite-element method, the thin-layer method takes its name from the requirement that individual elements (or layers) be much smaller than a wavelength to ensure numerical accuracy. As discussed later, we quantify the requirement of elements being much smaller than a wavelength in practice to mean that they are smaller than one-tenth of a wavelength.

In a Cartesian-coordinate system, GPR guided waves have the property of vanishing amplitude at infinite depth because past the critical angle, the wave in the substrate is evanescent. However, with a grid-based method, such as TLM, it is impossible to set the GPR guided waves to zero at an infinite depth. A good approximation assumes that the model extends sufficiently far in the vertical direction and sets the GPR guided waves to zero at the edges (Dirichlet condition). This approximation can be made without introducing significant error and is called the locked-mode approximation (Nolet et al., 1989). This approximation makes it possible to model GPR guided waves with high accuracy using TLM. Moreover, as we describe below, the main workhorse of TLM is an eigenvector and eigenvalue solver; such codes are widely available (e.g., *eigs* in MATLAB).

This is in contrast to the classical method of analyzing guided waves by finding the roots of a polynomial (Kausel, 2005). Kausel (2005, p. 19) points out that "in general, solving the transcendental eigenvalue problem ... for an arbitrarily layered system is very difficult." We avoid this difficulty here by applying TLM to the transverse electric case of electromagnetic wave propagation. We analyze the TE case because we process GPR data acquired in TE mode in a later section.

We used a model of GPR wave propagation that ignores material conductivity and takes permittivity $\epsilon$ and magnetic permeability $\mu$ to be dependent on depth: $\epsilon = \epsilon(z)$ and

$\mu = \mu(z)$. In this case, Maxwell's equations can be reduced to a single equation in terms of the transverse electric field component $e_y$ as

$$\epsilon \frac{\partial^2 e_y}{\partial t^2} = \frac{1}{\mu} \frac{\partial^2 e_y}{\partial x^2} + \frac{\partial}{\partial z}\left[\frac{1}{\mu}\frac{\partial e_y}{\partial z}\right]. \quad (5)$$

Note that no sources are present in equation 5 because we are considering normal mode solutions, such as guided waves, that exist without the presence of sources. Fourier transformation of equation 5 over time and lateral distance turns the partial differential equation into a real-valued, ordinary differential equation in terms of the Fourier transform of $e_y$ (denoted $E_y$) as

$$\epsilon \omega^2 E_y = \frac{k^2}{\mu} E_y - \frac{\partial}{\partial z}\left[\frac{1}{\mu}\frac{\partial E_y}{\partial z}\right]. \quad (6)$$

This equation can be shown to correspond directly with equations governing SH-component seismic waves (Carcione and Cavallini, 1995) and acoustic sound waves (Haney, 2009) in layered media.

We discretize the depth dependence of $E_y$ using a finite set of basis functions. This makes the ordinary differential equation into an algebraic equation. The transverse-electric field component $E_y$ is discretized as

$$E_y(z) = \sum_{W=1}^{N} E_y^W \phi_W(z). \quad (7)$$

Here, we use linear basis functions as in Lysmer (1970). For a nonuniform 1D spatial discretization with element thicknesses $h_W$ spanning the interval $[z_0, z_{N+1}]$, these basis functions are defined mathematically as

$$\phi_W(z) = \begin{cases} (z - z_{W-1})/h_{W-1} & \text{if } z_{W-1} \le z \le z_W, \\ (z_{W+1} - z)/h_W & \text{if } z_W \le z \le z_{W+1}, \\ 0 & \text{otherwise.} \end{cases} \quad (8)$$

We expand the depth-dependent material properties in terms of normalized boxcar functions $\Pi$

$$\epsilon(z) = \sum_{W=1}^{N+1} \epsilon_W \Pi_W(z). \quad (9)$$

$$\mu(z) = \sum_{W=1}^{N+1} \mu_W \Pi_W(z). \quad (10)$$

The normalized boxcar function is defined as

$$\Pi_W(z) = \begin{cases} 1 & \text{if } z_{W-1} \le z \le z_W, \\ 0 & \text{otherwise.} \end{cases} \quad (11)$$

Figure 6 shows the linear and boxcar functions graphically. Note that there are $N + 1$ elements and $N + 2$ nodes in the 1D finite-element mesh. However, the nodes at the upper and lower edges, $z_0$ and $z_{N+1}$, respectively, are fixed to be zero and therefore are not included in the vector of unknown nodal displacements. This yields a total of $N$ nodes for the numerical model.

An important difference between GPR waves and seismic waves evident in Figure 6 is that the numerical model for GPR waves includes nodes and elements in the air layer above the earth. In the seismic case, the earth's surface is treated as a stress-free surface, and no energy is exchanged between the earth and the atmosphere. This means that the numerical model for the GPR case must be approximately twice as large as for seismic waves. Although the air layer must be included for forward modeling GPR guided waves, the inversion step does not attempt to improve the model in the air layer, as described in a later section.

The Galerkin formulation of the finite-element method seeks to minimize the weighted average error induced by the incompleteness of the finite set of basis functions in equation 8. The Galerkin method weighs the errors with the same basis functions used in the approximation of $E_y$. This gives the following relation for the $W$th element located within the interval $z_W$ to $z_{W+1}$:

$$\int_{z_W}^{z_{W+1}} \left( \rho \omega^2 E_y - \frac{k^2}{\mu} E_y + \frac{\partial}{\partial z}\left[ \frac{1}{\mu} \frac{\partial E_y}{\partial z} \right] \right) \phi_J(z)\, dz = 0, \quad (12)$$

where $J = W$ and $W + 1$. Inserting equations 7, 9, and 10 into equation 12 leads to a matrix equation for the unknown coefficients $E_y^W$ and $E_y^{W+1}$. Note that integration by parts is used in the evaluation of the last term in the integrand of equation 12. For an interior element (one not at the upper or lower edge of the model), this boundary term gives the lateral, in-plane component of the magnetic field $H_x$ at the upper and lower edges of the element. Assuming $H_x$ and $E_y$ are continuous from element to element, the elemental matrices resulting from equation 12 are assembled over the entire model to form the complete matrices. This process is described in Appendix A and is illustrated in Figure A-1; it also is discussed by Lysmer (1970). For exterior elements, the vanishing Dirichlet boundary condition at the top and bottom of the model removes the boundary terms resulting from the integration by parts.

By organizing the vector of unknown nodal displacements with increasing depth as

$$\vec{v} = \begin{bmatrix} \dots & E_y^{W-1} & E_y^W & E_y^{W+1} & \dots \end{bmatrix}^T, \quad (13)$$

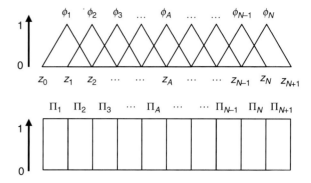

**Figure 6.** Linear and boxcar basis functions used for discretizing the eigenfunctions (linear) and material properties (boxcar). There are $N + 1$ total elements in the model and $N$ nodal points, with $A$ nodal points located in the air.

the complete system resulting from equation 12 gives a generalized linear eigenvalue problem in terms of the squared wavenumber $k^2$:

$$(k^2 \mathbf{B}_2 + \mathbf{B}_0)\vec{v} = \omega^2 \mathbf{M}\vec{v}. \quad (14)$$

The stiffness matrices $\mathbf{B}_2$ and $\mathbf{B}_0$ are dependent only on the magnetic permeability $\mu$, whereas the mass matrix $\mathbf{M}$ depends only on permittivity $\epsilon$. These matrices are discussed further in Appendix A. All three matrices are real valued and symmetric (Lysmer, 1970), a property that takes on an important role in the development of the inverse problem. Note that the adopted terminology referring to *mass* and *stiffness* matrices is a holdover from studies of elasticity in finite-element modeling. Once equation 14 has been solved for the eigenvalue and eigenvector corresponding to the fundamental mode, the group velocity $u$ can be calculated as

$$u = \frac{\delta \omega}{\delta k} = \frac{k \vec{v}^t \mathbf{B}_2 \vec{v}}{\omega \vec{v}^t \mathbf{M}\vec{v}}. \quad (15)$$

Appendix B gives a derivation of this equation.

We expect to observe the fundamental mode in studies of GPR guided waves. Efficiently finding the eigenvalue and eigenvector of the fundamental mode takes care in using the method described above. For fixed frequency, the fundamental mode corresponds to the largest eigenvalue in equation 14. This can be seen from the relation $k = \omega/c$: The fundamental mode has the lowest phase velocity and thus the largest value of $k$. One approach for a finite-element grid with $N$ nodes would be to solve for all $N$ eigenvalues and eigenvectors of equation 14 and search for the largest eigenvalue. In the interest of computational speed, it would be ideal if instead the eigenvalue/eigenvector solver could be asked to find only the fundamental mode. (In MATLAB, this can be accomplished using the function *eigs*, a solver based on the ARPACK linear solver [Lehoucq et al., 1998]). This solver can find an eigenvalue (or group of eigenvalues) closest to a particular value. Because the fundamental mode corresponds to the largest $k$ eigenvalue, this feature can be used once an upper bound on the fundamental mode eigenvalue is known.

The upper bound can be obtained by simply selecting the minimum value for wave speed in the model $c_{low}$. This gives an upper bound on the wavenumber, $k_{up} = \omega/c_{low}$. Asking the eigensolver to find the closest eigenvalue/eigenvector to this upper bound results in the eigensolver returning the fundamental mode. Assuming a linear scaling for finding any number of eigenvalues, this approach offers a factor of $N$ speedup compared to finding the maximum after calculating all eigenvalues and eigenvectors. Such a speedup is important given that the calculation of the eigenvalues/eigenvectors is the main workhorse for the inverse problem.

## Accuracy

The final consideration for forward modeling is one of accuracy. There are two factors important for the finite-element method described here: the extent of the model in the vertical direction $L$ away from the waveguide and the thickness of the elements $h_k$. The model must extend sufficiently far into the air layer and sufficiently deep into the subsurface such that the Dirichlet boundary conditions at the top and bottom of the model reasonably approximate the vanishing condition at infinity. This requires that the eigenvector of the fundamental mode be small at the upper and lower edges of the model. Lysmer (1970) discusses this issue and, given an estimate of the maximum desired wavelength for a model $\lambda_{max}$, the condition

$$L > \lambda_{max} \tag{16}$$

is sufficient. Regarding the thickness of the elements, the principle guiding accuracy is simply one of sufficiently sampling the eigenvector, similar to dispersion considerations for time-domain wave-propagation algorithms. A good rule of thumb, given a minimum desired wavelength $\lambda_{min}$, is

$$\lambda_{min} > 10 h_{max}. \tag{17}$$

In Figure 7, an example of the GPR guided-wave forward-modeling code is shown for a smooth model whose highest permittivities (lowest velocities) are located at the surface. The velocity model is plotted in Figure 7b. The smooth model is in fact used as the initial guess for the inversion of the GPR field data described in a later

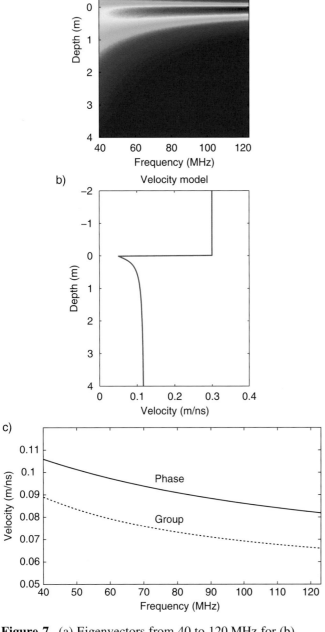

**Figure 7.** (a) Eigenvectors from 40 to 120 MHz for (b) velocity model with a high-permittivity surface layer. (c) Phase and group-velocity dispersion curves computed by the forward-modeling method described in the text.

section. Figure 7a shows all the eigenvectors modeled over the frequency range from 40 to 120 MHz. Note that the eigenvectors decay to a value close to zero near the edges of the model, at depths of $-2$ m (in the air layer) and 4 m. This ensures that the modeling adequately emulates a whole space, one in which the eigenvectors decay exponentially away from the waveguide. Figure 7c shows the modeled phase and group velocities over the frequency range, with lower velocities occurring at higher frequencies.

## Inversion of Group-velocity Dispersion Curves

The matrix-vector formulation of the forward problem for GPR guided waves in the previous section is well suited for developing the inverse problem using straightforward perturbation theory. As shown in Appendix C, the perturbation in phase velocity arising from perturbations in permittivity at fixed frequency is given by

$$\frac{\delta c}{c} = -\frac{\omega^2}{2k^2 \vec{v}^T \mathbf{B}_2 \vec{v}} \sum_{i=1}^{N+1} \vec{v}^T \frac{\partial \mathbf{M}}{\partial \epsilon_i} \vec{v} \delta \epsilon_i. \quad (18)$$

We have derived the inverse relation in equation 18 assuming only perturbations in permittivity; that is, we take the magnetic permeability in the subsurface to be the same as the initial guess. Note that for the application discussed here, the eigenvector $\vec{v}$ and wavenumber $k$ correspond to the fundamental mode; however, equation 18 applies individually to each mode and therefore is useful for the inversion of higher modes. The matrices appearing in equation 18 are the same as those in the forward problem, equation 14. Thus, the connection between the forward and inverse problems is clear using the matrix-vector notation.

Evaluated over many frequencies, equation 18 results in a linear matrix-vector relation between the perturbed phase velocities and the perturbations in permittivity, such that

$$\frac{\vec{\delta c}}{c} = \mathbf{K}_\epsilon^c \frac{\vec{\delta \epsilon}}{\epsilon}, \quad (19)$$

where $\mathbf{K}_\epsilon^c$ is the phase-velocity kernel for permittivity. Equation 19 is the linearized relation between phase velocity and permittivity. Because we measure group velocities, we require the linearized relation between group velocity and permittivity. The sensitivity kernel for group velocities is related to the phase-velocity kernel (Rodi et al., 1975) as

$$\mathbf{K}_\epsilon^u = \mathbf{K}_\epsilon^c + \frac{u\omega}{c} \frac{\partial \mathbf{K}_\epsilon^c}{\partial \omega}. \quad (20)$$

A linear relation similar to equation 19 thus can be set up for group velocity, as

$$\frac{\vec{\delta u}}{u} = \mathbf{K}_\epsilon^u \frac{\vec{\delta \epsilon}}{\epsilon}, \quad (21)$$

which is the basis for group-velocity inversion. For absolute perturbations (e.g., $\vec{\delta \epsilon}$) instead of relative perturbations (e.g., $\frac{\vec{\delta \epsilon}}{\epsilon}$), the kernel in equation 21 can be replaced with

$$\mathbf{G} = \mathrm{diag}(\vec{u}) \mathbf{K}_\epsilon^u \mathrm{diag}(\vec{\epsilon})^{-1}, \quad (22)$$

where $\mathrm{diag}(\vec{u})$ is a matrix with the vector $\vec{u}$ placed on the main diagonal and off-diagonal entries equal to zero.

### The air layer

A final detail of the inversion concerns perturbations in the air layer. Recall that the air layer is needed to accurately forward-model GPR guided waves. However, the properties of the air layer are well known ($\epsilon = 1$) and do not need to be taken into account in the inversion. Thus, kernel $\mathbf{G}$ and model $\vec{\delta \epsilon}$ in the linearized relation can be broken into their proportions within the air layer and subsurface as

$$\vec{\delta u} = \begin{bmatrix} \mathbf{G}^A & \mathbf{G}^E \end{bmatrix} \begin{bmatrix} \vec{\delta \epsilon}^A \\ \vec{\delta \epsilon}^E \end{bmatrix}. \quad (23)$$

Assuming no perturbations in the air layer ($\vec{\delta \epsilon}^A = 0$) yields

$$\vec{\delta u} = \mathbf{G}^E \vec{\delta \epsilon}^E, \quad (24)$$

which is the relation used to invert for subsurface permittivity from group-velocity dispersion curves.

The inversion of group-velocity dispersion curves uses data covariance and model covariance matrices $\mathbf{C}_d$ and $\mathbf{C}_m$, respectively, with the same form as for the group-velocity tomography, as shown in equations 2 and 3. With the covariance matrices so chosen, depth inversion proceeds using the algorithm of total inversion (Tarantola and Valette, 1982; Muyzert, 2007). We use an iterative algorithm for depth inversion because unlike group-velocity tomography, it is a nonlinear inverse problem. The $n$th model update $\vec{\epsilon}_n^E$ is calculated by forming the augmented system of equations (Snieder and Trampert, 1999; Aster et al., 2004), such that

$$\begin{bmatrix} \mathbf{C}_d^{-1/2} \\ 0 \end{bmatrix} (\vec{u}_0 - f(\vec{\epsilon}_{n-1}^E) + \mathbf{G}^E(\vec{\epsilon}_{n-1}^E - \vec{\epsilon}_0^E))$$

$$= \begin{bmatrix} \mathbf{C}_d^{-1/2}\mathbf{G}^E \\ \mathbf{C}_m^{-1/2} \end{bmatrix} (\vec{\epsilon}_n^E - \vec{\epsilon}_0^E), \qquad (25)$$

where $\vec{u}_0$ is the group-velocity data, $f$ is the (nonlinear) forward-modeling operator, and $n$ ranges from 1 to the value at which the stopping criterion is met or the maximum allowed number of iterations is reached. The stopping criterion used here is (Gouveia and Scales, 1998):

$$(f(\vec{\epsilon}_n^E) - \vec{u}_0)\mathbf{C}_d^{-1}(f(\vec{\epsilon}_n^E) - \vec{u}_0) \leq F, \qquad (26)$$

where $F$ is the number of measurements of local group velocity (number of frequencies where the group velocity has been measured). The augmented matrix-vector relation can be passed to a conjugate gradient solver, for instance LSQR. The inversion given by equation 25 then is iterated to convergence, except for the case when the maximum allowed number of iterations is reached.

## Case Study: Kuparuk River, Alaska

We analyzed a GPR data set collected over two stream channels in August 2005 on the North Slope of Alaska, near the Kuparuk River (see Figure 8). Beneath the streambeds, an area of thaw that varies seasonally and spatially in depth overlies permafrost (Figure 9). In addition to the two active stream channels, the site consists of a gravel bar to the west and a peat layer to the east. An estimate of the depth to the permafrost is known a priori to be deeper beneath the gravel bar and gradually shallowing toward the peat layer (Figure 9). We could not obtain a clearly interpretable image of the permafrost or reach the permafrost with a probe beneath the stream or on the gravel bank. As a result, the sketch is based on analogy with similar sites (Brosten et al., 2009a; Brosten et al., 2009b). The relatively high permittivity of the thawed moist gravel and peat were amenable to guided-wave propagation. The active stream channels also had the potential to act as waveguides. The GPR moveout survey was acquired using the TE configuration with acquisition parameters described in an earlier section. Each of the 62 shots in the survey had a maximum of 48 receivers; a diagram showing the source-receiver layout is given in Figure 10.

Guided waves occurred only from 20 to 38 m (Figure 10). This range was determined by interactively analyzing selected energy diagrams, such as those in Figures 3 and 4, over the entire survey. Over this range, we applied an automated procedure for picking group traveltimes, performing group-velocity tomography, and inverting local dispersion curves for depth structure. The automated process began by eliminating all traces with clipped amplitude. Clipping mostly affected the near-offset traces. We then scanned over all source-receiver pairs and formed

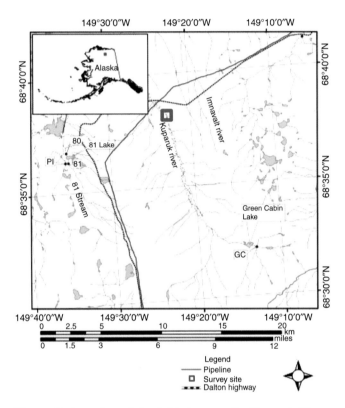

**Figure 8.** Location of the field site near the Kuparuk River, Alaska. After Brosten et al. (2006). Used by permission.

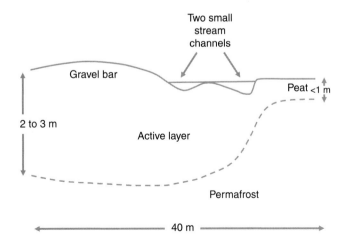

**Figure 9.** Conceptual sketch of the shallow structure and features at the field site.

energy diagrams using FTAN with frequencies from 20 to 120 MHz. For a single energy diagram from a source-receiver pair, we selected the maximum value in the energy diagram (the maximum the 2D surface shown in color in Figure 3b). From this point, the maximum at each frequency was found in the increasing and decreasing frequency direction away from the global maximum until the amplitude of the maxima was one-fourth of the global maximum value. This defines a possible dispersion curve, but to be acceptable, the dispersion curve must satisfy the following criteria:

1) The derivative of the dispersion curve with frequency never exceeds 0.002 m/ns/MHz.
2) The dispersion curve extends over at least 40 MHz.
3) The mean of the dispersion curve is less than 0.08 m/ns and greater than 0.04 m/s.

The dispersion curve then was transformed to group traveltime as a function of frequency and was saved for eventual input into group-traveltime tomography. The criteria reflect the properties of a desired dispersion curve: It is relatively smooth, extends over a broad frequency band, and has group velocities similar to those commonly observed over the entire survey during interactive data analysis. The group traveltimes are assumed to reflect the fundamental-mode GPR guided wave.

Group-velocity tomography requires a complete group-traveltime table as a function of source, receiver, and frequency. We performed the tomography from 20 to 120 MHz in steps of 1 MHz. Detailed parameters for the inversion are given in Table 1. The initial guess for the group-velocity structure was taken to be laterally homogeneous with a value equal to the average of all group-velocity measurements (computed from group traveltime) at a particular frequency. The model was broken into 60 cells laterally, each with a length of 0.3 m (1 ft). We chose the length of the cells based on the assumption that we could not resolve velocity variations on a scale smaller than the receiver spacing of the survey (1 ft). The output of the tomography is the pseudodepth section shown in Figure 11a — a matrix of local group velocity at all lateral locations as a function of frequency. By plotting the matrix with high frequencies on top, as in Figure 11a, the depthlike qualities of the section stand out because high frequencies are sensitive to shallower structure.

Because each vertical slice of the pseudodepth section represents a local group-velocity dispersion curve, a full 2D image of the relative permittivity in the subsurface is obtained by inverting all the dispersion curves. Figure 11b shows the predicted pseudodepth section based on the full 2D image in Figure 12. Excellent agreement, up to the

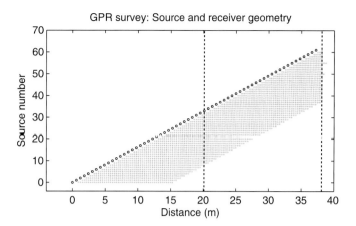

**Figure 10.** Acquisition geometry for the GPR survey. Source locations are open circles, with gray dots as receiver locations. Guided waves were excited between the vertical dotted lines.

**Table 1.** Parameters for the group-velocity tomography. The total number of traveltimes varies with frequency, from 106 at 20 MHz to 302 at 55 MHz.

| Parameter | Value |
|---|---|
| A priori data standard deviation $\sigma_d$ | 8 ns |
| A priori model standard deviation $\sigma_m$ | 1 ns/m |
| Smoothing length $\ell$ | 3.048 m |
| Number of data points | 106 to 302 |
| Number of cells | 60 |
| Lateral extent of model | $x = 19.8$ m to $x = 37.8$ m |
| Lateral cell length | .3048 m |

expected noise, is seen to exist between the two pseudo-depth sections in Figure 11. The detailed parameters for the depth-inversion step are given in Table 2. The initial guess for the permittivity structure with depth is the same smooth model shown in Figure 7b. The 2D image of relative permittivity in Figure 12 demonstrates that the GPR guided waves sense structure to depths of 1 to 2 m in the subsurface. Significant 2D structure is evident in the shallow parts of the permittivity model, with relative permittivities exceeding 50.

## Discussion

From the 2D permittivity profile shown in Figure 12, we observe the strongest waveguide structure at depths of less than 0.5 m and between lateral distances of 25 and 30 m. Based on the sketch of the survey area shown in Figure 9, this strong waveguide is interpreted to be the easternmost active stream channel. Relative permittivities in excess of five are on the same order of magnitude as the permittivity of 80 expected for freshwater (Davis and Annan, 1989). The lateral velocity contrast at a distance of approximately 30 m represents the transition from the active stream channel to the peat layer. Whereas water is the dominant waveguide above the stream channel, we interpret the moist peat above a shallow permafrost layer to be responsible for the guided waves to the east of the stream channel.

We add to the interpretation by analyzing a stacked GPR reflection profile for the same data set, shown in Figure 13. Whereas the guided-wave profile exists only over half of the survey (where the guided waves exist), the GPR reflection profile extends over the entire survey. From Figure 13, the quality of the reflection profile away from the two stream channels is evident. Between distances

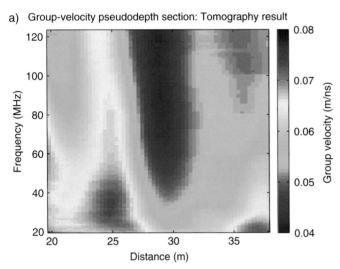

a) Group-velocity pseudodepth section: Tomography result

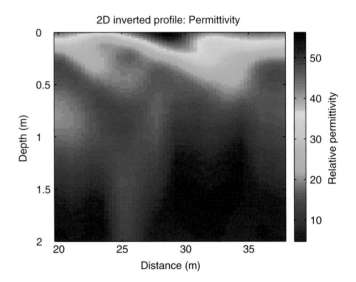

Figure 12. The 2D permittivity model obtained by inverting the dispersion curves shown in Figure 11a. The highest relative permittivities are associated with an active stream channel, and the structural change at a lateral distance of 30 m represents the transition from the stream channel to a peat layer. Laterally and vertically varying structure is resolved in the upper 1 to 2 m of the subsurface.

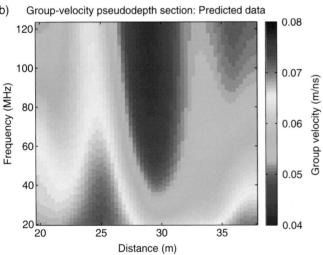

b) Group-velocity pseudodepth section: Predicted data

Figure 11. (a) The result of applying 1D group-velocity tomography in the lateral direction over the frequency band from 20 to 120 MHz. This pseudodepth section is a collection of local group-velocity dispersion curves that have had lateral propagation effects removed by tomography. (b) A comparison to data predicted by the inverted model in Figure 12. Excellent agreement exists between the tomography result and predicted data, to the degree of expected noise.

**Table 2.** Parameters for the dispersion-curve inversion.

| Parameter | Value |
|---|---|
| A priori data standard deviation $\sigma_d$ | 0.001 m/ns |
| A priori model standard deviation $\sigma_m$ | 1.8 |
| Smoothing length $\ell$ | 10 cm |
| Number of data points | 101 |
| Frequency interval of data | 20 to 120 MHz |
| Frequency spacing of data points | 1 MHz |
| Number of nodal points $N$ | 300 |
| Number of nodal points in the air layer $A$ | 100 |
| Depth extent of model | $z = -2$ m to $z = 4$ m |
| Grid spacing in depth $h$ | 2 cm |

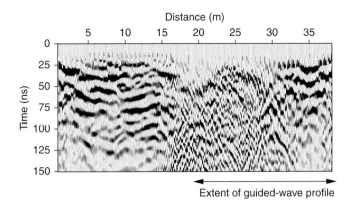

Distance (m)

Extent of guided-wave profile

**Figure 13.** A stacked GPR reflection profile for the same data as used for the guided-wave imaging.

of 17 and 30 m, where the stream channels exist, the presence of strong guided waves presents difficulties for reflection profiling. The guided waves are so strong that the stacking process inherent in reflection profiling is unable to dampen the amplitudes of the guided waves coming through the stack adequately. The result is a degraded profile beneath the stream channels, where high-amplitude and steeply dipping events crisscross the reflections from the layer boundaries. Therefore, the best areas for imaging with guided waves exist where the GPR reflection profile becomes most unreliable. In addition to the degraded quality of the reflection profile, the velocity "push-down" effect from the presence of a stream channel is evident at a time of approximately 50 ns for distances between 25 and 30 m.

The shallowest reflections observed in Figure 13 are still deeper than shallow depths ($< 0.5$ m) imaged by the guided waves shown in Figure 12. Therefore, the guided waves can resolve shallow structure within a wavelength of the surface, which is not possible to resolve in a GPR reflection profile. Knowledge of the shallow structure should be valuable for applying statics corrections to the GPR reflection data. However, we have demonstrated the value of analysis based on guided waves in its own right for mapping shallow 2D permittivity structure.

## Conclusions

The inversion of guided-wave group velocities for a 2D image of subsurface permittivity structure provides a starting point for further analysis of guided-wave properties. Because the GPR data we analyze are sampled densely in offset, phase velocities of the GPR guided waves should provide additional constraints on subsurface

structure. Phase velocities, once measured, can be combined with group velocities in a joint inversion. Full-waveform GPR modeling, such as time-domain finite-difference modeling, should be performed using the 2D permittivity image to gauge how well the model predicts other wave types, particularly the direct wave and any wide-angle refracted waves. Accuracy of the 2D permittivity image also can be tested in the field using the invasive method, which involves pressing a metal probe into the ground until it meets resistance from the frozen layer.

We plan to develop further the analysis of these guided waves to account for higher modes, conductivity, and TM configuration. The inclusion of higher modes improves the depth resolution of guided-wave inversions because higher modes have a greater depth of penetration. Producing 2D conductivity profiles, in addition to permittivity profiles, allows for better diagnostic tests to distinguish among different material types. Joint inversions of GPR data acquired in both TE and TM configurations have been shown to improve the uniqueness of the inverted models.

A major challenge for future research will focus on the inversion of leaky waves in addition to guided waves. The ability to treat leaky and guided waves on an equal footing would make the entire process more robust and model independent. In addition, our methodology can be extended easily to wide-azimuth GPR surveys to invert for 3D permittivity structure. Such an extension simply relies on the use of 2D tomography for sources and receivers over an area instead of 1D tomography for sources and receivers along a line. We have presented and applied successfully a methodology for imaging shallow permittivity structure based on guided waves in GPR data. The methodology is an extension of group-velocity techniques that have been developed within seismology for the analysis of Rayleigh and Love waves.

Although the extension is straightforward, some aspects (such as the importance of the air layer and the presence of leaky waves) are unique to GPR waves. The existence of guided waves allows more information to be derived from the recorded GPR wavefield, particularly information on shallow structure within a wavelength of the surface. This is an important zone because it often includes the interface between sediments and permafrost and is subject to strong seasonal variability.

## Acknowledgments

We appreciate the constructive comments of two anonymous reviewers.

# Appendix A

# Mass and Stiffness Matrices

Matrices $\mathbf{B}_2$, $\mathbf{B}_0$, and $\mathbf{M}$ (appearing in equation 14) are understood best as being assembled from fundamental $2 \times 2$ matrices known as elemental matrices. The elemental matrices are added together recursively in the manner shown in Figure A-1 to give the full matrices. For instance, from the discussion of Lysmer (1970), the elemental mass matrix associated with the $W$th element $\tilde{\mathbf{M}}_W$ is

$$\tilde{\mathbf{M}}_W = h_W \begin{bmatrix} \epsilon_W/3 & \epsilon_W/6 \\ \epsilon_W/6 & \epsilon_W/3 \end{bmatrix}. \tag{A-1}$$

The process known as "mass lumping" replaces this matrix by a diagonal matrix whose entries are equal to the row sum.

$$\tilde{\mathbf{M}}_W^L = h_W \begin{bmatrix} \epsilon_W/2 & 0 \\ 0 & \epsilon_W/2 \end{bmatrix}. \tag{A-2}$$

The full mass matrix $\mathbf{M}$ is assembled from this $2 \times 2$ matrix by adding individual $2 \times 2$ matrices in the manner shown in Figure A-1. A similar procedure applies for the stiffness matrices $\mathbf{B}_0$, and $\mathbf{B}_2$, although the $2 \times 2$ matrices in these cases are not lumped prior to assembly as in the case of mass matrix $\mathbf{M}$. Once these matrices are constructed, the phase velocity, group velocity, and eigenvector shapes of the guided waves can be forward-modeled by solving the eigenvalue/eigenvector problem in equation 14.

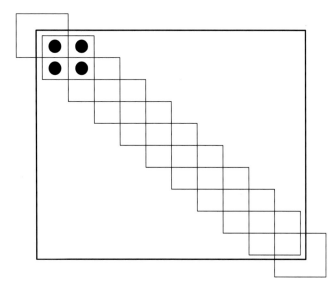

**Figure A-1.** The assembly of the complete mass and stiffness matrices from the elemental matrices. The individual $2 \times 2$ elemental matrices are added recursively to form the complete $N \times N$ matrices. The entries of the elemental matrices at the top and bottom edges, which extend past the complete matrix, are not used. Lysmer (1970) and Kausel (2005) discuss this procedure in the seismic case for Rayleigh waves.

# Appendix B

# Group Velocity

As shown by Lysmer (1970), the group velocity at a single frequency can be obtained without numerical differentiation of the phase-velocity dispersion curve once the phase velocity and eigenfunction are known. This result is given here but also, because it provides an introduction to perturbational techniques, it is developed further in Appendix C for the inverse problem.

From equation 14, the generalized quadratic eigenvalue problem for Rayleigh waves can be written as

$$(\mathbf{B}_k - \omega^2 \mathbf{M})\vec{v} = 0, \tag{B-1}$$

where $\mathbf{B}_k = k^2 \mathbf{B}_2 + \mathbf{B}_0$. To find an expression for group velocity, we perturb the wavenumber $k$ and frequency $\omega$ while keeping the material properties constant. This leads to the following perturbed equation:

$$\left( \mathbf{B}_k + \frac{\partial \mathbf{B}_k}{\partial k} \delta k - (\omega^2 + \delta\omega^2)\mathbf{M} \right)(\vec{v} + \delta\vec{v}) = 0. \tag{B-2}$$

Given the equality in equation B-1, this perturbed equation gives, to first order,

$$(\mathbf{B}_k - \omega^2 \mathbf{M})\delta\vec{v} + \frac{\partial \mathbf{B}_k}{\partial k} \delta k \vec{v} - \delta\omega^2 \mathbf{M}\vec{v} = 0. \tag{B-3}$$

We now multiply equation B-3 from the left by $\vec{v}^T$. The first term on the left-hand side of equation B-3 vanishes because $\mathbf{B}_k$ and $\mathbf{M}$ are both symmetric matrices and as a result yields

$$\delta\omega^2 \vec{v}^T \mathbf{M}\vec{v} = \vec{v}^T \frac{\partial \mathbf{B}_k}{\partial k} \delta k \vec{v}. \tag{B-4}$$

Given that $\delta\omega^2 = 2\omega\delta\omega$ and $\partial \mathbf{B}_k / \partial k = 2k\mathbf{B}_2$, an expression for the group velocity $u$ is

$$u = \frac{\delta\omega}{\delta k} = \frac{k\vec{v}^T \mathbf{B}_2 \vec{v}}{\omega\vec{v}^T \mathbf{M}\vec{v}}. \tag{B-5}$$

This is equation 15 in the text. It allows the group velocity $u$ to be computed once the eigenvalue/eigenvector problem in equation 14 has been solved.

# Appendix C

# Perturbation Theory

In Appendix B, we perturbed wavenumber and frequency while keeping the material properties constant to obtain an expression for the group velocity. Here, we perturb the wavenumber and material properties and fix the frequency. This approach leads to a first-order result relating perturbations in phase velocity to perturbations in the material properties. Such a formula forms the basis for the inversion of Rayleigh-wave dispersion curves.

The perturbation of permittivity and wavenumber in equation 14 is expressed as

$$((k + \delta k)^2 \mathbf{B}_2 + \mathbf{B}_0)(\vec{v} + \delta \vec{v})$$
$$= \omega^2 \left[ \mathbf{M} + \sum_{i=1}^{N+1} \frac{\partial \mathbf{M}}{\partial \epsilon_i} \delta \epsilon_i \right] (\vec{v} + \delta \vec{v}). \quad \text{(C-1)}$$

Because $\mathbf{B}_2$, $\mathbf{B}_0$, and $\mathbf{M}$ are symmetric, the following identity is valid:

$$\vec{v}^T (k^2 \mathbf{B}_2 + \mathbf{B}_0) \delta \vec{v} = \vec{v}^T \omega^2 \mathbf{M} \delta \vec{v}. \quad \text{(C-2)}$$

Substituting this relation along with equation 14 into equation C-1 gives, to first order,

$$2k \delta k \vec{v}^T \mathbf{B}_2 \vec{v} = \omega^2 \vec{v}^T \left[ \sum_{i=1}^{N+1} \frac{\partial \mathbf{M}}{\partial \epsilon_i} \delta \epsilon_i \right] \vec{v}. \quad \text{(C-3)}$$

Using the fact that $\delta k/k = -\delta c/c$, this first-order result leads to equation 18, the fundamental relation leading to the inversion of group-velocity dispersion curves.

# References

Abbott, R. E., L. C. Bartel, B. P. Engler, and S. Pullammanappallil, 2006, Surface-wave and refraction tomography at the FACT Site, Sandia National Laboratories, Albuquerque, New Mexico: Sandia National Laboratories Technical Report SAND2006Ð5098.

Aki, K., and P. G. Richards, 1980, Quantitative seismology: W. H. Freeman and Co.

Annan, A. P., 1973, Radio interferometry depth sounding: Part I — Theoretical discussion: Geophysics, **38**, 557–580.

Arcone, S. A., 1984, Field observations of electromagnetic pulse propagation in dielectric slabs: Geophysics, **49**, 1763–1773.

Arcone, S. A., A. J. Delaney, and P. R. Peapples, 2003a, GPR pulse attenuation in a fine-grained and partially contaminated formation: Journal of Environmental and Engineering Geophysics, **8**, 57–66.

Arcone, S. A., P. R. Peapples, and L. Liu, 2003b, Propagation of a ground-penetrating radar (GPR) pulse in a thin-surface waveguide: Geophysics, **68**, 1922–1933.

Aster, R., B. Borchers, and C. Thurber, 2004, Parameter estimation and inverse problems: Elsevier.

Becht, A., E. Appel, and P. Dietrich, 2006, Analysis of multi-offset GPR data: A case study in a coarse-grained aquifer: Near Surface Geophysics, **4**, 227–240.

Brosten, T. R., J. H. Bradford, J. P. McNamara, M. N. Gooseff, J. P. Zarnetske, W. B. Bowden, and M. E. Greenwald-Johnston, 2009a, Estimating 3D variation in active-layer thickness beneath arctic streams using ground-penetrating radar: Journal of Hydrology, **373**, 479–486.

Brosten, T. R., J. H. Bradford, J. P. McNamara, M. N. Gooseff, J. P. Zarnetske, W. B. Bowden, and M. E. Johnston, 2009b, Multi-offset GPR methods for hyporheic zone investigations: Near Surface Geophysics, **7**, 247–257.

Brosten, T. R., J. H. Bradford, J. P. McNamara, J. P. Zarnetske, M. N. Gooseff, and W. B. Bowden, 2006, Profiles of temporal thaw depths beneath two Arctic stream types using ground-penetrating radar: Permafrost and Periglacial Processes, **17**, 341–455.

Bussat, S. and S. Kugler, 2009, Recording noise — Estimating shear-wave velocities: Feasibility of offshore ambient-noise surface-wave tomography (answt) on a reservoir scale: 79th Annual International Meeting, SEG, Expanded Abstracts, 1627–1631.

Carcione, J. M., and F. Cavallini, 1995, On the acoustic-electromagnetic analogy: Wave Motion, **21**, 149–162.

Collin, R. E., 1960, Field theory of guided waves: McGraw-Hill.

Davis, J. L., and A. P. Annan, 1989, Ground-penetrating radar for high-resolution mapping of soil and rock stratigraphy: Geophysical Prospecting, **37**, 531–551.

Dorman, J., and M. Ewing, 1962, Numerical inversion of seismic surface wave dispersion data and crust-mantle structure in the New York–Pennsylvania area: Journal of Geophysical Research, **67**, 5227–5241.

Fisher, E., G. A. McMechan, and A. P. Annan, 1992, Acquisition and processing of wide-aperture ground-penetrating radar data: Geophysics, **57**, 495–504.

Gerstoft, P., K. G. Sabra, P. Roux, W. A. Kuperman, and M. C. Fehler, 2006, Green's functions extraction and surface-wave tomography from microseisms in southern California: Geophysics, **71**, no. 4, SI23–SI31.

Gouveia, W. P., and J. A. Scales, 1998, Bayesian seismic waveform inversion: Parameter estimation and uncertainty analysis: Journal of Geophysical Research, **103**, 2759–2779.

Haney, M. M., 2009, Infrasonic ambient noise interferometry from correlations of microbaroms: Geophysical Research Letters, **36**, L19808, doi:10.1029/2009GL040179.

Kausel, E., 2005, Wave propagation modes: From simple systems to layered soils: *in* C. G. Lai and K. Wilmanski, eds., Surface waves in geomechanics — Direct and inverse modelling for soil and rocks: CISM lecture notes: Springer-Verlag.

Lehoucq, R. B., D. C. Sorensen, and C. Yang, 1998, ARPACK users' guide: Solution of large scale eigenvalue problems with implicitly restarted Arnoldi methods: SIAM, Philadelphia.

Liu, L., and S. A. Arcone, 2003, Numerical simulation of the wave-guide effect of the near-surface thin layer on radar wave propagation: Journal of Environmental and Engineering Geophysics, **8**, 53–61.

Lysmer, J., 1970, Lumped mass method for Rayleigh waves: Bulletin of the Seismological Society of America, **60**, 89–104.

Masterlark, T., M. Haney, H. Dickinson, T. Fournier, and C. Searcy, 2010, Rheologic and structural controls on the deformation of Okmok volcano, Alaska: FEMs, InSAR, and ambient noise tomography: Journal of Geophysical Research, 115, B02409, doi:10.1029/2009JB006324.

Muyzert, E., 2007, Seabed property estimation from ambient-noise recordings: Part 2 — Scholte-wave spectral-ratio inversion: Geophysics, **74**, no. 4, U47–U53.

Nolet, G., R. Sleeman, V. Nijhof, and B. L. N. Kennett, 1989, Synthetic reflection seismograms in three dimensions by a locked-mode approximation: Geophysics, **54**, 350–358.

Paige C. C., and M. A. Saunders, 1982, LSQR: An algorithm for sparse linear equations and sparse least squares: ACM Transactions on Mathematical Software, **8**, 43–71.

Park, C. B., R. D. Miller, and J. Xia, 1999, Multichannel analysis of surface waves: Geophysics, **64**, 800–808.

Ritzwoller, M. H., and A. L. Levshin, 2002, Estimating shallow shear velocities with marine multicomponent seismic data: Geophysics, **67**, 1991–2004.

Rodi, W. L., P. Glover, T. M. C. Li. and S. S. Alexander, 1975, A fast, accurate method for computing group-velocity partial derivatives for Rayleigh and Love modes: Bulletin of the Seismological Society of America, **65**, 1105–1114.

Socco, L. V., and C. Strobbia, 2004, Surface-wave method for near-surface characterization: A tutorial: Near Surface Geophysics, **2**, 165–185.

Snieder, R., and J. Trampert, 1999, Inverse problems in geophysics, *in* A. Wirgin, ed., Wavefield inversion: Springer Verlag, 119–190.

Strobbia C., and G. Cassiani, 2007, Multilayer ground-penetrating radar guided waves in shallow soil layers for estimating soil water content: Geophysics, **72**, no. 4, J17–J29.

Tarantola, A., and B. Valette, 1982, Generalized nonlinear inverse problems solved using the least squares criterion: Reviews of Geophysics and Space Physics, **20**, 219–232.

van der Kruk, J., S. A. Arcone, and L. Liu, 2007, Fundamental and higher mode inversion of dispersed GPR waves propagating in an ice layer: IEEE Transactions on Geoscience and Remote Sensing, **45**, 2483–2491.

van der Kruk, J., C. M. Steelman, A. L. Endres, and H. Vereecken, 2009, Dispersion inversion of electromagnetic pulse propagation within freezing and thawing soil waveguides: Geophysical Research Letters, **36**, L18503, doi:10.1029/2009GL039581.

van der Kruk, J., R. Streich, and A. G. Green, 2006, Properties of surface waveguides derived from separate and joint inversion of dispersive TE and TM GPR data: Geophysics, **71**, no. 1, K19–K29.

van Wijk, K., J. A. Scales, W. Navidi, and L. Tenorio, 2002, Data and model uncertainty estimation for linear inversion: Geophysical Journal International, **149**, 625–632.

Chapter 11

# Sensitivity Studies of Fundamental- and Higher-mode Rayleigh-wave Phase Velocities in Some Specific Near-surface Scenarios

Carlos Calderón-Macías[1] and Barbara Luke[2]

## Abstract

Insufficient low-frequency data can pose a problem when inverting fundamental-mode Rayleigh-wave phase velocities using two-channel (SASW) or multiple-channel (MASW) active-source methods. Depth of penetration and accuracy of the inverted models are particularly sensitive to the low frequencies. Ambiguities might be reduced with supplemental geologic or geophysical information about the near surface, such as passive-source seismic data and well information. Dispersion data from higher modes might complement the fundamental mode in resolving shear-wave velocities in some cases because higher modes are more sensitive to shear-wave velocities at greater depths. However, interpretation of higher-mode phase velocities poses a challenge because the higher modes are not always easy to separate from the fundamental mode and possibly from other scattered energy. The energy partition into fundamental and higher modes results in different frequency-dependent signal-to-noise ratios for the different modes. Furthermore, convergence of the solution's data error to the hypothetical global minimum appears to be more complex when fundamental and higher modes are included in the data error to be minimized. Some of the difficulties encountered when inverting higher-mode surface waves are investigated for some simplified shallow (<10-m) earth models of engineering significance: a sediment profile with a constant stiffness gradient, a low-velocity sediment overlying shallow bedrock, and a high-velocity layer overlying and underlying softer sediment layers.

## Introduction

Inversion of dispersion curves (DCs) that reflect phase velocity in terms of temporal frequency or wavelength has become a standard method for deriving shallow shear-wave velocity profiles (Socco and Strobbia, 2004; O'Neill, 2005). The most common simplification used in these inversions corresponds to describing the medium as a stack of horizontal isotropic layers of unknown shear-wave velocity. Under this simplification, the measured phase velocities, up to a first order, depend solely on the distribution of shear-wave velocities ($V_S$) with depth. It is common then to assume density and acoustic velocity (or Poisson's ratio) as known and to invert only for $V_S$ (Xia et al., 1999). Typically, estimates of phase velocity in terms of frequency are obtained from two-channel acquisition in the spectral-analysis-of-surface-waves (SASW) method or from multiple-channel common-source profiling in the multichannel-analysis-of-surface waves (MASW) method.

In SASW data acquisition (Stokoe et al., 1994), a cross-correlation between the observed responses results in phase-difference data that when scaled by the receiver separation result in a phase-velocity measurement. Applying this process for several receiver separations results in an effective dispersion curve. Forming a dispersion curve from the data involves fitting measurements with a relatively smooth curve that edits outliers and rapid variations possible from ambient and artificial noise. Surface-wave energy partitions into different propagating modes, with the fundamental mode being slower and faster propagating

[1]*ION Geophysical, GXT-Imaging Solutions, Houston, Texas, U.S.A.*
[2]*University of Nevada, Applied Geophysics Center, Las Vegas, Nevada, U.S.A.*

modes appearing at frequencies higher than the cutoff frequency $f_n$, with $n$ denoting the $n$th higher mode (Aki and Richards, 1980). The SASW method provides particularly high accuracy in the shallowest depths, resulting from the localized nature of the measurements involved in building the dispersion curve.

MASW acquisition and data processing (Park et al., 1999; Xia et al., 1999) increase the signal-to-noise ratio (S/N) by transforming a multichannel record with many traces from space-time domain to frequency slowness or frequency–phase velocity. Resulting maps are used for manual or semimanual interpretation of phase-velocity peaks. Resolution of the transformed maps is a key for unambiguous identification of modes. Certainty or uncertainty of the picked velocities depends on field-acquisition parameters such as nature and strength of the seismic source, receiver spacing, receiver coupling, minimum and maximum source-receiver offset available, and heterogeneity of the shallow subsurface being tested. Often, low frequencies ($< \sim 10$ Hz) are sampled poorly because of limitations of the receiver-array length and the seismic source-and-receiver instrument characteristics (Ivanov et al., 2008). Optimizing field-acquisition parameters and increasing resolution when processing surface waves by data sorting and stacking (Hayashi and Suzuki, 2004; Ivanov et al., 2005; Grandjean and Bitri, 2006; Neducza, 2007) are subjects of great interest because the phase-velocity estimates correspond to the data that will be used to derive the $V_S$ profile.

Inversion of phase-velocity measurements is a nonlinear problem in which the measurements are fitted with synthetically computed phase velocities, through local minimization (Xia et al., 1999; Rix and Lai, 2007) and/or global minimization (Beaty et al., 2002; Calderón-Macías and Luke, 2007). The inversion requires some form of a priori information such as a reference model or expected $V_S$ ranges. Some form of regularization such as model smoothness (Rix and Lai, 2007) is important to prevent large deviations of the model parameters from a reference in the less constrained parts of the model, usually the deeper layers, and for avoiding unrealistic velocity contrasts.

Depth of penetration and accuracy of the inverted models at depth are particularly sensitive to the low frequencies (e.g., Casto et al., 2009). Low-frequency data can be collected in a surface-wave survey by using high-energy sources and extremely long arrays (e.g., Rosenblad et al., 2007; Luke et al., 2010), but the costs of such surveys are considerable. Alternatively, additional information about the near surface can help to resolve ambiguities in the inversion of phase velocities and simultaneously to improve resolution at depth. For example, active-source surface-wave measurements can be combined with seismic-refraction data (Ivanov et al., 2006) and/or passive-source surface-wave data (Luke et al., 2008) as well as lithologic data. Dispersion data from higher modes also might complement the fundamental mode in resolving shear-wave velocities in some cases because higher modes are more sensitive to shear-wave velocities at greater depths (e.g., Xia et al., 2003). However, interpretation of higher-mode phase velocities poses a challenge because the modes are not always easy to identify and to separate from one another and from other scattered energy.

We study the incorporation of higher modes for improving the resolution of inverted shear-wave velocities. We concentrate on three model scenarios: a sediment model in which velocity increases linearly with depth, a two-layer model representing sediment over hard bedrock, and a profile that has a high-velocity layer overlying and underlying slower sediments (Table 1). We first perform sensitivity studies on these three profiles to observe the relative sensitivity among propagating modes when the velocities of the layers are varied. We follow with an analysis of 2D error surfaces that result when two model parameters are varied systematically, for isolated modes and then when the modes are added together. Then we perform 2D elastic finite-difference (FD) modeling of the three models with the purpose of comparing the amplitudes of the spectra for the different propagating modes.

Finally, we provide an example in which we perform a staged linearized inversion for a field data set. The general intent is to gain understanding of the benefits and pitfalls of inverting higher modes, with the caveat that the models studied are an oversimplification of the real near surface.

## Base Models for Sensitivity and Synthetic Modeling Studies

We use three base models to study the importance of considering higher-mode data in the inversion of phase velocities. The models correspond to simplifications of problems we have found to be of interest in some engineering applications. Model properties are detailed in Table 1, and corresponding fundamental- and higher-mode dispersion curves (DCs), computed with surface-wave modal-inversion software (Rix and Lai, 2005), are shown in Figure 1. The maximum depth common to models in Table 1 is 10 m, and the maximum frequency of analysis corresponds to 50 Hz, which matches the highest frequency studied in the example data set presented later. The first model corresponds to a simple or regular profile in which velocity increases with depth following a constant

**Table 1.** Description of shallow 1D models used for the numeric studies; LVI stands for model with linear velocity increase, HVH for high-velocity half-space model, and HVL for high-velocity-layer model.

| Model | Layer no. | $V_S$ (m/s) | Thickness (m) | Poisson's ratio | Density (g/cm$^3$) |
|-------|-----------|-------------|---------------|-----------------|--------------------|
| LVI   | 1         | 200         | 2             | 0.3             | 1.7                |
|       | 2         | 271         | 3             | 0.3             | 1.7                |
|       | 3         | 385         | 5             | 0.3             | 1.7                |
|       | HS        | 600         | N/A           | 0.3             | 1.7                |
| HVH   | 1         | 300         | 10            | 0.3             | 1.7                |
|       | HS        | 1500        | N/A           | 0.25            | 2.2                |
| HVL   | 1         | 200         | 3             | 0.3             | 1.7                |
|       | 2         | 1500        | 1.5           | 0.25            | 2.2                |
|       | 3         | 400         | 5.5           | 0.3             | 1.7                |
|       | HS        | 600         | N/A           | 0.3             | 1.7                |

gradient. The second and third models are more complex because of large jumps in $V_S$, which are associated with partitioning of signal to higher mode and scattered energy (Stokoe et al., 1994; Supranata et al., 2007). A more detailed description and justification for the use of these models follow.

## Model with a linear velocity increase (LVI)

Model LVI in Table 1 represents a simple three-layer profile in which $V_S$ increases linearly with depth, starting with a relatively low velocity at the surface. The DCs up to the third higher mode are shown in Figure 1a. At high frequencies, dispersion curves asymptotically approach the velocity of the first layer (200 m/s), and at the low frequencies, phase velocities are constrained by the velocity of the half-space (600 m/s). (Rayleigh-wave velocities range from 89% to 95% of $V_S$ for values of Poisson's ratio between 0.1 and 0.49, a range that encompasses virtually all earthen materials [Graff, 1975]). Supranata et al. (2007) performs linearized inversion of multimode phase velocities for profiles that follow a linear increase in velocity and conclude from their tests that incorporating higher modes added little extra information to the estimation of $V_S$.

## Model with a high-velocity half-space (HVH)

In a case in which the active-source MASW method was applied to a site consisting of clay residuum overlying shallow basalt bedrock, inversion of MASW data for the fundamental-mode Rayleigh wave resulted in shear-wave velocities within the rock that were much lower than expected (Casto et al., 2009). Forward modeling revealed that the fundamental-mode dispersion curve was hardly sensitive to bedrock velocity perturbations above the minimum observed frequency. Casto et al. (2009) show that higher modes appeared to reach their high-velocity limits within the range of the recorded frequencies, making the higher modes of great interest for estimating the shear-wave velocity of the basalt and mapping the sediment-basalt interface. Figure 1b shows dispersion curves up to the fourth higher mode for the HVH base model described in Table 1, which represents a low-velocity clay residuum overlying shallow basalt bedrock.

## Model with a high-velocity layer (HVL)

In some arid regions where shallow groundwater is or was once present, it is common to encounter low-velocity sediments intermixed with rock-hard layers derived from precipitation of calcium carbonate from solution, and the transition between uncemented and cemented media can be abrupt (Werle and Luke, 2007). Inversion of Rayleigh-wave phase velocities in such settings has been shown to result in nonunique models, and hence the outcome of the inversion has been seen to depend heavily on a priori information about the site under investigation (Gucunski and Woods, 1991; O'Neill and Matsuoka, 2005; Calderón-Macías and Luke, 2007; Supranata et al., 2007). The ability to resolve an HVL depends on some knowledge of its depth, thickness, and/or velocity contrast with the surrounding material (O'Neill and Matsuoka,

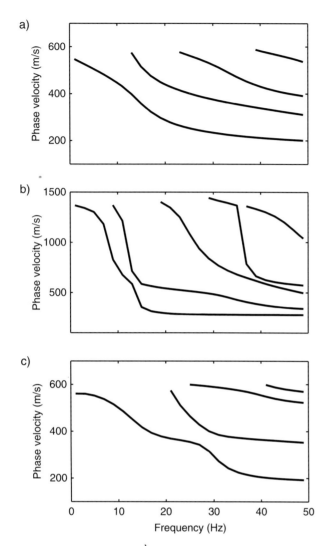

**Figure 1.** Dispersion curves of fundamental mode and higher modes for the models in Table 1: (a) LVI model, (b) HVH model, and (c) HVL model.

2005; Luke and Calderón-Macías, 2007). Supranata et al. (2007) show that using fundamental as well as higher-mode phase velocities results in an improvement toward developing profiles with shallow velocity inversions compared to using fundamental mode only.

The last model described in Table 1 corresponds to a model in which low-velocity sediments are imprinted with an HVL. Figure 1c shows the synthetically computed dispersion curves of this model. Characteristic to the presence of the HVL is a kink in the fundamental-mode dispersion curve.

## Sensitivity Studies

We track changes in phase velocities for the different modes, with respect to the reference models shown in

Figure 1, when perturbing the shear-wave velocities from their actual $V_S$ values in the range $V_{S_i} - V_{per_i}$ to $V_{S_i} + V_{per_i}$ in 25 constant increments, where $V_{per}$ is a maximum perturbation velocity and $i$ is the layer number. The synthetically computed dispersion curves resulting from the perturbations, made one layer at a time, are used to compute standard deviations in phase velocity with respect to frequency for each mode. The standard deviations indicate the magnitude by which the phase velocities vary because of the perturbations in $V_S$.

Figure 2 shows dispersion curves for the LVI model (identical to Figure 1a) with standard deviations resulting from velocity perturbations of each layer and the half-space independently. The maximum perturbation used for this case is 50 m/s for all the layers and half-space. As expected, the fundamental-mode DC becomes more sensitive to changes in $V_S$ of the first layer as frequency increases (Figure 2a), although at intermediate frequencies, the second and third layers of the system appear to dominate the dispersion curve (Figure 2a through 2c). Toward the low-frequency end of the spectrum, the half-space velocity dominates variations in the fundamental-mode DC. The first higher mode appears to be sensitive to the perturbations of the first and second layers but has little sensitivity to perturbations of the third layer and half-space. Interestingly, the second and third higher modes show sensitivity to changes in $V_S$ for all the layers. In particular, it is observed that these modes are sensitive to changes in half-space velocity at relatively high frequencies. Because of the frequency overlap, meaning that both fundamental and higher modes sense variations in velocity over a common frequency range, information provided by the higher modes might be redundant for estimating $V_S$ for the model studied. However, the redundant information provided by higher-mode data might benefit the inversion result of noisy data.

In the HVH model case, the low-velocity layer and half-space are perturbed by 50 and 250 m/s, respectively (Figure 3). Figure 3a shows that above 5 Hz, all modes are sensitive to changes in velocity in the first layer. In fact, the higher modes show a greater standard deviation than the fundamental mode. Perturbing $V_S$ of the half-space produces phase-velocity variations of the higher-mode data for frequencies approaching the cutoff frequency as well as for the fundamental mode at frequencies below 10 Hz (Figure 3b). This pattern is observed in the shallow bedrock study of Casto et al. (2009). Note also from Figure 3b that toward higher frequencies, sensitivity to changes in $V_S$ is large in the flatter part of the DC and then vanishes where the slope increases sharply.

Figure 4 shows a similar analysis for the HVL model case. Maximum perturbations for each layer are 50 m/s for layers 1, 3, and the half-space and 250 m/s for the HVL (layer 2). In Figure 4a, it is observed that the velocity of the first layer has a great influence on the first three modes, and similar to the previous case, its influence on the fundamental mode diminishes toward the low-frequency end. Variations in $V_S$ in the first layer have the strongest influence in the vicinity of the cutoff frequency for the first higher mode. The relatively large variations

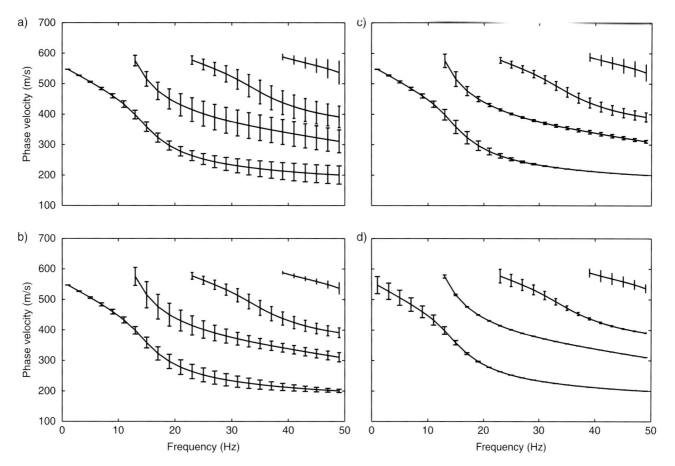

**Figure 2.** Dispersion curves of the LVI base model and standard deviations resulting from a parametric study where (a through c) $V_S$ of layers one to three and (d) half-space are varied by $\pm 50$ m/s from the base model shown in Table 1.

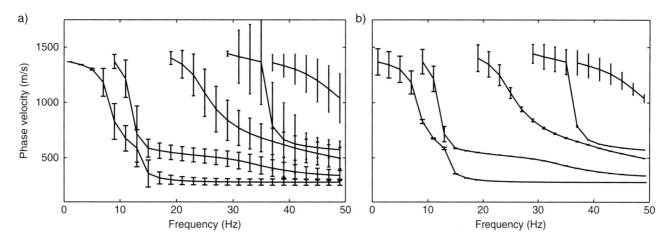

**Figure 3.** Dispersion curves of the HVH base model and standard deviations resulting from perturbing $V_S$ of (a) the first layer and (b) the half-space by $\pm 50$ m/s and $\pm 250$ m/s, respectively, from the base model shown in Table 1.

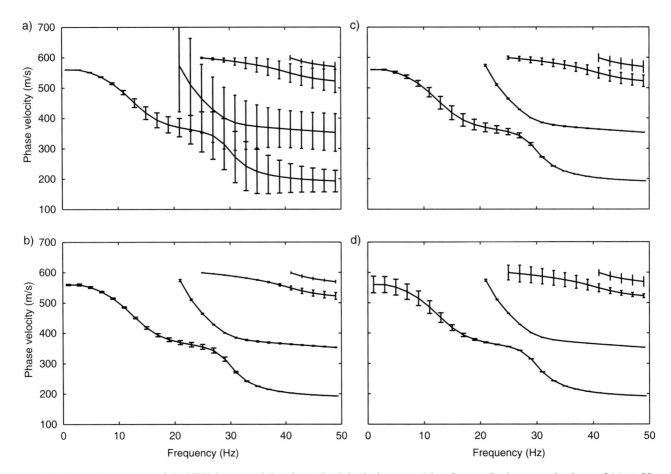

**Figure 4.** Dispersion curves of the HVL base model and standard deviations resulting from velocity perturbations of (a) $\pm 50$ m/s for layer 1, (b) $\pm 250$ m/s for the HVL, and (c) $\pm 50$ m/s for layer three and (d) the half-space.

observed in the first two modes point to the possibility that the modes potentially can coincide or even cross one another, resulting in erroneous or ambiguous picking.

Perturbing $V_S$ of the HVL (Figure 4b) produces almost no variations in the DCs for the range of frequencies studied here. The fundamental mode shows some sensitivity at intermediate frequencies, but it is small compared to the magnitude of the velocity perturbation in the layer. It is interesting to note that the second higher mode shows a higher sensitivity to changes in $V_S$ of the HVL than does the first higher mode, and likewise for the third layer and half-space (Figure 4c and 4d, respectively). In this experiment, the fundamental mode appears to be the most sensitive mode to velocity variations in the model, followed by the second higher mode.

## Analysis of 2D Error Surfaces

It is of particular interest to determine the importance of incorporating higher-mode data for mapping high

impedance contrasts. We look at error surfaces obtained from the root-mean-square (rms) difference between the DC of the base model and DCs from an ensemble of models that sample a 2D grid in which two model parameters are varied within a fixed range. For the HVH model case, depth and velocity of the half-space are discretized over the ranges of 4 to 20 m and 400 to 2200 m/s, respectively, with 51 samples in each dimension. These ranges are chosen according to a hypothetical uncertainty of the true parameters. We choose to perturb only these two parameters for the sake of simplicity. Dispersion curves for all model combinations, fixing $V_S$ of the first layer at 300 m/s, are computed, and the rms-error surface is obtained by comparing these curves with the base-model (target) DC. For the HVL model case, the depth and velocity of HVL are varied between 0.5 to 9.25 m and 400 to 2200 m/s, respectively.

The main objectives of the tests are twofold. First we want to compare the relative sensitivity to changes in depth and velocity in the rms-error surfaces of each mode and for all modes taken together, as measures of absolute and

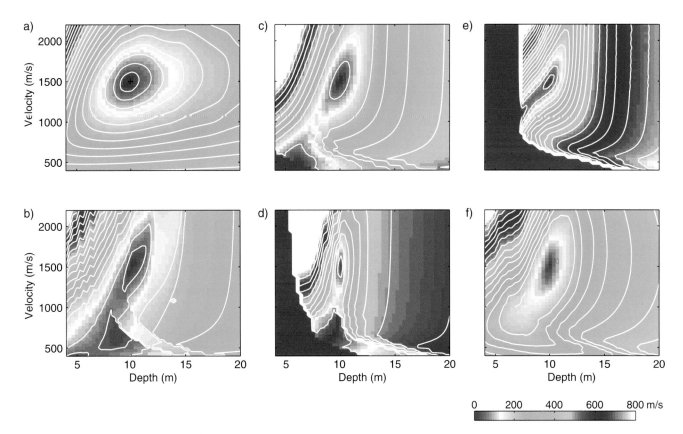

**Figure 5.** Surfaces of rms error (m/s) obtained by varying depth and velocity of the half-space for the HVH model within the plotted ranges. Frequency range considered is 0 to 50 Hz. Surfaces for (a) fundamental mode, (b through e) first to fourth higher modes, and (f) all modes combined.

relative parameter resolution. Second, we want to look at the complexity of the error surface because this has implications for error minimization.

Figure 5a-e shows error surfaces for the HVH model case of the fundamental and first to fourth higher modes, respectively, and Figure 5f displays the surface containing contributions of all modes obtained by summing surfaces in Figure 5a-e. Note the different shapes of the error surface among the modes, which translates to different sensitivities of each of the modes to the model parameters. Distinctive to the fundamental-mode error surface is a similar gradient observed in both axis directions. This implies that the fundamental mode has similar resolution for both parameters. Minimization of this error function is ideal in the sense that all starting models with parameters within the studied range would converge to the only minimum at a similar convergence rate. Error surfaces for the higher modes display a bananalike shape with preferential elongation along the velocity axis.

With respect to fundamental mode, higher modes have increased sensitivity to changes in depth of half-space and decreased sensitivity to changes in half-space velocity. Interestingly, the third higher mode shows the best

resolution in depth of all the modes. Error surfaces of the higher modes contain ripples caused by shifting of the cutoff frequency. A difference in the number of data points used for error evaluation from sample to sample in model-parameter space can cause spurious minima in the error surfaces, as observed, for instance, in the first higher mode surface for depths below 8 m (Figure 5b). Note also that for the second to fourth higher modes (Figure 5c and 5d), there is an area in the error surface that is not being sampled, the reason being that higher modes do not propagate for those model parameters within the frequency band of investigation. Also important to notice is the fact that a minimum develops for the higher-mode error surfaces away from the base model at an approximate depth of 7 m and relatively low velocities (<800 m/s). This will pose a problem when inverting higher-mode data because convergence to a minimum that represents a credible outcome is not guaranteed. Adding contributions of all modes results in a surface of a narrower shape in the depth axis compared to the fundamental-mode-only surface, but with a single minimum (Figure 5f).

In summary, this exercise suggests that inversion of the fundamental mode alone would converge to the true

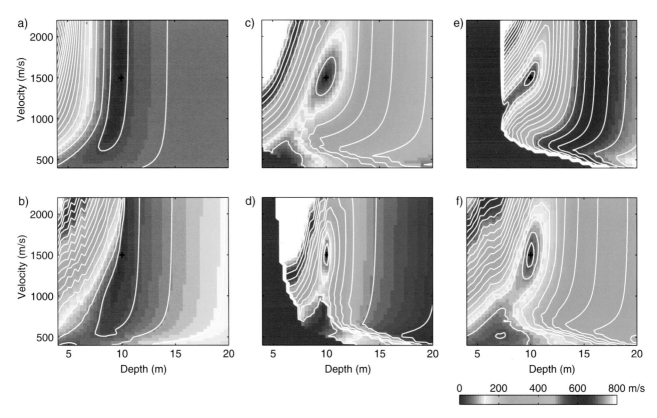

**Figure 6.** Surfaces of rms error (m/s) obtained by varying depth and velocity of the half-space for the HVH model within the plotted ranges. Frequency range considered is 12.5 to 50 Hz. Surfaces for (a) fundamental mode, (b through e) first to fourth higher modes, and (f) all modes combined.

model, but it is possible that some resolution could be gained if higher-mode data were incorporated. However, incorporation of higher modes could introduce pitfalls that result in a poorer solution than if the fundamental mode were inverted alone.

Figure 6 extends this exercise by repeating the experiment but now limiting the frequency range from 12.5 to 50 Hz. This experiment is of interest because low frequencies are lacking in surface-wave studies that use sledgehammer or accelerated weight-drop sources and are the most challenging part of the dispersion curve to populate in any case. Error surfaces of the fundamental and first higher modes (Figure 6a and 6b) provide some resolution in half-space depth but little information on the base layer velocity. Error surfaces of the second to fourth higher modes (Figure 6a and 6b) resemble those of Figure 5 because the cutoff frequencies of these modes are above the imposed low-frequency limit. Incorporating multiple higher-mode phase velocities in the inversion becomes essential for converging to the correct solution in this experiment. Notice from Figure 6f that combining all modes results in a more complex surface compared to the case in which low frequencies are included (Figure 5f).

Figure 7 contains error surfaces for the HVL case, in which depth and velocity of the HVL are varied, keeping the velocities of the other layers and half-space fixed to their base model values. The full frequency spectrum (0 to 50 Hz) is considered. Characteristic of these surfaces is the near absence of sensitivity in the direction of the velocity axis. Surfaces of the fundamental and first higher modes show a high sensitivity to changes in depth of the HVL (Figure 7a and b), with the first higher mode displaying slightly higher precision for situating the correct depth and some resolution in velocity for discerning a high velocity layer ($\sim > 700$ m/s). Error surfaces for the second and third higher modes show poorer resolution: one can only situate depth of the HVL to be below approximately 5.5 m. Adding the contribution of all modes results in a surface that is not superior to the error surface for fundamental mode. In this case, little is gained in terms of model-parameter resolution from the higher modes. Filtering the dispersion curves for the frequency range that is sampled commonly in the field, as in the previous exercise, produced surfaces that resemble the ones presented for the full frequency spectrum (not shown here for brevity).

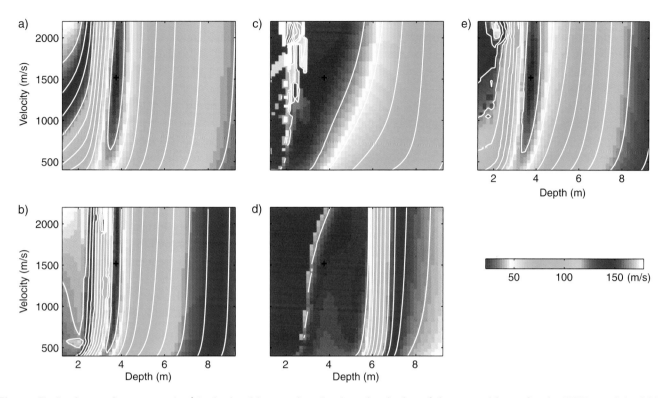

**Figure 7.** Surfaces of rms error (m/s) obtained by varying depth and velocity of the second layer for the HVL model within the plotted ranges. Frequency range considered is 0 to 50 Hz. Surfaces for (a) fundamental mode, (b through d) first to third higher modes, and (e) all modes combined.

The studies carried out so far assume that the higher modes can be resolved perfectly from the seismic data. In the following section, we carry out synthetic modeling tests to measure the relative amplitudes of the propagating modes.

## Synthetic Modeling

Synthetic seismograms are computed with the FD numerical method for the base models described in Table 1. Program E3D (Larsen and Schultz, 1995) is used for the simulations. Although the modeling is two dimensional, we expect similar relative amplitudes among the propagating modes for the 3D case. A vertical force source is applied at the free surface with a peak frequency of 20 Hz. Receivers are located on the surface of the model, spaced every 1 m for an offset range of 0 to 100 m. Note that all energy recorded by the sensors comes from depths shallower than the imposed half-space depth because absorbing boundaries are considered at the wall and bottom of the model. The seismograms were transformed to frequency–phase velocity or *f-c* domain by means of an *f-p* (frequency-slowness) transformation. Maps for the models of Table 1 are shown in Figures 8, 10, and 11. Dispersion curves shown in Figure 1 are plotted overlying the maps.

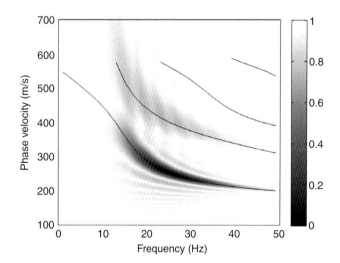

**Figure 8.** Phase velocity–frequency maps obtained from mapping FD synthetics for a vertical source and surface receivers for the LVI model shown in Table 1. Synthetically computed DCs (Figure 1a) overlie the plot.

For the LVI model case (Figure 8), the fundamental and first higher modes can be interpreted unambiguously for frequencies above 13 Hz. At about 20 Hz, the ratio of peak amplitudes between fundamental and first higher modes $A_0(20\ Hz)/A_1(20\ Hz)$, where $A_0$ refers to the amplitude in the *f-c* spectrum of the fundamental mode and $A_n$ to the amplitude of mode $n$, is approximately equal to four. Notice that resolution for picking phase velocities is highest for the fundamental mode. This can be seen better in the display of Figure 9, where the amplitudes for frequencies 20, 30, and 40 Hz have been extracted from the map in Figure 8. In our experience, it is most common that the energy of the fundamental mode overwhelms energy of higher modes (e.g., Calderón-Macías and Simmons, 2008; Jin et al., 2009), but this phenomenon depends on acquisition parameters and model properties.

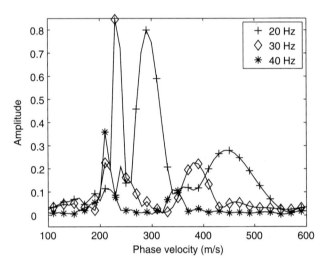

**Figure 9.** Amplitudes of constant frequency obtained from the map in Figure 8.

The HVH case is more interesting as well as complex because more higher-mode energy is discernable in the *f-c* map (Figure 10). The fundamental mode is observable in the frequency band of 13 to 50 Hz and the first higher mode in the range of 25 to 50 Hz. The second higher mode also can be identified, but with lower resolution, in a similar frequency band to that of the first higher mode. At high frequencies (>40 Hz), the second and third higher modes appear to overlap. At about 38 Hz, the third higher mode shows a band of high amplitude, which is related to the steepest part of the curve. The fourth higher mode can be identified at the high end of the spectrum, although with a poorer resolution than the lower propagating modes. Amplitude ratios between fundamental and higher modes at selected frequencies containing approximate amplitude maxima are $A_0(30\,\text{Hz})/A_1(30\,\text{Hz}) \approx 3.5$, $A_0(30\,\text{Hz})/A_2(30\,\text{Hz}) \approx 8$, $A_0(38\,\text{Hz})/A_3(38\,\text{Hz}) \approx 3$, and $A_0(45\,\text{Hz})/A_4(45\,\text{Hz}) \approx 15$ (amplitude curves are not shown for brevity).

Figure 11 shows the *f-c* amplitude spectrum for the HVL model case. The fundamental mode can be followed from a minimum frequency of approximately 12 Hz. The first higher mode is observable in the frequency range of 22 to 32 Hz. At relatively low frequencies (<28 Hz), the two modes have comparable amplitudes, although the fundamental mode dominates the map at higher frequencies. At about 30 Hz, the amplitude ratio between fundamental and first higher mode is $A_0(30\,\text{Hz})/A_1(30\,\text{Hz}) \approx 6$. Energy of the second higher mode and above is not discernable from the map.

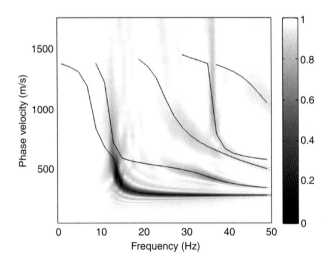

**Figure 10.** Phase velocity–frequency maps obtained from mapping FD synthetics for a vertical source and surface receivers for the HVH model in Table 1. Synthetically computed DCs (Figure 1b) overlie the plot.

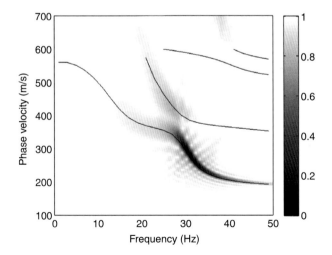

**Figure 11.** Phase velocity–frequency maps obtained from mapping FD synthetics for a vertical source and surface receivers for the HVL model shown in Table 1. Synthetically computed DCs (Figure 1c) overlie the plot.

## Inversion of Fundamental- and Higher-mode Data at a Site with Shallow Bedrock

The data set examined comes from a single MASW source gather, recorded during a seismic survey conducted in the southern Panama Canal region (Casto et al., 2009). The data were collected with a 24-channel system and 4.5-Hz geophones spaced at 1.5-m increments for a total array length of 34.5 m, with a minimum source-receiver separation of 6 m. An 80-lb elastic accelerated weight drop was used as the source. A geologic log from a boring near the seismic line showed clay overburden to a depth of 5.1 m above unweathered basalt, present to the total boring depth of 50 m. Seismic-refraction data collected along the same survey line (Casto et al., 2009) showed that the depth to top of rock is approximately 7.5 m at the location of the MASW gather, and it revealed bulk P-wave velocities ranging from approximately 440 to 2000 m/s within the overburden (including the top of weathered bedrock) and approximately 2400 to 5400 m/s within the indurated basalt bedrock. Assuming a $V_S/V_P$ ratio that is 0.4 within the unconsolidated materials and 0.6 within crystalline rocks (Press, 1966), shear-wave velocity is expected to range from 176 to 800 m/s within the overburden (including the top of weathered bedrock) and from 1440 to 3240 m/s within the basalt bedrock.

The fundamental-mode dispersion curve interpreted from the *f-c* transformation (Figure 12) is resolved in frequencies from approximately 13 to 50 Hz, with phase velocities ranging from 260 m/s at the higher frequencies to 700 m/s at the lowest reasonably discernible frequency. Picks of fundamental and higher-mode phase velocities are displayed in Figure 12. Phase-velocity picks lose resolution for decreasing frequency and increasing velocity. Linearized inversion of the fundamental mode performed by Casto et al. (2009) on the same data set produced velocities that were within the expected range for the sediments but that underestimated expected shear-wave velocities within the basalt by a large margin. Casto et al. (2009) conducted forward-modeling simulations using parameters similar to the expected site conditions, which showed little observable change in fundamental-mode phase velocity within the range of measured frequencies from the field-data case. The authors conclude that frequencies smaller than the minimum frequency sampled are required to characterize the bedrock velocity correctly in the fundamental-mode data.

We inverted this data set with the linearized inversion method for the following three cases: (1) fundamental mode alone; (2) fundamental and first higher mode; and

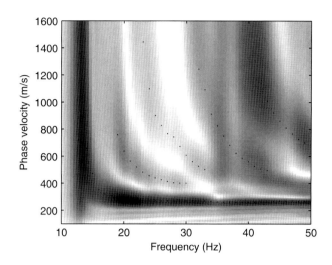

**Figure 12.** An *f-c* map of experimental data with phase-velocity picks (.) of fundamental and four higher modes.

(3) fundamental, first, and second higher modes. Shear-wave velocities for the reference model are derived directly from the observed dispersion curve (Casto et al., 2009). All inversions used four iterations of error minimization. Profile geometry is defined by eight model blocks (layers) with fixed thicknesses increasing 25% with depth over the half-space. The starting Poisson's ratio is fixed at 0.4. Density is fixed at 2 g/cm³. For inversion case 2, the reference model is the outcome of inverting the fundamental mode alone. For case 3, the result of case 2 is used. Inversions using all the data failed to converge to a meaningful answer, indicating possibly the lack of a reasonable starting model.

Figure 13 shows the extracted dispersion curves and calculated theoretical curves up to the second higher mode for the reference model and the fundamental-mode-only inversion. As one might expect, there is a good data fit between the picked and inverted fundamental-mode phase velocities but a poor fit between the picked and inverted higher-mode samples. Inversion for case 2, using the fundamental and first higher modes, produces an overall improvement in the fit of all modes, and inversion of all three modes (case 3) produces the best fit of all, taken as a whole (Figure 14). Incorporating the higher-mode data into the inversion comes at the cost of a larger misfit with the fundamental mode at the high-frequency end.

Figure 15 compares the final models. The model resulting from the fundamental-mode inversion (case 1) is similar to the reference profile for the first two layers, and then a velocity inversion is predicted. At depths greater than 7 m, the inversion predicts larger velocities than those from the reference profile. Incorporating the first higher mode into the inversion (case 2) results in a decrease in velocity at shallow depths associated with a poorer fit of

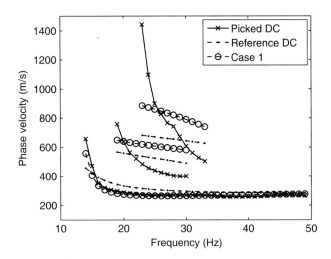

**Figure 13.** Picked, reference model, and inverted fundamental-mode (case 1) phase velocities of the fundamental and first two higher modes.

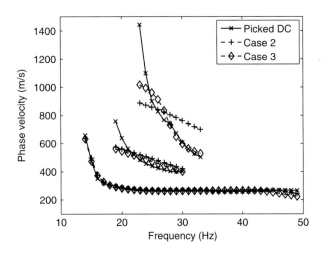

**Figure 14.** Picked, inverted fundamental mode plus first higher mode (case 2), and inverted fundamental plus first and second higher modes (case 3).

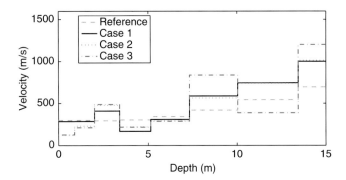

**Figure 15.** Starting $V_S$ (reference), fundamental-mode-only inversion (case 1), inversion of the fundamental mode plus first higher mode (case 2), and inversion of the fundamental mode plus first two higher modes (case 3).

the fundamental mode toward high frequency compared to the previous case. Higher velocity contrasts are observed in this inversion run with respect to inverting fundamental-mode-only data. At depths greater than 7 m, velocities for case 2 follow those of case 1 closely. Inversion of all three modes (case 3) leaves velocities of the first five layers unperturbed with respect to the reference model for this inversion test (case 2) and introduces larger velocity contrasts in layers six to eight with a second velocity inversion occurring in this depth range. There is a more pronounced jump in velocity at the expected overburden-bedrock interface ($\sim$7.5 m), but velocities overall are more erratic than for the other solutions.

In summary, phase-velocity picks from the first higher mode appear to be redundant for resolving the deeper layers because few differences are observed between case 1 and case 2 inversions. Near-surface resolution appears to be diminished by the addition of first-higher-mode data because of deterioration of the solution for the fundamental mode at high frequencies. This situation persists in the solution for the next higher modes. Adding the second higher mode results in a profile with a larger velocity contrast at the approximate expected depth from the refraction study but possibly at the cost of distorting velocities of other layers above and below. It might be necessary to extend the frequency range of the higher-mode picks to capture the velocity of the basalt. But as observed from the *f-c* spectrum (Figure 12), there is a high uncertainty for unambiguously identifying the higher modes and possibly a mode misidentification for the fundamental and first higher modes at relatively low frequencies. Attempts to include modes beyond the second higher mode (not presented) failed to converge on a meaningful solution.

## Discussion

Sensitivity studies carried out on the models illustrate that higher modes exhibit a higher sensitivity to changes in half-space velocities at intermediate to high frequencies than does the fundamental mode. Xia et al. (2008) show that by studying the Jacobian matrix (derivatives of phase velocities with respect to $V_S$ in this case), incorporation of higher-mode data into the inversion of phase velocities generally increases the resolution of the inverted $V_S$. In certain cases, such as the HVH model case shown here, higher modes might be the only information that can be extracted from surface-wave data to map a high stiffness contrast in the absence of low frequencies. On the other hand, the finite-difference modeling study carried out here showed that higher modes are more difficult to identify than the fundamental mode.

This problem can result in misidentification of modes and/or picks of phase velocities that have a large variance, with both having a potentially negative impact on inversion results (Zhang and Chan, 2003; Wathelet, 2005). Studies with synthetic data for a profile with abrupt velocity changes showed that picks of the fundamental and first higher mode are difficult to discern from one another (Jin et al., 2009). This problem also is shown in work by O'Neill and Matsuoka (2005), in which the dispersion curve that a user logically would pick as the fundamental mode actually transitions to the first higher mode at relatively low frequencies.

Resolution for picking higher-mode data might be improved by optimizing acquisition survey parameters and refining processing methods. In the first case, of particular interest are the selection of receiver density and spacing, minimum and maximum source-receiver offsets, dominant geophone frequency, and nature and strength of the seismic source. Although there are some general guidelines that might be followed, selection of optimum parameters is expected to be site dependent (e.g., Ivanov et al., 2008). Acquiring data at long offsets permits deeper penetration; however, lateral heterogeneity of the site of investigation can hamper resolution of the propagating modes.

In the second case, improving resolution by a more specialized processing flow that results in a higher-resolution mapping from the x-t domain to the f-c domain in MASW surveys might be achieved by selective windowing of the data (Hayashi and Suzuki, 2004; Neducza, 2007; Ivanov et al., 2009). In selective windowing, near- and far-offset traces are used to build different parts of the f-c spectrum (Neducza, 2007). The signal-to-noise ratio can be enhanced by combining (stacking) traces from source-receiver groups that span overlapping lateral positions (Hayashi and Suzuki, 2004; Neducza, 2007). In addition, enhanced mode identification can be achieved by applying mutes in the x-t domain that selectively remove interfering seismic energy traveling at certain ranges of velocity (Ivanov et al., 2009). Another approach to enhance resolution is the high-resolution radon transform, which searches for sparseness of energy in the frequency-slowness domain (Sacchi and Ulrych, 1995; Luo et al., 2008) through iterative minimization. All these approaches address the problem of reducing uncertainty in the data previous to inversion. A more numerically intensive solution that is receiving increasing attention (e.g., Pratt, 1999; Gelis et al., 2005) corresponds to performing a full-waveform inversion, thus bypassing the need to identify surface-wave modes.

The study of error surfaces points to the potential for a more complex error surface when higher-mode data are integrated for inversion. This is a topic that requires further scrutiny, particularly because the shape of the surface appears to be site dependent. Practical solutions that have been presented for reducing the impact that higher modes might have toward negatively biasing the solution include assigning a priori variances to the data so that lower modes are emphasized in the inversion (Beaty et al., 2002) and applying a staged inversion in which the fundamental mode provides the starting solution for inverting the higher modes (Supranata et al., 2007). This is the approach adopted here with the field experiment.

## Conclusions

Parametric studies of sensitivity of fundamental- and higher-mode dispersion curves to changes in $V_S$ are carried out for three near-surface geologic scenarios: a sediment model in which velocity increases linearly with depth (LVI model), a two-layer profile representing sediment over hard bedrock (HVH model), and a model representing a high-velocity layer imprinted in slower sediments (HVL model). Perturbing the velocity of a single layer produces changes in phase velocity over frequency ranges that vary among the modes. The deeper the perturbation is applied, the higher the frequencies at which the modes respond, with frequencies increasing toward increasing mode number.

Of particular interest are the HVH and HVL models because of problems encountered when inverting $V_S$ from fundamental-mode phase velocities alone in such geologic scenarios. The objective is to map shallow, high-shear-wave velocity contrasts. In the first case, lack of low frequencies results in unconstrained velocities and hence inability to map the interface. Our numerical studies suggest that higher-mode data can be used to retrieve depth and velocity of a shallow sediment/hard bedrock interface. The HVL model case is more complex. Fundamental- and higher-mode phase velocities provide constraints on the depth of the HVL, but velocity of the HVL appears to be unconstrained, with the higher modes providing little or no extra information.

Numerical modeling studies carried out on the three scenarios show that surface waves are predominantly of the fundamental mode and that subsurface properties have an influence on the occurrence of higher-mode energy. Furthermore, analysis of the f-c maps suggests that resolution for picking higher modes decreases with increasing mode number. Uncertainties in picking dispersion data in the presence of higher-mode energy need to be considered when fundamental and higher modes are combined for inversion of $V_S$.

Finally, inversion of a field-data test in which a thin layer of overburden covers hard bedrock shows that incorporation of higher modes produces changes in $V_S$ that are not predicted with fundamental-mode-only data. Even though a more realistic velocity trend might be predicted with higher-mode data, additional information is required to converge on a more credible outcome.

# Acknowledgments

We thank Ron Kaufmann for suggesting the study of shallow bedrock and Daniel Casto for discussions on the subject and for providing the data used in the section on multimode inversion. We thank Glenn Rix for use of the code for computing phase velocities. Coauthor Luke acknowledges financial support from the U. S. Department of Energy under contract number DE-FG52-03NA99204.

# References

Aki, K., and P. G. Richards, 1980, Quantitative seismology, 2nd ed.: University Science Books.

Beaty, K. S., D. R. Schmitt, and M. Sacchi, 2002, Simulated annealing inversion of multi-mode Rayleigh wave dispersion curves for geological structure: Geophysical Journal International, **151**, 622–631.

Calderón-Macías, C., and B. Luke, 2007, Improved parameterization to invert Rayleigh-wave data for shallow profiles containing stiff inclusions: Geophysics, **72**, no. 1, U1–U10.

Calderón-Macías, C., and J. Simmons, 2008, Constrained surface wave inversion from 9-component seismic reflection data: 78th Annual International Meeting, SEG, Expanded Abstracts, 1063–1067.

Casto, D. W., B. Luke, C. Calderón-Macías, and R. Kaufmann, 2009, Interpreting surface wave data for a site with shallow bedrock: Journal of Environmental and Engineering Geophysics, **14**, 115–129.

Gelis, C., J. Virieux, and G. Grandjean, 2005, Elastic full waveform inversion in a space frequency formulation for near-surface imaging: 67th Conference and Technical Exhibition, EAGE, Extended Abstracts, P274.

Graff, K. F., 1975, Wave motion in elastic solids: Oxford University Press.

Grandjean, G., and A. Bitri, 2006, 2M-SASW: Multifold multichannel seismic inversion of local dispersion of Rayleigh waves in laterally heterogeneous surfaces: Application to the Super-Sauze earthflow, France: Near Surface Geophysics, **4**, 367–375.

Gucunski, N., and R. D. Woods, 1991, Use of Rayleigh modes in interpretation of SASW test: Second International Conference on Recent Advances in Geotechnical Earthquake Engineering and Soil Dynamics, University of Missouri–Rolla, Proceedings, 1399–1408.

Hayashi, K., and H. Suzuki, 2004, CMP cross-correlation analysis of multi-channel surface-wave data: Exploration Geophysics, **35**, 7–13.

Ivanov, J., R. D. Miller, J. Xia, D. Steeples, and C. B. Park, 2006, Joint analysis of refractions with surface waves: An inverse solution to the refraction-traveltime problem: Geophysics, **71**, no. 6, R131–R138.

Ivanov, J., R. D. Miller, and G. Tsoflias, 2008, Some practical aspects of MASW analysis and processing: 2008 Symposium on the Application of Geophysics to Engineering and Environmental Problems, EEGS, Extended Abstracts, 1186–1198.

Ivanov, J., C. B. Park, R. D. Miller, and J. Xia, 2005, Analyzing and filtering surface-wave energy by muting shot gathers: Journal of Environmental and Engineering Geophysics, **10**, 307–322.

Jin, X., B. Luke, and C. Calderón-Macías, 2009, Role of forward model in surface-wave studies to delineate a buried high-velocity layer: Journal of Environmental and Engineering Geophysics, **14**, 1–14.

Larsen, S. C., and C. A. Schultz, 1995, ELAS3D: 2D/3D elastic finite-difference wave propagation code: Lawrence Livermore National Laboratory, Technical Report UCRL-MA-121792, Balkema, 31–38.

Luke, B., and C. Calderón-Macías, 2007, Inversion of seismic surface wave data to resolve complex profiles: Journal of Geotechnical and Geoenvironmental Engineering, **133**, 155–165.

Luke, B., H. Murvosh, P. Kittipongdaja, A. Karasa, P. Tamrakar, and W. J. Taylor, 2010, Rayleigh-wave dispersion curves for long, linear arrays at a predominantly-gravel site: 2010 Symposium on the Application of Geophysics to Engineering and Environmental Problems, EEGS, Extended Abstracts, 742–750.

Luke, B., W. Taylor, C. Calderón-Macías, X. Jin, H. Murvosh, and J. Wagoner, 2008, Characterizing anomalous ground for engineering applications using surface-based seismic methods: The Leading Edge, **27**, 544–1549.

Luo, Y., J. Xia, R. D. Miller, Y. Xu, J. Liu, and Q. Liu, 2008, Rayleigh-wave dispersive imaging using a high-resolution linear Radon transform: Pure and Applied Geophysics, **165**, 1–20.

Neducza, B., 2007, Stacking of surface waves: Geophysics, **72**, no. 2, V51–V58.

O'Neill, A., 2005, Seismic surface waves special issue guest editorial: Journal of Environmental and Engineering Geophysics, **10**, 67–86.

O'Neill, A., and T. Matsuoka, 2005, Dominant higher surface-wave modes and possible inversion pitfalls:

Journal of Environmental and Engineering Geophysics, **10**, 85–201.

Park, C. B., R. D. Miller, and J. Xia, 1999, Multichannel analysis of surface waves (MASW): Geophysics, **64**, 800–808.

Pratt, R. G., 1999, Seismic waveform inversion in the frequency domain, Part 1: Theory and verification in a physical scale model: Geophysics, **64**, 888–891.

Press, F., 1966, Seismic velocities, *in* S. P. Clark Jr., ed., Handbook of physical constants: Geological Society of America Memoir 97, 97–173.

Rix, G. J., and C. G. Lai, 2005, SWAMI v.1.2.0 — Surface wave modal inversion software: Georgia Institute of Technology.

———, 2007, Model-based uncertainty in surface wave inversion: 2007 Symposium on the Application of Geophysics to Engineering and Environmental Problems, EEGS, Extended Abstracts, 969–975.

Rosenblad, B. L., J. Li, F. Y. Menq, and K. H. Stokoe II, 2007, Deep shear wave velocity profiles from surface wave measurements in the Mississippi Embayment: Earthquake Spectra, **23**, 791–808.

Sacchi, M. D., and T. J. Ulrych, 1995, High-resolution velocity gathers and offset space reconstruction: Geophysics, **70**, 1199–1177.

Socco, L. V., and C. Strobbia, 2004, Surface wave method for near-surface characterization: A tutorial: Near Surface Geophysics, **2**, 165–185.

Stokoe, K. H. II, G. W. Wright, J. A. Bay, and J. M. Roësset, 1994, Characterization of geotechnical sites by SASW method, *in* R. D. Woods, ed., Geophysical characterization of sites: ISSMFE Technical Committee No. 10, Oxford & IBH Publishing, 15–25.

Supranata, Y. E., M. E. Kalinski, and Q. Ye, 2007, Improving the uniqueness of surface wave inversion using multiple-mode dispersion data: International Journal of Geomechanics, **7**, 333–343.

Werle, J., and B. Luke, 2007, Engineering with heavily cemented soils in Las Vegas, Nevada, *in* A. J. Puppala, N. Hudyma, and W. J. Likos, eds., Problematic soils and rocks and in situ characterization: Proceedings of Geo Denver 2007: Geotechnical Special Publication 162 (CD-ROM), ASCE.

Wathelet, M., 2005, Array recordings of ambient vibrations: Surface-wave inversion: Ph.D. dissertation, University of Liège, Belgium.

Xia, J., R. D. Miller, and C. B. Park, 1999, Estimation of near surface shear-wave velocity by inversion of Rayleigh waves: Geophysics, **64**, 691–700.

Xia, J., R. D. Miller, and Y. Xu, 2008, A trade-off between model resolution and variance with selected Rayleigh-wave data: 78th Annual International Meeting, SEG, Expanded Abstracts, 1293–1297.

Zhang, S. X., and L. S. Chan, 2003, Possible effects of misidentified mode number on Rayleigh wave inversion: Journal of Applied Geophysics, **53**, 17–29.

Chapter 12

# Void Detection Using Near-surface Seismic Methods

Steven D. Sloan[1], Shelby L. Peterie[2], Julian Ivanov[2], Richard D. Miller[2], and Jason R. McKenna[1]

## Abstract

Detection of anomalies such as voids in the shallow subsurface using noninvasive geophysical techniques has proved to be challenging at best. Three near-surface seismic methods are introduced, including diffracted body waves, backscattered surface waves, and changes in reflection moveout velocities to detect voids directly or their effects on surrounding material properties using different parts of the wavefield. Examples are presented, including modeled and field data sets to demonstrate each technique. Body-wave diffractions were used to identify and locate man-made tunnels in multiple geologic settings. Variations in shear-wave reflection velocities are shown to correlate to changes in stress over known void locations; backscattered surface waves are shown to correlate with a known void location. Results of the studies show that the field data correlate well with the synthetic, and these methods show promise in furthering the ability to locate subsurface voids and their effects on the surrounding media.

## Introduction

Detecting and imaging near-surface anomalies (such as voids, cavities, and tunnels) using noninvasive geophysical methods has been pursued for a long time and has proved to be challenging. Such voids could be a product of natural processes such as dissolution (an indirect result of intentional means such as room-and-pillar or dissolution mining) or completely intentional, as is the case with clandestine tunneling. These voids potentially can lead to surface collapse in the form of sinkholes — creating public-safety hazards around roadways, rail lines, or buildings. Cross-border clandestine tunneling compromises na-

tional borders and security, providing conduits for the transportation of illegal drugs, arms, or people. We address three near-surface seismic methods, including surface-wave backscatter, body-wave diffraction enhancement, and the use of shear-wave velocity profiles to image near-surface anomalies (such as voids) directly or their effects on subsurface properties.

Researchers have attempted to identify subsurface voids using an array of geophysical methods, including gravity (Butler, 1984), ground-penetrating radar (Fenner, 1995), and electrical methods (Militzer et al., 1979; Ogilvy et al., 1991), to name a few. Numerous seismic techniques have been used also, including the common-midpoint (CMP) method (Steeples and Miller, 1987; Branham and Steeples, 1988), surface waves (Miller et al., 1999), and refraction (Turpening, 1976). Some studies have had varying degrees of success. However, often the interpretation is nonunique (applicable to all geophysical techniques) or the method lacks the necessary resolution, depth of penetration, or applicability to a wide range of geologic environments (or all of the above). Advancements in seismic-reflection methods in the past decade have opened the door for a wide range of applications. However, the techniques often lack the resolution necessary to image small-scale features in the shallow subsurface, which becomes increasingly difficult with increased depth.

Unconventional seismic methods, such as the use of body-wave diffractions and backscattered surface waves, have demonstrated potential in detecting subsurface anomalies beyond the resolution of traditional techniques. Three seismic methods are discussed here, each focusing on a different part of the wavefield. Body-wave diffractions are used to detect subsurface voids too small to be detected reliably by reflection methods; backscattered surface waves

[1]*U. S. Army Corps of Engineers, Engineer Research & Development Center, Vicksburg, Mississippi, U.S.A.*
[2]*Kansas Geological Survey, Lawrence, Kansas, U.S.A.*

are used to identify anomalous changes in seismic properties or boundaries; shear-wave velocity variations are used to identify increased failure potential associated with dissolution-mining voids.

# Diffracted Body Waves

Diffracted components of the seismic wavefield typically have been considered noise, and the usefulness of diffracted energy was limited to its ability to facilitate the interpretation of reflections. Numerous authors recognize that diffracted waves contain valuable information about local structure and suggest that information from diffractions from horizontal plane terminations (edge diffractions) should be used routinely to supplement reflection data for accurate interpretation of pinch-outs, faults, and other geologic discontinuities (Khaidukov et al., 2004; Fomel et al., 2007). However, some diffracted energy originates from pointlike subsurface anomalies (diffractions) that are unrelated to reflecting interfaces. Examples of these types of anomalies include buried pipes, man-made tunnels, karst features, and abandoned mines. Authors suggest that processing algorithms designed specifically for diffraction enhancement and imaging can be used to detect these types of small anomalies (relative to the seismic wavelength) and various types of plane terminations (Landa et al., 1987; Belfer et al., 1998; Berkovitch et al., 2009).

Diffraction imaging has proved to be successful for detection of relatively small subsurface anomalies that are difficult to detect with reflection imaging. Kanasewich and Padke (1988) propose a method to detect fault planes whereby a fault is assumed to be located and produce an edge diffraction beneath each common midpoint (CMP) in the data set. Common-fault-point (CFP) gathers of traces that contribute to diffracted energy are created for each assumed fault location. Amplitude and phase corrections are applied to each CFP gather based on the dynamic properties of edge diffractions. Diffraction moveout correction is applied using normal-moveout (NMO) velocities (separately determined during CMP processing), and select traces are stacked. The result is a CFP section that indicates fault locations in both synthetic and real data.

Belfer et al. (1998) use a phase-correlation procedure developed by Landa et al. (1987) to calculate diffraction sections (D-sections) that are analogous to common-offset gathers whose diffraction apexes have been enhanced. They successfully detect the presence of karst features and faults beneath an artificial water reservoir. Landa and

Keydar (1998) calculate D-sections for data from baseline and monitor surveys acquired at the same location with the same acquisition parameters. The D-sections are differenced to produce a differential D-section that indicates changes in the scattering characteristics of the subsurface between surveys. A cylindrical tunnel dug beneath the seismic line after the baseline survey was detected successfully as a high-amplitude event on the differential D-section.

We propose that single diffractions observed on shot records can be used to characterize a subsurface anomaly and the overlying medium. We begin with a brief overview of the theory associated with diffractions, the diffraction traveltime equation, and the relationship between the diffraction arrival pattern on shot records and physical properties of the anomaly and the overlying subsurface. Models and synthetic data are provided to demonstrate the characteristics of seismic signatures anticipated for small anomalies relative to the seismic wavelength. Real examples of diffractions observed in field data are presented from three recent studies. Characteristics of diffractions are used to approximate the lateral location, average velocity of the overlying subsurface, and depth of the anomaly.

## Theory

The Huygens-Fresnel principle states that every unobstructed point on a primary wavefront is the source of a secondary wave, and the primary wavefront at a later time can be regarded as the superposition of the secondary waves (Hecht, 2002). A diffraction occurs when a seismic wavefront encounters a subsurface anomaly that is small relative to the seismic wavelength, which acts as a secondary source. Energy from the primary seismic wavefront is redistributed and diffracted energy from the anomaly is returned to the surface (Sheriff, 2002), where it is recorded by receivers (Figure 1a). The arrival time of diffracted energy ($t_d$) is defined as

$$t_d = \frac{1}{v_1}\sqrt{x_s^2 + z^2} + \frac{1}{v_2}\sqrt{x_r^2 + z^2}, \qquad (1)$$

where $v_1$ is the average velocity of the primary seismic wavefront as it travels from the source to the anomaly, $x_s$ is the lateral distance between the source and anomaly, $z$ is the depth of the anomaly, $v_2$ is the average velocity of the diffracted seismic wave as it travels from the anomaly to receivers, and $x_r$ is the lateral distance between the anomaly and receiver. This equation is hyperbolic, and the associated arrival pattern on seismic shot records is unique to diffracted energy.

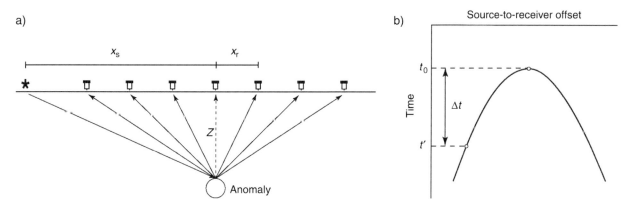

a)

b)  Source-to-receiver offset

**Figure 1.** (a) Diagram of energy from a seismic source diffracted by an anomaly and recorded by receivers. (b) Illustration of the hyperbolic diffraction-arrival pattern on a shot record.

Observations about equation 1 lead to important insights about diffraction energy on shot records and the type of information that can be derived from a recorded diffraction hyperbola (Figure 1b). The shortest traveltime of diffracted energy arrives when $x_r$ is equal to zero. Therefore, the apex of the diffraction always arrives at the receiver directly over the anomaly at time $t_0$, and the lateral location of the anomaly can be determined directly from shot records. In general, $v_1$ and $v_2$ are both equal to the same velocity $v$, and the average velocity can be determined directly from the hyperbolic curvature by

$$v = \sqrt{\frac{x_r^2 t_0 + x_s^2 \Delta t}{t_0 t' \Delta t}}, \qquad (2)$$

where $t'$ is the arrival time of the diffraction hyperbola at the receiver located at $x_r$, and $\Delta t$ is the difference in arrival times $t'$ and $t_0$. Once $v$ is determined, the depth of the anomaly can be calculated by setting $v_1$ and $v_2$ in equation 1 equal to $v$ and solving for $z$:

$$z = \frac{t_0^2 v^2 - x_s^2}{2 t_0 v}. \qquad (3)$$

There are some instances when $v_2$ is not equal to $v_1$. For example, if a mode conversion occurs at the interface of the anomaly, the velocity of the diffracted wave will not be equal to the velocity of the primary wave. In this case, information derived from the diffraction hyperbola using equations 2 and 3 will not characterize accurately the depth of the anomaly or the seismic velocity of the overburden.

## Model data

Synthetic shot records were calculated using 2D elastic finite-difference modeling (Virieux, 1986; Levander, 1988)

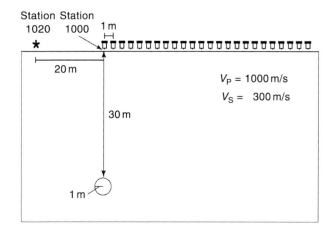

**Figure 2.** Field layout of the first shot record simulated using the model.

with SeisSyn, a proprietary software package from Kansas Geological Survey, to determine the characteristic seismic signatures associated with a small, shallow void. Model data were generated to simulate acquisition of roll-along-style surveys above an air-filled circular void with a depth of 30 m and a 1-m radius. The source wavelet is the first derivative of the Gaussian function with a central frequency of 100 Hz. The source offset is 20 m from the nearest of 24 receivers spaced at intervals of 1 m. The first shot was acquired with the first receiver directly over the void located beneath station 1000 (Figure 2). The source and the receiver spread were advanced four station intervals, and another shot was acquired. This procedure was repeated for a total of six shots.

Four types of diffractions are observed on both compressional (P) and vertically polarized shear ($S_V$) shot records: P-wave, mode-converted S- to P-wave (S-P), mode-converted P- to S-wave (P-S), and S-wave diffractions (Figure 3). The apex of each diffraction type is

**Figure 3.** Synthetic shot records produced for the (a) compressional and (b) $S_V$ components of the modeled wavefield. The Rayleigh wave has been muted to enhance the relative amplitudes of diffractions.

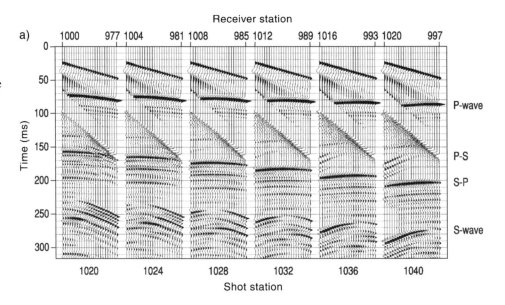

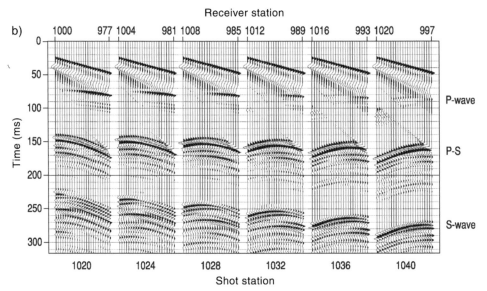

recorded by the receiver at station 1000, located directly above the anomaly in the model. Snapshots of the wavefield reveal amplitude and phase relationships characteristic of each type of diffraction (Figures 4 and 5). Each diffraction type has a distinguishing attribute that can be used to help identify it on shot records. The P-wave diffraction is the only type that undergoes a polarity reversal symmetric about the apex on $S_V$ records. The P-S diffraction is the only type that undergoes a polarity reversal symmetric about the apex on compressional records. The S-P diffraction amplitude is generally much weaker than the other diffraction types because of the general preference for energy to diffract as an S-wave (Korneev and Johnson, 1996). Therefore, it is unlikely that an S-P diffraction will be observed without another diffraction type on the same record. The S-wave diffraction exhibits resonant emission

(Korneev, 2008) and therefore appears as a cyclic diffraction late in the record.

Depth of the anomaly and average velocity of the overburden can be derived for each type of diffraction using equations 2 and 3. The velocity derived from P-wave diffractions is approximately 1000 m/s, and the calculated depth is 30 m for each record. S-P diffractions arrive with the same hyperbolic curvature as P-wave diffractions. However, the velocity derived from these diffractions is approximately 600 m/s. The depth calculated for each shot gather ranges from 42.8 m at a source offset of 20 m to 52.2 m at a source offset of 40 m. The velocity derived from S-wave diffractions is approximately 300 m/s, and the calculated depth is 30 m. The P-S diffractions arrive with the same hyperbolic curvature as S-wave diffractions. However, the velocity derived from these diffractions is approximately 400 m/s, and the depth ranges from 23.5 m to 16.7 m with source offsets of 20 m and 40 m, respectively.

The lateral location of an anomaly can be determined accurately from any diffraction type. Depth of the anomaly and velocity of the overburden can be determined from a single shot record only if the recorded diffraction is a pure P- or S wave diffraction. The apparent depth and velocity calculated from S-P diffractions are, respectively, too deep and somewhat slower than the actual P-wave velocity. The apparent depth and velocity calculated from P-S diffractions are, respectively, too shallow and somewhat faster than the actual S-wave velocity. For mode-converted diffractions, information from shot records acquired with different source offsets would be required to estimate the depth and average velocity of the overburden.

## Field examples

### Man-made clandestine tunnel in Arizona, U.S.A.

Data were acquired over a suspected tunnel in Arizona near the United States–Mexico border. The surveys were acquired using a roll-along style of acquisition approximately perpendicular to the suspected tunnel axis. Line 1 was acquired from west to east on an all-weather road. Line 2 was acquired from east to west in native sandy sediment adjacent to the road at a subset of stations from line 1. The seismic source was a rubber-band-accelerated weight drop (RAWD), and each of 24 receivers was a 40-Hz geophone mounted to steel plates with 1.2-m station spacing. The source was located 41.4 m from the nearest active receiver. Automatic gain control (AGC) and frequency-wavenumber (*f-k*) filtering were applied to each record to scale amplitudes and to attenuate ground-roll energy.

Only one diffraction was observed clearly on shot records from line 1 (Figure 6). The apparent velocity of the diffraction is approximately 275 m/s. The P-wave velocity determined from first arrivals is approximately 520 m/s; therefore, the observed event is not a P-wave diffraction. Calculated depth is approximately 8.2 m. Because no diffraction was observed at this station on adjacent shot records, this event cannot be interpreted definitively as a diffraction.

On line 2, diffractions were observed on multiple records at approximately the same locations as the diffraction observed on line 1 (Figure 7). The apparent velocity and depth calculated from each recorded diffraction are 190 m/s and 8.8 m, respectively. Velocity of the diffrac-

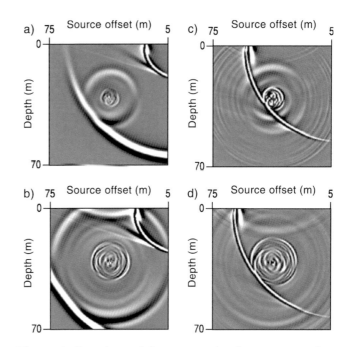

**Figure 4.** Snapshots of the compressional component of the modeled wavefield velocity produced with a source offset of 40 m from the void. The seismic events are the (a) P-wave diffraction at 75 ms, (b) P-S diffraction at 95 ms, (c) S-P diffraction at 195 ms, and (d) S-wave diffraction at 220 ms.

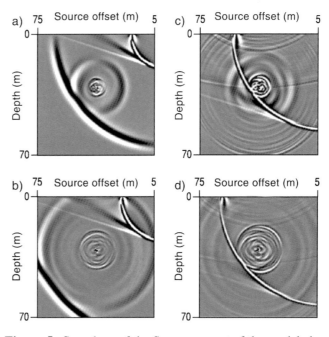

**Figure 5.** Snapshots of the $S_V$ component of the modeled wavefield velocity produced with a source offset of 40 m from the void. The seismic events are the (a) P-wave diffraction at 75 ms, (b) P-S diffraction at 95 ms, (c) S-P diffraction at 195 ms, and (d) S-wave diffraction at 220 ms.

tions observed on line 2 is consistent with the S-wave velocity derived from multichannel analysis of surface waves (MASW) at this site. Because of the velocity, late arrival time, strong amplitude, ringing character, and consistent depth, these diffractions are interpreted to be the S-wave type. The lateral location and depth of this anomaly are consistent with intelligence information from local law-enforcement officials.

Apparent velocity of the diffraction observed on line 1 is slightly faster than the actual S-wave velocity and much slower than the actual P-wave velocity, and the depth is shallower than the depth calculated from S-wave diffrac-

tions on line 2. Therefore, the event recorded on line 1 is likely a P-S diffraction. A polarity reversal symmetric about the apex is not observed. However, this event arrived at the same time as strong Rayleigh waves, which masked much of the signal. The *f-k* filtering applied to attenuate ground roll and residual ground-roll energy might affect the phase characteristics of the diffraction and their use as an interpretive aid. The change in diffraction type from line 1 to line 2 likely is caused by changes in the transmission characteristics of the all-weather road and the undisturbed sediment.

## Railroad tunnel in Colorado, U.S.A.

Data were acquired on a hillside near Winter Park, Colorado. Moffat Tunnel is a man-made railroad tunnel 7 × 5 m constructed in rhyolitic granite that cuts through the Continental Divide. The granite was overlain by a thick organic layer at the survey site. One line of data was acquired with a roll-along style of acquisition at an estimated distance of 5 to 10 m above and perpendicular to Moffat Tunnel. The seismic source was a silenced surface .30–06 rifle, and the 24 receivers were 100-Hz geophones with station spacing of 0.6 m. The source was located 9.1 m from the nearest active receiver. A bandpass filter and AGC were applied to shape the data spectra, to enhance the signal-to-noise ratio (S/N), and to scale amplitudes.

Diffractions from Moffat Tunnel were observed on each record acquired with the spread of receivers directly over the tunnel (Figure 8). The apparent velocity is approximately 530 m/s. Calculated depth varies from one record to another but is approximately 3.5 m on average. These diffractions undergo a polarity reversal that is symmetric about the apex, and they have an amplitude of

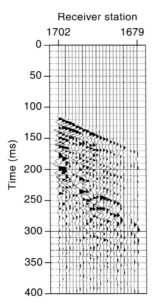

**Figure 6.** Processed shot record from line 1 of the Arizona site. The diffraction apex arrives at approximately 230 ms.

**Figure 7.** Processed shot records from line 2 of the Arizona site.

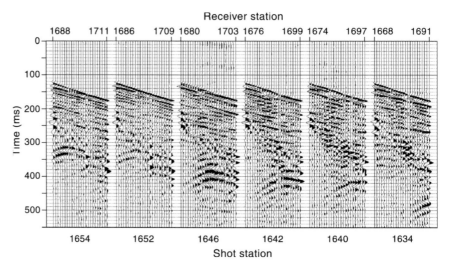

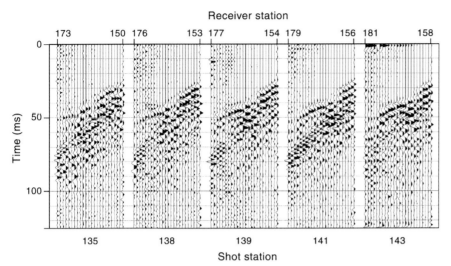

Receiver station

Shot station

nearly zero at the apex on each record. These observations suggest that these events are P-S diffractions and that the tunnel is at a depth greater than 3.5 m. The lateral location of the anomaly derived from the apexes is correct for Moffat Tunnel.

## Man-made clandestine tunnel in California, U.S.A.

Data were acquired over an area in California near the United States–Mexico border suspected to contain numerous clandestine tunnels. One line of data was acquired with a roll-along style of acquisition on an all-weather road approximately perpendicular to the axis of suspected man-made tunnels. The seismic source was a RAWD, and receivers were 40-Hz geophones mounted to steel plates with station spacing of 1.2 m. The source was located 26.8 m from the nearest active receiver. Prior to acquisition, several test shots were recorded with a source offset of 5.5 m. A bandpass filter, AGC, and *f-k* filter were applied to shape the data spectra, to scale amplitudes, and to attenuate ground roll and guided-wave energy.

Numerous potential diffractions were observed at this site, but not all were interpreted to have been caused by a subsurface anomaly. Some potential diffractions lacked trace-to-trace consistency because of strong ambient noise caused by nearby traffic (Figure 9a), inconsistent receiver coupling because of grating and potholes in the all-weather road (Figure 9b), or near-surface statics. In general, poor trace-to-trace consistency either resulted in only a fragment of a potential diffraction observed on shot records or arrival times that were not consistent for a true diffraction. Other potential diffractions lacked record-to-record consistency because of changes in ambient noise conditions, changes in receiver coupling, and changes in the travel path of the

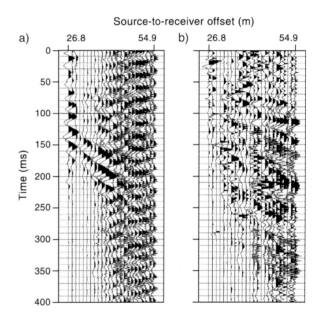

Source-to-receiver offset (m)

**Figure 9.** Shot records with (a) strong ambient noise and (b) inconsistent receiver coupling.

diffracted energy through an inhomogeneous subsurface. In general, poor record-to-record consistency resulted in potential diffractions on multiple records but at different apex locations or traveltimes and different apparent velocities. Some true diffractions might have been interpreted falsely as noise from lack of consistency.

Diffractions from three anomalies were observed with certainty at this site. A diffraction centered at station 1405 appears on five records (Figure 10a). The apparent velocity is approximately 1050 m/s. Depth calculated from each diffraction is approximately 13.7 m. The apparent velocity is approximately equal to the shallow P-wave velocity determined from the direct wave recorded during testing. Therefore, these events are likely P-wave diffractions.

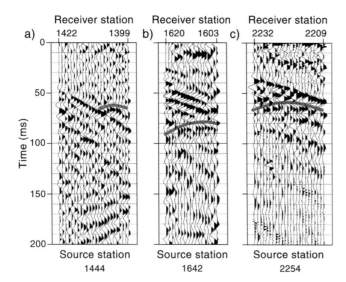

**Figure 10.** Processed shot records acquired at the California site. Diffraction apexes are located at stations (a) 1405 at 60 ms, (b) 1606 at 80 ms, and (c) 2220 at 60 ms. Diffractions are interpreted in gray.

A diffraction centered at station 1606 was observed on one record with an apparent velocity and depth of approximately 915 m/s and 21 m, respectively (Figure 10b). Because the diffraction is observed on only one shot record and only on one side of the apex, it is difficult to determine whether this event is a P-wave or P-S diffraction. However, the depth calculated assuming it is a P-wave diffraction is consistent with the range of suspected tunnel depths, and the velocity is not unexpected for the P-wave velocity because of lateral variations in velocity at this site. Therefore, this event is determined to be a P-wave diffraction.

A diffraction centered at station 2220 is observed on one record with an apparent velocity and depth of 990 m/s and 16 m, respectively (Figure 10c). Because the calculated depth is consistent with suspected tunnel depths and the velocity is approximately equal to the P-wave velocity derived from the direct wave, this event is determined to be a P-wave diffraction. The anomaly at station 1606 is consistent with intelligence information from local law-enforcement agencies.

## Summary

Detection of a diffraction on one shot record does not always lead to an unambiguous interpretation. Diffraction type can vary within a single survey depending on the energy-transmission characteristics of the ground surface. Uncertainty of the diffraction type because of identifica-

tion of a diffraction on only one shot record with unclear phase characteristics might result in inaccurate calculation of the average velocity of the overburden and depth of the anomaly. Diffractions with inconsistent characteristics might be misinterpreted as noise. Lack of trace-to-trace consistency can be caused by ambient noise, inconsistent receiver coupling, statics, and other near-surface conditions. Lack of record-to-record consistency can be caused by changes in ambient noise, receiver coupling, and changes in the raypath of seismic energy. Therefore, it is vital that minimal ambient noise and optimum consistency are maintained during acquisition for the successful detection of diffractions and correct interpretation of small anomalies.

Information about the lateral location and wave type of the energy diffracted by a subsurface anomaly can be inferred directly from a diffraction observed on as little as one shot record. The average velocity of the overlying medium can be determined from the hyperbolic curvature of P- and S-wave diffractions. For P-S diffractions, the diffraction must occur on multiple shot records to determine the average P- and S-wave velocities of the overlying medium. Using the known velocities, the depth of the anomaly can be calculated from one shot record using the apex arrival time and source offset.

# Surface-wave Backscatter Analysis

When surface-wave energy propagates away from the source and meets an object, part of its energy scatters back toward the source (Figure 11). The surface-wave backscatter-analysis technique focuses on enhancing backscatter energy and attenuating/filtering the usual body- and surface-wave energy to pinpoint the location of the anomalous object. The method is sensitive to changes in subsurface seismic properties such as voids and boundaries. Anomalies such as air-filled cavities or voids are especially suitable for this approach because of the abrupt change in material properties (such as velocity and stiffness) between the geologic medium and the air.

## Model data

Synthetic seismic data were calculated with SeisSyn using 2D elastic finite-difference modeling specifically tuned for the estimation of surface-wave propagation. Seismic model parameters were selected using a two-layer model (Figure 12) containing a void starting at a depth of 4.5-m that is 1.4 m wide and 2 m high. Synthetic seismic shot records were calculated using a 25-Hz Ricker wavelet with a 14-m source offset from the nearest receiver, 30 receivers spaced at 1.2 m, and the acquisition system moving (rolling) above the void with 2.4 m from shot to shot. For seismic data-processing convenience, the left edge of the void is assigned to station number 8036 (a horizontal location number) at $x = 39.2$ m. In this example, station numbers increase to the left toward the source every 1.2 m (i.e., the geophone interval).

Backscatters can be identified on shot records as seismic events originating from the surface-wave energy trend at the horizontal location of an anomalous object (e.g., void) and having a slope in a direction opposite to that of the surface wave (Figure 13). The middle synthetic seismic record is calculated using source and receiver locations indicated on the seismic model (Figure 12).

Modeled surface-wave data were analyzed using the backscatter technique. The backscatter-analysis method first applies an *f-k* filter on shot records to remove the dominant, forward-propagating surface-wave energy (from source to receivers), resulting in enhanced backscatter energy propagating backward from a backscatter location (if such exists) toward the source. Linear-moveout (LMO) corrections then are applied to the shot records (Figure 14) using estimated surface-wave velocities derived from dispersion curves. Such a correction positions all remnant forward-propagating surface-wave energy and the peak of any backscatter events to time zero. As a result, remnant surface-wave energy (after *f-k* filtering) appears as horizontal events at time zero. Backscatter energy would appear as the strongest-dipping linear events on a shot record with a slope from trace to trace opposite to that of the forward-propagating surface wave and crossing time zero at the horizontal location of the void, i.e., at station 8063 (Figure 14).

The last step is obtaining a common-receiver stack (CRS) by stacking data from all shot records using receivers positioned at identical locations (i.e., stations). Such a CRS processing scheme enhances backscatter imaging by constructively stacking backscatter signal from different shot records and attenuating other types of signals (Figure 15). As a result, all coherent dipping signals converge to the horizontal location of the anomaly by crossing time zero at station 8063.

Closer inspection of the backscatter signal shows that its frequency range is approximately 13 to 52 Hz, which roughly corresponds to wavelengths between 4.5 and 18 m (using $V_{S1} = 230$ m/s for simplicity). These modeled data indicate that wavelengths shorter than 4.5 m did not reach the top of the void (which is at 4.5 m) and, as a result, did not scatter back. These data also show that the void

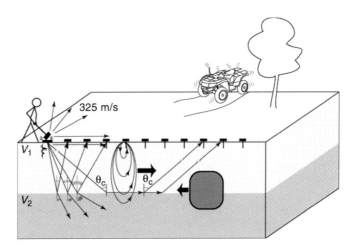

**Figure 11.** Seismic body waves (straight arrows), surface waves (curved arrows and thick arrow to the right showing direction of propagation), and surface-wave backscatter (thick arrow to the left showing direction of propagation) from an object located between layers 1 and 2.

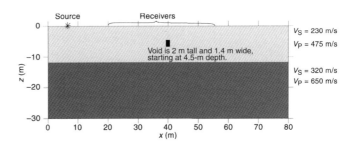

**Figure 12.** Seismic model image used for calculating synthetic seismic shot records showing the location of the void (black rectangle).

managed to reflect/diffract back signal wavelengths significantly longer than its size ($18 \gg 2$ m), suggesting that its main cause is the contact between void and native material rather than the size of the void, which is significant for the backscatter process.

## Field example

Data were acquired for the 2D full-wavefield surveys at a field site visited during this campaign using state-of-the-art near-surface imaging equipment and techniques. Four Geometrics StrataView R60 seismographs were interfaced to a Geometrics StrataVisor NZC to allow the flexibility necessary to record one- to 240-channel configurations with 24-bit A/D conversion. Because of the broad spectral requirements of full-wavefield measurements, it was necessary to use a low-frequency source and matched low-frequency receivers. Receivers were deployed in both a conventional spike-coupled format and a towed spread. In the spike-coupled format, all receivers were Geospace GS-11D 4.5-Hz vertical geophones. For the towed array, both Geospace GS-11D 4.5-Hz vertical geophones and Geospace GS-11 14-Hz horizontal geophones were towed in a continuous pressure-coupled streamer. Receiver stations were spaced 1.2 m apart, with the source impact points separated by 2.5 m. Three ground impacts from a rubber-band-accelerated weight drop were stacked vertically in the seismograph at each shot station.

## Data processing

Processing concentrated on data recorded from 30 vertical receiver stations optimally offset from the source by a range of distances determined during preliminary testing at the known void site with confirmation of the selected range of offsets at each site. All processing was centered around wavefield anomalies caused by small voids in otherwise relatively uniform earth materials less than 30 m below the ground surface. Traditional approaches to surface-wave processing were used to estimate dispersion characteristics

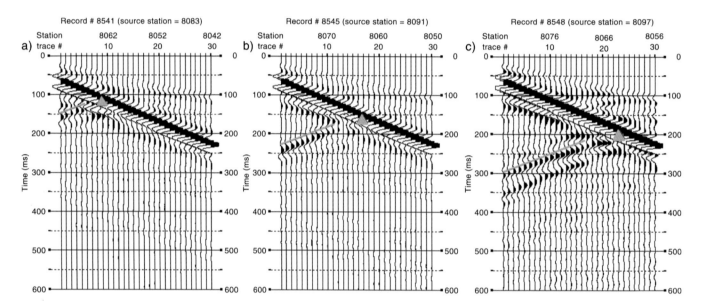

**Figure 13.** Synthetic seismic shot records showing backscatters (below light gray lines) from the void located below station 8063 (light gray triangle).

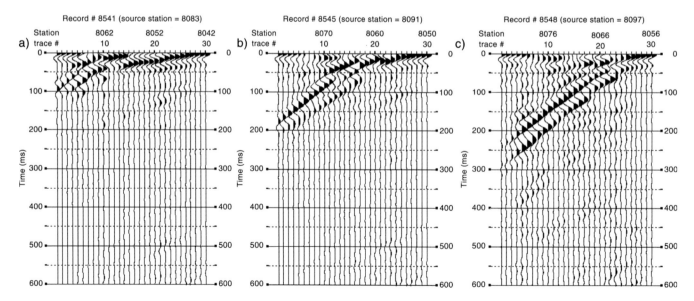

**Figure 14.** Synthetic seismic shot records after *f-k* filter and LMO corrections, making backscatter from the void (at station 8063) the strongest-dipping event converging to station 8063 at time zero.

and to designate the acquisition geometries and operational procedures.

Analysis of dispersion curves (frequency versus phase velocity) and shot gathers (scattered energy, nonlinearity of wave propagation, and frequency versus source offset) provided a variety of approaches to enhance the empirical and model-derived seismic characteristics of voids. Data were analyzed and processed using SurfSeis 3.0 and beta versions of both WinSeis Turbo 1.9 and KGS SeisUtilities 1.0 (proprietary software packages from Kansas Geological Survey, facilitating use of MASW as well as general seismic data processing). These data were run through a large and diverse set of processing routines with the most notable listed below.

- offset-dependent estimations of dispersion curves
- inversion of dispersion curves to establish velocity-depth dependencies
- linear moveout as a function of phase velocity in the *f-k* domain
- filtering in the *f-k* domain of all energy consistent with normal surface-wave propagation in a homogeneous earth in both shot and receiver domains
- common-receiver stacking after LMO and receiver gathers for different offsets
- digital filtering pre- and post-LMO and receiver stack

## Backscatter processing results

As is the case with all geophysical methods, no single data set can provide a unique answer or solution, and all seismic data are nonunique. To best provide a defensible, confident interpretation, it is important to first construct numerical models and synthetic seismic data based on best estimates of target characteristics and second (if possible) to acquire data at a site with similar geology and a known feature closely resembling in depth and size the target feature of the survey. By establishing a "seismic template" in this manner (Figures 13, 14, and 15), pattern recognition and associated assignment of candidate features with confidence ratings can be achieved effectively.

Following the above approach, seismic data were analyzed with the MASW method (Miller et al., 1999) to obtain $V_S$ estimates for this site (Figure 16). Such estimates are essential for modeling surface-wave seismic data. These estimated MASW $V_S$ results were used as a basis for the creation of the model (Figure 12) used for calculating the synthetic seismic data (Figure 13), discussed earlier. After close examination of raw shot gathers (using very high gain), hardly noticeable hints of backscatter energy can be observed on the second and third shot records at

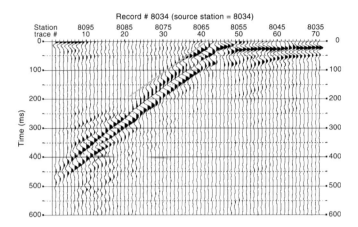

**Figure 15.** Common-receiver stack of synthetic seismic shot records after *f-k* filter and LMO corrections. Backscatter signal crossing time 0 at station 8063 (below the gray line) marks the horizontal location of the void.

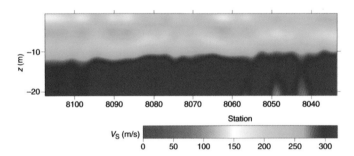

**Figure 16.** MASW $V_S$ cross-sectional plot.

near- and far-offset traces (Figure 17). After *f-k* filtering was applied, these events became stronger and more coherent, suggesting that the origin of the backscatter is at station 8063. Next, LMO correction was applied to all shot records using dispersion-curve estimates from the MASW analysis. Then all traces recorded at the same receiver station from different shot records were stacked to obtain a CRS (Figure 18).

Comparisons of backscatter processing results of real (Figure 18) and modeled data (Figure 15) show close resemblance in both geometric appearance and wavelet signature. This observation implies that the suggested depth of the void in the model used to calculate synthetic data is a possible explanation for the backscatter anomaly in the real data. Close resemblance between the modeled and real data can be used as supporting evidence that a void is a possible explanation for an observed backscatter anomaly in seismic data. In this work, we show that such resemblance can be achieved. Still, ground truth is critical. If invasive confirmation is the next step, enhanced processing and modeling of the specific anomaly to be confirmed

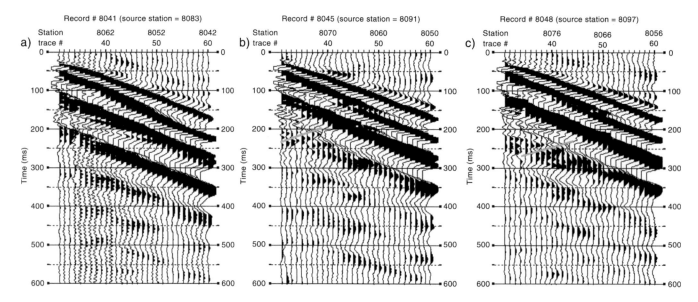

**Figure 17.** Raw seismic shot records showing hints of backscatter energy (below light gray lines).

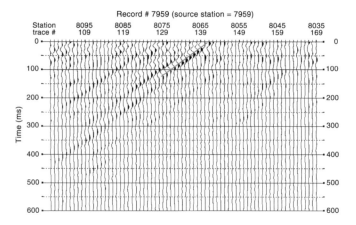

**Figure 18.** Seismic shot records after *f-k* filtering and backscatter processing of shot records.

could reduce greatly the number of boreholes necessary to confirm or refute these interpretations.

## Stress and Shear-wave Velocity

Formation of sinkholes through catastrophic failure of subsurface voids represents a public-safety hazard in Kansas where mining for resources such as coal, lead, zinc, and salt have left voids whose location, extent, or condition are often unknown. Dissolution mining of the Hutchinson Salt Member over the last century in central Kansas has left abandoned wells and brine-filled voids in the subsurface that eventually can lead to sinkholes. The current condition of many of these "jugs" (as they are commonly called) is unknown, warranting investigation to assure public safety.

Geophysical methods represent the only viable approach to characterizing the condition and location of abandoned mine voids over a sizable area prior to drilling out the well plugs for direct interrogation. Of the various geophysical approaches with potential to detect these voids, seismic methods have the greatest theoretical resolution and sensitivity. Previous studies have demonstrated the ability of seismic-reflection methods to image the subsurface expression of subsidence features and bedding geometries within the collapse zone using compressional waves (Miller et al., 1993; Miller et al., 2005) and changes over time (Lambrecht and Miller, 2006). However, no work has been done to evaluate void-induced changes in the localized stress field for the potential of near-term roof failure.

An abandoned dissolution minefield in Hutchinson, Kansas, was chosen to evaluate the use of shear-wave reflection methods to discriminate locations with increased potential for roof failure based on documented subsurface conditions, confirmed through drilling and the use of sonar, acoustic televiewer, and modern borehole logs. We show that changes in shear-wave velocity ($V_S$) correlate to known anomalous subsurface conditions and appear indicative of relative changes in effective stress in rock layers overlying subsurface voids. The $V_S$ calculated using normal-moveout (NMO) estimates of reflection curvature might serve as an indicator of risk for vertical migration through roof failure.

Sinkholes resulting from subsidence within dissolution minefields in and around Hutchinson, Kansas, have been reported for more than 90 years (Walters, 1978). Documented sinkholes related to dissolution mining of the Hutchinson Salt have formed as a result of roof failure associated with jugs formed from the single-well mining method. In general, the single-well method involves injecting fresh water near the base of the salt and recovering the brine solution near the top of the salt through a multi-plumbed borehole (Figure 19a).

Problems with the method occur when the volume of salt removed reaches a point at which the void beneath the unsupported roof span exceeds a size that the roof strength will endure (Figure 19a). Failure near the wellbore results in rupture of the freshwater tubing and upward movement of the dissolution zone (Figure 19b). Once the failure and collapse cycle reaches the caprock, the process becomes much more horizontal (Figure 19c). When two wells with developing voids are in proximity, horizontal void migration might lead to the formation of a gallery, where the voids become interconnected (Figure 19d).

When caprock fails, vertical growth continues progressively upward through the overburden as stress exceeds the strength in each successive unsupported span of new roof rock. The void space available within the salt interval to accommodate roof-collapse breccia is key to whether void movement through the entire overburden rock column proceeds as a relatively continuous process or in intermittent segments.

Near-surface materials at the test site are comprised of 20 m of Quaternary alluvium and unconsolidated Plio-Pleistocene Equus beds. Bedrock is defined as the Ninnescah Shale, which overlies the 5-m-thick Three Finger Dolomite at a depth of 70 m. The top of the Upper Wellington Shale is at 75 m, followed by the top of the Permian Hutchinson Salt Member at 125 m. Caprock characteristics are a very important component of subsidence and the formation of sinkholes. The Permian shales (Wellington and Ninnescah) that overlay the Hutchinson Salt are about 100 m thick in this area and are characterized as unstable when exposed to freshwater, being susceptible to sloughing and collapse (Swineford, 1955). The basal contact of the Wellington Shale is key to the general strength of roof rock if dissolution-mined voids reach the top of the salt zone. Directly above the salt-shale contact is a shale layer with halite-filled joint and bedding cracks (Walters, 1978). Once unsaturated brine comes in contact with this layer, joints and bedding planes are leached rapidly, leaving a structurally weak layer.

## Stress-velocity relationship

Changes in $V_S$ are a key indicator of either previous subsidence activity (low-strength setting) or areas where void growth has elevated the stress field and increased the potential for roof collapse. The $V_S$ is related directly to the ratio of stress to strain. Because $V_S$ is dependent on the rock matrix more so than pore conditions, monitoring changes in velocity should represent a highly sensitive method of detecting failure potential arising from nonlinear changes in strain relative to stress. Changes in $V_S$ for a particular rock are related to differential stress and associated nonlinearity in the stress-strain curve (Dvorkin et al., 1996).

Shear-wave velocity can be expressed as

$$V_S = \sqrt{\frac{\mu}{\rho}}, \qquad (4)$$

where $\mu$ represents the shear modulus and $\rho$ is the bulk density. Assuming that density remains constant, $V_S$ is controlled by the shear modulus, which is the change in force across a unit area (stress) divided by the lateral change in

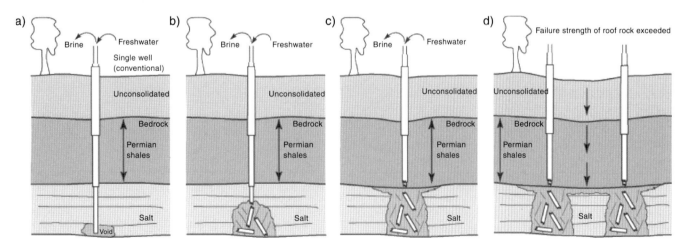

**Figure 19.** (a-c) Illustration of void migration with the single-well mining method and (d) the formation of a gallery by two adjacent wells and associated voids.

cross section for a given length (strain), expressed as

$$\mu = \frac{\sigma}{\gamma} = \frac{\Delta F/A}{\Delta L/L}. \qquad (5)$$

The shear modulus is assumed to remain constant under static pressure and within the elastic deformation portion of the stress-strain curve.

Laboratory measurements of compressional and shear-wave velocities show marked nonlinear increases with increased confining stress (Eberhart-Phillips et al., 1989; Khaksar et al., 1999; Siggins, 2006). These measured velocity changes suggest that variations in effective stress, such as variations present above a void with a large roof span or a reservoir under production pressures (Sayers, 2004; Herwanger and Horne, 2005), could cause significant changes in velocity of the affected rocks. Carrying the confining stress to the logical extreme introduces plastic deformation and failure. At this end of the stress-strain curve, the strain response to small changes in stress is substantial. $V_S$ is reduced when mechanical damage occurs in the rock as a result of stress-induced plastic deformation (Winkler, 2005).

The strength of individual rock layers can be described qualitatively in terms of stiffness or rigidity and empirically estimated from measurements of $V_S$. $V_S$ is directly proportional to stress and inversely related to strain. Figure 20 depicts the results of a simple model showing the changes in effective stress in the media surrounding a void. Because the $V_S$ of earth materials changes when the stress on those materials becomes relatively large, it is reasonable to suggest that load-bearing roof rock above dissolution voids might experience elevated $V_S$ because of loading between load-bearing sidewalls (Figure 20). This localized increase in $V_S$ is not related to increased strength but is caused by increased load. High $V_S$ "halos" are likely key indicators of the potential for near-term roof failure.

Seismic survey lines were selected based on known borehole and subsurface conditions selected at well locations within the minefield. Wells were drilled out and interrogated using acoustic televiewer, sonar, and modern borehole logs to determine void conditions. We compared seismic-reflection data acquired over areas representative of native subsurface conditions and where the void has migrated into the overlying shale caprock with the potential for further roof failure and migration.

## Data acquisition

Downhole seismic data were acquired to obtain a 1D velocity profile that ties reflectors to two-way traveltime on surface-seismic sections and therefore reflection NMO velocity. The downhole-velocity profile and borehole data were used to generate synthetic seismograms representative of the site geology. The forward model aided in the design of data-acquisition parameters and interpretation of common-source gathers.

Maximum 120-fold CMP, horizontally polarized shear-wave ($S_H$) reflection profiles were acquired at well locations with known subsurface conditions. A fixed spread of 240 14-Hz horizontal-component geophones was planted on 1-m spacing along each line. The source was an IVI Mini-Vibe I with a horizontally oriented mass and waffle-style base plate inputting three 10-s-long linear upsweeps ranging from 15 to 150 Hz at each source location. The source walked through the spread at intervals of 2 m. Data were recorded using four networked 60-channel Geometrics StrataView seismographs with 24-bit A/D conversion. The sampling interval was 1 ms, producing record lengths of 12 s. Line locations were restricted by surface obstacles and access constraints, necessitating less than ideal locations for some receiver spreads and overall line lengths. Lines were bulldozed to remove brush and the upper few inches of soil. This line preparation dramatically improved geophone and source coupling and subsequent data quality.

CMP data were processed using techniques commonly applied to near-surface seismic-reflection data (Steeples and Miller, 1990). Large vertical changes in NMO velocity, ranging from ~200 m/s in the unconsolidated overburden

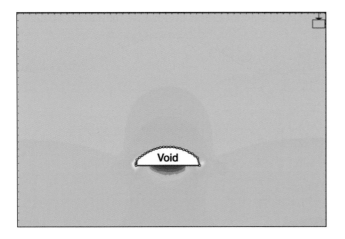

**Figure 20.** Simple model depicting a subsurface void and corresponding increases in stress around the roof and walls and decreased stress at the base. Areas of increased stress are indicated by warm colors, and lower stress values are represented by cool colors.

to ~550 to 650 m/s in the Ninnescah Shale, necessitated the use of offset-dependent subsets to achieve optimal imaging. Steeply dipping low-velocity reflections require severe stretch muting to avoid stacking low-frequency stretched portions of the wavelet after NMO corrections (Miller, 1992), which eliminates valuable reflection information from deeper events, leading to poorer S/N and degraded images (Miller and Xia, 1998). The data were divided into near- and far-offset subsets, separating the bedrock reflection (near offsets) from the dolomite and salt reflections (far offsets). Each subset was NMO-corrected independently and then recombined into a single data set prior to CMP sorting and stacking.

## Results

Comparison of the synthetic seismogram with a raw common-source gather from the reflection survey provides conclusive correlation of reflections to reflectors and the match between borehole-measured velocity and NMO velocity (Figure 21). Reflections from the tops of the Ninnescah Shale, Three Finger Dolomite, and Hutchinson Salt are indicated by the arrows and correlate very well with the synthetic data and drill records. The P-S converted energy also is observable between the bedrock and dolomite reflections at ~350 ms.

Line 1 (Figure 22a) was collected over an area representing native conditions with no known voids above the salt/shale contact. Line 5 (Figure 22b and 22c) was acquired over an area where voids are present above the salt interval and vertical migration has progressed upward through the Upper Wellington Shale. Both the presence and extent of the voids would be expected to cause a relative increase in the effective stress manifested as changes in the $V_S$ of the overlying materials.

Decreasing lateral coverage of the image with depth arises from the fixed receiver spread, where longer offsets required to image deeper reflectors were not recorded when the source was in the middle of the spread. Normal-moveout velocities were picked at every tenth CMP location using constant velocity stacks with 10 m/s steps and hyperbola fitting. Figure 22d shows a plot of NMO velocities for the bedrock (at ~200 ms) and dolomite (at ~450 ms) reflections along lines 1 and 5. Of particular interest is the more than 20% differential between the NMO velocities of the dolomite reflection on lines 1 and 5. Velocity of the unconsolidated overburden remains relatively constant between the two lines, suggesting that this velocity difference is contained fully within the Ninnescah Shale with no influence from overlying unconsolidated sediments.

One void along line 5 is known from drilling and sonar data to have migrated into the Upper Wellington Shale, constrained temporarily by the Three Finger Dolomite. Several other dissolution jugs located adjacent to the seismic line are of the same construction and vintage. Figure 22c shows the relative positions of adjacent wells with overlays of inferred void geometries. Dashed lines indicate approximate geometry based on salt tonnage removed and historical borehole data. Solid lines are based on modern sonar, modern geophysical logs, tonnage, and historical borehole data. The overlays were not derived from seismic data. Elevated velocity along the entire line suggests that more than one void could be influencing the velocity measurements. A localized velocity increase is observed in the area between locations 275 and 300, which lies between wells 25 and 40. This increase in velocity might be related to either well (well 25 is offset to the north by about 18 m and well 40 to the south by about 9 m) or might be a gallery between the two. These

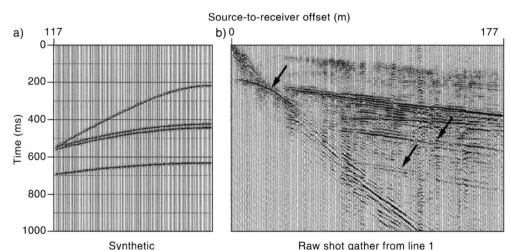

**Figure 21.** Comparison of (a) a synthetic gather created by using downhole velocities and (b) a raw shot gather from line 1. Reflections from the Ninnescah Shale, Three Finger Dolomite, and the Hutchinson Salt are indicated by arrows from top to bottom, respectively.

particular wells have not been interrogated to determine the location or condition of the jug, but this seismic investigation suggests elevated stress conditions indicative of an upward-migrating void, and further investigation could be warranted.

Theoretically, lateral variability in the velocity gradient can be related to changes in the stress field as defined by principles governing the elastic moduli in rocks. Reflection data from this study are the first known to provide empirical evidence relating the changes in stress associated

**Figure 22.** Stacked reflection sections from (a) line 1 and (b and c) line 5, acquired over (a) native subsurface conditions and (b) a void that has migrated into the overlying shale caprock. (c) Void geometry is indicated by overlays. Dashed lines indicate that geometry and size are based approximately on tonnage removed and historical borehole data. Solid lines are derived from modern sonar, modern geophysical logs, tonnage, and historical borehole data. (d) Graph shows shear-wave NMO velocities for the dolomite and bedrock reflections for lines acquired over native conditions (diamonds) and known voids (squares).

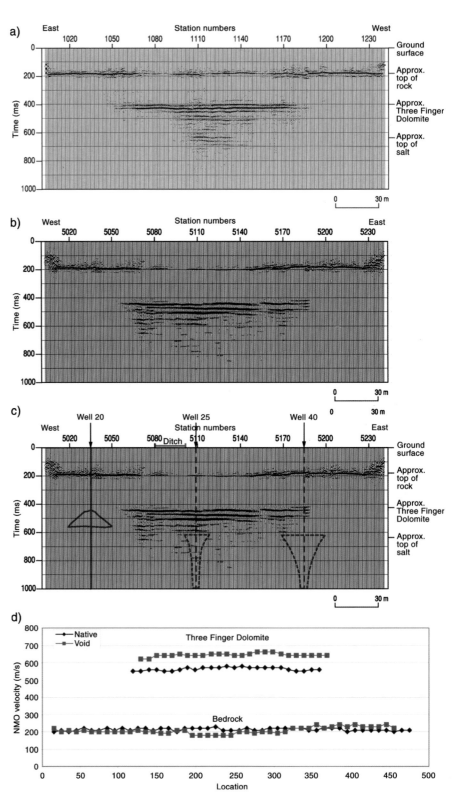

with the tensional dome concept of Davies (1951) to non-invasively measured shear-wave seismic-reflection velocities. These reflection data provide some of the most conclusive evidence to date for direct detection of salt jugs with the potential for upward migration using surface geophysics. Elevated $V_S$ values correlate to areas where jugs have migrated above the Upper Wellington Shale–Hutchinson Salt contact. These changes are likely diagnostic of changes in rock properties or layer geometry related to voids. For shear-wave seismic reflection to be a viable method to characterize the degree of change in stress associated with a single cavern at this site, resolution must be improved and velocity-discrimination techniques must be developed that can isolate changes in velocity to suboptimum offset distances.

## Discussion

Three near-surface seismic techniques have been discussed here, including body-wave diffractions, surface-wave backscatter, and $V_S$ variations using multiple examples of both modeled and real data sets. Each technique is based on different portions of the seismic wavefield and can be applied to direct detection of anomalous subsurface features such as voids or their effects on surrounding material properties.

To summarize, diffraction analysis uses relatively high-frequency body waves and is responsive to relatively small features in comparison to wavelengths. There are four types of diffractions: P, S, P-S, and S-P. Each of their signatures is unique and is based on observable characteristics such as polarity and velocity. Accurate depth and velocity measurements can be made using only pure P- or S-wave diffractions. S-P diffractions will produce depths and velocities that are too deep and slow, respectively, although P-S diffraction analysis is just the opposite. Although there are dependencies related to velocity and depth, the lateral location can be determined from any type of diffraction directly from shot gathers and theoretically is limited only by the receiver spacing used, where smaller intervals increase the lateral resolution.

Surface-wave backscatter analysis is based on low-frequency surface-wave components of the wavefield. The method is sensitive to anomalous changes in seismic properties, most notably at contact boundaries such as the change from a semiconsolidated medium to air. The size of the void is less important than the contrast across those boundaries. The lateral resolution of the technique is limited to several stations but is a valuable tool for discerning subsurface contrasts, especially when used in

conjunction with other methods such as diffraction enhancement.

The use of shear-wave NMO velocity variations is not intended as a direct imaging technique but as a way to observe the effects that the presence of a void imposes on the surrounding media. This technique can be used to identify voids with potential of roof failure through the buildup of stress in the roof rock and walls, which is linked directly to Vs. In its current state, the method is applicable to larger voids at depths that are not feasible for MASW, but it is not well suited for small shallow voids because the overlying materials are less likely to exhibit observable increases in stress.

## Conclusions

The three near-surface seismic methods described here have demonstrated the potential of identifying subsurface voids or their effects on surrounding media using field data and validated by modeling. These methods can be used independently in many applications or in conjunction with other seismic or geophysical methods to increase interpretation confidence and to decrease ambiguity. Applications of the methods include the detection of voids and cavities, karst features, tunnels, abandoned mine workings, and dissolution features such as salt jugs. Further research and development are warranted to continue to develop these methods and to increase their applicability to a wider range of geologic and cultural environments and gain a better fundamental understanding of the underlying physics.

## Acknowledgments

The authors wish to thank Brett Wedel, Tony Wedel, Joe Anderson, Owen Metheny, Justin Schwarzer, Birgit Leitner, and Arno Rech for their assistance in the field and Mary Brohammer for her editorial and graphics work. Permission to publish this paper was granted by the director of the Geotechnical and Structures Laboratory, U. S. Army Engineer Research and Development Center.

## References

Belfer, I., I. Bruner, S. Keydar, A. Kravtsov, and E. Landa, 1998, Detection of shallow objects using refracted and diffracted seismic waves: Journal of Applied Geophysics, **38**, 155–168.

Berkovitch, A., I. Belfer, Y. Hassin, and E. Landa, 2009, Diffraction imaging by multifocusing: Geophysics, **74**, no. 6, WCA75–WCA81.

Branham, K. L., and D. W. Steeples, 1988, Cavity detection using high-resolution seismic reflection methods: Mining Engineering, **40**, 115–119.

Butler, D. K., 1984, Microgravimetric and gravity gradient techniques for detection of subsurface cavities: Geophysics, **49**, 1084–1096.

Davies, W. E., 1951, Mechanics of cavern breakdown: Bulletin of the National Speleological Society, **13**, 36–43.

Dvorkin, J., A. Nur, and C. Chaika, 1996, Stress sensitivity of sandstones: Geophysics, **61**, 444–455.

Eberhart-Phillips, D., D.-H. Han, and M. D. Zoback, 1989, Empirical relationships among seismic velocity, effective pressure, porosity, and clay content in sandstone: Geophysics, **54**, 82–89.

Fenner, T., 1995, Ground penetrating radar for identification of mine tunnels and abandoned mine stopes: Mining Engineering, **47**, 280–284.

Fomel, S., E. Landa, M. T. Taner, 2007, Poststack velocity analysis by separation and imaging of seismic diffractions: Geophysics, **72**, no. 6, U89–U94.

Hecht, E., 2002, Optics, 4th ed.: Addison Wesley.

Herwanger, J., and S. Horne, 2005, Predicting time-lapse stress effects in seismic data: The Leading Edge, **24**, 1234–1242.

Kanasewich, E. R., and S. M. Padke, 1988, Imaging discontinuities on seismic sections: Geophysics, **35**, 334–345.

Khaidukov, V., E. Landa, and T. J. Moser, 2004, Diffraction imaging by focusing-defocusing: An outlook on seismic superresolution: Geophysics, **69**, 1478–1490.

Khaksar, A., C. M. Griffiths, and C. McCann, 1999, Compressional- and shear-wave velocities as a function of onfining stress in dry sandstones: Geophysical Prospecting, **47**, 487–508.

Korneev, V., 2008, Resonant seismic emission of subsurface objects: Geophysics, **74**, no. 2, T47–T53.

Korneev, V. A., and L. R. Johnson, 1996, Scattering of P and S waves by a spherically symmetric inclusion: Pure and Applied Geophysics, **147**, 675–718.

Lambrecht, J. L., and R. D. Miller, 2006, Catastrophic sinkhole formation in Kansas: The Leading Edge, **25**, 342–347.

Landa, E., and S. Keydar, 1998, Seismic monitoring of diffraction images for detection of local heterogeneities: Geophysics, **63**, 1093–1100.

Landa, E., V. Shtivelman, and B. Gelchinsky, 1987, A method for detection of diffracted waves on common-offset sections: Geophysical Prospecting, **35**, 359–373.

Levander, A. R., 1988, Fourth-order finite-difference P-SV seismograms: Geophysics, **58**, 1425–1436.

Militzer, H., H. Rosler, and W. Losch, 1979, Theoretical and experimental investigations for cavity research with geoelectrical resistivity methods: Geophysical Prospecting, **27**, 640–652.

Miller, R. D., 1992, Normal moveout stretch mute on shallow-reflection data: Geophysics, **57**, 1502–1507.

Miller, R. D., J. Ivanov, D. W. Steeples, L. W. Watney, and T. R. Rademacker, 2005, Unique near-surface seismic reflection characteristics within an abandoned salt-mine well field, Hutchinson, Kansas: 75th Annual International Meeting, SEG, Expanded Abstracts, 1041–1044.

Miller, R. D., D. W. Steeples, L. Schulte, and J. Davenport, 1993, Shallow seismic-reflection feasibility study of the salt dissolution well field at North American Salt Company's Hutchinson, Kansas, facility: Mining Engineering, October, 1291–1296.

Miller, R. D., and J. Xia, 1998, Large near-surface velocity gradients on shallow seismic-reflection data: Geophysics, **63**, 1348–1356.

Miller, R. D., J. Xia, C. B. Park, and J. M. Ivanov, 1999, Multichannel analysis of surfaces waves to map bedrock: The Leading Edge, **18**, 1392–1396.

Ogilvy, R. D., A. Cuadra, P. D. Jackson, and J. L. Monte, 1991, Detection of an air-filled drainage gallery by the VLF resistivity method: Geophysical Prospecting, **39**, 845–859.

Sayers, C. M., 2004, Monitoring production-induced stress changes using seismic waves: 74th Annual International Meeting, SEG, Expanded Abstracts, 2287–2290.

Sheriff, R. E., 2002, Encyclopedic dictionary of applied geophysics, 4th ed.: SEG Geophysical References Series No. 13.

Siggins, A. F., 2006, Velocity-effective stress response of $CO_2$-saturated sandstones: Exploration Geophysics, **37**, 98–103.

Steeples, D. W., and R. D. Miller, 1987, Direct detection of shallow subsurface voids using high-resolution seismic-reflection techniques; *in* B. F. Beck, W. L. Wilson, and A. A. Balkema, eds., Karst hydrogeology: Engineering and environmental applications: Balkema, 179–183.

———, 1990, Seismic reflection methods applied to engineering, environmental, and groundwater problems, *in* S. H. Ward, ed., Geotechnical and environmental geophysics, v. 1.: Review and tutorial: SEG Investigations in Geophysics Series No. 5, 1–30.

Swineford, A., 1955, Petrography of upper Permian rocks in south-central Kansas: State Geological Survey of Kansas Bulletin 111.

Turpening, R. M., 1976, Cavity detection by means of seismic shear and compressional wave refraction techniques: Environmental Research Institute of Michigan, Report 116400-1-F.

Virieux, J., 1986, P-SV wave propagation in heterogeneous media: Velocity-stress finite-difference method: Geophysics, **56**, 889–901.

Walters, R. F., 1978, Land subsidence in central Kansas related to salt dissolution: Kansas Geological Survey Bulletin 214.

Winkler, K. W., 2005, Borehole damage indicator from stress-induced velocity Variations: Geophysics, **70**, no. 1, F11–F16.

Chapter 13

# Inversion Methodology of Dispersive Amplitude and Phase versus Offset of GPR Curves (DAPVO) for Thin Beds

Jacques Deparis[1, 2] and Stéphane Garambois[1]

## Abstract

The presence of a thin layer embedded in any formation creates complex reflection patterns because of interferences within the thin bed. Amplitude variation with offset (AVO) is used increasingly in seismic interpretation and has been tested more recently on ground-penetrating radar (GPR) data to characterize nonaqueous-phase liquid contaminants. In those analyses, phase and dispersion properties of the reflected signals generally are omitted, although they contain useful information. An inversion methodology to examine thin-bed properties — dispersive amplitude and phase versus offset (DAPVO) — combines all reflectivity properties (amplitude, phase, and dispersion) of the reflected GPR signal generated by a thin bed embedded within a homogeneous material. A brief description of electromagnetic (EM) phenomena is presented. The dispersive properties of the dielectric permittivity of investigated materials can be described using a Jonscher parameterization, which allows the study of the dependency of amplitude and phase versus offset (APVO) curves on the frequency of thin-bed properties (filling nature, aperture). Simplifying assumptions and using careful corrections are necessary to convert raw common-midpoint (CMP) reflected data into DAPVO curves and to study the propagation and radiation-pattern corrections. The inversion methodology is explained and validated to a synthetic set of CMP GPR data and can be illustrated with a real CMP data set acquired along a vertical cliff. This allows for extraction of the characteristics of a subvertical fracture while simultaneously satisfying resolution and confidence. Such a study
motivates interest in combining the dispersion dependency of the reflection-coefficient variations with classical AVO analysis for thin-bed characterization.

## Introduction

Electromagnetic (EM) reflection properties have been studied widely for purposes of ground-penetrating radar (GPR) (Annan, 2002; Grégoire et al., 2003; Carcione et al., 2006). The properties are sensitive to several contributing properties of the ground medium, such as dielectric permittivity, electrical conductivity, and magnetic permeability (Annan, 2002). Furthermore, they depend on instrument setting and use, such as antenna polarization (Lehmann, 1996; Lutz et al., 2003), incident angle of the GPR wave (Annan, 2002; Bradford and Deeds, 2006), and spectrum of the studied wave (Grégoire et al., 2003; Deparis and Garambois, 2009). In seismic interpretation, amplitude variation analysis of the seismic reflectivity as a function of amplitude variation with offset (AVO) proved to be a useful tool to assess the contrasts in elastic properties (Ostrander, 1984; Castagna, 1993). Those variations were related to changes in lithology (Kindelan et al., 1989) and in fluid content (Simmons and Backus, 1994; Hall and Kendall, 2003; Mahob and Castagna, 2003; Stovas et al., 2006).

AVO tests were performed successfully on GPR data to qualitatively characterize the presence of nonaqueous-phase liquid (NAPL) contaminants in the subsurface (Baker, 1998; Deeds and Bradford, 2002; Jordan and

[1]Laboratoire de Géophysique Interne et Tectonophysique (LGIT), Université Joseph Fourier & Centre National de la Recherche Scientifique, Grenoble, France.
[2]Currently at BRGM, Orléans, France. E-mail: j.deparis@brgm.fr.

Baker, 2002; Jordan et al., 2004). This work was supplemented by the numerical analyses of Lehmann (1996) and of Carcione et al. (2006), who studied transverse electric (TE) and transverse magnetic (TM) reflection variations versus offset for different contrasts in EM properties such as NAPL concentrations.

In the presence of a thin bed, i.e., when a layer is thin compared to the wavelength of the GPR signal propagating through it, it is not possible to detect and separate reflections coming from the upper and lower surfaces. Consequently, classical approaches to velocity interpretation or AVO analysis are impossible to apply. In that case, multiple reflections coming from the two sides of the bed create interferences and generate complex reflection patterns. For seismic applications, the pioneering work of Widess (1973) shows how the composite reflection amplitudes from a thin bed vary as a function of its aperture for a cosine wavelet. That has been generalized and studied more thoroughly by Schoenberger and Levin (1976), Koefoed and de Voogd (1980), and Stephens and Sheng (1985).

More recently, AVO response of a single thin bed was discussed by Liu and Schmitt (2003) in terms of its capability to characterize a seismic reservoir. For GPR data, Grégoire and Hollender (2004) compared the spectral ratio between measured reflected wavelets and a reference wavelet for the case of thin-bed reflectors to estimate the dispersive dielectric permittivity of the reflectors and their apertures. This frequency-sensitive approach was applied only to constant offset sections and not to AVO data. Bradford and Deeds (2006) analyzed AVO curves using an analytical solution of the thin-bed reflectivity and successfully applied their modeling to two case studies dealing with NAPL-contaminated zones. In that case, however, the authors did not account for the dispersive properties of constitutive parameters. New developments in GPR not only use amplitude and phase versus offset (APVO) or

dispersion characteristics but also aim to perform full-waveform inversion or to use attributes (e.g., Sassen and Everett, 2009).

This study is a complement to previous work (Deparis and Garambois, 2009), adding explication and illustration on the inversion scheme. We focus on the global scheme of the inversion process by illustrating its efficiency with different apertures and filling materials and by testing noise effect. In addition, we present a new case study. After an introduction on thin-bed reflectivity theory and dispersion, we illustrate the inversion strategy using synthetic data computed into various contexts and considering various geologic settings. An overview of conversion processes which must be applied to raw data to derive dispersive amplitude and phase versus offset (DAPVO) curves as input of the inversion scheme is presented and illustrated with synthetics. Finally, the global procedure is applied successfully to an original set of data acquired for rock-fall hazard assessment on a vertical cliff.

# Basic Theory

## Reflection coefficient

Fresnel reflection and transmission equations quantify amplitude and phase distribution of EM waves on both sides of an interface for two polarization planes: the TE mode (electric-field vector is parallel to the strike of a dipping fracture plane) and the TM mode (electric-field vector is perpendicular to the strike of a dipping fracture plane).

In the case of thin-bed problems (Figure 1), the corresponding reflection coefficient results from interferences between reflections coming from both sides of the thin bed and presents a more complex equation, where $k_m$ and $k_f$ denote the wavenumbers of the matrix and filling materials, respectively, (Liu and Schmitt, 2003):

$$R(\omega, \theta_i) = \frac{r_{m,f}(\omega, \theta_i) - r_{m,f}(\omega, \theta_i)e^{-i\phi(\omega)}}{1 - r_{m,f}^2(\omega, \theta_i)e^{-i\phi(\omega)}}, \tag{1}$$

where $\phi(\omega) = 2.k_f d.\cos(\omega)$, $k_{m,f} = \omega(\mu_{m,f}\,\varepsilon_{m,f})^{1/2}$ are the wavenumbers and $r_{m,f}$ is the Fresnel reflection coefficient.

Fresnel reflection coefficients no longer are valid to describe offset-dependent reflectivity when EM waves cross a thin bed. The validity limit of the Fresnel equation was discussed previously by Deparis and Garambois (2009), using a comparison between analytical and finite-difference numerical simulations performed in the time domain (Giannopoulos, 2005). They showed that equation 1 is

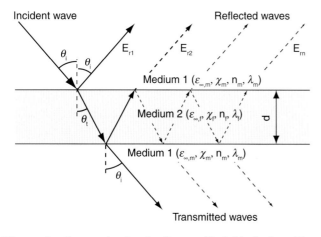

**Figure 1.** Geometric sketch of a stratified thin-bed problem.

valid if the aperture of the fracture remains lower than $\lambda_f/2$. Figure 2 illustrates the effect of the aperture on wavelet reflectivity when considering a normalized thin-bed aperture $(d/\lambda_f)$ varying from 0.1 to 1 and filled by dry sand $(\varepsilon_f/\varepsilon_0 = 4.25, \sigma_f = 0.0027 \text{ S/m})$. The geometric model is made of a thin bed at a depth of 3 m, embedded in homogeneous limestone $(\varepsilon_m/\varepsilon_0 = 9, \sigma_m = 0.001 \text{ S/m})$. Data were computed in the H-plane (with a central frequency of about 100 MHz). The grid resolution $(N = \lambda/\delta_x)$ is larger than 10, with a cell size $(\delta_x)$ of 0.33 cm to keep numerical dispersion low (Holberg, 1987; Bergmann et al., 1996). The effect of the aperture on the fracture is clearly visible.

When the normalized thickness $(d/\lambda_f)$ is lower than 0.5, the reflection generates a complex wave with a complex shape. This limit depends on the bandwidth of the signal. For the useful GPR antennae, the bandwidth of the signal is tight, so we can use the central frequency to calculate the length of waves. For a normalized thickness greater than 0.5, two distinct waves were generated, the first on the top of the fracture and the second on the bottom.

## Frequency dependence

Here, the frequency content of the reflected signal is used with amplitude and phase variation with offset (APVO) properties to improve the convergence of the inversion process and to decrease nonuniqueness in potential solutions. Relaxation mechanisms influencing the constitutive properties must be described correctly in the forward and inverse problems. In our application, only the dielectric polarization mechanism has been included, thus restricting the applications to materials which present weak dispersive characteristics for current conduction (electrical conductivity) and have a constant magnetic permeability (absence of any magnetic particles). The dielectric permittivity, which characterizes redistribution of the locally connected charge under the action of an electric field, is a complex frequency-dependent quantity, with the real component representing the "instantaneous" energy (in phase) and the imaginary component representing the energy dissipation. The different individual relaxation mechanisms combine into bound-charge effects (electronic, atomic, and dipolar polarization) and free-charge effects.

In the GPR frequency range, the most important relaxation phenomenon is the dipolar polarization of bound and/or free charges, whereas the others tend to fall (Cassidy, 2009). Some authors (Debye, 1929; Cole and Cole, 1941; Jonscher, 1977) have proposed different models to account for the dielectric dispersion effects, which are thoroughly presented and discussed by Bano (2004). Among them, Jonscher parameterization (Jonscher,

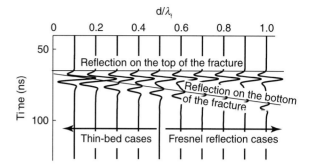

**Figure 2.** Reflected wavelets determined using a captured GPR wavelet and the thin-layer reflection model (equation 1) as a function of the ratio of the thin-layer thickness to the dominant wavelength of the wavelet in the thin layer. Dashed lines show the thickness of the thin layer. The character of the background medium is $\varepsilon_m/\varepsilon_0 = 9$, $\sigma_m = 0.001$, and the thin bed is $\varepsilon_f = 4.25/\varepsilon_0$, $\sigma_f = 0.0027$. After Widess, 1973. Used by permission.

1977) is well adapted to precisely describe complex permittivity $(\varepsilon_e)$ in the frequency range used for GPR surveys (Hollender and Tillard, 1998). This consistency is particularly effective when the signal frequency is higher than the relaxation frequencies; therefore, it is necessary to pay attention to the lower part of the GPR spectrum, where interfacial polarization is present. The original law proposed by Jonscher uses four parameters, but as suggested by Hollender and Tillard (1998), the complex permittivity dispersion can be simplified with three parameters $(\varepsilon_\infty, \chi_r$ and n) in many cases as

$$\varepsilon_e(\omega) = \varepsilon_0 \chi_r \left(\frac{\omega}{\omega_r}\right)^{n-1} \left(1 - i\cot\frac{n\pi}{2}\right) + \varepsilon_\infty, \quad (2)$$

where $\varepsilon_e$ is complex permittivity, $\varepsilon_0$ is permittivity of free space, and $\omega_r$ is the reference frequency (equal to $2\pi\,100$ MHz in our case).

To illustrate the reflectivity sensitivity of a thin bed, analytical APVO curves were computed for the TE mode according to frequency (Figure 3a and 3b) for a vertical incidence and as a function of incidence angle at a frequency of 100 MHz (Figure 3c and 3d). Those computations were performed considering a sand-filled thin bed of different apertures. The thin bed was considered to be embedded within a limestone formation. Those examples highlight that the reflection coefficient is highly sensitive to frequency, aperture, and incidence angle. The frequency variations show destructive interferences for a critical normalized aperture of the thin bed $(d = \lambda_f/2)$, and the phase variation exhibits large fluctuations around that value. The amplitude of the reflection coefficient is maximal for

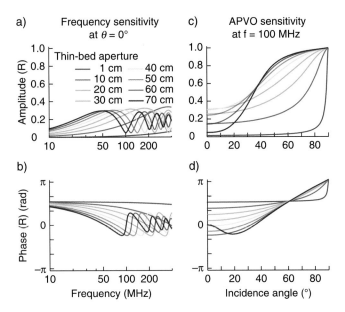

**Figure 3.** Amplitude and phase of the thin-bed reflection coefficient as function of (a and b, respectively) frequency and (c and d, respectively) incidence angle for TE plane waves (equation 1). The considered background medium is limestone ($\varepsilon_{\infty,m} = 8.14$, $\chi_{r,m} = 0.94$, $n_m = 0.82$), and the thin bed is dry sand ($\varepsilon_{\infty,f} = 2.5$, $\chi_{r,f} = 2$, $n_f = 0.85$) with various apertures.

a thin-bed aperture of $\lambda_f/4$, whereas the phase exhibits a smooth variation around that value. The variation of amplitude of the reflection coefficient according to the incidence angle increases more or less quickly (depending on the considered aperture) to a value of one. On the contrary, the phases present weak fluctuations according to incidence angle and aperture. Those basic computations illustrate well the potential of the proposed global approach, which aims to characterize the thin bed and surrounding material properties by combining and inverting all reflection-coefficient variations (dispersive amplitude and phase versus offset).

## Deriving Dispersive APVO Curves

Conversion of field GPR data into reflectivity properties (APVO, dispersion) is a very delicate operation which requires several corrections, discussed by Bradford and Deeds (2006) and by Deparis and Garambois (2009). In the Fourier domain, the global reflected electric field $E_{mes}(\omega, x)$ was recorded at an offset x (incidence angle $\theta_i$) and at a discrete angular frequency $\omega$

$$E_{mes}(\omega, x)$$
$$= E_0(\omega) \frac{D(\omega, \theta_i).C(\omega, \theta_i) \cdot T(\omega, \theta_i) \cdot e^{-ik_m(\omega)r}}{r} R(\omega, \theta_i). \quad (3)$$

In this expression, $E_0(\omega)$ denotes the wavelet originally generated by the GPR system and $k_m$ the matrix material wavenumber, $D(\omega, \theta_i)$ describes the influence of the transmitter and receiver antenna radiation patterns, $C(\omega, \theta_i)$ the influence of transmitter and receiver/soil couplings, $T(\omega, \theta_i)$ potential energy losses resulting from possible interfaces between the source and the studied thin bed, r the length of the travel path studied ($r = x/\cos(\omega_i)$, and x the offset between the emitting and receiving antennae. The factor $e^{-ik_m(\omega)r}$ is the propagation term, which includes the intrinsic attenuation and dephasing when plane-wave propagation is assumed.

Considering this expression, access to the reflection-coefficient ($R(\omega, \theta_i)$) properties is not straightforward, and the raw recorded signal has to be corrected from several terms presented in equation 3, which presents complex trade-offs. Fortunately, as proposed originally by Bradford and Deeds (2006), simplifying assumptions can decrease those difficulties. For example, it is reasonable to assume that the coupling factor does not vary significantly over the offset range for a uniform surface with laterally homogeneous material and that the transmission losses do not vary significantly with offset. Bradford and Deeds (2006) conclude that the error generated by omitting those phenomena is less than 5% and that it might generate problems only for large angles.

## Propagation corrections

As shown in the propagation equation (equation 3), the reflected signal has to be corrected for propagation effects. The process primarily depends on the permittivity of the homogeneous material and on the propagation distance. Rock materials generally show low dispersion of the real part of $k_m$ in the GPR frequency range, and consequently, that dispersion can be considered negligible. The imaginary part of $k_m$ can be deduced from electrical measurements acquired at the test site using electrical tomography or EM methods or laboratory measurements on rock samples. Those electrical acquisitions generally are performed at low frequency (quasi-static conditions) and can be extended to GPR frequencies using the complex permittivity models, such as the Jonscher. However, as noted by Bradford (2007), low-frequency EM measurements provide fundamentally different information than the dispersion parameter D which might be extracted directly from GPR frequency, although nonuniqueness in the solution also could raise problems.

From those considerations, one can conclude that propagation corrections can be roughly performed when the real part of $k_m$ is derived from velocity analysis.

However, when phase properties are studied, the drawback of such an approach is that there is a trade-off between phase shifts because of propagation, relaxation, and reflection; consequently, large errors can be generated in the phase estimation of the reflected signal. Here, the chosen approach consists of inverting the properties of the homogeneous material during a global inversion process. In this way, the better solutions overcome the trade-off problem.

## Radiation-pattern corrections

The second correction concerns radiation patterns of the antennae. For the RAMAC unshielded GPR antennae used in this study, radiation-pattern studies were carried out for antennae placed in free space (Streich and van der Kruk, 2007). One of the observations of Streich and van der Kruk (2007) was that the field results were consistent with the model proposed by Lampe and Holliger (2003) for bow-tie antennae with a Wu and King (1965) resistivity profile. In this study, the radiation pattern was computed using numerical finite-difference time-domain (FDTD) simulations originally presented by Lampe and Holliger (Lampe and Holliger, 2003, 2005). Figure 4 presents a simulation of the radiation pattern obtained for bow-tie GPR antennae with a scaled Wu and King (1965) profile for amplitude and phase as a function of frequency and radiated angle. The numerical FDTD radiation pattern is included in the inversion algorithm. It was extracted from modeled data on a circle around the source at all distances provided by all solutions of the depth of the interface within the inversion process and corrected by a propagation term. In the example shown in Figure 4, the computation used a cell size of 3 cm to achieve a grid resolution ($N = \lambda/\delta_x$) above 10 for a frequency as high as 300 MHz. These figures show that amplitude and phase present weak frequency dependency in the range of 100 to 200 MHz.

## Inversion of Dispersive APVO Curves

Because the corrections needed to extract dispersive APVO curves are highly sensitive to the properties of the homogeneous medium, difficulties can arise when using an AVO or APVO approach. One solution is to try to estimate the properties of the homogeneous medium, apply those corrections, and then extract dispersive APVO curves. However, it is almost impossible to estimate Jonscher parameters from such an approach, and moreover, phase properties of reflectivity might be estimated too inaccurately or might be wrong. To overcome this problem, we propose a global inversion methodology in which all Jonscher

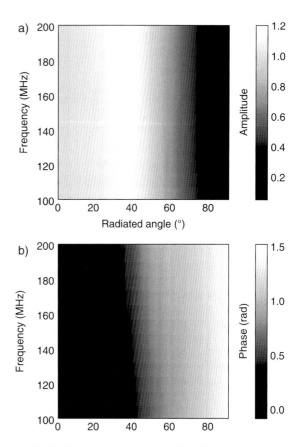

**Figure 4.** Radiation pattern normalized by the propagation term $e^{-ik_m(\omega)r}$ between 100 and 200 MHz obtained from numerical modeling for (a) amplitude and (b) phase.

parameters (thin-bed and homogeneous material) are inverted along with thin-bed depth and aperture. Using that approach, all necessary corrections are inverted considering each solution of the parameter space before inversion.

## Inversion strategy of DAPVO and scheme

The inversion strategy is based on the neighborhood algorithm (Sambridge, 1999a, 1999b) which consists of searching a set of best models inside a given parameter space that fits the experimental data well. The search is performed using the nearest neighbor regions defined under a suitable distance norm. The algorithm requires solving many forward problems, initialized by a random seed number. The advantage of the neighborhood algorithm compared with classical random inversion methods (e.g., Monte Carlo) lies in the speed of solving for space parameters, especially when the number of unknowns is greater than three, which is the case here.

The different inversion steps are presented in Figure 5. We computed a thin-bed dispersive APVO response using equation 3 for CMP offset varying from 0 to 10 m, with a

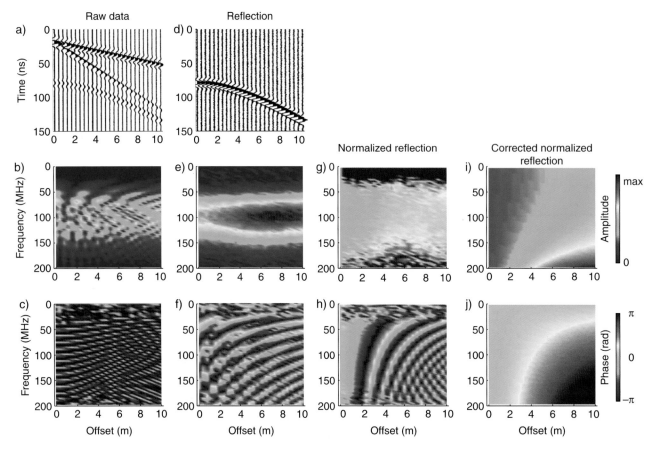

**Figure 5.** Inversion procedure. (1) Preprocessing: raw data in (a) time domain and frequency domain for (b) amplitude and (c) phase, reflected studied event in (d) time domain and frequency domain for (e) amplitude and (f) phase. (2) Inversion: normalized studied event in frequency domain for (g) amplitude and (h) phase, normalized reflectivity model derived from inversion in frequency domain for (i) amplitude and (j) phase.

step of 40 cm and 10% of Gaussian noise in the reflected hyperbola. The thin-bed reflectivity was modeled with the following parameters: limestone, ($\varepsilon_{\infty,\mathrm{m}} = 8.14$, $\chi_{\mathrm{r,m}} = 0.94$, $n_{\mathrm{m}} = 0.82$); filling properties, ($\varepsilon_{\infty,\mathrm{f}} = 2.5$, $\chi_{\mathrm{r,f}} = 2$, $n_{\mathrm{f}} = 0.85$); thickness, 28.3 cm; depth, 3 m.

The raw synthetic CMP acquisition is presented in Figure 5a, 5b, and 5c for TE acquisition mode, respectively, for the time domain, frequency domain, and amplitude and phase of the reflected wave as a function of offset. To test the robustness of this methodology, we added 10% of Gaussian noise to the reflected signal. The CMP initially was processed to mute the direct airwave and ground wave (Figure 5d). The Fourier transform was performed and is presented in Figure 5e and 5f for the amplitude and phase, respectively. The first step of the inversion algorithm is to normalize the signal in the frequency domain for the amplitude and phase (Figure 5g and 5h, respectively). Assuming that the coupling factor and the transmission losses do not vary significantly with offset (Bradford and Deeds, 2006) and the signal source ($E_0$) remains constant

for each acquisition permits the following simplification of the obtained electromagnetic signal (equation 3)

$$E_{\mathrm{mes}}(\omega, x) / E_{\mathrm{mes,ref}}(\omega, x_{\mathrm{ref}})$$
$$= \frac{D(\omega, \theta_{\mathrm{i}}) \cdot e^{-ik_{\mathrm{m}}(\omega)r} r_{\mathrm{ref}}}{D(\omega, \theta_{\mathrm{i,ref}}) \cdot e^{-ik_{\mathrm{m}}(\omega)r_{\mathrm{ref}}} r} \frac{R(\omega, \theta_{\mathrm{i}})}{R(\omega, \theta_{\mathrm{i,ref}})}. \tag{4}$$

Whereas the choice of offset selected for normalization is flexible, preliminary tests show that the inversion algorithm is most efficient when low offsets are used.

The inversion process then directly compares the normalized field data to synthetic normalized data (generated using equation 4) in the frequency domain.

As discussed before, the main advantage of this strategy is that the algorithm takes into account the trade-off among the radiation pattern, the propagation effect, and the complex dispersive permittivity of the matrix and filling properties for each investigated solution. The normalized reflection coefficient obtained after correcting for propagation effects and the radiation pattern is presented in the frequency domain in Figure 5i and 5j for amplitude and phase, respectively.

## Parameter space

The normalized recorded reflectivity (equation 4) shows that the thin-bed reflection coefficient depends on the Jonscher parameters that describe the matrix (homogeneous material) and the thin-bed properties, i.e., the aperture of the fracture and the incidence angle $\theta_I$ (a function of the depth of the thin bed and the studied offset). The radiation pattern is a function only of $\theta_i$, whereas the propagation effect depends on the depth of the fracture and the Jonscher parameters of the matrix. Consequently, the parameter space is an 8D space. It is, possible, however, to reduce the investigation range of some parameters, depending on the application and materials encountered. For example, the real part of the dielectric permittivity can be approximated from velocity analysis of the CMP, thus restricting its possible variations and the possible variations of the depth of the thin bed. In highly resistive formations such as limestone, the potential variations of the dielectric permittivity also can be reduced significantly. Finally, under the thin-bed assumption, the aperture of the thin layer cannot exceed $\lambda_f/2$.

To sum up, if eight unknowns are investigated during the inversion process, four present a broad variation range ($\varepsilon_{\infty,f}$, $\chi_{r,f}$, $n_f$, d), whereas four present a reduced variation range ($\varepsilon_{\infty,m}$, $\chi_{r,m}$, $n_m$, z) in highly resistive formations.

## Algorithm validation: Analytical studies

The inversion program first was tested on analytical signals with 10% of Gaussian noise, presented in Figure 6 in a 50–150-MHz frequency range with the transverse electric reflectivity mode. Gaussian noise has been computed depending on the amplitude of the reflected event and using the properties presented in Figure 5. For each test, inversion was started from three distinct initial random seeds, and 400 iterations were performed successively, resulting in 20,000 models. The parameter space was reduced, as presented in Table 1.

Figure 6 shows all inverted parameters for all models generated by the inversion process in the parameter range displayed in Table 1 as a function of rms error. Each generated model is represented by a dot, with a color scale depending on the misfit value. Dashed lines correspond to the solution. The minimum achieved misfit (red) is about 1.8%. Shapes for the lowest misfit values in Figure 6a

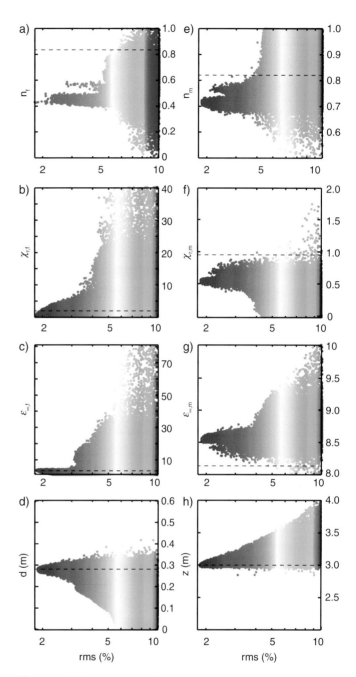

**Figure 6.** Convergence of each parameter for all models generated by the inversion process as a function of rms error.

through 6h give valuable information about the posterior marginal uncertainties of one parameter and show the efficiency of the convergence process in this case. For example, if we accept a level of error of 2.5%, all values of d between 0.24 and 0.32 m ensure a good fit of the data curve. For the parameters, each additional run might bring new solutions, improving the global sampling of the parameter space. Even with only eight parameters, the complexity of the parameter space is such that an exhaustive sampling would be prohibitive.

It is clear from Figure 6 that the inversion process retrieved the thickness and depth of the fracture but was inefficient in deriving the correct Jonscher parameterization for the filling and matrix, particularly when considering Figure 6a. Figure 7 compares the best model results (found by the inversion process) with the true solution for the real and imaginary parts of the permittivity of the filling material (respectively, Figure 7a and 7b) and the matrix (respectively, 7c and 7d) as a function of frequency. The agreements between the theoretical and the inverted complex permittivity are more satisfying now. That highlights the problem of nonuniqueness of the solution for Jonscher parameters, which does not propagate on complex permittivity or constitutive properties. For that reason, only the complex permittivity results are plotted for a given frequency.

**Table 1.** Investigation range for the parameter space during inversion of synthetic data.

| Parameters | Investigated range | Solution |
|:---:|:---:|:---:|
| $\varepsilon_{\infty,f}$ | $1-81$ | 2.5 |
| $\chi_{r,f}$ | $0-25$ | 2 |
| $n_f$ | $0-1$ | 0.85 |
| $\varepsilon_{\infty,m}$ | $5.5-7$ | 8.14 |
| $\chi_{r,m}$ | $0-1$ | 0.94 |
| $n_m$ | $0.5-1$ | 0.82 |
| d (normalized) | $0.01-\lambda_f/2$ | $\lambda_f/5$ (28.3 cm) |
| z (m) | $2.5-3.5$ | 3 |

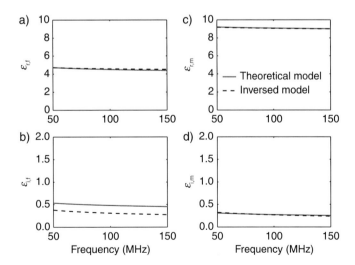

**Figure 7.** Comparison between the best-model solution and the theoretical solution for complex permittivity as a function of frequency. (a and b) Filling properties. (c and d) Matrix properties.

This numerical study also highlights the low sensitivity to noise of the inversion methodology.

## Generalization

To generalize the validation presented before for a single case, different inversion processes were computed considering various normalized apertures (ranging from $\lambda_f/100$ to $\lambda_f/2$) with the TE reflectivity mode. This part was performed without noise. Here, the Jonscher parameters of the filling material and matrix were converted directly into complex permittivity for a frequency of 100 MHz with the following variables:

- $\varepsilon_{r,f}/\varepsilon_0$: real part of the effective permittivity of the filling
- $\varepsilon_{i,f}/\varepsilon_0$: imaginary part of the effective permittivity of the filling
- $\varepsilon_{r,m}/\varepsilon_0$: real part of the effective permittivity of the matrix
- $\varepsilon_{i,m}/\varepsilon_0$: imaginary part of the effective permittivity of the matrix

Figure 8 displays the results of the inversions, the dotted lines corresponding to the theoretical solutions. The best solutions (values shown in red) are near the modeled solutions for all apertures and all parameters except for the complex permittivity of the filling when the normalized aperture is lower than $\lambda_f/10$.

The successful generalization to different thicknesses with wet sand as filling material has been extended to four other fillings typical of those encountered in the context of fractured media (air, saturated sands, water, and saturated clays). They can be described by the Jonscher parameters displayed in Table 2.

For the air case, the effective permittivity is lower than the matrix permittivity and does not exhibit any frequency dispersion, whereas the saturated sand is a weakly dispersive material which presents a real part of permittivity higher than the real part of the matrix permittivity. For water, the real part of permittivity is higher than the real part of the matrix permittivity, whereas saturated clays present a high dispersion with a real part of permittivity higher than the real part of the matrix permittivity.

Figure 9 provides a synthesis of the generalized inversion results for the TE mode, the aperture of the thin bed, and the complex permittivity of the different filling materials. It shows that the aperture is well determined for all fillings. The real and imaginary parts of the permittivity of the matrix also are well determined when the normalized aperture is greater than $\lambda_f/10$ for nondispersive or low-dispersive materials (i.e., air, water, and saturated sands).

For wet clays, the inverted complex permittivity of the filling material is in agreement with the theoretical solution but presents higher variations.

Those results show the potential of this proposed inversion strategy to accommodate a wide range of filling properties and apertures.

## Application to Field Data

Application of the inversion strategy to field tests was presented by Deparis and Garambois (2009), who successfully retrieved the characteristics of a near-vertical fracture (aperture, filling material, and depth) and of the homogeneous limestone in the context of cliff investigations. We present another set of data acquired on a higher vertical cliff, Rocher du Midi, in the context of rock-fall hazard (Deparis et al., 2008). GPR data were acquired using 200-MHz unshielded antennae in the TE mode, directly inside an open limestone fracture 40 m high (see Deparis et al, 2008). Acquisition was performed by increasing the distance between the transmitting and receiving antennae in increments of 20 cm. The central location of CMP acquisition is 8 m from the top of the cliff. GPR profiles (vertical and horizontal) conducted on the cliff face show a succession of fractures subparallel to the cliff wall.

The CMP section first was filtered using a 50- to 300-MHz band-pass Butterworth filter and was processed using automatic gain-control time equalization for visualization purposes, which showed a good signal-to-noise ratio (Figure 10a). Only the highlighted reflection wave (t = 50 ns) is analyzed, and consequently, a muting process was applied to the data to keep only this hyperbolic event. A raw Fourier transform of this reflected event is presented within a 50- to 200-MHz frequency range in Figure 10b and 10c for amplitude and phase, respectively. As displayed in Figure 10b, a large decrease in the amplitude of the reflectivity is observed for all frequencies for offsets larger than 4.5 m. That observation cannot be explained fully by a 1D reflectivity model because it is generated by 2D effects (presence of a weathered zone at the edge of the cliff, airwave reflections, scattering). For this reason, only the first 4.5-m offsets have been selected for the DAPVO inversion process. The first step of the inversion has consisted of normalizing in the frequency domain the hyperbolic event by the second trace.

The inversion results are plotted in Figure 11. Jonscher parameters were transformed into complex permittivity parameters at a frequency of 130 MHz, which correspond to the maximum energy displayed in Figure 10b.

Figure 11 shows a satisfying convergence of solutions toward a minimal error solution focusing in a unique region of the different parameter spaces. Inversion shows that (1) (Figure 11a) complex filling permittivity ranges between 18 and 40 for the real part and between 8 and 16 for the imaginary part; (2) (Figure 11b) depth of the fracture ranges between 2.1 and 2.3 m with a fracture aperture ranging between 0.15 and 0.23 m; (3) (Figure 11c) complex matrix permittivity exhibits less clustered results, ranging between

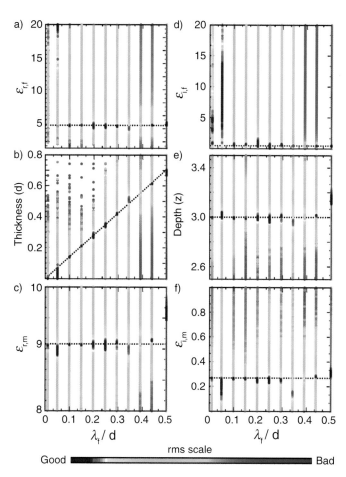

**Figure 8.** Generalization of inversion process for various thin-layer thicknesses as a function of $\lambda/d$. The background medium is limestone ($\varepsilon_{\infty,m} = 8.14$, $\chi_{r,m} = 0.94$, $n_m = 0.82$), and the thin bed is wet sand ($\varepsilon_{\infty,f} = 2.5$, $\chi_{r,f} = 2$, $n_f = 0.85$).

**Table 2.** Jonscher parameters for different modeled filling materials.

| Filling | Air | Saturated sands | Water | Saturated clays |
|---|---|---|---|---|
| $\varepsilon_{\infty,f}$ | 1 | 29 | 81 | 55 |
| $\chi_{r,f}$ | 0 | 4 | 0 | 30 |
| $n_f$ | 1 | 0.5 | 1 | 0.25 |

**Figure 9.** Synthesis of inversion results considering different apertures and filling materials. The background medium is limestone ($\varepsilon_{\infty,m} = 8.14$, $\chi_{r,m} = 0.94$, $n_m = 0.82$). The different thin-bed properties are provided in Table 2.

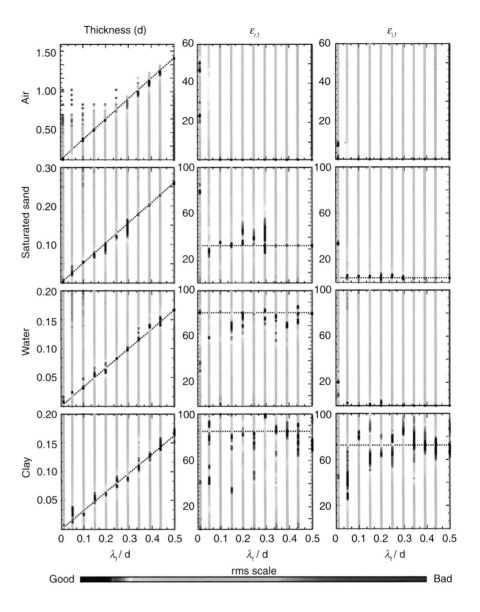

**Figure 10.** (a) Field CMP data acquired using a 200-MHz unshielded antenna in time domain. (b) Reflected event in the spectral domain as a function of offset for amplitude and (c) phase.

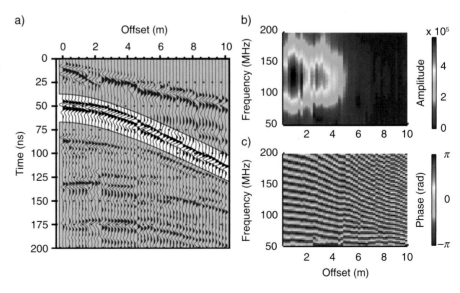

7 and 8 for the real part and between 0.5 and 1 for the imaginary part. The best model (lowest rms) found by the inversion process corresponds to a thin bed located at a depth of 2.21 m with an aperture of 22 cm and filling Jonscher parameters of $n_f = 0.31$, $\chi_{r,f} = 3.5$ and $\varepsilon_{\infty,f} = 15.2$. For the matrix material, the inversion proposes a solution with the parameters $n_m = 0.62$, $\chi_{r,m} = 1.2$, and $\varepsilon_{\infty,m} = 7.0$.

This best solution indicates a GPR velocity of the filling centered around 7 cm/ns and an electrical resistivity of about 32 $\Omega$m (deduced from the Jonscher model). Those values indicate that the filling appears to be relatively dry clays. To test the best solution, Figure 11d displays the time waveform of the reflected wave after it was corrected for propagation effects and radiation pattern using the best model delivered by the inversion process for the homogeneous material. The variation of the traces presents coherent attributes for amplitude and phase as a function of reflectivity and underlines the efficiency of the inversion.

To go further with the post-process validation, Figure 12 presents a comparison between normalized field data (after propagation effects and radiation-pattern corrections, dotted) and three normalized theoretical thin-bed reflections of the following apertures: the best-model solution (22 cm), 17 cm, and 27 cm. All curves were normalized for each frequency with respect to the trace acquired at an incidence angle of 2.6°. Figure 12a and 12b compare, respectively, the amplitude and phase obtained at an incidence angle of 26° as a function of frequency, whereas Figure 12c and 12d compare, respectively, the amplitude and phase at a frequency of 130 MHz as a function of incidence angle. For the four representations, it is clear that the best model derived from the inversion correlates well and presents the lowest rms value with the field data. The other two models show divergence from our data, especially for amplitude. However, it must be noted that contrary to the case presented by Deparis and Garambois (2009), the phase varies weakly here as a function of aperture and seems to be of little use in this study.

Figure 13 generalizes the comparison between normalized theoretical reflectivity modes, obtained using the best

model deduced from the inversion process, and field data after corrections were applied. The figure shows the consistency between the model for (12a) amplitude and (12b) phase and (12c) data for amplitude and (12d) phase for all frequencies and incidence angles. Although the comparison is satisfying, again, the case presented by Deparis and

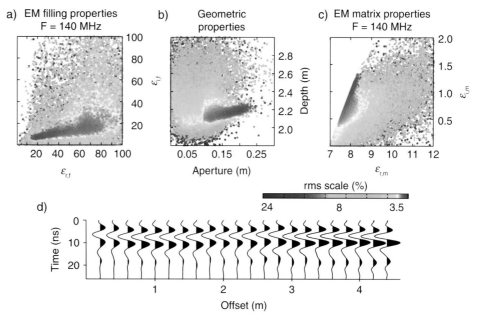

**Figure 11.** Results of the inversion. (a) Thin-bed complex permittivity, (b) depth of the fracture as a function of its aperture, and (c) limestone complex permittivity. Best models are shown in red. (d) Time waveform of the reflected event after propagation, reflectivity, and radiation-pattern corrections.

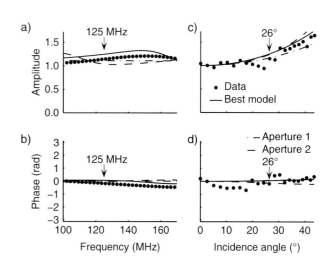

**Figure 12.** Comparison between normalized reflectivity results derived from field data and those deduced from three thin-bed models. (a) Amplitude and (b) phase as a function of frequency derived at an incidence angle of 26°.(c) Amplitude and (d) phase as a function of frequency derived at a frequency of 130 MHz.

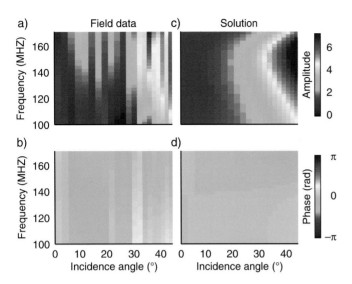

**Figure 13.** Comparison of DAPVO curves derived from corrected field data and for the best solution obtained after the inversion process, for all frequencies and incidence angles.

Garambois (2009) was most efficient in terms of amplitude and phase sensitivities, as a result of the larger variations in DAPVO curves resulting from filling properties (wet sands) and thin-bed aperture.

These two examples show the benefits of using DAPVO data rather than only AVO or APVO data because the inversion problem is constrained better and robust solutions are derived.

## Conclusion

This study was dedicated to the evaluation of the potential of dispersive amplitude and phase versus offset (DAPVO) curves deduced from GPR data to characterize thin-bed properties (aperture, filling material) when embedded within an homogeneous formation. For this, an original algorithm was proposed which exhibits the interesting property that all necessary corrections to derive correct DAPVO curves are made during the inversion process for all tested models. We showed with different analytical data and in one case with noise that the inversion process was efficient in providing not only the properties of the thin bed but also its depth and the properties of the surrounding homogeneous soil. This approach was successful mainly because of the introduction of the dispersion characteristics, which restrict the nonuniqueness problems potentially encountered when only AVO curves are used. In one case, synthetic data with 10% of Gaussian noise were inverted successfully. It must be noted that the inversion

technique is valid only if the thin-bed aperture remains lower than half of the filling wavelength. The advantage of this global inversion method is that the process is quasi-automatic.

The inversion strategy was tested successfully on field CMP data acquired along the wall of a vertical cliff for rockfall hazard issues. Results showed a satisfying convergence toward a unique solution for the depth and material properties of the thin bed and the properties of the surrounding material (limestone). The inversion also permitted identification of the clayey nature of the filling without any direct observation of the fracture.

The inversion strategy is restricted to a single thin-bed reflection event within a homogeneous formation. To develop this approach to other applications of thin-bed problems (glacier-bedrock for example), extensions will be performed for a three-bed scenario, i.e., when the deeper formation is different from the surface material. It also must be developed to allow for multiple thin-bed events when a succession of thin beds (fractures, for example) is present within a homogeneous formation. In that case, different problems could be encountered, notably because of propagation of errors as multiple corrections must be made for each interface and because of the decrease of the signal-to-noise ratio for late reflection events.

## Acknowledgments

We are grateful to M. Sambridge, who allows the free distribution of his inversion code for research purposes. We are grateful to K. Hollinger, who provided the FDTD modeling code to study the radiation pattern of GPR antennae. We also thank Robin Bruce (MEEES student) for comments about English. The constructive comments of the reviewers are greatly appreciated.

## References

Annan, A., 2002, Ground-penetrating radar workshop notes: Sensors and Software Inc., Ontario, Canada.

Baker, G. S., 1998, Applying AVO analysis to GPR data: Geophysical Research Letters, **25**, 397–400.

Bano, M., 2004, Modelling of GPR waves for lossy media obeying a complex power law of frequency for dielectric permittivity: Geophysical Prospecting, **52**, 11–26.

Bergmann, T., J. O. A. Robertsson, and K. Holliger, 1996, Numerical properties of staggered finite-difference solutions of Maxwell's equations for ground-penetrating radar modeling: Geophysical Research Letters, **23**, 45–48.

Bradford, J. H., 2007, Frequency-dependent attenuation analysis of ground-penetrating radar data: Geophysics, **72**, no. 3, J7–J16.

Bradford, J. H., and J. C. Deeds, 2006, Ground-penetrating radar theory and application of thin-bed offset-dependent reflectivity: Geophysics, **71**, no. 3, K47–K57.

Carcione, J. M., D. Gei, M. A. B. Botelho, A. Osella, and M. de la Vega, 2006, Fresnel reflection coefficients for GPR-AVO analysis and detection of seawater and NAPL contaminants: Near Surface Geophysics, **4**, 253–263.

Cassidy, N. J., 2009, Electrical and magnetic properties of rocks, soils and fluids, *in* H. M. Jol, ed., Ground penetrating radar: Theory and applications: Elsevier, 41–72.

Castagna, J. P., 1993, AVO analysis — Tutorial and review, *in* J. P. Castagna and M. M. Backus, eds., Offset-dependent reflectivity — Theory and practice of AVO analysis, Chapter 1: Principles: SEG Investigation in Geophysics Series No. 8, 3–36.

Cole, K. S., and R. H. Cole, 1941, Dispersion and absorption in dielectrics: I. Alternating current characteristics: Journal of Chemical Physics, **9**, 341–351.

Debye, P., 1929, Polar molecules: Dover Press.

Deeds, J., and J. H. Bradford, 2002, Characterization of an aquitard and direct detection of LNAPL at Hill Air Force Base using GPR AVO and migration velocity analyses: Ninth International Conference on Ground Penetrating Radar, 323–329.

Deparis, J., B. Fricout, D. Jongmans, T. Villemin, L. Effendiantz, and A. Mathy, 2008, Combined use of geophysical methods and remote techniques for characterizing the fracture network of a potential unstable cliff site (the "Roche du Midi," Vercors Massif, France): Journal of Geophysics and Engineering, **5**, 147–157.

Deparis, J., and S. Garambois, 2009, On the use of dispersive APVO GPR curves for thin-bed properties estimation: Theory and application to fracture characterization: Geophysics, **74**, no. 1, J1–J12.

Giannopoulos, A., 2005, Numerical modeling of ground penetrating radar using GprMax: Third International Workshop on Advanced Ground Penetrating Radar, 10–15.

Grégoire, C., L. Halleux, and V. Lukas, 2003, GPR abilities for the detection and characterisation of open fractures in a salt mine: Near Surface Geophysics, **1**, 139–147.

Grégoire, C., and F. Hollender, 2004, Discontinuity characterization by the inversion of the spectral content of ground-penetrating radar (GPR) reflections — Application of the Jonscher model: Geophysics, **69**, 1414–1424.

Hall, S. A., and J.-M. Kendall, 2003, Fracture characterization at Valhall: Application of P-wave amplitude variation with offset and azimuth (AVOA) analysis to a 3D ocean-bottom data set: Geophysics, **68**, 1150–1160.

Holberg, O., 1987, Computational aspects of the choice of operator and sampling interval for numerical differentiation in large-scale simulation of wave phenomena: Geophysical Prospecting, **35**, 629–655.

Hollender, F., and S. Tillard, 1998, Modeling ground-penetrating radar wave propagation and reflection with the Jonscher parameterization: Geophysics, **63**, 1933–1942.

Jonscher, A. K., 1977, The "universal" dielectric response: Nature, **267**, 673–679.

Jordan, T. E., and G. S. Baker, 2002, Field testing amplitude and phase variation with offset (APVO) analysis of ground penetrating radar data: 72nd Annual International Meeting, SEG, Expanded Abstracts, 1516–1518, doi: 10.1190/1.1816954.

Jordan, T. E., G. S. Baker, K. Henn, and J.-P. Messier, 2004, Using amplitude variation with offset and normalized residual polarization analysis of ground penetrating radar data to differentiate an NAPL release from stratigraphic changes: Journal of Applied Geophysics, **56**, 41–58.

Koefoed, O., and N. de Voogd, 1980, The linear properties of thin layers, with an application to synthetic seismograms over coal seams: Geophysics, **45**, 1254–1268.

Kindelan, M., G. Seriani, and P. Sguazzero, 1989, Elastic modelling and its application to amplitude versus angle interpretation: Geophysical Prospecting, **37**, 3–30.

Lampe, B., and K. Holliger, 2003, Effects of fractal fluctuations in topographic relief, permittivity and conductivity on ground-penetrating radar antenna radiation: Geophysics, **68**, 1934–1944.

———, 2005, Resistively loaded antennas for ground-penetrating radar: A modeling approach: Geophysics, **70**, no. 3, K23–K32.

Lehmann, F., 1996, Fresnel equations for reflection and transmission at boundaries between two conductive media, with applications to georadar problems: Sixth International Conference on Ground Penetrating Radar, 555–560.

Liu, Y., and D. R. Schmitt, 2003, Amplitude and AVO responses of a single thin bed: Geophysics, **68**, 1161–1168.

Lutz, P., S. Garambois, and H. Perroud, 2003, Influence of antenna configurations for GPR survey: Information from polarization and amplitude versus offset measurements: Geological Society [London] Special Publication 211, 299313.

Mahob, P. N., and J. P. Castagna, 2003, AVO polarization and hodograms: AVO strength and polarization product: Geophysics, **68**, 849–862.

Ostrander, W. J., 1984, Plane-wave reflection coefficients for gas sands at nonnormal angles of incidence: Geophysics, **49**, 1637–1648.

Sambridge, M., 1999a, Geophysical inversion with a neighbourhood algorithm — I. Searching a parameter space: Geophysical Journal International, **138**, 479–494.

———, 1999b, Geophysical inversion with a neighbourhood algorithm — II. Appraising the ensemble: Geophysical Journal International, **138**, 727–746.

Sassen, D. S., and M. E. Everett, 2009, 3D polarimetric GPR coherency attributes and full-waveform inversion of transmission data for characterizing fractured rock: Geophysics **74**, no. 3, J23–J34.

Schoenberger, M., and F. K. Levin, 1976, Reflected and transmitted filter functions for simple subsurface geometries: Geophysics, **41**, 1305–1317.

Simmons, J. L. Jr., and M. M. Backus, 1994, AVO modeling and the locally converted shear wave: Geophysics, **59**, 1237–1248.

Stephens, R. B., and P. Sheng, 1985, Acoustic reflections from complex strata: Geophysics, **50**, 1100–1107.

Stovas, A., M. Landrø, and P. Avseth, 2006, AVO attribute inversion for finely layered reservoirs: Geophysics, **71**, no. 3, C25–C36.

Streich, R., and J. van der Kruk, 2007, Characterizing a GPR-antenna system by near-field electric field measurements, Geophysics, **72**, no. 5, A51–A55.

Widess, M. B., 1973, How thin is a thin bed?: Geophysics, **38**, 1176–1180.

Wu, T. T., and R. W. P. King, 1965, The cylindrical antenna with nonreflecting resistive loading: IEEE Transactions on Antennas and Propagation, **AP-13**, 369–373.

Chapter 14

# Characterizing the Near Surface with Detailed Refraction Attributes

Derecke Palmer[1]

## Abstract

The tomographic inversion of near-surface seismic-refraction traveltime data is fundamentally nonunique. It is possible to generate tomograms, which range from the geologically improbable to the very detailed, all of which accurately satisfy the traveltime data. A comparison of the starting models with the final tomograms demonstrates that refraction tomography usually does not improve lateral resolution significantly. Therefore, if important geotechnical features are to be delineated, it is essential that they be included in the starting model, especially zones with low seismic velocities. Suitable detailed starting models for both traveltime and full-waveform inversion can be derived using a suite of parameters, generally known as seismic attributes. Refraction attributes can be computed readily from all near-surface seismic-refraction traveltimes, amplitudes, and waveforms, using the generalized reciprocal method (GRM) and the refraction convolution section (RCS). Furthermore, refraction attributes can be employed as a priori information to resolve nonuniqueness before the acquisition of any a posteriori information, such as borehole or other geophysical data. Narrow and wide zones with low seismic velocities are delineated with detailed attribute-based tomograms and are consistent with other refraction attributes derived from head-wave amplitudes and the RCS. Those zones are not detected with refraction tomograms which use low-resolution starting models, such as the smooth vertical velocity gradient. Additional models of the near surface, such as scaled density ratios and the P-wave modulus, can be computed from combinations of the refraction attributes. The use of a suite of attributes and combined attributes as well as the seismic velocity facilitates derivation of more comprehensive quantitative models of the near surface and thus more effective integration of seismic with borehole and other geotechnical data, using either multivariate geostatistics or full-waveform inversion.

## Introduction

Seismic velocities can be a useful measure of several petrophysical properties, and accordingly, the delineation of their lateral variations in the subweathering zone can be important in many geotechnical and environmental investigations (McCann et al., 1997). Such variations can range from moderate reductions caused by local increases in fracturing and jointing to major reductions caused by shear zones.

Shear zones and fractured bedrock usually are considered to be the product of major tectonic activity. However, normal weathering processes also can result in differential isostatic uplift (Simpson, 2004), and it is not uncommon for increased fracturing to be associated with areas of low elevation, especially in regions of rugged topographic relief, typical of many dam sites (Zaruba, 1956; Nichols, 1980; Fell et al., 2005). Accordingly, fracture-induced lowering of seismic velocities in the subweathering zone could be more widespread than often is recognized.

Determining lateral variations in seismic velocities in the subweathering zone is an ill-posed problem: Small variations in processed traveltimes generated with standard inversion algorithms can result in large variations in the computed velocities. Palmer (2010b) demonstrates that seismic velocities that are both lower and higher by approximately 30% of the true values, in addition to the true value, are consistent with the traveltime data for a

[1]*University of New South Wales, Sydney, Australia. E-mail: d.palmer@unsw.edu.au.*

syncline model. Other model studies show that even modest departures from plane layering can result in artifacts in the seismic velocities of more than 70% (Palmer et al., 2005).

Ill-posed problems are synonymous with nonunique solutions in model-based inversion, in which an initial starting model is updated systematically through iteratively comparing the modeled response with the observed data (Treitel and Lines, 1988; Oldenburg and Li, 2005). It is recognized generally now that nonuniqueness is a fundamental reality with the inversion of virtually all sets of geophysical data, including near-surface seismic-refraction data (Ivanov et al., 2005a; Ivanov et al., 2005b). A common source of nonuniqueness is the starting model and in turn, the inversion algorithm selected to produce the starting model (Oldenburg, 1984).

Refraction tomography (Stefani, 1995; Zhu et al., 1992; Lanz et al., 1998; Zhang and Toksöz, 1998), which is an example of model-based inversion, is used widely to invert seismic-refraction data recorded for geotechnical and environmental investigations. In this study, several starting models are used to generate refraction tomograms with wavepath-eikonal-traveltime (WET) tomography (Schuster and Quintus-Bosz, 1993) using traveltime data recorded across a major shear zone and a narrow, steeply dipping massive sulfide orebody.

The starting models include smooth velocity gradients, which are the default for many implementations of refraction tomography and which generally emphasize the vertical resolution of many layers. Other starting models are generated with the inversion algorithms of the generalized reciprocal method (GRM) (Palmer, 1980, 1981, 1986, 1991, 2006, 2008a, 2008b, 2009a, 2010b, 2010c), which emphasize the lateral resolution of individual layers. This study demonstrates that it is possible to generate quite different tomograms, all of which are consistent with the traveltime data.

Two major conclusions are reached in this study. The first is that the final tomogram usually is very similar to the starting model. Therefore, if any important feature is not included in the starting model, then it is unlikely that that feature will be generated in the final refraction tomogram. This study demonstrates that the 50-m-wide zone in the first case study and the narrow 10-m-wide zone of massive sulfides in the second case study (both of which exhibit low seismic velocities) are detected with all GRM tomograms but not with the smooth velocity-gradient tomograms.

The second conclusion is that refraction attributes, such as inverted head-wave amplitudes and spectral analysis of the refraction convolution section (RCS) (Palmer, 2001a, 2001c; de Franco, 2005; Palmer, 2008a, 2009b), often can

resolve many important ambiguities associated with nonuniqueness, such as the existence or absence of zones with low seismic velocities. Furthermore, the use of the head coefficient can facilitate more representative measures of the rock strength in which variations in the density (such as with the massive sulfides) can affect seismic velocities.

Finally, this study questions the usefulness of simplistic comparisons of the misfit errors of tomographic inversion. Instead, this study proposes that the three tomograms generated with the GRM parameters for the optimum *XY* value and plus/minus half the station spacing constitute a more practical measure of uncertainty.

## Seismic-refraction Attributes

This study proposes the routine computation of a suite of parameters, generally known as seismic attributes. A seismic attribute is *any* measure that helps to better visualize or quantify features of interest in seismic data (Chopra and Marfurt, 2007). The refraction attributes can be computed readily from all near-surface seismic-refraction data using the GRM and RCS.

Two major roles are proposed for refraction attributes. The first is to compute detailed starting models for tomographic inversion and as a priori information to resolve nonuniqueness before the acquisition of any a posteriori information such as borehole or other geophysical data. This study demonstrates the use of refraction attributes derived from head-wave amplitudes and the RCS, to validate narrow and wide zones with low seismic velocities detected with refraction tomograms derived from detailed starting models generated with the GRM. By contrast, the regions with low seismic velocities are not detected with refraction tomograms that use smooth velocity-gradient starting models.

The second role is to employ various combinations of the refraction attributes to compute additional models of the near surface, such as scaled density ratios and the P-wave modulus. The aim is to generate more comprehensive quantitative models of the near surface, to facilitate more effective integration of seismic with borehole and other geotechnical data. With the acquisition of any borehole data, the task becomes one of integrating rather than comparing and contrasting the seismic and borehole data. Danbom (2005), for example, describes an anecdote in which the results of a detailed geophysical survey were rejected in favor of a more limited geotechnical drilling program.

The refraction attribute most commonly used in geotechnical investigations is seismic velocity because it is often a useful measure of rock strength and fracturing. With the standard approach, a physical relationship using the single attribute of seismic velocity is employed to predict the petrophysical property of rock strength or fracturing over a 2D line or 3D volume (McCann et al., 1990; McCann and Fenning, 1995). Frequently, the physical relationship is semiquantitative, with lower seismic velocities indicating lower rock strength or increased fracturing.

An alternative approach proposed by Schultz et al. (1994a, 1994b) and by Ronen et al. (1994) is to derive statistical rather than deterministic relationships. These techniques, generally known as multivariate geostatistics, encompass those methods that use more than one variable to predict some other parameter of interest. Multiattribute transforms (Hampson et al., 2001) facilitate the more effective integration of borehole data (which has superior vertical resolution) with seismic data (which has superior spatial resolution).

This study proposes that the application of multivariate geostatistics with refraction attributes, along with their various combinations, can facilitate the more effective integration of disparate sets of data acquired by engineering and geoscience investigators, irrespective of their proficiency in the geosciences (Danbom, 2005). Furthermore, this strategy offers the opportunity for expanding the application of seismic-refraction methods from an initial reconnaissance or semiquantitative role prior to the acquisition of any borehole data to an integral component for the entire lifetime of each site-investigation program, during which the detailed quantitative characterization of the site is updated as additional geotechnical data are acquired.

Finally, refraction attributes and the various combinations can be used to generate detailed starting models for full-waveform inversion (Smithyman et al., 2009; Virieux and Operto, 2009). Full-waveform inversion is a deterministic approach for extracting several parameters in addition to the seismic velocity from seismic data, whereas multivariate geostatistics can be viewed as a pragmatic approach to obtaining comparable information inherent within the refraction attributes.

# The Mount Bulga Shear Zone

## Traveltime data

The first set of traveltime data to be examined in this study was recorded at Mount Bulga, near Orange in southeastern Australia. It is the site of a small massive sulfide orebody in steeply dipping to vertical Silurian metasediments. The site underwent extensive investigations

several decades ago but subsequently was abandoned because the orebody was considered to be uneconomic. The data were recorded on a relatively flat area away from the orebody and near the now demolished core shed (Palmer, 2001a, 2001c, 2008a, 2009a, 2010c).

Traveltime data (Figure 1 in Palmer, 2010c) allow three layers to be recognized. They are a surface soil layer with a seismic velocity of 300 to 500 m/s, a weathered layer with a seismic velocity of approximately 1000 to 1200 m/s for which the upward concave graphs indicate cycle skipping and thus the occurrence of a velocity reversal, and a main refractor at the base of the weathered layer with a seismic velocity of 2500 to 6000 m/s.

## Smooth velocity-gradient starting model

Figure 1 presents the smooth velocity-gradient starting model and the WET tomogram. The mean unsigned error is 6.44 ms for the starting model and 1.86 ms for the WET tomogram. A comprehensive summary of all errors,

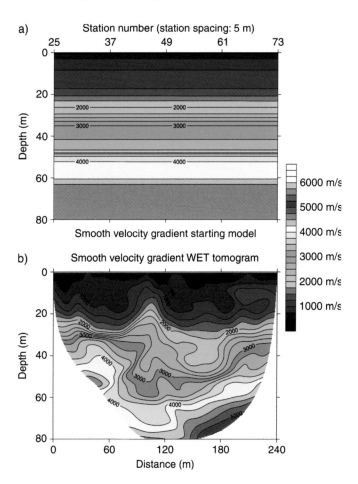

**Figure 1.** (a) Smooth velocity-gradient starting model and (b) WET tomogram. From Palmer, 2010a. Used by permission of EAGE.

which are defined in Schuster and Quintus-Bosz (1993), is shown in Table 1.

The seismic velocities in the lowest layers of the WET tomogram are greater than 4000 m/s for the entire traverse, and there are no regions with seismic velocities that would be indicative of shear zones. The region between stations 43 and 61 is interpreted to be a local increase in the depth of the weathered layer and not a shear zone because the seismic velocities in the lower part of the tomogram are still 4000 m/s or higher.

Alternatively, the base of the weathered layer can be taken as the region where the vertical velocity gradients are a maximum. Regions in the vicinity of the 2000 m/s and 3000 m/s contours would constitute "interfaces" using this criterion. However, a similar region of increased gradients occurs in the starting model, which demonstrates the critical importance of the starting model in most implementations of refraction tomography. Furthermore, these seismic velocities are probably too low for unweathered Silurian metasediments.

An inspection of the ray-coverage diagram (not shown here) shows that coverage in the region of the 2500 m/s contour is considerably less than that for the region where velocities are 4000 m/s or more. Therefore, if the base of the weathered layer were taken to be the region where

seismic velocities are 2000 to 3500 m/s, then it could be concluded that the region of higher seismic velocities in the vicinity of station 61 with a seismic velocity of approximately 5000 m/s is either an artifact (Palmer, 2010b) or an indication of no shear zone.

These results demonstrate that there is no clear consistency between the velocity gradients and the magnitudes of the seismic velocities and that identification of the base of the weathered layer (Figure 1) is ambiguous. It is concluded that there can be pitfalls in intuitive approaches to interpreting refraction tomograms.

## Smooth GRM starting model

A simple starting model is obtained by converting the GRM time model (Palmer, 2010c, Figure 4) to a depth model using an approximate average seismic velocity of 1000 m/s in the weathered layer and the "smoothed" average seismic velocities in the subweathering zone, namely, 6000 m/s (stations 25 to 51 and stations 61 to 68) and 2500 m/s (stations 51 to 61 and stations 68 to 73). The smooth GRM velocity model is equivalent to that obtained with manual curve fitting to the GRM velocity-analysis function (Palmer, 2010c, Figure 3).

**Table 1.** Misfit errors — Mount Bulga shear zone.

| Starting models and WET tomograms | Mean unsigned error (ms) | Relative misfit function (ms) | Maximum unsigned error (ms) |
|---|---|---|---|
| Smooth velocity-gradient starting model: vertical velocity gradients | 6.44 | 33.94 | 24.57 |
| Smooth velocity-gradient WET tomogram: vertical velocity gradients | 1.86 | 3.07 | 8.29 |
| Smooth GRM starting model $XY = 2.5$ m: no velocity reversal | 5.53 | 23.81 | 22.42 |
| Smooth GRM WET tomogram $XY = 2.5$ m: no velocity reversal | 1.24 | 1.57 | 8.52 |
| Smooth GRM starting model $XY = 2.5$ m: velocity reversal | 5.53 | 23.81 | 22.42 |
| Smooth GRM WET tomogram $XY = 2.5$ m: velocity reversal | 1.95 | 3.92 | 20.66 |
| Detailed GRM starting model $XY = 2.5$ m: no velocity reversal | 10.93 | 83.89 | 29.35 |
| Detailed GRM WET tomogram $XY = 2.5$ m: no velocity reversal | 2.01 | 4.13 | 20.67 |
| Detailed GRM starting model $XY = 0$: velocity reversal | 8.60 | 46.12 | 18.10 |
| Detailed GRM WET tomogram $XY = 0$: velocity reversal | 2.15 | 4.88 | 12.85 |
| Detailed GRM starting model $XY = 2.5$ m: velocity reversal | 9.3 | 55.42 | 20.74 |
| Detailed GRM WET tomogram $XY = 2.5$ m: velocity reversal | 2.14 | 5.10 | 14.23 |
| Detailed GRM starting model $XY = 5$ m: velocity reversal | 9.29 | 55.42 | 20.74 |
| Detailed GRM WET tomogram $XY = 5$ m: velocity reversal | 2.14 | 5.11 | 15.02 |

The WET tomogram shown in Figure 2 confirms that two regions in the refractor with low seismic velocities can be recognized between stations 51 and 61 and between stations 68 and 73. These seismic velocities, which are less than 2500 m/s and 40% lower than the lowest seismic velocity in the refractor indicated in the smooth velocity-gradient tomogram (Figure 1), provide clear evidence for one or more shear zones. The results of the seven crosslines from a later 3D survey (Palmer, 2001b; Figure 14 in Palmer, 2010c) confirm that a 50-m-wide zone with a low seismic velocity extends for at least 60 m along strike.

There are very few significant visual differences between the two images in Figure 2, despite the fact that the mean unsigned error in the starting model of 5.53 ms is larger than the mean unsigned error in the final WET tomogram of 1.24 ms. These errors are smaller than but of a comparable magnitude to those obtained with the smooth velocity-gradient model. The differences in the misfit errors in Figure 2 can be attributed largely to the updating of the approximate seismic velocity in the weathered layer with values that are more consistent with the traveltime data and to the elimination of gridding artifacts. It can be concluded that the use of refraction tomography is not essential for achieving acceptable accuracy with the GRM.

There is little ambiguity in determining the interface between the weathered layer and the subweathering zone in either the starting model or the WET tomogram (Figure 2). There is both a significant change in the seismic velocities and a narrow region of large gradients in the seismic velocities. Furthermore, the seismic velocities in the weathered layer exhibit considerably less variability or complexity than is the case with the smooth velocity-gradient WET tomogram shown in Figure 1.

## Velocity reversal in the weathering

There is clear evidence for the presence of a reversal in the seismic velocities in the weathered layer, in which a layer with a higher seismic velocity occurs above a layer with a lower seismic velocity. Velocity reversals are recognized readily by rapid attenuation (Figure 9 in Palmer, 2010c) and cycle skipping in the shot records, which result in concave upward segments in the traveltime graphs.

The likely occurrence of the velocity reversal is supported by a comparison of the horizontal seismic velocities in the weathered layer of approximately 1000 m/s computed from direct arrivals in the traveltime graphs, with the average vertical velocity of 500 m/s (average time model, $t_G = 30$ ms; refractor seismic velocity, $V_R = 6000$ m/s) computed with equation 1 and an optimum $XY$

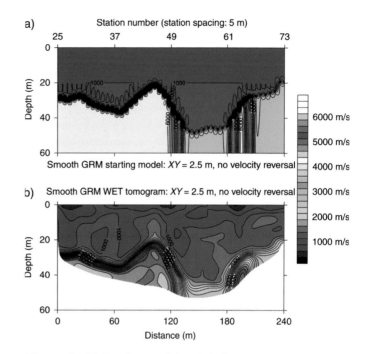

**Figure 2.** (a) Starting model and (b) WET tomogram obtained with the smooth GRM seismic velocities in the refractor. From Palmer, 2010c. Used by permission of EAGE.

value of 2.5 m. Palmer (2001a, 2001c) uses an average vertical velocity of 700 m/s, based on an optimum $XY$ value of 5 m:

$$\overline{V} = \sqrt{\frac{XY_{\text{optimum}} V_R^2}{(XY_{\text{optimum}} + 2t_G V_R)}}. \quad (1)$$

A simple measure of the errors in the average vertical velocity in equation 1, resulting from errors in the determination of the optimum $XY$ values, is shown in equation 2, after differentiation of equation 1 and a little algebraic manipulation:

$$\frac{\Delta \overline{V}}{\overline{V}} = \frac{\Delta XY}{2XY} \cos^2 \overline{i}, \quad \text{where} \quad \sin \overline{i} = \frac{\overline{V}}{V_R}. \quad (2)$$

Therefore,

$$\frac{\Delta \overline{V}}{\overline{V}} \approx \frac{\Delta XY}{2XY}. \quad (3)$$

The error in determining the optimum $XY$, value, $\Delta XY$, can be taken as half the station spacing, that is, 2.5 m (Palmer, 2010c). The error from equation 3 is 50%. Therefore, the average velocity is not well determined, principally because the optimum $XY$ value of 2.5 m is less than the station spacing. Nevertheless, the average vertical velocity of $500 \pm 250$ m/s is still consistent with a velocity reversal rather than uniform velocities or vertical velocity gradients in the weathered layer, as is assumed in Figure 1. The error in the optimum *XY* value using the analysis of Leung (2003) is 4%, which demonstrates that the station spacing is usually a more significant source of errors than changes in dip at the base of the weathered layer.

## Smooth GRM starting model with a velocity reversal

The starting model shown in Figure 3 was generated with seismic velocities of 500 m/s in the weathered layer and with the smooth seismic velocity model in the sub-weathering zone and the average time model. Furthermore, traveltimes for the direct arrivals were replaced with values that were representative of a seismic velocity of 500 m/s in the weathered layer.

The WET tomogram shows the two regions with the low seismic velocities as representative of shear zones between stations 51 and 61 and between stations 68 and 73. However, the seismic velocity between stations 51 and 61 is approximately 1750 m/s, which is 30% less than the 2500 m/s shown in the starting model. Furthermore, there is an indication of lower seismic velocities near station 37.

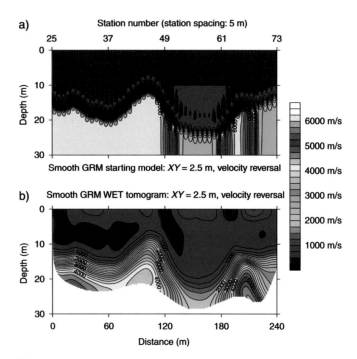

**Figure 3.** (a) Starting model and (b) WET tomogram for the smooth GRM starting model. An average seismic velocity in the weathered layer of 500 m/s has been used to accommodate the velocity reversal in the weathered layer. From Palmer, 2010c. Used by permission of EAGE.

## Detailed GRM starting model

Detailed seismic velocities in the subweathering zone can be generated with a novel application of the Hilbert transform described in Chopra and Marfurt (2007, p. 102–105) (Palmer, 2010c, equation A8). This algorithm computes an average of the reciprocal of the seismic velocities derived over a range of distances, namely 5, 10, 15, 20, 25, and 30 m, which are centered on the reference station (Palmer, 2010c, Figure A5). The benefit of this and similar approaches (Palmer, 2009a, equation 8) is that they provide objective methods for computing detailed lateral variations in the seismic velocities in the sub-weathering, as shown in Figure 4.

Figure 5 presents the starting model generated with an average seismic velocity in the weathered layer of 1000 m/s and a detailed model of the seismic velocities in the refractor (Palmer, 2010c, Figure A5). The unrealistically high seismic velocity of approximately 9000 m/s at station 49 is inferred to represent sideswipe, and therefore, it has a valid geophysical explanation. Alternative more realistic seismic velocities can be obtained simply by increasing the length of the windows over which the Hilbert transform is applied to the averaged GRM refractor velocity-analysis function. Nevertheless, these values have been retained to generate more lateral variability in the final tomogram. Furthermore, it was anticipated that the regularization operations of inversion would moderate any extreme values.

The WET tomogram in Figure 5, which is very similar to the final tomogram generated with the smooth starting

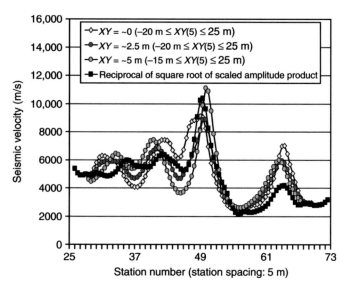

**Figure 4.** The detailed seismic velocity determined with *XY* values of 2.5 ± 2.5 m and the inverted amplitude product.

model in Figure 2, shows that all seismic velocities are less than 6000 m/s and that regularization has been effective. Although there is some indication of lower seismic velocities in the refractor below stations 25 and 37 shown in Figure 5, it is not definitive. In addition, the seismic velocities in the weathered layer are considerably simpler than those shown in Figure 1.

## Detailed GRM starting models with a velocity reversal

The WET tomograms were generated with seismic velocities of 500 m/s in the weathered layer to accommodate the velocity reversal in the weathered layer and with the seismic velocity models for $XY$ values of zero, 2.5 m, and 5 m, i.e., the optimum $XY$ value of 2.5 m $\pm$ 2.5 m, in the subweathering zone.

The three WET tomograms in Figures 6 through 8 are consistent with the seismic velocities in the subweathering zone shown in Figure 2. For example, the low-velocity region between stations 51 and 61 is reproduced on all three tomograms. The lower seismic velocities between stations 25 and 30 occur on all three tomograms, and they update the higher seismic velocities shown in Figure 2. However, the seismic velocities below station 37 increase

with increasing $XY$ value, indicating a possible limit in the lateral resolution of the seismic velocities in the subweathering zone and the need to use a priori information.

The misfit errors for the three WET tomograms in Figures 6 through 8 are of a comparable magnitude. This study proposes that the three tomograms generated with the GRM parameters for the optimum $XY$ value and plus/minus half the station spacing constitute as useful a measure of the uncertainties of inversion as the standard error analyses.

## Head-coefficient attributes

Head-wave amplitudes constitute half the volume of refraction data recorded with all seismic surveys. Regrettably, these data usually are underused. A simple 1D analysis of the shot amplitudes usually can indicate whether uniform seismic velocities, vertical velocity gradients, or even velocity reversals are appropriate (Palmer, 2010c).

Palmer (2001a) demonstrates that a 2D analysis of shot amplitudes with the multiplication of the forward and reverse amplitudes largely compensates for geometric spreading and that the resulting amplitude products are approximately proportional to the square of the head coefficient. Palmer (2001c) also demonstrates that the head coefficient $K$, which is the refraction analogue of the Zoeppritz transmission coefficient in reflection seismology, is approximately proportional to the ratio of the specific acoustic

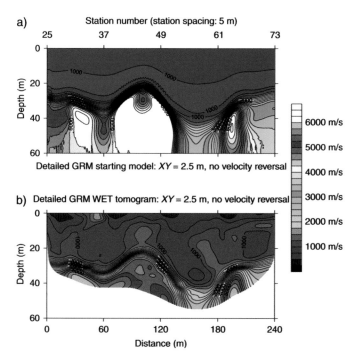

**Figure 5.** (a) Starting model and (b) WET tomogram obtained with the detailed GRM starting model. The velocity reversal in the weathered layer has been ignored. From Palmer, 2010c. Used by permission of EAGE.

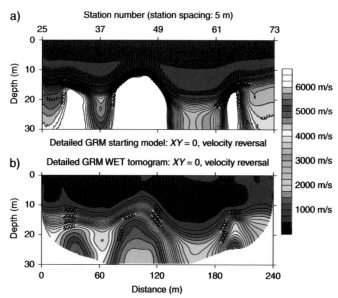

**Figure 6.** (a) Starting model and (b) WET tomogram obtained with the detailed GRM starting model for $XY = 0$. An average seismic velocity of 500 m/s has been used to accommodate the velocity reversal in the weathered layer. From Palmer, 2010a. Used by permission of EAGE.

impedance in the overburden to that in the refractor:

$$\sqrt{\text{amplitude product}} \propto K \propto \frac{\rho_1 V_1}{\rho_2 V_2}. \qquad (4)$$

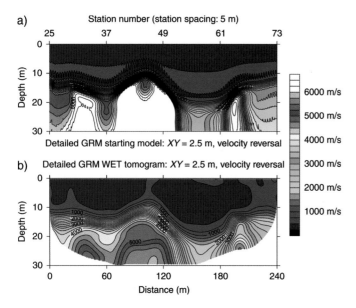

**Figure 7.** (a) Starting model and (b) WET tomogram obtained with the detailed GRM starting model for $XY =$ 2.5 m. An average seismic velocity of 500 m/s has been used to accommodate the velocity reversal in the weathered layer. From Palmer, 2010a. Used by permission of EAGE.

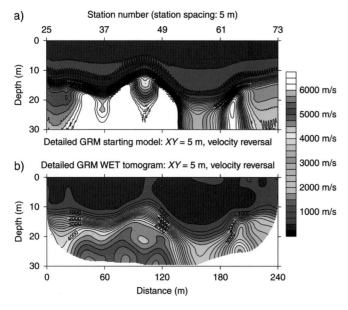

**Figure 8.** (a) Starting model and (b) WET tomogram obtained with the detailed GRM starting model for $XY =$ 5 m. An average seismic velocity of 500 m/s has been used to accommodate the velocity reversal in the weathered layer. From Palmer, 2010a. Used by permission of EAGE.

Layer 1 overlies layer 2, $\rho$ is density, and $V$ is the P-wave velocity. Because the amplitude product, which has been corrected for near-surface irregularities (Palmer, 2006; Palmer, 2010c, Figure A3), is approximately proportional to the square of the head coefficient, it follows that

$$V_2 \propto \frac{1}{\sqrt{\text{amplitude product}}}. \qquad (5)$$

Accordingly, amplitude products inverted with equation 5 often can provide an additional useful attribute, which is a measure of the seismic velocities in the subweathering zone (provided, of course, that there are no significant lateral variations in the seismic velocities in the weathered layer). Figure 4 presents the square root of the reciprocal of scaled amplitude products that have been corrected for the residual geometric-spreading component (Palmer, 2009a) and seismic velocities computed at each station with the Hilbert transform for the optimum $XY$ value of 2.5 m $\pm$ 2.5 m. The correlation between seismic velocities and the inverted amplitude products is generally very good.

Equation 4 can be rearranged to form

$$\frac{\rho_2}{\rho_1} = \frac{V_1}{K V_2}. \qquad (6)$$

Figure 9 is a cross section which presents the scaled density ratio computed as $V_1/KV_2$. This density model would be suitable for model-based inversion of any coincident gravity data, in which a starting model is updated systematically until the gravity response is comparable to the field data.

The velocity of P-waves is related to elastic moduli in isotropic media by

$$V = \sqrt{\frac{k + 4\mu/3}{\rho}} = \sqrt{\frac{\lambda + 2\mu}{\rho}}, \qquad (7)$$

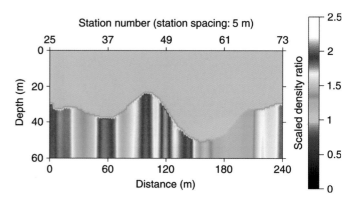

**Figure 9.** The scaled density contrast computed from the seismic velocities and the head coefficient. From Palmer, 2010a. Used by permission of EAGE.

where $k$ is the bulk modulus, $\mu$ is the shear modulus, $\lambda$ is Lamé's lambda constant, and $\rho$ is density (Sheriff and Geldart, 1995, Table 2). The numerator under the square-root sign is the P-wave modulus. Equations 6 and 7 can be rearranged to form

$$\text{P-wave modulus} = \rho_2 V_2^2 \propto \frac{\rho_1 V_1 V_2}{K}. \qquad (8)$$

Figure 2 shows that the P-wave velocity and, by inference, the density in the weathered layer do not vary greatly. Therefore, to a first approximation, the ratio of the subweathering velocity to the head coefficient $V_2/K$ is proportional to the P-wave modulus. Figure 10 is a cross section that presents the P-wave modulus.

An alternative representation of the P-wave modulus would be the square of the subweathering velocity, assuming that the density of the subweathering zone does not vary greatly. Therefore, it is possible to derive three attributes, which are a measure of the P-wave modulus. They are the square of the seismic velocity, the reciprocal of the square of the head coefficient, and the seismic velocity divided by the head coefficient. Although each of these transformed attributes is useful in its own right, they can be more effective when integrated with multivariate geostatistics.

## Spectral analysis of the RCS

The RCS is generated by the convolution of individual pairs of forward and reverse traces. The convolution integral adds traveltimes and multiplies amplitudes and hence generates the same time-structural model of the refractor as that obtained with the time-model algorithm of the GRM (Figure 12 in Palmer, 2010c).

The inelastic attenuation through the weathered layer at each source point and through the subweathering zone is the same for all convolved traces for a given pair of forward and reverse sources. Therefore, any variations in the waveform of the convolved traces can be attributed to variations in the weathered and subweathered regions below each receiver.

Spectral analysis of the first few cycles of the RCS in Figure 11 shows that the region in the subweathering zone between stations 51 and 61 with the low seismic velocity in Figures 2 through 8 has significantly attenuated the high-frequency components. This reduction in the high-frequency response is consistent with increased fracturing and a reduction in mechanical rigidity in the shear zone. There are also smaller reductions in the high-frequency content between 68 and 73 and in the vicinity of stations 25 and 37. These regions correspond with lower seismic velocities in the GRM WET tomogram generated with detailed seismic velocities with $XY = 2.5$ m in Figure 7.

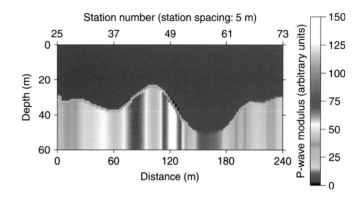

**Figure 10.** The P-wave modulus in arbitrary units computed as the subweathering velocity divided by the head coefficient. From Palmer, 2010a. Used by permission of EAGE.

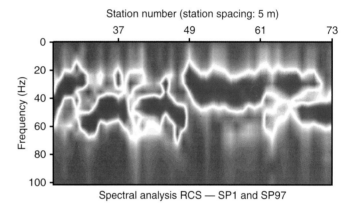

Spectral analysis RCS — SP1 and SP97

**Figure 11.** Spectral analysis of the RCS generated with SP1 and SP97. From Palmer, 2010a. Used by permission of EAGE.

Furthermore, there is generally good correlation between the spectral analysis of the RCS in Figure 11 and the P-wave modulus in Figure 10.

The simple attribute derived from Figure 11 is bandwidth in octaves. This attribute can be used in conjunction with seismic velocity and the head coefficient as a measure of rock strength or fracturing.

## The Mount Bulga Massive Sulfide Orebody

### Traveltime data

The Mount Bulga orebody consists of narrow (5 to 10 m) syngenetic, fine-grained, banded massive pyrite-galena-sphalerite-chalcopyrite in steeply dipping unfolded Silurian altered metasediments. Several major shear zones are associated with the orebody. A 2D seismic-refraction profile was recorded across a small ridge, which also marks the approximate location of the massive sulfide orebody (Palmer, 2006). A more extensive profile was recorded earlier in the same approximate location by Whiteley et al. (1984).

There were considerable variations in surface conditions along the profile. Outcrop occurs between stations 31 and 49, where there were difficulties in planting receivers and auguring shot holes. Gossan and recently restored historical mine workings occur in the vicinity of stations 49 to 53. Soil occurs between stations 53 and 70, where planting receivers and auguring shot holes were considerably easier. At each end of the profile, there were areas of significant disruption to the surface layers, caused by ongoing forestry activities. In those areas, the soil contained a large biogenic component.

Sections of the orebody were subjected to mining of the enriched supergene layer in the historical past. These historical workings were restored in the recent past. The seismic profile was offset by about 10 m from the areas disturbed by mining and restoration.

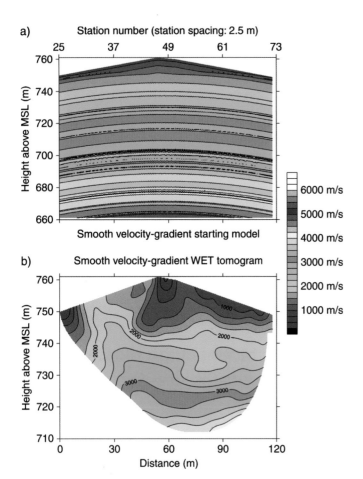

**Figure 12.** (a) Smooth velocity-gradient starting model and (b) WET tomogram for the massive sulfide orebody.

## Smooth velocity-gradient starting model

Figure 12 presents the smooth velocity gradient starting model (mean unsigned error of 3.94 ms) and WET tomogram (mean unsigned error of 1.23 ms) (Table 2). The seismic velocities of approximately 1800 m/s in the vicinity of station 37 correspond with weathered outcrop, the values of approximately 750 m/s in the vicinity of station 49 correspond with gossan, and the values of approximately 1000 m/s in the vicinity of stations 53 to 70 correspond with soil and completely weathered rock.

The seismic velocities systematically increase with depth to a maximum value of approximately 4000 m/s below station 61. There are no regions with large vertical velocity gradients that would provide a clear indication of the base of the weathered layer. Furthermore, there is no indication of any distinctive features that might correspond with the steeply dipping massive sulfide orebody or any associated shear zones.

## Detailed GRM starting model

Figure 13 presents the detailed GRM starting model (mean unsigned error of 3.44 ms) and WET tomogram (mean unsigned error of 1.19 ms) for the massive sulfide orebody. As with the smooth GRM study of the shear zone shown in Figure 2, the starting model and the WET tomogram are very similar after allowances are made for the effects of the gridding.

Contrary to statements by Sheehan et al. (2005), both the GRM starting model and the GRM WET tomogram can accommodate and define lateral variations in the weathered layer and subweathering zone successfully at this site. In fact, there is a better correlation between the high seismic velocities in the weathered layer in the vicinity of station 37 and the high seismic velocities in the subweathering zone than is the case with the smooth velocity-gradient tomogram in Figure 12. Similar observations concerning correlations between the seismic velocities can be made with both tomograms in the vicinity of stations 49 and 61. It can be concluded that the GRM tomogram at this site demonstrates superior geologic verisimilitude because it is more consistent with the predominantly vertically dipping geologic structure.

## Detailed GRM starting model with a velocity reversal

The application of equations 1 and 3 (average time model of 17 ms, average refractor velocity of 3500 m/s

**Table 2.** Misfit errors — Mount Bulga orebody.

| Starting models and WET tomograms | Mean unsigned error (ms) | Relative misfit function (ms) | Maximum unsigned error (ms) |
|---|---|---|---|
| Smooth velocity-gradient starting model: vertical velocity gradients | 3.94 | 11.17 | 12.50 |
| Smooth velocity-gradient WET tomogram: vertical velocity gradients | 1.23 | 1.51 | 6.74 |
| Detailed GRM starting model $XY = 3.75$ m: no velocity reversal | 3.44 | 10.31 | 15.12 |
| Detailed GRM WET tomogram $XY = 3.75$ m: no velocity reversal | 1.20 | 1.30 | 5.70 |
| Detailed GRM starting model $XY = 3.75$ m: velocity reversal | 14.92 | 180.69 | 46.86 |
| Detailed GRM WET tomogram $XY = 3.75$ m: velocity reversal | 1.64 | 2.45 | 9.98 |

and optimum $XY$ value of 3.75 m) yields an average vertical seismic velocity in the weathered layer of approximately $600 \pm 100$ m/s. This value, which is comparable to that obtained several hundred meters away with the shear-zone traverse (approximately $500 \pm 250$ m/s), is unexpectedly low, but nevertheless, it shows that vertical velocity gradients are unlikely.

The tomogram in Figure 13 shows that there are significant lateral variations in seismic velocities in the weathered layer. These lateral variations also can be accommodated in the GRM average vertical velocity simply by evaluating equation 1 at each station. The starting model presented in Figure 14 has been computed with the GRM average vertical velocity using the optimum $XY$ value of 3.75 m. As with the shear-zone case study above, the traveltimes have been altered to reflect the revised seismic velocities in the weathered layer.

The WET tomogram in Figure 14 shows similar features to the GRM WET tomogram in Figure 13. Although seismic velocities in the subweathering zone are lower by approximately 500 m/s, nevertheless, the regions at both ends and in the center of the profile with low seismic velocities are confirmed. In addition, the lateral variations in the seismic velocities in the weathered layer show similar trends.

## Head-coefficient attributes

The head-wave amplitudes can be affected by variations in the composition of the surface soil layers. Palmer (2006) shows that the amplitudes increase in accordance with the transmission coefficient of the Zoeppritz equations where an additional soil layer occurs. Because the seismic velocities and densities of the uppermost few centimeters of surface soil layers can be very low, it is not unusual for the transmission coefficient to be almost two. Accordingly, the computed head coefficient for the orebody profile has

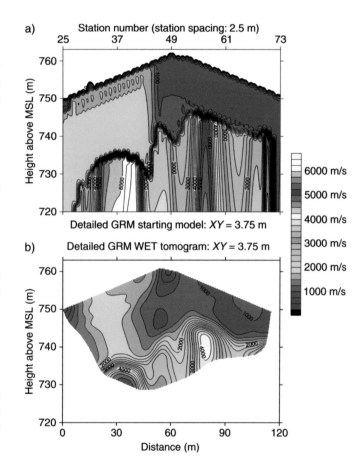

**Figure 13.** (a) Detailed GRM starting model and (b) WET tomogram for the massive sulfide orebody.

a "noise" envelope of a factor of two, arising from the occurrence and disappearance of soil and completely weathered rock.

A model of the scaled density ratios is presented in Figure 15. In this model, the density of the weathered layer has been assumed to be uniformly one. In the absence of any density values measured on discrete samples, no account has been taken of the likely higher densities of the outcropping rocks between stations 31 and 49, the lower densities over the recently restored historical mine workings in the vicinity of stations 49 to 53, or the intermediate den-

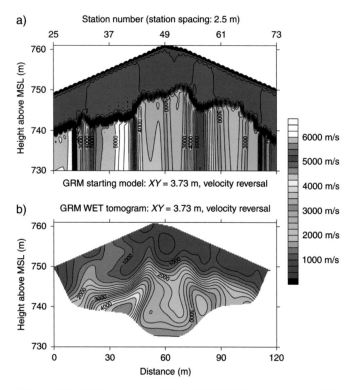

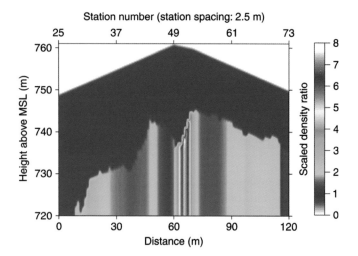

**Figure 15.** The scaled density ratios computed from the seismic velocities and the head coefficient.

**Figure 14.** (a) Starting model and (b) WET tomogram, which accommodates a possible velocity reversal in the weathered layer for the massive sulfide orebody.

sities between stations 53 and 70, where soil and completely weathered rock occur.

Nevertheless, the density model clearly shows the increased density of the orebody, and it provides a useful starting model for inversion. Despite the uncertainties in the densities in the weathered layer, the density model shows there is an increase in thickness of the low-density weathered layer over the high-density orebody between stations 49 and 53. This might explain anecdotal evidence that there is no significant gravity anomaly over the orebody. Accordingly, the density model of the weathered layer derived from the seismic-refraction data can be used to strip away the gravity effects of the weathered layer in a manner similar to the application of static corrections with seismic-reflection data, to enhance the detectability of the deeper gravity target — in this case, the orebody. Elsewhere, other targets might include karsts or clandestine tunnels.

The P-wave modulus computed with equation 8 is shown in Figure 16. This parameter appears less sensitive to lateral variations in densities in the weathered layer, and there is generally quite good correlation with the seismic velocities.

The P-wave modulus is quite high for the region of high density between stations 49 and 53, where the massive sulfides occur. In this case, the inclusion of the density with the head coefficient has counteracted the effect of the low seismic velocity effectively, and it has generated a more realistic measure of the rock strength of the massive sulfides. It demonstrates that accommodation of density with the head coefficient often can result in more representative estimates of rock strength.

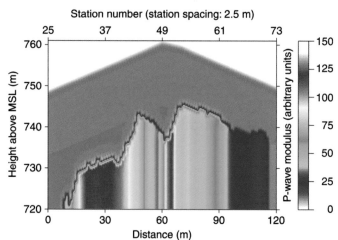

**Figure 16.** The P-wave modulus in arbitrary units computed as the subweathering velocity divided by the head coefficient.

## Spectral analysis of the RCS

Figure 17 shows the spectral analysis for the first few cycles of the RCS for the shots at stations 1 and 97. As with head-wave amplitudes, the spectral content of the RCS is affected by the large lateral variations in the weathered layer. The high-frequency response is reduced significantly between stations 25 and 31 and between stations 56 and 70, which correspond with intervals of low seismic velocity. The correlation is better with the detailed GRM starting model than with the GRM WET tomogram.

The region between stations 49 and 54 exhibits low seismic velocities but good high frequency response. This region corresponds with the increase in the density in the subweathering zone, as shown in Figure 15. Accordingly, both head-wave amplitudes and the RCS facilitate the differentiation of shear zones from massive sulfides where both exhibit low seismic velocities.

# Discussion

## Refraction attributes as a priori information

This study demonstrates that the use of a suite of refraction attributes, derived from traveltimes, head-wave amplitudes, and the RCS, as a priori information often can confirm the most likely detailed models of the seismic velocities. The major advantage of using refraction attributes as a priori information is that they are different components of the same seismic signal. Therefore, they sample the same volumes and relate to the same layers, interfaces, and petrophysical properties. Consistency between the seismic model and the refraction attributes constitutes internal consistency, which validates reliability of the inversion of traveltime data.

The shear-zone traverse represents favorable conditions for the application of the GRM and the RCS and the derivation of refraction attributes. The topography is essentially flat, and there are only moderate lateral variations in the weathered layer, which facilitates effective use of head-wave amplitudes. With this case study, there is overwhelming evidence for the existence of a major shear zone, based on a priori refraction attributes and a posteriori orthogonal 3D profiles.

By contrast, the traverse across the orebody represents considerably less favorable conditions. In particular, the large lateral variations in the surface soil layers, especially at each end of the traverse, complicate the use of head-wave amplitudes. The massive sulfide orebody does not constitute a common geotechnical target. However, it is a nar-

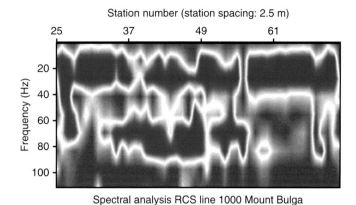

Station number (station spacing: 2.5 m)

**Figure 17.** Spectral analysis of the RCS for SP1 and SP97.

row vertical feature with quite distinctive petrophysical properties. Accordingly, it provides a crucial test of the lateral resolution of refraction tomography. This study demonstrates that the GRM and RCS can generate good estimates of the lateral extent and various useful petrophysical and geotechnical parameters. By contrast, the orebody and associated shear zones remain undetected, undefined, and undifferentiated with the smooth velocity-gradient tomogram.

## Detailed seismic models in the subweathering zone

There is a common expectation that refraction tomography can reveal details that are not apparent in either the traveltime data or the starting model, that is, tomography can improve the resolution of the final tomogram. However, the role of inversion is to provide information about the unknown numerical parameters that go into the model, *not to provide the model itself* (Menke, 1989, p. 2). As a result, usually there is a close similarity between the starting models and the respective final tomograms. Accordingly, the choice of the starting model and in turn the inversion algorithm used to generate the starting model are critical, whether it be for traveltime or full-waveform inversion.

If important features, such as the 50-m-wide, low-velocity shear zone between stations 51 and 61, are not included in the starting model, as is the case with the smooth velocity-gradient model in Figure 1, then it is unlikely that tomographic inversion will recover them. By contrast, even tomograms generated with the conventional reciprocal method, which represents nonoptimum application of the GRM with a zero *XY* value, clearly demonstrate the existence of major shear zones at Mount Bulga and elsewhere (Palmer, 2007, 2008b).

Further improvements in the resolution of the seismic velocities in the subweathering zone are generated readily and objectively with a novel application of the Hilbert transform to the velocity analysis function of the GRM. This method can recover major lateral variations in seismic velocities, such as those which are representative of low-velocity shear zones between stations 51 and 61 and stations 68 and 73. Furthermore, this method can indicate the likely occurrence of more moderate variations, such as those in the vicinity of stations 25 and 37, which also can be significant in geotechnical investigations.

In Figures 5 through 8, the detailed GRM starting models intentionally included unrealistic values of the seismic velocities in the subweathering zone. However, the major effect of tomography has been to moderate the unrealistically high seismic velocities. The regions with the low seismic velocities remain relatively unaffected. Furthermore, there are few significant differences in the depths to the base of the weathered layer between the starting models and the WET tomograms.

Therefore, it can be concluded that refraction tomography essentially smoothes unrealistic or inconsistent starting models. Refraction tomography rarely extracts additional lateral resolution that is not apparent in the starting model.

## Seismic velocities in the weathered layer

It is not widely recognized that the accommodation of any velocity reversal in the weathered layer can affect the detailed resolution of the seismic velocities in the subweathering zone. For the shear zone between stations 51 and 61, the seismic velocity varies from 2500 m/s with the two tomograms that uncritically accept the traveltime data from the weathered layer to 1750 m/s and 2000 m/s for the tomograms that accommodate the velocity reversal.

Although there is no ambiguity in recognizing that these very low velocities represent a major shear zone, this is not the case with more moderate variations. In the vicinity of stations 25 and 37, the use of a lower seismic velocity in the weathered layer is critical for their validation in the final WET tomogram in Figure 7 as regions of lower seismic velocities. It can be concluded that the use of unrealistically high seismic velocities in the weathered layer, such as the default use of vertical gradients with most refraction tomography programs, can result in a significant loss of lateral resolution of the seismic velocities in the subweathering zone. In fact, it can be argued that overfitting in the weathering and underfitting in the subweathering zone are inevitable consequences of the use of vertical velocity gradients.

This study demonstrates the usefulness of generating revised tomograms where updated models of the seismic velocities in the weathered layer are available. The average vertical velocity of the GRM provides an objective approach for accommodating velocity reversals, vertical velocity gradients and seismic anisotropy, in the absence of suitable a posteriori information, such as uphole surveys, sonic logs, or borehole tomography.

## The usefulness of error analyses

This study questions the usefulness of simplistic comparisons of the misfit errors of refraction tomography. The numerous tomograms obtained in this study, which present quite different models of the subsurface, are generally of comparable accuracy. Furthermore, the errors of inversion usually can be reduced simply by increasing the number of iterations, without any significant changes in the major features of the final tomogram.

Because a comparison of the misfit errors of tomographic inversion does not prove that a given result is either correct, defensible, or even geologically reasonable, alternative measures of the uncertainties with refraction inversion are required. A commonly advocated strategy is to generate several refraction tomograms using a variety of starting models, and then to select those features that are common to all tomograms (Oldenburg, 1984).

This strategy can be implemented usefully where the GRM is used because the GRM algorithms honor the traveltime data independently of the *XY* value selected (Palmer, 1980, p. 50; Palmer, 1986, p. 106–107). Because lateral resolution of the GRM time model does not vary significantly with the *XY* value, the major source of nonuniqueness is usually in the determination of the lateral variations in the seismic velocities.

It is proposed that the three tomograms generated with detailed GRM time and velocity models for the optimum *XY* value and plus/minus half the station spacing are a more useful measure of uncertainty. The Welcome Reef studies (Palmer, 2007, 2008b), in which the station spacing is 10 m, have used even greater variations in *XY* values (30 ± 30 m, 60 ± 30 m) to illustrate uncertainty.

## Starting models for routine geotechnical investigations

The Mount Bulga shear zone is 50 m or 10 stations wide and exhibits a large contrast in the seismic velocities with the surrounding rock (about 2000 m/s compared with approximately 6000 m/s). It is a major geologic feature that would be of considerable importance in any geotechnical or environmental investigation. This wide zone with a low seismic velocity is detected with the WET tomograms generated with both smooth and detailed GRM starting models but not with WET tomography using a smooth velocity-gradient starting model.

It can be concluded from these data that refraction tomograms generated from low resolution starting models are not able to provide a reliable measure of the occurrence or otherwise of even major lateral variations in seismic velocities. Therefore, it can be concluded that inversion algorithms (such as those of the GRM), which generate detailed starting models, are essential for the majority of routine geotechnical investigations.

## Conclusions

The tomographic inversion of near-surface seismic-refraction traveltime data is fundamentally nonunique. It is possible to generate tomograms, which range from the geologically improbable to the very detailed, all of which satisfy the traveltime data to sufficient accuracy. Therefore, unless specific measures are taken to address nonuniqueness explicitly with both a priori and a posteriori information, then the production of a single refraction tomogram which fits the traveltime data to sufficient accuracy does not necessarily demonstrate that the result is either accurate, defensible, or even geologically reasonable.

A comparison of the starting models with the final tomograms does not demonstrate that tomography can improve the resolution significantly. If important geotechnical features, such as both narrow and wide zones with low seismic velocities, are not included in the starting model, as is the case with smooth velocity-gradient starting models, then it is unlikely that the inversion process will recover them. Accordingly, the choice of the inversion algorithm used to generate the starting model is critical.

There are usually few significant differences between the smooth GRM starting models and the WET tomograms, where the traveltime data are accepted uncritically. These results indicate that the use of refraction tomography is not essential for achieving acceptable accuracy with the GRM. However, in those cases where there is an obvious difference, either because of deficiencies with the starting models (as is the case with the smooth velocity-gradient tomograms) or by intent to create greater variability (as is the case with the use of the Hilbert transform), then the application of tomography can be useful. Therefore, it can be concluded that refraction tomography essentially smoothes unrealistic or inconsistent starting models

and that refraction tomography rarely improves lateral resolution.

In addition to the use of low-resolution starting models, unrealistically high seismic velocities in the weathered layer, such as those generated with vertical velocity gradients, also can limit the lateral resolution of seismic velocities in the subweathering zone. The GRM average vertical velocity can provide representative seismic velocities in the weathered layer, irrespective of the occurrence of constant velocities, vertical velocity gradients, velocity reversals, or seismic anisotropy.

A simplistic comparison of the errors of tomographic inversion does not prove that a given result is correct, defensible, or even geologically reasonable. Instead, it is proposed that the three tomograms generated with detailed GRM time and velocity models for the optimum *XY* value and plus/minus half the station spacing constitute a more useful measure of the uncertainties of refraction inversion.

Refraction attributes can be useful during all stages of the geotechnical characterization of any site. In the initial or reconnaissance stage, refraction attributes can be used to compute detailed starting models and to resolve nonuniqueness with refraction tomography. With the acquisition of borehole and other geotechnical data, refraction attributes can be used to compute more detailed quantitative models of the geotechnical parameters, especially with multivariate geostatistics.

## References

Chopra, S., and K. J. Marfurt, 2007, Seismic attributes for prospect identification and reservoir characterization: SEG Geophysical Developments Series No. 11.

Danbom, S. H., 2005, Special challenges associated with the near surface, *in* D. K. Butler, ed., Near-surface geophysics, Chapter 2: SEG Investigations in Geophysics Series No. 13, 7–29.

de Franco, R., 2005, Multi-refractor imaging with stacked refraction convolution section: Geophysical Prospecting, **53**, 335–348.

Fell, R., P. MacGregor, and D. Stapledon, 2005, Geotechnical engineering of dams: CRC Press.

Hampson, D. P., J. S. Schuelke, and J. A. Quirein, 2001, Use of multiattribute transforms to predict log properties from seismic data: Geophysics, **66**, 220–236.

Ivanov, J., R. D. Miller, J. Xia, and D. Steeples, 2005a, The inverse problem of refraction travel times, Part II: Quantifying refraction nonuniqueness using a three-layer model: Pure and Applied Geophysics, **162**, 461–477.

Ivanov, J., R. D. Miller, J. Xia, D. Steeples, and C. B. Park, 2005b, The inverse problem of refraction travel times, Part I: Types of geophysical nonuniqueness through minimization: Pure and Applied Geophysics, **162**, 447–459.

Lanz, E., H. Maurer, and A. G. Green, 1998, Refraction tomography over a buried waste disposal site: Geophysics, **63**, 1414–1433.

Leung, T. K., 2003, Controls of traveltime data and problems of the generalized reciprocal method, Geophysics, **68**, 1626–1632.

McCann, D. M., M. G. Culshaw, and P. J. Fenning, 1997, Setting the standard for geophysical surveys in site investigations, in D. M. McCann, M. Eddleston, P. J. Fenning, and G. M. Reeves, eds., Modern geophysics in engineering geology: Geological Society [London] Engineering Geology Special Publication 12, 3–34.

McCann, D. M., M. G. Culshaw, and K. J. Northmore, 1990, Rock mass assessment from seismic measurements, in F. G. Bell, M. G. Culshaw, J. C. Cripps, and J. R. Coffey, eds., Field testing in engineering geology: Geological Society [London] Engineering Geology Special Publication 6, 257–266.

McCann, D. M., and P. J. Fenning, 1995, Estimation of rippability and excavation conditions from seismic velocity measurements: Geological Society [London] Engineering Geology Special Publication 10, 335–343.

Menke, W., 1989, Geophysical data analysis: Discrete inverse theory: Academic Press.

Nichols, T. C., 1980. Rebound — Its nature and effect on engineering works: Quarterly Journal of Engineering Geology, **13**, 133–152.

Oldenburg, D. W., 1984, An introduction to linear inverse theory: IEEE Transactions on Geoscience and Remote Sensing, **GE-22**, 665–674.

Oldenburg, D. W., and Y. Li, 2005, Inversion for applied geophysics: A tutorial, in D. K. Butler, ed., Near-surface geophysics, Chapter 5: SEG Investigations in Geophysics Series No. 13, 89–150.

Palmer, D., 1980, The generalized reciprocal method of seismic refraction interpretation: SEG.

———, 1981, An introduction to the generalized reciprocal method of seismic refraction interpretation: Geophysics, **46**, 1508–1518.

———, 1986, Refraction seismics: The lateral resolution of structure and seismic velocity: Geophysical Press.

———, 1991, The resolution of narrow low-velocity zones with the generalized reciprocal method: Geophysical Prospecting, **39**, 1031–1060.

———, 2001a, Imaging refractors with the convolution section: Geophysics, **66**, 1582–1589.

———, 2001b, Measurement of rock fabric in shallow refraction seismology: Exploration Geophysics, **32**, 907–914.

———, 2001c, Resolving refractor ambiguities with amplitudes: Geophysics, **66**, 1590–1593.

———, 2006, Refraction traveltime and amplitude corrections for very near-surface inhomogeneities: Geophysical Prospecting, **54**, 589–604.

———, 2007, Is it time to re-engineer geotechnical seismic refraction methods?: 19th Conference and Exhibition, ASEG, Extended Abstracts, doi: 10.1071/ASEG2007ab106.

———, 2008a, Is it time to re-engineer geotechnical seismic refraction methods?: First Break, **26**, 69–77.

———, 2008b, Non-uniqueness in near-surface refraction inversion, in Y. X. Xu, and J. H. Xia, eds., Proceedings of the 3rd International Conference on Environmental and Engineering Geophysics: Science Press, 42–54.

———, 2009a, Maximising the lateral resolution of near-surface seismic refraction methods: Exploration Geophysics, **40**, 85–90.

———, 2009b. Integrating long and short wavelength time and amplitude statics: First Break, **27**, 57–65.

———, 2010a, Are refraction attributes more useful than refraction tomography?: First Break, **28**, 43–52.

———, 2010b, Non-uniqueness with refraction inversion — A synclinal model study: Geophysical Prospecting, **58**, 203–218.

———, 2010c, Non-uniqueness with refraction inversion — The Mt. Bulga shear zone: Geophysical Prospecting, **58**, 561–575.

Palmer, D., R. Nikrouz, and A. Spyrou, 2005, Statics corrections for shallow seismic refraction data: Exploration Geophysics, **36**, 7–17.

Ronen, S., P. S. Schultz, M. Hattori, and C. Corbett, 1994, Seismic guided estimation of log properties: Part 2: Using artificial neural nets for nonlinear attribute calibration: The Leading Edge, **13**, 674–678.

Schultz, P. S., S. Ronen, M. Hattori, and C. Corbett, 1994a, Seismic guided estimation of log properties: Part 3: A controlled study: The Leading Edge, **13**, 305–310.

———, 1994b, Seismic guided estimation of log properties: Part 1: A data-driven interpretation methodology: The Leading Edge, **13**, 770–776.

Schuster, G. T., and A. Quintus-Bosz, 1993, Wavepath eikonal traveltime inversion: Theory: Geophysics, **58**, 1314–1323.

Sheehan, J. R., W. E. Doll, and W. A. Mandell, 2005, An evaluation of methods and available software for seismic refraction tomography analysis: Journal of Engineering and Environmental Geophysics, **10**, 21–34.

Sheriff, R. E., and L. P. Geldart, 1995, Exploration seismology, 2nd ed.: Cambridge University Press.

Simpson, G., 2004, Role of river incision in enhancing deformation: Geology, **32**, 341–344.

Smithyman, B., R. G. Pratt, J. Hayles, and R. Wittebolle, 2009, Detecting near-surface objects with seismic waveform tomography: Geophysics, **74**, no. 6, WCC119–WCC127.

Stefani, J. P., 1995, Turning-ray tomography: Geophysics, **60**, 1917–1929.

Treitel, S., and L. Lines, 1988, Geophysical examples of inversion (with a grain of salt): The Leading Edge, **7**, 32–35.

Virieux, J., and S. Operto, 2009, An overview of full-waveform inversion in exploration geophysics: Geophysics, **74**, no. 6, WCC1–WCC26.

Whiteley, R. J., L. V. Hawkins, and G. J. S. Govett, 1984, The seismic, electrical, and electrogeochemical character of the Mount Bulga massive sulfide orebody, N.S.W., Australia: 54th Annual International Meeting, SEG, Expanded Abstracts, 310–314.

Zaruba, Q., 1956, Bulged valleys and their importance for foundations of dams: Transactions of the Sixth International Congress for Large Dams, 509–515.

Zhang, J., and M. N. Toksöz, 1998, Nonlinear refraction traveltime tomography: Geophysics, **63**, 1726–1737.

Zhu, X., D. P. Sixta, and B. G. Andstman, 1992, Tomostatics: Turning-ray tomography + static corrections: The Leading Edge, **11**, 15–23.

Chapter 15

# Direct Determination of Electric Permittivity and Conductivity from Air-launched GPR Surface-reflection Data

Evert Slob[1] and Sébastien Lambot[2, 3]

## Abstract

Knowledge of the spatial distribution of surface-soil water content and its dynamics is important on many scales for research and applications in agriculture, hydrology, meteorology, and climatology. At the field scale of one hectare, it is difficult at present to obtain reliable estimates in an efficient way. A versatile method for determining the electric permittivity of a volume just below the earth surface uses off-ground monostatic ground-penetrating radar (GPR) reflection data. This volume average can be called the surface-soil permittivity, and the permittivity and conductivity of the ground are determined as close to the surface as possible given the bandwidth used in acquisition. It can be demonstrated that under reasonable conditions, in which the time-domain surface-reflection method works with satisfactory accuracy, the extended surface-reflection method performs better when used in the frequency domain. This method allows for accurate conductivity estimates when these conductivities can be obtained from full-waveform inversion.

## Introduction

Surface water-content estimates can be obtained accurately and efficiently at the field scale using ground-penetrating radar (GPR) (Huisman et al., 2003). This complements existing and well-established methods such as time-domain reflectometry (TDR) and gravimetric methods at scales below 1 m (Huisman et al., 2001), and air-

borne/spaceborne remote sensing or radiometry, which are used at scales above 100 m (Schneeberger et al., 2004a). From GPR data, estimates on electric permittivity, and possibly conductivity, can be obtained. The electric permittivity and conductivity can be frequency-dependent but often are assumed constant over the limited frequency band used for this application. A petrophysical model is required to obtain an estimate of surface water content from measured electrical properties. The property of water-content transform will not be discussed here in any detail and we refer the reader to Rhoades et al. (1976), Topp et al. (1980), and Ledieu et al. (1986). Several applications can be found in Allred et al. (2008).

Investigations of the influence of the topsoil structure on the estimated surface electric permittivity have shown that when the depth variation of permittivity near the surface is smooth relative to the wavelength, we can assume a single half-space model to determine the surface electric permittivity (Schneeberger et al., 2004b; Lambot et al., 2006a; Lambot et al., 2006b). A more sophisticated model is required for situations in which it is not reasonable to assume a homogeneous half-space below the surface. Assuming the homogeneous half-space model is adequate, the surface electric permittivity can be obtained using the direct ground wave with ground-coupled antennae or using the reflection from the surface using elevated antennae.

From the ground wave, electric permittivity can be obtained by fitting a straight line through the ground-wave arrival times as a function of transmitter-receiver antenna distance (Huisman et al., 2002; Galagedara et al., 2003;

[1]*Delft University of Technology, Department of Geotechnology, Delft, the Netherlands.*
[2]*Université Catholique de Louvain, Earth and Life Institute, Louvain, Belgium.*
[3]*Forschungszentrum Jülich, Institute of Chemistry and Dynamics of the Geosphere, Agrosphere ICG IV, Jülich, Germany.*

Grote et al., 2003; Galagedara et al., 2005b). This can require a large offset range and some difficulties occur when electric permittivity is not constant over the offset range and when the near-surface conductivity is high. Galagedara et al. (2005a) perform a modeling study to investigate this depth of influence for the direct ground wave. They find that water content estimated from the ground wave depends primarily on the depth range of influence, which depends on the frequency bandwidth used. This means that the depth range of influence depends itself on moisture content, making the method unreliable. Other disadvantages are that it is a relatively labor-intensive geometry and that identification of the direct ground wave can be difficult or impossible. A recent experimentally developed method that seems robust is presented by Pettinelli et al. (2007) and shows good correlations between surface-soil permittivity and the early time response.

The second and possibly more appropriate method relies on the surface reflection. Hence reflection amplitude is used, not traveltime information. Initiated by Maser and Scullion (1991) for GPR data analysis for road-pavement thickness determination, it is known as the surface-reflection method (SRM). Studies include Chanzy et al. (1996), Redman et al. (2002), and Serbin and Or (2003), but the method has not been explored fully.

These SRM studies assume that the surface reflection can be isolated using the reflection from a metallic plate as a calibration signal. This requires that the height of the antenna above the metallic sheet remains constant for all following surface-reflection measurements. Small errors in antenna height lead to large errors in estimated permittivity (Lambot et al., 2006b). A more general approach employs full-waveform inversion on monostatic GPR data with the aid of a carefully chosen antenna-antenna and antenna-ground interaction model (Lambot et al., 2004). This model requires the determination of the characteristic antenna transfer functions, obtained from a number of calibration measurements at different heights above a metal sheet that need to be carried out once for a particular antenna. The main advantage for practical applications is the ability to keep the antenna height relatively small, because 3D wave propagation and antenna-soil interaction are modeled. The antenna height does not have to be known a priori, nor does it have to be kept constant. This method also allows including a small topsoil layer that is different from the layer just below it. The method has been tested successfully for estimating soil-surface water content (Lambot et al., 2006b).

Here we demonstrate that frequency-domain implementation of the surface-reflection model allows for estimation of the surface-soil electric conductivity. It is much

simpler and faster than the full-waveform inversion method (Lambot et al., 2008; Lambot et al., 2009) and performs better than SRM. When surface roughness (Lambot et al., 2006a) is larger than one-eighth of the dominant wavelength over the first Fresnel zone for pulse radars, both methods will fail, and a lower dominant frequency has to be used. Using the same criterion, the upper frequency limit can be determined where stepped-frequency continuous-wave (SFCW) radar will work. Of course, sensitivity to larger depth levels will increase with decreasing the dominant frequency of pulse radar, or with decreasing the maximum frequency (reducing the bandwidth) for SFCW radar. In case there is near-surface layering at a scale much smaller than the vertical resolution, both surface methods fail. This situation is investigated in Minet et al. (2010).

We show that the capabilities of this method on numerical data and on field data obtained in irrigated areas in the Gabes region of southern Tunisia are different from those presented in Lambot et al. (2008).

## Surface-reflection Models

The estimation of surface-soil water content from air-launched monostatic GPR data from SRM assumes that the surface-reflection coefficient can be obtained by the ratio of the received field that reflects from the surface and the received field that reflects from a metallic sheet. The metallic sheet should be a very good conductor, and its size should be larger than the antenna footprint on the surface in the whole frequency bandwidth or for pulse radars when the edge effects can be windowed out in the time domain. This assumption also implies sufficient but minimal height of the antenna above the ground surface and that the normalization with the reflected field from the metallic sheet eliminates all propagation and antenna effects.

Suppose there is a sufficient distance between the antenna and the ground surface, such that the antenna's internal reflection is separated in time from the surface reflection. Let the first-order multiple between antenna and ground surface arrive sufficiently separated in time from the direct ground-surface reflection. Then it is clear that the measured direct ground-surface reflection is related linearly to the field that is incident on the ground surface. This incident field is unknown, but let us assume that it is emitted by a point source with unknown strength. The same antenna is used as receiver (monostatic mode) so that by normalizing the surface-reflection response by the metallic-sheet response, we eliminate all propagation and antenna effects. Then, we have a model similar to the one given in Lambot et al. (2004). This model has a Green's function that describes the reflection process at the surface of a heterogeneous half-space.

The model and calibration procedure are described in more detail in the next section. Because we perform the analysis on a single measurement, a plane-layered half-space is assumed. The SRM (Maser and Scullion, 1991; Chanzy et al., 1996) assumes the subsurface can be approximated by a homogeneous, nonconductive half-space and that the reflection coefficient can be approximated by the normal-incidence plane-wave reflection coefficient. Combining these assumptions, we obtain the relative electric permittivity just below the surface in the SRM as

$$\varepsilon_{\rm r} = \left(\frac{E^{\rm PEC}(h, t) + E(h, t)}{E^{\rm PEC}(h, t) - E(h, t)}\right)^2, \quad (1)$$

where $E(h, t)$, $E^{\rm PEC}(h, t)$ denote the surface-reflection data and the measurement over the metallic sheet, respectively, and $t$ is time. Parameter $h$ in the argument of both functions denotes the two-way height between the antenna phase center and the ground surface, which must be constant. Equation 1 is in principle valid for any time point in the whole time window of the surface reflection. One can optimize the solution in a certain time window, but of course, the simplest way to compute the reflection coefficient is by using the peak-to-peak amplitude. This is the original implementation of the method by Maser and Scullion (1991); see Jol (2009) for a more detailed description of this method.

A slight modification of SRM employs the frequency dependence of the reflection coefficient to improve the permittivity estimate and allows for an estimation of conductivity. We still approximate the reflection coefficient by its normal-incidence plane-wave value, but keep its frequency dependence to allow for nonzero conductivity or more general relaxation mechanisms. Then, it is more useful to write an extended form of equation 1 in the frequency domain including the frequency dependence of the relative electric permittivity:

$$\hat{\varepsilon}_{\rm r} = \varepsilon_{\rm r}' - i\varepsilon_{\rm r}'' = \left(\frac{\hat{E}^{\rm PEC}(h, \omega) + \hat{E}(h, \omega)}{\hat{E}^{\rm PEC}(h, \omega) - \hat{E}(h, \omega)}\right)^2. \quad (2)$$

In the case that a constant permittivity and conductivity can be assumed, the complex relative permittivity is given by $\hat{\varepsilon}_{\rm r} = \varepsilon_{\rm r} - i\sigma/(\omega\varepsilon_0)$. This extended surface-reflection model (XSRM) is similar to the SRM equation 1, but it gives a better estimate of surface permittivity and does not require picking extreme values of the surface reflection.

Equation 2 can be used over the whole frequency band where the radar has sufficient energy levels, which allows for possible dispersion estimates. In addition, we can work with a point source in air above a homogeneous half-space and leave the real constant electric permittivity and conductivity of the half-space as unknowns. We match the model to the data using the normalized root-mean-square (rms) error function

$$\Phi = \left(\frac{\sum_\omega |\hat{E}^{\rm d}(h, \omega) - \hat{E}^{\rm m}(h, \omega, \varepsilon_{\rm r}, \sigma)|^2}{\sum_\omega |\hat{E}^{\rm d}(h, \omega)|^2}\right)^{1/2}, \quad (3)$$

where the recorded data is $\hat{E}^{\rm d}(h, \omega)$ and modeled data is $\hat{E}^{\rm m}(h, \omega, \varepsilon_{\rm r}, \sigma)$, as described in Lambot et al. (2004).

## Antenna Model and Calibration

When the vertical two-way travel distance from antenna to surface is $4/3$ of the dominant wavelength or more, the signal-to-noise ratio is decreased, hyperbolic moveout of scattered events becomes flat, and side-looking ability is reduced (Bloemenkamp and Slob, 2003). These last two aspects occur because a larger part of the total travel path occurs in the air, which is the fast medium, and rays are bent toward the vertical axis for waves that penetrate the subsurface. At sufficient height, this can be used advantageously by assuming a 1D earth model and monostatic mode of acquisition. The antenna effectively becomes decoupled from the earth, which allows for removing antenna effects by a simple calibration procedure.

The statement that the antenna effectively is decoupled from the earth does not mean that it does not interact with the earth, but only that its impedance does not depend on the earth impedance. We assume the antenna is connected to the radar by a coaxial transmission line and it is positioned such that the surface is in the far field of the antenna. Then the antenna can be approximated by an interactive point source that can be characterized by its scattering parameters, as depicted in Figure 1. The full scattering matrix consists of four terms: the reflection and transmission coefficients from both sides of the antenna. When a unit-amplitude signal enters the antenna at the input side, the antenna will partially reflect (represented by $S_{11}^{\rm ant}$) and transmit (represented by $S_{12}^{\rm ant}$) the incoming signal when there are no other scattering domains. Similarly, when a unit-amplitude signal enters the antenna at the opening where it would act as a receiving antenna, the antenna will partially reflect (represented by $S_{22}^{\rm ant}$) and transmit (represented by $S_{21}^{\rm ant}$) the incoming signal when there are no other scattering domains. Only the product $S_{12}^{\rm ant}S_{21}^{\rm ant}$

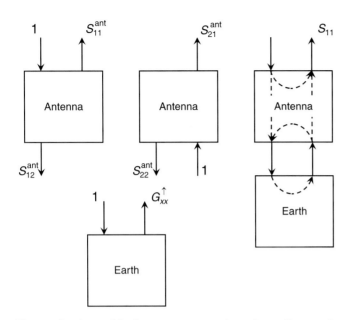

**Figure 1.** A graphic S-port representation of an off-ground antenna. When a unit-amplitude signal enters the antenna at the input side, the antenna will partially reflect (represented by $S_{11}^{ant}$) and transmit (represented by $S_{12}^{ant}$) the incoming signal when there are no other scattering domains. Similarly, when a unit-amplitude signal enters the antenna at the opening where it would act as a receiving antenna, the antenna will partially reflect (represented by $S_{22}^{ant}$) and transmit (represented by $S_{21}^{ant}$) the incoming signal when there are no other scattering domains. The earth is assumed to be below the antenna and its reflection response to a unit-amplitude incoming wave is the earth impulse-reflection response $G_{xx}^{\uparrow}$. When the antenna is located above the earth, the antenna and the earth will interact as represented in the cartoon on the right. The result is a feedback system that is described mathematically by a linear fractional transformation as given in equation 4.

needs to be known for reflection measurements, and it is given by $H = S_{12}^{ant} S_{21}^{ant}$. The earth is assumed to be below the antenna and its reflection response to a unit-amplitude incoming wave is the earth impulse-reflection response $G_{xx}^{\uparrow}$. When the antenna is located above the earth, the earth and antenna will interact as depicted on the right of Figure 1. The result is a feedback system that is described mathematically by a linear fractional transformation (Redheffer, 1961)

$$S_{11} = S_{11}^{ant} + \frac{H G_{xx}^{\uparrow}}{1 + S_{22}^{ant} G_{xx}^{\uparrow}}. \quad (4)$$

See Lambot et al. (2004) for further details about this model. The true earth reflection response is obtained from equation 4 after the three antenna scattering parameters are determined. This can be achieved by performing a

number of independent measurements with a known earth reflection response, e.g., a perfect electric conductor, for different antenna heights.

Computing the Green's function for the reflection of the interface between two homogeneous half-spaces is done optimally using a fast integration technique. The Green's function for a point source in air over a half-space is known in closed form in the radial wavenumber domain, which leads to the following Fourier-Bessel integral given by

$$G_{xx}^{\uparrow} = \frac{1}{8\pi i\omega\varepsilon_0} \int_{\kappa=0}^{\infty} (\Gamma_0^2 r^{TM} + k_0^2 r^{TE}) \frac{\exp(-\Gamma_0 h)}{\Gamma_0} \kappa d\kappa, \quad (5)$$

where vertical wavenumbers $\Gamma$, $\Gamma_0$ are given by $\Gamma = \sqrt{\kappa^2 - \hat{\varepsilon}_r k_0^2}$, $\Gamma_0 = \sqrt{\kappa^2 - k_0^2}$ with nonnegative real parts, the free-space wavenumber $k_0 = \omega/c_0$, and $c_0$, $\varepsilon_0$ are the wave speed and electric permittivity, respectively, in free space. The transverse magnetic (TM) mode and transverse electric (TE) mode reflection coefficients are given by

$$r^{TM} = \frac{\hat{\varepsilon}_r \Gamma_0 - \Gamma}{\hat{\varepsilon}_r \Gamma_0 + \Gamma}; \quad r^{TE} = \frac{\Gamma_0 - \Gamma}{\Gamma_0 + \Gamma}. \quad (6)$$

Numerical evaluation of the integral in equation 5 is carried out most efficiently by going into the upper half of the complex $\kappa$ plane and ensuring that $\Im(\Gamma_0) = $ constant. We derive a new and more efficient method by deforming the integration path along the steepest descent contour by demanding that

$$\Gamma_0 = \xi + ik_0 \Rightarrow \kappa = \sqrt{\xi^2 + 2i\xi k_0}, \quad (7)$$

where $\xi$ is real and increasing monotonically, and it can serve as the new variable of integration, with $0 < \xi < \infty$. Because we must deform the contour in the upper half of the complex $\kappa$ plane, the sign of the square root for $\kappa$ in equation 7 must be taken positive. The Jacobian of this transformation is found as

$$\frac{\partial \kappa}{\partial \xi} = \frac{\Gamma_0}{\kappa}. \quad (8)$$

Using the results of equations 7 and 8 in equation 5 gives the reflected Green's function as a steepest descent integral over $\xi$:

$$G_{xx}^{\uparrow} = \frac{\exp(-ik_0 h)}{8\pi i\omega\varepsilon_0} \int_{\xi=0}^{\infty} ((\xi + ik_0)^2 r^{TM} + k_0^2 r^{TE}) \exp(-\xi h) d\xi.$$

$$(9)$$

This expression of equation 9 is new and leads to a more efficient computation of the integral than the expressions used in Michalski and Butler (1987) and Lambot et al. (2007). This method can be extended to include subsurface layering by inserting in equation 9 the global reflection coefficients given in Slob and Fokkema (2002) instead of the local reflection coefficients from equation 6.

## Numerical Results

Effectiveness of the two surface-reflection models to estimate the surface electric parameters from GPR reflection data has been shown to depend on two main assumptions that we investigate here numerically. Both assume the plane-wave reflection coefficient is an accurate representation. The other assumption is to either neglect or incorporate subsurface conductivity. Therefore, we use a suite of two half-spaces as possible earth models. The reflection from a model with two half-spaces is computed (see Figure 2). In the model, we choose for the lower half-space all possible electric permittivity and conductivity combinations that can be expected in areas varying from arid to humid and from clean sands to loamy soils. Conductivity values range from $\sigma = 0.1$ mS/m to $\sigma = 1$ S/m, giving 17 values on a logarithmic scale, and relative electric permittivity ranges independently from $\varepsilon_r = 2.5$ to $\varepsilon_r = 25$, yielding 10 values on a linear scale. This mimics 170 possible earth-surface conditions that can be encountered. The source point is 24 cm above the earth surface and data is modeled according to equations 6 and 7, and the fast integration method of equation 9. Source height is taken to be large enough to eliminate possible antenna-ground interactions and to ensure the effectiveness of SRM.

Considering a factor 30 between the far-field and intermediate-field terms in a homogeneous medium, the far-field region starts at approximately five times the wavelength away from the source. For the given source height of 24 cm, the incident wave at the earth surface can be considered as coming from a point source located at the phase center of the antenna for frequencies higher than 650 MHz. For this reason, we take a lower frequency limit of 800 MHz for the frequency-domain analysis of the extended surface-reflection model.

The SRM as given in equation 1 is investigated for this suite of permittivity-conductivity combinations. The XSRM as given in equation 2 is compared to the SRM results to show the range of permittivity-conductivity combinations where the plane-wave reflection assumption is accurate. The normalized (or relative) error (in %) in

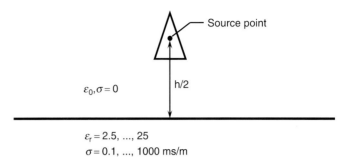

**Figure 2.** The single reflector model with many permittivity-conductivity pairs.

permittivity is used for the SRM and XSRM methods. We define the normalized error as

$$\text{ERR}^\varepsilon = 100 \times \frac{\varepsilon_r^{\text{retrieved}} - \varepsilon_r^{\text{model}}}{\varepsilon_r^{\text{model}}}. \quad (10)$$

where the retrieved values can be obtained from SRM or XSRM. Figure 3a shows the obtained real part of the electric permittivity values obtained with the SRM method and Figure 3b shows that obtained with the XSRM method, both as a function of all model values. From Figure 3a, it can be seen that for low values of conductivity, the permittivity is underestimated slightly. This underestimation becomes stronger (as much as −10%) with increasing conductivity, and finally goes quickly to very large overestimations (far above 50%) when conductivity becomes large. On the other hand, it can be observed from Figure 3b that the plane-wave reflection model itself is quite accurate for the whole permittivity range and almost independent of conductivity values used with XSRM. The error is small, about −2% for most permittivity-conductivity combinations, and remains below −5% for all values used.

Figure 4 shows that, using XSRM, the relative error in conductivity is quite low when conductivity has a measurable effect in the reflection strength of the surface. For conductivity values above 10 mS/m, the errors are below 1 mS/m for all combinations used. When the conductivity is below 1 mS/m, there is no information in the reflection data to retrieve the conductivity, as can be observed by the large errors in that zone in Figure 4, where they rise quickly to above 50%. From numerical analysis of full-waveform inversion (Lambot et al., 2006b), we know that full-waveform inversion is preferable over the surface-reflection method for soil-surface water-content determination. In that study, results focused on permittivity estimation to compare full-waveform-inversion results with SRM. Lambot et al. (2006b) found that forcing the conductivity to be zero did not deteriorate their permittivity

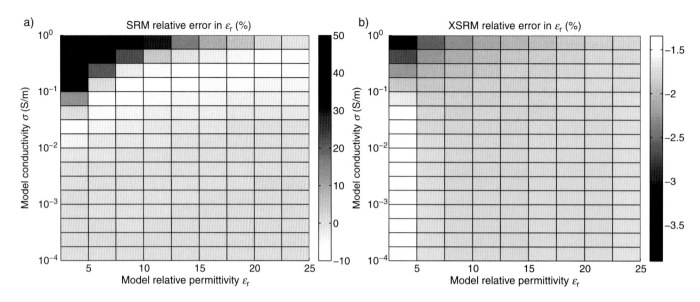

**Figure 3.** Normalized error in electric permittivity values (a) obtained with SRM, and (b) obtained with XSRM, both as a function of model permittivity and conductivity values. Grayscale bars indicate the error in percent.

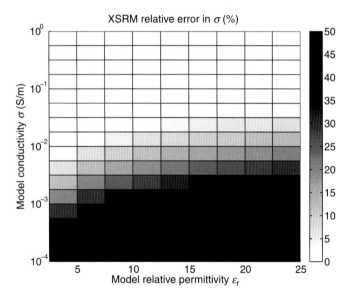

**Figure 4.** The normalized error in electric conductivity values obtained with XSRM as a function of model permittivity and conductivity values. Grayscale bar indicates the error in percent.

estimates when $\sigma < \sim 10$ mS/m, but that for higher values of the conductivity, the estimated permittivity values were less accurate and even unrealistic for $\sigma \sim > 100$ mS/m.

This implies the information content related to conductivity in the reflection data is very low for small conductivity values of $\sigma < \sim 10$ mS/m and increases in the range up to $\sigma \sim > 100$ mS/m, although information content is high above these values. This corresponds to our results, with which we demonstrate that the extended surface-

reflection model allows retrieving both electric permittivity and conductivity to a similar degree of accuracy obtained through full-waveform inversion. Full-waveform inversion using a global-optimization method is very time-consuming, but local methods work fast for this simple problem. The extended surface-reflection model is the preferable model for surface permittivity and conductivity estimations because it comes at no computational cost and performs with the same accuracy as full-waveform inversion.

It is conceivable that electric properties show frequency dependence and the implementation of equation 2 allows retrieving the complex relative permittivity for each frequency separately. To test how well the method works, we have modeled a dispersive electric response function for the lower half-space. The dispersive model shows a decreasing real part of the relative permittivity from $\hat{\varepsilon}_r(f = 0.8 \text{ GHz}) \approx 11.8$ to $\hat{\varepsilon}_r(f = 2.5 \text{ GHz}) \approx 10$, and increasing conductivity from $\hat{\sigma}_r(f = 0.8 \text{ GHz}) = 0.012$ S/m to $\hat{\sigma}_r(f = 2.5 \text{ GHz}) = 0.018$ S/m. The result of XSRM in this frequency range is shown in Figure 5 along with the exact values. Figure 5a shows that relative error in the real part of the permittivity is about $-4\%$ at 800 MHz, and it is below 1% at 2.5GHz.

It is interesting to observe that the surface-reflection model finds a reasonable value for the real part of the relative permittivity as some kind of average over all frequencies. This is shown in Figure 5a only for the frequencies used for the extended surface-reflection model, but we have computed the estimate using equation 1 and are taking the peak-to-peak values as the best estimate. The time-domain response was computed using all frequencies

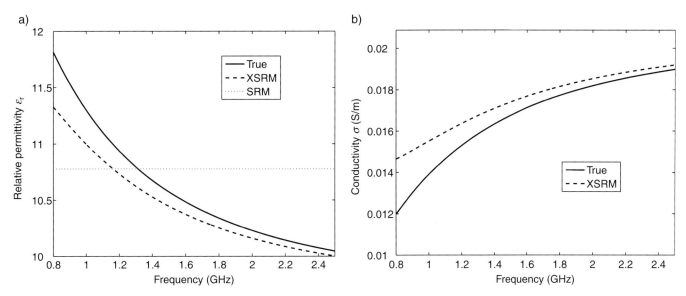

**Figure 5.** (a) True values and XSRM and SRM estimates of relative electric permittivity of a dispersive model as a function of frequency and (b) true values and XSRM estimates of conductivity of a dispersive model as a function of frequency.

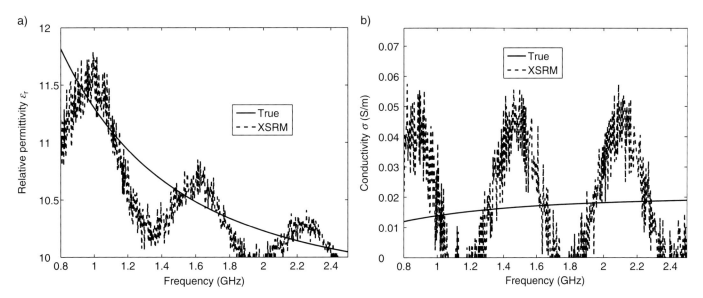

**Figure 6.** (a) True values and XSRM and SRM estimates obtained from data with only 50 dB precision of relative electric permittivity of a dispersive model as a function of frequency and (b) true values and XSRM estimates obtained from data with only 50 dB precision of conductivity of a dispersive model as a function of frequency.

and with a 1-GHz center-frequency Ricker wavelet. Figure 5b illustrates that errors in conductivity for the extended surface-reflection model are 22% at 800 MHz and 1% at 2.5 GHz.

Another interesting test is to investigate the influence of uncertainty. Most measurement systems have a limited dynamic range and because of all kinds of operational differences, the true reliable range in the data is the signal that falls in the first 40 to 80 dB of the receiver. We truncate the signal for the dispersive model for signals below

−50 dB, where 0 dB is the maximum of the received signal. Results are shown in Figure 6a for estimates of the real part of relative permittivity and in Figure 6b for estimated conductivity over the whole frequency range. It can be observed that the solutions become oscillating functions of frequency because of limited precision of the data. The smoothed version of the relative electric permittivity that would be obtained by curve fitting with a Debye-type model still would be quite accurate and would follow the ideal curve of Figure 5a. Hence, data-precision degradation is not a severe limitation in estimations of permittivity. In Figure 6b, we find strong oscillating behavior of conductivity, including nonphysical negative values. Moreover, the smoothed curve that would be obtained from data fitting using a Debye-type model leads to a conductivity estimate two times higher than it should be. It is clear that high data precision is necessary for reliable estimations of conductivity from surface-reflection data.

## Experimental Results

The radar system consisted of a handheld coherent impedance analyzer (Rohde & Schwarz FSH6). We used a linear polarized double-ridged broadband horn antenna with a height of 22 cm and an aperture area of $14 \times 24$ cm$^2$. The antenna operational frequency range is from 800 MHz to 5 GHz and its isotropic gain ranges from 6 to 18 dBi. The relatively small beam width of the antenna (3-dB beam width of $45°$ in the E-plane and $30°$ in the H-plane at 1 GHz) makes it suitable for using off the ground. The antenna was connected to the reflection port of the FSH6 via a high-quality N-type coaxial cable with an impedance of 50 ohms. We calibrated the FSH6 using the standard calibration procedure. The frequency-dependent complex ratio $S_{11}$ between the returned signal and the emitted signal was measured sequentially at 301 stepped operating frequencies over the range of 800 to 2500 MHz with a frequency step of approximately 5.667 MHz. However, only data in the range of 1000 to 2000 MHz, for which the highest quality was observed,

**Figure 7.** One of the measurement locations in Tunisia. From Lambot et al., 2008. Photograph by S. Lambot. Courtesy of EAGE. Used by permission.

were used in the inversion process. Measurements were taken with the antenna aperture situated at determined heights varying from 0.23 to 0.27 m above the soil surface.

In total, 36 measurements were taken at 17 field locations in the area of the Gabes region in Tunisia (see Figure 7). At every location, we first established that the soil surface satisfies the Rayleigh criterion for surface smoothness, which defines a maximum difference in surface height of one-eighth of the wavelength over the first Fresnel zone of the antenna. For our setup, this corresponds to a maximum allowable roughness of 2 cm, and measurements were restricted to areas satisfying this condition. The Fresnel zone is approximately 37 cm at 1.5 GHz and the depth resolution for the final 1 GHz bandwidth is approximately $10 \, \mathrm{cm}/\sqrt{\varepsilon_r}$, which is then better than 5 cm. This depth resolution is important because it implies that vertical heterogeneities at smaller scales cannot be identified as such and will lead to erroneous estimates in permittivity and conductivity.

The volume of soil contributing to the estimated soil properties is determined by the area of the inner half of the first Fresnel zone and the resolution depth. The sampling volume defined in this way is less than 400 cm$^3$ for this experiment. This volume will increase rapidly on fields where the surface roughness is more than 3 cm, because side scattering might become important and the method will fail in this frequency range. A lower frequency band has to be used in such cases, with the disadvantage of increasing the sampling volume and sensitivity to subsurface layering.

The XSRM is applied directly on frequency-domain data, while SRM is used directly on time-domain data because it exploits only the maximum peak-to-peak values. The combined permittivity-conductivity results of XSRM are shown in Figure 8, where the results are sorted in an increasing permittivity sequence and the corresponding conductivity values are shown with a vertical scale given in the right side of the graph. Conductivity values are rather high, but mostly below 100 mS/m. The question is still if the estimated conductivity values have significance. Unfortunately this cannot be validated, because as shown below, the comparison between TDR and GPR cannot be made for this site. There are large variations in conductivity for small changes in permittivity. This could indicate that the precision with which measurements are taken with the FSH6 is insufficient to obtain reliable estimates for conductivity.

From the conductivity values obtained, it is expected that the SRM results also will be accurate. This comparison is shown in Figure 9. Numerical results predict that SRM produces appreciable differences in permittivity estimates

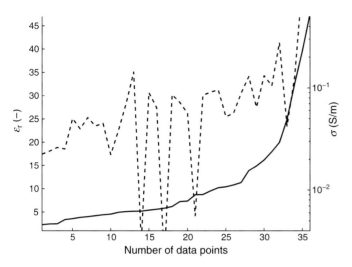

**Figure 8.** The sorted retrieved permittivity values (solid line, left axis) and conductivity values (dashed line, right axis) obtained by applying XSRM to the data.

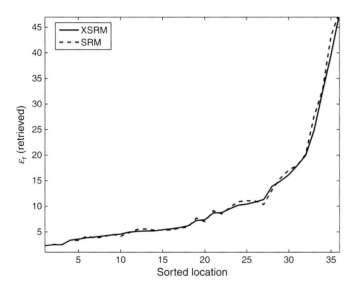

**Figure 9.** The sorted retrieved permittivity values with XSRM (solid line) and with SRM (dashed line).

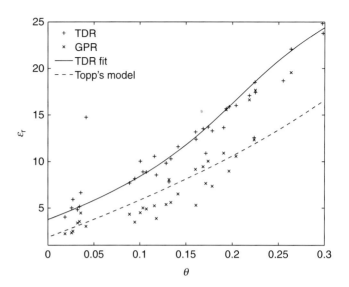

**Figure 10.** TDR and GPR results of permittivity as a function of measured water content along with a third-order polynomial fit for the TDR results and Topp's model as a fit for the GPR results.

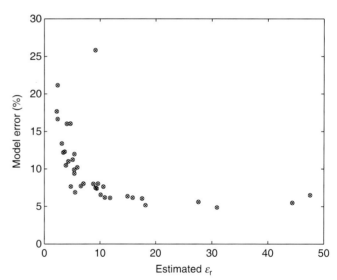

**Figure 11.** Values of data misfit for the GPR estimated permittivity and conductivity values shown in Figure 8 that have been used to compute model data.

for conductivity values higher than 100 mS/m only for permittivity values below 20. By comparing the numerical and experimental results, it can be concluded that higher conductivity values occur for permittivity values above 20 and both methods show similar results. Lambot et al. (2008) show that the full-waveform inversion results and results from the surface-reflection model match very well. Here we show that results from the extended surface-reflection model also match those from the surface-reflection model.

In addition to GPR measurements, three 100-cm³ soil samples were taken in the footprint of the antenna, up to a

depth of 5 cm below the surface. These samples were used for direct water-content determination using the standard oven-drying method at 105°C for more than 24 hours. Additional TDR measurements were performed with a TDR Hydra Probe system (Stevens Water Monitoring Systems, Inc.), operating at a center frequency of 50 MHz. At each sampling site, the probe with 5.7-cm-long stainless-steel rods was inserted vertically into the ground at three places within the footprint of the antenna. From these results, negligible variation was found within the antenna footprint.

Comparing TDR and GPR results is notoriously difficult as explained in Serbin and Or (2004) and Weihermüller et al. (2007). This is because of different frequencies used, different sampling volumes of sensitivity, and varying vertical water-content profiles. Relative electric permittivity values obtained with TDR and GPR are shown in Figure 10 as a function of the water-content values as determined from the oven-drying method. Both methods show a good correlation between their respective obtained real part of permittivity and water content, but TDR results lie consistently above GPR results. For TDR data fit, we used a third-order polynomial with a correlation coefficient of 0.91. For GPR, we used Topp's model, resulting in a correlation coefficient of 0.8. If an arbitrary third-order polynomial had been used, a better value for the correlation coefficient would have been obtained, but Topp's model is already adequate.

The relative permittivity values obtained from GPR that lie above $\varepsilon_r = 15$ are outliers with respect to Topp's model, which corresponds to total water content of $\theta = 0.2$. At high permittivity and/or conductivity values where the surface-reflection coefficient approaches unity and the difference with the reflection from a conducting plate becomes small, we would expect a larger data misfit than at lower permittivity values. However, in Figure 11, where the data misfit is shown as a function of estimated relative permittivity, it can be observed that model error is generally higher for lower values of permittivity than for higher values. This means that the data corresponding to high permittivity values is well modeled using those values. Yet, for these high permittivity values, and hence high water content, full-waveform inversion seems to fail as well as surface-reflection and extended surface-reflection models. It remains unclear why this is the case.

## Conclusions

A simple extension of the surface-reflection method has been presented to obtain accurate estimates for relative electric permittivity and conductivity from GPR surface-reflection data in a fast and robust way. This has been achieved by generalizing the known surface-reflection model and allowing for frequency dependence in the permittivity values below the surface. Relative electric permittivity can be obtained directly from the data in real time through an explicit expression, very similar to the surface-reflection model. This new model combines the accuracy of the full-waveform-inversion method and the speed from the surface-reflection method. A lower frequency limit has been found, above which the new method works to a high degree of accuracy. In the examples here, an average permittivity was given over the whole frequency band used, but an estimate is obtained for each frequency, which allows us to determine frequency dependence of the material parameters. This was demonstrated using a dispersion model for the complex permittivity. In practice, frequency dependence can be physical or caused by near-surface heterogeneities. This adds a check feature to the XSRM method. Together with the discussed calibration scheme for antenna effects and antenna-soil interaction, the approximation can be used in an inversion scheme in case a layered earth must be assumed instead of a mere surface reflection. We believe that this provides a versatile method that might have applications in areas where surface-soil water is the governing factor.

## References

Allred, B. J., M. R. Ehsani, and J. J. Daniels (eds.), 2008, Handbook of agricultural geophysics: CRC Press, LLC.

Bloemenkamp, R., and E. Slob, 2003, The effect of the elevation of GPR antennas on data quality: Proceedings of the 2nd International Workshop on Advanced Ground Penetrating Radar, 201–206.

Chanzy, A., A. Tarussov, A. Judge, and F. Bonn, 1996, Soil water content determination using digital ground penetrating radar: Soil Science Society of America Journal, **60**, 1318–1326.

Galagedara, L., G. Parkin, and J. Redman, 2003, A derivation of the seismic representation theorem using seismic reciprocity: Hydrological Processes, **17**, 3615–3628.

Galagedara, L., J. Redman, G. Parkin, A. Annan, and A. Endres, 2005a, Numerical modeling of GPR to determine the direct ground wave sampling depth: Vadose Zone Journal, **4**, 1096–1106.

Galagedara, L., G. Parkin, J. Redman, P. von Bertoldi, and A. Endres, 2005b, Field studies of the GPR direct ground wave method for estimating soil water content during irrigation and drainage: Journal of Hydrology, **301**, 182–197.

Grote, K., S. Hubbard, and Y. Rubin, 2003, Field-scale estimation of volumetric water content using GPR ground wave techniques: Water Resources Research, **39**, 1321.

Huisman, J., S. Hubbard, J. Redman, and A. Annan, 2003, Measuring soil water content with ground penetrating radar: A review: Vadose Zone Journal, **2**, 476–491.

Huisman, J., C. Sperl, W. Bouten, and J. Verstraten, 2001, Soil water content measurements at different scales: Accuracy of time domain reflectometry and ground-penetrating radar: Journal of Hydrology, **245**, 48–58.

Huisman, J., A. Weerts, T. Heimovaara, and W. Bouten, 2002, Comparison of travel time analysis and inverse modeling for soil water content determination with time domain reflectometry: Water Resources Research, **38**, 1224.

Jol, H. M., ed., 2009, Ground penetrating radar theory and applications: Elsevier, 410.

Lambot, S., M. Antoine, M. Vanclooster, and E. Slob, 2006a, Effect of soil roughness on the inversion of off-ground monostatic GPR signal for non-invasive quantification of soil properties: Water Resources Research, **42**, W03403.

Lambot, S., E. C. Slob, I. van den Bosch, B. Stockbroeckx, and M. Vanclooster, 2004, Modeling of ground-penetrating radar for accurate characterization of subsurface electric properties: IEEE Transactions on Geoscience and Remote Sensing, **42**, 2555–2568.

Lambot, S., E. Slob, D. Chavarro, M. Luczynski, and H. Vereecken, 2008, Measuring soil surface water content in irrigated areas of Southern Tunis using full-waveform inversion of proximal GPR data: Near Surface Geophysics, **6**, 403–410.

Lambot, S., E. C. Slob, J. Rhebergen, O. Lopera, K. Z. Jadoon, and H. Vereecken, 2009, Remote estimation of the hydraulic properties of a sand using full-waveform inversion of time-lapse, off-ground GPR data: Vadose Zone Journal, **8**, 743–754.

Lambot, S., E. Slob, and H. Vereecken, 2007, Fast evaluation of zero-offset Green's function for layered media with application to ground-penetrating radar: Geophysical Research Letters, **34**, L21405.

Lambot, S., L. Weihermüller, J. Huisman, H. Vereecken, M. Vanclooster, and E. Slob, 2006b, Analysis of air-launched ground-penetrating radar techniques to measure the soil surface water content: Water Resources Research, **42**, W11403.

Ledieu, J., P. de Ridder, P. De Clercq, and S. Dautrebande, 1986, A method of measuring soil moisture by time domain reflectometry: Journal of Hydrology, **88**, 319–328.

Maser, K., and T. Scullion, 1991, Use of radar technology for pavement layer evaluation: Second International Symposium on Pavement Response Monitoring Systems for Roads and Airfields, CD-ROM.

Michalski, K., C. Butler, 1987, Evaluation of Sommerfeld integrals arising in the ground stake antenna problem: IEE Proceedings-H Microwaves Antennas and Propagation, **134**, 93–97.

Minet, J., S. Lambot, E. C. Slob, and M. Vanclooster, 2010, Soil surface water content estimation by full-waveform GPR signal inversion in the presence of thin layers: IEEE Transactions on Geoscience and Remote Sensing, **48**, 1138–1150.

Pettinelli, E., G. Vannaroni, B. Di Pasquo, E. Mattei, A. Di Matteo, A. De Santis, and P. A. Annan, 2007, Correlation between near-surface electromagnetic soil parameters and early-time GPR signals: An experimental study: Geophysics, **72**, no. 2, A25–A28.

Redheffer, R., 1961, Difference equations and functional equations in transmission-line theory: McGraw-Hill.

Redman, J., J. Davis, L. Galagedara, and G. Parkin, 2002, Field studies of GPR air launched surface reflectivity measurements of soil water content, *in* S. Koppenjan and K. Lee, eds., Proceedings of the Ninth International Conference on GPR: SPIE 4758, 156–161.

Rhoades, J. D., P. A. C. Raats, and R. J. Prather, 1976, Effects of liquid-phase electrical conductivity, water content, and surface conductivity on bulk soil electrical conductivity: Soil Science Society of America Journal, **40**, 651–655.

Serbin, G., and D. Or, 2003, Near surface water content measurements using horn antenna radar: Methodology and overview: Vadose Zone Journal, **2**, 500–510.

———, 2004, Ground-penetrating radar measurement of soil water content dynamics using a suspended horn antenna: IEEE Transactions on Geoscience and Remote Sensing, **42**, 1695–1705.

Schneeberger, K., M. Schwank, C. Stamm, P. de Rosnay, C. Mätzler, and H. Flühler, 2004a, Topsoil structure influencing soil water retrieval by microwave radiometry: Vadose Zone Journal **3**, 1169–1179.

Schneeberger, K., C. Stamm, C. Mätzler, and H. Flühler, 2004b, Ground-based dual-frequency radiometry of bare soil at high temporal resolution: IEEE Transactions on Geoscience and Remote Sensing, **42**, 588–595.

Slob, E., and J. Fokkema, 2002, Coupling effects of two electric dipoles on an interface: Radio Science, **37**, 1073.

Topp, G., J. L. Davis, and A. P. Annan, 1980, Electromagnetic determination of soil water content: Measurements in coaxial transmission lines: Water Resources Research, **16**, 574–582.

Weihermüller, L., J. A. Huisman, S. Lambot, M. Herbst, and H. Vereecken, 2007, Mapping the spatial variation of soil water content at the field scale with different ground penetrating radar techniques: Journal of Hydrology, **340**, 205–216.

Chapter 16

# Theory of Viscoelastic Love Waves and their Potential Application to Near-surface Sensing of Permeability

Paul Michaels[1] and Vijay Gottumukkula[2]

## Abstract

In computing Love-wave solutions, the choice of constitutive model depends on the domain of application. In the domain of global earthquake seismology, the search for solutions in the complex plane began in the vicinity of the elastic solutions. In the case of near-surface engineering work, damping levels can be large, and elastic stiffness can be much less than in global seismology. Furthermore, the choice of representation should depend on the permeability and degree of water saturation. The study of dry or impermeable soils and rock, where viscous effects are largely absent, has led to an alternative representation for the Kelvin-Voigt damping property. Under that alternative of *effective viscosity*, the damping ratio is a frequency-independent soil constant. Permeable, water-saturated soils, on the other hand, have shown *viscous* behavior. A method to solve for Love waves can be used under a truly viscous assumption. Applications would include near-surface remote sensing of either water content or permeability.

## Introduction

Most of the shallow-engineering surface-wave methods use Rayleigh waves, a mixture of compressive P-waves and vertically polarized shear SV-waves. Some recent examples of geotechnical applications applied to soils and pavements are provided by Nazarian et al. (1983), Stokoe and Nazarian (1985), Gucunski and Woods (1991), Luke and Stokoe (1998), Park et al. (1999), Yuan et al. (2006), and Park et al. (2007). Love waves, on the other hand, are much simpler and involve only horizontally polarized shear SH-waves.

Kelly (1983) provides a numerical study of how elastic Love waves might be used in common geologic settings. Kim (1992) studies the effect on Love waves of a viscous fluid over an elastic layered medium. A field study on the use of Love waves can be found in Eslick et al. (2008). Li (1997) explores the elimination of higher mode Love waves as noise. Anderson (1962) provides a study of Love waves in anisotropic media. Anderson concludes that as long as the wavelengths are long enough, one might replace fine layering with a single layer.

The viscous damping of Rayleigh waves involves both compressional and shear contributions. Compressional-wave damping can occur for water-saturated and dry media because even a low-density fluid will be squeezed through pore spaces. On the other hand, shear-wave damping is inherently inertial. Shear viscous damping requires a dense fluid and permeable pore spaces so that shaking will result in relative fluid-frame motion. A low-density fluid, like air, produces low damping in shear (Michaels, 2006a). The dynamic behavior of soils is of interest in some areas, and the relevance of water content has been discussed in the context of earthquake engineering (Michaels, 2008). Our interest in Love waves began with the recognition that fluid-frame motion can explain viscous damping. The observation of damping in Love waves might be used to assess properties of water content or permeability, if interpreted in the context of a Biot (1956) two-phase medium.

To begin, we needed to compute the forward problem. Given a soil profile, what will be the Love-wave velocity dispersion and attenuation? An elastic layered medium produces configurational dispersion. The addition of viscosity

[1]*Boise State University Center for Geophysical Investigation of the Shallow Subsurface, Boise, Idaho, U.S.A.*
[2]*ExxonMobil Exploration Company, Houston, Texas, U.S.A.*

can introduce other trends in dispersion, different from those produced by the wave-guide effect. Summaries of the relevant viscoelastic mathematics are provided in Appendix A (body waves), Appendix B (Love waves), and Appendix C (wave guides). For a more complete discussion, the reader is referred to Michaels (1998) for body waves and to Gottumukkula (2008) for Love waves.

## Theoretical background

Pioneering work on surface waves was done by Haskell (1953). We have chosen to make our extension to viscosity following a more recent exposition for elastic Love waves found in Aki and Richards (1980). The key elements of the elastic problem are Hooke's law (stress proportional to strain) and Newton's second law (acceleration proportional to stress). These elements are applied in the context of two boundary conditions, vanishing stress at the free surface and radiation boundary conditions in the half-space. Extending this derivation to viscoelasticity requires adopting a specific viscous representation for the constitutive properties of the medium. Our choice of viscous representation is to add strain rate to the elastic Hooke's law.

## Viscous representations

Many possible ways exist to introduce viscosity, each producing its own specific Love-wave solution. We have chosen the Kelvin-Voigt (KV) representation (spring in parallel with dashpot). Here, we assume that the viscosity is a constant property of the medium, just like the shear modulus and the density. This choice makes the damping ratio of a vibrating element dependent on its natural frequency (see equation A-11 in Appendix A). In addition, the attenuation of a propagating wave predicts a frequency variable quality factor $Q$ (see equation A-12 in Appendix A). A propagating shear body wave will exhibit dispersion, with wave speed increasing with increasing frequency. Laboratory experience justifies this representation for saturated soils (Stoll, 1978, 1985). Field measurements of body-wave dispersion and attenuation also support this model for permeable, water-saturated soils (Stoll and Bryan, 1970; Michaels, 1998, 2001). A summary of changes to the wave equation using KV viscosity is given in Appendix A. We must emphasize that our chosen *viscous* KV representation is different from a very similar representation, one that uses *effective viscosity*. The existence of two KV viscous concepts can lead to confusion.

Effective viscosity has origins in a constant-$Q$ experience, founded in part by resonant column studies such as those of Hardin (1965). In those studies, samples were dry (saturated with air, not water). Air is of such low density that it is difficult to inertially decouple the fluid motion from the frame when strained in shear. A feature of this representation is that the damping ratio becomes a constant property for the medium, regardless of the frequency of vibration. This is achieved by assuming that material viscosity is not constant, but decreases with frequency (Hardin, 1965). It appears to be a way of retaining the convenience of a KV model while representing coulomb damping (force independent of frequency). Our view is that using effective viscosity in a Love-wave solution could be appropriate for dry or impermeable soils, but perhaps less appropriate for water-saturated, permeable soils (Michaels, 2008). Impermeable, water-saturated soils are expected to present like dry soils because fluid-frame interaction would be minimal in both cases. For those with an interest in Love-wave solutions using the effective viscosity representation, the reader is referred to Rix et al. (2000) and Lai and Rix (2002).

## Search for Love-wave solutions

Early constant-$Q$ work was done by global seismologists who expected modest levels of damping for crustal and mantle materials (Schwab and Knopoff, 1971). Not having information about the frequency dependence of damping, Schwab and Knopoff used a constant value for the imaginary component of shear velocity (effective viscosity assumption). Their search was biased toward elastic solutions as a starting point. Our work focuses on near-surface problems for which we believe a truly viscous model is appropriate. We use constant values for stiffness and damping, the coefficients $C_1$ (m²/s²) and $C_2$ (m²/s) of the viscoelastic wave equation A-4 in Appendix A. The shallowest layer of the crustal Schwab and Knopoff model was 8000 m thick. If one were to convert their constitutive properties to our notation at a frequency of 10 Hz, then they would be using $Q = 609$, or relaxation time of $T_R = 0.0000261$ s. When this relaxation time is compared to our near-surface experience, we see much larger relaxation times in the near surface (Table 1).

The search for Love-wave modes requires finding zeros of an objective function defined by equation B-13 in Appendix B. We scan a region of the complex velocity plane approximately bounded by the viscoelastic properties of the model layers. A grid search determines points close to solutions. Brackets are established. Then solutions are determined by a steepest descent algorithm that uses a sequence of bisection searches. Details on the algorithms and source code are in Gottumukkula (2008).

# Love Waves in a Viscoelastic Wave Guide

We consider the problem of a viscoelastic layer over an elastic half-space. This type of soil profile often appears in the context of bridge-foundation surveys (Michaels, 2001). In Figure 1, we consider a permeable gravel formation above a far less permeable siltstone. We choose to characterize our media properties in terms derived from the viscoelastic wave equation A-4 in Appendix A. The upper layer stiffness of $C_1 = 100^2$ $(\text{m}^2/\text{s}^2)$ is referenced by its square root $\beta_1 = 100$ (m/s), the zero-frequency limit of complex body-wave velocity (see equation A-9 in Appendix A). We specify the upper layer damping in terms of the wave-equation coefficient itself, $C_2 = 10$ $(\text{m}^2/\text{s})$. Relaxation time is found from the ratio $T_r = C_2/C_1$ (equation A-10 in Appendix A), and has units of time (in this case, $T_r = 1$ ms).

Relaxation time expresses the balance between damping and stiffness. For example, consider the analogy of a sponge squeezed under water and then released. It would take time for the sponge to restore its shape (depending on the "spring" stiffness of the fibers and the viscous resistance to fluid flowing into the sponge's pores, the "dashpot"). Table 1 shows some relaxation times determined from field downhole surveys (Michaels, 1998, 2001). The third parameter for any medium is mass density $\rho$, in units of $\text{kg/m}^3$.

Figure 2 illustrates how viscosity in the upper layer alters the Love-wave solution. For viscosity to have an effect, there must be significant motion if friction is to be produced. We have found that, for the layer over half-space models, the result is most sensitive to the upper layer viscosity. Viscosity in the half-space has far less influence because most of the motion is in the upper layer. In comparing the elastic higher modes with the viscoelastic higher modes, we note that the phase velocities of viscoelastic solutions are not bound by the same limits.

Viscoelastic Love waves exist in two forms, normal modes and leaky modes. Love-wave phase velocities are apparent velocities. They are an expression of the angle of emergence. Normal modes (such as the fundamental mode in Figure 2a) are supercritical, just like normal modes in an elastic wave guide. In addition, valid leaky-mode solutions become possible with a viscoelastic layer. These modes emerge from the imaginary axis and satisfy the differential equation and boundary conditions of Love waves. A key difference is that leaky modes have smaller angles of emergence and leak motion into the half-space. Their small angle of emergence produces large phase velocities because the wavefront becomes nearly parallel to the recording surface.

Viscoelastic Love-wave solutions are located in the complex velocity plane at the zeros of the objective function

**Table 1.** Downhole measurements of stiffness and damping compared to shallow layer at 10 Hz from Schwab and Knopoff (1971). Relaxation times $T_R$ and damping $C_2$ are constant for Michaels (1998, 2001) but vary with frequency for Schwab and Knopoff.

| Measured near-surface soil properties Boise and Logan (Michaels, 1998); Horseshoe Bend (Michaels, 2001) | | | | |
|---|---|---|---|---|
| **Soil** | $C_1$ (m$^2$/s$^2$) | $C_2$ (m$^2$/s) | $T_R$ (ms) | **Depth (m)** |
| Logan, Utah, silt | $25567 \pm 218$ | $1 \pm 1$ | 0.0391 | $2 \leq Z \leq 7$ |
| Logan, Utah, sand | $51343 \pm 375$ | $14 \pm 1$ | 0.2727 | $8 \leq Z \leq 13$ |
| Boise, Idaho, sand | $182751 \pm 4860$ | $69 \pm 17$ | 0.3776 | $10 \leq Z \leq 15$ |
| Boise, Idaho, gravel | $94917 \pm 2913$ | $255 \pm 9$ | 2.6866 | $5 \leq Z \leq 10$ |
| Horseshoe Bend, Idaho, siltstone | $234356 \pm 6685$ | $297 \pm 13$ | 1.2673 | $9 \leq Z \leq 18$ |
| Horseshoe Bend, Idaho, gravel | $92755 \pm 6633$ | $369 \pm 17$ | 3.9782 | $4 \leq Z \leq 9$ |
| Crustal model: effective viscosity (Schwab and Knopoff, 1971) | | | | |
| Crustal model 10 Hz | 8352083 | 218 | 0.02610 | $0 \leq Z \leq 8000$ |

Viscoelastic layer over elastic half-space 1D model

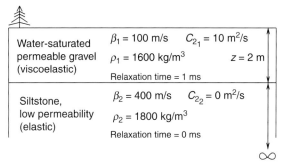

**Figure 1.** Model of a viscoelastic layer over a half-space. Upper layer damping, $C_{2_1} = 10$ m$^2$/s, corresponds to a relaxation time of 1 ms. The half-space medium is elastic as a result of low permeability.

**Figure 2.** Love-wave (a) velocity dispersion and (b) attenuation of a viscoelastic soil profile (Figure 1) compared to (c, d) an elastic overburden. Only damping $C_2$ of the upper layer is different between the viscoelastic and elastic case. Higher mode velocities and attenuation of the fundamental are significantly impacted by viscosity. These higher modes are leaky-mode solutions with small angles of emergence and large apparent velocities.

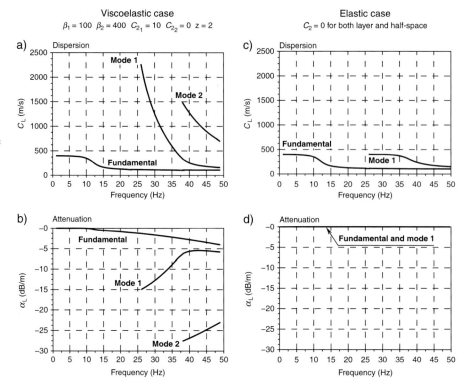

as given by equation B-13 in Appendix B. Figure 3 shows the modulus of the complex objective function, with contours clipped at 0.5 amplitude. This particular case is for 30 Hz and corresponds to the fundamental (normal) and first mode (leaky) of Figure 2a. The horizontal and vertical axes of the complex plane are given by real and imaginary parts of the complex phase velocity $\tilde{C}$, respectively. Because the modulus of the objective function can soar to extremely large values away from solutions, clipping contours at 0.5 amplitude permits one to see the roots, which are needlelike depressions.

Shown also in Figure 3 are blowups of the region around the solutions. Note that as one zooms in on a root, the contours become circular. The fundamental solution converged in six iterations (from the initial grid search) with the modulus of the objective function $|\tilde{B}_{21}| = 4.8573496399743431E - 16$. The first mode converged in five iterations with $|\tilde{B}_{21}| = 4.4154165088704674E - 16$. Confidence is high in the software because each author developed an independent code base (one in MATLAB, the other in C-language). Solutions were found to be in agreement when the different code results were compared.

Figure 4 shows the path of the Figure 1 model solutions. Shown are locations for solutions from 5 to 35 Hz computed at a 1-Hz interval. Shown also are the phase velocities for the endpoints, computed by equation B-16 in Appendix B. These paths correspond to the dispersion curves in Figure 2a. For a layer over a half-space, elastic Love waves

are supercritical reflected plane waves. This means that the angle of emergence is larger than the critical angle. The phase velocity is limited by the lower half-space body-wave velocity. This is understood by considering the phase delays caused by propagation and reflection, which result in constructive interference.

Figure 5 illustrates the constructive interference view of Love waves in an elastic wave guide. Constructive interference occurs when the multiple reflections become aligned in phase. For normal modes, this is illustrated by a phase delay of $n2\pi$ between points A and C. This integer multiple of $2\pi$ is formed from the difference between the propagation delay from A to C and the phase shift at point A, reflection at the bottom of the layer. Appendix C gives a derivation of the complex reflection coefficient extended to viscoelastic layers forming the reflective boundary. Constructive interference explains only the normal-mode viscoelastic Love waves. The leaky modes are quite different.

Consider normal modes. The propagation delay for the path ABC in Figure 5 is given by the following:

$$\phi_{ABC} = K_R[2H\cos(\theta_1)], \tag{1}$$

where the real part of the complex wavenumber $\tilde{k} = \omega/\tilde{\beta}_1$ is given by

$$K_R = \frac{\omega\beta_R}{\beta_R^2 + \beta_I^2}. \tag{2}$$

The phase shift resulting from the reflection coefficient is computed from equation C-18 in Appendix C:

$$\phi_{RC} = \tan^{-1}\left(\frac{imag(RC)}{real(RC)}\right). \quad (3)$$

The total phase shift must be an integral multiple of $2\pi$ for constructive interference,

$$\phi = \phi_{ABC} - \phi_{RC} = n2\pi. \quad (4)$$

Table 2 gives the computations for the elastic and viscoelastic versions of the soil profile of Figure 1. Note that the reflection coefficient for a leaky mode is less than unity, permitting motion to leak into the half-space.

Figure 6a shows the fundamental, normal-mode motion-stress vectors for our example soil profile of Figure 1, computed at a frequency of 30 Hz. Note that most of the motion is in the upper layer. Damping effects are predicted by that stronger upper layer motion, which shakes the pore

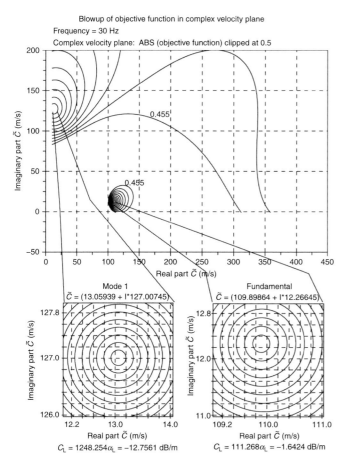

**Figure 3.** Love-wave solutions located at zeros of the complex objective function. This figure shows a blowup of the fundamental (normal mode) and first leaky mode in the complex velocity plane. The model is that of Figure 1. Solutions are for 30 Hz (see Figure 2a).

fluids relative to the soil frame. If one considers the alternative of an elastic layer over a viscous half-space, the result should be reduced damping effects when compared to our example soil profile of Figure 1. Figure 6b shows the first leaky-mode motion-stress vectors. Note that motion leaks into the half-space, and considerably more stress is computed in the half-space than is the case for the normal mode. Note also that the validity of this solution is confirmed by the vanishing stress at the free surface and the radiation boundary conditions.

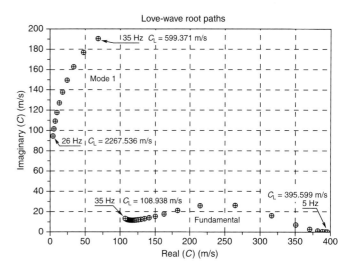

**Figure 4.** Path of solutions in the complex velocity plane for the fundamental and first mode. Solutions shown for a 1-Hz step size. The model is that of Figure 1.

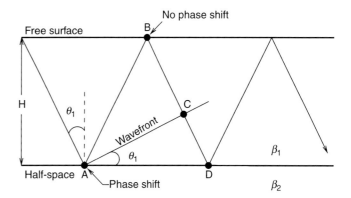

**Figure 5.** Wave guide formed by a layer over a half-space. Constructive interference occurs when the phase shift caused by propagation from point A to C minus the phase shift caused by reflection at point A total to a multiple of $2\pi$. Under those conditions, the motion is in phase at both points A and C, forming a constructive wavefront. This view can be used to explain supercritical normal modes, but it is inappropriate for the leaky modes.

**Table 2.** Comparison of elastic ($C_2 = 0$) and viscoelastic ($C_2 = 10$) waveguide properties for soil profile of Figure 1. The constructive interference view of Love waves is valid only for normal modes.

| | Waveguide computations for fundamental and mode 1 at 30 Hz | | | |
| Property | Elastic | | Viscoelastic | |
| | Fundamental (normal) | Mode 1 (normal) | Fundamental (normal) | Mode 1 (leaky) |
|---|---|---|---|---|
| $C_L$ (m/s) | 109.6272 | 397.1781 | 111.26778 | 1248.2537 |
| $\theta_1$(deg.) | 65.808705 | 14.582644 | 65.582811 | 4.6555994 |
| $\theta_2$ (deg.) | 90. + 112.769311i | 90. + 6.8258824i | 90. + 111.8838 i | 18.690001 |
| $|RC|$ | 1.00 | 1.00 | 1.00 | 0.6197916 |
| $\phi_{RC}$ (deg.) | 177.02688 | 58.083328 | 176.93883 | 177.3396 |
| $n = \frac{\phi}{2\pi}$ (deg.) | 0 | 1 | 0 | 0.6878958 |

These leaky modes should not be considered as the result of constructive interference. Instead, they satisfy the differential equation and boundary conditions. Leaky modes can exist in other contexts. For those with an interest, see the discussion on a fluid layer over a fluid half-space in Aki and Richards (1980). The period equation is similar to that of Love waves with the exception that all modes have a cutoff frequency. Our viscoelastic Love-wave leaky modes appear to share that same kind of cutoff frequency behavior.

## Using fundamental-mode Love waves

The most recordable viscoelastic Love-wave mode is the fundamental mode. It can be seen from Figure 2 that higher modes suffer large attenuation effects when viscosity is present, even at modest levels. This appears to be the combined result of viscosity and their leaky nature. For the first higher mode, just above 25 Hz, the decay is computed to be $-15$ dB/m. Thus, at a mere 10-m offset, the mode will be attenuated by 150 dB, making recording of that mode unlikely. Mode 2 is even more heavily attenuated, and is projected to be down by 250 dB at 45 Hz, 10 m from the source.

Figure 7 illustrates how upper layer damping can affect Love-wave solutions. It appears that the decay of the propagating wave is more sensitive to variations in damping than the phase velocity. We suggest that one might remotely probe damping $C_2$ by making observations of wave decay. Spectral observations of fundamental-mode decay at different offsets would benefit any inversion strategy that seeks both stiffness $C_1$ and damping $C_2$. However, this strategy will require an appropriate bandwidth to sample the attenuation effects. For a 2-m layer thickness presented here, one should have frequencies that extend beyond about 12 Hz, at which the curves become separated.

## Mapping Damping $C_2$ to Permeability

In engineering, a distinction is made between *absolute permeability* (units of area, m$^2$) and *permeability*, which is synonymous with the term *hydraulic conductivity* (units m/s) that is favored by hydrogeologists. For example, the American Society for Testing and Materials (ASTM) standard test for permeability in granular soils yields a value in units of m/s (American Society for Testing and Materials [ASTM], 1996). Our mapping is also of permeability (m/s). Laboratory or field methods need to sense the friction developed between the solid grains and whatever pore fluid is present. This friction results from the combined effect of pore fluid viscosity and pore diameters. Absolute permeability can be computed if the pore fluid is known. For water as a pore fluid, the formula is given by Das (1993):

$$\bar{K} = K_d \frac{\eta_w}{\gamma_w}, \qquad (5)$$

where $\bar{K}$ is absolute permeability (m$^2$), $K_d$ is permeability (m/s), $\eta_w$ is the absolute viscosity of water (Pa · s), and $\gamma_w$ is the unit weight of water (N/m$^3$).

Another unit of absolute viscosity commonly used in the oil business is the millidarcy (1 mD = 9.86923E − 16 m$^2$). However, the actual definition of the darcy is in terms of permeability for a specific fluid viscosity. In other words, a porous medium has a permeability of 1 darcy when differential pressure of 1 atmosphere across a sample 1 cm long and 1 cm$^2$ in cross section will force a liquid of 1 centipoise (cp) of viscosity through the sample at the rate of 1 cm$^3$ per s. For reference, a centipoise is a unit of absolute viscosity (1 cp = 0.001 Pa · s).

Michaels (2006b) provides a mapping between damping and permeability (m/s). Working from basic principles, the mapping is conducted in the context of vibrations and combines Biot (1956) theory with a Darcy-Weisbach representation of viscous friction for fluid flowing through cylindrical pore spaces. The constitutive model that permits the mapping is a blend of Kelvin-Voigt, Maxwell, and Biot representations. Thus, the representation is named KVMB (see Figure 8). The KVMB model requires an estimate of porosity to proportion the solid and fluid phases properly. Solid and fluid masses are connected by a dashpot. Depending on the frequency of shaking, the fluid and frame become inertially decoupled, leading to friction and a characteristic decay of a vibrating system. The link between wave propagation and vibrational points of view is through the damping ratio (see equation A-11 in Appendix A).

Two solutions map a Kelvin-Voigt (KV) damping ratio to a KVMB damping ratio. One is the largely coupled solution, the other the largely uncoupled solution. For most soils and rocks, the largely coupled solution appears to be relevant, and leads to a stable mapping that varies little with frequency. It should be noted that the transition from coupled to uncoupled conditions depends jointly on porosity, permeability, and the frequency of shaking. Thus, for any given porosity and permeability, uncoupled conditions will result if the shaking is done at a high-enough frequency (Michaels, 2006b). The design of an SH-source spectrum will need to take into account the range of expected porosity and permeability.

Field testing of the mapping has been conducted using downhole SH-wave recordings, porosity logs, and flow testing in instrumented boreholes. The results have been encouraging (Pool, 2007). The next step would be to replace downhole measurements with Love-wave recordings, resulting in a method to sense permeability remotely.

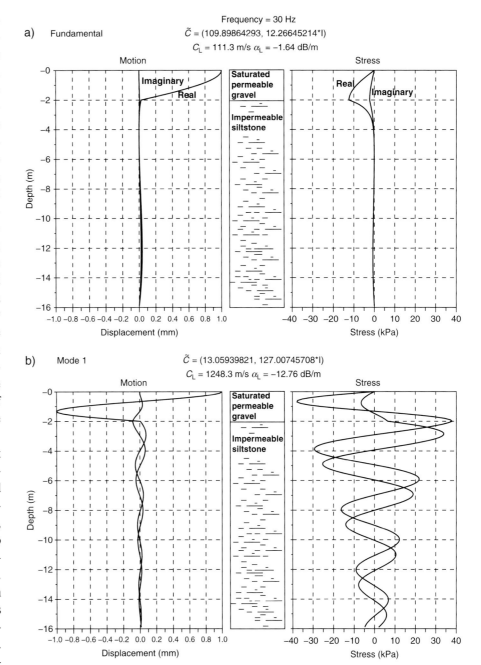

**Figure 6.** Motion-stress vectors at 30 Hz for the viscoelastic soil profile of Figure 1: (a) is for the fundamental mode, (b) is for the first (leaky) mode. The motion in the upper layer is much larger than in the half-space. This dominance of upper layer motion makes the upper layer viscosity more important than any viscosity in the half-space. The leaking of mode 1 energy into the half-space results in more rapid amplitude decay.

## Proposed method to sense permeability with Love waves

The procedure would be to determine stiffness $C_1$ and damping $C_2$ from observations of SH-wave propagation. The forward problem is given in Appendix B. This can be

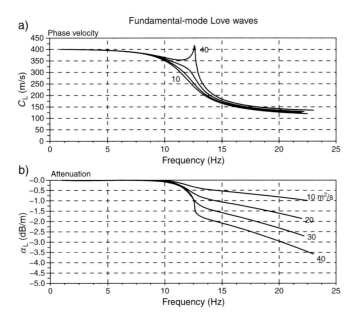

**Figure 7.** (a) Dispersion and (b) attenuation for different levels of damping in the upper layer. The Figure 1 model is modified by changing the damping of the upper layer to the following values: $C_2 = 10, 20, 30$, and $40$ m$^2$/s. The decay of the wave appears to be more sensitive than the phase velocity to variations in damping.

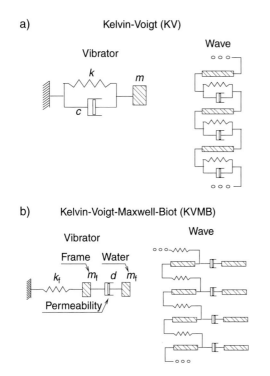

**Figure 8.** Schematic spring, mass, and dashpot representations of (a) KV and (b) KVMB representations. The KVMB model requires an independent measure of porosity, which then is used to split the mass into two components, fluid and solid. The wave representation is formed from an assemblage of vibrators.

inverted using standard inverse-method techniques. The data would be measurements of phase velocity and decay as a function of frequency. From these measurements, one would jointly invert dispersion and decay for viscoelastic SH-wave-equation coefficients ($C_1$ and $C_2$). These coefficients then could be used to compute a KV damping ratio at a frequency representative of the wave spectra (equation A-11 in Appendix A). Permeability follows from the intersection of that damping ratio with the KVMB curve for a given porosity.

Figure 9 shows an example of how a constant KV damping ratio intersects the convex down-mapping curve for a specified frequency and porosity. Details on computing this map can be found in Michaels (2006b). Intersection with the branch on the left side of the apex provides the largely coupled solution (fluid moving with the frame, but slightly out of phase). The higher permeability solution is on the right side of the apex, and is referred to as the largely uncoupled solution (fluid and frame motions essentially separate).

Fortunately, most geologic materials are largely coupled at frequencies commonly generated by impact hammer sources (from 5 to 50 Hz). As can be seen from the figure, this means that the solution is almost completely insensitive to the choice of frequency for selecting a mapping curve. As the frequency decreases, the mapping curve moves to the right, and the KV damping ratio decreases, resulting in an intersection that remains constant on a permeability value. Although the choice of vibration frequency

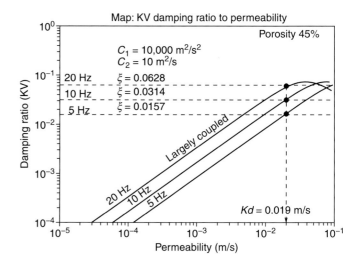

**Figure 9.** Mapping KV damping ratio $\xi$ to permeability for an uncompacted sand or gravel (porosity 45%). Note that as long as one is on the largely coupled side of the mapping peak, the frequency used to convert wave-determined damping ratio $\xi = (C_2\omega)/(2C_1)$ (equation A-11 in Appendix A) produces essentially the same permeability $K_d$.

to map wave determinations to permeability is fairly arbitrary, one should avoid using too high a frequency because that would move one to an intersection near the peak. Choosing a frequency that is too high could result in no intersection at all.

## Choosing a frequency band

Depending on the type of soils present, one should give consideration to shaking the soil in a band that produces largely coupled behavior. Figure 10 shows how the mapping shifts with different soil types. Coarse-grained granular soils such as a gravel should be shaken at lower frequencies than less permeable fine-grain soils. If one started with some downhole sampling of the subsurface, the proposed Love-wave technique would be designed to extend control by shaking at appropriate frequencies.

Another consideration will be the thickness of the upper layer. As was noted with the 2-m-layer case of Figure 7, the attenuation curves separate beyond some frequency. We suggest that normalization of the frequency axis provides a guide. The separation of curves occurs about at the point where $\frac{FH}{\beta_1} = 0.25$, where $F$ is frequency, $H$ is the upper layer thickness, and $\beta_1 = \sqrt{C_1}$. Combining this guide with the permeability guide above will limit the useful range of this proposed technique. Depending on the soil types expected, Figure 10 can be used to select a bandwidth needed to produce measureable KV damping ratios that are expressions of largely coupled motion. Then the range of useful upper layer thicknesses would follow from the 0.25 rule above. In other words, the minimum layer thicknesses would be

$$H_{\min} \geq \frac{0.25\beta_1}{F_{\min}}. \qquad (6)$$

A layer thinner than this limit would shift the dispersion curves to higher frequencies. Depending on the available bandwidth of the source, this could result in the attenuation curves being squeezed together over the source band. Such a condition would not be helpful.

## Viscosity, pore diameter, and dashpot

One might think that changing the fluid viscosity would impact the mapping of the KV damping ratio to permeability. It turns out that the choice of fluid viscosity is not important; it merely predicts a different pore diameter, which would result in the same level of friction. If one were to construct an alternative map that replaces permeability with the pore diameter on the Figure 9 mapping, the pore-

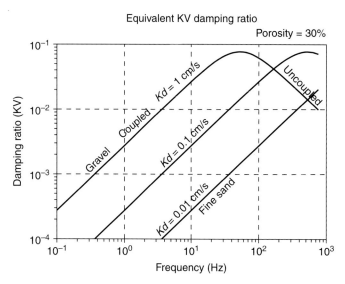

**Figure 10.** The mapping between KV and KVMB representations as a function of frequency. The largely coupled domain can be excited by choosing an appropriate frequency of vibration. Shown are three typical cases, gravel to fine-sand permeabilities ($Kd = 1$ to $Kd = .01$ cm/s, respectively). Coarse-grain materials, such as a gravel, should be shaken at lower frequencies than a fine-grain sand to keep the behavior in the largely coupled region.

diameter solution would depend strongly on the fluid viscosity. However, because the cylindrical pore representation is merely a device to produce a functional relationship that honors a Biot two-phase model, the implied pore diameter is of less value than the actual permeability (which is generally the object of inquiry). Note that the map to permeability has an apex. This means that KV damping ratio is limited, and increasing the dashpot value will not increase the damping without limit. Neither very large pores nor very small pores will produce much friction. Fluid moves easily in large pores, and fluid might not move at all in very small pores, hence causing an apex in the map.

## Porosity and the vertical position of the map

Of importance to the mapping is an independent measure of porosity. For example, a scenario might be to use Love waves to extend downhole control. One could use neutron logging of the borehole to provide porosity values. The model assumes saturated conditions, and thus porosity splits the mass into two components, fluid and solid. For the fluid to provide a strong, viscous resistance to the frame motion, the fluid must be both dense and constitute a significant proportion of a unit volume of soil or rock. A low porosity will translate the map downward, placing the apex at a lower KV damping ratio. A high porosity will produce the opposite effect.

## Error analysis

The uncertainty in a determined permeability involves several components. The proposed measurements of phase velocity and decay will have associated variances at each frequency. These uncertainties can be mapped using standard inverse-theory methods to uncertainties in the SH-wave-equation coefficients for each layer (Menke, 1989). Uncertainties in $C_1$ and $C_2$ for a layer of interest would be used to compute an uncertainty in the KV damping ratio when determined by equation A-11 in Appendix A. This could be done using the error propagation equation

$$\sigma_\xi^2 = \left(\frac{\partial \xi}{\partial c_1}\right)^2 \sigma_{c_1}^2 + \left(\frac{\partial \xi}{\partial c_2}\right)^2 \sigma_{c_2}^2 + 2\sigma_{c_1 c_2}^2 \left(\frac{\partial \xi}{\partial c_1}\right)$$
$$\times \left(\frac{\partial \xi}{\partial c_2}\right), \qquad (7)$$

where $\xi$ is damping ratio and the $\sigma^2$ are respective variance or covariance estimates from the $C_1$ and $C_2$ determinations (Bevington and Robinson, 2003). The uncertainty in the damping ratio, combined with the uncertainty in porosity, then would be used to compute the uncertainty in the permeability. An example of this type of error analysis can be found in Pool (2007). In Pool, the wave equation coefficients $C_1$ and $C_2$ were computed from downhole measurements of direct body-wave dispersion and amplitude decay. The only difference would be to use Love-wave measurements in place of downhole measurements.

## Conclusions

We are at an early stage and have presented viscoelastic Love waves in a way that relates viscosity to soil permeability in a saturated two-phase medium. We predict that a saturated permeable soil layer over a half-space should produce measureable decay that depends on the damping coefficient of the viscoelastic wave equation. We have proposed a procedure to extract the stiffness and damping coefficients ($C_1$ and $C_2$) from Love-wave measurements, which can be used to estimate a KV damping ratio if one chooses a mapping frequency. As long as the mapping frequency places one on the largely coupled side of the mapping apex, then a stable estimate of permeability might be obtained. Implementation of this proposed procedure would require an appropriate design of the survey, as follows:

- The seismic-source spectral band should be chosen according to the expected range of soil porosity and permeability. The criteria would be to keep the shaking in the largely coupled region. Highly permeable soils will require lower frequencies than tight soils. One then could use equation 6 to define a minimum layer thickness. A maximum layer thickness would depend on the available bandwidth.
- A goal should be to choose a source spectrum that provides a measurable viscous friction effect. This can be done by shaking closer to the apex of the mapping curve (as opposed to down on the extreme limb).
- An independent measure of porosity is required for the mapping to permeability. Downhole neutron logging would be one way to do this, with Love waves being used to fill in control between borehole determinations of permeability.
- Higher leaky-mode Love waves are predicted to be highly attenuated, and difficult to measure. The best opportunity appears to be with the fundamental mode. The forward problem provides theoretical predictions of phase-velocity dispersion and decay. These estimates then can be used to determine a useful range of offsets for a given dynamic range of recording.

## Acknowledgments

We wish to thank Boise State University for supporting this work, and Dr. Warren Barrash for providing boreholes, logs, and flow-test results at the Boise Hydrogeophysical Research Site (BHRS). Downhole seismic surveys, porosity logs, and flow testing conducted at the BHRS have provided the field evidence that lead to the development of the KVMB model and the theory that Love waves also might be used to estimate permeability.

## Appendix A

## Viscoelastic Body Waves

Extending the elastic wave equation to KV viscosity is done by first adding a strain rate term to Hooke's law. Thus, Hooke's law for spatial components $ab$ becomes

$$\tau_{ab} = \lambda \delta_{ab} \epsilon_{cc} + \psi \delta_{ab} \dot{\epsilon}_{cc} + 2G\epsilon_{ab} + 2\eta \dot{\epsilon}_{ab}, \qquad (A\text{-}1)$$

where $\tau_{ab}$ is stress, $\lambda$ is Lamé's constant, $\epsilon_{ab}$ is strain, $\epsilon_{cc}$ is cubical dilatation (summation convention of repeated subscripts), $\delta_{ab}$ is the delta function (unity if $a = b$ and zero if $a \neq b$), $\psi$ is compressional viscosity, $\eta$ is shear viscosity,

$G$ is the shear modulus, and $\dot{\epsilon}$ is strain rate. For an S-wave, there are no normal stresses, and the KV Hooke's law reduces to

$$\tau_{ab} = 2G\epsilon_{ab} + 2\eta\dot{\epsilon}_{ab}. \qquad \text{(A-2)}$$

For an S-wave propagating in the $x$-direction with particle motion in the horizontal $y$-direction (right-hand system, $z$ positive down), Newton's law of motion reduces to

$$\rho\ddot{u}_y = \frac{\partial\tau_{xy}}{\partial x}, \qquad \text{(A-3)}$$

where $\rho$ is mass density.

For the case of a 1D planar S-wave, substituting shear stress into Newton's second law leads to (Kramer, 1996)

$$\frac{\partial^2 u_y}{\partial t^2} = \left(\frac{G}{\rho}\right)\frac{\partial^2 u_y}{\partial x^2} + \left(\frac{\eta}{\rho}\right)\frac{\partial^3 u_y}{\partial x^2 \partial t}$$
$$= (C_1)\frac{\partial^2 u_y}{\partial x^2} + (C_2)\frac{\partial^3 u_y}{\partial x^2 \partial t}, \qquad \text{(A-4)}$$

where $C_1$ represents stiffness and $C_2$ represents damping, both constants of the medium. Following Michaels (1998), a solution of equation A-4 can be written as an exponentially decaying sinusoid

$$u(x, t) = e^{-\alpha x}\cos(\beta x - \omega t), \qquad \text{(A-5)}$$

where the complex wavenumber is $\tilde{k} = (\beta + i\alpha)$. In this study, we use the tilde over variables to indicate complex quantities. Substituting A-5 into A-4 leads to a decay $\alpha$:

$$\alpha = \frac{4\sqrt{D}\omega^2 C_2}{(2\omega C_2)^2 + D^2},$$
$$D = 2\left(C_1 + \sqrt{C_1^2 + \omega^2 C_2^2}\right) \qquad \text{(A-6)}$$

and a body-wave phase velocity $V_S$:

$$V_S = \frac{\omega}{\beta} = \frac{2\omega^2 C_2}{D\alpha}. \qquad \text{(A-7)}$$

Note that decay and phase velocity are functions of frequency $\omega$. Only if damping vanishes, $C_2 = 0$, does phase velocity become constant, $V_S = \sqrt{G/\rho} = \sqrt{C_1}$.

Other useful expressions include the complex shear modulus

$$\tilde{G} = G + i\omega\eta = G_R + iG_I, \qquad \text{(A-8)}$$

the complex shear velocity

$$\tilde{C} = \sqrt{\frac{\tilde{G}}{\rho}} = \sqrt{\frac{G + i\eta\omega}{\rho}} = \sqrt{C_1 + i\omega C_2}, \qquad \text{(A-9)}$$

and the loss tangent

$$\tan(\delta) = \frac{G_I}{G_R} = \omega\frac{C_2}{C_1} = \omega T_R, \qquad \text{(A-10)}$$

where $T_R$ is relaxation time. If we take the perspective of a vibrating element of specified dimension and oscillatory frequency, the damping ratio is given by

$$\xi = \frac{\eta\omega}{2G_R} = \frac{C_2\omega}{2C_1}. \qquad \text{(A-11)}$$

The quality factor $Q$ is given by

$$Q = \frac{1}{2\xi} = \frac{C_1}{\omega C_2}. \qquad \text{(A-12)}$$

## Comparison to effective viscosity

As mentioned in the text, an alternative formulation derived from dry soils is effective viscosity. Effective viscosity is not constant with that view, but varies as

$$\eta = \frac{2G_R}{\omega}\xi, \qquad \text{(A-13)}$$

where $\xi$ is a constant damping ratio. This is a "nonviscous" viscosity, and it is a mathematical convenience to represent contact friction. The use of this device results in a complex shear modulus given by

$$\tilde{G} = G_R(1 + i2\xi). \qquad \text{(A-14)}$$

We do not use this concept, which treats the damping ratio $\xi$ as a frequency-independent soil property. Instead, we use viscosity $\eta$ as the constant soil property.

# Appendix B

# Viscoelastic Love Waves

We extend the elastic derivation found in Aki and Richards (1980) by replacing the elastic Hooke's law with the viscoelastic extension given by equation A-1 in Appendix A. The resultant formulas are essentially the same for the propagator matrix and motion-stress vectors. The chief difference is that many quantities become complex.

A tilde over a symbol indicates a complex variable. For a detailed discussion, see Gottumukkula (2008).

Thus, our coordinate system and particle displacements are written

$$u_x = 0$$

$$\tilde{u}_y = \tilde{l}_1(\tilde{k}, z, \omega)e^{i(\omega t - \tilde{k}x)}.$$

$$u_z = 0 \qquad \text{(B-1)}$$

The components of the motion-stress vector are given by $\tilde{l}_1$ (displacement) and $\tilde{l}_2 = \tilde{G}\frac{\partial \tilde{l}_1}{\partial z}$ (stress). Applying Hooke's law to Newton's second law leads to the differential equation for the motion-stress vector in matrix form,

$$\frac{d}{dz}\underbrace{\begin{bmatrix} \tilde{l}_1 \\ \tilde{l}_2 \end{bmatrix}}_{\frac{d\tilde{\mathbf{l}}}{dz}} = \underbrace{\begin{bmatrix} 0 & 1/\tilde{G} \\ \tilde{k}^2\tilde{G} - \omega^2\rho & 0 \end{bmatrix}}_{\tilde{\mathbf{A}}} \cdot \underbrace{\begin{bmatrix} \tilde{l}_1 \\ \tilde{l}_2 \end{bmatrix}}_{\tilde{\mathbf{l}}} \qquad \text{(B-2)}$$

or

$$\frac{d\tilde{\mathbf{l}}}{dz} = \tilde{\mathbf{A}} \cdot \tilde{\mathbf{l}}. \qquad \text{(B-3)}$$

The complex propagator-matrix solution to equation B-2 is (Gottumukkula, 2008)

$$\underbrace{\begin{bmatrix} \tilde{l}_1 \\ \tilde{l}_2 \end{bmatrix}}_{\tilde{\mathbf{l}}(z)} = \underbrace{\begin{bmatrix} \cosh[\tilde{v}_j\Delta z] & (\tilde{v}_j\tilde{G}_j)^{-1}\sinh[\tilde{v}_j\Delta z] \\ \tilde{v}_j\tilde{G}_j\sinh[\tilde{v}_j\Delta z] & \cosh[\tilde{v}_j\Delta z] \end{bmatrix}}_{\tilde{\mathbf{P}}_j(z,z_{j-1})} \cdot \underbrace{\begin{bmatrix} \tilde{l}_1 \\ \tilde{l}_2 \end{bmatrix}}_{\tilde{\mathbf{l}}(z_{j-1})},$$

$$\text{(B-4)}$$

or

$$\tilde{\mathbf{l}}(z) = \tilde{\mathbf{P}}(z, z_{j-1}) \cdot \tilde{\mathbf{l}}(z_{j-1}), \qquad \text{(B-5)}$$

where the eigenvalues of matrix $\mathbf{A}$ of equation B-2 are given by

$$\tilde{v}_j = \pm\sqrt{\tilde{k}^2 - \frac{\omega^2}{\tilde{C}_j^2}} \qquad \text{(B-6)}$$

for the $j$th layer and $\Delta z = z - z_{j-1}$. Here, $z_{j-1}$ is the depth of the top of a layer where the material properties (subscripted "j") are constant, and $z$ is some point deeper within that layer, limited to the bottom of the layer, where a change in properties would make a new matrix required to go farther. At the bottom of the layer, $z = z_j$. In short, one

forms a $\Pi$-product of propagator matrices to handle a multi-layered medium, connecting the surface to a point within a stack of layers.

## Applying boundary conditions

Following the program of Aki and Richards (1980), the layer matrix formulation is used to express the solution in terms of up- and downgoing waves. This permits the radiation boundary condition to be imposed by an absence of upgoing wave content in the bounding half-space at the bottom of a stack of 1D layers. For SH-waves in a homogeneous body, we have

$$\underbrace{\begin{bmatrix} \tilde{l}_1 \\ \tilde{l}_2 \end{bmatrix}}_{\tilde{\mathbf{l}}(z)} = \underbrace{\begin{bmatrix} e^{-\tilde{v}z} & e^{\tilde{v}z} \\ -\tilde{v}\tilde{G}e^{-\tilde{v}z} & \tilde{v}\tilde{G}e^{\tilde{v}z} \end{bmatrix}}_{\tilde{\mathbf{F}}(z)} \cdot \underbrace{\begin{bmatrix} \grave{S} \\ \acute{S} \end{bmatrix}}_{\mathbf{w}} \qquad \text{(B-7)}$$

or

$$\tilde{\mathbf{l}}(z) = \tilde{\mathbf{F}}(z) \cdot \mathbf{w}. \qquad \text{(B-8)}$$

Solving for the amplitude of the downgoing $\grave{S}$ and the upgoing wave $\acute{S}$ in the bounding half-space requires inverting the layer matrix $\mathbf{F}$:

$$\underbrace{\begin{bmatrix} \grave{S} \\ \acute{S} \end{bmatrix}}_{\mathbf{w}} = \underbrace{\left(\frac{1}{2\tilde{v}\tilde{G}}\right)\begin{bmatrix} (\tilde{v}\tilde{G})e^{\tilde{v}z} & -e^{\tilde{v}z} \\ (\tilde{v}\tilde{G})e^{-\tilde{v}z} & e^{-\tilde{v}z} \end{bmatrix}}_{\tilde{\mathbf{F}}^{-1}(z)} \cdot \underbrace{\begin{bmatrix} \tilde{l}_1(z) \\ \tilde{l}_2(z) \end{bmatrix}}_{\tilde{\mathbf{l}}(z)}. \qquad \text{(B-9)}$$

For a stack of layers with a half-space at the bottom, we choose to number the free surface as depth $z_0$, the top of the first layer. The bottom of the first layer is at $z_1$. The first layer has the properties velocity $V_1 = \sqrt{C_{1_1}}$, density $\rho_1$, and damping $C_{2_1}$. For an $n$-layer problem, the half-space is designated with the properties velocity $V_{n+1} = \sqrt{C_{1_{n+1}}}$, density $\rho_{n+1}$, and damping $C_{2_{n+1}}$. For a stack of layers, the $\Pi$-product of propagator matrices connects the surface to the top of the half-space:

$$\tilde{\mathbf{P}}_n(z_n, z_0) = \tilde{\mathbf{P}}_n(z_n, z_{n-1})\tilde{\mathbf{P}}_{n-1}(z_{n-1}, z_{n-2})$$

$$\ldots\tilde{\mathbf{P}}_2(z_2, z_1)\mathbf{P}_1(z_1, z_0). \qquad \text{(B-10)}$$

The up- and downgoing amplitudes in the half-space then are given in terms of the half-space inverse-layer matrix $\tilde{\mathbf{F}}_{n+1}^{-1}$ and the propagator matrix result of equation B-10,

$$\mathbf{w}_{n+1} = \tilde{\mathbf{F}}_{n+1}^{-1}\mathbf{P}_n(z_n, z_0)\tilde{\mathbf{l}}(z_0) = \tilde{\mathbf{B}} \cdot \tilde{\mathbf{l}}(z_0), \qquad \text{(B-11)}$$

where the complex matrix $\tilde{\mathbf{B}}$ is the product of the inverse-layer matrix and the propagator matrix. Boundary conditions are applied by setting the upgoing wave in the half-space to zero, $\acute{S} = 0$, and the stress at the surface to zero, $\tilde{l}_2(z_0) = 0$:

$$\begin{bmatrix} \grave{S}_{n+1} \\ 0 \end{bmatrix} = \begin{bmatrix} \tilde{B}_{11} & \tilde{B}_{12} \\ \tilde{B}_{21} & \tilde{B}_{22} \end{bmatrix} \begin{bmatrix} \tilde{l}_1(z_0) \\ 0 \end{bmatrix}. \qquad \text{(B-12)}$$

A solution for a Love-wave mode is found by searching the complex plane for complex wavenumbers $\tilde{k}$, which cause the modulus of the $\tilde{B}_{21}$ element to vanish:

$$|\tilde{B}_{21}| = 0. \qquad \text{(B-13)}$$

Instead of searching in the complex wavenumber plane, we conduct our search in the complex velocity plane. This plane has a real axis corresponding to $C_R$, the real component of complex phase velocity. The imaginary axis is $C_I$, the complex component. Thus, we scan for $\tilde{C}$, which satisfies equation B-13:

$$\tilde{C} = \frac{\omega}{\tilde{k}} = C_R + iC_I = \sqrt{C_1 + i\omega C_2}. \qquad \text{(B-14)}$$

Velocity and decay are calculated from the real and imaginary parts of the complex wavenumber $\tilde{k} = k_R + ik_I$ at each frequency:

$$\tilde{k} = \frac{\omega}{C_R + iC_I} = \frac{\omega C_R}{(C_R^2 + C_I^2)} + \frac{-i\omega C_I}{(C_R^2 + C_I^2)}. \qquad \text{(B-15)}$$

Thus, Love-wave phase velocity $C_L$ (m/s) is given by

$$C_L = \frac{\omega}{k_R} = \frac{(C_R^2 + C_I^2)}{C_R}, \qquad \text{(B-16)}$$

and the Love wave decay $\alpha_L$ (1/m) is found from

$$\alpha_L = k_I = \frac{-\omega C_I}{(C_R^2 + C_I^2)}. \qquad \text{(B-17)}$$

Early work in global seismology assumed small damping levels and limited the search near the elastic solutions on the real axis (Schwab and Knopoff, 1971). Near-surface damping levels in soils are much larger, requiring the search to extend much deeper into the complex plane.

# Appendix C

# Viscoelastic Reflections and Wave Guides

Under elastic assumptions, an alternative view explains Love waves as constructive interference in a wave guide. Central to this view is an understanding of the reflection coefficient and any associated phase shift at layer boundaries. We begin by extending the elastic case for SH-plane waves at a layer boundary found in Aki and Richards (1980). The principal change is to replace the elastic Hooke's law with the viscoelastic extension given by equation A-1 in Appendix A.

Consider a downgoing plane SH-wave in layer 1 that reflects off a planar interface separating the two viscoelastic media of complex SH-velocities $\tilde{\beta}_1$ and $\tilde{\beta}_2$. Let the ray normal to the wavefront make an angle $\theta_1$ with the normal to the interface. Further, let the displacement of the incident and reflected wave in the upper layer 1 be written as

$$\tilde{u}_2^{(1)} = A \cdot \exp\left[i\omega\left(t - \frac{\sin(\theta_1)}{\tilde{\beta}_1}x_1 + \frac{\cos(\theta_1)}{\tilde{\beta}_1}x_3\right)\right]$$
$$+ B \cdot \exp\left[i\omega\left(t - \frac{\sin(\theta_1)}{\tilde{\beta}_1}x_1 - \frac{\cos(\theta_1)}{\tilde{\beta}_1}x_3\right)\right], \quad \text{(C-1)}$$

where the superscript indicates layer and the subscript indicates that the motion is in the $x_2$-direction, orthogonal to the propagation direction $x_1$. The vertical axis is $x_3$ and is positive downward. The transmitted wave-particle displacement in the lower layer 2 is given by

$$\tilde{u}_2^{(2)} = C \cdot \exp\left[i\omega\left(t - \frac{\sin(\theta_2)}{\tilde{\beta}_2}x_1 + \frac{\cos(\theta_2)}{\tilde{\beta}_2}x_3\right)\right]. \quad \text{(C-2)}$$

For a viscoelastic medium, we will need to consider particle velocity also. These particle velocities are given for layer 1 as

$$\dot{\tilde{u}}_2^{(1)} = +i\omega\tilde{u}_2^{(1)} \qquad \text{(C-3)}$$

and

$$\dot{\tilde{u}}_2^{(2)} = +i\omega\tilde{u}_2^{(2)} \qquad \text{(C-4)}$$

for layer 2. The relevant shear stress follows from the viscoelastic Hooke's law (equation A-1 in Appendix A) and simplifies to

$$\sigma_{23} = G\frac{\partial u_2}{\partial x_3} + \eta\frac{\partial \dot{u}_2}{\partial x_3}. \tag{C-5}$$

There is no loss of generality by setting $x_3 = 0$ at the layer boundary. Further, by Snell's law, the ray parameter is constant at any given frequency,

$$p = \frac{\sin(\theta_1)}{\beta_1} = \frac{\sin(\theta_2)}{\beta_2} = constant, \tag{C-6}$$

where $\beta_1$ and $\beta_2$ are real SH-velocities at any single frequency. Thus, after taking derivatives, we will evaluate the results at $x_3 = 0$ and use the following compact notation for the exponential at the boundary:

$$\exp(*) = \exp[i\omega(t - px_1)]. \tag{C-7}$$

Substituting equations C-1 through C-4 into equation C-5, we write the shear stresses as

$$\tilde{\sigma}_{23}^{(1)} = [G_1 + (\eta_1\omega)i]\left(\frac{i\omega\cos(\theta_1)}{\tilde{\beta}_1}\right)[A - B] \cdot \exp(*),$$
$$\tag{C-8}$$

and

$$\tilde{\sigma}_{23}^{(2)} = [G_2 + (\eta_2\omega)i]\left(\frac{i\omega\cos(\theta_2)}{\tilde{\beta}_2}\right)[C] \cdot \exp(*), \tag{C-9}$$

in layers 1 and layer 2, respectively. We now apply the boundary conditions of continuity of displacement and stress

$$\tilde{u}_2^{(1)} = \tilde{u}_2^{(2)}, \tag{C-10}$$

and

$$\tilde{\sigma}_{23}^{(1)} = \tilde{\sigma}_{23}^{(2)}. \tag{C-11}$$

Evaluated at the boundary, the exp(*) terms will divide out, simplifying the boundary conditions to the following two equations:

$$A + B = C, \tag{C-12}$$

and

$$\left[\frac{G_1 + (\eta_1\omega)i}{\tilde{\beta}_1}\right](i\omega\cos(\theta_1))[A - B]$$
$$= \left[\frac{G_2 + (\eta_2\omega)i}{\tilde{\beta}_2}\right](i\omega\cos(\theta_2))[C]. \tag{C-13}$$

We recognize the complex shear moduli in equation C-13. For example,

$$\tilde{G} = [G + (\eta\omega)i] = \tilde{\beta}^2\rho. \tag{C-14}$$

Applying equation C-14 to equation C-13, the second boundary condition becomes

$$[\tilde{\beta}_1\rho_1](i\omega\cos(\theta_1))[A - B]$$
$$= [\tilde{\beta}_2\rho_2](i\omega\cos(\theta_2))[C]. \tag{C-15}$$

We divide both sides of equations C-12 and C-15 by $A$ to obtain a system of two equations in terms of the unknown reflection coefficient $\frac{B}{A}$ and unknown transmission coefficient $\frac{C}{A}$. In matrix notation, this is given as

$$\begin{bmatrix} (-1) & (+1) \\ (+1) & D \end{bmatrix} \cdot \begin{bmatrix} \frac{B}{A} \\ \frac{C}{A} \end{bmatrix} = \begin{bmatrix} (1) \\ (1) \end{bmatrix}, \tag{C-16}$$

where

$$D = \frac{\rho_2\tilde{\beta}_2\cos(\theta_2)}{\rho_1\tilde{\beta}_1\cos(\theta_1)}. \tag{C-17}$$

The solution gives the following reflection and transmission coefficients:

$$RC = \frac{B}{A} = \frac{\rho_1\tilde{\beta}_1\cos(\theta_1) - \rho_2\tilde{\beta}_2\cos(\theta_2)}{\rho_1\tilde{\beta}_1\cos(\theta_1) + \rho_2\tilde{\beta}_2\cos(\theta_2)}, \tag{C-18}$$

and

$$TC = \frac{C}{A} = \frac{2\rho_1\tilde{\beta}_1\cos(\theta_1)}{\rho_1\tilde{\beta}_1\cos(\theta_1) + \rho_2\tilde{\beta}_2\cos(\theta_2)}. \tag{C-19}$$

This solution compares well with the elastic equivalent, equation 5.32, found in Aki and Richards (1980).

# References

Aki, K., and P. Richards, 1980, Quantitative seismology: W. H. Freeman and Co.

Anderson, D. L., 1962, Love wave dispersion in heterogeneous anisotropic media: Geophysics, **27**, 445–454.

American Society for Testing and Materials (ASTM), 1996, Standard test method for permeability of granular soils (constant head): ASTM D2434-68.

Bevington, P. R., and D. K. Robinson, 2003, Data reduction and error analysis for the physical sciences: McGraw-Hill.

Biot, M., 1956, Theory of propagation of elastic waves in a fluid-saturated porous solid: Part 1 — Low-frequency range: Journal of the Acoustical Society of America, **28**, 168–178.

Das, B., 1993, Principles of geotechnical engineering, 3rd ed.: PWS Publishing Co.

Eslick, R., G. Tsoflias, and D. Steeples, 2008, Field investigation of Love waves in near-surface seismology: Geophysics, **73**, no. 3, G1–G6.

Gottumukkula, V. R. C., 2008, Love wave propagation in viscoelastic media: Ph.D. thesis, Boise State University.

Gucunski, N., and R. D. Woods, 1991, Instrumentation for SASW testing, *in* S. K. Bhatia and G. W. Blaney, eds., Recent advances in instrumentation, data acquisition and testing in soil dynamics: American Society of Civil Engineers Geotechnical Special Publication No. 29, 1–16.

Hardin, B. O., 1965, The nature of damping in sands: Journal of the Soil Mechanics and Foundations Divison, Proceedings of the American Society of Civil Engineers, **91**, 63–97.

Haskell, N., 1953, The dispersion of surface waves in multilayered media: Bulletin of the Seismological Society of America, **43**, 17–34.

Kelly, K. R., 1983, Numerical study of Love wave propagation: Geophysics, **48**, 833–853.

Kim, J. O., 1992, The effect of a viscous fluid on Love waves in a layered medium: Journal of the Acoustical Society of America, **91**, 3099–3103.

Kramer, S., 1996, Geotechnical earthquake engineering: Prentice-Hall.

Lai, C. G., and G. J. Rix, 2002, Solution of the Rayleigh eigenproblem in viscoelastic media: Bulletin of the Seismological Society of America, **92**, 2297–2309.

Li, X. P., 1997, Elimination of higher modes in dispersive in-seam multi-mode Love waves: Geophysical Prospecting, **45**, 945–961.

Luke, B. A., and K. H. Stokoe, 1998, Application of SASW method underwater: Journal of Geotechnical and Geoenvironmental Engineering, **124**, 523–531.

Menke, W., 1989, Geophysical data analysis, discrete inverse theory: Academic Press.

Michaels, P., 1998, In situ determination of soil stiffness and damping: Journal of Geotechnical and Geoenvironmental Engineering, **124**, 709–719.

——, 2001, Use of engineering geophysics to investigate a site for a bridge foundation, *in* T. L. Brandon, ed., Foundations and ground improvement: American Society of Civil Engineers Geotechnical Special Publication No. 113, 715–727.

——, 2006a, Comparison of viscous damping in unsaturated soils, compression and shear, *in* Proceedings of the 4th International Conference on Unsaturated Soils: American Society of Civil Engineers Geotechnical Special Publication No. 147, 565–576.

——, 2006b, Relating damping to soil permeability: International Journal of Geomechanics, **6**, 158–165.

——, 2008, Water, inertial damping, and the complex shear modulus, *in* Proceedings of the 4th Geotechnical Earthquake Engineering and Soil Dynamics Conference: American Society of Civil Engineers Geotechnical Special Publication No. 181, CD-ROM.

Nazarian, S., K. Stokoe, and W. Hudson, 1983, Use of spectral analysis of surface waves method for determination of moduli and thickness of pavement systems: Transportation Research Record, **980**, 38–45.

Park, C. B., R. D. Miller, and J. Xia, 1999, Multichannel analysis of surface waves: Geophysics, **64**, 800–808.

Park, C. B., R. D. Miller, J. Xia, and J. Ivanov, 2007, Multichannel analysis of surface waves (MASW) — Active and passive methods: The Leading Edge, **26**, 60–64.

Pool, L., 2007, Determination of hydraulic conductivity from SH-waves: Master's thesis, Boise State University.

Rix, G. J., C. G. Lai, and A. W. Spang, 2000, In situ measurement of damping ratio using surface waves: Journal of Geotechnical and Geoenvironmental Engineering, **126**, 472–480.

Schwab, F., and L. Knopoff, 1971, Surface waves on multilayered anelastic media: Bulletin of the Seismological Society of America, **61**, 893–912.

Stokoe, K. H., and S. Nazarian, 1985, Use of Rayleigh waves in liquefaction studies, *in* R. D. Woods, ed., Measurement and use of shear wave velocity for evaluating dynamic soil properties: American Society of Civil Engineers, 1–17.

Stoll, R., 1978, Damping in saturated soil, *in* Proceedings of the Earthquake Engineering and Soil Dynamics Conference: American Society of Civil Engineers, 960–975.

——, 1985, Computer-aided studies of complex soil moduli, *in* R. Woods, ed., Measurement and use of shear wave velocity for evaluating dynamic soil properties: American Society of Civil Engineers, 18–33.

Stoll, R., and G. Bryan, 1970, Wave attenuation in saturated sediments: Journal of the Acoustical Society of America, **47**, 1440–1447.

Yuan, D., S. Nazarian, and I. Abdallah, 2006, Surface wave method as applied to geotechnical site investigation and pavement evaluation, *in* B. Huang, R. Meier, J. Prozzi, and E. Tutumluer, eds., Pavement mechanics and performance: American Society of Civil Engineers Geotechnical Special Publication No. 154, 110–116.

# Section 3

# Integrative Approaches

Chapter 17

# High-resolution Seismic Imaging of Near-surface Fault Structures within the Upper Rhine Graben, Germany

Patrick Musmann[1] and Hermann Buness[1]

## Abstract

High-resolution reflection seismic profiles were acquired at two study sites in the Upper Rhine Graben, Germany, to image fault zones in the near-surface domain. The profiles fill the gap left by large-scale 3D seismic imaging focused on targets several kilometers deep. The survey design comprises a very dense sampling of reflection points with frequencies as high as 360 Hz emitted by a small hydraulic vibrator. Dip-moveout processing with poststack migration is required to image the fault systems properly. At the first study site, a broad normal fault zone with a width of about 300 m was imaged. It shows two major faults accompanied by numerous smaller parallel and subparallel faults. At the second study site, the survey reveals a horst structure with two bounding normal faults that branch into several smaller ones with depth. Thicker sedimentary units in the hanging walls of the faults suggest synsedimentary fault growth. Significantly, the high-resolution, 2D near-surface seismic measurements provide deeper insight into the architecture and kinematics of the fault systems than is possible from 3D measurements that are focused on deeper targets.

## Introduction

Fault systems are important targets in fields of exploration geophysics because they have a significant impact on hydrocarbon migration and trapping (Knipe et al., 1998; Manzocchi et al., 1998; Walsh et al., 1998; Krawczyk et al., 2007), ore mineralization (Blundell et al., 2002; Zhang et al., 2003), and regional hydrogeology (Haneburg and Hawley, 1996; Bense and van Balen, 2003; Hess et al., 2009). More recently, fault systems also have been considered to be valuable hydrogeothermal reservoirs for heat and energy extraction. In the latter case, exploration is focused on open and high-permeable fracture zones, where fluid flow is enhanced compared to the surrounding host rock (fracture-dominated aquifers). Naturally, fault systems are far from being simple planar interfaces. Instead, they form zones of deformed material, with complex structures of smaller faults developed within the damage zone around a fault core (Wibberley et al., 2008).

Getting detailed structural information about fault systems requires geophysical methods. Three-dimensional seismic surveying is a well-established exploration tool in the hydrocarbon industry (Greenlee et al., 1994; Aylor, 1995); it has been used increasingly in recent years for geothermal exploration, especially in Germany. Because the depths for economically valuable, low-enthalpy hydrogeothermal reservoirs range from 2 to 5 km (Schellschmidt et al., 2007), acquisition parameters have to be set accordingly. As a consequence, the shallow subsurface, that is, down to about 500 m, is not imaged adequately or at all.

High-resolution near-surface seismic measurements have the potential to illuminate this gap (Shtivelman et al., 1998; Steeples, 1998; Buness et al., 2008). They can be designed to map the continuation of fault systems, already recognized at greater depths, to the surface with a much higher resolution compared to deep seismic exploration. Hence, the internal structure of a fault system and its kinematic history can be derived in more detail. Knowing the kinematic history is a key to modeling the amount and quantity of fractures and hence to understanding and estimating the permeability distribution and its heterogeneity

[1]Leibniz Institute for Applied Geophysics (LIAG), Hannover, Germany.

a)

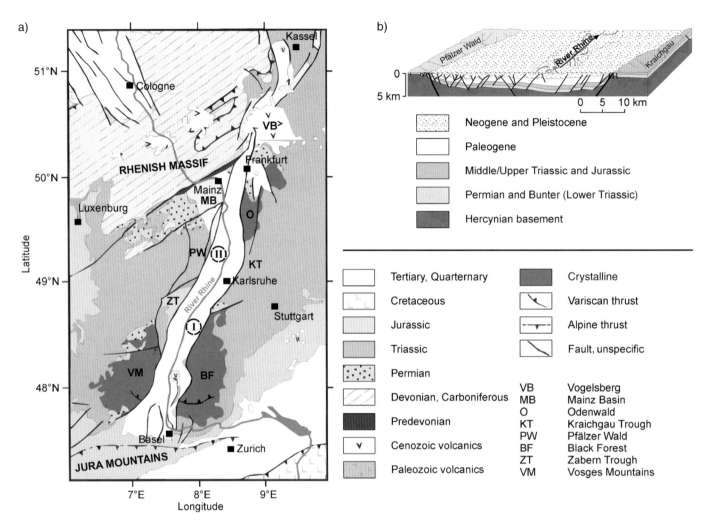

Figure 1. (a) Geologic map of the Upper Rhine Graben (modified after Peters, 2007). Study areas are marked by dashed circles: (I) Neuried and (II) Speyerdorf. The profile locations are given in more detail in Figure 2. (b) Schematic cross section of the Upper Rhine Graben directly north of Karlsruhe. After Illies, 1974.

along the fault (Krawczyk et al., 2006; Krawczyk et al., 2007; Endres et al., 2008; Lohr et al., 2008). Furthermore, information about the last active period of a fault system allows more reliable estimation of fault-sealing processes, in which the geologic time of activity is one of many factors (Knipe et al., 1998; Schleicher, 2005; Kurz et al., 2008).

In the context of the research project titled "Risk Mitigation for Geothermal Projects Using 3D Seismics," the Leibniz Institute for Applied Geophysics is investigating the capabilities of contemporary 2D seismic surveys for near-surface exploration of fault structures (Buness et al., 2010). High-resolution seismic lines were measured across local fault structures in the Upper Rhine Graben (URG), Germany. We discuss two of them here. These profiles are located in the middle and the northern part of the URG, where 3D reflection seismic surveys

were carried out recently for deep hydrogeothermal exploration (Figure 1a). With the combination of 2D shallow and 3D reservoir-scale reflection seismics, we intend to test the potential to improve fault tracing from the depth to the surface, thereby determining fault activity in the URG area.

## Geologic Settings at Study Sites

### Upper Rhine Graben (URG)

The URG belongs to the central part of the European Cenozoic Rift System, which crosses the Northwestern European Plate (Ziegler, 1992). With a width of approximately 30 to 40 km and a length of 300 km, the graben runs from the foreland of the Alpine Jura Mountains

(Basel, Switzerland) to the south to the Rhenish Massif (Frankfurt, Germany) in the north (Figure 1a).

The URG is considered to be a passive rift system, evolved in conjunction with the Alpine orogeny and started during the Eocene (Illies, 1977; Sissingh, 1998; Froitzheim et al., 2008). The main rifting phase across the entire URG occurred with an east-west crustal extension during the Oligocene. Since the late Oligocene, a major reorientation of the stress field changed the extensional to a sinistral transtensional system (Schumacher, 2002). In the Quaternary, the URG is dominated by sinistral strike-slip kinematics with two major pull-apart basins in its northern and southern parts (Schumacher, 2002; Dèzes et al., 2004).

The Tertiary graben infill consists of several successions ranging from terrestrial to brackisch, evaporitic, and marine environments (Derer, 2003) with a thickness of as much as 3300 m (Doebl and Olbrecht, 1974). Quaternary deposits, dominated by fluvial sands and gravels with intercalated limnic fine clastic deposits, are found with a thickness of as much as 350 m (Bartz, 1974).

The main tectonic lineaments are characterized by synthetic and antithetic faulting, mostly parallel or sub-parallel to the margin of the URG. However, repeated extensional and strike-slip movements have led to intensive fracturing and block faulting of the Mesozoic sediments, including the basement (Figure 1b).

Because of its tectonic history, a geothermal gradient of $4°$ to $6°C/km$ prevails in the URG, which is much higher than that found throughout most other regions in Germany (Hänel and Staroste, 1988). Triassic units of Muschelkalk and Buntsandstein constitute the main target horizons for deep hydrogeothermal exploration. The primary permeability of the rock matrix is too low for economic energy extraction (Münch et al., 2005), but fracture systems might allow this problem to be overcome (Schellschmidt et al., 2007).

## Study site I: Neuried

The study site Neuried (Figure 2a) is in the middle part of the URG in the vicinity of the Black Forest (Figure 1a). This area was explored with reflection seismic profiles and several boreholes in the 1970s and 1980s by the hydrocarbon industry, but the results are published sparsely (Mauthe et al., 1993). Since 2005, seismic activities have started again with the acquisition of 2D and 3D data sets for geothermal exploration (Stober and Jodocy, 2005). Including older unpublished borehole information, consisting of several deep boreholes at about 10 km south and

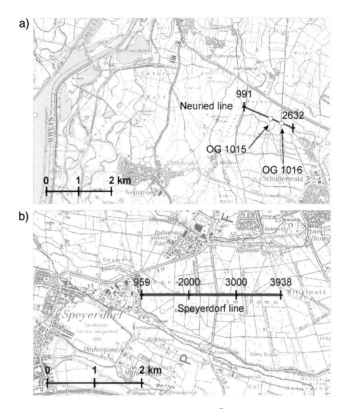

**Figure 2.** Topographic maps (ATKIS®, 2004) with seismic reflection profiles (labeled red lines) recorded at (a) Neuried and (b) Speyerdorf. Numbers refer to CMP numbers and equal the profile distances in meters. Circles mark the location of boreholes at the Neuried site.

two shallow boreholes nearby (Figure 3), they form the state of knowledge.

The main tectonic feature in the area of Neuried is the Kehler Mulde, which constitutes a local basin inside the URG approximately 10 km wide. It is characterized by numerous steeply dipping synthetic and antithetic faults (most with incident angles of $60°$ to $70°$), confining the basin at its eastern and western borders. At least two subparallel normal faults mark its eastern border with a total throw of more than 1000 m. The western border is less well defined; it is a series of synthetic stepped faults with minor displacements forming a broader transition zone.

The Quaternary and Pliocene strata consist of an approximately 200-m-thick series of unconsolidated fluvial deposits, overlying unconformably consolidated sediments from Oligocene to Eocene age. The Quaternary strata consist mainly of two sandy gravel units, with thin, intercalated units of sand and silt. The Pliocene consists of alternating gravel, sand, and silt (Stober et al., 2002). Thickness of the Tertiary fill reaches approximately 2000 m in the center of the basin, followed by another

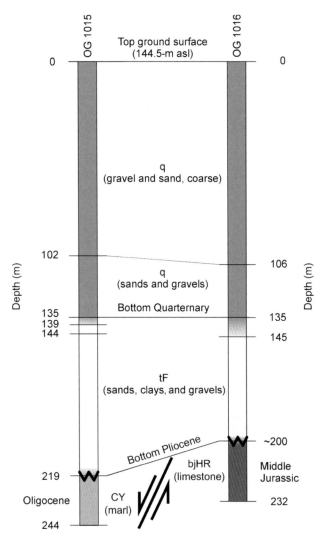

**Figure 3.** Sediment succession of the boreholes Offenburg (OG) 1015 and 1016, located at the Neuried study site: Quaternary (q), fluvial upper Tertiary (tF) (formerly Pliocene), Cyrene marl (CY), Hauptrogenstein formation (bjHR) (German for "master oolite formation"). The distance between boreholes is only 760 m, but they are located on different tectonic blocs.

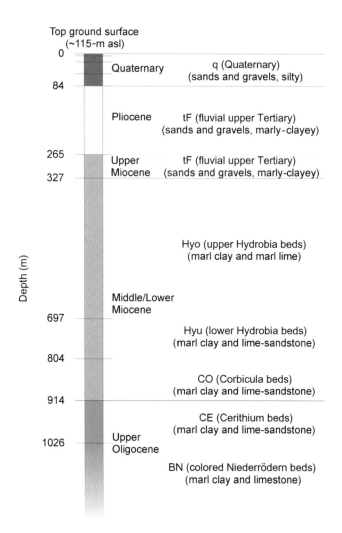

**Figure 4.** Succession of Quaternary and Tertiary sediments at the Speyerdorf study site. The Quaternary and Pliocene strata are derived from hydrogeologic investigations (Kärcher, 1987). Information on the Miocene to Oligocene layers is deduced from an average of four deep drillings at approximately 5 km west of our seismic line. These depths were adjusted according to the depth trend of the Pliocene-Miocene boundary from Kärcher (1987).

1000 m of Mesozoic sediments. No Tertiary sediments occur east of the main boundary fault.

## Study site II: Speyerdorf

The study site Speyerdorf (Figure 2b) is in the northern part of the URG at the frontier to the Pfälzerwald (Figure 1a). The structural setting in this area was largely unknown until geothermal exploration started in 2006. This most recent seismic exploration imaged a prominent horst structure continuing downward into the basement. The horst is bound by

steeply dipping faults (60° or more) and is accompanied by extensional structures. These constitute the target of current exploration activities (Lotz and Greiner, 2009) and the near-surface exploration presented here.

The next boreholes were drilled directly at the western border fault of the URG at a distance of about 5 km from our profile. Along with shallow hydrogeologic investigations (Kärcher, 1987), the structural situation can be summarized as follows: The Quaternary and Pliocene units consist of a series of unconsolidated fluvial deposits of interbedded gravel, sand, and silt (Figure 4) that are approximately 265 m thick. The strata are bedded almost parallel.

They are underlain by approximately 2300-m-thick consolidated deposits from Miocene to Eocene ages that rest unconformably on Triassic sediments, and these constitute the target horizons for geothermal exploration. The tectonic blocks are generally slightly dipping toward the center of the graben, but their local dip does not exceed 15°.

# Methods

## Survey design and data acquisition

When designing a high-resolution survey, the most essential decisions have to be made about the spread layout and sources used. Regarding spread layout, there is always a trade-off between the benefits of small sampling intervals and the maximum offset, when the number of channels is limited. A well-known estimation of the maximum CMP bin size $b$ to avoid spatial aliasing is given by Yilmaz (2001):

$$b \le v_{min}/(4f_{max}\sin\alpha_{max}).$$

Here, $v_{min}$ is the minimum velocity in the subsurface, $f_{max}$ is the maximum frequency of the seismic signal, and $\alpha_{max}$ is the maximum expected subsurface dip. These parameters must be defined on a site-specific basis. Test shots prior to seismic data acquisition revealed a minimum velocity of approximately 1900 m/s in the Neuried area and 1500 m/s in the Speyerdorf area, that is, the velocity immediately below the water level at depths between 2 and 5 m.

The choice of maximum frequency leads us to another trade-off between a desirable broad bandwidth and the field effort to achieve it. Using a vibratory source, we can choose the bandwidth within the technical limits of the vibrator. The vibrator used in this survey was designed especially for high-resolution surveying (Figure 5). It can generate frequencies between 16 and 500 Hz using a peak force of 27 kN (Buness, 2007). The lower limit was set to 30 Hz, a value at which the vibrator generates the full output power. Setting the upper limit is more difficult because it depends on the maximum desired depth penetration. Sweeping very high frequencies for deep targets is a useless effort because the earth's attenuation prevents their recording. Moreover, increasing the upper frequency limit is very expensive in terms of vibration time when linear sweeps are used. Unfortunately, judgment of an optimum frequency bandwidth cannot be made on the basis of the signal-to-noise ratio (S/N) of single shots. Instead, it must be determined by reviewing the data after full processing, as discussed by Buness (2007). Thus, based on test shots, we fixed upper frequencies of 360 Hz for the Neuried area and 300 Hz for the Speyerdorf area. This resulted in sweeps wider than three octaves.

Because we intended to image steeply dipping fault zones, we let $\alpha_{max} = 65°$. Having done this, we could determine the minimum CMP size $b$ to be less than 1.2 m. Using the symmetric sampling principle (Vermeer, 1998), a shot-and-receiver distance of 2 m would follow. However, a shot distance of only 2 m was judged to be too expensive; hence, it was set to 4 m. Available channel numbers were 264 at Neuried and 288 at Speyerdorf, resulting in maximum spread lengths of 526 m and 574 m, respectively. These values are well suited to image the desired depth range of about 500 m. The actual maximum offset range varied because the Geode recording system has no continuous roll-along capabilities. Acquisition parameters are given in Table 1.

Because of the high upper-frequency limit, the sample interval was set to 0.5 ms. Nyquist criteria would allow a larger sample interval of, for example, 1 ms, but the corresponding possible mismatch in sampling peak-amplitude values suggests a finer temporal sampling. Assuming a sinusoidal waveform, the maximum error of a sampled peak-amplitude value relative to the true peak value is given by

$$e_{rel} \le 1 - \cos(\pi f_{max}\Delta t),$$

where $\Delta t$ is the sampling interval (Schwetlick, 1997). Setting $\Delta t = 1$ ms and $f_{max} = 360$ Hz, the maximum rela-

**Figure 5.** The LIAG's small high-frequency MHV 2,7 vibrator (left) used in our studies and a M12 Mertz vibrator (right), which is a typical vibrator used by industry for deeper exploration. The MHV weighs 2.7 t and can radiate frequencies of 16 to 500 Hz; the M12 weighs 18.6 t and usually emits sweeps from 10 to 100 Hz. Photo by Rüdiger Thomas. Used by permission.

**Table 1.** Data-acquisition parameters for seismic lines measured at Neuried and Speyerdorf.

| | Neuried line | Speyerdorf line |
|---|---|---|
| Seismic source type | Vibrator MHV 2,7 | Vibrator MHV 2,7 |
| Sweep parameters | Linear, 30 to 360 Hz, 12 s | Linear, 30 to 300 Hz, 12 s |
| Number of source points | 414 | 745 |
| Source spacing | 4 m | 4 m |
| Geophone type | SM 4/7, 20 Hz | SM 4/7, 20 Hz |
| Geophone spacing | 2 m | 2 m |
| Seismograph | Geometrics/Geode | Geometrics/Geode |
| Number of channels | 264 + pilot sweep | 288 + pilot sweep |
| Sample interval | 0.5 ms | 0.5 ms |
| Record length | 14 s (uncorrelated) | 2 s (correlated) |
| CMP spacing | 1 m | 1 m |
| CMP profile length | 1.641 km | 2.979 km |
| Offsets<br>Left spread<br>Right spread | End-on and split-spread<br>0 to 526 m<br>0 to 94 m | End-on and split-spread<br>0 to 574 m<br>0 to 142 m |
| Nominal fold | 66 | 72 |

tive error amounts to 58%; setting $\Delta t = 0.5$ ms, it is only 16%. The finer sampling results in less amplitude variation along seismic events observable in the shot gathers. However, the high fold cancels out random sampling mismatches.

In total, three high-resolution profiles with lengths between 1.6 and 3.0 km were recorded across projected outcrop locations of faults identified by means of industry 3D and 2D seismic data sets. Two principal shallow seismic profiles are discussed in this article (Figure 2). Both profiles were recorded on forest tracks consisting of highly compacted gravel and soil with only slight variations in topographic relief (<5 m) along the entire profile. Quiet weather conditions and low anthropogenic noise caused excellent preconditions for seismic measurements.

Unprocessed shot gathers (Figures 6a and 7a) show numerous reflection events with excellent S/N throughout the optimum window. The lower bound of the optimum window for reflection events is marked by the air-coupled wave (Speyerdorf area; Figure 7a) or by shear waves (Neuried area; Figure 6a), presumably converted near to the vibrator source. The upper bound is marked by direct and refracted waves. Spectral analysis of the reflected signal reveals a strong decrease of spectral energy toward

higher frequencies (Figures 6c and 7c). Therefore, one is tempted to restrict sweep frequencies to a maximum of 240 Hz. However, the recording of frequencies near the upper limit is confined to near offsets and short recording times; hence, they appear increasingly weak with larger analysis windows.

## Data processing

To a large extent, the processing sequence follows a standard for vibroseis data (described in detail, e.g., by Yilmaz, 2001, or Sheriff and Geldart, 1995) that includes a conventional normal-moveout/dip-moveout (NMO/DMO) stack with subsequent time migration and depth conversion (Table 2). Here, we used commercial ProMAX™ software (Landmark Corporation), and we paid special attention to fault imaging within the uppermost 500 m of depth.

We took special care to determine proper static corrections because traveltime variations caused by topographic relief and by thickness and velocity variations of the weathering layer influence shallow seismic data much more than data acquired for deeper targets because of higher-frequency content (Steeples et al., 1990). Refraction statics for sources and receivers were calculated from first-break picks by applying Gulunay's (1985) diminishing-residual-matrix method. Datum elevation was set a few meters above the highest level of topography to ensure that no shallow reflection events would get truncated. The first refractor in both areas is interpreted as water level at depths of 2 to 5 m; it shows average velocities of ~2000 m/s at the Neuried site and ~1600 m/s at the Speyerdorf site.

Reconstruction of the high-frequency spectrum of the data within the sweep range improves the vertical resolution. Tests recommended different optimization techniques for the profiles. At Speyerdorf, a spiking deconvolution followed by band-pass filtering yielded the best results, whereas at Neuried, spectral whitening gave better overall results, that is, a higher S/N with comparable or better resolution at targets of interest. A flat whitening

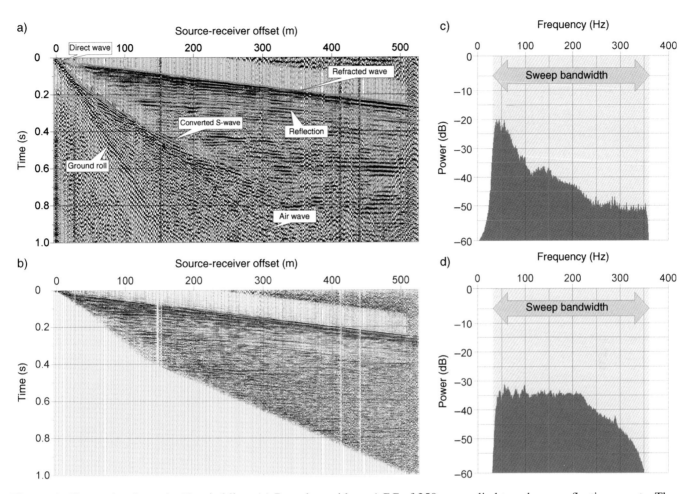

**Figure 6.** Shot gather from the Neuried line. (a) Raw data with an AGC of 250 ms applied to enhance reflection events. The first-break analysis reveals a velocity of 450 m/s for the direct wave (short red line); the strong refracted events (blue line) have a nearly constant velocity of 2150 m/s. This results from the refraction at the water table at ~3 m depth. The green hyperbola marks a reflection event that is associated with a strong impedance contrast at the Pliocene-Oligocene unconformity. (b) Data after processing steps 2 through 8 (see Table 2). Frequency spectra of (c) raw and (d) processed shot gathers. The analysis is restricted to the offset-traveltime range of the optimum window.

of the spectra above 240 Hz results in obscured reflections for deeper targets but reveals more details for shallow reflections. As a compromise, frequencies above 240 Hz are damped linearly to the maximum sweep frequency. The resulting spectra are shown in Figures 6d and 7d along with their corresponding shot gathers.

A bottom mute was applied to remove the noise cone below the optimum window (Figures 6 and 7), although the reflection signal was enhanced inside this area after deconvolution or spectral whitening, respectively. Muting the noise cone leads to a better S/N in the stacked data, down to ~500 ms, but it decreases the stack quality below. The same holds true for other filters that were tested to attenuate noise or to enhance signal, such as *f-k*. An explicit top-mute function to eliminate first-break

events of direct and refracted waves was not applied because the chosen stretch mute of 40% during NMO correction was considered sufficient for removing the first-break energy. Velocity models (Figure 8) were derived by an iterative process of velocity analysis, NMO/DMO, and stack (Table 2, items 9 through 13); no additional information was available (e.g., VSP data). The partial prestack-migration process of dip moveout was incorporated into processing because of the presence of steeply dipping events. Thereby, DMO ideally makes the stacking velocities independent of dip and corrects stacking of simultaneous events with conflicting dips (Deregowski, 1986). The process has proved to be essential here, especially at the Neuried line, where dipping reflection events in the fault zone are present. Iterations were stopped if significant

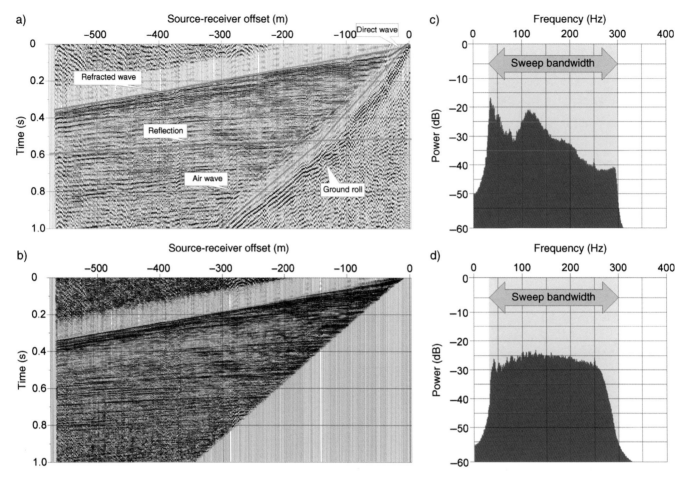

**Figure 7.** Shot gather from the Speyerdorf line. (a) Raw data with an AGC of 250 ms applied to enhance reflection events. The first-break analysis reveals a velocity of ~450 m/s for the direct wave (short red line); the strong refracted events (blue line) show an increasing velocity from 1650 to 1850 m/s. This results from the water table at ~3 m depth. Distorted events at an offset of ~100 m are from a fault coming close to the surface. (b) Data after processing steps 2 through 8 (see Table 2). Frequency spectra of (c) raw and (d) processed shot gathers. The analysis is restricted to the offset-traveltime range of the optimum window.

changes in the velocity model were no longer observed. Dip moveout preserves diffractions well through the stacking process so that an improved definition of *discontinuities* is given after both stacking (Figure 9) and migration.

Poststack time migration was carried out using an explicit finite-difference migration algorithm. Slight smoothing of stacking velocities prior to conversion into interval velocity improved the migration result and the results of depth conversion. This was noticeable in that there were fewer undulant horizons (viz., fewer lateral velocity variations) and better agreement with borehole data at the Neuried site. Prestack depth migration (PSDM) also has been approved as a valuable tool for shallow seismic imaging (Bradford et al., 2006). However, its results depend on seismic velocities, and we were not able to improve the imaging results using PSDM with respect to the NMO/DMO processing.

# Results

## Study site I: Neuried

The most distinct features of the time-migrated depth section (Figure 10; see Figure 2a for location) are (1) a high-amplitude reflector at a depth of approximately 200 m that can be followed across the entire section and (2) steeply dipping events below this reflector, located between profile distances of 1900 and 2200 m. Interpretation of these features is enabled by two boreholes only a few meters beside the line (Figures 2a and 3):

1) The strong reflector marks the base of unconsolidated Quaternary and Pliocene units overlying discordantly older consolidated units. The unconsolidated units show a high reflectivity and a low to moderate continuity of reflectors, which express the fluvial character of

**Table 2.** Processing sequence applied to seismic data gathered at Neuried and Speyerdorf.

| | Neuried line | Speyerdorf line |
|---|---|---|
| 1 | Vibroseis correlation | (Done during acquisition) |
| 2 | Trace editing | Trace editing |
| 3 | Vertical stack | Vertical stack |
| 4 | Trace/amplitude normalization (AGC, 250 ms) | Trace/amplitude normalization (AGC, 250 ms) |
| 5 | First-break picking and refraction statics to floating datum | First-break picking and refraction statics to floating datum |
| 6 | Spectral whitening: 30/45 to 240/360 Hz | Surface-consistent spiking deconvolution |
| 7 | | Bandpass filtering: 30/40 to 240/300 Hz |
| 8 | Bottom mute (air/surface/shear waves) | Bottom mute (air/surface waves) |
| 9 | Velocity analysis | Velocity analysis |
| 10 | NMO correction (stretch mute 40%) | NMO correction (stretch mute 40%) |
| 11 | Common offset *f-k* DMO (five iterations of velocity analysis) | Common offset *f-k* DMO (two iterations of velocity analysis) |
| 12 | | Residual statics |
| 13 | CMP stack | CMP stack |
| 14 | Static correction to final datum (150-m asl) | Static correction to final datum (115-m asl) |
| 15 | Trace/amplitude normalization (AGC) | Trace/amplitude normalization (AGC) |
| 16 | Finite-difference migration | Finite-difference migration |
| 17 | Depth conversion | Depth conversion |

these sediments and their frequent and small-sized changes of different sands and gravel units. They can be divided further by a slight improvement in the continuity of the reflection pattern at ∼130 m into Quaternary and Pliocene deposits.

2) The steeply dipping events (30° to 45°) are located between two major parallel normal faults, cutting the Pliocene-Quaternary boundary at profile distances of 2100 m and 2220 m, respectively. Both faults, dipping approximately 65° to 75°, also are interpreted on deep seismic lines. The eastern one constitutes the eastern-basin bounding fault (EBBF) of the Kehler Mulde with a throw of more than 1000 m, whereas the western one is of minor throw with regard to the uppermost Tertiary and Quaternary units (Stober and Jodoky, 2005).

West and east of these two major faults, numerous parallel or subparallel faults are recognizable. At least two faults clearly are cutting the Pliocene basis and the lowermost portion of the Pliocene strata, but their continuation into Quaternary units is equivocal.

Total width of the entire fault zone amounts to approximately 300 m. Keeping in mind the complex 3D structures of fault zones, reflecting elements located between the two major faults are not interpreted to be fault surfaces themselves but rather lithologic boundaries, probably interfering with side reflections and imaging artifacts. East of the fault zone, Pliocene sediments rest directly on Middle Jurassic oolite units. To the west, they rest unconformably on Upper Oligocene units that dip increasingly (4° to 15°) toward the center of the basin.

## Study site II: Speyerdorf

The time-migrated depth section at Speyerdorf (Figure 11; see Figure 2b for location) images reflective sedimentary structures with a very high resolution. Strata are generally dipping east (2° to 4°), with slightly increasing thicknesses to the center of the Rhine Graben. A distinctive reflector at a depth of ∼80 m is interpreted as the base of the Quaternary. The underlying Pliocene, which has a thickness of ∼170 m, shows significantly higher amplitudes than the

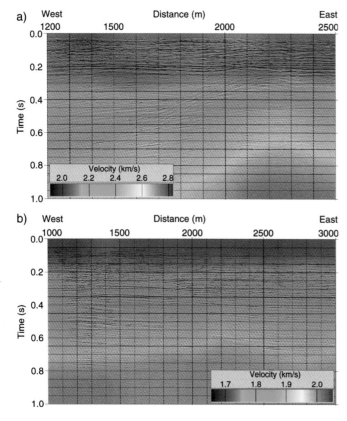

**Figure 8.** Stacking velocities derived from DMO processing for the (a) Neuried and (b) Speyerdorf lines. The color-coded velocities are underlain by their NMO/DMO stack sections.

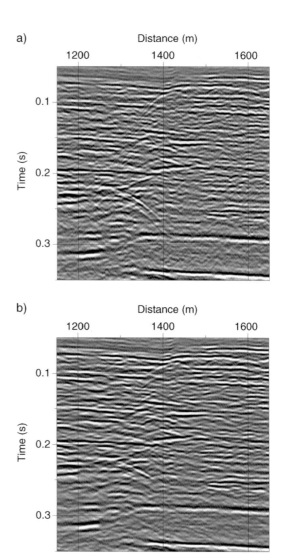

**Figure 9.** Comparison of the (a) NMO/DMO stack and (b) NMO stack for a cutout of the Speyerdorf seismic line. The DMO processing enhances diffractions, improving the interpretation of the discontinuities and faults.

Quaternary. It is followed by Miocene units. At about 300-m depth, continuity of reflectors increases. We interpret this as a transition from a fluvial to a brackish-marine depositional environment, which occurred during the Miocene (Derer, 2003).

The most prominent feature on this section is a horst structure in the eastern part, bounded by two normal faults, with apparent dip between 50° and 70°. With increasing depth, the faults seem to branch into a complex fault system. The throw along the main normal faults increases with depth, ranging from several meters near the surface to about 100 m at a depth of 1000 m. Synsedimentary growth of the main normal faults can be deduced because the thicknesses of strata outside the horst generally are larger than those on the horst itself. This is most obvious for the Pliocene and Upper Miocene units across the eastern fault of the horst. Thus, main fault movement can be assumed during these epochs. The faults can be followed upward to a depth of only 20 m, which proves at least sub-recent activity of the horst structure.

## Discussion

### Study site I: Neuried

The surveys aimed to clarify the age of latest activity of the fault systems because this is one factor for estimation of potential permeability. In the Neuried area, activity of the fault system can be inferred confidently to mid-Pliocene times (Figure 10), but a continuation to the base Quaternary is equivocal. Thickness of Pliocene deposits varies across the fault zone. However, a distinct fault surface or a displacement of strata along a shear surface cannot be recognized because of the high-energy fluvial character of the Pliocene and Quaternary sediments. A continuous movement along the fault during the Pliocene and Quaternary cannot be

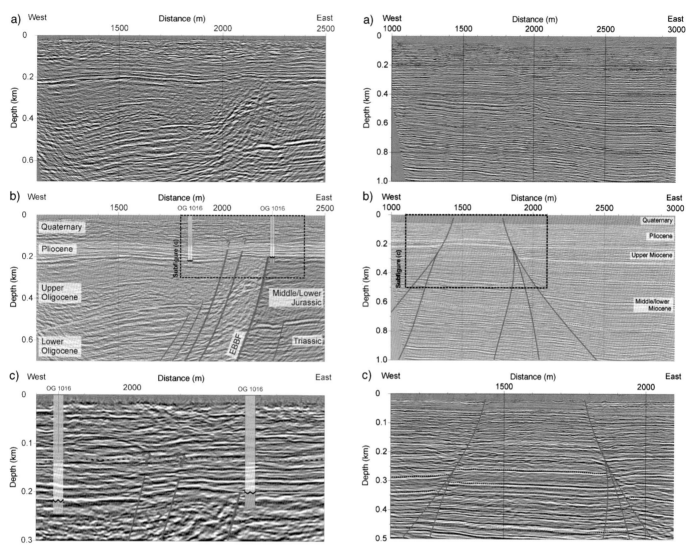

**Figure 10.** (a) Time-migrated and depth-converted seismic section of the Neuried line. (b) Interpretation based on borehole information and deep seismic data sets (EBBF = eastern-basin-bounding fault). (c) Enlargement showing the Plio-Quaternary sediments above the fault zone.

**Figure 11.** (a) Time-migrated and depth-converted seismic section of the Speyerdorf line. (b) Interpretation based on borehole information and deep seismic data sets. (c) Enlargement showing the upper part of the horst structure with its bounding faults.

ruled out. In principle, thickness variations in Pliocene and Quaternary sediments could result from compaction of the large pile of Tertiary sediments at the western side of the fault zone also. A compaction analysis could help to answer this question. A first estimation (D. Tanner, personal communication, 2008) amounts to 12 m of differential compaction, compared to the approximately 20-m throw observed at the base Pliocene.

## Study site II: Speyerdorf

Continuity of the reflections and hence the fault imaging are much better at Speyerdorf than at Neuried

because of a lower-energy depositional environment. The image allows one to follow the fault displacements in great detail as they split into several branches with increasing depth. Branching of the fault at both sides of the horst at depths of 300 to 1000 m seems to be well defined. Seismic interpretation must be done with care because fault-zone imaging can be affected by artifacts. This issue is known as the fault shadow problem. Thereby, lateral velocity variations across the fault can disrupt, pull up, or pull down reflectors in the footwall of the fault and can be misinterpreted, for example, as fault bifurcation (Fagin, 1996; Trinchero, 2000; Couples et al., 2007). Usually, this problem is addressed by prestack depth migration, which

**Figure 12.** (a) Transect cutout of the 3D industry seismic data set at Speyerdorf along the same position as the high-resolution seismic line shown in Figure 11a and b. Note the lower frequency content in this image and the missing information at above a depth of 300 m. The course of the host bounding faults can be determined only crudely. (b) Density plot of data shown above with overlaid interpretation from 2D high-resolution data (Figure 11b).

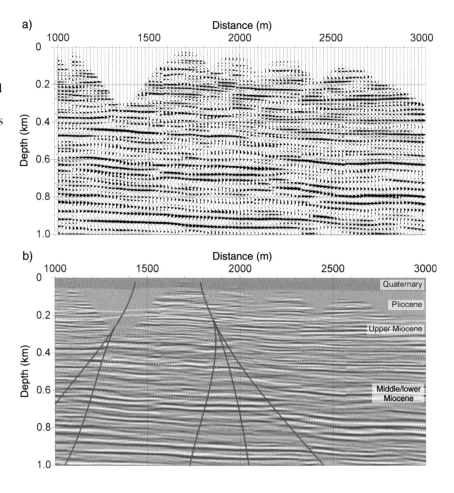

can correct local velocity variations and therefore yields a truer picture of the subsurface. However, the result of pre-stack depth migration depends on reliability of the velocity model. Therefore, verification of the interpretation results by seismic modeling techniques is advisable. Nevertheless, the image of the faults down to a depth of more than 1000 m allows reconstruction of the kinematic history of the horst and graben structure from the Oligocene to the present time with a high resolution.

## Survey site II: Comparison of shallow 2D and deep 3D seismic data at Speyerdorf

To compare the data discussed here with industry-scale 3D exploration seismic data, a transect has been cut out of the volume (Figure 12) in coincidence with the 2D line at Speyerdorf (Figure 11a and b). The 3D seismic data had been recorded in 2008 for geothermal exploration. According to the target depth of 3 to 5 km, source- and receiver-line spacing of 400 m, source and receiver spacing of 40 m, and five vibrators with a sweep of 12 to 96 Hz were used. The geometry setup results in a CMP bin size of 20 m × 20 m.

Although the main structural elements are observable in the 3D data set as well, its lower-frequency content and the larger CMP spacing contribute to blurring of the seismic image (Figure 12). In consequence, the horst structure is still recognizable, but the course of the fault zones becomes vague. In particular, only the two faults bounding the horst can be identified, and the branching of fault splays can be inferred only from the supplemental high-resolution seismic profile (Figure 12b).

The two data sets show different reflection characteristics. In the 3D data set, only the low-frequency reflectors with medium-high amplitudes can be tracked continuously along the profile. However, the 2D section reveals more high-frequency reflection events. Only the latter enable a detailed interpretation of the seismic facies units. For example, the increase of reflector continuity at a depth of about 300 m, marking the change of the depositional environment (as discussed in the "Results" section) can be deduced from the high-resolution section only.

Moreover, the chosen source-and-receiver line spacing at 3D acquisition prevent full imaging of the uppermost 300 m so that no conclusion about the fault activity beyond the Miocene could be drawn (Figure 12). This gap is bridged with our high-resolution 2D seismic line up to a depth of 20 m only.

Each of our data sets was recorded with a crew of four people in one week. Comparing this field effort with the acquisition of 3D seismic data, the 2D high-resolution data gave valuable additional information to the deep seismic data at low costs. A further increase in field efficiency could be possible with use of the landstreamer technique (Van der Veen et al., 2001; Inazaki, 2004; Pugin et al., 2004; Beilecke et al., 2006; Polom et al., 2007).

## Seismic resolution and shallow surveying

Both surveys image stratigraphic layers and fault zones with high resolution. The theoretical limit of the vertical resolution often is cited as a quarter of the wavelength, the $\lambda/4$ or Rayleigh criterion (Sheriff and Geldart, 1995; Yilmaz, 2001). This implies a limit of $\sim$2.5 m when using the data of the surveys presented here ($v = 2000$ m/s, $f = 200$ Hz). The lateral limit of resolution is more difficult to define; it depends on depth and is bound to the radius of the Fresnel zone before migration. The numbers given above result in values of 14 m and 44 m for traveltimes of 40 ms and 400 ms, respectively. Migration tends to collapse the Fresnel zone, at least in the direction of the profile, and hence the lateral resolution is given by $\lambda$ (Stolt and Benson, 1986) or by $\lambda/2$ (Lindsey, 1989), giving smaller values of $\sim$5 m or $\sim$10 m, respectively. Compared to the deep 3D seismic data presented above (in which $f_{\max} = 96$ Hz), the resolution of our 2D profiles is more than doubled.

Tracking of faults farther up to the surface, as done for the Speyerdorf line (viz., <20 m) is difficult to achieve. Although seismic recording is reported from as shallow as 2 m (Bachrach and Nur, 1998; Baker et al., 1999; Bachrach and Mukerij, 2004a, 2004b; Schmelzbach et al., 2005), the methodology differs from the kind of seismic recording used in this study. Using ultrashallow seismic methods, it would not be possible to reach the depth necessary to connect the data reliably to industry-type 3D or 2D data. Moreover, a survey of a fault zone with a 300-m width, such as the one imaged at Neuried, would be difficult to carry out. With the methodology presented here, one has to consider the physical-resolution limit because the wavelength is on the order of 10 m, as stated above.

An increase in resolution eventually could be achieved by the adoption of shear-wave techniques, which have gained new popularity, especially in near-surface seismic (Carr et al., 1998; Simmons and Backus, 2001; Polom et al., 2008; Pugin et al., 2009a). Wavelengths to an order of magnitude shorter than that found in P-wave data have been reported for unconsolidated sediments (Pugin et al.,

2009b). Thus, we carried out a shear-wave reflection seismic survey at the Neuried site, using a small hydraulic vibrator generating horizontal polarized shear waves. Although good results were reported from other surveys using the same source (Polom et al., 2009; Polom et al., 2010), the results of the survey at our study site were disappointing, underlining the need for further research.

## Conclusions

Reflection seismic measurements carried out in the Upper Rhine Graben resulted in high-resolution images of fault zones and sedimentary strata in a depth range of 20 to 1000 m. They gave new insights into the structural architecture of the fault zones and, in combination with stratigraphic information, into their kinematic history. In both cases, the age of the last activity could be stated more precisely than before. However, the fidelity of this assignment depends on the sedimentary environment. In the Speyerdorf area, the fault zones can be inferred quite precisely because of continuous reflectors imaged to very shallow depths, which is the result of relatively low-energy deposits. In contrast, the situation in Neuried is not as favorable because the sedimentation changes to a high-energy fluvial deposition. Therefore, the course of the fault zones is lost when reaching the Quaternary.

Conventional processing including the NMO/DMO stack followed by migration and depth conversion was proved to image the near-surface fault system. Dip-moveout processing is necessary for improvement of the velocity model and interpretation of the fault zones. However, because of the complex structure of fault zones, processing or imaging artifacts cannot be excluded, especially concerning the footwall of the fault zone. Thus, further improvements could be achieved by shallow-seismic 3D acquisition. In addition, seismic-modeling techniques might be useful for investigating imaging artifacts and verifying results.

Once the position of a fault is known, for example, from conventional 3D surveying, the acquisition and processing of our high-resolution 2D data set exhibit a very effective method for imaging faults at shallow depths, especially if compared to the effort necessary for 3D exploration. Thus, both surveys complement each other, with the shallow seismics giving additional value to deep 3D data.

## Acknowledgments

We thank our dedicated field team of Stefan Cramm, Eckhardt Großmann, Siegfried Grüneberg, and Walter

Rode for their effort to record high-quality data. Furthermore, we thank Charlotte Krawczyk, Susanne Wölz, and various reviewers for their constructive comments on this manuscript. GeoEnergy GmbH kindly provided the 3D seismic data of the Speyerdorf site. The project is funded by the Federal Ministry of Environment, Nature Conservation, and Nuclear Safety of Germany (BMU).

# References

ATKIS®, 2004, DTK50-V: © Vermessungsverwaltungen der Länder und BKG 2004, http://www.geodatenzentrum.de, accessed 10 May 2010.

Aylor, W. K., 1995, Business performance and value of exploitation 3-D seismic: The Leading Edge, **14**, 797–801.

Bachrach, R., and A. Nur, 1998, High-resolution shallow seismic experiments in sand, Part I: Water table, fluid flow and saturation: Geophysics, **63**, 1225–1233.

Bachrach, R., and T. Mukerji, 2004a, Portable dense geophone array for shallow and very shallow 3D seismic reflection surveying — Part 1: Data acquisition, quality control, and processing: Geophysics, **69**, 1443–1455.

Bachrach R., and T. Mukerji, 2004b, Portable dense geophone array for shallow and very shallow 3D seismic reflection surveying — Part 2: 3D imaging tests: Geophysics, **69**, 1456–1469.

Baker, G. S., C. Schmeissner, D. W. Steeples, and R. G. Plumb, 1999, Seismic reflections from depths of less than two meters: Geophysical Research Letters, **26**, 279–282.

Bartz, J., 1974, Die Mächtigkeit des Quartärs im Oberrheingraben, *in* J. H. Illies, and K. Fuchs, eds., Approaches to taphrogenesis: E. Schweizerbart, 78–87.

Beilecke, T., U. Polom, and S. Hoffmann, 2006, Efficient high resolution subsurface shear wave reflection imaging in sealed urban environments: 12th European Meeting of Environmental and Engineering Geophysics, EAGE, Extended Abstracts, A033.

Bense, V., and R. van Balen, 2003, Hydrogeologic aspects of fault zones on various scales in the Roer Valley Rift System: Journal of Geochemical Exploration, **78–79**, 317–320.

Blundell, D. J., F. Neubauer, and A. von Quadt, 2002, The timing and location of major ore deposits in an evolving orogen: Geological Society [London] Special Publication 204.

Bradford, J. H., L. M. Liberty, M. W. Lyle, W. P. Clement, and S. Hess, 2006, Imaging complex structure in shallow seismic-reflection data using prestack depth migration: Geophysics, **71**, no. 6, B175–B181.

Buness, H., 2007, Improving the processing of vibroseis data for very shallow high-resolution measurements: Near Surface Geophysics, **5**, 173–182.

Buness, H., G. Gabriel, and D. Ellwanger, 2008, The Heidelberg Basin drilling project: Geophysical pre-site surveys: Quaternary Science Journal (Eiszeitalter und Gegenwart), **57**, 338–366.

Buness, H., H. von Hartmann, H.-M. Rumpel, T. Beilecke, P. Musmann, and R. Schulz, 2010, Seismic exploration of deep hydrogeothermal reservoirs in Germany, World Geothermal Congress, Proceedings, 1346.

Carr, B. J., Z. Hajnal, and A. Prugger, 1998, Shear wave studies in glacial till: Geophysics, **63**, 1273–1284.

Couples, G., J. Ma, H. Lewis, P. Olden, J. Quijano, T. Fasae, and R. Maguire, 2007, Geomechanics of faults: Impacts on seismic imaging: First Break, **25**, 83–90.

Deregowski, S. M., 1986, What is DMO?: First Break, **4**, 7–24.

Derer, C. E., 2003, Tectono-sedimentary evolution of the northern Upper Rhine Graben (Germany), with special regard to the early syn-rift stage: Ph.D. thesis, University of Bonn.

Dèzes, P., S. M. Schmid, and P. A. Ziegler, 2004, Evolution of the European Cenozoic Rift System: Interaction of the Alpine and Pyrenean orogens with their foreland lithosphere: Tectonophysics, **389**, 1–33.

Doebl, F., and W. Olbrecht, 1974, An isobath map of the Tertiary base in the Rhinegraben, *in* J. H. Illies, and K. Fuchs, eds., Approaches to taphrogenesis: E. Schweizerbart, 71–72.

Endres, H., T. Lohr, H. Trappe, R. Samiee, P. O. Thierer, C. M. Krawczyk, D. C. Tanner, O. Oncken, and P. A. Kukla, 2008, Quantitative fracture prediction from seismic data: Petroleum Geoscience, **14**, 369–377.

Fagin, S., 1996, The fault shadow problem: Its nature and elimination: The Leading Edge, **15**, 1005–1013.

Froitzheim, N., D. Plasienka, and R. Schuster, 2008, Alpine tectonics north of the Alps and Western Carpathians, *in* T. McCann, ed., The Geology of Central Europe, v. 2: Mesozoic and Cenozoic: Geological Society [London], 1141–1232.

Greenlee, S. M., G. M. Gaskin, and M. G. Johnson, 1994, 3-D seismic benefits from exploration through development: An Exxon perspective: The Leading Edge, **13**, 730–734.

Gulunay, N., 1985, A new method for the surface-consistent decomposition of statics using diminishing residual matrices (DRM): 55th Annual international Meeting, SEG, Expanded Abstracts, 293–295.

Haneburg, W. C., and J. W. Hawley, 1996, Characterization of hydrogeological units in the northern Albuquerque Basin, New Mexico: Bureau of Mines and Mineral Resources, Open-file Report 402C.

Hänel, R., and E. Staroste, 1988, Atlas of geothermal resources in the European community: Th. Schäfer.

Hess, S., J. P. Fairley, J. H. Bradford, M. Lyle, and W. Clement, 2009, Evidence for composite hydraulic architecture in an active fault system based on 3D seismic reflection, time domain electromagnetic, and temperature data: Near Surface Geophysics, **7**, 341–352.

Illies, J. H., 1974, Taphrogenesis, introductory remarks, *in* J. H. Illies, and K. Fuchs, eds., Approaches to taphrogenesis: E. Schweizerbart, 1–13.

———, 1977, Ancient and recent rifting in the Rhinegraben: Geologie en Mijnbouw, **56**, 329–350.

Inazaki T., 2004, High-resolution seismic reflection surveying at paved areas using S-wave type land streamer: Exploration Geophysics, **35**, 1–6.

Kärcher, T., 1987, Beiträge zur Lithologie und Hydrogeologie der Lockergesteinsablagerungen (Pliozän, Quartär) im Raum Frankenthal, Ludwigshafen-Mannheim, Speyer, *in* K. Hinkelbein, ed., Jahresberichte und Mitteilungen des Oberrheinischen Geologischen Vereines: N. F. 69: E. Schweizerbart, 279–320.

Knipe, R. J., G. Jones, and Q. J. Fisher, 1998, Fault sealing and fluid flow in hydrocarbon reservoirs: Geological Society [London] Special Publication 147.

Krawczyk, C. M., T. Lohr, D. C. Tanner, H. Endres, R. Samiee, H. Trappe, O. Oncken, and P. Kukla, 2006, Structural architecture and deformation styles derived from 3-D reflection seismic data in the north German basin: 68th Conference and Technical Exhibition, EAGE, Extended Abstracts, P104.

Krawczyk, C. M., T. Lohr, D. C. Tanner, H. Endres, H. Trappe, O. Oncken, and P. A. Kukla, 2007, A workflow for sub-/seismic structure and deformation quantification of 3-D reflection seismic data sets across different scales, *in* Presentations from the DGMK/ÖGEW spring meeting, CD-ROM.

Kurz, W., J. Imber, C. A. J. Wibberley, R. E. Holdsworth, and C. Collenetti, 2008, The internal structure of fault zones: Fluid flow and mechanical properties, *in* C. A. J. Wibberley, W. Kurz, J. Imber, R. E. Holdsworth, and C. Collenetti, eds., The internal structure of fault zones: Implications for mechanical and fluid-flow properties: Geological Society [London] Special Publication 299, 1–3.

Lindsey, J. P., 1989, The Fresnel zone and its interpretive significance: The Leading Edge, **8**, 33–39.

Lohr, T., C. M. Krawczyk, D. C. Tanner, R. Samiee, H. Endres, P. O. Thierer, O. Oncken, H. Trappe, R. Bachmann, and P. A. Kukla, 2008, Prediction of subseismic faults and fractures: Integration of three-dimensional seismic data, three-dimensional retrodeformation, and well data on an example of deformation around an inverted fault: AAPG Bulletin, **92**, 473–485.

Lotz, U., and G. Greiner, 2009, Großräumige geologisch-seismische Exploration als wesentliches Tool zur Risikominimierung bei Projekten der Tiefengeothermie: Der Geothermiekongress 2009, GTV, Expanded Abstracts, TF01, http://www.geothermie.de/fileadmin/useruploads/aktuelles/Geothermiekongress/vortraege/TF01_Lotz.pdf, accessed 19 May 2010.

Manzocchi, T., P. S. Ringrose, and J. R. Underhill, 1998, Flow through fault systems in high-porosity sandstones, *in* M. P. Coward, T. S. Daltaban, and H. Johnson, eds., Structural geology in reservoir characterization: Geological Society [London] Special Publication 127, 65–82.

Mauthe, G., H.-J. Brink, and P. Burri, 1993, Kohlenwasserstoffvorkommen und-potential im deutschen Teil des Oberrheingrabens: Bulletin der Vereinigung Schweizerischer Petroleum-Geologen und-Ingenieure, **60**, 15–29.

Münch, W., H. P. Sistenich, C. Bücker, and T. Blanke, 2005, Feasibility of geothermal power generation in the German Upper Rhine Graben: VGB PowerTech, **10**, 58–66.

Peters, G., 2007, Active tectonics in the Upper Rhine Graben: Integration of paleoseismology, geomorhology and geomechanical modeling, Ph.D. thesis, Vrije Universiteit Amsterdam.

Polom, U., I. Arsyad, and H.-J. Kümpel, 2008, Shallow shear-wave reflection seismics in the tsunami struck Krueng Aceh River Basin, Sumatra: Advances in Geosciences, **14**, 135–140.

Polom, U., I. Arsyad, and S. T. Wiyono, 2007, Shallow reflection seismic shear-wave velocity analysis on paved soils using a land streamer unit: 69th Conference and Technical Exhibition, EAGE, Extended Abstracts, 19.

Polom, U., L. Hansen, J. S. L'Heureux, O. Longva, I. Lecomte, and C. Krawczyk, 2009, High-resolution shear wave reflection seismic in the harbour area of Trondheim, Norway: 71st Conference and Technical Exhibition, EAGE, Extended Abstracts, T008.

Polom, U., L. Hansen, G. Sauvin, J. S. L'Heureux, I. Lecomte, C. Krawczyk, O. Longva, and M. Vanneste, 2010, High-resolution, shear-wave seismics for characterization of onshore ground conditions in the Trondheim Harbor area, central Norway, this volume.

Pugin, A. J. M., T. H. Larson, S. L. Sargent, J. H. McBride, and C. E. Bexfield, 2004, Near-surface mapping using SH-wave and P-wave seismic land-streamer data acquisition in Illinois, U. S.: The Leading Edge, **23**, 677–682.

Pugin, A. J. M., S. E. Pullan, and J. A. Hunter, 2009a, Multicomponent high-resolution seismic reflection profiling: The Leading Edge, **28**, 1248–1261.

Pugin, A. J. M., S. E. Pullan, J. A. Hunter, and G. A. Oldenborger, 2009b, Hydrogeological prospecting using P- and S-wave landstreamer seismic reflection methods: Near Surface Geophysics, **7**, 315–327.

Schellschmidt, R., B. Sanner, R. Jung, and R. Schulz, 2007, Geothermal energy use in Germany: Proceedings of the European Geothermal Congress: 292, http://www.sanner-geo.de/media/final$20paper$20Germany$20update.pdf, accessed 19 May 2010.

Schleicher, A., 2005, Clay mineral formation and fluid-rock interaction in fractured crystalline rocks of the Rhine rift system: Ph.D. thesis, University of Heidelberg.

Schmelzbach, C., A. G. Green, and H. Horstmeyer, 2005, Ultra-shallow seismic reflection imaging in a region characterized by high source-generated noise: Near Surface Geophysics, **3**, 33–46.

Schumacher, M. E., 2002, Upper Rhine Graben: Role of preexisting structures during rift evolution: Tectonics, **21**, 1006–1022.

Schwetlick, H., 1997, PC-Messtechnik: Grundlagen und Anwendungen der rechnergestützten Meßtechnik: Vieweg and Teubner.

Sheriff, R. E., and L. P. Geldart, 1995, Exploration seismology: Cambridge University Press.

Shtivelman, V., U. Frieslander, and E. Zilberman, 1998, Mapping shallow faults at the Evrona playa site using high-resolution reflection method: Geophysics, **63**, 1257–1264.

Simmons, J., and M. Backus, 2001, Shear waves from 3-D–9-C seismic reflection data: Have we been looking for signal in all the wrong places?: The Leading Edge, **20**, 604–612.

Sissingh, W., 1998, Comparative Tertiary stratigraphy of the Rhine Graben, Bresse Graben and Molasse Basin: Correlation of Alpine foreland events: Tectonophysics, **300**, 249–284.

Steeples, D. W., 1998, Shallow seismic reflection section — Introduction: Geophysics, **63**, 1210–1212.

Steeples, D. W., R. D. Miller, and R. A. Black, 1990, Static corrections from shallow-reflection surveys: Geophysics, **55**, 769–775.

Stober, I., and M. Jodocy, 2005, Projektstudie Tiefe Geothermie (Hydrogeothermie) über die Optimierung geophysikalischer Untersuchungen am Beispiel Neuried (Oberrheingraben) — Teil 1: Technical report, Landesamt für Geologie, Rohstoffe und Bergbau Baden-Württemberg, 1–20.

Stober, I., O. Wendt, and R. Traub, 2002, Tiefenabhängige hydrogeologische Untersuchungen im Quartär und Pliozän des Oberrheingrabens — Ergebnisse der Erkundungs-und Messstellenbohrung Marlen bei Kehl: Abhandlungen des Landesamt für Geolgie, Rohstoffe und Bergbau Baden-Württemberg, **15**, 255–301.

Stolt, R. H., and A. K. Benson, 1986, Seismic migration — Theory and practice: Geophysical Press.

Trinchero, E., 2000, The fault shadow problem as an interpretation pitfall: The Leading Edge, **19**, 132–135.

Van der Veen, M., R. Spitzer, A. G. Green, and P. Wild, 2001, Design and application of a towed landstreamer for cost effective 2D and pseudo-3D shallow seismic data acquisition: Geophysics, **66**, 482–500.

Vermeer, G. J. O., 1998, 3-D symmetric sampling: Geophysics, **63**, 1629–1647.

Walsh, J. J., J. Watterson, A. Heath, P. A. Gillespie, and C. Childs, 1998, Assessment of the effects of sub-seismic faults on bulk permeabilities of reservoir properties, *in* M. P. Coward, T. S. Daltaban, and H. Johnson, eds., Structural geology in reservoir characterization: Geological Society [London] Special Publication 127, 99–114.

Wibberley, C. A. J., G. Yielding, and G. di Torro, 2008, Recent advances in the understanding of fault zone internal structure: A review, *in* C. A. J. Wibberley, W. Kurz, J. Imber, R. E. Holdsworth, and C. Collettini, eds., The internal structure of fault zones: Implications for mechanical and fluid-flow properties: Geological Society [London] Special Publication 299, 5–33.

Yilmaz, Ö., 2001, Seismic data analysis: Processing, inversion, and interpretation of seismic data: SEG Investigations in Geophysics Series No. 10.

Zhang, Y., B. E. Hobbs, A. Ord, A. Barnicoat, C. Zhao, J. L. Walshe, and G. Lin, 2003, The influence of faulting on host-rock permeability, fluid flow and ore genesis of gold deposits: A theoretical 2D numerical model: Journal of Geochemical Exploration, **78–79**, 279–284.

Ziegler, P. A., 1992, Euopean Cenozoic Rift System: Tectonophysics, **208**, 91–111.

Chapter 18

# High-resolution SH-wave Seismic Reflection for Characterization of Onshore Ground Conditions in the Trondheim Harbor, Central Norway

Ulrich Polom[1], Louise Hansen[2, 3], Guillaume Sauvin[2], Jean-Sébastien L'Heureux[2, 3], Isabelle Lecomte[2, 4], Charlotte M. Krawczyk[1], Maarten Vanneste[2, 5], and Oddvar Longva[2, 3]

## Abstract

The area around Trondheim Bay in central Norway is affected by landslides, both onshore and within the fjord, with several events documented to have occurred in the last century. As urban development, including land reclamation, is taking place in the harbor, assessing in situ soil conditions is paramount for infrastructure and operational safety. To obtain better insight into the harbor setting in terms of subsurface structures and potential coastal geohazards, a high-resolution multichannel SH-wave seismic-reflection land survey was carried out during summer 2008, which complements a dense network of high-resolution, single-channel marine seismic profiles over the deltaic sediments in the fjord. The SH-wave seismic reflection was chosen because the resulting interval shear-wave velocity provides a nearly direct proxy for in situ soil stiffness, a key geotechnical parameter. In total, 4.2 km of 2.5D SH-wave profiles was acquired along roads and parking places. Highly resolved images of the sediments were obtained, overlying the bedrock at a depth of about 150 m. The high quality of the data is ascribed to the quieter ambient noise conditions of the nighttime data collection and an efficient suppression of Love waves arising from the presence of a high-velocity layer at the surface. Five main stratigraphic units were identified based on reflection patterns and amplitudes. Distinct SH-wave reflection events enabled detailed S-wave velocity determination down to the bedrock. Subsequently, interval velocities were remapped into soil stiffness. Low S-wave velocities of about 100 m/s occurring in the upper 50 m of the fjord-deltaic sediment succession suggest low sediment stiffness (50 to 100 MPa) directly below the stiffer man-made fill that is 10 to 15 m thick. The results indicate that SH-wave seismic reflection is well suited for urban ground investigation.

## Introduction

A good understanding of soil conditions is a prerequisite for safe development of urban areas. This is particularly true for areas prone to landslides or other critical ground movements. Traditionally, geotechnical investigations are performed through invasive techniques such as drillings or in situ tests, after which soil properties are interpolated between the individual drill holes. Geophysical techniques, such as P-wave seismic reflection and electromagnetic and electric methods, are useful to outline the spatial variability of sediments and their history. However, they do not provide soil parameters for geotechnical applications. In addition, their resolution in the upper soil units is often relatively poor, and their use in urban areas might be hindered by various noise sources.

In contrast, a range of geophysical techniques directly provides S-wave velocity, an elastic parameter related to

[1]*Leibniz Institute for Applied Geophysics (LIAG), Hannover, Germany.*
[2]*International Center for Geohazards (ICG), Oslo, Norway.*
[3]*Geological Survey of Norway (NGU), Trondheim, Norway.*
[4]*NORSAR, Kjeller, Norway.*
[5]*Norwegian Geotechnical Institute (NGI), Oslo, Norway.*

soil stiffness and hence of importance for engineering purposes. S-wave velocity of the shallow subsurface can be obtained from the geometric dispersive character of surface waves (Rayleigh-Scholte, Love) propagating within the soil units, e.g., using multichannel analysis of surface waves (MASW) (Nazarian and Stokoe, 1984; Park et al., 1999; Stokoe and Santamarina, 2000; Long and Donohue, 2007). The maximum penetration depth of MASW is determined by the longest wavelength, which is limited mostly by the seismic source. This implies that for deep targets, heavy and cumbersome hardware is necessary (Rosenblad et al., 2007).

Therefore, surface-wave techniques typically are used for shallow site-specific geotechnical studies down to 30 m (high-frequency approach). Such limitations led to the development of shallow-target S-wave seismic-reflection systems providing high horizontal and vertical resolutions using high-frequency seismic sources (Ghose et al., 1996). The use of landstreamers for near-surface seismic-reflection acquisition is a developing technology with the ability to acquire a large amount of multifold data efficiently (van der Veen and Green, 1998; Inazaki, 1999; van der Veen

et al., 2001; Beilecke et al., 2006; Arsyad et al., 2007; Pugin et al., 2007; Polom et al., 2008a). The SH-wave reflection profiling works particularly well when a high-velocity layer overlies the soils, e.g., paved surface, because this essentially inhibits the generation of coupled Love surface waves (Inazaki, 2004; Pugin et al., 2004; Arsyad et al., 2007).

Trondheim — Norway's third-largest city — has seen substantial urban development along its coast over the last century. A large part of the harbor has extended into the fjord through land reclamation (Figure 1). During this period, the historical record documents at least three major landslides along the shoreline of the bay with severe consequences (Figure 1) (Emdal et al., 1996; L'Heureux et al., 2007; L'Heureux et al., 2010). During the last 15 years, further development of the harbor area has raised concerns about soil conditions and stability of the shoreline slopes that might pose a threat to society. Unlike in the Trondheim city center, SAR interferometry data indicate that active subsidence is taking place in the harbor area that extends into the fjord (Dehls, 2005), further emphasizing why this area in particular is under scrutiny. To shed light on the causes and possible consequences of landslides, detailed

**Figure 1.** Map showing the bathymetry of the bay of Trondheim, Norway, with location of the study area (harbor zone in a reclaimed area delimited on land by the 1875 coastline — blue line), location of the S-wave seismic-survey profiles (red lines), and historical shoreline landslides (black lines with year of event). Green triangles indicate location of CPTu piezocone tests. The black line offshore is the location of one reference subbottom profile.

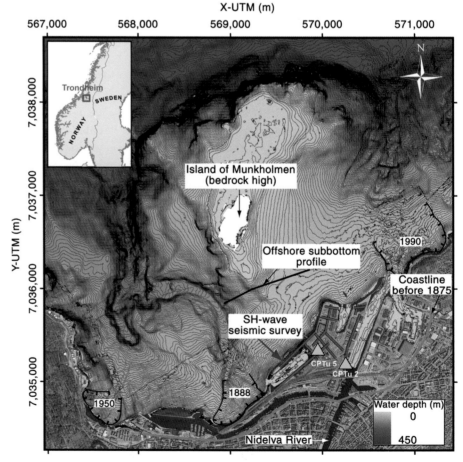

*Chapter 18: High-resolution SH-wave Seismic Reflection* **299**

investigations were carried out in the fjord, essentially through a combination of sediment core analysis and very high-resolution marine seismic profiling (L'Heureux, 2009).

Onshore investigations in Trondheim's harbor were limited to sparse geotechnical drillings, laboratory testing, and some early geophysical measurements (gravimetry). However, none of the previously acquired geophysical data provided relevant and detailed information for geotechnical purposes. Therefore, we started to consider acquisition of SH-wave seismic-reflection data to improve the onshore data coverage. During a preliminary site investigation in 2007, some 50-MHz ground-penetrating radar (GPR) profiles were acquired in an attempt to map the top part of the ground, i.e., the man-made fill and a few meters below. However, penetration was 5 to 10 m (allowing the bottom of the man-made fill to possibly be identified at some locations), and the data were contaminated heavily by air diffractions caused by buildings, trucks, and so on (Sauvin, 2009). Spectral analysis of Rayleigh waves from a small P-wave refraction seismic test (hammer blow and 24 channels) revealed low to very low S-wave velocities. Hence, the site conditions for high-resolution imaging and S-wave velocity mapping using the planned SH-wave seismic-reflection data were considered appropriate. Acoustic noise estimation in the harbor showed that tide noise could be filtered out easily. However, seismic data could be acquired only without noise from the nearby train station, boats, and trucks/cars in the harbor area.

The SH-wave seismic-reflection survey was carried out in 2008, using an SH-wave vibrator and a landstreamer system developed at the Leibnitz Institute for Applied Geophysics (LIAG) in Germany. The goals of the SH-wave seismic investigation were to obtain high-resolution information on the internal structure of the sediments, to map the depth to bedrock, and to determine elastic properties (i.e., S-wave velocity) of the sediments as a key proxy for their stiffness. Our objective is to present the acquisition technology and seismic poststack results of this SH-wave seismic-reflection survey, including 2D depth sections and S-wave velocity profiles down to the bedrock. We illustrate the ability and importance of this type of seismic surveying method in urban areas for the integration of geologic and geotechnical site characterization.

## Geologic Setting

Bedrock around Trondheim is dominated by westward-dipping low-grade metamorphic, volcanic rocks of Precambrian-Silurian age (Wolff, 1979). It is exposed locally on land and at the seabed. During repeated glaciations, bedrock was eroded along weakness zones, shaping the landscape of valleys and fjords, including the present valley at Trondheim.

In the bay of Trondheim, core data combined with high-resolution seismic profiles show that a 125-m-thick succession of fjord-deltaic deposits on top of marine and glaciomarine clays covers the bedrock (Figure 2). The glaciomarine and marine sediments were deposited during and after the end of the last glaciation (Reite, 1995; Rise et al., 2006; Lyså et al., 2007). After deglaciation, the area was subject to glacioisostatic rebound, causing rapid fall of relative sea level from a maximum of 175 m above sea level at Trondheim. As a consequence, glaciomarine deposits became exposed locally above sea level, and the river outlet in the fjord moved continuously northward in phase with Holocene delta progradation (Reite, 1995; L'Heureux et al., 2009). The shallow near-shore areas were reclaimed from the sea during urbanization in the last century (see Figure 1). Landslide activity in Trondheim Bay is

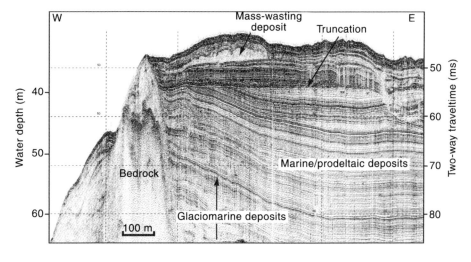

**Figure 2.** High-resolution offshore P-wave seismic section (subbottom profiling; see location in Figure 1). Modified from L'Heureux et al., 2010. Used by permission.

witnessed by distinct escarpments of both prehistoric and known historical age (Figure 1). Near-shore landsliding started with translational failure along specific weak layers (L'Heureux, 2009; L'Heureux et al., 2010).

# Method

Data acquisition was carried out using the vibroseis method (Crawford et al., 1960), which is common practice in onshore hydrocarbon exploration in which explosive charges cannot be used. The advantages of this method are (1) control of the resulting source signal in terms of frequency bandwidth and energy, (2) control of the initial energy applied to the subsurface to prevent damage of the

**Figure 3.** MHV4S hydraulic S-wave vibrator in operation. (a) The shaker system is included in a light-weight baseplate of tensile-strength aluminum alloy mounted below a special truck frame. The hold-down force to support ground coupling by the truck weight was balanced to the best fit of 3800 kg at the center of the shaker unit by optimizing the truck frame design. Photograph courtesy of L. Hansen. (b) S-wave landstreamer during data acquisition on a detached mode in the Trondheim Harbor area. Arrows indicate the orientation of the SH-wave particle motion of source (white arrow closest to the land streamer) and receivers (black arrow near bottom of image). Photograph courtesy of U. Polom.

paved surface or buried structures, and (3) excellent repeatability of the seismic source signal. However, this method requires a sophisticated control of the vibratory source and the use of a recording system that can perform vibroseis-correlation processing for field quality control. To lower the initial energy applied, the total amount of source signal energy is stretched in time. Therefore, the recording time for each seismic record is several times longer than required for impulsive seismic sources.

The S-wave source was combined with an S-wave receiver-array system (landstreamer) designed for operation on paved or compacted surfaces. Extension of the application toward SH-waves minimizes body-wave-type conversions between P- and S-waves. Furthermore, this wave type is decoupled widely from Rayleigh surface waves (ground roll). However, operating in SH-wave mode introduces another surface-wave type, i.e., Love waves. In contrast to Rayleigh waves, Love-wave propagation requires a low-velocity layer at the surface above a half-space of higher velocity. In such cases, Love waves introduce strong distortions of the reflected waves along the receiver spread. This complicates S-wave reflection analysis because Love-wave velocities fall in a similar range as body S-wave velocities in the shallow subsurface. In contrast, the propagation of Love waves is hindered by a high-velocity surface layer above a half-space of lower velocity. Typically, this is found for, e.g., all-weather roads constructed above soft soil, even if the surface is not paved. Such a high-velocity layer is present in the area of Trondheim Harbor, i.e., the man-made fills on fjord-deltaic sediments. Therefore, the area is well suited for SH-wave investigations using a modified landstreamer system.

## The S-wave source

The S-wave source consists of a hydraulically driven shaker unit installed below the modified frame of a small truck originally designed for urban application (Figure 3). The seismic-source concept was developed by LIAG (Hannover), Kiefer GmbH (Dorfen), and Prakla-Bohrtechnik GmbH (Peine). The source system was designed for shallow high-resolution applications and to meet the technical requirements of the European Union (EU) concerning environmental safety and traffic requirements. Beyond the integration of a S-wave vibrator unit, the buggy is designed conceptually for integration of a vertical vibrator as well. The most important technical data on the source system are summarized in Table 1.

Mechanical vibrations of the shaker unit are decoupled from the truck in three dimensions by an array of eight air-bag devices. The typical shaking orientation is

**Table 1.** Technical data of the MHV4S shear-wave vibrator.

| Buggy unit | Oscillation generator |
|---|---|
| Two-stage hydraulic four-wheel drive | Full hydraulic drive |
| Four-cylinder diesel injection (IVECO) | Hydraulic suspended reaction mass |
| 2800 ccm, 92 kW at 3600 rpm, 290-Nm torque at 1800 rpm | Hydraulic rotation and lock device |
| Meets Euro 3 emission requirements, additional cleaning by particle filter | Reaction mass weight 240 kg |
| Climbing performance 45% | Oscillation house weight 160 kg, cast from special aluminium alloy |
| Top speed 62 km/h | Baseplate diameter 1200 mm, quick-change ground-coupling pads |
| Tandem pump 2×70 l/min (drive), 117 l/min at Pmax 280 bar (shaking) | Theoretical peak force 30 kN |
| Full biodegradable hydraulic fluid | Frequency range 16 to 300 Hz |
| Offroad tires 315/55-16 | 3D vibrational decoupling |
| Rollover protection frame | Servo valve: Mannesmann Rexroth |
| Length 4.5 m, width 1.5 m, height 2.12 m | Drip-free quick-change clutch |
| | Oscillator control: Pelton VibPro |

perpendicular to the frame of the truck, to generate SH-waves. A hydraulically driven rotation mechanism allows changing to vertically polarized (SV) mode or other desired horizontal-shaking polarization if required. This adjustment can be made by rotating the shaker mass within the inner frame of the baseplate. After rotation, a hydraulic coupling mechanism leads to a force-locked contact of the shaking mass and the baseplate unit. This construction principle enables a low center of gravity of the whole unit to reduce tilt moving during shaking operation. Ground coupling is achieved by an easily exchangeable plate below the shaker casing, attached with small spikes, rubber pads, or similar friction-enhancement material, depending on specific surface conditions. For flat asphalt surfaces in the area of Trondheim Harbor, a rubber pad was preferred to prevent damage to the surface.

A Pelton VibPro vibrator-control system was used for radio-controlled operation, sweep-signal generation, hydraulic-shaker control, and phase locking. To obtain optimal results and signal-to-noise ratio (S/N), numerous adjustments of the control system were carried out prior to the surveying. Ultimately, two 10-s sweep signals were used at each source location, differing only by a $180°$ phase shift (i.e., opposite polarity) to enable the plus-minus operation and subtractive-stacking method for the

enhancement of S-waves (Polom, 2005a, 2005b). Frequencies of the upsweep ranged from 25 to 100 Hz, with 100-ms tapers at either end, after several test sequences with the intention to record a distinct response from the bedrock and to minimize airwave reflections from surrounding buildings. The pilot-sweep signal was recorded in geophone channel 121.

## The receiver system

The landstreamer consists of 120 single geophone units at 1-m intervals (split into five subsets with 24 channels). The ground coupling of the landstreamer geophones is achieved solely by gravitation force using a static, stable three-point contact, attached by tough metal vats for road application or plastic vats if damage to the road needs to be prevented. The streamer unit was reeled on a special trailer attached with an electrically driven reeling unit for a fast roll-on, roll-off application. The horizontal geophones were oriented perpendicular to the profiling direction (pure SH mode). During data acquisition, the landstreamer was attached to a Geometrics Geode system included in the tracking and recording car. The sampling rate was 1 kHz.

**Figure 4.** Profile location (white lines). (a) Side view showing the coastline of Trondheim and the fjord-reclaimed harbor. Photograph courtesy of Trondheim Port Authority. Used by permission. (b) View above the site with indication of the profile numbers (white text). Black triangles indicate the location of two CPTu piezocone tests, and the gray circle represents a reference position along profile 4 to be used later. Orthophoto source © Norwegian Mapping Authority. Used by permission.

**Table 2.** Acquisition parameters of SH-wave profiles after initial parameter tests.

| | |
|---|---|
| Period | 19–30 June 2008, 10 nights of data acquisition |
| Instrument | Geometrics Geode |
| Channels per record | 120 + 1 auxiliary |
| Seismic source | LIAG MHV4S shear-wave vibrator, adjusted to 60% peak force for friction contact by rubber pad |
| Sweep type | 25- to 100-Hz linear, 10 s, 10-ms taper at front and end |
| Recording | 14 s, 4 s after correlation |
| Sampling interval | 1 ms |
| Recording filter | Out |
| Spread type | 2D variable-split spread, SH-SH configuration, roll-on operation using a 40-m streamer shift interval |
| Geophone type | SM6 H (10 Hz), single units attached to landstreamer unit |
| Receiver interval | 1 m |
| Source interval | 4 m |
| Vertical stack | Twofold [+Y]–[−Y] alternated vibrations, stored separately |

## Seismic field operation

To minimize ambient noise level and for crew safety, the seismic experiment was conducted only at night (11:30 P.M. to 5:00 A.M) during 10 nights, 19 June to 30 June 2008. Eleven 2D profiles were acquired along roads and car parks with a total length of 4.2 km (Figure 4). Local companies and the public authorities granted permission for fieldwork and periodic road closures, either for the whole harbor area or in the immediate vicinity of the survey. Detailed planning led to efficient acquisition with little downtime, e.g., for railway traffic or bad weather (one night went without acquisition because of strong wind and heavy rain).

The general seismic acquisition parameters used are listed in Table 2. Seismic data were recorded and stored uncorrelated to enable editing and fine-tuning prior to full-scale processing. The operator carried out quality control during surveying using the display-correlation adjustment of the Geode system. In addition, seismic data-quality control of each production night was done during the day using the GEDCO-VISTA processing system to validate or modify recording parameters for optimal seismic-reflection processing.

Independent handling of source and receiver units allows much more variability in the spread layout, including the undershooting of problematic zones (e.g., buildings, gateways). This is in contrast to the most often used fixed configuration, in which the landstreamer is attached to the source (e.g., Pugin, 2007). In general, the spread configuration shown in Figure 5 was used. The landstreamer move-up was done only after 40 m of source move-up by an ingoing and outgoing ramp of 80 m each. Thereby, the acquisition scheme results in a zigzag-formed CMP coverage after data processing.

The positions of every fortieth receiver and offset shot were marked on the pavement during the survey. Unfortunately, GPS positioning did not work in the area, possibly because of electromagnetic noise. Therefore, geodetic surveying was carried out during the day, using a terrestrial-based

Trimble S6 DR300+ system referenced to three topographic points (TP).

## Processing

The general processing scheme of SH-SH reflections is similar to conventional processing of onshore P-P seismic-reflection data (Kurahashi and Inazaki, 2006; Pugin et al., 2006). Examples of on-site preprocessed single shot series from profile 4 (location in Figure 4) are shown in Figure 6. The source-receiver configuration used for the SH-wave

survey in the harbor yields little conversion to P- and SV-waves and surface waves. The inverse velocity contrast from the surface pavement and man-made fills on top of softer soils in the Trondheim Harbor area also suppresses the generation of Love waves and refracted SH-waves. However, Figure 6d shows some refractionlike waves (red line; velocity of about 1000 m/s) and surface waves (green line; velocity of about 100 m/s). The former could correspond to S-to-P converted waves, especially if heterogeneities are below the source; the latter are probably weak channeled Love waves traveling in the low-velocity zone

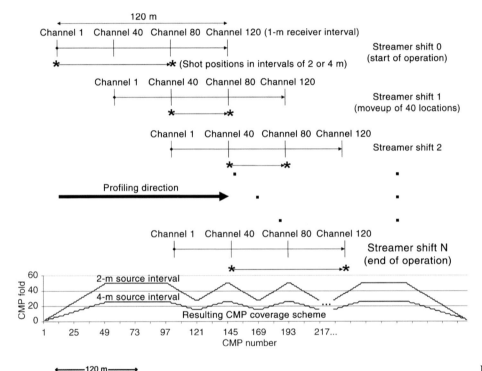

**Figure 5.** General source-to-receiver spread scheme used for operation with decoupled source and receiver units. This spread enables high flexibility in dense building and traffic areas.

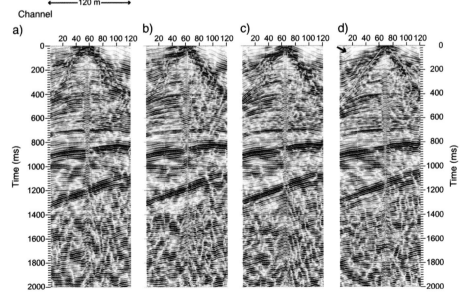

**Figure 6.** Shot-record examples of profile 4 at the location given in Figure 2b. The preprocessing applied on site consisted of vibroseis correlation, stacking of the opposite-polarity records, 300-ms AGC, and no filters.

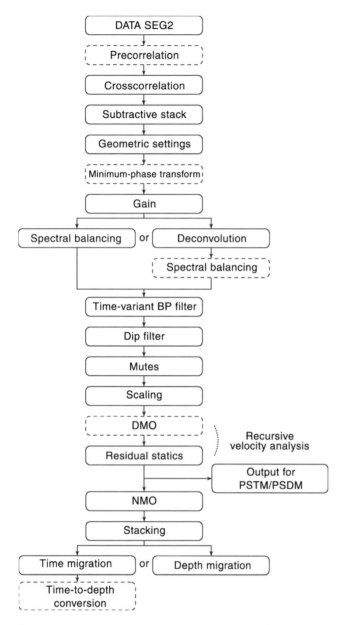

**Figure 7.** Actual processing sequence. Solid-border boxes are attached to processing applied systematically to all profiles. Dashed-border boxes represent optional processing applied to some profiles only.

below the pavement. Source airwave attenuation, commonly used in P-wave seismic-processing sequences to reduce source noise, was found to be unnecessary because SH-SH source-and-receiver configuration is affected less by direct airwaves than are common P-P source-and-receiver configurations. Some coherent noise remaining in the data (e.g., black arrow in Figure 6), occurring before and after the first arrivals, is explained by ringing because of crosscorrelation with the sweep signal and is emphasized

further by amplitude scaling. This noise and the refraction-like arrivals are eliminated later by muting. The Love channel waves, although weak, are removed efficiently by *f-k* filters.

During data acquisition, a preliminary processing sequence using only raw geometry parameters was developed to perform on-site quality control and obtain an initial idea of subsurface structure and S-wave velocity distribution. This sequence allows for better profiling control in the area. Next, the preliminary processing sequence was extended in a first run conducted after the survey, using crooked-line geometry and an amplitude-preserving processing sequence, and was applied to all profiles, followed by a first combined interpretation (Polom et al., 2008b, 2009a; Polom et al., 2009b). The high-quality, well-resolved structure observed on the processed data led to further enhanced imaging of the shallow parts, i.e., the actual targets for geotechnical applications (L'Heureux, 2009). Then the processing sequence also was reviewed and extended, and parameter settings were fine-tuned (Figure 7).

In addition, reviewing and testing of each individual processing parameter were carried out carefully for each profile individually, to obtain optimal results in amplitude-preserving processing and to derive the best possible subsurface S-wave velocity information. The most important aspects of the processing are outlined below. For further details and profiles illustrating the results of all processing steps performed in GEDCO-VISTA, see Sauvin (2009). Profile 4 of Sauvin (2009) is used in the following to illustrate the major results.

## Vibroseis signal contraction

Because the data were stored as uncorrelated time series, precorrelation data-quality analyses were performed. Time-frequency domain analysis (Gabor or Fourier transform) shows generally good coherency of transmitted signals and low noise levels with few spike and burst distortions. The amplitudes of harmonics, caused by nonideal vibrator-to-ground coupling, are also low relative to the primary signal, indicating that no corrections were necessary prior to crosscorrelation. Next, signal contraction was performed by common vibroseis correlation using the pilot sweep. For some shots from profiles 1 and 2, the pilot-sweep signal was distorted because of a temporary malfunctioning of the pilot signal generator in the recording car. These pilot sweeps then were reconstructed from the opposite polarity recorded at the same location. Ultimately, the trace length was 4 s after correlation (10-s sweep length, 4-s listening period).

## Subtractive stack

After compression of the vibroseis signal, stacking of two records was applied by summing correlated records with opposite sweep-signal polarities. This process enhances the S/N of the S-waves and suppresses P-wave energy inherently present. For further details, we refer to Edelman and Helbig (1983), Tatham and McCormack (1991), Hasbrouck (1993), and Polom (2005a, 2005b).

## Geometry

Common-midpoint (CMP) geometry binning was carried out using crooked-line tracking because of the local requirements of nonlinear profiling, which results in some midpoint scattering.

## Amplitude corrections and balancing

Geometric spreading and preserving of amplitudes were carried out by an analytical spherical-divergence correction of $t^2$ followed by common-trace energy normalization. An offset-dependent noise removal (radial transform) also was tested but showed only marginal effects.

## Deconvolution

Subsequent to gain correction, a surface-consistent predictive deconvolution with an operator length varying between 100 and 180 ms, a prediction lag of 5 ms, and a prewhitening of 0.1% was applied. To fine-tune the spectral content, this was followed by spectral balancing using frequency windows of 30 or 40 Hz and a slope of 10 Hz. Surface-consistent spiking deconvolution yielded generally poorer results. Hence, only spectral balancing was applied. Tests of minimum-phase transformations prior to deconvolution led to insufficient results and will be readdressed in later investigations.

## Filtering and muting

The signal-to-noise ratio was improved further by time-variant bandpass filtering (zero-phase Ormsby filter) with parameters determined for each individual profile, after screening the results of numerous filter settings using normal moveout (NMO) and brute-stack data. Typically, different filters were applied to four to seven windows, with 50-ms overlap, for the different profiles. The bandwidth changed from 38-42-90-100 Hz for the uppermost window to approximately 20-24-55-60 Hz for the last window. Subsequently, a unique dip filter was designed for each profile in the frequency-wavenumber domain, with the aim of further suppressing Love-wave energy (see discussion about Figure 6) which might mask the shallower horizons (our target). Finally, position- and offset-dependent top and bottom mutes were designed at several positions and were interpolated. Muting was necessary to eliminate converted P-waves and reverberations at the top of the recorded energy and to suppress source-generated noise close to the source (near field).

## Velocity analyses, NMO, and DMO

Velocity analyses were performed recursively with dip moveout (DMO) and residual-statics corrections, giving rise to a very consistent S-wave velocity field from about 5 to 10 m in depth (bottom of man-made fill) and down to the bedrock. The S-wave velocity was not properly estimated underneath the bedrock because there were no laterally coherent reflections to pick (possibly offline reflections) and an insufficient maximum offset. Semblance velocity analysis was performed for every fiftieth or eightieth CMPs and more regularly in the presence of pronounced structures for which lateral velocity variations were anticipated. At least three velocity analyses were performed for each line, i.e., one after dip moveout correction and one after each of the two residual-static corrections. The procedure works as follows. An initial semblance velocity analysis that best flattens the horizons is used for NMO correction. Then the data are sorted into common-offset gathers for DMO correction, using a DMO log stretch. After inverse NMO using the same velocity field, a new semblance velocity analysis is performed. Then, using the best-fitting velocity field, NMO corrections are applied, followed by residual statics computing and application. After inverse NMO, a new velocity field is determined, and the procedure with residual statics is repeated. Finally, the data were stacked. An example of the brute stack is presented in Figure 8.

Despite the fact that the seismic records do not show steeply dipping events or conflicting dips with different stacking velocities (with the exception of the bedrock), the applied DMO corrections clearly improved the velocity model and hence the quality of the seismic data. This is most evident for the undulating horizon corresponding to the top of the bedrock and to a lesser degree for reflections between different sedimentary units. The effect of the second residual-static corrections was modest compared to the first run, but nevertheless, it improved the final velocity field. Overall, the iterative velocity-analysis process, including DMO and residual statics, gave the most significant improvement compared with preliminary processing.

**Figure 8.** Profile 4: Brute-stack section with the reference position given in Figure 2b noted in blue, i.e., corresponding to the shot profiles of Figure 6.

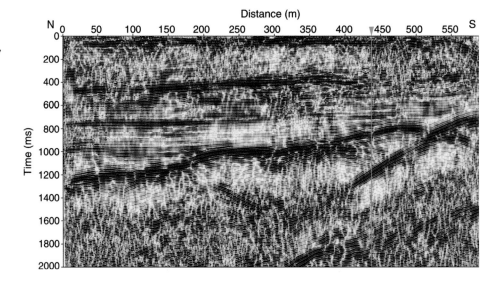

**Figure 9.** Profile 4: Final poststack time-migrated section (45° to 65° FD).

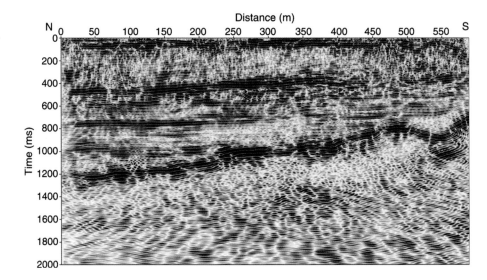

## Poststack migration

For further improvement, essential for interpretation, the poststack data were migrated in time and depth. Different methods were tested for poststack time migration, including finite-difference (FD), Kirchhoff, and frequency-wavenumber (*f-k*) migration. Finite-difference migration (filtered 45° to 65° algorithm) using an rms velocity field yielded the best results. Prior to application, the velocity field was smoothed slightly to prevent migration artifacts. Figures 9 and 10 show an example of the time-migrated section of profile 4, with and without the interval velocity field in the background, respectively. Similarly, several algorithms were tested for poststack depth migration using Seismic Unix instead of VISTA and including FD, Fourier-FD, Gazdag phase shift plus interpolation (PSPI; Gazdag and Sguazerro, 1984), and split-step Fourier mi-

gration (Stoffa et al., 1990). The latter two migration schemes seem to give the best results (visual inspection only). As for time migration, slight velocity-field smoothing was required. The PSPI depth-migrated section of profile 4 is given in Figure 11 with superimposed interpretation as explained below.

## Results

In total, 4.2 km of SH-wave seismic reflection profiles was gathered during 10 nights of time-restricted data acquisition, including experimental testing. The survey area proved to be highly suited for the methods used, and high-quality recordings were obtained. All profiles show a high-resolution image of the fjord-fill sediments and bedrock down to a depth of at least 200 m (vertical res-

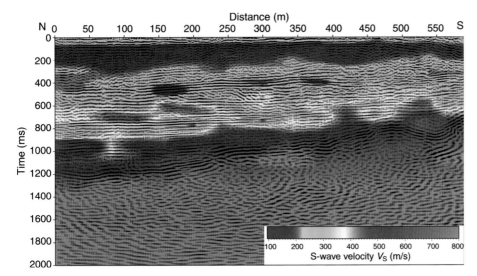

**Figure 10.** Profile 4: Final poststack time-migrated section (45° to 65° FD) with superimposed S-wave interval velocity ($V_S$).

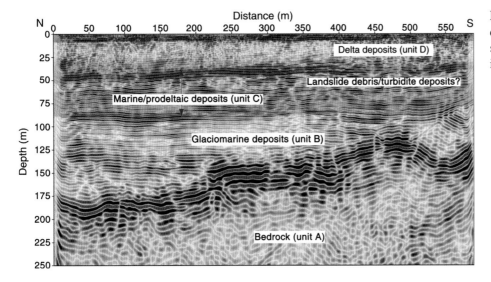

**Figure 11.** Profile 4: Final poststack depth-migrated section (Gazdag phase shift plus interpolation) with geologic interpretation.

olution of 1 to 5 m). In addition, MASW acquisition was tested briefly during the 2008 campaign, but along only a part of profile 2 (Figure 2), with 48 vertical geophones every 1 m (streamer) and a distance of 12 m to the source (5-kg sledgehammer). The resulting S-wave velocity profile (Sauvin, 2009) reached only 10 m in depth at that site. This prevents a proper comparison with the velocity profiles obtained with seismic reflection (see below), which are not sufficiently constrained so close to the surface.

## Geologic interpretation

The high resolution of the SH-wave seismic data allows for a detailed geologic interpretation in depth (Figure 11). Five main stratigraphic units are identified by clear reflection patterns in the seismic-reflection sections. The units make up a typical fjord-fill succession above bedrock and

correspond to the stratigraphic succession documented in the fjord (e.g., Figure 2) and general models of fjord-fill successions (Corner, 2006). Detailed features can be observed within some units. All profiles have been interpreted, correlating the picked events from one profile to the other at the crossing points in the time-migrated domain (Figure 12).

Unit A, interpreted as bedrock, forms the base of the seismic sections and contains irregular high-amplitude reflections and an undulating upper boundary at depths of 90 to 160 m. Structures within bedrock are observed, but further investigation is needed for verification. Unit B, interpreted as mainly glaciomarine deposits, is 20 to 70 m thick and is characterized by continuous subparallel reflections draping and onlapping the bedrock (unit A). The onlapping infill pattern reflects an influence from sediment gravity flows during initial sedimentation after retreat of

**Figure 12.** Three-dimensional view of interpreted poststack time-migrated SH-wave seismic sections. Five main stratigraphic units are indicated by letters A through E.

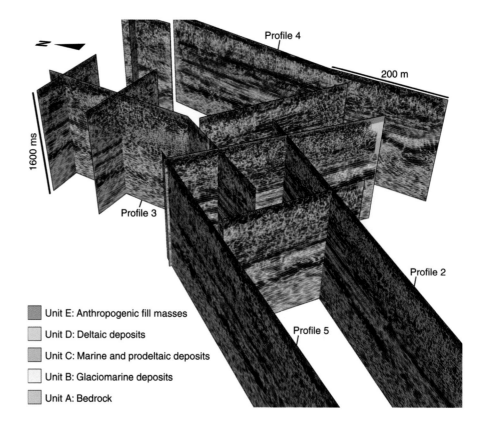

Unit E: Anthropogenic fill masses
Unit D: Deltaic deposits
Unit C: Marine and prodeltaic deposits
Unit B: Glaciomarine deposits
Unit A: Bedrock

the glacier from the fjord basin. Distinct high-amplitude reflections at the top might reflect a change in the depositional environment at the transition to the Holocene. However, more data are needed for confirmation.

Unit C, interpreted as consisting of fjord marine and prodeltaic deposits, conformably overlies unit B and is less than 50 m thick. It is characterized by partially continuous reflections with local lenses displaying a more irregular reflection pattern. Unit D, interpreted as deltaic deposits, displays wavy to gently fjordward-dipping continuous-to-discontinuous reflections and is 10 to 50 m thick. The fjordward-dipping reflections are interpreted as delta foresets (Figure 11). Topsets also are present. Unit E displays continuous to irregular reflections and is less than 15 m thick. It is interpreted as anthropogenic fill, which lies on the previous tidal flat/submarine part of the Nidelva River delta (Figure 11).

## Estimation of S-wave velocity and geotechnical properties

Because of the distinct and continuous reflection events, the S-wave interval velocity could be determined with good accuracy down to bedrock in the survey area. Interval velocities were calculated by the smoothed-gradient method to prevent artifacts typically introduced by

the Dix algorithm. Estimated velocity errors by the rms velocity analysis were 10% to 20% on average but could increase to as much as 30% because of the strong 3D surface topography of the bedrock. This was reduced again by the DMO and the subsequent fitting of the 2.5D profile network. Velocity error after conversion to interval velocities might be somewhat more for some intervals. In general, S-wave velocity ($V_S$) increases from 100 m/s in top deltaic deposits (unit D, Figure 12) to 600 m/s at the bedrock interface (unit A, Figure 12). An average velocity of 300 m/s is recorded in the anthropogenic fills.

The measurement of dynamic or small-strain shear modulus $G_{max}$ of a soil is important for a range of geotechnical applications (e.g., Hight and Leroueil, 2003). According to elastic theory, $G_{max}$ can be calculated from the S-wave velocity ($V_S$) using

$$G_{max} = \rho V_S^2, \qquad (1)$$

where $G_{max}$ is the shear modulus (in Pa), $V_S$ is the shear-wave velocity (in m/s), and $\rho$ is the density of the soil (in kg/m$^3$). Numerous density logs exist in the area of interest, offshore and onshore, but they are very limited in depth (a few meters), and no attempts have been made so far to use them to generate spatially varying density models. Soil density measured in several samples from

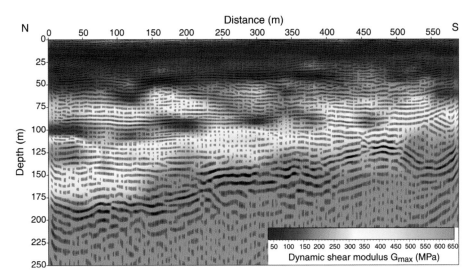

**Figure 13.** Profile 4: Final poststack depth-migrated section (as in Figure 11) with superimposed dynamic shear modulus ($G_{max}$), which was derived from the interval velocity using a constant density of 1.9 g/cm³.

cores range from 1.7 to 2.4 kg/m³, but an average constant density of 1.9 kg/m³ was used for calculation of $G_{max}$, considering that the major variability of that parameter comes from $V_S^2$. The result for profile 4 is given in Figure 13 as a function of depth. The delta deposits show values of 50 to 100 MPa down to 50-m depth, whereas the marine/prodeltaic deposits below exhibit higher values (200 to 300 MPa). Approaching bedrock depths gives a shear modulus of 350 MPa and higher (Figure 13).

Shear modulus $G_{max}$ also can be derived empirically from cone-penetration-test (CPT) data in silty, sandy soils using the measured cone-tip resistance ($q_c$) and the following empirically derived formula (Rix and Stokoe, 1992):

$$G_{max} = 1634 \times q_c \left(\frac{q_c}{\sqrt{\sigma'_{v0}}}\right)^{0.75}, \qquad (2)$$

where $\sigma'_{v0}$ is the vertical effective stress. A comparison between $G_{max}$ values determined from interval S-wave velocities using the SH-wave profiling and empirically derived values from two CPTu tests is shown in Figure 14. As expected, $G_{max}$ derived from CPTu data is much more detailed. With the exception of the anthropogenic fill (unit E), the match between the two data sets is good. The differences likely are caused by several factors, including the overconsolidation ratio (OCR), strain levels, fine contents, particle angularity, layering, and soil plasticity.

## Discussion

The SH-wave data presented above allow for a detailed geologic interpretation of sediments down to the bedrock. The vertical resolution is estimated as being as much as

10 times greater than what would be obtained from on-shore P-wave seismic reflection. These data yield information on S-wave velocities and therefore soil stiffness ($G_{max}$), which is important for geotechnical applications. However, the resolution of the resulting S-wave velocities is still too low for detection of very thin, decimeter-size weak layers, detected in cores and in offshore data. Therefore, additional drillings and borehole geophysics, including high-resolution VSP, are required for detailed site investigations and calibration of the seismic data. Nevertheless, the results from 2.5D mapping of S-wave velocity reveals where $G_{max}$ values appear to be low and helps to correlate low-velocity layers to the geology.

The SH-wave acquisition system with a landstreamer is highly flexible and can be applied efficiently in urban areas on paved surfaces. Therefore, the method is excellent for spatial mapping of soil conditions near the shoreline and on land. Ideally, it would have been interesting (1) to use 3C source/geophones to control wave-conversion behavior and to improve splitting of P-, SV-, and SH-wavefields and (2) to acquire P-wave data as well. For the first point, we refer to Pugin et al. (2009) because the equipment is novel technology currently at the research level. For the second point, expected P-wave velocities at target depths probably would not differ significantly from water velocity, i.e., about 1500 m/s. This would imply a much lower vertical resolution (4 to 5 m) compared to the SH data (1 m) and would not provide the needed geotechnical parameters, but it would help to discriminate fluid content and lithology and detect gas accumulations because of the sensitivity of P-waves to pore-space content.

In terms of processing, further data-quality enhancement will be done by vibroseis deconvolution and minimum-phase transformations prior to surface-consistent predictive

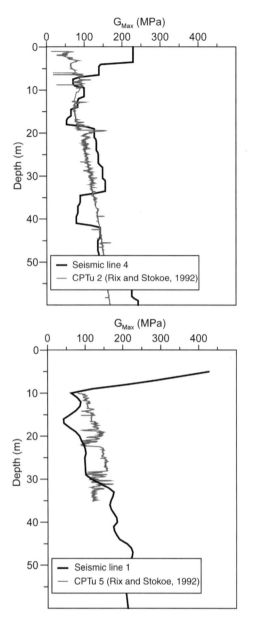

**Figure 14.** Comparison of $G_{max}$ estimated from S-wave seismic and derived from CPTu data. (see Figures 1 and 2b for the locations of seismic profile 4, CPTu 2, and CPTu 5).

deconvolution. Prestack depth migration (PSDM) is the next step for further quality control of the interval velocity field. Residual moveout of common-image-point gathers (CIG) can be used to improve the velocity model and thus possibly refine the seismic imaging. Seismic modeling also should be used to better understand the complex wave propagation at very shallow depths (early arrivals in Figure 6 and Love channel waves) and to control the velocity model and seismic interpretation. Finally, a preserved-amplitude approach of PSDM could allow for extraction of more information from seismic data via various attributes, e.g., amplitudes.

The focus of future investigations will be to refine the velocity model using additional CPTu or preferably other geophysical methods, e.g., crosshole tests or seismic CPT (Ghose, 2008). In addition, at least one borehole down to bedrock is necessary for ground truth. Another focus is the correlation of the onshore SH-wave data with the P-wave marine data, including P-to-S converted waves acquired by Statoil and the Norwegian Geotechnical Institute (NGI) a few years ago. The latter data set indeed can provide a link between P- and S-wave velocity models offshore. Finally, an ideal case would be to carry out SH-wave seismic reflection offshore as well, but marine S-wave technology is not sufficiently mature, although potentially interesting results have been obtained in recent years (Westerdahl, 2004; Vanneste et al., 2007).

## Conclusion

Full SH-wave seismic acquisition using an S-wave vibrator and a landstreamer system equipped with SH-geophones is highly suited for ground investigations in urban areas with a paved surface. The handling of this system in a flexible manner enables advanced field techniques and allows for sophisticated processing tools that make use of distinct wavefields. At this point, future work also should encompass multicomponent acquisition to provide the full wavefield. In addition, the 1-m vertical resolution of the seismic sections achieved here allowed us for the first time to reasonably combine land seismics with marine data sets. Furthermore, the high-resolution SH-wave data yield detailed geologic information that can be integrated into a complete 3D ground characterization. Nonetheless, a drill hole for calibration or additional VSP data should constrain the subsurface model further. Determination of small-scale structure and the translation of seismic velocities to soil stiffness bear great potential for geotechnical applications. The dynamic shear modules obtained in this study are comparable to those obtained using conventional geotechnical engineering techniques. Especially in urban areas, this noninvasive technique is one of the few that provides important types of information delineating the spatial extent of low-stiffness zones.

## Acknowledgments

The authors acknowledge the Trondheim Port Authority and local municipalities and companies for their permission and support during the field campaign. The high-resolution SH-wave seismic-reflection survey was sponsored kindly by Statoil. The authors are most thankful to the German acquisition crew (Stefan Cramm, Siegfried Grueneberg, and Eckhardt Grossmann) and to the students involved in field testing (Karl-Magnus Nielsen, University of Oslo, for preliminary site investigations in fall 2007 and to Eugene Morgan, TUFTS University, for MASW acquisition). Trafikkvakta and Det FINNs Trafikkhjelp AS helped with road closures. The Norwegian Public Road Administration and Rambøll gave permission to use CPTu data. The authors thank André Pugin, Robert A. Williams, and one anonymous reviewer for their great help in improving the manuscript. An academic license (ICG) of GEDCO VISTA software was used for most processing, and Seismic Unix (Colorado School of Mines) was used for depth-migration tests. This is ICG contribution no. 298.

# References

Arsyad, I., U. Polom, and S. T. Wiyono, 2007, Shallow reflection seismic shear-wave velocity analysis on paved soils using a land streamer unit: 69th Conference and Technical Exhibition, EAGE, Extended Abstracts, P19.

Beilecke, T., U. Polom, and S. Hoffmann, 2006, Efficient high resolution subsurface shear wave reflection imaging in sealed urban environments: 68th Conference and Technical Exhibition, EAGE, Extended Abstracts, A033.

Corner, G. D., 2006, A transgressive-regressive model of fjord-valley fill: Stratigraphy, facies and depositional controls, *in* R. W. Dalrymple, D. Leckie, and R. Tillman, eds., Incised valleys in time and space: Society for Sedimentary Geology (SEPM) Special Publication 85, 161–178.

Crawford, J. M., W. Doty, and M. R. Lee, 1960, Continuous signal seismograph: Geophysics, **25**, 95–105.

Dehls, J. F., 2005, Subsidence in Trondheim, 1992–2003: Results from PSInSAR analysis: Geological Survey of Norway, Report 2005.082, 12.

Edelmann, H. A. K., and K. Helbig, 1983, Seismisches Aufschlußverfahren, Offenlegungsschrift: German Patent DE 3212357 A1 (in German).

Emdal, A., N. Janbu, and K. Sand, 1996, The shoreline slide at Lade, *in* K. Senneset, ed., Landslides: Proceedings of the 7th International Symposium on Landslides, v.1: Balkema, 533–538.

Gazdag, J., and P. Sguazzero, 1984, Migration of seismic data by phase shift plus interpolation: Geophysics, **49**, 124–131.

Ghose, R. J., 2008, Quantitative integration of seismic and CPT for soil property estimation — Sensitivity analyses on field and lab data: 70th Conference and Exhibition, EAGE, Extended Abstracts, D025.

Ghose, R., J. Brouwer, and V. Nijhof, 1996, A portable S-wave vibrator for high-resolution imaging of the shallow subsurface: 59th Conference and Exhibition, EAGE, Extended Abstracts, MO37.

Hasbrouck, J. C., 1993, Final report, TTP AL921102: An integrated geophysics program for non-intrusive characterization of mixed-waste landfill sites: U. S. Department of Energy, GJPO-GP-7.

Hight, D. W., and S. Leroueil, 2003, Characterisation of soils for engineering purposes, *in* T. S. Tan, K. K. Phoon, D. W. Hight, and S. Leroueil, eds., Characterisation and engineering properties of natural soils, v. 2: Taylor and Francis, 255–362.

Inazaki, T., 1999, High-resolution S-wave reflection survey in urban areas using a woven belt type land streamer: 68th Conference and Exhibition, EAGE, Extended Abstracts, A016.

Inazaki, T., 2004, High resolution reflection surveying at paved areas using S-wave type land streamer: Exploration Geophysics, **35**, 1–6.

Kurahashi, T., and T. Inazaki, 2006, A seismic reflection survey using S-wave vibrator for an active fault: Civil Engineering Journal, **91**, 52–57.

L'Heureux, J.-S., 2009, A multidisciplinary study of shoreline landslides: From geological development to geohazard assessment in the bay of Trondheim, mid-Norway: Ph.D. thesis, Norwegian University of Science and Technology.

L'Heureux, J.-S., L. Hansen, and O. Longva, 2009, Development of the submarine channel in front of the Nidelva River, Trondheimsfjorden, Norway: Marine Geology, **260**, 30–44.

L'Heureux, J.-S., L. Hansen, O. Longva, A. Emdal, and L. Grande, 2010, A multidisciplinary study of submarine slides at the Nidelva fjord delta, mid-Norway — Implications for hazard assessments: Norwegian Journal of Geology, **90**, 1–20.

L'Heureux, J.-S., O. Longva, L. Hansen, and G. Vingerhagen, 2007, The 1990 submarine slide outside the Nidelva River mouth, Trondheim, Norway, *in* V. Lykousis, D. Sakellariou, and J. Locat, eds., Submarine mass movements and their consequences: Kluwer, 259–267.

Lyså, A., L. Hansen, O. Christensen, J.-S. L'Heureux, O. Longva, H. Olsen, and H. Svian, 2007, Landscape evolution and slide processes in a glacio-isostatic rebound area: A combined marine and terrestrial approach: Marine Geology, **248**, 53–73.

Long, M., and S. Donohue, 2007, In situ S-wave velocity from multichannel analysis of surface waves (MASW) tests at eight Norwegian research sites: Canadian Geotechnical Journal, **44**, 533–544.

Nazarian, S., and K. H. Stokoe, 1984, In-situ shear-wave velocities from spectral analysis of surface waves:

Proceedings of the 8th World Conference on Earthquake Engineering, v. 3: Prentice Hall, 31–38.

Park, C. B., R. D. Miller, and J. Xia, 1999, Multichannel analysis of surface waves (MASW): Geophysics, **64**, 800–808.

Polom, U., 2005a, Verfahren zur Erzeugung und Verarbeitung seismischer Anregungen zur geophysikalischen Erkundung des Untergrundes, Offenlegungsschrift: German patent DE 10332972 A1 (in German).

———, 2005b, Vibration generator for seismic applications: U. S. Patent Application 0200 5010 2105 A1.

Polom, U., I. Arsyad, and H.-J. Kümpel, 2008a, Shallow shear-wave reflection seismics in the tsunami struck Krueng Aceh River Basin, Sumatra: Advances in Geosciences, **14**, 135–140.

Polom, U., L. Hansen, J.-S. L'Heureux, O. Longva, I. Lecomte, and C. M. Krawczyk, 2008b, S-wave reflection seismic surveying in the Trondheim Harbor area — Imaging of landslide processes: 2008 fall meeting, AGU, Abstract S11C–1755.

———, 2009a, High-resolution S-wave reflection seismic in the harbor area of Trondheim, Norway: 71st Conference and Exhibition, EAGE, Extended Abstracts, T008.

Polom, U., J. S. L'Heureux, L. Hansen, I. Lecomte, O. Longva, and C. M. Krawczyk, 2009b, Joint land and shallow-marine seismic investigations of landslide processes in the bay of Trondheim, mid-Norway: Near Surface 2009 — 15th European Meeting of Environmental and Engineering Geophysics, Extended Abstracts (CD-ROM), A17.

Pugin, A. J., S. E. Pullan, and J. A. Hunter, 2009, Multicomponent high-resolution seismic reflection profiling: The Leading Edge, 1248–1261.

Pugin, A. J. M., A. J. Hunter, D. Motazedian, G. R. Brooks, and K. B. Kasgin, 2007, An application of S-wave reflection land streamer technology to soil response evaluation of earthquake shaking in an urban area, Ottawa, Ontario: 20th Symposium on the Application of Geophysics to Engineering and Environmental Problems, Annual Meeting, EEGS, Proceedings, 885–896.

Pugin, A. J. M., T. H. Larson, S. L. Sargent, J. H. McBride, and C. E. Bexfield, 2004, Near-surface mapping using SH-wave and P-wave seismic land-streamer data acquisition in Illinois, U. S.: The Leading Edge, **23**, 677–682.

Pugin, A. J. M., S. L. Sargent, and L. Hunt, 2006, SH- and P-wave seismic reflection using landstreamers to map shallow features and porosity characteristics in Illinois: 19th Symposium on the Application of Geophysics to Engineering and Environmental

Problems, Annual Meeting, EEGS, Proceedings, 1094–1109.

Reite, A., 1995, Deglaciation of the Trondheimsfjord area, central Norway: Geological Survey of Norway Bulletin, **427**, 19–21.

Rise, L., R. Bøe, H. Sveian, A. Lyså, and H. Olsen, 2006, The deglaciation history of Trondheimsfjorden and Trondheimsleia, central Norway: Norwegian Journal of Geology, **86**, 419–437.

Rix, G. J., and K. H. Stokoe, 1992, Correlation of initial tangent modulus and cone resistance: Proceedings of the International Symposium on Calibration Chamber Testing: Elsevier, 351–362.

Rosenblad, B. L., J. Li, F.-Y. Menq, and K. H. Stokoe, 2007, Deep shear wave velocity profiles from surface wave measurements in the Mississippi Embayment: Earthquake Spectra, **23**, 791–808.

Sauvin, G., 2009, S-wave seismic for geohazards: A case study from Trondheim Harbor: Diploma engineering thesis, University of Strasbourg, 2009-6-1.

Stoffa, P. L., J. T. Fokkema, R. M. de Luna Freire, and W. P. Kessinger, 1990, Split-step Fourier migration: Geophysics, **55**, 410–421.

Stokoe, K. H., and J. C. Santamarina, 2000, Seismic-wave based testing in geotechnical engineering: Proceedings of the International Conference on Geotechnical and Geological Engineering, GeoEng 2000, v. 1: Technomic Publishing, 1490–1536.

Tatham, R. H., and R. D. McCormack, 1991, Multicomponent seismology in petroleum exploration: SEG Investigations in Geophysics Series No. 6.

van der Veen, M., and A. G. Green, 1998, Land streamer for shallow seismic data acquisition: Evaluation of gimbal-mounted geophones: Geophysics, **63**, 1408–1413.

van der Veen, M., R. Spitzer, A. G. Green, and P. Wild, 2001, Design and application of a towed land-streamer for cost-effective 2D and pseudo-3D shallow seismic data acquisition: Geophysics, **66**, 482–500.

Vanneste, M., H. Westerdahl, P. M. Sparrevik, C. Madshus, I. Lecomte, and L. Zühlsdorff, 2007, S-wave source for offshore geohazard studies: A pilot project to improve seismic resolution and better constrain the shear strength of marine sediments: Offshore Technology Conference, SPE **18756**.

Westerdahl, H., P. M. Sparrevik, C. Madshus, L. Amundsen, and J. P. Fjellanger, 2004, Development and testing of a prototype seabed coupled shear wave vibrator: 74th Annual International Meeting, SEG, Expanded Abstracts, 929–932.

Wolff, F. C., 1979, Beskrivelse til de berggrunngeologiske kart i Trondheim og Østersund 1:250,000: Geological Survey of Norway, **353** (in Norwegian).

Chapter 19

# Integrated Hydrostratigraphic Interpretation of 3D Seismic-reflection and Multifold Pseudo-3D GPR Data

John H. Bradford[1]

## Abstract

To map the 3D distribution of major hydrologic boundaries in a shallow aquifer near Boise, Idaho, 3D seismic-reflection data and multifold, pseudo-3D ground-penetrating-radar (GPR) data were analyzed. The seismic data covered a 75- × 70-m area and imaged horizons from 18 to 150 m deep. The 10-fold, 50-MHz GPR data were acquired on a 20- × 30-m grid using a multichannel GPR system and offsets ranging from 2 m to 20 m. By correlating the well-resolved GPR depth image with a clay-aquitard seismic reflection, the seismic-velocity model was improved substantially and the accuracy of the final interpretation was improved. The resulting clay-aquitard surface differed by 0.12 ± 0.46 m from the depth to clay measured in wells. By integrating the interpretations of the GPR and seismic data, a 3D map of major hydrostratigraphic boundaries was produced.

## Introduction

It is useful to think of the similarities and differences between electromagnetic (EM) and elastic waves in terms of kinematics and dynamics. The dynamics describe the mechanisms of a physical interaction, and clearly we cannot draw analogies between EM dynamics and elastic-wave dynamics. The kinematics of a system describes the changes in time without regard to the physical mechanisms. Through kinematics, we can find similarities among all types of wave propagation, and we can identify mathematical analogies. Indeed, several authors have used a unified mathematical framework to describe elastic and EM wave propagation

(Szaraneic, 1976, 1979; Ursin, 1983). Because of the kinematic analogy between ground-penetrating-radar (GPR) and seismic techniques, often many of the processing tools and interpretation techniques that we use to analyze the data are similar or the same.

In conventional reflection imaging, reflected waves are recorded at the surface and are used to produce a 2D or 3D reflector map of the subsurface. The map indicates the location of contrasts in subsurface properties. Seismic reflections are generated at acoustic-impedance contrasts (i.e., at contrasts in the product of density and velocity), whereas GPR reflections are a function primarily of contrasts in dielectric permittivity and electric conductivity (magnetic permeability also plays a role but is approximately constant in many applications). Often, elastic and dielectric contrasts coincide. For example, in a thorough petrophysical study of a low-water-content sandstone, Koesoemadinata and McMechan (2002) found an imperfect but positive correlation ($R^2 = 0.55$) between dielectric permittivity and compressional-wave velocity. Because of this relationship, GPR-reflection images and seismic-reflection images provide complementary information that sometimes reveals the same boundaries. The equivalent boundaries then can be used to integrate seismic and GPR data sets and may help to improve the overall interpretation.

Seismic- and GPR-reflector maps are subject to the resolution limits of the data, and scale differences can complicate interpretation and data integration. For both seismic reflections and GPR reflections, the resolution limit often is taken as one-quarter of the wavelength at the dominant frequency of the reflected wavelet. That is, two boundaries must be separated vertically by greater than that amount for

[1]Center for Geophysical Investigation of the Shallow Subsurface, Boise State University, Boise, Idaho, U.S.A. E-mail johnb@cgiss.boisestate.edu.

the adjacent wavelet peaks or troughs to be differentiated in the reflection image. Ground-penetrating radar operates in the $10^7$- to $10^9$-Hz range, whereas seismic operates in the $10^1$- to $10^3$-Hz range. Similarly, GPR velocities are on the order of $10^8$ m/s, and seismic velocities are on the order of $10^3$ m/s. Although the magnitudes of GPR and seismic frequencies and velocities differ by five to six orders of magnitude, the ratios scale proportionally, and seismic and GPR wavelengths often are of the same order.

Consider a seismic signal with a dominant frequency of 200 Hz and a 50-MHz GPR signal. In a typical unsaturated sand, both the seismic and GPR wavelengths are about 2 m. Once the sand is saturated, the seismic velocity increases by a factor of about five, whereas the EM velocity decreases by a factor of about two and the respective wavelengths become 10 m and 1 m. The resolution potential of each method can vary dramatically and differently within a given survey, and this possibility must be considered when one is interpreting coincident data sets jointly. Table 1 lists ranges of compressional-wave and electromagnetic velocities and provides a comparison of wavelengths from typical near-surface studies.

Given the similarities and complementary information provided by GPR- and seismic-reflection methods, it may be surprising that the two methods are not used together often for site investigation. To understand this,

we return to the dynamics of wave propagation. Seismic waves are generated and recorded most efficiently where cohesive materials are found at the surface. In unconsolidated sediments, this means moist, fine-grained material. However, GPR signals are strongly attenuated by electrically conductive material and therefore propagate most efficiently through dry, coarse-grained material. Therefore, the ideal conditions for the two methods are in very different environments.

Despite the differences in ideal conditions for seismic and GPR methods, there is a substantial region of overlap in suitable material types so that in some cases, both tools can be deployed effectively. Several studies have exploited that overlap to investigate joint analysis and to improve interpretation for applications ranging from neotectonic characterization of faults to hydrogeophysical and waterborne applications.

In one of the first published studies of coincident GPR- and seismic-reflection data, Cardimona et al. (1998) imaged the top of a 10- to 15-m-deep aquitard at the base of a shallow unconfined aquifer. A shallow clay lens prevented GPR signal penetration to the aquitard at some locations, but the complementary information provided by the two data sets substantially improved interpretation.

Baker et al. (2001) acquired an ultrashallow (<3 m) seismic-reflection survey using a .22-caliber-rifle source

**Table 1.** Typical velocity ranges ($V_P$ and $V_{EM}$) and wavelengths ($\lambda_P$ and $\lambda_{EM}$) for a 200-Hz seismic P-wave (subscript "P") and a 50-MHz electromagnetic wave (subscript $_{EM}$) in common, unconsolidated near-surface (low-pressure) materials. No velocity is reported for the EM wave in clay because the GPR signal typically is attenuated very rapidly and does not propagate effectively in clay. Data were compiled from a variety of sources, including Annan (2005), Bertete-Aguirre and Berge (2002), Bertete-Aguirre et al. (2003), Carmichael (1982), Hamilton (1971), Han et al. (1986), and Santamarina et al. (2005).

| Material | $V_P$ (m/s) | $V_{EM}$ (m/ns) | $\lambda_P$ (m) @ 200 Hz | $\lambda_{EM}$ (m) @ 50 MHz |
|---|---|---|---|---|
| **Sand/gravel** | | | | |
| Vadose | 100–600 | 0.10–0.16 | 0.5–3 | 2–3.2 |
| Saturated | 1700–3000 | 0.06–0.09 | 8.5–15 | 1.2–1.8 |
| **Silt** | | | | |
| Vadose | 300–800 | 0.09–0.14 | 1.5–4 | 1.8–2.8 |
| Saturated | 1500–1800 | 0.05–0.07 | 8–11 | 1–1.4 |
| **Clay** | | | | |
| Vadose | 300–800 | — | 1.5–4 | — |
| Saturated | 1500–1800 | — | 8–11 | — |
| Water | 1500 | 0.033 | 10 | 0.66 |

that produced a signal with a 400-MHz dominant frequency. Sediments at the site consisted of medium- and coarse-grained sand and gravel. Multiple horizons in the seismic data were coincident with horizons imaged using a 225-MHz GPR system. The somewhat higher resolution of the GPR signal provided greater detail in the interpretation. However, because of a diffuse transition from unsaturated to saturated conditions, a GPR reflection was not generated at the water table, whereas that horizon was delineated clearly in the seismic image.

Bachrach and Rickett (1999) acquired a similarly high-resolution (dominant frequency of 300 Hz) seismic data set and were able to correlate reflections with 100-MHz GPR data above the water table. In this case, both methods imaged the water table. Bachrach and Rickett (1999) noted difficulty in interpreting some details of the images because of changes in the support volumes of GPR measurements and seismic measurements caused by opposite and substantial changes in EM and seismic wavelengths in response to water-content variation. This study highlights the need to consider scaling issues when one is integrating seismic and GPR measurements.

Sloan et al. (2007) merged 190-Hz seismic-reflection and 75-MHz GPR-reflection data sets by summing in the depth domain. This approach provided a single coherent reflection section. The GPR data provided a high-resolution image of the unsaturated zone ($<6$ m deep), whereas the seismic data imaged deeper stratigraphic features as well as bedrock at a depth of approximately 30 m. Because the radar signal penetrated only to the water table and the wavelengths of the seismic and GPR signals were similar in the vadose zone, summation worked well. As Sloan et al. (2007) noted, the opposite change in wavelengths would have made summation ineffective for radar reflectors in the saturated zone. In studies of active faults, Rashed and Nakagawa (2004) and Chow et al. (2001) also supplemented deeper seismic images by using GPR for high-resolution imaging of the upper few meters.

In freshwater environments, GPR can be effective for imaging subbottom stratigraphy. Typically, seismic signals can be generated and recorded in the marine environment at an order of magnitude higher than on land. Because of this, where GPR and seismic data can be acquired coincidentally in a waterborne environment, a similar scale of resolution can be maintained. That similarity is illustrated by Haeni (1996), who summarizes coincident seismic and GPR case studies from pond and river environments. Schwamborn et al. (2002) describe the study of a lake in northern Siberia in which seismic subbottom images were acquired in as much as 13 m of open water and coincident GPR images were acquired in the winter when the surface of

the lake was frozen. It also should be noted that in addition to their use in subbottom imaging, both acoustic and GPR methods can be used to image thermal stratigraphy and other features within the water column (Thorpe and Brubaker, 1983; Imberger, 1985; Bradford et al., 2007).

All of the studies mentioned above used qualitative comparisons of seismic and GPR reflections, but quantitative, coupled analysis of the two types of data can improve estimates of material properties. Ghose and Slob (2006) show that by using a common earth model to formulate the solution for both seismic and GPR analysis, joint inversion of offset-dependent reflectivity curves can produce a unique solution for water saturation and porosity.

For completeness, I mention here the seismoelectric effect, which provides a different sort of symmetry between elastic waves and electromagnetic fields. It is well established now that a propagating elastic wave causes a temporary charge dislocation in subsurface materials and that this dislocation produces an electromagnetic field that can be measured. The symmetry arises because charge dislocation caused by a propagating electric field produces a mechanical disturbance that can be measured as a seismic wave. The electromagnetic field produced by the seismoelectric effect is in the same frequency range as that of the seismic signal that generated it ($10^0 - 10^2$ Hz for a typical surface study). This is well below the electromagnetic wave-propagation regime of GPR ($10^6 - 10^9$ Hz), but such a low-frequency electric field also can be used in tandem with GPR to image boundaries. Dupuis et al. (2007) give an excellent example of a horizon imaged by both the seismoelectric field and GPR.

In this paper, I present the results of coincident 3D seismic-reflection and pseudo-3D multifold-GPR imaging in the study of an aquifer system located near Boise, Idaho. The primary objective of the work was to image the upper surface of a clay aquitard at a depth of approximately 20 m and to identify beneath the aquitard significant stratigraphic horizons that might have hydrologic significance.

The dry, coarse-grained surface sediments at the site are ideal for GPR data acquisition. However, at first glance, the environment appears to be ill suited to the surface-seismic-reflection method. Despite this apparent adversity, a shallow water table enabled acquisition of high-quality seismic data. I take advantage of the high resolution afforded by the GPR data and a common seismic and GPR horizon to improve the seismic-velocity model. I then combine the interpretations of both data sets to produce a 3D image of significant hydrostratigraphic horizons.

## Field-data Acquisition and Processing

### Site description

Data for this study were acquired as part of an ongoing effort to characterize the Boise Hydrogeophysical Research Site (BHRS). The BHRS is an experimental well field located on a gravel bar adjacent to the Boise River, 15 km from downtown Boise, Idaho, U.S.A. (Barrash et al., 1999) (Figure 1). The surface aquifer is composed of the youngest in a series of Pleistocene to Holocene coarse fluvial deposits that mantle a sequence of successively older and higher terraces. Core surveys at the BHRS show coarse, unconsolidated and unaltered fluvial deposits underlain by a red clay at a depth of 18–20 m below the surface (Barrash and Clemo, 2002; Barrash and Reboulet, 2004). Analogous deposits that are exposed in the Boise area show massive coarse-gravel sheets; sheets with weak subhorizontal layering and with planar and trough-cross-bedded coarse-gravel facies; and sand channels, lenses, and drapes that are similar to classic deposits such as the Rhine gravels (Jussel et al., 1994; Klingbeil et al., 1999; Heinz et al., 2003).

The central well field (approximately 20 m in diameter) comprises 13 wells, arranged in two concentric rings of six wells each around a central well (Barrash et al.,

1999). In addition, there are five boundary wells approximately 10–35 m from the central area (Figure 1). This design has supported a wide variety of borehole-based hydrologic and geophysical tests and analysis (Barrash et al., 2006; Clement and Knoll, 2006; Clement et al., 2006; Moret et al., 2006; Ernst et al., 2007; Irving et al., 2007; Johnson et al., 2007; Cardiff et al., 2009; Jardani et al., 2009; Mwenifumbo et al., 2009). The saturated thickness of the shallow, unconfined aquifer at the BHRS ranges between 16 m and 18 m, depending on seasonal variation in river stage. Similarly, the vadose zone above the aquifer varies from <1 m to 2.5 m in thickness, depending on surface topography and river stage.

The stratigraphy of the shallow, unconfined aquifer comprises five distinct units that are based on stratigraphic position in the sediment column and on differentiation by porosity and lithology (Figure 2) (Barrash and Clemo, 2002; Barrash and Reboulet, 2004). The major characteristics of each unit are listed in Table 2.

In a 3D multifold GPR study, Bradford et al. (2009) used reflection tomography to derive a 3D GPR velocity volume for the central BHRS. They show that the surface-derived velocities agree with velocities determined from 1D inversion (Clement and Knoll, 2006) of vertical radar

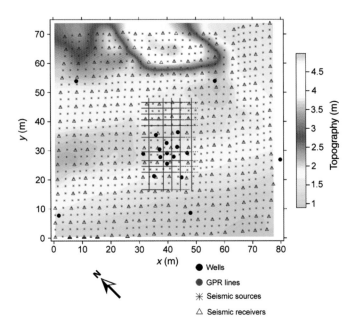

**Figure 1.** Map of the BHRS, showing topographic variation, well positions, 3D seismic coverage, and GPR profiles. The sand berm on the northeast side of the site created a significant statics problem. Well casings and other surface obstructions prevented true 3D GPR data acquisition.

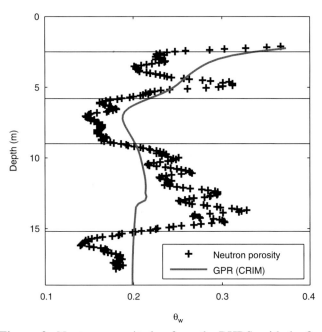

**Figure 2.** Neutron porosity log from the BHRS, with the five major hydrostratigraphic units defined by Barrash and Clemo (2002). Unit 5 is a medium- to coarse-grained sand channel, whereas Units 1–4 are composed of poorly sorted sand, gravel, and cobbles and are differentiated primarily by porosity (Table 2). The water content derived from GPR velocity inversion approximates a running average of the porosity log and follows the large-scale trends in porosity.

**Table 2.** Properties of the five primary lithologic units at the BHRS, interpreted from cores and neutron porosity logs (see Figure 2). Adapted from Barrash and Clemo (2002).

| Unit | Thickness, approximate (m) | Mean porosity | Porosity variance | Dominant composition |
|---|---|---|---|---|
| 5 | 0–4 | 0.429 | 0.003 | Coarse sand |
| 4 | 1–5 | 0.232 | 0.002 | Pebble/cobble dominated |
| 3 | ~3 | 0.172 | 0.0006 | Pebble/cobble dominated |
| 2 | ~6 | 0.243 | 0.002 | Pebble/cobble dominated |
| 1 | ~2 | 0.182 | 0.0006 | Pebble/cobble dominated |

profiles (VRP) to within 2% when averaged over a few wavelengths. Bradford et al. (2009) went on to estimate the porosity distribution using the complex refractive index method (CRIM) equation (Wharton et al., 1980) and found that the GPR-estimated porosities agree with neutron-porosity-log measurements to within 8% of the mean porosity. With the exception of the Unit 2/Unit 3 transition, the porosity contrasts across unit boundaries result in dielectric permittivity contrasts that generate well-defined radar reflectors observed in surface GPR profiles (Barrash and Clemo, 2002; Barrash and Reboulet, 2004; Clement et al., 2006; Bradford et al., 2009). The transition from Unit 2 to Unit 3 is irregular and gradational and does not generate a distinct GPR reflection.

Borehole seismic data also have proved useful at the BHRS. Using 1D inversion of vertical seismic profiles (VSP), Moret et al. (2004) show that compressional seismic slowness values correlate strongly with porosity in the unconfined aquifer. Prior to the present exposition, no surface-seismic-reflection data at the site have been published.

Although the unconfined aquifer is well characterized, little is known about the system below the shallow clay aquitard, nor is the surface of the aquitard well constrained outside the central well field. The Boise State University hydrogeophysical research group has an interest in extending the experimental well field laterally as well as deeper into the underlying confined aquifer. The purpose of the present study is to map the clay aquitard over the entire well field and to identify and map the distribution of major units in the underlying confined aquifer system.

## GPR data

### Acquisition

The GPR survey was designed to image the central well field and covered an area of 20 × 30 m (Figure 1).

A Sensors and Software PulseEKKO Pro system, with a multichannel adapter, 1000-V transmitter, and 50-MHz antennae, was used to acquire multifold data along five inline (*y*-direction) profiles and 11 crossline (*x*-direction) profiles. Data were acquired in three passes in an off-end geometry, with one transmitter and four receivers for the first two passes and one transmitter and two receivers for the final pass. Offsets varied from 2 m to 20 m, with 2-m spacing between receivers. The antennae were attached to wheeled carts, and the entire system was towed by hand. Traces were acquired every 15 cm, with position control maintained by an odometer wheel trigger, and 16 traces were stacked vertically on each trigger. Because of surface obstructions, including well casings, trees, and bushes, it was not possible to avoid spatial aliasing between profiles; spacing between lines varied from 2 m to 6 m. Because of spatial aliasing between lines, true 3D processing was impossible, and therefore I term this a pseudo-3D survey.

Acquisition of these data coincided with and used the same GPR system as did the 100-MHz 3D survey described by Bradford et al. (2009). The 100-MHz survey provided a 3D image of the major stratigraphic boundaries to a depth of 15 m (Units 2–5) but failed to image the aquifer-aquitard boundary at a depth of 18–20 m. Imaging that clay surface was the primary objective of the 50-MHz survey.

### Processing

Data preprocessing consisted of (1) 3D binning onto a 0.5- × 0.5-m common-midpoint (CMP) binning grid, (2) channel-dependent time-zero correction as described by Bradford et al. (2009), (3) band-pass filtering (6–12–60–120 MHz) to attenuate the low-frequency transient and high-frequency random noise, and (4) automatic gain control (AGC) with a 40-ns window. After this processing stream, CMP gathers show coherent reflections over the full range of offsets throughout the survey area (Figure 3). The deepest reflection is from the clay aquitard and has a near-offset arrival time of approximately 450 ns.

Although the clay reflection is evident in the CMP gathers, it is obscured almost totally by surface scatter in a conventional single-offset GPR image (Figure 4a). The clarity of this reflection is enhanced significantly simply by NMO correcting and stacking of the data (Figure 4b). To further enhance the subsurface reflections, I applied an NMO correction at air velocity in the CMP domain. After this correction, surface-scattered events had infinite apparent velocity, whereas subsurface reflections maintained positive residual moveout. I used an *f-k* filter to attenuate events with infinite apparent velocity and followed this with an inverse NMO correction at air velocity. The *f-k* filtered data were then NMO-corrected and stacked at the subsurface velocity. This procedure effectively eliminates surface scatter and has minimal impact on the subsurface reflectors. In the final CMP stack, the surface scatter is strongly attenuated and the clay reflection is revealed throughout the survey area (Figure 4c).

To derive a depth-velocity model, I used the reflection-tomography method of Stork (1992) to invert for velocity along all profiles oriented in the inline (*y*) direction

(Figure 1). This orientation is approximately perpendicular to the prominent dip of the Unit 4/Unit 5 stratigraphic boundary, in which case the 2D velocity-model assumption is reasonable. For the starting velocity model, I used the velocity model previously derived by Bradford et al. (2009). The Unit 1 velocity was not constrained by the earlier 100-MHz survey because the 100-MHz signal did not reach the clay aquitard. I inverted for the Unit 1 velocity while holding the velocity constant in the region above the Unit 2/Unit 1 boundary. The final GPR data volume is a CMP stack that is depth-converted using the tomographic depth-velocity model (Figure 5). This new velocity model now forms a complete picture of the shallow aquifer and correlates with the distribution of Units 1–5 (Figures 2 and 5). Note that whereas the Unit 2/Unit 3 boundary does not produce a coherent reflection, it is possible to map all five units on the basis of GPR velocity distribution alone (Figures 2 and 5).

## Seismic data

### Acquisition

The seismic survey grid was deployed to span the outer ring of wells with dimensions of approximately 75 × 70 m

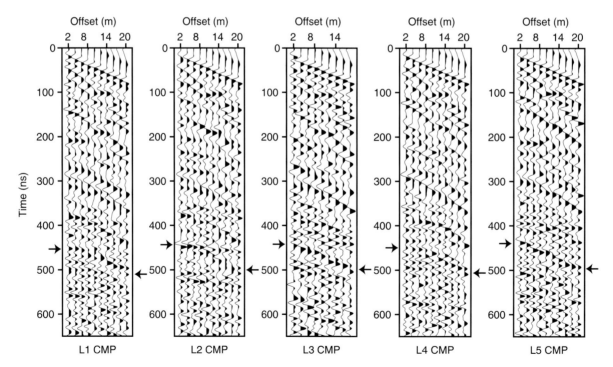

**Figure 3.** Representative CMPs from each of the five inline GPR profiles. The clay reflection has a near-offset arrival time of about 450 ns and is indicated by arrows. Heavy contamination of late arrivals by surface scatter resulted in a signal-to-noise ratio of less than 1 at the clay reflection.

(Figure 1). The acquisition system consisted of four 60-channel seismographs recording separately but with a common trigger cable. Receivers consisted of single 40-Hz geophones planted on a 5- × 5-m grid, with 15 stations in the inline ( $y$ ) direction and 16 in the crossline ( $x$ ) direction. The receiver grid was held static while shooting took place through the source grid. Spatial density was increased by locating source points on a 2.5- × 2.5-m grid. The source was an 8-kg sledgehammer on a steel plate, with eight hammer blows stacked vertically at each shot location. The acquisition procedure resulted in good data quality, with the reflection from the clay aquitard present at a near-offset traveltime of 30–40 ms as well as a series of deeper reflections at traveltimes as long as 180 ms (Figure 6).

## Processing

For CMP processing, I binned the data onto a 1.25- × 1.25-m CMP grid. This resulted in a maximum CMP fold of 240 and an average fold of 55, with the highest coverage near the central portion of the well field (Figure 7). I applied a conventional data-processing scheme to the data that consisted of (1) time-variant band-pass filtering that varied from 100–200–500–1000 Hz at 50 ms to 60–120–500–1000 Hz at 80 ms and greater, (2) AGC (50-ms time gate), (3) elevation and residual statics (described in greater detail below), (4) NMO velocity analysis, (5) inside and top muting to remove the noise cone and first-arrival refraction, respectively, and (6) CMP stacking and depth conversion.

A 2-m-high sand berm on the northeastern side of the site (Figure 1) resulted in substantial statics anomalies (Figure 6a), and application of residual statics was particularly important. The residual-statics procedure consisted of an initial NMO velocity analysis and moveout correction followed by residual-statics calculation using a modified form of the surface-consistent, stack-power maximization algorithm described by Ronen and Claerbout (1985). After application of residual statics, NMO velocity analysis was repeated to produce the final stacking-velocity model. The shot records after static corrections show a substantial improvement in reflector coherence (Figure 6b). That improvement led to high-quality CMP gathers for velocity analysis (Figure 8), with the dominant frequency of the reflections being controlled by the low shoulder of the band-pass filter (200 Hz for shallow reflections, grading to 120 Hz for reflections at 80 ms or greater). The importance of statics is evident in the inline stack slices that cross the sand berm (Figure 9). Prior to residual statics, the clay-aquitard reflection is imaged poorly below the

sand berm (Figure 9a), whereas after statics, the reflection is coherent throughout the data volume, and the coherence of deeper reflections is improved substantially (Figure 9b).

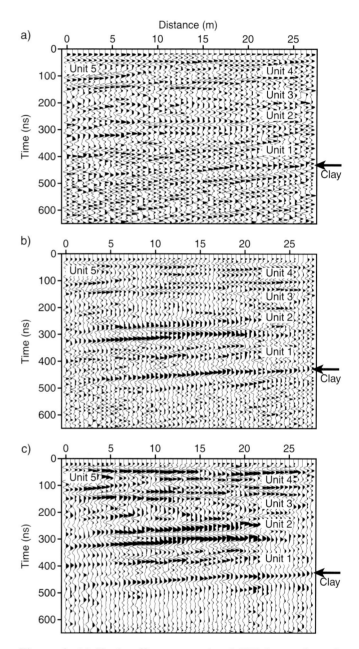

**Figure 4.** (a) Single-offset, conventional GPR image through the center of the well field. Reflections are interpretable to about 300 ns, but the clay reflection between 450 and 500 ns is obscured almost totally by surface scatter. (b) CMP stacking significantly improves the signal-to-noise ratio. (c) Final stacked image after prestack *f-k* filtering in the CMP domain followed by CMP stacking. In the final section, all major hydrostratigraphic boundaries down to the clay aquitard are imaged clearly.

**Figure 5.** Inline GPR velocity model through the center of the well field, produced using reflection tomography. Interpretation of the clay surface after prestack depth migration with this velocity model agrees with clay depths measured in the wells to within 0.03 m ± 0.3 m. The portion of the velocity model that is constrained by the full range of offsets is shown. The velocity model is clipped at 0.11 m/ns but reaches 0.15 m/ns in the vadose zone. The water table is 2 m deep.

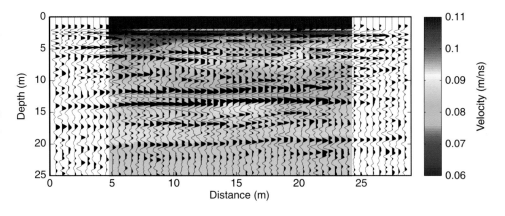

**Figure 6.** A representative shot gather (a) before residual statics and (b) after residual statics. A 2-m-high sand berm on the northeast side of the site (Figure 1) produced a significant statics anomaly. Residual statics were effective at improving the coherence of reflecting horizons; this is particularly evident between channels 100 and 240.

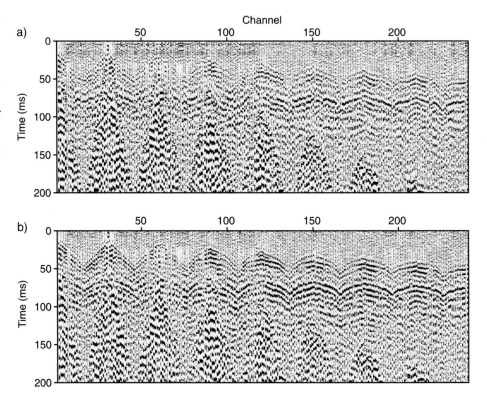

To determine the seismic-interval-velocity model, I first modified the stacking-velocity functions to include the low-velocity surface layer. I picked the traveltime of the direct arrival from 20 shot records distributed throughout the survey, and then I fitted a line to all the picks and found a velocity of 495 m/s ± 18 m/s. I computed the mean two-way traveltime to the water table (8 ms) on the basis of the mean water-table depth of 2 m measured in the wells, and then I inserted the velocity of 495 m/s into the stacking-velocity functions. Finally, I computed the interval-velocity model using the Dix inversion (Figure 10), and I used this interval-velocity model to depth-convert the stacked data.

# Data Integration and Interpretation
## Improving the seismic-velocity model

With stacked, depth-converted data volumes complete, I picked the clay-aquitard reflection in the seismic and GPR data. I first picked the central peak of the reflection, then I shifted the picks upward so that the horizon would correspond to the first motion of the reflected wavelet — the true position of the reflecting boundary. The upward shift was 4 m for the seismic data and 1 m for the GPR data. When I subtracted these initial clay horizon picks from the clay depths measured in well data,

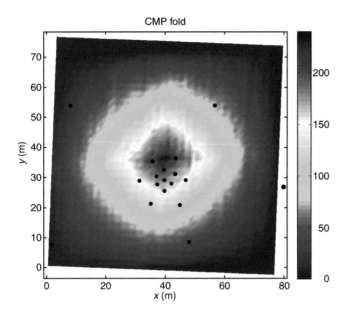

**Figure 7.** A CMP fold map for the 3D seismic survey, with well positions shown as black circles. The maximum fold is 240, and the mean fold over the entire grid is 55. The survey was designed to provide high fold near the center of the well field.

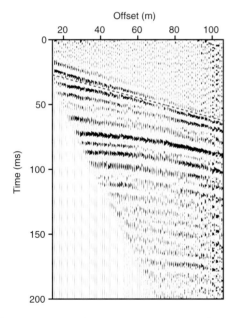

**Figure 8.** A representative CMP supergather after residual statics. The supergather was formed by summing traces within a 3×3 CMP bin area and was used for the final round of velocity analysis. All processing steps have been applied to this gather except the top mute, which is shown as a red dashed line. The data above the top mute show the relation of the clay reflection to the water-table refraction. The top mute was applied prior to producing the final stack. Note that the supergathers were used only for velocity analysis and that individual CMP bins were used to produce the stacked volumes.

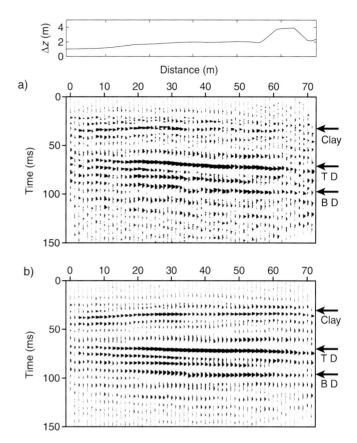

**Figure 9.** Inline seismic CMP stack that crosses through the center of the well field, showing the clay aquitard dipping toward the left and deeper reflecting horizons dipping toward the right. TD and BD are top and bottom of the deep unit, respectively. The surface topography is shown at the top. (a) The clay surface in the stack before statics corrections is evident on the left but decreases in coherence on the right. (b) After statics corrections, both the clay and deeper reflections are imaged clearly across the entire profile.

I found that the difference was 1.5 ± 0.5 m for the seismic horizon and 0.03 ± 0.32 m for the GPR horizon. Relative to the dominant signal wavelengths (12 m for seismic, 2.8 m for GPR), these estimates are remarkably accurate. However, the difference in the seismic horizon is greater than the uncertainty, which suggests that there may be some bias in the result.

In the saturated zone above the aquitard, the initial seismic-interval-velocity estimate was 2155 m/s. This value is 10.6% lower than the mean value of 2400 m/s determined by Moret et al. (2004) from VSP analysis. The underestimate of the interval velocity explains the initial underestimate of clay depth. The noise cone and the first-arrival refraction limit the offset aperture that is available for velocity analysis to between 18 m and 35 m,

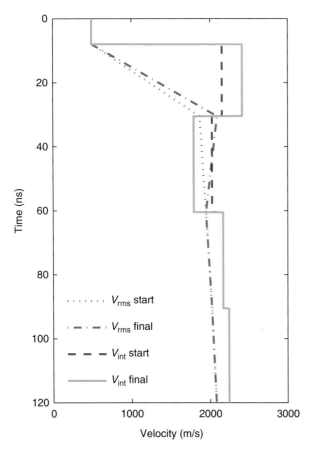

**Figure 10.** Site-averaged seismic-velocity models before (start) and after (final) adjustment, based on correlation of the clay interface reflection with the GPR interpretation. After adjustment, the velocity of 2417 m/s in the surface aquifer is consistent with velocity estimates from borehole measurements. The velocity of 1800 m/s below the clay interface is consistent with expected values for fine-grained lake sediments. $V_{rms}$ is rms velocity, and $V_{int}$ is interval velocity.

and this small offset aperture limits the accuracy of the stacking-velocity measurement.

To test how well I could improve the seismic velocity model and clay depth estimate using only information derived from the seismic-reflection and GPR data, I first computed the mean difference between the seismic and GPR clay horizons ($1.4 \pm 0.3$ m). I then computed the mean change in seismic rms velocity that would be required to shift the seismic clay horizon downward by 1.4 m and added this value to my stacking-velocity functions. Finally, I used these updated velocity functions to compute a new interval-velocity function (Figure 10). The updated mean interval velocity in the shallow aquifer is 2417 m/s, which agrees well with Moret et al.'s (2004) VSP-determined value of 2400 m/s. After the velocity model

is updated, the difference between seismic and GPR clay horizons is $0.05 \pm 0.31$ m, and the difference between the seismic clay horizon and clay depths measured in wells is $0.12 \pm 0.46$ m. In both of these comparisons, the mean difference now is substantially less than the variability, thereby indicating that the bias in the seismic surface has effectively been removed.

Note that a thin basalt ($<0.5$ m thick) has been encountered in some wells at the aquifer-aquitard boundary. Although the lateral distribution of basalt is unknown, surprisingly, there is no clear manifestation of this anomaly in the GPR amplitudes or seismic amplitudes (prior to AGC), in their reflection traveltimes, or in their reflection surfaces. This suggests that the basalt is present only in isolated distributions. The mean basalt thickness is less than uncertainty in the depths to clay, so the variability probably includes scattering from the irregular distribution of basalt.

Better resolution of velocities in the GPR data and the resulting improvement in accuracy of clay-depth estimates enabled me to improve the estimates of seismic velocity and depth significantly. The uncertainty in depth to clay determined from the GPR and seismic interpretations is nearly equivalent ($\pm 0.32$ m for GPR and $\pm 0.46$ m for seismic), but the wavelengths differ by a factor of greater than four. It is surprising that the large difference in wavelengths between the seismic data and the GPR data was not manifested as a greater difference in uncertainty of depth to clay. This observation suggests that little lateral variability exists in the nature of the Unit 1/clay interface.

## Combined 3D interpretation

With accurate velocity and depth models for both the GPR- and seismic-data volumes, it now is possible to derive an integrated 3D interpretation that describes the hydrologic system more fully (Figures 11 and 12). In the shallow aquifer, coherent GPR reflections are generated at the base of the Unit 5 sand channel, the Unit 4/Unit 3 boundary, and the Unit 2/Unit 1 boundary. These horizons and interpretation are discussed in greater detail by Bradford et al. (2009). Because of a gradational and irregular distribution of porosity, the Unit 3/Unit 2 boundary does not produce a coherent GPR reflection, but the porosity decrease associated with Unit 3 is manifest as a velocity increase that is well resolved through reflection tomography (Figures 2 and 5). Furthermore, the Unit 1/aquitard boundary is imaged clearly with the 50-MHz GPR data; multifold acquisition and processing were critical to producing this high-quality image. The clay surface in the interpretation of the GPR data appears as a relatively flat surface that begins to dip downward at the southwestern extent of the survey area.

The seismic data volume reveals that the aquitard surface continues to trend to greater depths toward the southwest. The depression roughly parallels the crossline (*x*) direction of the 3D surveys (Figures 11 and 12). The depression parallels the course of the Boise River, and I interpret this feature as a paleochannel that is cut deeper into the clay. The base of the channel reaches a maximum depth of 23 m and is deeper than the mean clay surface by as much as 5 m. This depression is a large-scale feature that likely has a significant impact on flow dynamics in the shallow aquifer. Toward the northeast, the clay surface is variable but generally shallows and reaches a minimum depth of just 13 m (Figures 11 and 12). The shallowest depth occurs in an area where no borehole control was available, and previously, the clay was not thought to reach this near the surface at the BHRS.

There is no well control below the clay aquitard, but the seismic data provide new information that helps to constrain our understanding of the deeper aquifer system. From the aquitard to a depth of 60 m, no strong reflections are present. It is unlikely that the clay aquitard is 20 m thick, so the thickness of the clay likely is below the seismic resolution. The interval from 20- to 60-m depth is associated with a substantial velocity inversion that decreases from approximately 2400 m/s in the surface aquifer to approximately 1800 m/s in the deeper interval (Figure 10). A series of lakes filled the western Snake River Valley during late Miocene-Pliocene time, and the sediments from those lakes typically have a substantially higher concentration of fine-grained materials than the BHRS surficial aquifer. Compressional-wave velocity tends to decrease with increasing concentration of fines (Table 1) (Han et al., 1986). Therefore, the velocity inversion below the aquitard is consistent with a transition to fine-grained lacustrine sediments. Note that of course there is overlap among the velocities of sand-rich and fine-rich materials, so the interpretation is not unique.

The top of a 25-m-thick unit of strong reflectivity is present at a depth of 60 m. This unit dips gently toward the northeast, and its upper and lower surfaces are nearly parallel. However, the lower surface dips more steeply toward the north, which causes the package to thicken in

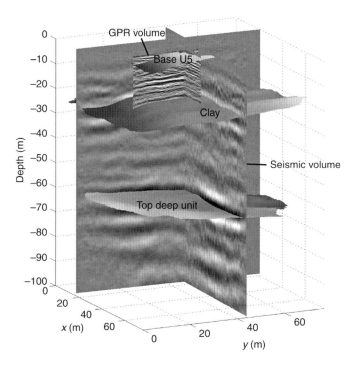

**Figure 11.** Combined GPR and seismic volumes, which show excellent correlation of the clay interface. Note that this horizon has been shifted down by 1.5 m from the true position so that the GPR and seismic reflections are visible. The high resolution of the GPR data enables detailed interpretation of the surface-aquifer stratigraphy, whereas the seismic data image deeper strata that may have significance for the underlying confined aquifer.

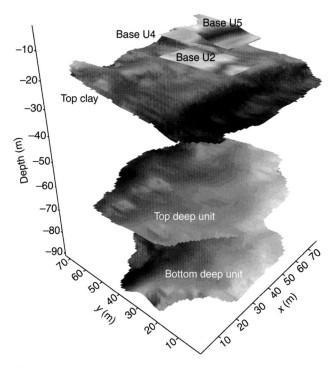

**Figure 12.** Shaded-relief surface images show the combined interpretation of all major hydrostratigraphic interfaces at the BHRS. Horizons in color were interpreted from the GPR data volume, and grayscale surfaces were interpreted from the 3D seismic volume.

that direction. The deep unit is associated with an increase in velocity to 2050 m/s. In this environment, it would not be surprising to find a layer of basalt, but the measured velocity is too low to support that interpretation. The 2050 m/s value is closer to the velocity of the alluvium found in the surface aquifer, and the unit likely comprises coarser-grained materials than those that make up the overlying low-velocity unit. This deep unit may well comprise a deeper confined aquifer.

## Conclusions

Despite apparently adverse conditions for surface seismic data acquisition, the BHRS proved to be a suitable environment for acquisition of high-quality seismic and GPR data. Because of interference by energy scattered from surface objects, acquisition of multifold GPR data coupled with prestack dip filtering proved critical to producing a coherent image of the clay aquitard at a depth of 18–20 m. Similarly, the presence of a 2-m-high sand berm produced statics anomalies that made residual statics corrections a critical step in producing a coherent seismic-data volume.

Both GPR reflection and seismic reflection were effective for delineating the clay aquitard. Using a quantitative comparison of the clay surface interpreted from the GPR and seismic data, I improved the seismic velocity estimates for the surficial aquifer by about 10%, which resulted in their differing by less than 1% from the average velocity measured in previously analyzed vertical seismic profiles. The resulting seismic-data volume provided a laterally continuous image of the clay surface over the entire well field that is within $\pm 0.46$ m of the depths to clay measured in wells. In addition, I interpreted the upper and lower surfaces of what may be a 25-m-thick confined aquifer at a depth of 60 m. This new information provides an improved understanding of the hydrologic system at the BHRS that can be used in future planning, development, and research at the site.

The relatively simple quantitative integration of GPR and seismic data presented here suggests the significant advantages that can be gained where it is possible to acquire both types of data simultaneously. In addition, by improving estimates of depth to the water table, for example, the use of shallow GPR images to develop and improve seismic statics corrections in arid environments is likely to be a fruitful area for future research. The complementary data provided by the seismic and GPR methods are a rich source of information that can lead to improved site characterization.

## Acknowledgments

The U. S. Environmental Protection Agency funded this work under Grant # X-97008501-0. Boise State University acknowledges support of this research by Landmark Graphics Corporation via the Landmark University Grant Program. Graduate students Josh Nichols, Joel Brown, Dylan Mikesell, and Leah Steinbronn acquired the multifold GPR data. Graduate students Scott Hess and Vijaya Raghavendra and research faculty member Bill Clement acquired the 3D seismic-reflection data. Warren Barrash provided the porosity logs and insight to the BHRS geologic setting and hydrology.

## References

Annan, A. P., 2005, Ground-penetrating radar, *in* D. K. Butler, ed., Near-surface geophysics, Chapter 11: SEG Investigations in Geophysics Series No. 13, 357–438.

Bachrach, R., and J. Rickett, 1999, Ultra shallow seismic reflection in depth: Examples from 3D and 2D ultra shallow surveys with applications to joint seismic and GPR imaging: 69th International Meeting, SEG, Expanded Abstracts, 488–491.

Baker, G. S., D. W. Steeples, C. Schmeissner, M. Pavlovic, and R. Plumb, 2001, Near-surface imaging using coincident seismic and GPR data: Geophysical Research Letters, **28**, 627–630.

Barrash, W., and T. Clemo, 2002, Hierarchical geostatistics and multifacies systems: Boise Hydrogeophysical Research Site, Boise, Idaho: Water Resources Research, **38**, 1196, 10.1029/2002WR001436.

Barrash, W., and E. C. Reboulet, 2004, Significance of porosity for stratigraphy and textural composition in subsurface coarse fluvial deposits, Boise Hydrogeophysical Research Site: Geological Society of America Bulletin, **116**, 1509–1073, doi:10.1130/B25370.1.

Barrash, W., T. Clemo, J. J. Fox, and T. C. Johnson, 2006, Field, laboratory, and modeling investigation of the skin effect at wells with slotted casing, Boise Hydrogeophysical Research Site: Journal of Hydrology, **326**, 181–198.

Barrash, W., T. Clemo, and M. D. Knoll, 1999, Boise Hydrogeophysical Research Site (BHRS): Objectives, design, initial geostatistical results: SAGEEP '99, Symposium on the Application of Geophysics to Environmental and Engineering Problems, Environmental and Engineering Geophysical Society, 713–722.

Bertete-Aguirre, H., and P. Berge, 2002, Recovering soil distributions from seismic data using laboratory velocity measurements: Journal of Environmental and Engineering Geophysics, **7**, 1–10.

Bertete-Aguirre, H., P. Berge, and J. J. Roberts, 2003, A method for using laboratory measurements of electrical and mechanical properties to assist in the interpretation of field data from shallow geophysical measurements: Journal of Environmental and Engineering Geophysics, **8**, 23–29.

Bradford, J. H., W. P. Clement, and W. Barrash, 2009, Estimating porosity via ground-penetrating radar reflection tomography: A controlled 3D experiment at the Boise Hydrogeophysical Research Site: Water Resources Research, **45**, W00D26, doi:10.1029/2008WR006960.

Bradford, J. H., C. R. Johnson, T. R. Brosten, J. P. McNamara, and M. N. Gooseff, 2007, Imaging thermal stratigraphy in freshwater lakes using georadar: Geophysical Research Letters, **34**, L24405, doi:10.1029/2007GL032488.

Cardiff, M., W. Barrash, P. Kitanidis, B. Malama, A. Revil, S. Straface, and E. Rizzo, 2009, A potential-based inversion of unconfined steady-state hydrologic tomography: Ground Water, **47**, 259–270.

Cardimona, S. J., W. P. Clement, and K. Kadinsky-Cade, 1998, Seismic reflection and ground-penetrating radar imaging of a shallow aquifer: Geophysics, **63**, 1310–1317.

Carmichael, R. S., 1982, CRC Handbook of physical properties of rocks II: CRC Press.

Chow, J., J. Angelier, J. J. Hua, J. C. Lee, and R. Sun, 2001, Paleoseismic event and active faulting: From ground penetrating radar and high-resolution seismic reflection profiles across the Chihshang Fault, eastern Taiwan: Tectonophysics, **333**, 241–259.

Clement, W. P., and M. D. Knoll, 2006, Traveltime inversion of vertical radar profiles: Geophysics, **71**, no. 3, K67–K76.

Clement, W. P., W. Barrash, and M. D. Knoll, 2006, Reflectivity modeling of ground-penetrating radar: Geophysics, **71**, no. 3, K59–K66.

Dupuis, J. C., K. E. Butler, and A. W. Kepic, 2007, Seismoelectric imaging of the vadose zone of a sand aquifer: Geophysics, **72**, no. 6, A81–A85.

Ernst, J. R., A. G. Green, H. Maurer, and K. Holliger, 2007, Application of a new 2D time-domain full-waveform inversion scheme to crosshole radar data: Geophysics, **72**, no. 5, J53–J64.

Ghose, R., and E. C. Slob, 2006, Quantitative integration of seismic and GPR reflections to derive unique estimates for water saturation and porosity in subsoil: Geophysical Research Letters, **33**, L05404.

Haeni, F. P., 1996, Use of ground-penetrating radar and continuous seismic-reflection profiling on surface-water bodies in environmental and engineering studies: Journal of Environmental and Engineering Geophysics, **1**, 27–35.

Hamilton, E. L., 1971, Elastic properties of Marine Sediments: Journal of Geophysical Research, **76**, 579–604.

Han, D., A. Nur, and D. Morgan, 1986, Effects of porosity and clay content on wave velocities in sandstones: Geophysics, **51**, 2093–2107.

Heinz, J., S. Kleineidam, G. Teutsch, and T. Aigner, 2003, Heterogeneity patterns of Quaternary glaciofluvial gravel bodies (SW-Germany): Applications to hydrgeology: Sedimentary Geology, **158**, 1–23.

Imberger, J., 1985, The diurnal mixed layer: Limnology and Oceanography, **30**, 737–770.

Irving, J. D., M. D. Knoll, and R. J. Knight, 2007, Improving crosshole radar velocity tomograms: Geophysics, **72**, no. 4, J31–J41.

Jardani, A., A. Revil, W. Barrash, A. Crespy, E. Rizzo, S. Straface, M. Cardiff, B. Malama, C. R. Miller, and T. C. Johnson, 2009, Reconstructrion of the water table from self-potential data during dipole pumping/injection test experiments: Ground Water, **47**, 213–227.

Johnson, T. C., P. S. Routh, W. Barrash, and M. D. Knoll, 2007, A field comparison of Fresnel zone and ray-based GPR attenuation-difference tomography for time-lapse imaging of electrically anomalous tracer or contaminant plumes: Geophysics, **72**, no. 2, G21–G29.

Jussel, P., F. Stauffer, and T. Dracos, 1994, Transport modeling in hetergeneous aquifers: 1. Statistical description and numerical generation: Water Resources Research, **30**, 1803–1817.

Klingbeil, R., S. Kleineidam, U. Asprion, T. Aigner, and G. Teutsch, 1999, Relating lithofacies to hydrofacies: Outcrop-based hydrogeological characterization of Quaternary gravel deposits: Sedimentary Geology, **129**, 299–310.

Koesoemadinata, A. P., and G. A. McMechan, 2002, Correlations between seismic parameters, EM parameters, and petrophysical/petrological properties for sandstone and carbonate at low water saturations: Geophysics, **68**, 870–883.

Moret, G. J. M., W. P. Clement, M. D. Knoll, and W. Barrash, 2004, VSP traveltime inversion: Near-surface issues: Geophysics, **69**, 245–351.

Moret, G. J. M., M. D. Knoll, W. Barrash, and W. C. Clement, 2006, Investigating the stratigraphy of an alluvial aquifer using crosswell seismic traveltime tomography: Geophysics, **71**, no. 3, B63–B73.

Mwenifumbo, C. J., W. Barrash, and M. D. Knoll, 2009, Capacitive conductivity logging and electrical stratigraphy in a high-resistivity aquifer, Boise Hydrogeophysical Research Site: Geophysics, **74**, no. 3, E125–E133.

Rashed, M., and K. Nakagawa, 2004, High-resolution shallow seismic and ground penetrating radar investigations revealing the evolution of the Uemachi Fault system, Osaka, Japan: The Island Arc, **13**, 144–156.

Ronen, J., and J. F. Claerbout, 1985, Surface-consistent residual statics estimation by stack-power maximization: Geophysics, **50**, 2759–2767.

Santamarina, J. C., V. A. Rinaldi, D. Fratta, K. A. Klein, Y.-H. Wang, G. C. Cho, and G. Cascante, 2005, A survey of elastic and electromagnetic properties of near-surface soils, *in* D. K. Butler, ed., Near-surface geophysics, Chapter 4: SEG Investigations in Geophysics Series No. 13, 71–87.

Schwamborn, G. J., J. K. Dix, J. M. Bull, and V. Rachold, 2002, High-resolution seismic and ground penetrating radar — Geophysical profiling of a thermokarst lake in the Western Lena Delta, Northern Siberia: Permafrost and Periglacial Processes, **13**, 259–269.

Sloan, S. D., G. P. Tsoflias, D. W. Steeples, and P. D. Vincent, 2007, High-resolution ultra-shallow subsurface imaging by integrating near-surface seismic reflection and ground-penetrating radar data in the depth domain: Journal of Applied Geophysics, **62**, 281–286.

Stork, C., 1992, Reflection tomography in the postmigrated domain: Geophysics, **57**, 680–692.

Szaraneic, E., 1976, Fundamental functions for horizontally stratified earth: Geophysical Prospecting, **24**, 528–548.

———, 1979, Towards unification of geophysical problems for horizontally stratified media: Geophysical Prospecting, **27**, 576–583.

Thorpe, S. A., and J. M. Brubaker, 1983, Observations of sound reflection by temperature microstructure: Limnology and Oceanography, **28**, 601–613.

Ursin, B., 1983, Review of elastic and electromagnetic wave propagation in horizontally layered media: Geophysics, **48**, 1063–1081.

Wharton, R. P., G. A. Hazen, R. N. Rau, and D. L. Best, 1980, Electromagnetic propagation logging: Advances in technique and interpretation: SPE 9267.

Chapter 20

# Refraction Nonuniqueness Studies at Levee Sites Using the Refraction-tomography and JARS Methods

Julian Ivanov[1], Richard D. Miller[1], Jianghai Xia[1], Joseph B. Dunbar[2], and Shelby Peterie[1]

## Abstract

The utility of two varied approaches to first-arrival time analysis of seismic data acquired at several unique levee sites is demonstrated by solving the inverse refraction-traveltime problem (IRTP). These data were evaluated using conventional refraction tomography and joint analysis of refractions with surface waves (JARS). The JARS approach uses a reference model, derived from surface-wave-calculated shear-wave velocity estimates, as a constraint in reducing refraction nonuniqueness. At those levee sites, conventional refraction-tomography and JARS methods provided different solutions, equally matching the observed data. This observation suggests both approaches are equally possible from a numerical perspective. The JARS images reveal horizontal layering patterns, laterally uniform velocity trends, mild velocity variations, and channel-like features consistent with geologic expectations. In addition, the JARS approach demonstrated the capability for imaging low-velocity layers/zones, something not seen using conventional refraction or refraction-tomography techniques. As a result of these qualitative observations, without ground truth to support an earth model (e.g., from wells), the JARS approach can be viewed as an additional method for finding solutions to the IRTP. However, from all evidence in those studies, the JARS approach represents a possible solution and an example of the potential adverse affect of nonuniqueness. These empirical results support the understanding that for a given refraction data set, significantly different and equally possible velocity-model solutions can exist, resolving which is truly best using invasive ground truth.

## Introduction

There are many possible approaches to formulating solutions to the inverse refraction-traveltime problem (IRTP) (Slichter, 1932; Healy, 1963; Ackerman et al., 1986; Burger, 1992; Lay and Wallace, 1995). Different refraction and refraction-tomography algorithms can converge on different equally possible solutions (Sheehan and Doll, 2003). Ivanov et al. (2005b) note that even using a simple three-layer model and assuming exact data, the IRTP has a continuous range of possible solutions. Such a continuous range of solutions can be viewed as a valley (Figure 1) in the surface of the prediction error function (i.e., objective function) between the measured and calculated (function of model parameters) data (Menke, 1989; Ivanov et al., 2005b). This observation explains the wide range of solutions offered by conventional inversion algorithms for specific data sets. Each method converges toward a solution in a valley of nonuniqueness based on assumptions about the geologic model and the type and degree of numerical regularization used (Ivanov et al., 2005b).

The joint analysis of refractions with surface waves (JARS) method was proposed to overcome preferential algorithm behavior based on initial model and regularization characteristics (Ivanov et al., 2006). Key to the effectiveness of JARS is the use of a reference compressional-wave velocity ($V_P$) model, which is derived from surface-wave shear-wave velocity ($V_S$) estimates. The JARS method involves two key steps: obtaining a $V_S$ model from surface-wave analysis of seismic data, and obtaining a pseudo-$V_P$ model by rescaling the $V_S$ estimates using assumptions about the $V_P/V_S$ trend.

[1]*Kansas Geological Survey, Lawrence, Kansas, U.S.A.*
[2]*U. S. Army Engineer Research and Development Center, Vicksburg, Mississippi, U.S.A.*

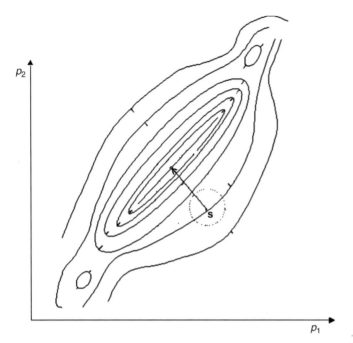

**Figure 1.** A 2D map of the mismatch error (objective) function. Conceptual continuous nonuniqueness visualized using 2D (i.e., two-parameter, $p_1$ and $p_2$) continuous nonuniqueness, which can be viewed as a valley in the 2D map. In the absence of accurate reference information about $p_1$ or $p_2$, an inaccurate reference parameter $s$ can be used to choose a point in the valley that is closest.

These critical assumptions can be a constant, linear, or nonlinear function based on any available information about the distribution of the $V_P/V_S$ ratio across the specific site. In the absence of any supporting information, the general trend of $V_P$ is selected to follow $V_S$. This general rule of thumb is based on observation by many researchers (e.g., Lay and Wallace, 1995) and the commonality $V_p$ and $V_S$ have with the elastic moduli (Sheriff, 1994). The overall match between the observed first arrivals and those calculated from the derived pseudo-$V_P$ model determine the degree of $V_S$ rescaling necessary. The pseudo-$V_P$ model is used as an initial model, as well as a reference model during the inversion process.

Traditional geophysical deterministic inversion seeks a solution to the least-square system:

$$\begin{bmatrix} \mathbf{L} \\ \beta\mathbf{D}_d \\ \lambda\mathbf{D}_s \end{bmatrix} [\mathbf{s}^{est}] = \begin{bmatrix} \mathbf{t}^{obs} \\ \beta\mathbf{s}^a \\ \lambda\mathbf{h} \end{bmatrix}, \qquad (1)$$

where matrix $\mathbf{L}$ represents the ray lengths through the earth model, $\mathbf{s}^{est}$ is the model vector of the estimated velocity field, and $\mathbf{t}^{obs}$ is a vector of observed first-arrival times.

Regularization is present in the form of weighted smoothing ($\lambda$; not to be confused with Lamé's constant or wavelength) and damping ($\beta$) constraints. Damping constrains the solution to the neighborhood of the reference a priori model $\mathbf{s}^a$. The matrix containing weights for the reference model (usually set to a value of 1) is $\mathbf{D}_d$, and $\mathbf{D}_s$ is the matrix containing the smoothing constraints (first, second, or higher derivative). Vector $\mathbf{h}$ is usually set to 0, resulting in maximum degree of smoothness.

The JARS method expands the system in equation 1 to include the reference pseudo-$V_P$ model in equation 2:

$$\begin{bmatrix} \mathbf{L} \\ \beta\mathbf{D}_d \\ \beta_2\mathbf{D}_d \\ \lambda\mathbf{D}_s \end{bmatrix} [\mathbf{s}^{est}] = \begin{bmatrix} \mathbf{t}^{obs} \\ \beta\mathbf{s}^a \\ \beta_2\mathbf{s}^{aa} \\ \lambda\mathbf{h} \end{bmatrix}, \qquad (2)$$

where $\mathbf{s}^{aa}$ is the reference pseudo-$V_P$ model and $\beta_2$ is the corresponding weighting coefficient. The damping weight $\beta$ is used to control the influence of hard evidence data about the model (e.g., data from wells) and is given greater weight, although $\beta_2$ is used to control the influence of the reference pseudo-$V_P$ model and is given smaller weight. The smaller the weight, the greater variance the solutions can have from the reference model (Ivanov et al., 2006). Such smaller weight reflects the understanding that the reference pseudo-$V_P$ model of the JARS method is inaccurate for a variety of reasons, including possible errors in $V_S$ estimates and $V_P/V_S$ ratio trend approximations. As a result, the reference model is viewed to be off to the side of the nonuniqueness valley, and therefore it is used as a guide (clue) for selecting a solution, i.e., a point in the nonuniqueness valley that is closest to the reference model (Figure 1).

An important element of the JARS algorithm is the removal of conventional regularization (e.g., smoothing) in the vertical direction so the velocity gradient with depth would be influenced mainly by the $V_S$ trend derived from surface-wave analysis (through the pseudo-$V_P$ reference model), and not by any mathematical assumptions used by conventional algorithms (Ivanov et al., 2006). Comparisons using seismic data acquired in the Sonoran Desert, Arizona, U.S.A., clearly suggest the JARS method can provide more realistic appearing results relative to standard IRTP algorithms (Ivanov et al., 2006). The qualitative term *realistic* reflects the observation of traits, such as horizontal layering and channel features, which are typical for the Sonoran Desert.

We prefer to use the multichannel analysis of surface waves (MASW) method (Song et al., 1989; Miller et al., 1999; Park et al., 1999; Xia et al., 1999), developed to estimate near-surface S-wave velocity from high-frequency ($\geq 2$ Hz) Rayleigh-wave data to obtain the $V_S$ model of the subsurface that can be used by the JARS algorithm. Shear-wave velocities estimated using MASW have correlated with drill data reliably and consistently. Using the MASW method, Xia et al. (2000) noninvasively measure $V_S$ within 15% of $V_S$ measured in wells. Miller et al. (1999) map bedrock with 0.3-m accuracy at depths of about 4.5 to 9 m, as confirmed by numerous borings. The Rayleigh-wave MASW imaging method includes three steps: fundamental-mode dispersion-curve estimation from compressional-wave shot-record data, dispersion-curve inversion into a 1D vertical $V_S$ profile, and calculation of a pseudo-2D $V_S$ profile using an interpolation algorithm (Miller et al., 1999).

## Texas and New Mexico test sites

Refraction-tomography and JARS algorithms were tested at several levee sites along the Rio Grande in the San Juan quadrangle, Texas, U.S.A. and the La Mesa quadrangle, New Mexico, U.S.A. (Figure 2). Simple two-layer solutions that would have been appropriate for the data were not explored because such simple models could not meet the resolution objectives of the surveys for these sites (Ivanov et al., 2006). The near-surface sediments of the Rio Grande Valley consist of river deposits sequentially laid down during and after the last major episode of valley entrenchment. At the base of these deposits are pebble- to cobble-sized gravels, although sands with little to no gravel are near the ground surface (Gile et al., 1981). Local deposits of fine-grained sediment resulted from complex channel shifts that caused slack-water conditions (Hawley and Lozinsky, 1992).

Surface-wave data were acquired as part of a wide-ranging investigation effort intended to evaluate the applicability of several seismic techniques to identify, delineate, and estimate changes in physical characteristics or properties of materials within and below levees during high-water events. Several surface seismic measurements were taken using state-of-the-practice equipment and were analyzed using both well-established and new research methods. Analysis methods included compressional and shear-wave refraction tomography, surface-wave propagation, surface-wave dwell tuning, and MASW of both Rayleigh and Love waves. Multi-channel surface-wave techniques (e.g., Song et al., 1989; Park et al., 1999) have proven capable of detecting anomalous shear-wave velocity zones within and below fill materials (Miller et al., 1999; Xia et al., 1999; Xia et al., 2004) and have provided

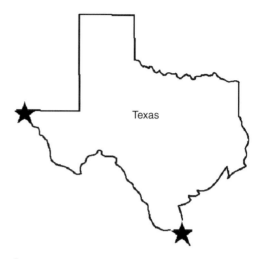

**Figure 2.** The star at bottom indicates location of the San Juan quadrangle, Texas, U.S.A. The star on the left indicates location of the La Mesa quadrangle, New Mexico, U.S.A.

shear-wave-velocity maps reliably within and beneath levees.

We compare solutions using the JARS method with those obtained from conventional refraction tomography (Zhang and Toksöz, 1998) using several seismic data sets. At all sites, refraction-tomography and JARS methods provided different solutions that match observed data equally well; therefore, both approaches were considered equally possible from a numerical perspective.

# Data Acquisition
## Southern Texas site

Data were acquired using two 2D, two-component (2C) survey lines with receiver lines deployed along adjacent edges of the levee road (Figure 3). The crest of the 5-m-high levee was approximately 6 m wide, with a 1:3-ratio slope on each side. Receiver station spacing was 0.9 m with two receivers at each location (10-Hz vertical geophones and one 14-Hz horizontal geophone oriented perpendicularly to the profile). The total spread length was 108 m with 120 channels recording compressional data and 120 channels recording shear data. Source spacing through the spread was 1.8 m with off-end shooting extending out to a distance equivalent to the maximum depth of investigation. Sources tested included sledgehammers of various sizes and a mechanical weight drop, each impacting appropriately sized striker plates. A 7.25-kg sledgehammer was used to acquire three records per station. Each data set (compressional and shear) was processed using a variety of methods and flows (Ivanov et al., 2005a).

## Southern New Mexico site

This work on this site was a continuation of an applied research project designed to evaluate the applicability of several seismic techniques to identify, delineate, and estimate the changes in physical characteristics or properties of materials within levees during a simulated flood event. A pond, approximately 30 m wide, was designed and built using earth material on the south side of a levee segment, which was of interest. A 2D vertical-geophone survey line was deployed along the edge of the levee road toward the pond. This 3-m-high levee was approximately 6 m wide at the crest with a 1:3-ratio slope on each side. The seismic-survey line was on the pond side of the levee (Figure 4). Receivers were single 10-Hz vertical geophones

**Figure 3.** South Texas site, on top of the levee road. Photograph by R. Miller. Used by permission.

**Figure 4.** Southern New Mexico site, on top of the levee road, next to the pond. Photograph by R. Miller. Used by permission.

spaced 0.6 m apart. Sources tested included sledgehammers of various sizes and a mechanical weight drop, each impacting appropriately sized striker plates. A 7.25-kg sledgehammer was used to acquire three records per station. A 120-channel seismograph system was used to record the compressional energy, resulting in a total spread length of 72 m. Source spacing through the spread was 2.4 m with shot stations extending beyond the 120-channel spread a distance equivalent to the maximum depth of investigation.

# Results

## Southern Texas site

Compressional-wave first arrivals were picked from data acquired along the crest. Two distinctively different apparent first-arrival velocity trends are evident based on trace-to-trace comparison (Figure 5). The numerical velocity models were made of $125 \times 26$ cells with a square cell size of 0.91 m. Comparing a JARS IRTP 2D $V_P$ solution (Figure 6a) with a conventional refraction-tomography solution (Figure 6b) estimated using the preferred second-order smoothing regularization (Depprat-Jannaud and Lailly, 1993; Zhang and Toksöz, 1998) demonstrates the increased apparent detail possible using JARS. The JARS algorithm uses a reference pseudo-2D $V_P$ initial model derived from a 2D $V_S$ model. Justification for this approach at this site was based on the general assumption (Ivanov et al., 2006) that the $V_P$ trend follows that of $V_S$, and the most effective way to estimate $V_S$ using compressional data is with the MASW method (Figure 6c). It was difficult to find a simple $V_P/V_S$ trend at this site. We tested constant, linearly decreasing, and increasing-with-depth $V_P/V_S$ trends, searching for the one that would provide a pseudo-2D $V_P$ model, which would produce first arrivals that would fit the observed data best overall. We managed to find a trend with distinct upper and lower portions. The selected $V_P/V_S$ function had a constant value of 6 for the bottom and middle layers of cells followed by a step drop down to 2.8 at 9 m and gradually decreased to 2.1 (2.8, 2.7, 2.5, 2.4, 2.3, 2.1) at 4-m depth. The top 3 m of the $V_P$ model were estimated from the direct wave observed on the few traces nearest to the source. We speculate that such a trend at the top 9 m might be a result of the levee-related construction activities at the site. The reference pseudo-2D $V_P$ model is identical in appearance to the 2D $V_S$ profile with the scaling function applied related to $V_P/V_S$ (Figure 6c).

Blank areas within the tomography images intentionally were left without interpolation to highlight areas that lacked ray coverage. Those coverage holes are an indica-

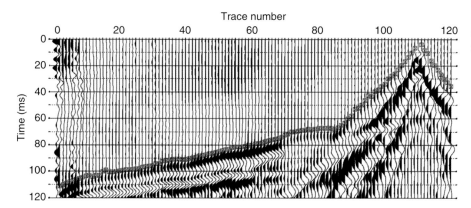

**Figure 5.** First-arrival picks from a compressional-wave seismic-shot record recorded at levee crest in southern Texas.

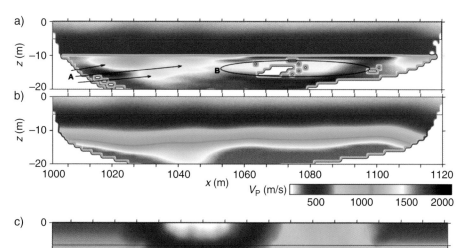

**Figure 6.** Southern Texas 2D images from compressional-wave seismic data acquired at the levee crest. (a) JARS compressional-wave solution with channel-like features indicated by A (black arrows), with B (black ellipse) indicating low-velocity intervals. (b) Conventional compressional-wave refraction-tomography solution. (c) Rayleigh-wave MASW shear-wave estimates. Red rectangles at the very top indicate the 5-m-high levee. Blank areas within the JARS image indicate lack of ray coverage and are retained deliberately.

tion of the instability and nonuniqueness of the inverse problem. Any velocity values within the blank cells lower than the surrounding cells can represent a valid solution.

The JARS solution is consistent with the expected geology along the Rio Grande Valley based on published geologic information in this area (Gile et al., 1981; Hawley and Lozinsky, 1992). Channel-like features are evident in the left half of the line below the base of the levee. A low-velocity zone sandwiched between high-velocity layers is distinguishable below the base of the levee across the right half of the section.

Data from the levee toe was picked (Figure 7) and processed to obtain a JARS 2D $V_P$ solution (Figure 8a) as well as a conventional refraction-tomography 2D $V_P$ image (Figure 8b). Consistent with previous observations, the JARS image appears to be a more geologically plausible and detail-oriented representation at this setting. The JARS

image reveals horizontal layering patterns, laterally uniform velocity trends, and mild velocity variations, which are expected for the Rio Grande Valley. In comparison, the conventional tomography image exhibits two very-high-velocity anomalies at depth at both ends and a very-low-velocity, vertically unbound wide channel following the very high velocity on the left. The latter features do not match any geologic expectation, and therefore appear unrealistic. A relatively low-velocity zone under a high-velocity lens also was imaged by the JARS method between depths of 10 to 13 m and a length of about 20 m — not evident on conventionally processed data.

## Southern New Mexico site

Compressional-wave seismic data were recorded on the levee crest at the Southern New Mexico site. First arrivals were picked (Figure 9) and processed through the same series of steps previously described for the South Texas site. The numerical velocity models were made of $125 \times 24$ cells using a square cell size of 0.61 m. We tested $V_P/V_S$ trends that were constant, linearly and stepwise decreasing, and increasing with depth, searching for one that would provide a pseudo-2D $V_P$ model that would produce first arrivals fitting the observed data best overall.

**Figure 7.** First-arrival picks from a compressional-wave seismic-shot record recorded at the levee toe in southern Texas.

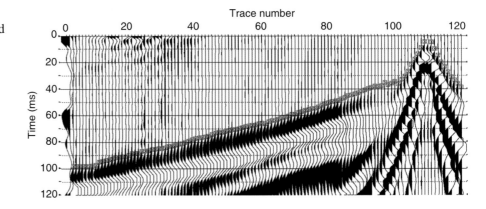

**Figure 8.** Southern Texas 2D images from compressional-wave seismic data acquired at the levee toe. (a) The JARS compressional-wave solution. Black ellipse encircles an interval A, which has low velocity under high velocity. (b) Conventional compressional-wave refraction-tomography solution. (c) Rayleigh-wave MASW shear-wave estimates.

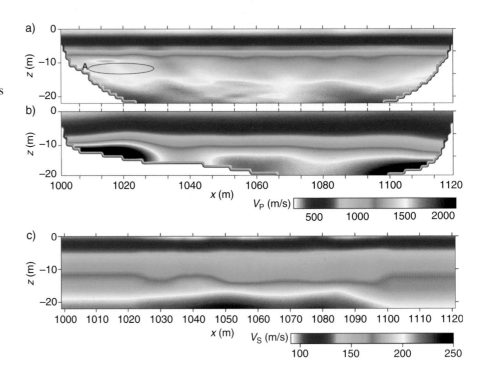

**Figure 9.** First-arrival picks from a compressional-wave seismic-shot record recorded at levee crest in southern New Mexico.

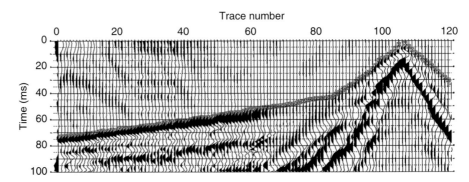

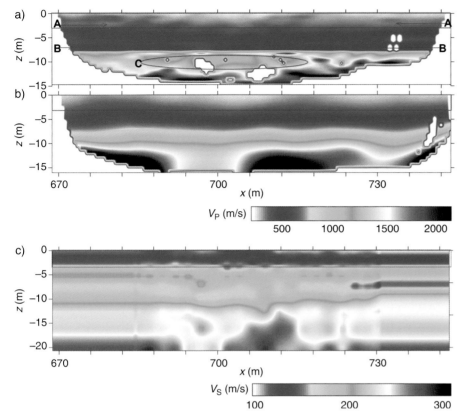

**Figure 10.** Southern New Mexico 2D images from compressional-wave seismic data acquired at the levee crest. (a) The JARS compressional-wave solution, with low-velocity intervals, A, B (red arrows), and C (red ellipse). (b) Conventional compressional-wave refraction-tomography solution. (c) Rayleigh-wave MASW shear-wave estimates. Red rectangles at the very top indicate the 3-m-high levee. Blank areas within the JARS image indicate lack of ray coverage and are retained deliberately.

The selected $V_P/V_S$ function had a constant value of 6 for the bottom layers of cells, decreased to 3 from a depth of 8 m to 6 m, and to 2.6 for the top 6 m. We speculate that such a trend at the top 8 m might result from levee-related construction activities. Comparing the JARS 2D $V_P$ solution (Figure 10a) with the conventional refraction-tomography 2D $V_P$ estimates (Figure 10b), once again, it is apparent that the JARS solution possesses more detail and has a texture consistent with the expected depositional geometries. After close examination of the JARS image, three distinct low-velocity zones appear layered between high-velocity layers. The velocity inversion with the greatest contrast is between 9 and 11 m below ground surface. This feature appears similar to the one interpreted on JARS images from below the south Texas levee. Another low-contrast, low-velocity layer is evident between 7 and 8 m of depth. Velocity inversions of these types are consistent with the depositional settings along the Rio Grande Valley, where alternating sandy and clayey sand deposits are encountered in drill holes. The top 3 m imaged by the JARS method is consistent with the known levee construction methods where a higher-velocity levee core at a depth of 1 to 2 m sits a top a lower-velocity native base (Hawley and Lozinsky, 1992).

Additional comparisons of specific features on crest and toe data processed equivalently through JARS and

conventional tomography consistently possessed similar patterns and characteristics across all levees interrogated during this study.

All solutions provided a good fit between the measured and calculated first arrivals, converging to below 2-ms rms error, which is less than 4% for most of the long-offset first arrivals ranging from 50 to 110 ms (Figures 3, 7, and 9). Such a misfit error was estimated to be an appropriate threshold for these data because a lower rms error could be achieved only at the expense of lesser regularization but greater instability.

## Discussion

The successful application of the JARS method requires the use of an adequate $V_P/V_S$ ratio trend. In the absence of such information, the JARS method assumes that the general $V_P$ trend is consistent with the $V_S$ trend following a constant or a simple (e.g., linear) function. For these particular levee-site data sets, this assumption was valid for two subsets of the velocity model: the upper and lower portions of the sections. We arrived at such $V_P/V_S$ ratio trends because it was not possible to find simpler trends that would fit the observed first-arrival data acceptably well.

We realize that simple assumptions about the $V_P/V_S$ trend might not be adequate or valid for other sites as well. In some settings with extreme lateral variability in deposition, a variable $V_P/V_S$ trend might be needed to optimize a JARS solution. From a practical perspective, however, the $V_P/V_S$ ratio trend utilized by JARS (regardless of how accurate it is) appears to be an improvement for selecting the initial geophysical model and constraining the inversion compared with the purely mathematical assumptions used by conventional refraction (tomography) algorithms. That perspective has been supported empirically in comparisons of JARS results with those from conventional methods, in this manuscript and in all previously published results.

For these levee case studies, the JARS IRTP $V_P$ solutions appear equally possible with those obtained from conventional refraction tomography based on depositional settings, local drilling information, and past JARS studies. Additionally, the JARS method was able to provide IRTP $V_P$ solutions that included low-velocity layers between high-velocity layers. Such solutions are not possible using conventional IRTP algorithms. The ability of the JARS method to image low-velocity layers comes from the establishment of an overall vertical-velocity gradient from the site-specific reference physical model (derived from $V_S$). Establishment of such a trend is a factor that can improve the vertical resolution of the model. This can be inferred from the three-layer model nonuniqueness research (Ivanov et al., 2005b). In that study, a priori knowledge of the depth of the third high-velocity layer uniquely defined the thickness and velocity of the second hidden (or blind zone) layer. The JARS approach produced images with low-velocity layers and accurate, high-resolution lithologic representations consistent with local geologic understanding. This manuscript uses seismic data acquired using a standard approach that addresses real-world problems.

From a numerical perspective, JARS solutions provide models that address the nonuniqueness problem that all conventional refraction-tomography solutions suffer. Conventional refraction algorithms use mathematically based assumptions in defining their initial model (i.e., number of layers, type and degree of regularization, etc.). The JARS method uses physical measurements of $V_S$ with only assumptions or a priori information about its relationship to $V_P$ to condition the initial model. Thus, from a geophysical perspective, the JARS algorithm offers $V_P$ solutions that are numerically sound, physically plausible, and consistent with available ground truth and $V_S$ trends calculated using MASW.

The best way to address nonuniqueness is to use hard evidence about the earth model (e.g., well logs), which is expensive and very often an impractical approach. When there is no hard evidence for such reasons, the JARS method could serve as a possible way to address the first-arrival-time inverse-problem nonuniqueness.

Two-layer (or three-layer) refraction solutions defined by using data with only two (or three, respectively) unique, first-arrival apparent-velocity trends can match the observed data reasonably. However, in many cases, this simplistic approach will not meet survey objectives at levee sites such as the ones studied. In view of the wide range of refraction nonuniqueness, when mathematical or physical assumptions used by these methods to define an initial model (velocity gradient, $V_P/V_S$ ratio trend, etc.) are not accurate, we speculate that the refraction two- or three-layer approach might provide realistic velocity models in comparison to either the refraction-tomography or JARS methods.

## Conclusions

Comparisons of experimental results from application of the joint analysis of refractions with surface-waves method to conventional refraction-tomography methods demonstrate that this new inversion method has the potential to analyze first arrivals equally well for velocity-field estimation. Solutions obtained from JARS in all published studies appear possible and geologically realistic. Therefore, JARS can be considered an advancement in the struggle with the inverse refraction-traveltime problem.

The JARS expansions principally lie in the qualitative nature in the development of the initial model. This statement requires quantitative support. Hard evidence or ground truth is required to define and resolve the refraction nonuniqueness problem. The JARS method results previously discussed provide a viable option for investigating and possibly narrowing the range of refraction nonuniqueness at a specific site. The main value of this work is a heightened awareness of how different refraction solutions can be while still being equally possible from a numerical perspective. Also, it is important to note how explicit and implicit assumptions about the velocity model can affect the final results.

## Acknowledgments

Without the support and assistance of the International Boundary and Water Commission, this research would not have been possible. We are deeply regretful that Bob Ballard, U. S. Army Corps of Engineers, passed away before his ideas and visions about interrogating levees with seismic could be developed fully. We will continue to explore his many creative and truly ingenious ideas for years to come. He was a true visionary and a southern gentleman. The U. S. Border Patrol provided a safe working environment for our field crew. We appreciate the discussions with Öz Yilmaz on the JARS method. We are very thankful to John Brandford and two anonymous reviewers for their helpful comments and suggestions. We also appreciate Mary Brohammer for her assistance in manuscript preparation. Permission to publish this paper was granted by the commissioner of the U. S. IBWC, and by the director of the Geotechnical and Structures Laboratory, ERDC.

# References

Ackerman, H. D., L. W. Pankratz, and D. Dansereau, 1986, Resolution of ambiguities of seismic refraction traveltime curves: Geophysics, **51**, 223–235.

Burger, H. R., 1992, Exploration geophysics of the shallow subsurface: Prentice Hall, Inc.

Delprat-Jannaud, F., and P. Lailly, 1993, Ill-posed and well-posed formulations of the reflection traveltime tomography problem: Journal of Geophysical Research, **98**, 6589–6605.

Gile, L. H., J. W. Hawley, and R. B. Grossman, 1981, Soils and geomorphology in a basin and range area of southern New Mexico — Guidebook to the desert project: New Mexico Bureau of Mines and Mineral Resources, Memoir 33.

Hawley, J. W., and R. P. Lozinsky, 1992, Hydrogeologic framework of the Messilla Basin in New Mexico and western Texas: New Mexico Bureau of Geology and Mineral Resources, Open-file report 323.

Healy, J. H., 1963, Crustal structure along the coast of California from seismic-refraction measurements: Journal of Geophysical Research, **68**, 5777–5787.

Ivanov, J., R. D. Miller, J. B. Dunbar, and S. Smullen, 2005a, Time-lapse seismic study of levees in southern Texas: 75th Annual International Meeting, SEG, Expanded Abstracts, 1121–1124.

Ivanov, J., R. D. Miller, J. Xia, and D. Steeples, 2005b, The inverse problem of refraction traveltimes, part II: Quantifying refraction nonuniqueness using a three-layer model: Pure and Applied Geophysics, **162**, 461–477.

Ivanov, J., R. D. Miller, J. Xia, D. Steeples, and C. B. Park, 2006, Joint analysis of refractions with surface waves: An inverse solution to the refraction-traveltime problem: Geophysics, **71**, no. 6, R131–R138.

Lay, T., and T. Wallace, 1995, Modern global seismology: Academic Press, Inc.

Menke, W., 1989, Geophysical data analysis: Discrete inverse theory: Academic Press, Inc.

Miller, R. D., J. Xia, C. B. Park, and J. M. Ivanov, 1999, Multichannel analysis of surface waves to map bedrock: The Leading Edge, **18**, 1392–1396.

Park, C. B., R. D. Miller, and J. Xia, 1999a, Multichannel analysis of surface waves: Geophysics, **64**, 800–808.

Sheehan, J., and W. Doll, 2003, Evaluation of refraction tomography codes for near-surface applications: 73rd Annual International Meeting, SEG, Expanded Abstracts, 1235–1238.

Sheriff, R. E., 1994, Encyclopedic dictionary of exploration geophysics, 4th ed.: SEG Geophysical References Series No. 13, 100.

Slichter, L. B., 1932, The theory of interpretation of seismic traveltime curves in horizontal structures: Physics, **3**, 273–295.

Song, Y.Y., J. P. Castagna, R. A. Black, and R. W. Knapp, 1989, Sensitivity of near-surface shear-wave velocity determination from Rayleigh and Love waves, 59th Annual International Meeting, SEG, Expanded Abstracts, 509–512.

Xia, J., C. Chen, P. H. Li, and M. J. Lewis, 2004, Delineation of a collapse feature in a noisy environment using a multichannel surface wave technique: Geotechnique, **54**, 17–27.

Xia, J., R. D. Miller, and C. B. Park, 1999, Estimation of near-surface shear-wave velocity by inversion of Rayleigh wave: Geophysics **64**, 691–700.

Xia, J., R. D. Miller, C. B. Park, J. A. Hunter, and J. B. Harris, 2000, Comparing shear-wave velocity profiles from MASW with borehole measurements in unconsolidated sediments, Fraser River Delta, B. C., Canada: Journal of Environmental and Engineering Geophysics, **5**, 1–13.

Zhang, J. and M. N. Toksöz, 1998, Nonlinear refraction traveltime tomography: Geophysics, **63**, 1726–1737.

# Section 4

# Case Studies

Chapter 21

# Near-surface Shear-wave Velocity Measurements for Soft-soil Earthquake-hazard Assessment: Some Canadian Mapping Examples

J. A. Hunter[1], D. Motazedian[2], H. L. Crow[1], G. R. Brooks[1], R. D. Miller[3], A. J.-M. Pugin[1], S. E. Pullan[1], and J. Xia[3]

## Abstract

The shear-wave velocity profile of a near-surface soil and/or rock site is one of the most important parameters for geotechnical estimation of earthquake shaking response at the ground surface. Downhole and surface seismic methods for measuring shear-wave velocities and mapping subsurface impedance boundaries include seismic cone penetrometer, downhole shear-wave vertical seismic profiling, surface-geophone array sites using refraction and reflection methods including array-to-source reversals, multichannel analysis of surface waves, seismic-reflection profiling, and horizontal-to-vertical spectral analyses of ambient noise. A suite of seismic shear-wave measurement methods has been tested in two Canadian cities with relatively high seismic hazard. Regional maps of National Earthquake Hazard Reductions Program (NEHRP) seismic site classifications (following the National Building Code of Canada) and fundamental site periods were created with the data. These maps indicate that broadband amplification effects and fundamental resonance periods can be extremely variable over short lateral distances within both survey areas. Such information needs to be considered by land-use planners and engineers working in such areas. Shear-wave velocity techniques constitute the most versatile approaches to earthquake-hazard mapping and site investigations.

## Introduction

The nature of earthquake seismic waves radiating through the earth is strongly dependent on the source mechanism, the source location at depth, and the character of rock types along the travel path to a particular surface site. However, the character of the shaking at the ground surface (amplitude, frequency, and duration) is affected strongly by the materials through which the waves travel over the last few hundred meters (or less). It is known that damage from earthquake shaking tends to be concentrated at locations where soft soils are present. Damage from earthquake events, such as those in Niigata, Japan (1964), Alaska, U.S.A. (1964), Tangshan, China (1976), and more recently, Mexico City, Mexico (1985), Loma Prieta (near San Francisco), California, U.S.A. (1989), Northridge, California, U.S.A. (1994), and Kobe, Japan (1995) exemplify this phenomenon (Anderson et al., 1986; Holzer, 1994; Choi and Stewart, 2005).

Soft soil conditions are correlated with low near-surface shear-wave velocities, usually with strong positive

[1]*Geological Survey of Canada, Ottawa, Ontario, Canada.*
[2]*Carleton University, Ottawa, Ontario, Canada.*
[3]*Kansas Geological Survey, Lawrence, Kansas, U.S.A.*

velocity gradients at depth (Kramer, 1996). Shear-wave velocity gradients as well as seismic impedance boundaries (e.g., the overburden-bedrock interface) can result in strong earthquake amplification effects such as

1) velocity gradient amplification across a significant impedance boundary, which can result in a shortening of shear-wave wavelengths and an increase in shear-wave amplitudes over a wide frequency band of earthquake shaking as the seismic energy passes from a high-velocity medium (e.g., rock) to a lower-velocity medium (e.g., soil) (Shearer and Orcutt, 1987)

2) resonance amplification, forming when seismic shear waves that have traveled up through the crust reflect back and forth between the free surface of the ground and the underlying impedance boundary at the soil-bedrock interface. The resonance amplification effect at the fundamental frequency and higher harmonics can be significantly larger than broadband amplification effects.

3) focusing or defocusing effects, much like a concave or convex mirror modifying light beams (Bard and Bouchon, 1985)

4) basin-edge effects, in which upcoming seismic waves impinging on the edges of the buried bedrock valley might generate surface waves that can interfere constructively within the buried valley feature, resulting in anomalously large-amplitude horizontal and vertical energy (Lomnitz et al., 1999)

The key to unlocking the complexities of such ground-motion effects lies in detailed delineation of the fundamental geotechnical and geophysical properties of soils and the underlying bedrock, including shear-wave velocity-depth functions and the identification and mapping of significant shear-wave velocity boundaries within the unconsolidated overburden sequence or at the overburden-bedrock contact. To this end, we have directed research efforts since 1985 toward development of cost-effective geophysical measurement techniques for the estimation of shear-wave velocities of soils and rock and their 3D structure (e.g., Hunter et al., 2002). Similar studies have been conducted in the United States (e.g., Williams et al., 1994; Williams et al., 1997), Italy (Cardarelli et al., 2008), Mexico (Ramos-Martínez et al., 1997), Turkey (Ulusay et al., 2004), Venezuela (Gonzalez et al., 2003), Romania (Mandrescu et al., 2006), and many other countries worldwide.

The need for reliable measurements of shear-wave velocities has been highlighted by the system developed by the National Earthquake Hazard Reduction Program (NEHRP) in the 1990s for the United States (Building Seismic Safety Council, 1994, 1995). Canada also adopted this system in

the 2005 National Building Code of Canada (NBCC) (Finn and Wightman, 2003; National Research Council, 2005). The definitions of NEHRP soil and rock classes as defined in the 2005 NBCC are given in Table 1 (NRC, 2005). Five of the six site conditions (classes A through E) are defined by either average seismic shear-wave velocities or average geotechnical properties such as undrained shear strength or values for standard blow count (N). The sixth site classification (F) requires further on-site geotechnical investigations and is defined on specific properties and thicknesses of poor soils. From Table 1, it can be seen that the most versatile measurement parameter that can be used in classes A through E is the thickness-weighted average shear-wave velocity of the upper 30 m below ground surface ($V_S30$). As a result, it is apparent that $V_S30$ should be considered as the "front-line" measurement, and near-surface geophysical techniques for $V_S$ measurements should be considered first in geotechnical site evaluations.

These provisions of the NBCC are designed to address all aspects of seismic amplification of soils and rock, including the above-mentioned soft-soil phenomena. As an example, Table 2 gives the amplification factors (previously known as factors of safety) for accelerations in the period of 0.2 s (5 Hz), which structural engineers must use to account for near-surface site conditions when assessing seismic safety factors. Similar tables and definitions form an envelope of upper-boundary shearing-force conditions recommended for a 1:2500 return-period event (the design earthquake). Note that the input ground-force values (e.g., Adams and Halchuk, 2003, 2004) are given for "firm ground" (NEHRP zone C soil conditions) rather than "bedrock surface" conditions (NEHRP A).

In Canada, we have focused our geophysical research on two soft-soil areas of the country where the earthquake hazard is considered moderate to high: the Fraser River Delta of the lower mainland of British Columbia and the Ottawa–St. Lawrence lowlands of eastern Canada (provinces of Ontario and Quebec). The locations of these sites are shown on a seismic-hazard map of Canada in Figure 1. Both sites have large areal extents of geologically young unconsolidated materials (Holocene and Pleistocene) that can have considerable thicknesses and lateral variability of material types. We have examined geophysical techniques that characterize the soils and rocks not only to a depth of 30 m (per current NEHRP definitions) but rather down to and including firm bedrock (in some cases, to depths in excess of 1000 m). We attempt to summarize the application and testing of near-surface geophysical techniques along with specific examples of the regional mapping that can be produced from such measurements.

**Table 1.** Definition of NEHRP soil and rock classes. Reprinted with permission from NBCC, 2005.

| Site class | Ground profile name | Average properties in top 30 m | | |
|---|---|---|---|---|
| | | Traveltime-weighted average shear-wave velocity $V_S30$ (m/s) | Average standard penetration resistance $N_{60}$ | Soil undrained shear strength $s_u$ (kPa) |
| A | Hard rock | $V_S30 > 1500$ | n/a | n/a |
| B | Rock | $760 < V_S30 \leq 1500$ | n/a | n/a |
| C | Very dense soil and soft rock | $360 < V_S30 \leq 760$ | $N_{60} > 50$ | $s_u > 100$ |
| D | Stiff soil | $180 < V_S30 \leq 360$ | $15 \leq N_{60} \leq 50$ | $50 \leq s_u \leq 100$ |
| E | Soft soil | $V_S30 \leq 180$ | $N_{60} < 15$ | $s_u < 50$ |
| | | Any profile with more than 3 m of soil with the following characteristics: <br> • plasticity index: PI > 20 <br> • moisture content: w ≥ 40%, and <br> • undrained shear strength: $s_u < 25$ kPa | | |
| F | Other soils[4] | Site-specific evaluation required | | |

[4]Other soils include: (1) liquefiable soils, quick and highly sensitive clays, collapsible weakly cemented soils, and other soils susceptible to failure or collapse under seismic loading; (2) beat and/or highly organic clays greater than 3 m in thickness; (3) highly plastic clays (PI > 75) more than 8 m thick; (4) soft to medium-stiff clays more than 30 m thick.

**Table 2.** Example of amplification factors ($F_a$; previously known as factors of safety) required to multiply earthquake shearing forces at any site in Canada for the 0.2-s period of shaking (5 Hz). After Finn and Wightman, 2003. Reprinted with permission from NRC, 2005.

| Site class | Values of $F_a$ | | | | |
|---|---|---|---|---|---|
| | $S_a(0.2) \leq 0.25$ | $S_a(0.2) \leq 0.50$ | $S_a(0.2) \leq 0.75$ | $S_a(0.2) \leq 1.00$ | $S_a(0.2) \leq 1.25$ |
| A | 0.7 | 0.7 | 0.8 | 0.8 | 0.8 |
| B | 0.8 | 0.8 | 0.9 | 1.0 | 1.0 |
| C | 1.0 | 1.0 | 1.0 | 1.0 | 1.0 |
| D | 1.3 | 1.2 | 1.1 | 1.1 | 1.0 |
| E | 2.1 | 1.4 | 1.1 | 0.9 | 0.9 |
| F | (5) | (5) | (5) | (5) | (5) |

[5]Other soils include: (1) liquefiable soils, quick and highly sensitive clays, collapsible weakly cemented soils, and other soils susceptible to failure or collapse under seismic loading; (2) beat and/or highly organic clays greater than 3 m in thickness; (3) highly plastic clays (PI > 75) more than 8 m thick; (4) soft to medium-stiff clays more than 30 m thick.

## Shear-wave Velocity-measurement Techniques

Hunter et al. (2002) provide an overview of shear-wave seismic methods used for thick-soil site investigations. Here, we summarize some of that material but also provide an update on several new methods (shear-wave reflection profiling, spectral motion measurements of ambient seismic noise) and developments in data processing

(MASW). The techniques are summarized in Table 3. First we will discuss downhole surveys (seismic cone-penetrometer testing and downhole vertical seismic profiling) because these are the most direct measurements of shear-wave velocity. Following that, surface techniques are presented, including surface refraction/reflection site surveys, MASW (multichannel analysis of surface waves), seismic reflection profiling, and passive measurements of the spectral ratio of horizontal-to-vertical motion.

One-dimensional earthquake models suggest that seismic energy might travel vertically through the soil column from depth. It is possible that vertical-to-horizontal shear-wave velocity anisotropy could exist in unconsolidated surficial sediments, particularly those that might be deposited through waterborne means (e.g., lacustrine or deltaic sediments) resulting from quasihorizontal stratigraphy or possibly grain-fabric orientation. Hence, techniques that measure the vertical traverse of shear waves are deemed to be most applicable to the assessment of earthquake hazards. Such methods include borehole vertical seismic profiling (VSP), seismic cone-penetration testing (SCPT), and downhole acoustic source and sensor arrays. On the other hand, if the investigation of a soil site requires delineation of possible azimuthal shear stresses, techniques that favor horizontal measurements of velocities with depth might be preferred.

## Seismic cone-penetrometer testing

The seismic cone penetrometer (SCPT) is a device with geophone(s) or accelerometer(s) embedded in the tip of a standard cone penetrometer (Figure 2; e.g., Robertson et al., 1986). It provides an excellent method for measuring in-situ shear-wave velocities in soft soils, with minimal disturbance of local sediments.

Finn et al. (1989) give a detailed description of the use of this tool in the Fraser River Delta. Hunter and Woeller (1990) compare downhole VSP and SCPT measurements in Holocene sediments in adjacent locations. Both surveys use a horizontally polarized surface source consisting of a 7.5-kg hammer and plate and similar horizontal geophone detectors. Shear-wave velocities measured by the two techniques were matched closely (Figure 3). However, the running least-squares fits of velocities from traveltime measurements suggested slightly less picking error with the SCPT. Such comparisons might vary with location and the skill of the SCPT operator or (for a cased borehole) the quality of the grouted casing. It is suggested that the close contact of the seismic cone with the formation materials results in better signal-to-noise ratio (S/N) than that of a downhole clamped geophone in a cased and grouted borehole. In addition, seismic tube-wave interference could limit the accuracy of some borehole measurements (see below). Thus, where soft soils exist to considerable depth and the seismic cone penetrometer can be pushed easily (with or without drillout), the SCPT would be a preferred method for determining shear-wave velocities. A major advantage of the SCPT approach is the acquisition of shear-wave velocities in combination with other geotechnical data.

## Downhole VSP

At locations where firm ground is encountered within 30-m depth or where the requirement is detection of firm

**Figure 1.** Locations of the two survey sites discussed in this paper, shown on a map that is a simplification of the seismic-hazard map from the National Building Code of Canada for spectral acceleration at a 0.2-s period (five cycles per second). This map shows the ground motions that might damage one- and two-story buildings. From http://earthquakes-canada.nrcan.gc.ca/hazard-alea/simphaz-eng.php; see also Adams and Halchuk, 2003, 2004. Reproduced with the permission of Natural Resources Canada 2010, Courtesy of Earthquakes Canada.

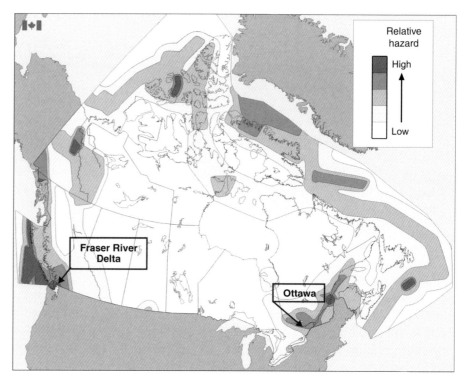

Relative hazard

High

Low

Fraser River Delta

Ottawa

ground at any depth, SCPT surveys are not an option. In those locations, it is suggested that a VSP survey in a cased and grouted borehole is the preferred method for measuring shear-wave velocities (e.g., Hunter et al., 1998a). This method can determine and characterize significant S-wave velocity boundaries at depth, which could cause resonance effects during earthquake shaking.

One potential problem with this technique is that poor grouting of the casing in some areas of the borehole can result in interfering "tube-wave" arrivals that might mask or deteriorate the quality of the shear-wave arrival-time pick. Tube-wave radiation also can occur as a result of

significant voids in the formation behind the casing or occasionally from significant shear-wave velocity contrasts in the formation (e.g., a boundary between soft porous material and firm ground) despite good-quality grouting.

Figure 4 shows an example downhole VSP from the Fraser River Delta of variable data quality resulting from "bridging" of the grout around a 3-inch casing in some areas of the borehole. However, the data still show clearly the distinct velocity increase (<2:1) associated with the silt-till interface at 53-m depth.

Figure 5 shows an example suite of borehole horizontal-component traces in soft soil (marine sediments) in the

**Table 3.** Shear-wave velocity-measurement techniques discussed in this chapter.

| Technique | Description | Advantages | Disadvantages | Application areas |
|---|---|---|---|---|
| Seismic cone penetrometer | Horizontal geophone(s) installed in a penetrometer pushed into soft soil, surface shear-wave source | Very good contact with soil, minimal signal-to-noise, additional geotechnical data collected, range can be extended with drillout | Usually limited to near-surface soft soils only, refusal in stiff soils | Zones containing thick soft (Quaternary) soils |
| Downhole shear-wave VSP | Well-locking horizontal geophone(s) in a PVC-cased borehole, surface shear-wave source | Good contact with soil depending on quality of casing grout, repeat measurements in preserved BH, ancillary geophysical logs, geologic and geotechnical samples | Cost of drilling and casing, possible poor-quality grouting resulting in tube-wave interference | All zones containing soft or stiff soils and rock |
| Site-specific refraction and/or reflection | Surface array of horizontal geophones, surface shear-wave source | Refraction (interval) velocities, reflection average velocities, for significant impedance contrasts | Refraction model requires $V_S$ to increase with depth, "hidden" layers missed | Most soils and rock where significant seismic-impedance boundaries occur |
| MASW | Surface array of geophones, inversion of dispersion curves | Economical, velocity inversions mapped | Depth limited | All zones containing soft or stiff soils and rock |
| Seismic-reflection profiling | Moving array of horizontal geophones, surface shear-wave source | Detailed subsurface shear-wave stratigraphic depths and velocities, velocity inversions mapped | Operational costs are high, significant impedance contrasts are required | Best results in soft soils overlying bedrock |
| Site resonance (HVSR) | Surface-mounted three-component low-frequency seismometers | Simple acquisition procedure, minimal equipment | Data quality strongly affected by local noise | All zones where subsurface seismic-impedance boundaries occur; best results in soft soils overlying bedrock |

a)

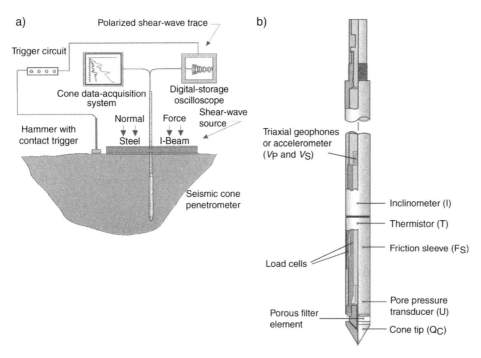

b)

Figure 2. (a) Schematic representation of the operation of a seismic cone penetrometer. (b) The components of the penetrometer instrument. Used by permission of Conetec, Ltd.

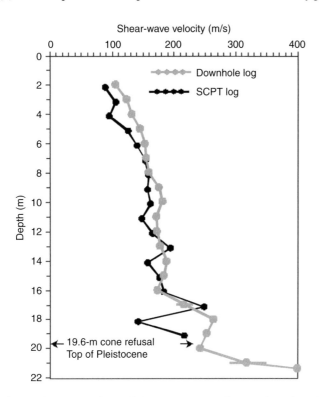

Figure 3. Comparison of shear-wave velocity as a function of depth obtained in the Fraser River Delta using SCPT and downhole VSP techniques. From Hunter et al., 1998b. Reproduced with the permission of Natural Resources Canada 2010, courtesy of the Geological Survey of Canada (Bulletin 525).

Ottawa, Ontario, area. These records were acquired using a well-locking 14-Hz three-component sonde in a cased borehole at sonde spacings of 0.5 m. The source was a Minivib Mark I swept-frequency horizontal vibratory source located on the surface, 4 m to one side of the well. In this case, the PVC casing was well grouted to the formation, and the signal-to-noise ratio was very high. The objective of the borehole and the well survey was to delineate and confirm the presence of a coarse-grained (Pleistocene?) aquifer beneath the (Holocene) fine-grained marine sediments, as first indicated by a surface shear-wave reflection-seismic survey (Pugin et al., 2007; Pugin et al., 2009a), but the resultant shear-wave velocity-depth profile also provides valuable information for earthquake-hazard analyses.

The fine-grained marine sediments of the St. Lawrence Lowlands of eastern Canada commonly exhibit a surface overconsolidated zone, which can reach a few meters in depth and is characterized by higher shear-wave velocities than the clay or silt at depth. These sediments also commonly show slight but well-defined velocity gradients with depth (Hunter et al., 2007; Motazedian and Hunter, 2008). Figure 5a shows these effects clearly through the curvature of the first shear-wave arrival times. In addition, reflections from the top of the aquifer at the bottom of the borehole and from the bedrock beneath the bottom of the hole are clearly visible. These reflections can be traced to surface as later arrivals. The curvature of these late arrival times duplicates the velocity gradients shown from the first arrivals. The shear-wave velocity-depth function in Figure 5b was derived from the first arrival times, using reflection-analysis techniques described by Hunter et al. (1998a). It clearly shows the velocity gradient through the marine-sediment sequence from 20- to 90-m depth and the beginning of the velocity increase associated with coarser sediments at the bottom of the borehole. Although these data cannot provide information on the velocities of these lower units, the high amplitudes of the reflections from the top of the aquifer and underlying bedrock surface suggest that these are large shear-wave impedance boundaries.

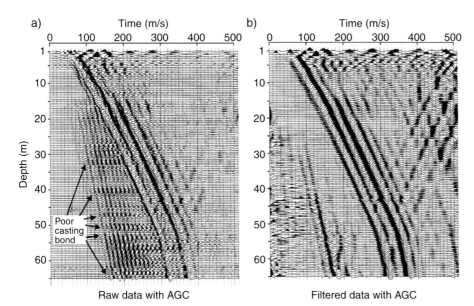

**Figure 4.** Downhole VSP data collected using a three-component geophone sonde, Fraser River Delta, British Columbia. (a) Raw and (b) filtered downhole record suites from one of the horizontal geophones show the interfering effects of poor casing bond in some areas of the borehole. The interference can be reduced by filtering out the higher frequencies, but it is at the expense of some resolution. A long automatic-gain-control (AGC) window has been used to normalize trace amplitudes. From Hunter et al., 1998a.

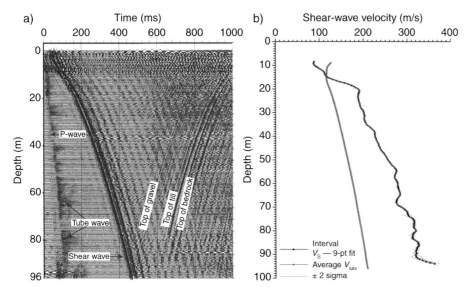

**Figure 5.** Downhole VSP data collected using a three-component geophone sonde and a Minivib vibratory source, Ottawa, Ontario, Canada. (a) Downhole record suite from one of the horizontal geophones showing a strong shear-wave arrival and reflections from below the base of the borehole interpreted to be from the underlying coarse-grained sediments and the bedrock surface. (b) Downhole shear-wave average and interval velocity profiles determined from the first-arrival picks of the data in (a). Running least-squares fits of traveltime picks yield shear-wave velocities at 0.5-m intervals; in this example, the fit is over nine points, or 4 m.

## Surface shear-wave refraction-reflection surveys

Surface shear-wave refraction-reflection surveys were conducted at sites throughout the Fraser River Delta (e.g., Hunter et al., 2002) and Ottawa (Crow et al., 2007; Motazedian and Hunter, 2008) survey areas to determine a shear-wave velocity-depth function in areas where boreholes were not available. At each site, a surface array of horizontally oriented geophones was laid out and records were obtained using shear-wave sources off both ends of the array.

Array lengths and source positions were varied depending on the anticipated layer velocities and depth to bedrock. Such arrays were positioned in available open spaces such as parks, schoolyards, edges of parking lots, and roadside rights-of-way when ambient noise conditions permitted.

Seismic-refraction methods rely on the fundamental assumption that (shear) velocities increase with depth, so the arrival time of refracted energy on surface at the geophone array can be used to estimate the velocity-depth function. In both project areas, this assumption is basically correct, and refraction methodology was applied (despite a thin, 1- to 5-m-thick, surface high-velocity screening layer encountered throughout the Ottawa area). In both the Fraser River Delta and Ottawa, low shear-wave velocity surface materials were associated with unconsolidated Holocene sediments, and these overlie Pleistocene glacial deposits. In the Fraser River Delta area, the basal bedrock unit consists of Tertiary sandstones, whereas bedrock in the Ottawa area consists of firm, lower Paleozoic limestones and shales and Precambrian gneiss. Within the Fraser River Delta survey area, the Holocene thickness range between 0 and

300 m, and the depth to the base of the Pleistocene ranges from approximately 400 to 1000 m. In contrast, in the Ottawa area, postglacial Champlain Sea sediments reach thicknesses of 120 m, and the underlying Pleistocene materials are relatively thin (about 3 m), thickening to about 15-m depth in buried bedrock topographic depressions.

Because of the differing subsurface lithologic structure, seismic-refraction array designs differed between the two survey areas. However, in both locations, the array lengths were designed to measure shear-wave refractions to a minimum depth of 30 m, so as to yield an estimate of $V_S30$. For the Fraser Delta, the standard array consisted of 210-m length with horizontal geophones spaced at 3 m. In the Ottawa region, geophone spacing was commonly 3 m, and the array length was 72 m. Both off-end forward and reversed seismic source locations were occupied, and where possible, 30-m step-off source locations also were recorded.

These same arrays were used for seismic-reflection measurements if there were significant velocity-density contrasts at the boundaries between the various infraoverburden lithologic units, including the top of bedrock. A least-squares fit of the hyperbolic, wide-angle reflection-event arrival times yielded the average shear-wave velocity from surface down to the boundary. The most significant reflections were associated with the Holocene-Pleistocene boundary in the Fraser River Delta and with the overburden-bedrock boundary in the Ottawa area. Impedance contrasts at these boundaries range from 7 to 45; however, where high-quality data were recorded, reflections from even minor changes in seismic impedance also could be detected and measured. Hence, by varying the geophone and source array geometry, data obtained from surface refraction-reflection sites yielded accurate and complete shear-wave velocity-depth functions, even at depths exceeding 100 m. Computation of $V_S30$ constitutes only a subset of the complete seismic-site data collected at most sites.

Figure 6 shows an example reversed-refraction shear-wave record and interpretation from the Fraser River Delta where the top of the Pleistocene surface is relatively deep. The 24-channel array of horizontal SH-polarized geophones was moved over a distance of 600 m along a roadside for successive records between the forward and reverse source locations. The source consisted of an eight-gauge "buffalo-gun" impulsive source that generated sufficient S-wave energy to produce clearly identifiable shear-wave arrivals out to 600-m offset. The first-arrival shear-wave event indicates a significant velocity increase from 120 m/s to 370 m/s through the Holocene sequence (Figure 6b). A significant velocity discontinuity is interpreted to correlate with the top of Pleistocene ($V_S = 540$ m/s) dipping between depths of 165 to 190 m across the spread.

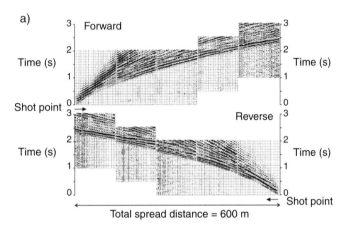

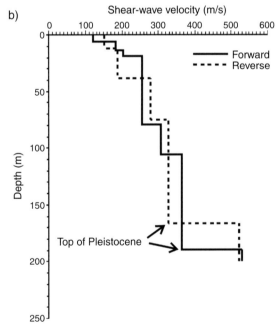

**Figure 6.** (a) Composite forward and reverse seismic records from the Fraser River Delta covering source-receiver offsets from 5 to 600 m at 5-m geophone spacings; the source was an eight-gauge, in-hole "buffalo gun." (b) Velocity-depth interpretations from layered-model refraction analysis of the first-arrival data. From Hunter et al., 2002. Copyright 2002. Used by permission of Elsevier.

Figure 7 shows a combined refraction-reflection shear-wave record from the Ottawa area. A polarized hammer-and-plate source was used to acquire data from a 24-channel array of SH-geophones. The time-distance data (Figure 7b) indicate the presence of a thin high-velocity (overconsolidated) layer (first arrivals on near traces). The main refractions through the Holocene marine sediments and the refraction from the top of bedrock are clearly visible as the first arrival at offsets of greater than 10 m. A significant wide-angle reflection associated with the bedrock surface also is clearly visible. However, filtering and the application of automatic gain control (not shown) make it possible to identify weak infraoverburden reflections to obtain several independent measurements of average velocity down through the section. Figure 8 shows the combined refraction-reflection interpretation of shear-wave velocity as a function of depth using the complete suite of data acquired from the forward, reverse, and offset source locations.

The cost of combined refraction-reflection site investigations can be relatively low compared to SCPT or borehole shear-wave velocity determinations. The technique can be used for regional reconnaissance mapping purposes and in urban areas where there is adequate space for the deployment of seismic arrays.

## Multichannel analyses of seismic waves (MASW)

Multichannel analysis of surface waves (MASW) is a noninvasive and environmentally friendly method, keenly suited for estimations of shear-wave velocity as a function of depth. The method consists of three unique steps: acquisition of time-versus-distance shot gathers, transformation of time-versus-distance data to the domain of phase velocity versus frequency, and finally, inversion of dispersion curves for shear-wave velocity as a function of depth. With an overall difference between the direct borehole measurements and inverted S-wave velocities of approximately 15% or less (Xia et al., 2002), MASW can be a cost-effective and noninvasive method of determining shear-wave velocity-depth functions or $V_S30$ where SCPT or borehole measurements are not feasible.

Traditionally, surface waves have been viewed as noise on multichannel seismic data collected to investigate targets for shallow-engineering, environmental, and groundwater purposes (Steeples and Miller, 1990). Advances in the use of surface waves for near-surface imaging have combined spectral-analysis techniques (SASW), developed for civil-engineering applications (Nazarian et al., 1983),

with multitrace reflection technologies developed for near-surface (Schepers, 1975) and petroleum applications (Glover, 1959). MASW is a combination of these two

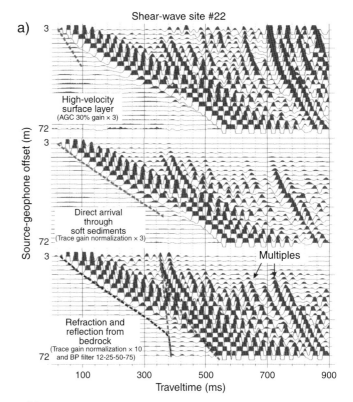

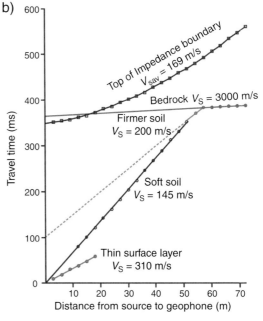

**Figure 7.** (a) One off-end shear-wave record from a site in the Ottawa area showing first arrivals though the high-velocity surface layer, postglacial materials, bedrock, and the associated bedrock reflection and multiples. (b) Plot of arrival time versus offset for the events indicated in (a).

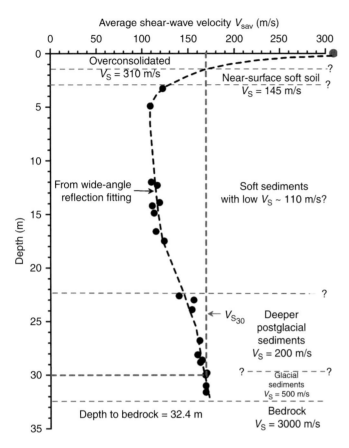

**Figure 8.** Plot of average shear-wave velocity versus depth from combined analysis of reflection and refraction data for a site in the Ottawa area. As shown by the medium-gray lines, such plots can be used to determine $V_S30$ (which equals 170 m/s at this site).

uniquely different approaches to seismic imaging of the shallow subsurface that permits noninvasive estimation of shear-wave velocities and delineation of horizontal and vertical variations in near-surface material properties based on changes in these velocities (MASW) (Xia et al., 1999; Park et al., 1999).

Economic advantages of the MASW method to measure S-wave velocity in near-surface material over other approaches are significant. Time needed for data acquisition at each station for the MASW method is approximately a few minutes; the time required for the SASW method is about one hour. The cost of acquiring borehole velocities can be an order of magnitude greater than using MASW to estimate the S-wave velocity (several thousand dollars versus several hundred dollars). Health, safety, and environmental concerns routinely play a major role in drilling and can increase the cost of a borehole measure significantly.

Multichannel surface-wave data were acquired at sites in the Fraser Delta with existing borehole measurements,

using 4.5-Hz vertical geophones and a 60-channel Geometrics StrataView seismograph. Geophones were deployed at the eight sites on either an interval of 0.6 or 1.2 m (depending on target-depth range) with the nearest source-to-geophone offset ranging from 1.2 to 90 m, again depending on target-depth considerations. Geophone spreads were placed as close as possible to the measurement well and never more than 50 m from it. Three to 10 impacts were stacked vertically at each offset using an accelerated weight drop designed and built by Kansas Geological Survey (KGS). To ensure that the entire surface-wave train was recorded and appropriately sampled, data for all sites were recorded with a 1-ms sample interval for 2048 ms. Once the geophone spread was deployed at a site, it took only two to three minutes to acquire all the data necessary to estimate the S-wave velocity profile confidently as a function of depth.

The MASW velocity profile at a "blind" test site in the Fraser River Delta matches borehole measurements extremely well (Figure 9). The relative difference between the results was only 9%. This almost insignificant difference is primarily the result of random noise and/or reflected ground roll contaminating portions of the Rayleigh-wave signals. An overall difference between the direct borehole measurements and inverted S-wave velocities of approximately 15% or less was observed at seven well locations in the Fraser River Delta area. Using blind testing as the ultimate verification of computational accuracy, the 9% differences between S-wave velocities using MASW and velocities measured in the blind-test borehole confirm the legitimacy and accuracy of this method.

Advancement in the accuracy of the MASW method has come from inclusion of higher modes into the inversion process. Higher modes are independent from the fundamental-mode phase velocities, and they exist under a specific frequency condition (Aki and Richards, 1980). Stokoe et al. (1994) note that generation of higher modes has been associated with the presence of a velocity reversal (a lower S-wave velocity layer between high S-wave velocity layers) and that higher-mode surface waves, when trapped in a layer, are much mores sensitive to the fine structure of the S-wave velocity field (Kovach, 1965). Unlike early exploitation of surface waves, multichannel surface-wave recording allows reliable observation of higher modes. Direct construction of high-resolution, multimodal dispersion curves from a multichannel record with a relatively small number of traces (e.g., 30 traces) spaced across a small lateral distance (20 m) (Figure 10) will produce notable improvements in the resolution and accuracy of shear-wave velocity estimates as a function of depth.

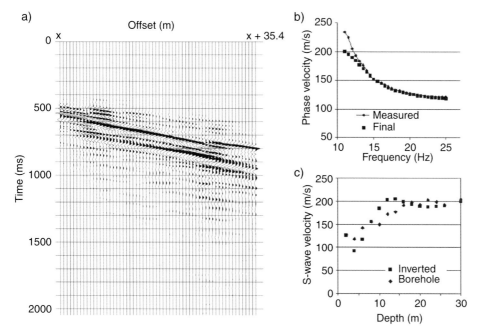

a) Offset (m)

b)

c)

**Figure 9.** (a) Field shot gather with 60 traces recorded close to a Fraser River Delta borehole. (b) Rayleigh-wave phase velocities measured from field data shown in (a) and calculated based on inverted $V_S$ model shown in (c); (c) inverted $V_S$ model compared to measured velocities in a borehole. The seismograph was triggered manually because of interference of an electro-magnetic field nearby. First arrivals are not related correctly to source-geophone offsets.

It must be noted, however, that applicability of the MASW method could be dependent on site and source; in some situations, it might be difficult to measure the low frequencies required to reach 30-m depth with standard sources and geophones. In those cases, it might be required to use lower-frequency geophones and larger-impact sources or to resort to passive-array monitoring of distance low-frequency ambient noise.

## Seismic-reflection profiling

Land-based seismic-reflection methods use measurements of the time taken for elastic energy to travel from a source on the surface through the subsurface and back to a series of receivers on the ground. Energy is reflected or refracted at boundaries where there is a change in acoustic impedance (the product of material density and seismic velocity). Because contrasts in acoustic impedance generally are associated with lithologic boundaries, seismic-reflection techniques can be used to obtain subsurface structural information (bedrock surface and infraoverburden stratigraphy) that is important in assessing resonance amplification, focusing, and basin-edge effects in earthquake-hazard analyses.

Early shallow seismic-reflection work was based almost exclusively on the analyses of compressional (P-) waves (e.g., Pullan and Hunter, 1990; Steeples and Miller, 1990; Steeples, 1998). These methods have produced very useful information in the Fraser River Delta area, as discussed below. There has been an ongoing interest in using shear-wave reflection methods for shallow applications (e.g.,

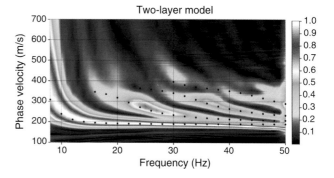

Two-layer model

**Figure 10.** Higher modes of Rayleigh waves appear shingled above the fundamental mode. Black dots represent analytical results calculated by the Knopoff method (Schwab and Knopoff, 1972).

Helbig and Mesdag, 1982; Woolery et al., 1993), but shear-wave methods have been applied more routinely in the last decade only. This has been associated largely with the development of landstreamers or towed receiver arrays (e.g., Inazaki et al., 1999, 2004; Pugin et al., 2006; Pugin et al., 2007; Pugin et al., 2009a; Pugin et al., 2009b).

More than 125 line km of conventional large-scale (related to oil and gas) P-wave reflection profiling was carried out in the Fraser River Delta by Dynamic Oil and Gas, Inc., and its partners. Although of relatively low frequency, the results were sufficient to identify the Holocene-Pleistocene boundary and the top of the buried Tertiary bedrock surface between 400 and 1000 m below surface (Britton et al., 1995). In addition, the 183 CMP (common-midpoint) velocity analyses done along these lines were sufficiently detailed to be used for estimation of the shear-wave velocity-depth function using an empirical compressional-shear-wave velocity relation given by Hunter et al. (1998b). The resulting deltawide velocity-depth function could be used to estimate the shear-wave velocity of the upper 500 m within the bedrock.

Higher-resolution subsurface information was obtained in the southern Fraser River Delta from more than 60 line km of high-resolution "optimum-offset" P-wave reflection profiles (Pullan et al., 1998). Hunter et al. (1998b) describe the results of high-resolution CMP profiling in the delta, where permitted by the good transmissive qualities of the soil. Initial CMP reflection tests of shear-wave velocity structure in young unconsolidated sediments were carried out in the Fraser River Delta using horizontal geophone arrays in SH mode and both polarized-hammer and swept-frequency (Minivib) sources (Hunter et al., 2002). In those cases, using planted geophone arrays and a moving source, productivity was limited to approximately 0.5 to 1.0 line km per day with a relatively large field crew.

Our recent research has focused on design of a landstreamer with geophones mounted on metal sleds and the mating of this with a variable swept-frequency vibrator source (e.g., Pugin et al., 2007; Pugin et al., 2009a; Pugin et al., 2009b). With our current minivib-landstreamer system (Figure 11), we routinely collect 3 to 5 line km

**Figure 11.** Minivib/landstreamer system from the Geological Survey of Canada in operation. In this survey, the landstreamer consists of three-component geophones mounted on 48 sleds at a spacing of 0.75 m. Photograph by Andre Pugin. Reproduced with the permission of Natural Resources Canada 2010, courtesy of the Geological Survey of Canada (Open File 6273).

per day of P- and S-wave reflection data along paved or gravel roads. For shallow applications, source spacings of 3 to 6 m are used commonly. Receiver spacings can be varied, depending on the target depths, but they typically range between 0.75 and 3 m. The vibrator source transmits a coded swept-frequency wave (usually 5 to 300 Hz over 7 s) into the ground, in either horizontal or vertical mode in a manner similar to but at higher frequency than the large vibrator sources used in oil exploration. The use of a vibrator source permits data to be obtained in noisy environments (e.g., urban areas, windy conditions, heavy traffic), which used to preclude the application of these techniques. Landstreamer signal-to-noise quality is as good as that acquired by conventional buried geophone arrays with impulsive point sources yet with a fivefold increase in the rate of data acquisition.

Figure 12 shows an example seismic section shot over a buried valley infilled with Champlain Sea postglacial sediments in the Ottawa area. Both P- and SV-wave sections were obtained using the vertical and inline horizontal components from a multicomponent data set obtained with the vibrator source oriented in horizontal mode (e.g., Pugin et al., 2009a). Most landstreamer work is conducted over firm roadbeds, which offer a thin surface layer with a high shear-wave velocity immediately below the source and receivers. The net effect is improved S/N through generation of only a narrow set of surface-wave noise (in both frequency and offset distance) that is removed by CMP stacking.

In the Ottawa area, shear-wave reflection sections indicate prominent seismic-impedance contrasts at depth and significant lateral stratigraphic changes within the postglacial and glacial sediments. Commonly, the most prominent shear-wave reflector on the section is related to the bedrock surface or the top of the overlying glacial sediments in areas where they exist (e.g., Figures 12, 13a). If the glacial sediments are thick enough, the top and bottom of the layer can be identified as two separate reflectors. However, because glacial deposits are commonly only a few meters thick in this area, the top of glacial and top of bedrock are often one reflection package. Hence, the earliest-arriving, large-amplitude reflector is considered to be the significant seismic impedance boundary associated with the fundamental site period.

The P-wave sections are lower in resolution but might better indicate reflections at depth or beneath thick layers of coarse-grained sediments where the shear-wave reflection energy tends to be attenuated. From the point of view of wavelength resolution, shear-wave-reflection vertical resolution is usually greater than that of the P wave by a factor of as much as 3:1 (e.g., P-wavelength = period × velocity =

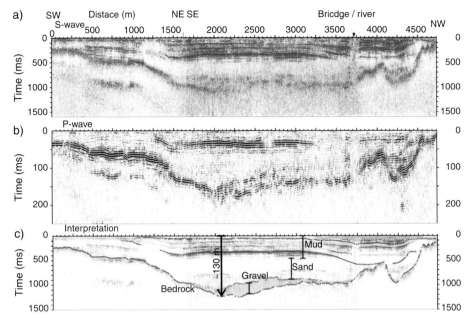

**Figure 12.** High-resolution seismic-reflection profile (in two-way traveltime) obtained over a buried valley west of Ottawa, Ontario, Canada. The entire profile is more than 4 km in length. (a) The SV-wave section. (b) The P-wave section, derived from data recorded by the vertical receivers but with the source oriented horizontally. The very high resolution of the SV-section is related to the low shear-wave velocity of the Holocene marine sediments (about 150 m/s). (c) The interpreted SV-section, which is supported by a logged borehole drilled to about 96 m in the center of the buried valley.

5 ms $\times$ 1.5 m/ms = 7.5 m; S-wavelength = 20 ms $\times$ 0.1 m/ms = 2 m).

During field surveys, we attempt to ensure that the receiver array (landstreamer) is both long enough and has a small enough receiver spacing to obtain high accuracy in average and interval shear-wave velocity determination. An example of average and interval velocity determinations in Champlain Sea sediments is shown in Figure 13b and 13c. Figure 14 shows an example semblance velocity analysis from this section; the estimated standard error on NMO velocities to the base of the unconsolidated overburden area, based on the semblance maxima, is commonly $\pm$ 5%. From correlation with borehole shear-wave measurements, the associated error on interval velocities between successive reflections is on the order of $\pm$ 10%. Hence, as an aid to determine the primary stratigraphic units, the shear-wave velocity structure of the overburden can be contoured with a relatively large degree of accuracy if significant reflectors are present.

This example is typical of the Ottawa area in that the upper portion of the Champlain Sea sediments is characterized by low shear-wave velocity material, whereas the lower portion of the postglacial unit yields somewhat higher velocities. The postglacial unit overlies glacially derived material with considerably higher interval velocity. If the depth of overburden exceeds 30 m, then it is possible to estimate $V_S30$ by interpolation between the reflecting horizons using the average velocities. Where overburden thickness is less than 30 m, average overburden shear-wave velocities and bedrock shear-wave velocities can be combined to compute $V_S30$.

## Site-resonance (HVSR) measurements

For vertically traveling earthquake energy, in areas where substantial soft soil (characterized by low shear-wave velocities) overlies competent bedrock (high shear-wave velocity), there is a possibility of resonance with very high amplitudes; resonance frequency is governed by the shear-wave velocity of the soft soil and thickness of the layer. Nakamura (1989) introduces a method to estimate the fundamental site period ($T_0$) or site frequency ($f_0$) using ambient seismic noise in the same frequency range as earthquake energy. The spectral ratio of horizontal-to-vertical spectral ratio (HVSR) can indicate a peak frequency equivalent to the resonant frequency, the argument being that only horizontal components of motion are amplified through a surface layer with low shear-wave velocity.

The ambient seismic noise was thought to be a combination of vertically traveling teleseismic body-wave noise and the contribution of Rayleigh and Love surface waves from local sources. Current research suggests that the predominant energy is that of surface waves; however, the peak spectral ratio occurs at the same values as that of resonance from vertically incident shear waves at the soil-bedrock boundary (SESAME, 2004). Fundamental site periods computed using horizontal-to-vertical spectral ratios (HVSR) correlate closely to those computed from the shear-wave transfer function for numerous soil-rock structures. The estimated resonant peak amplitudes might represent a lower bound to broadband amplification for small-strain (linear soil response) seismic events.

Because of the low cost of data acquisition, the HVSR technique is adopted frequently in seismic microzonation investigations and has proved useful to estimate the fundamental period of soil deposits. It also has helped to constrain the geologic and geotechnical models used for numerical

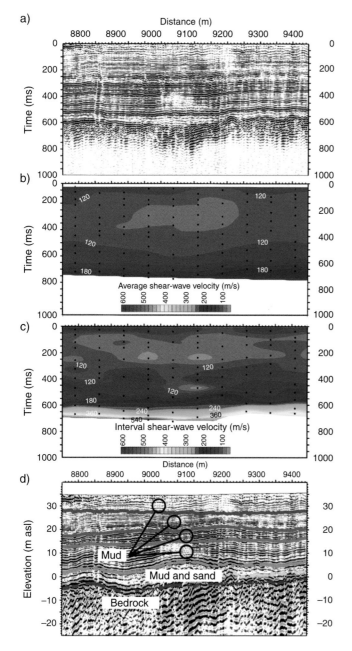

**Figure 13.** (a) Detail of about 700 m of SV-wave section (in time) acquired in Champlain Sea sediments. (b) Contoured plot of average shear-wave velocity as a function of depth as determined from velocity analyses of the CMP gathers (black dots). (c) Contoured plot of interval shear-wave velocity as a function of depth, calculated from (b). (d) Interpreted section after conversion to depth and correction for surface topography.

computations of ground motions. In addition, this technique is useful in calibrating site-response studies at specific locations. In particular, the method is recommended in areas of low and moderate seismicity where significant earthquake recordings are rare or are even nonexistent (SESAME, 2004).

As part of our analyses of all surface refraction-reflection sites in the Ottawa area, we have computed the resonant frequencies for those sites where sufficient shear-wave velocity contrast and thickness of low-velocity layer occurred (generally $> 10$ m of overburden). We assumed a simple two-layer model, consisting of a layer with low shear-wave velocity overlying high-shear-wave-velocity bedrock. For this model, the fundamental period is equivalent to $T_0 = 4H/V_S$ (or fundamental resonant frequency of $f_0 = V_S/[4H]$), where H is the thickness of the low-velocity layer (overburden) and $V_S$ is its average shear-wave velocity. To ground-truth these estimates, we used the Micromed Tromino electrodynamic three-component data-acquisition system, designed specifically for HVSR measurements, to acquire passive seismic data at more than 300 locations in the Ottawa area where thick Champlain Sea sediments had been recognized. Figure 15 shows a typical HVSR record.

To compare the computed fundamental site-period values with those measured by the Tromino HVSR, we selected surface refraction/reflection sites where we had observed excellent signal-to-noise data from the surface refraction/reflection array for all events and high-quality H/V ratios from passive noise measurements. Figure 16 shows a comparative plot of the site periods computed by both approaches, illustrating a systematic variation between the measured HVSR site-period measurements and the $4H/V_S$ estimates. At long periods (1.5 to 2.5 s, approximately equivalent to the natural period of a 15- to 25-story building), the deviation between the two methods reaches 30% to 40%.

The variance between the two methods is not yet fully understood. We suggest that the estimated site period from direct measurement of the two-way traveltime of the reflection from firm ground should be very precise, but we note that these types of reflection measurements are made at 30 to 100 Hz, whereas the ambient noise measurements HVSR are made at frequencies between 0.1 and 10 Hz. Velocity dispersion possibly might explain the differences.

Alternative explanations for the differences between the estimated fundamental site period and HVSR include the possibility of error from a simplified model consisting of a single soft-sediment layer (postglacial) overlying rigid bedrock. Dobry et al. (1976) and Hadjian (2002) suggest that slight velocity gradients within the soil layer could alter the computed fundamental site period substantially. Such conditions have been observed within the postglacial sed-

iments (e.g., a linear increase in the average $V_S$ leading to a nonlinear increase of interval $V_S$), but commonly, this velocity-increase effect appears to be minimal or is overlooked on plots of velocity versus depth where extremely large velocity contrasts occur at the postglacial-bedrock boundary.

Until the discrepancy is resolved, we continue to be guided by the 2005 NBCC (NRC, 2006) and offer the fundamental site-period map as estimated from a database of single-layer average shear-wave velocities and depths, consisting of surface, borehole, and landstreamer measurements of shear-wave reflection traveltime. However, the reader is cautioned that calculated values from the single-layer approximation as shown on the maps below should be used only as a guide to areas where significant long-period fundamental resonance can occur; the exact values could be systematically in error.

## Mapmaking Using Seismic Data

The systematic collection of shear-wave velocity data using the methods described above has allowed us to develop maps of parameters related to the ground response to earthquake shaking for our two study areas in Canada.

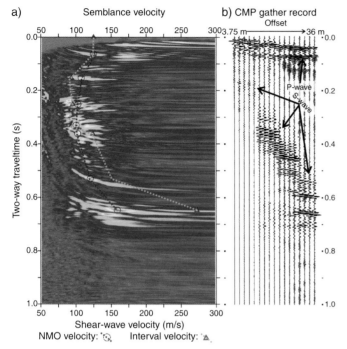

**Figure 14.** (a) Semblance velocity analysis of (b) the CMP gather, showing the very high degree of resolution in the NMO velocity estimates from these data for the fine-grained marine sediments.

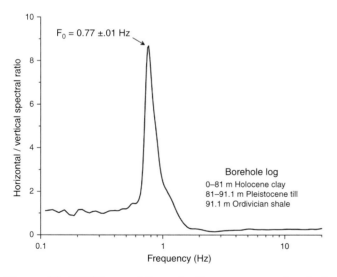

**Figure 15.** HVSR record from the Ottawa area at a borehole site where thick Champlain Sea sediments overlie bedrock. The horizontal-to-vertical spectral ratios are obtained using three-component low-frequency geophones (0.1 to 60 Hz). The fundamental resonance peak is defined clearly in this geologic environment.

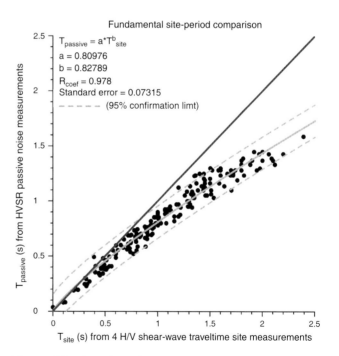

**Figure 16.** Plot of the site periods computed from HVSR measurements (*y*-axis) against those computed assuming a two-layer model ($T_0 = 4H/V_S$). Above periods of about 0.5 seconds, the HVSR measurements are systematically and significantly lower than the $4H/V_S$ estimates.

## Fraser River Delta

Shear-wave velocity data have been collected from approximately 425 seismic-refraction, reflection, and MASW sites, 50 downhole seismic sites, and 20 SCPT sites in the Fraser River Delta (Figure 17a). Even though direct measurements of shear-wave velocities at some sites were as shallow as 30-m depth, best estimates of velocity-depth functions were made down to and including the Tertiary bedrock from available adjacent seismic data. From these estimates, it was possible to produce regional maps of the lateral variation of NEHRP zones (Figure 17b), fundamental site period (Figure 17c), and impedance amplification using the one-quarter-wavelength approach of Joyner et al. (1981) (assuming an average shear-wave velocity of a surface low-velocity layer with a thickness equal to the one-quarter wavelength of a particular frequency, e.g., 1 Hz). Most of the thick Holocene and Pleistocene unconsolidated sediments in the survey area yield low shear-wave velocities, and consequently, the NERHP zones D and E are prevalent. Because the sediments are water saturated and might consist of thick sands in places, it is possible that these might liquefy in a significant earthquake event; hence, these zones or parts of them should be considered NEHRP F. Seismic measurements alone cannot differentiate these special cases. Surrounding the Fraser Delta (including the southern Tsawwassen uplands), the ground surface consists of very firm glacial till overlying bedrock. This is reflected in the NEHRP zones C and B. The fundamental site periods calculated using the average shear-wave velocity down to the top of the Pleistocene layer in the delta mirror the topography of the buried surface. Very long-period resonance is mapped throughout the west side of the project area, which includes the city of Richmond.

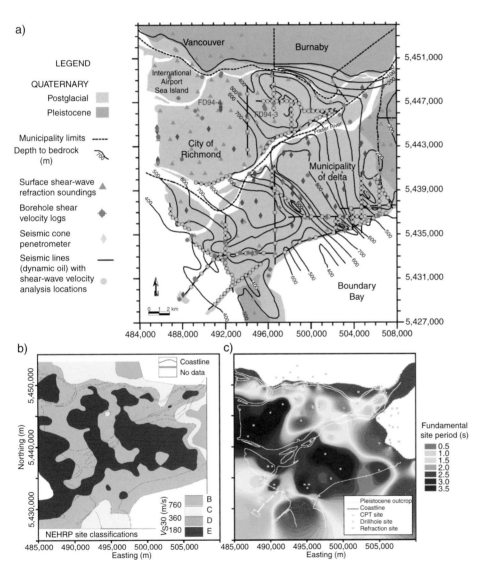

**Figure 17.** (a) Location map of shear-wave velocity-versus-depth site data acquired in the Fraser River Delta. (b) Derived $V_S30$ NEHRP site zonations (Table 1). (c) Fundamental site resonance for the Holocene-Pleistocene seismic-impedance boundary $T_0 = 4H/V_S$, where H is the thickness of the Holocene layer and $V_S$ is its thickness-weighted average shear-wave velocity.

## City of Ottawa

Within the city of Ottawa, the Geological Survey of Canada (GSC) previously had compiled geologic descriptions of soils and bedrock from more than 20,700 water-well and geotechnical drilling records (Belanger, 1998). From these, we could determine variations in thickness of the three basic geotechnical units: postglacial sediments (primarily fine-grained marine silts), glacial deposits, and bedrock. This distinction was made so we could calculate a shear-wave velocity profile at each borehole location.

Data were collected at 685 surface-reflection/refraction sites across the city between the 2004 and 2008 field seasons, and this allowed us to determine velocity estimates for the three basic stratigraphic units. Refraction breakovers interpreted at 505 seismic sites were used to determine average velocities for the various Precambrian and Paleozoic bedrock types underlying the city (e.g., Figure 18). Using GIS software, locations of the 685 seismic sites and 20,700 boreholes were intersected through the bedrock geology map to determine the type of bedrock below each site or borehole. From limited measurements of the thin glacial unit, we assigned an average shear-wave velocity of 580 m/s as a single representative value. For the postglacial Champlain Sea sediments, we compiled a citywide average shear-wave velocity-depth function using downhole and surface refraction/reflection data.

Based on this average velocity-depth function for postglacial soils and interval velocities for glacial materials and bedrock, we calculated an average shear-wave velocity down to 30-m depth for all boreholes, as shown by the example in Figure 19. To further refine the estimation of shear-wave velocity as a function of depth, the reflection data from the seismic sites were analyzed one site at a time. This created 150 site-specific average velocity-depth equations in areas of the thickest Champlain Sea deposits in the city. Using these 150 relations, a unique velocity-depth function was determined for numerous borehole locations based on their spatial location from the geophysical site (or sites) within a specified search radius. This led to an improved estimate of variation in average $V_S$ with depth in soft soil across the city. The citywide equation was used where no geophysical sites were found within the borehole's search radius.

In all, 21,700 average velocity-depth functions were developed from surface to bedrock. From these, a subset consisting of $V_S30$ NEHRP zones and the fundamental site periods (using a single postglacial layer defined by $T = 4H/V_S$) were calculated. Preliminary maps were developed using nearest-neighbor gridding and contouring techniques. Manual editing of contour positions was done to respect surficial geologic boundaries (Figure 20a). A detailed description of the process can be found in Hunter et al. (2010).

The resulting seismic site-classification map is presented as Figure 20b. The NEHRP zones A through E can be found within city limits, in some places within a few hundred meters of each other. The NEHRP zone A is

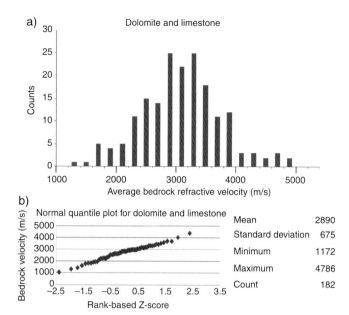

**Figure 18.** (a) Velocity distribution of interbedded dolomite and limestone bedrock, based on 182 refraction analyses. (b) A relatively linear normal quantile plot indicates a normal distribution of velocities. The resultant velocity assigned to this bedrock type is 2890 m/s.

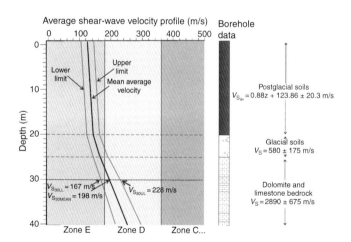

**Figure 19.** Typical average shear-wave velocity profile for the Ottawa area, calculated using stratigraphic thicknesses provided by the borehole database. Average velocity-depth functions for the postglacial materials are combined with interval velocities from the glacial and bedrock materials to calculate a traveltime-weighted $V_S30$ value. This calculation is repeated three times using the mean, upper, and lower velocity limits of the materials to provide a measure of error on the $V_S30$ value. It is not uncommon to have the upper and lower ranges straddle two seismic site categories, as shown in this figure.

**Figure 20.** Maps showing: (a) simplified surficial geology of the Ottawa area; (b) NEHRP seismic site classes that are defined based on $V_S30$ as specified in the 2005 NBCC (Hunter et al., 2010); and (c) the fundamental site period. Reproduced with the permission of Natural Resources Canada 2010, courtesy of the Geological Survey of Canada (Open File 6273).

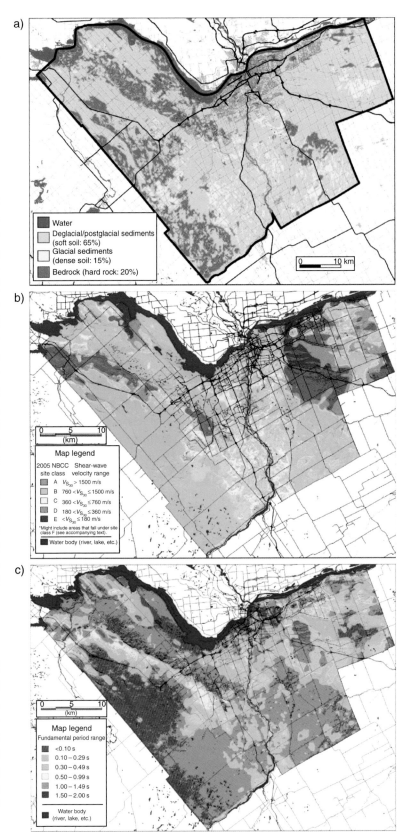

associated with firm bedrock outcrop and is subject to deamplification effects (amplification factors <1) for the "design" earthquake under the National Building Code of Canada (see examples in Table 2). On the other hand, NEHRP Zone E commonly is associated with very thick deposits of postglacial Champlain Sea sediments and is subject to amplification. As mentioned, zones defined by special case F cannot be distinguished by seismic methods only. Such zones commonly contain thick, low-velocity, geotechnically sensitive clay or organics with low shear-wave velocities similar to those found in zones of NEHRP D and E. Figure 20c presents the fundamental site-period variations calculated from a single-layer average $V_S$ for the unconsolidated postglacial overburden map. Most of the city falls within the range of 0.1 to 0.4 s, which coincides with the areas of thin to negligible overburden. Ranges of 0.8 to 0.12 s, 1.21 to 1.6 s, and 1.61 to 2.0 s coincide with the seismic site class E, which reflects the occurrences of greater overburden thicknesses. In urban locations, longer-period areas generally are observed in eastern parts of the city.

## Conclusions

We have presented various geophysical methods of determining shear-wave velocity as a function of depth, assessing its lateral variation, and mapping significant shear-wave velocity boundaries at depth. Because there is a clear link between ground-motion amplification effects and these properties of the subsurface, such information is critical to understanding the response of soft soils to earthquake shaking. We have examined and assessed downhole and surface techniques.

We have conducted testing and application of velocity-measurement techniques in two study areas in Canada where earthquake hazards are relatively high: the Fraser River Delta near Vancouver in British Columbia, and the city of Ottawa in eastern Ontario. Survey design and the applicability of some techniques varied between these two areas in response to the differing subsurface Quaternary stratigraphy. In both survey areas, results have shown that considerable variation in NEHRP zones and fundamental site periods can occur over relatively short distances; this suggests that there will be considerable variation in the shaking response to significant earthquake events.

Such information is needed for informed land-use planning by city planners and the geotechnical-engineering community. It is cautioned that such regional maps should be used only as a guide and do not replace the need for detailed site-specific assessment. However, this work has shown the utility of geophysical methods for earthquake-hazard mapping, and it is suggested that geophysical techniques can be used more extensively for site-specific characterization of geotechnical parameters required for earthquake-hazard assessment.

## Acknowledgments

The authors wish to acknowledge the financial support and direction of the Reducing Risk from Natural Hazards Program (2006–2009; Chris Tucker, program manager) and Public Geoscience Program (2009–2010; Calvin Klatt, program manager) of the Geological Survey of Canada and National Research Council of Canada funding through the Canadian Seismic Research Network for Carleton University researchers. The GSC technical staff and many students were instrumental in collecting the field data, and we thank them for their help in making this work possible. J. A. H. wishes to thank W. D. L. Finn (University of British Columbia) for urging him to initiate studies of shear-wave velocities in soft sediments after the Mexico City earthquake of 1985.

## References

Adams, J., and S. Halchuk, 2003, Fourth generation seismic hazard maps of Canada: Values for over 650 Canadian localities intended for the 2005 National Building Code of Canada; Geological Survey of Canada, Open-file Report 4459.

———, 2004, A review of NBCC 2005 seismic hazard results for Canada — The interface to the ground and prognosis for urban risk mitigation: 57th Canadian Geotechnical Conference–5th Joint CGS/IAH-CNC Conference, G33-296, CD-ROM.

Aki, K., and P. G. Richards, 1980, Quantitative seismology: W.H. Freeman and Company.

Anderson, J. G., P. Bodin, J. N. Brune, J. Prince, S. K. Singh, R. Quass, and M. Onate, 1986, Strong motion from the Michoacan, Mexico, earthquake: Science, **233**, 1043–1049.

Bard, P. Y., and M. Bouchon, 1985, The two-dimensional resonance of sediment-filled valleys: Bulletin of the Seismological Society of America, **75**, 519–541.

Belanger, J. R., 1998, Urban geology of Canada's national capital area, in P. F Karrow and O. W. White, eds., Urban geology of Canadian cities: Geological Association of Canada Special Paper 42, 365–384.

Britton, J. R., J. B. Harris, J. A. Hunter, and J. L. Luternauer, 1995, The bedrock surface beneath the Fraser River Delta in British Columbia based on seismic measurements, in Current Research 1995-E: Geological Survey of Canada, 83–89.

Building Seismic Safety Council (BSSC), 1994, NEHRP recommended provisions for seismic regulations of new buildings: Part 1, Provisions: FEMA 222A, Federal Emergency Management Agency.

Building Seismic Safety Council (BSSC), 1995, A non-technical explanation of the 1994 NEHRP recommended provisions: FEMA 99, Federal Emergency Management Agency.

Cardarelli, E., M. Cercato, R. deNardis, G. Di Filippo, and G. Milana, 2008, Geophysical investigations for seismic zonation in municipal areas with complex geology: The case study of Celano, Italy: Soil Dynamics and Earthquake Engineering, **28**, 950–963.

Choi, Y., and J. P. Stewart, 2005, Nonlinear site amplification as function of 30 m shear wave velocity: Earthquake Spectra, **21**, 1–30.

Crow, H. L., J. A. Hunter, A. J.-M. Pugin, G. Brooks, D. Motazedian, and K. Khaheshi-Banab, 2007, Shear wave measurements for earthquake response evaluation in Orleans, Ontario: 60th Canadian Geotechnical Conference & 8th Joint CGS/IAH-CNC Groundwater Conference, Proceedings, 871–879, CD-ROM.

Dobry, R., I. Oweis, and A. Urzua, 1976, Simplified procedures for estimating the fundamental period of a soil profile: Bulletin of the Seismological Society of America, **66**, 1293–1321.

Finn, W. D. L., and A. Wightman, 2003, Ground motion amplification factors for the proposed 2005 edition of the National Building Code of Canada: Canadian Journal of Civil Engineering, **30**, 272–278.

Finn, W. D. L., D. J. Woeller, M. P. Davies, J. L. Luternauer, J. A. Hunter, and S. E. Pullan, 1989, New approaches for assessing liquefaction potential of the Fraser Delta, British Columbia: Geological Survey of Canada Current Research, Part E, Paper 89-1E, 221–231.

Glover, R. H., 1959, Techniques used in interpreting seismic data in Kansas, in W. W. Hambleton, ed., Symposium on Geophysics in Kansas: Kansas Geological Survey Bulletin 137, 225–240.

Gonzalez, J., M. Schmitz, F. Audemard, R. Contreras, A. Mocquet, J. Delgado, and F. De Santis, 2003, Site effects of the 1997 Cariaco, Venezuela earthquake: Engineering Geology, **72**, 143–177.

Hadjian, A. H., 2002, Fundamental period and mode shape of layered soil profiles: Soil Dynamics and Earthquake Engineering, **22**, 885–891.

Helbig, K., and C. S. Mesdag, 1982, The potential of shear-wave observations: Geophysical Prospecting, **30**, 413–431.

Holzer, T. L., 1994, Loma Prieta damage largely attributed to enhanced ground shaking: Eos Transactions, **75**, 299–301.

Hunter, J. A., B. Benjumea, J. B. Harris, R. D. Miller, S. E. Pullan, R. A. Burns, and R. L. Good, 2002, Surface and downhole shear wave seismic methods for thick soils site investigations: Soil Dynamics and Earthquake Engineering, **22**, 931–941.

Hunter, J. A., R. A. Burns, R. L. Good, J. M. Aylsworth, S. E. Pullan, J. B. Harris, A. Skvortzov, and N. N. Goriainov, 1998a, Downhole seismic logging for high-resolution reflection surveying in unconsolidated overburden: Geophysics, **63**, 1371–1384.

Hunter, J. A., R. A. Burns, R. L. Good, J. M. Aylsworth, S. E. Pullan, D. Perret, and M. Douma, 2007, Borehole shear wave velocity measurements of Champlain Sea sediments in the Ottawa-Montreal region: Geological Survey of Canada, Open-file Report 5345, CD-ROM.

Hunter, J. A., H. L. Crow, G. R. Brooks, M. Pyne, M. Lamontagne, A. J.-M. Pugin, S. E. Pullan, T. Cartwright, M. Douma, R. A. Burns, R. L. Good, D. Motazedian, K. Kaheshi-Banab, R. Caron, M. Kolaj, I. Folahan, L. Dixon, K. Dion, A. Duxbury, A. Landriault, V. Ter-Emmanuil, A. Jones, G. Plastow, and D. Muir, 2010, Seismic site classification and site period mapping in the Ottawa area using geophysical methods: Geological Survey of Canada, Open File 6273.

Hunter, J. A., H. L. Crow, G. Brooks, M. Pyne, M. Lamontagne, A. J.-M. Pugin, S. E. Pullan, T. Cartwright, M. Douma, R. A. Burns, R. L. Good, D. Motazedian, K. Kaheshi-Banab, R. Caron, M. Kolaj, D. Muir, A. Jones, L. Dixon, G. Plastow, K. Dion, A. Duxbury, A. Landriault, V. Ter-Emmanuil, and I. Folahan, 2009, City of Ottawa seismic site classification map from combined geological/geophysical data: Geological Survey of Canada, Open-file Report 6191, 1 map sheet.

Hunter, J. A., M. Douma, R. A. Burns, R. L. Good, S. E. Pullan, J. B. Harris, J. L Luternauer, and M. E. Best, 1998b, Testing and application of near-surface geophysical techniques for earthquake hazards studies, Fraser River Delta, British Columbia, in J. J. Clague, J. L. Luternauer, and D. C. Mosher, eds., Geology and natural hazards of the Fraser River Delta, British Columbia: Geological Survey of Canada Bulletin 525, 123–145.

Hunter, J. A., and D. J. Woeller, 1990, Comparison of surface, borehole and seismic cone penetrometer methods of determining the shallow shear-wave velocity structure of the Fraser River Delta, British Columbia: 60th Annual International Meeting, SEG, Expanded Abstracts, 385–388.

Inazaki, T., 1999, Land streamer: A new system for high-resolution S-wave shallow reflection surveys: 1999 Symposium on the Application of Geophysics to Engineering and Environmental Problems, EEGS, Proceedings, 207–216.

———, 2004, High resolution reflection surveying at paved areas using S-wave type land streamer: Exploration Geophysics, **35**, 1–6.

Joyner, W. B., R. W. Warrick, and T. E. Fumal, 1981, The effect of Quaternary alluvium on strong ground motion in the Coyote Lake, California, earthquake of 1979: Bulletin of the Seismological Society of America, **71**, 1333–1349.

Kovach, R. L., 1965, Seismic surface waves: Some observations and recent developments, in L. H. Ahrens, S. K. Runcorn, and H. C. Urey, eds., Physics and chemistry of the earth, v. 6: Pergamon Press, 251–314.

Kramer, S. L., 1996, Geotechnical earthquake engineering: Prentice Hall.

Lomnitz, C., J. Flores, O. Novaro, T. H. Seligman, and R. Esquivel, 1999, Seismic coupling of interface modes in sedimentary basins: A recipe for disaster: Bulletin of the Seismological Society of America, **89**, 14–21.

Mandrescu, N., M. Radulian, and G. Marmureanu, 2006, Geological, geophysical and seismological criteria for local response evaluation in Bucharest urban area: Soil Dynamics and Earthquake Engineering, **27**, 367–393.

Motazedian, D., and J. A. Hunter, 2008, Development of an NEHRP map for the Orleans suburb of Ottawa, Ontario: Canadian Geotechnical Journal, **45**, 1180–1188.

Nakamura, Y., 1989, A method for dynamic characteristics estimation of subsurface using microtremor on the ground surface: Quarterly Report of Railway Technical Research Institute, **30**, no.1, 25–33.

National Research Council Canada (NRC), 2005, National building code of Canada 2005, v. 1, Division B, 4-22.

National Research Council Canada (NRC), 2006, User's guide — NBC 2005 structural commentaries: Canadian

Commission on Buildings and Fire Codes, Division B, Part 4, Commentary J.

Nazarian, S., K. H. Stokoe II, and W. R. Hudson, 1983, Use of spectral analysis of surface waves method for determination of moduli and thicknesses of pavement systems: Transportation Research Record No. 930, 38–45.

Park, C. B., R. D. Miller, and J. Xia, 1999, Multichannel analysis of surface waves (MASW): Geophysics, **64**, 800–808.

Pugin, A. J.-M., J. A. Hunter, D. Motazedian, G. R. Brooks, and K. Khaheshi-Banab, 2007, An application of shear wave reflection landstreamer technology to soil response of earthquake shaking in an urban area, Ottawa, Ontario: 2007 Symposium on the Application of Geophysics to Environmental and Engineering Problems, EEGS, Proceedings, 885–896.

Pugin, A. J.-M., S. E. Pullan, and J. A. Hunter, 2009a, Multicomponent high resolution seismic reflection profiling: The Leading Edge, **28**, 1248–1261.

Pugin, A. J.-M., S. E. Pullan, J. A. Hunter, and G. A. Oldenborger, 2009b, Hydrogeological prospecting using P- and S-wave landstreamer seismic reflection methods: Near Surface Geophysics, **7**, 315–327.

Pugin, A. J.-M., S. L. Sargent, and L. Hunt, 2006, SH and P-wave seismic reflection using landstreamers to map shallow features and porosity characteristics in Illinois: 2006 Symposium on the Application of Geophysics to Engineering and Environmental Problems, EEGS, Proceedings, 1094–1109.

Pullan, S. E., and J. A. Hunter, 1990, Delineation of buried bedrock valleys using the optimum offset shallow seismic reflection technique, *in* S. H. Ward, ed., Geotechnical and Environmental Geophysics, v. 3: Geotechnical: SEG Investigations in Geophysics Series No. 5, 75–87.

Pullan, S. E., J. A. Hunter, H. M. Jol, M. C. Roberts, R. A. Burns, and J. B. Harris, 1998, Seismostratigraphic investigations of the southern Fraser River Delta, *in* J. J. Clague, J. L. Luternauer, and D. C. Mosher, eds., Geology and natural hazards of the Fraser River Delta, British Columbia: Geological Survey of Canada Bulletin 525, 91–121.

Ramos-Martínez, J., G. J. Chavez-García, E. Romaero-Jimenez, J. L. Rodrigues-Zuniga, and J. M. Gomez-Gonzalez, 1997, Site effects in Mexico City: Constraints from surface wave inversion of shallow refraction data: Journal of Applied Geophysics, **36**, 137–165.

Robertson, P. K., R. G. Campanella, D. Gillespie, and A. Rice, 1986, Seismic CPT to measure in situ shear wave velocity: Journal of Geotechnical Engineering Division, ASCE, **112**, 791–803.

Schepers, R., 1975, A seismic reflection method for solving engineering problems: Journal of Geophysics, **41**, 267–284.

Schwab, F. A. and L. Knopoff, 1972, Fast surface wave and free mode computations, *in* B. A. Bolt, ed., Methods in computational physics: Academic Press, 87–180.

SESAME, 2004, Guidelines for the implementation of the H/V spectral ratio technique on ambient vibrations — Measurements, processing and interpretation, http://sesame-fp5.obs.ujf-grenoble.fr/SES_Reports.htm, accessed 1 February 2010.

Shearer, P. M., and J. A. Orcutt, 1987, Surface and near-surface effects on seismic waves — Theory and borehole seismometer results: Bulletin of the Seismological Society of America, **77**, 1168–1196.

Steeples, D. W., 1998, Shallow seismic reflection section — Introduction: Geophysics, **63**, special issue, 1210–1212.

Steeples, D. W., and R. D. Miller, 1990, Seismic reflection methods applied to engineering, environmental, and groundwater problems, *in* S. H. Ward, ed., Geotechnical and environmental geophsyics, v. 1: Review and tutorial: SEG Investigations in Geophysics Series No. 5, 1–30.

Stokoe, K. H. II, S. G. Wright, J. A. Bay, and J. M. Roësset, 1994, Characterization of geotechnical sites by SASW method, *in* R. D. Woods, ed., Geophysical characterization of sites: ISSMFE Technical Committee No. 10, Oxford & IBH Publishing, 15–25.

Ulusay, R., O. Aydan, A. Erkin, E. Tuncay, H. Kumsar, and Z. Kaya, 2004, An overview of geotechnical aspects of the Cay-Eber (Turkey) earthquake: Engineering Geology, **73**, 51–70.

Williams, R. A., E. Cranswick, and K. W. King, 1994, Site response models from high-resolution seismic reflection and refraction data, Santa Cruz, California, *in* R. D. Borcherdt, ed., The Loma Prieta, California, earthquake of October 17, 1989 — Strong ground motion: U. S. Geological Survey Professional Paper 1551-A, 217–242.

Williams, R. A., W. J. Stephenson, J. K. Odum, and D. W. Worley, 1997, High-resolution surface-seismic imaging techniques for NEHRP soil profile classifications and earthquake hazard assessments in urban areas: U. S. Geological Survey Open-file Report 97-501.

Woolery, E. W., R. L. Street, Z. Wang, and J. B. Harris, 1993, Near-surface deformation in the New Madrid seismic zone as imaged by high resolution SH-wave seismic methods: Geophysical Research Letters, **20**, 1615–1618.

Xia, J., R. D. Miller, and C. B. Park, 1999, Estimation of near-surface shear-wave velocity by inversion of Rayleigh waves: Geophysics, **64**, 691–700.

Xia, J., R. D. Miller, C. B. Park, J. A. Hunter, J. B. Harris, and I. Ivanov, 2002, Comparing shear-wave velocity profiles from multichannel analysis of surface wave with borehole measurements: Soil Dynamics and Earthquake Engineering, **22**, 181–190.

# Chapter 22

# Integrating Seismic-velocity Tomograms and Seismic Imaging: Application to the Study of a Buried Valley

Femi O. Ogunsuyi[1] and Douglas R. Schmitt[1]

## Abstract

The architectural complexity of a paleovalley ~350 m deep has been revealed by acquisition and conventional processing of a high-resolution seismic-reflection survey in northern Alberta, Canada. However, processing degraded much of the high quality of the original raw data, particularly with respect to near-surface features such as commercial methane deposits, and that motivated use of additional processing algorithms to improve the quality of the final images. The additional processing includes development of a velocity model, via tomographic inversion, as the input for prestack depth migration (PSDM); application of a variety of noise-suppression techniques; and time-variant band-pass filtering. The resulting PSDM image is of poorer quality than the newly processed time-reflection profile, thus emphasizing the importance of a good velocity function for migration. However, the tomographic velocity model highlights the ability to distinguish the materials that constitute the paleovalley from the other surrounding rock bodies. Likewise, the reprocessed seismic-reflection data offer enhanced spatial and vertical resolution of the reflection data, and they image shallow features that are newly apparent and that suggest the presence of gas. This gas is not apparent in the conventionally processed section. Consequently, this underscores the importance of (1) ensuring that primarily high-frequency signals are kept during the processing of near-surface reflection data and (2) experimenting with different noise-suppression and elimination procedures throughout the processing flow.

## Introduction

Buried valleys are exactly what their name implies: valleys that have been filled with unconsolidated sediments and covered so that their existence is not apparent at the earth's surface. They are abundant in recently glaciated areas in North America and Europe (e.g., ÓCofaigh, 1996; Fisher et al., 2005; Hooke and Jennings, 2006; Jørgensen and Sandersen, 2008). Their internal structure is complex, with a heterogeneous mix of fluid-saturated porous and permeable sands and gravels mixed with low-porosity and low-permeability diamicts and clays.

Surface geologic mapping often cannot locate or delineate the extent of such buried valleys because they are masked at the surface by recently deposited glacial sediments; invasive methods such as boreholes have been employed to characterize them (e.g., Andriashek and Atkinson, 2007). Moreover, the physical properties of these glacially derived sediments often differ significantly from the surrounding bedrock into which the valleys had been cut; that contrast allows use of many complementary geophysical methods. Geophysical techniques also can provide laterally continuous information about the subsurface, so they might be preferred over intrusive methods.

Geophysical methods have been used widely to investigate near-surface targets (Hunter et al., 1984; Miller et al., 1989; Clague et al., 1991; Belfer et al., 1998; Nitsche et al., 2002; Benjumea et al., 2003; Sharpe et al., 2004; Chambers et al., 2006), and specifically, buried valleys (Greenhouse and Karrow, 1994; Jørgensen et al., 2003a; Jørgensen

[1]*Institute for Geophysical Research, Department of Physics, University of Alberta, Edmonton, Alberta, Canada. E-mail: ogunsuyi@phys.ualberta.ca; dschmitt@ualberta.ca.*

et al., 2003b; Steuer et al., 2009). In particular, refraction- and reflection-seismic methods have been used extensively in various glacial environments to image subsurface structures (e.g., Roberts et al., 1992; Wiederhold et al., 1998; Büker et al., 2000; Juhlin et al., 2002; Schijns et al., 2009) and to study buried valleys (Büker et al., 1998; Francese et al., 2002; Fradelizio et al., 2008). In such studies, seismic inversion often is limited to first-arrival traveltimes for the input (e.g., Lennox and Carlson, 1967; Deen and Gohl, 2002; Zelt et al., 2006), rather than inputting both refracted and reflected traveltimes (e.g., De Iaco et al., 2003).

The fresh groundwater within buried valleys is usually their most important resource. Consequently, in the last 15 years, numerous and varied geophysical investigations have been undertaken in northern Europe (e.g., Gabriel et al., 2003; Sandersen and Jørgensen, 2003; Wiederhold et al., 2008; Auken et al., 2009) and North America (e.g., Sharpe et al., 2003; Pullan et al., 2004; Pugin et al., 2009) to locate and define such features for their exploitation and protection. Further, the porous sands and gravels in buried valleys also can contain local biogenic gas (Pugin et al., 2004) or leaked thermogenic methane. Such gas sometimes exists in modest commercial quantities, but it also can be a significant safety hazard for drillers.

Ahmad et al. (2009) recently described an integrated geologic, well-log, direct-current electrical, and seismic-reflection study of one such large buried valley in northern Alberta, Canada. Using the same seismic data set, we extend Ahmad et al.'s (2009) work by exploring development of a seismic tomographic velocity model to characterize the paleovalley on the basis of the material/interval velocities. The model was generated by performing an inversion of the traveltimes of both refracted and reflected waves.

Refracted, guided, air, and surface waves are examples of source-generated noise (coherent noise) that presents considerable problems during seismic data processing (Büker et al., 1998; Montagne and Vasconcelos, 2006). Accurate separation of refracted and guided waves from shallow reflections is difficult, and such linear events can stack coherently on reflection profiles, causing misinterpretation (Steeples and Miller, 1998). Performing a noise-cone muting might prove useful in eliminating some of the source-generated noise, but reflections also are muted in the process (Baker et al., 1998). Surgical noise-cone muting was employed to remove the coherent noise in the previous study (Ahmad, 2006; Ahmad et al., 2009). Apart from the possibility that this procedure might not have removed the guided waves located outside the noise-cone

zone on the shot gathers, it inadvertently also might have eliminated some true reflections during muting. Either way, the resulting seismic profile is not as satisfactory as one might expect from the quality of the raw shot gathers. Therefore, in our study, we instead performed radial and slant-stack noise-suppression procedures on the data set to retain and enhance shallow reflections that might have been removed by muting or masked by source-generated noise.

Because the dominant frequency of a seismic wave controls the separation of two close events (Yilmaz, 2001), our new processing scheme was aimed also at improving the spatial and vertical resolution of the data set because the seismic-processing sequence of the previous study did not account for adequately filtering out the low frequencies. We subsequently attempted to use the tomographic velocity model to perform a prestack depth migration (PSDM) after the noise-suppression strategies had been used on the data set. To our knowledge, PSDM algorithms and the radial and $\tau$-$p$ noise-suppression techniques that often are employed in more conventional and deeper petroleum exploration have not been applied to such near-surface seismic data.

The goal here is not to provide a new geologic interpretation of Ahmad et al.'s (2009) study, but instead to share the experiences gained in applying several tools, some of which to our knowledge have been employed heretofore only in deeper petroleum exploration. One special improvement over the earlier work is that this reprocessing has permitted imaging of shallow methane deposits within the glacial materials, and as such this work has implications for both resource exploration and safety enhancement for drillers.

## Background

Details and maps of the location of the survey in the northwestern corner of Alberta, centered at approximately 58° 35' N and 118° 31' W, are found in Ahmad et al. (2009). The near-surface geology of northeastern British Columbia and northwestern Alberta has been studied extensively in the last decade (e.g., Best et al., 2006; Hickin et al., 2008; Levson, 2008). The surface geology of the area immediately over the profile has been investigated by Plouffe et al. (2004) and Paulen et al. (2005), and it is established to be blanketed primarily with glacial, lacustrine, and glaciolacustrine sediments of variable depths; those authors also have produced numerous complementary maps of the region's general surface geology.

A brief explanation of the bedrock geology is necessary to assist the reader's understanding of later geophysical responses. The consolidated bedrock sediments beneath the Quaternary cover in the region lie nearly flat, and when they are undisturbed they consist of ~250 m of Cretaceous siliclastic sands and shales underlain, across a sharp unconformity (sub-Cretaceous), by more indurated Paleozoic carbonates and shales. Hickin et al. (2008) and references therein offer more detailed descriptions of the region's bedrock geology. Ahmad et al. (2009) also provide representative well logs that, to first order, approximately categorize the sediments on the basis of sonic velocity and density.

## Seismic Field Program

A high-resolution 2D seismic profile was acquired over a survey length of approximately 9.6 km in an east-west direction (Figure 1). The seismic survey line straddles the surface over the large buried valley to the east and the out-of-valley region to the west, as determined from the maps of bedrock topography and surface geology (Plouffe et al., 2004; Paulen et al., 2005; Pawlowicz et al., 2005a, 2005b) (Figure 1). The purpose of the survey was to image the formation above the sub-Cretaceous unconformity and hence to delineate the buried channel and obtain important information about its internal structure.

A summary of the acquisition parameters is provided in Table 1. The P-wave seismic source used was the University of Alberta's IVI Minivib™ unit, operated with linear sweeps of a 7-s period, from 20 Hz to 250 Hz, at a force of approximately 26,690 N (6000 lb). The seismic traces were acquired with high-frequency (40-Hz) geophone singles (to attenuate some of the ground-roll noise) at a 4-m spacing using a 240-channel semidistributed seismograph that consists of 10 24-channel Geode™ field boxes connected via field intranet cables to the recording computer. Approximately five to eight sweeps per shotpoint were generated by the seismic source at a 24-m spacing. Cross-correlation of the seismic traces with the sweep signal (to generate a spike source), as well as vertical stacking, were carried out in the field. The final stacked records were saved in SEG2 format and later were combined with survey header information into a single SEGY format file for processing.

Generally, good coupling between the vibrator plate and the frozen, snow-covered ground was achieved, as determined by the constant nature of the force over time during the sweep period and by the transmission of high seismic frequencies (Ahmad et al., 2009). In a similar manner, there was good coupling between the ground and the geophones, which were frozen in place overnight, thus improving the signal-to-noise ratio (S/N). The average

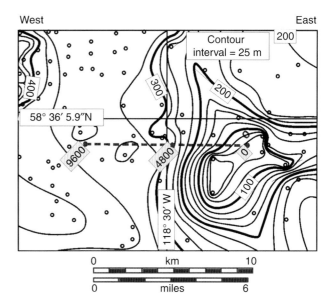

**Figure 1.** Contours of bedrock-topography elevation in meters above sea level (asl), with the location of the 2D survey line shown as a broken line. Black labels inside gray boxes indicate distance in meters from east to west, with the origin at 0 m. The small unfilled circles are wellbore locations used to generate the bedrock-topography map. After Pawlowicz et al. (2005a). Used courtesy of Alberta Energy and Utilities Board/Alberta Geological Survey.

**Table 1.** Acquisition parameters for the 2D seismic survey.

| Parameter | Value |
|---|---|
| 2D line direction | East-west |
| Length of profile | ~9.6 km |
| Source | 6000-lb IVI Minivib™ unit |
| Source frequency | 20–250 Hz |
| Source type | Linear |
| Source length | 7 s |
| Source spacing | 24 m |
| Vertical stacks | 5–8 |
| Number of unique shotpoints | 399 |
| Receivers | 40-Hz single geophones |
| Receiver spacing | 4 m |
| Recording instrument | Geometrics Geode™ system |
| Number of channels | 192–240 |
| Sampling interval | 0.5 ms |
| Record length | 1.19 s |
| Nominal fold | ~40 |

a)

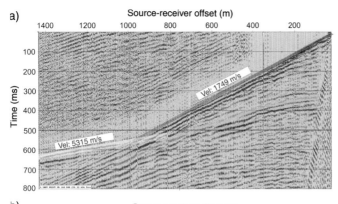

b)

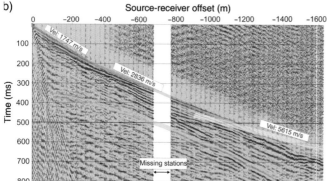

**Figure 2.** Raw shot gathers acquired at different locations on the survey line. The apparent velocities of different refractors, obtained from a simple intercept-time refraction method, are displayed. Positive-offset values are to the west of the respective shotpoint, whereas negative offsets are to the east. The gray highlight shows a direct wave through the Quaternary fill, the blue highlight is a refracted wave turning through the Cretaceous rock, the yellow line denotes the refracted wave through the Devonian, and the reflection from the top of the Devonian is in green. (a) An eastern shot gather (shotpoint 360) acquired over the valley on the seismic line. (b) A western shot record (shotpoint 1848) acquired outside the buried valley.

common-midpoint (CMP) fold for the survey was approximately 40. Two representative shot gathers from either end of the seismic profile (Figure 2) highlight the evolution of the traveltimes (and hence of velocity structure) from east to west.

## Seismic Traveltime Inversion

A linearized traveltime-inversion procedure, developed primarily for modeling 2D crustal refraction and wide-angle data (comprehensively described in Zelt and Smith, 1992; and applied in Song and ten Brink, 2004), was used. The inversion incorporates traveltimes of the direct, refracted, and reflected events. The geometry of the model is outlined

by boundary node points that are connected through linear interpolation, whereas the velocity field is specified by velocity value points at the top and base of each layer. The velocity within each block varies linearly with depth (between the upper and lower velocities in a layer) and laterally across the velocity points along the upper and lower layer boundaries.

Examination of the refracted events on the shot gathers offers an important insight into the apparent velocity structure of the survey line. Figure 2a shows a shot gather acquired at the east end of the profile immediately over the buried valley. Performing a simple intercept-time refraction analysis on this gather suggests that a simple two-layer model could be sufficient (see Ahmad et al., 2009). The wave passage through the top layer (gray highlight) might be linked to a lower material velocity of the Quaternary fill compared with the higher velocity of the Devonian carbonates (yellow highlight). The velocity of the Quaternary rock is expected to be much lower than that of the Paleozoic rocks because of its weak consolidation, which resulted from minimal overburden pressures. On the other hand, the shot gather obtained at the west end and outside of the buried valley indicates three layers (Figure 2b): (1) a direct wave (gray highlight) that passes through a thin low-velocity Quaternary rock, (2) a refracted event, of intermediate velocities, from the top of the Cretaceous rock, and (3) another wave, with higher velocities, refracted from the top of the Paleozoic.

The Vista® processing package (GEDCO, Calgary) was used to pick the traveltimes within the shot gathers of both the refracted and reflected arrivals. Approximately 72,000 traveltime picks were made from 143 shot gathers, and these were assigned an average error of ± 5 ms to account for far-offset ranges and large depths (Ogunsuyi et al., 2009). The program RayGUI (Song and ten Brink, 2004) was used for the forward modeling and inversion.

After a series of tests was conducted to determine the optimal initial model, the starting model chosen was constructed with six layers, with the velocities varying linearly in each layer. The starting model was made up of coinciding locations of velocity and boundary nodes, which were almost equally spaced laterally at a distance of about 330 m for the top of the second and third layers, and were spaced irregularly for the other layers. Short-offset (<100-m) direct arrivals were inverted in the first layer to account for the near-surface velocity variations and the relatively flat surface topography. The second layer was defined for the inversion of the rest of the direct waves (the gray highlights in Figure 2), which constituted the greater part of the seismic waves through the Quaternary deposits. The third layer was defined for the refracted events turning through the Cretaceous rock (i.e., the blue highlight in Figure 2b), whereas the base of the fourth layer represented the waves that reflected prominently from the top of the sub-Cretaceous unconformity. Technically, the third and fourth layers are supposed to be one and the same, but they were separated to give a measure of the validity of the final tomographic model, as demonstrated by the degree to which the gap (in depth) between the base of the third layer and base of the fourth layer will be reduced subsequent to the inversion process. The refracted events (yellow highlights in Figure 2) through the Devonian rock were inverted for the fifth layer. A reflected event that could not be picked successfully for the whole length of the survey line was inverted for layer six, so less confidence is placed on the tomographic results at elevations below ~200 m below sea level.

The model was inverted layer by layer, from the top down, in a layer-stripping method intended to speed up and simplify the process (Zelt and Smith, 1992). This method involved the following steps: (1) simultaneously inverting the model parameters (both velocity and boundary nodes) of the topmost layer, (2) updating the model with the calculated changes, (3) repeating steps 1 and 2 until the stopping criteria are satisfied for the layer, (4) holding the model parameters of the layer constant for the subsequent inversion of all the parameters of the next layer in line, and (5) repeating steps 1 through 4 for all the other underlying layers in sequence. The uncertainties in the depths of the boundary nodes and velocity values were 10 m and 200 m/s, respectively.

The average values for the root-mean-square (rms) traveltime residual between the calculated and observed times for all layers was 16.4 ms after seven iterations. Notably, adding more model parameters generally reduces the traveltime residual, but it does so at the expense of reducing the spatial resolution of the final model parameters (Zelt and Smith, 1992). Subsequent to the inversion, the difference in depth between the bases of the third and fourth layers was reduced acceptably on the west but was not reduced adequately on the east end inside the valley area. Merging the third and fourth layers before the inversion scheme, however, produced low ray coverage, thus violating one of the conditions for choosing a final model. Consequently, the six-layer model was chosen as the optimal model for the subsurface of the area under investigation.

## Processing of Seismic-reflection Data

Ahmad et al.'s (2009) earlier study involves a conventional 2D CMP processing scheme (Table 2) whereas the

**Table 2.** Previous seismic processing sequence, as performed by Ahmad et al. (2009).

| Processing step | Parameters |
|---|---|
| Geometry | — |
| Editing of bad trace | — |
| Offset limited sorting | −500 m to −12 m and 12 m to 500 m |
| Surgical mute | Auto bottom mute; manual surgical mute |
| CMP sorting | 4-m bin size |
| Velocity analyses | — |
| NMO corrections | 15% stretch mute |
| Elevation/refraction statics corrections | 400-m asl datum; 1500 m/s replacement velocity |
| Residual statics corrections | Stack power algorithm |
| Inverse NMO corrections | — |
| Final velocity analyses | — |
| NMO corrections | 15% stretch mute |
| Final residual statics corrections | Stack power algorithm |
| CMP stack | — |
| Band-pass filtering | 45 to 240 Hz |
| Mean scaling | — |
| *f-x* prediction | — |
| Automatic gain control | 150 ms |

new processing sequence (adapted from Spitzer et al., 2003) followed in this study is complemented with noise-suppression procedures that involve transformation of time-space (*t-x*) data into other domains (Table 3). The motivation for designing a new processing scheme for the seismic data was the need to determine whether reflections were eliminated or degraded by the muting functions adapted in the previous study or were covered up by source-generated noise. If either was the case, we wished to recover the affected reflections and additionally to improve the lateral and vertical resolutions of the reflection profile.

Low-quality traces resulting from noisy channels, high amplitude, and frequency spikes are problematic to a final image. Starting with the data set that already has geometry information assigned, the bad traces, with abnormally high amplitudes and frequencies, were identified by computing amplitude and frequency statistics on all the traces and subsequently were removed. To adjust for the lateral changes in the thickness and velocity of the shallow depths and for the small elevation variations of sources and receivers, elevation/refraction statics corrections were conducted. A model of the shallow subsurface was established by inverting the first-break picks. Using a weathering velocity of 500 m/s, the average velocity of the first refractor was approximately 1700 m/s. The computed statics corrections

**Table 3.** Time-processing sequence for the 2D seismic survey.

| Processing step | Justification |
|---|---|
| Trace editing | Removal of spurious traces |
| First-break picking | |
| Elevation/refraction statics corrections | Correction for shallow lateral variations |
| Spiking deconvolution | Compression of wavelet |
| Time-variant band-pass filtering | Suppression of low-frequency noise |
| Trace equalization | |
| CMP binning | |
| Initial velocity analyses | Determination of stacking velocities |
| NMO corrections | |
| Residual statics corrections | Correction for near-surface velocity changes |
| Inverse NMO corrections | |
| Radial domain processing | Removal of guided waves |
| Linear $\tau$-$p$ processing | Suppression of source-generated noise |
| Predictive deconvolution | Elimination of multiples |
| Dip-moveout (DMO) corrections | Preservation of conflicting dips |
| Final velocity analyses | Determination of stacking velocities |
| Final residual statics corrections | Correction for near-surface velocity changes |
| NMO corrections | |
| CMP stack | |
| $f$-$x$ prediction | Reduction of incoherent noise |
| 2D Kirchhoff time migration | Placing reflections in their true positions |

were applied to a flat datum of 385 m above sea level, which was slightly above the highest elevation of the survey line. Total elevation/refraction statics corrections ranged from approximately −6.5 ms to +21 ms.

Predictive deconvolution was not successful in removing some of the multiples in the data at this stage, so it was carried out in later processing. To compress the wavelet to a spike and thereby to increase the temporal resolution, spiking deconvolution was applied. After testing with different operator lengths for optimal results, a 20-ms operator length finally was employed. Low frequencies in the amplitude spectrum of the seismic data were dominated by direct and surface waves, but application of a low-cut frequency filter to the data set for the purpose of suppressing the noise might also inadvertently remove some deeper reflections that are characterized by low frequencies. To avoid this, time-variant band-pass filtering was applied to the data (80–300 Hz for a 0- to 380-ms time interval and 65–150 Hz for a 380- to 800-ms interval).

In addition, the band-pass-filtering step provides a means of enhancing temporal resolution of the seismic profile. Following the spiking deconvolution step, the amplitude spectra of the shot gathers were equalized adequately. To appreciate the value of these processing steps, a raw shot gather (shot-point number 660), affected by surface waves after elevation/refraction statics corrections (Figure 3a) and after application of spiking deconvolution, band-pass filtering, and trace equalization (Figure 3b), is displayed for comparison. Although most of the source-generated noise has not been eliminated, most of the surface waves have been suppressed and the reflection wavelet improved. Moreover, this processing step appears to have exposed additional guided waves that were concealed in the initial shot gather (Figure 3b). Details about the properties of these common seismic arrivals, which form the basis of our interpretation, can be found in Robertsson et al. (1996) and in Yilmaz (2001).

Choosing the best possible CMP bin size is imperative for minimizing spatial aliasing when one is processing seismic data for moderately to steeply dipping reflections (Spitzer et al., 2003). To determine the maximum CMP bin size $b$ to use (Yilmaz, 2001), we evaluated

$$b \leq \frac{V_{min}}{4f_{max}\sin\theta}, \tag{1}$$

where $V_{min}$ is the minimum velocity, $f_{max}$ is the maximum frequency, and $\theta$ is the maximum expected dip of structures. We arrived at a value of 3 m as the appropriate bin size for our data. Initial velocity analyses to determine the stacking velocities were carried out on CMP supergathers by creating a panel of offset sort/stack records, constant-velocity stacks, and semblance output.

Residual statics are needed to correct for short-wavelength changes in the shallow velocity underneath each source-and-receiver pair. Surface-consistent residual statics by a stack power-maximization algorithm (Ronen and Claerbout, 1985) were estimated from the data after applying normal-moveout (NMO) corrections on the basis of the initial velocity analyses. The resulting average time shifts

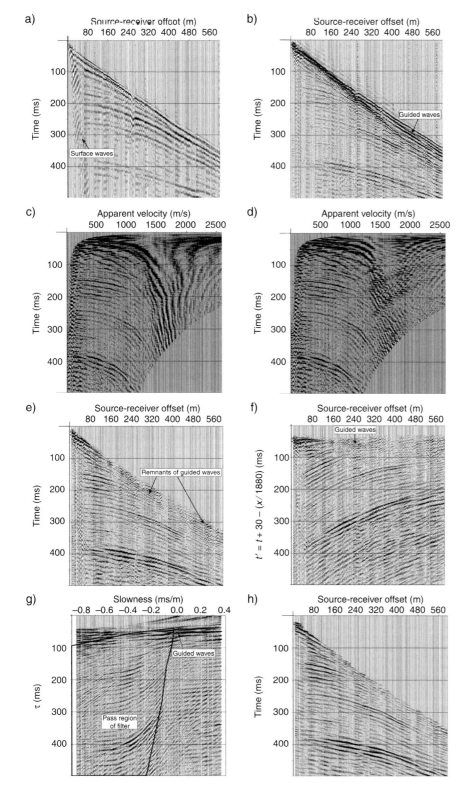

**Figure 3.** (a) A typical raw shot gather (shotpoint number 660) after elevation/refraction static corrections. (b) The same shot gather as in (a), after spiking deconvolution, time-variant band-pass filtering, and trace equalization. (c) The same shot gather as in (b), but after transformation to the *r-t* domain. (d) The same shot gather as in (c), after applying a low-cut filter of 45 Hz to suppress the linear events. (e) The same shot gather as in (d), after radial processing that involved transforming the shot gather from *r-t* to *t-x* coordinates. (f) The same shot gather as in (e), except that the time axis is reduced: traveltimes $t' = t + 30 - (x/1880)$, where 1880 m/s is the average velocity for the first arrival (as obtained from intercept-time refraction analyses). (g) The same shot gather as in (f), after linear $\tau$-$p$ transformation. The pass region of the $\tau$-$p$ filter is shown by a solid black line. (h) Result of linear $\tau$-$p$ processing obtained from filtering (g) and applying the inverse $\tau$-$p$ transformation. The records were scaled in relation to the rms amplitude of their respective gathers.

of about 4 ms were applied subsequently to the inverse NMO-corrected data.

As was noted earlier, most of the source-generated noise was not eliminated by band-pass filtering. Linear events (e.g., guided waves), which can affect the interpretation of shallow seismic adversely if they are not suppressed, still can be observed (Figure 3b). Some of this coherent noise can be reduced by mapping the data from a normal *t-x* domain into an apparent-velocity versus two-way-traveltime (radial or *r-t*) domain. The basis of this noise-attenuation process is that linear events in the *t-x* gather transform into a relatively few radial traces, with apparent frequencies shifting from the seismic band to subseismic frequencies (Henley, 1999). After transformation to the *r-t* domain (Figure 3c), a low-cut filter of 45 Hz was applied to the radial traces (Figure 3d) to eliminate the coherent noise mapped by an *r-t* transform to low frequencies. Subsequently, the data were transformed back to the *t-x* domain (Figure 3e). With regard to removal of some of the guided waves, the improvement of the data passed through radial processing (Figure 3e) compared with the quality of the original (Figure 3b) is quite noticeable.

To further reduce the source-generated noise (direct waves, surface waves, and remnants of guided waves) in the data, linear time-slowness (*τ-p*) processing was applied next. Linear and hyperbolic events in the *t-x* domain are mapped into points and ellipses, respectively, in the linear *τ-p* (or slant-slack) domain during linear *τ-p* transformation (Yilmaz, 2001). Hence, it is possible to separate these events in slant-slack gathers, to facilitate noise suppression. The steps involved in the linear *τ-p* processing are outlined below (modified from Spitzer et al., 2001).

1)  The shot gathers were converted to reduced-traveltime format (linear-moveout terms) using velocities derived from intercept-time refraction analyses. To generate the gathers,

$$t' = t + 30 - (x/V_{av}) \qquad (2)$$

was applied to each trace, where $t'$ is the reduced time in milliseconds, $t$ is the original time in milliseconds, $x$ is the source-receiver offset in meters, and $V_{av}$ is the average velocity in kilometers per second. Generally, the average velocities change across the survey line. A bulk shift of 30 ms was applied to the data to accommodate possible overcorrections of linear moveout. As seen in Figure 3f, the first arrivals and related source-generated noise have been converted to horizontal or nearly horizontal events.

2)  Because the recording direction is not preserved during *τ-p* mapping (Spitzer et al., 2001), we separated the positive source-receiver offsets from negative offsets before processing them further.

3)  The reduced-traveltime shot gathers were transformed into the linear *τ-p* domain using a range of *p* (slowness) values from −0.9 to 0.4 ms/m for positive offsets and −0.4 to 0.9 ms/m for negative offsets, to exclude surface waves and other low-velocity coherent noise, and with *τ* being intercept time. Although minor aliasing of the surface waves was observed in the slant-stack gathers (as observed in the frequency-wavenumber or *f-k* domain) of the data, it does not seem to pose a major problem to our data.

4)  The reflected events (i.e., ellipses) in the *τ-p* domain are quite distinguishable from the source-generated noise (mapped to points around $p = 0$ ms/m). We defined a 2D pass filter (as illustrated in Figure 3g) around the elliptical events for data on each side of the split source-receiver offset, and we set the amplitudes of the regions outside the area to zero. A 5-ms taper was applied in the *τ* direction to the data, to minimize artifacts.

5)  We then performed inverse linear *τ-p* transformation on the filtered *τ-p* data. Subsequently, the data for the positive and negative offsets were recombined, and the linear moveout terms and the time bulk shift were removed. The results show that most of the linear source-generated noise has been reduced with no adverse effect on the reflections (Figure 3h).

Some linear events still remain in the data. This could be because spatially aliased events in the *t-x* domain might spread over a range of slowness values, including the pass region of the filter, in the *τ-p* domain (Spitzer et al., 2001). These remnant linear events were removed carefully by surgical muting. Applying predictive deconvolution with 150-ms operator length and a prediction distance of 15 ms at this stage appeared to remove some of the multiples at greater depths.

Stacking velocities are dip-dependent, so in the case of an intersection between a flat event and a dipping event, one can choose a stacking velocity in favor of only one of these events, not both (Yilmaz, 2001). Dip-moveout (DMO) correction preserves differing dips with dissimilar stacking velocities during stacking. We applied DMO corrections to the NMO-corrected gathers (using velocities from the initial conventional CMP velocity analyses) and then performed an inverse NMO on the resultant data. Subsequently, a final velocity analysis was carried out on CMP supergathers (made up of 15 adjacent composite CMPs). The stacking velocities of the *τ-p*-processed and DMO-corrected data can be picked with greater assurance compared with the data that were not passed through those processing steps. The final stacking velocities from conventional CMP velocity analyses (Figure 4a) and the corresponding interval velocities after conversion (Figure 4b) show the lateral variation in the velocities from the buried valley to the Cretaceous bedrock, as does the result of the tomographic inversion (Figure 4c).

Using statics estimates that were computed after NMO corrections based on the final stacking velocities had been done, residual statics again were carried out. As a result of NMO correction, a frequency distortion occurs, particularly for shallow events and large offsets (Yilmaz, 2001). A stretch mute of 60% was applied to the data to get around that problem. The data later were stacked and frequency-space (*f-x*) prediction was performed on the data to reduce incoherent noise (Canales, 1984). For display purposes, an automatic gain control of 300 ms was applied to the final stacked section (Figure 5a). To place the reflections

in the true subsurface positions, 2D Kirchhoff poststack time migration also was performed on the seismic data (Figure 5b).

Although it is possible to make an interpretation about the structure of the buried valley from the time-stack section, correlation with depth values cannot be made without having a time-depth relationship. We conducted a simple depth conversion of the seismic section, using average velocities from the generated tomographic model and from conventional CMP velocity analyses. However, the results showed a prominent reflection (the sub-Cretaceous unconformity) being pulled down substantially at the west end of the profile. This probably is the result of the considerable lateral variation in velocity. Instead, prestack depth-migration (PSDM) processing was performed on the data, after the noise-suppression procedures (as outlined in Table 4), by using the velocity distribution derived from the traveltime tomography of refracted and reflected events

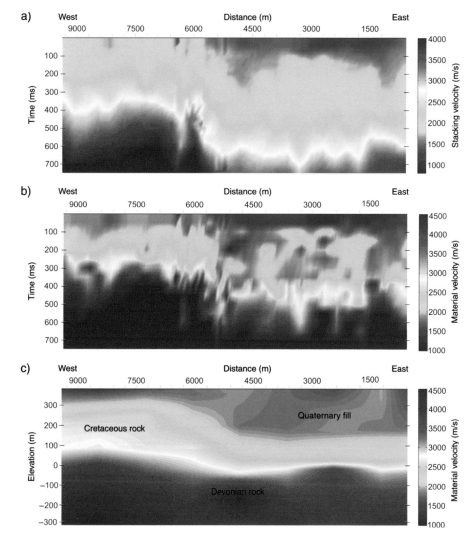

**Figure 4.** (a) Final stacking velocities as picked during traditional velocity analyses on CMP supergathers. The vertical axis is two-way time in milliseconds. (b) for the same data as in (a), after conversion to interval velocities. (c) The interval velocities of the subsurface, as acquired from the traveltime inversion with a vertical exaggeration of about five. The vertical axis is elevation (m) above sea level (asl). The different rock bodies, as labeled, can be distinguished on the basis of their respective material velocities.

**Figure 5.** (a) Unmigrated time seismic profile with interpretative tags. (b) Post-stack time-migrated seismic section. (c) Final prestack depth-migrated seismic. (d) The same data as in (c), overlaid here with the traveltime inversion velocities. The interpretation tags in (a) and (d) are D: Devonian rock; C: Cretaceous rock; Q: Quaternary fill; w: washed-out regions; Dt: top of Devonian unconformity horizon; Qa: possible top of Cretaceous bedrock; Qb: strong Quaternary event with possible gas presence; Qc: Quaternary dipping reflectors; Qd: strong dipping event in the Quaternary; and Qe: flat-lying reflectors in the Quaternary fill. An automatic gain control of 300 ms was applied to each of the stacked seismic images and they were scaled in relation to the mean amplitude of their respective section.

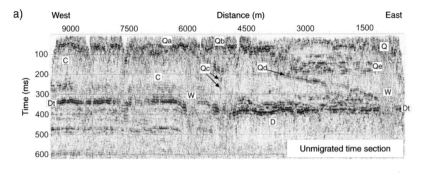

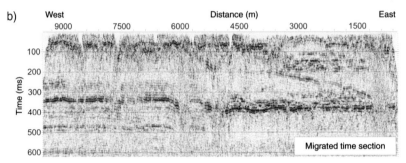

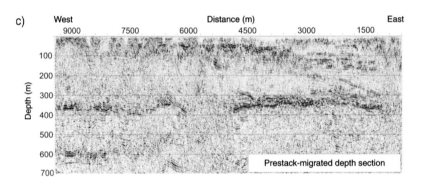

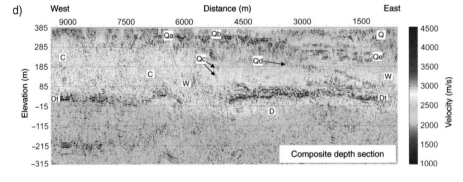

(Figure 4c). Although the results from the PSDM procedure showed poorer quality than had been expected, the depth to the prominent reflector at the top of the unconformity agrees acceptably with available wellbore information. The resulting PSDM image obtained with a split-step (Stoffa et al., 1990) shot-profile (Biondi, 2003) migration algorithm is displayed on its own (Figure 5c) and is overlaid with a velocity field from traveltime tomography and some interpretation tags (Figure 5d).

## Results and Discussion

The tomographic velocity distribution of the subsurface (Figure 4c) that was created from inversion of refracted and reflected traveltimes is an improvement over the simple refraction analysis carried out in the previous study. Delineating the paleovalley on the basis of differences in the material velocities of the rocks in the area is the foremost reason for generating the tomography. The Quaternary sediments in the east are distinguished readily from the Cretaceous bedrock in the west with material velocities, which vary from ~1700 m/s in the buried valley to ~2800 m/s in the bedrock (Figure 4c). This is not surprising because the Quaternary-fill deposits are expected to be loosely consolidated (thereby giving rise to lower compressional-wave velocities) compared with the stiffer Cretaceous bedrock. Accordingly, the edge of the valley is defined by rapid changes in the material velocities, as can be observed in the inversion result at distances between 4800 and 6000 m. Vertically, the transition in the velocities from the inversion (Figure 4c) is not abrupt, but the geologically sharp unconformity (see Ahmad et al., 2009) at the base of both the Cretaceous and Quaternary sediments still is noticeable. The velocities rise rapidly in the tomographic result to values >3500 m/s, which is typical of the deeper Devonian carbonates. The stacking velocities that were converted to interval velocities (Figure 4b) are comparable to the final material velocities derived from the results of the traveltime inversion (Figure 4c). The similarities, both in magnitude and features, between the two velocity images (Figure 4b and 4c) are apparent.

However, it is noteworthy that the edge of the valley (at an approximate distance of 5700 m), as deduced from the interval velocities (Figure 4b) derived from the picked stacking velocities, appears more abrupt than is observed in the tomographic velocity model (Figure 4c). This difference in velocity transition from the valley to the Cretaceous rock could be related to challenges encountered in the course of picking the stacking velocities on washed-out CMP supergathers, because reflections were difficult to detect in the washout zones.

It also is observed that the top of the sub-Cretaceous unconformity (Ahmad et al., 2009), which is characterized by a noticeable jump in velocities, appears to be more uneven in the converted velocities (Figure 4b) than in the results of the traveltime inversion (Figure 4c). Minor errors in the stacking-velocity picking could account for that contrast. It is clear that the tomographic data are successful in delineating the paleovalley.

However, the low-resolution inversion results (Figure 4c) could not image details of the structure within the valley that are evident in the reflection profiles (Figure 5a and 5b). Some of the features include a variety of dipping reflectors Qc at the edge of the valley; a strong, dipping reflector Qd that is unconformable with the other reflectors; and the numerous flat-lying reflectors Qe. Nonetheless, because of inadequate well information within the valley area, we cannot ascertain whether substantial material-velocity differences exist in the various sediments that constitute the buried valley. Further, if there are material-

**Table 4.** PSDM processing sequence for the 2D seismic survey.

| Processing step |
| --- |
| Trace editing |
| First-break picking |
| Elevation/refraction-statics corrections |
| Spiking deconvolution |
| Time-variant band-pass filtering |
| Trace equalization |
| CMP binning |
| Initial velocity analyses |
| NMO corrections |
| Residual statics corrections |
| Inverse NMO corrections |
| Radial domain processing |
| Linear $\tau$-$p$ processing |
| Predictive deconvolution |
| Final velocity analyses |
| Final residual statics corrections |
| Prestack depth migration using tomographic velocity model |
| Stack |
| $f$-$x$ prediction |

velocity differences, we are not sure whether they can be observed clearly on sonic logs.

To convert the time section to depth, we used the tomographic velocity model to perform a PSDM on the data (Figure 5c and 5d). The quality of the results, however, was not as good as anticipated, as can be observed from the degraded reflection continuities, mostly in the western part of the line (i.e., distances > 5000 m). This could be related to minor problems in the velocity model, which for improved results might require iterative refinement with the aim of serving as input to the PSDM algorithm (see Bradford and Sawyer, 2002; Morozov and Levander, 2002; Bradford et al., 2006). Seismic anisotropy also might play a role here because the tomographic image, which includes refracted head and turning waves, could be biased by these more horizontal propagation paths. In addition, because a 2D migration can only collapse the Fresnel zone in the migration direction (Liner, 2004), the discontinuous nature of the reflections in the PSDM data could result from the fact that we are imaging an irregular 3D structure into the 2D profile.

Most of the wellbores in the immediate vicinity of the 2D line are for shallow gas production; hence, they are not deep enough to reach the top of the unconformity. Nonetheless, two wells to the south at a distance of <3 km from the survey line penetrated the unconformity at an elevation of about 28 m above sea level, which is approximately 7 m deeper than the depth to the unconformity that is clearly observable on the PSDM image (Figure 5d). Considering the uneven topography on top of the unconformity (Figure 5a), this minor discrepancy in depth is not unreasonable. However, uncertainty is involved in estimating the top of the unconformity — which is known from core and well logs to be abrupt — using the "smeared" results of the traveltime inversion. As mentioned earlier, the initially separated third and fourth layers of the tomographic velocity

model are supposed to be one and the same. However, subsequent to the inversion they were merged adequately only for the western part of the profile line and not for the eastern side (i.e., for the distance of 1200 to 5100 m, see the section on seismic traveltime inversion). Hence, either the top or the base of the fourth layer can be picked as the top of the unconformity. If the top of layer four is selected as the top of the unconformity, the results of the traveltime inversion and the PSDM stacked section agree, but if the base of layer four is picked, a depth discrepancy of approximately 43 m occurs. In addition, estimating the depth of the unconformity from the tomography, on the basis of an interpretation of the colors, is quite subjective and easily biased.

The result of the reflection data processed previously using conventional steps (without radial and linear $\tau$-$p$ processing) (Ahmad et al., 2009) is displayed in Figure 6a, and the result from the processing steps presented in this con-

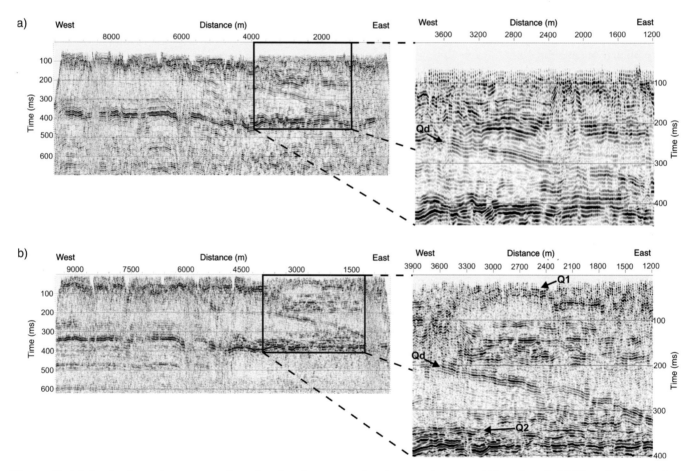

**Figure 6.** (a) A seismic section processed previously using conventional processing steps (Ahmad et al., 2009) and not optimized to reduce source-generated noise. (b) Newly processed seismic section, optimized to reduce noise. An automatic gain control of 300 ms was applied to both stacked seismic sections and they were scaled with respect to the mean amplitude of the respective section. The images on the right-hand sides are magnified to highlight better resolution of the newly processed reflection seismic profile. The interpretation tags are: Qd: strong dipping event within the Quaternary; Q1: shallow event within the Quaternary; and Q2: resolved Quaternary horizons lying above the Devonian.

tribution is shown in Figure 6b for comparison. In the magnified images beside each profile, better temporal and lateral resolution in the newly processed seismic is observed. Using the quarter-wavelength limit of vertical resolution (Widess, 1973) and a velocity of 2000 m/s, a vertical resolution of 16.6 to 6.25 m was obtained for the previously processed seismic data, from dominant frequencies typically ranging from 30–80 Hz.

Similarly, a vertical resolution of 10 to 3.3 m was obtained for the newly processed seismic profile, from dominant frequencies primarily in the 50- to 150-Hz range. Evaluating the equation for the threshold of lateral resolution — i.e., for the radius of the first Fresnel zone (Sheriff, 1980) — with 1000 m/s as velocity at 150-ms two-way traveltime, the lateral resolution of the seismic section from the previous study ranged from 35.4 to 21.6 m (30- to 80-Hz dominant frequencies). That of the new seismic profile ranged from 27.4 to 15.8 m (50- to 150-Hz dominant frequencies) at about the same two-way traveltime. Thus, it is evident from these values that the resolution of the data has been enhanced in the new seismic profile.

It is unclear from the previous seismic profile (Figure 6a) whether the low-amplitude horizons immediately below strong, dipping event Qd are flat lying or at an incline. On the other hand, the new profile (Figure 6b) shows clearly that the above-mentioned horizons dip from west to east. Clarification of the dipping nature of these events was perhaps a result of the improved resolution of the new data, but one cannot rule out the possibility that the new processing scheme recovered some eliminated parts of the reflections. Those parts could have been removed by the mute functions used in the previous processing, thus making them less coherent. At the distance of about 2400 to 3000 m and the time 30 ms, it also is possible to identify a nearly horizontal feature Q1 on the new seismic data (Figure 6b).

To avoid misinterpreting as reflections what really are coherent events and artifacts from various processing steps for enhancement (e.g., Steeples and Miller, 1990; Steeples et al., 1997; Sloan et al., 2008), we attempted to validate the true nature of the Q1 feature directly from the filtered shot records (Figure 7a). However, this shallow feature exhibited a lower frequency than did deeper reflections on shot gathers, and it could not be correlated with certainty to any true reflection. Thus, without additional supporting evidence, the Q1 feature cannot be considered to be any more than a stacked coherent event.

Clearly noticeable on the new seismic profile (Figure 6b), between distances 3000 and 4200 m, are horizons Q2, lying directly on top of the unconformity. On the old seismic section (Figure 6a), these horizons appear to be merged with the sub-Cretaceous unconformity and cannot

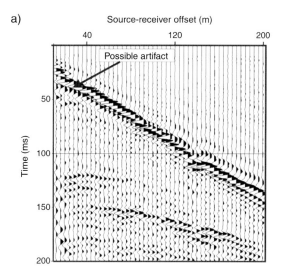

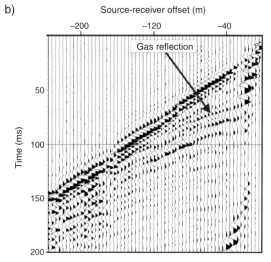

**Figure 7.** Raw shot gathers after elevation/refraction static corrections, spiking deconvolution, time-variant band-pass filtering, and trace equalization, located near (a) the poorly constrained shallow, 30-ms Q1 feature from Figure 6b, and (b) a strong reflection that is interpreted to be the top of a near-surface methane-saturated sand shown in Figure 8b.

be distinguished easily. On the new profile, these high-amplitude reflections seem to cover almost the entire extent of the bottom of the valley, from distance 4200 m on the west to distance 1200 m on the east — at which point they become incoherent because of the smeared zone. Considering the fact that these flat-lying reflectors were not observed outside the paleovalley (distances >6000 m), they are likely Quaternary sediments that were deposited immediately after the erosion caused by glacial meltwaters; alternatively, they might be remnants of the erosion of the valley itself.

Washout/smeared zones were problematic to the imaging (Figure 5). In particular, the western edge of the

a)

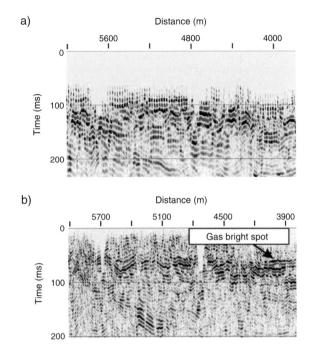

b)

**Figure 8.** Detailed comparison of the topmost section of a profile over a known near-surface gas reservoir. (a) The previously processed seismic section of Ahmad et al. (2009), and (b) the reprocessed seismic-reflection profile, showing a strong reflector that indicates the presence of free gas.

valley (at approximately distance 5700 m) was not well imaged because of a large washout zone at that location. The washout zones are attenuated regions where continuous reflections are not observed. As can be seen from the unprocessed shot gathers, it is not possible to make out any strong reflections in these zones. Although the exact cause of this attenuation is not known, it likely is associated with thicker zones of muskeg (bogs filled with sphagnum moss). The most conspicuous event in the reflection sections is the strong reflector Dt located approximately halfway down the vertical axis of the profiles. Aside from the washout zones W, this reflector, which is the unconformity above the Devonian rock, spans the entire survey line. Above the unconformity lie Cretaceous bedrock C to the west of distance 6000 m and Quaternary sediments Q to the east of distance 4800 m. The edge of the paleovalley, dipping from west to east, lies between distances 4800 and 6000 m.

There is a shallow high-amplitude reflection Qa from approximately distance 6000 to 9300 m at a two-way traveltime of ~50 ms (Figure 5a). This event could be the top of the Cretaceous bedrock. It is interesting to note that the reflection is not continuous across the entire survey line to the valley area on the eastern side. A possible explanation for this could be the minimal impedance contrast between

the glacial sediment that blankets the whole area and the deposits that constitute the buried valley. This event could not be seen clearly on the previously processed seismic profile, which points to the fact that the new data are improved relative to the old.

Shallow features that were not apparent in Ahmad et al.'s (2009) previous processing (Figure 8a) now are visible in Figure 8b and are particularly noteworthy. As also can be seen on the PSDM seismic profile (Figure 5d), there is a strong reflector Qb at an elevation of approximately 345 m above sea level (at an ~40-m depth), inside but at the edge of the valley (between distances 3900 and 5100 m). This strong reflector, also clearly visible in the raw shot gathers (e.g., Figure 7b), likely indicates the presence of free gas. Such gas has been produced in commercial quantities from this site (Rainbow and Sousa fields in northern Alberta) in the last decade, at depths of less than 100 m (see Pawlowicz et al., 2004; Kellett, 2007), and it still is being produced. Considering that the shallow gas in the Rainbow field has a chemical signature that indicates a deeper, thermogenic origin and given the high electrical resistivities recorded in our survey area, it has been suggested that gases migrated from the Cretaceous bedrock formations and were trapped in the porous Quaternary sediments (Ahmad et al., 2009).

These shallow gas deposits had been found serendipitously during previous drilling of shallow water or deeper petroleum boreholes, and on numerous occasions they have led to dangerous releases of flammable methane gas that sometimes have destroyed rigs. Ground-based electrical-resistivity tomography (ERT) studies have been used in the past to indicate free, dry gas on the basis of high electrical resistivities. However, separating gas-saturated zones from freshwater-saturated zones can be difficult. The result of reprocessing this current data set suggests that with sufficient care, such shallow gas-filled zones can be distinguished. Conducting high-resolution seismic surveys over areas already targeted for drilling on the basis of ERT could add confidence and warn drillers about potential shallow blowout hazards.

## Conclusions

We have presented the results of reanalyses of a near-surface seismic data set acquired over a paleovalley in northern Alberta, Canada. Our study includes (1) generation of a traveltime inversion to better delineate the buried valley, (2) reprocessing of the reflection data to enhance their resolution and recover any muted or degraded horizon from the previous processing, and (3) employment of prestack depth migration, using velocities generated from the traveltime inversion, to obtain proper presentation of the data in depth scale.

Construction of tomographic velocity from inversion of direct, refracted, and reflected waves aimed to determine accurate representation of the velocity distribution of the area over the buried valley. That accomplishment would be an improvement over the results of the simple refraction analysis performed in the previous study. The results from the traveltime inversion show clearly that on the basis of the material velocities, a buried valley can be delineated from the surrounding rock bodies. The interval velocity of the loosely consolidated Quaternary-fill valley was observed to be approximately 1700 m/s, whereas that of the more competent Cretaceous bedrock was approximately 2800 m/s; the edge of the valley was defined by rapid changes in the material velocities. The interpretation made from the tomographic model is important because of the significant knowledge it provides about velocity contrasts of different materials. On the other hand, such detailed information about exact contrasts in the physical properties that produce reflections might not be deduced readily from seismic-reflection profiles alone, particularly given the absence of appropriate sonic and density well logs in the area. It is noteworthy, however, that the resolution of the traveltime inversion was insufficient to image the different rock features within the valley clearly, perhaps because of the structural complexities of the valley. It is suggested that waveform inversion with the tomographic results as the input model might be a viable option in imaging a paleovalley with complex architecture.

The seismic-reflection data were processed with a strategy optimized to enhance the lateral and vertical resolutions of the profile. An additional goal was to recover any muted or degraded reflection from the old study by employing noise-suppression techniques in other domains as opposed to the total muting of the noise cone in *t-x* coordinates. The processing steps included radial and linear *τ-p* processing used to reduce noise, and time-variant band-pass filtering to improve resolution.

Better resolution and more continuous events characterize the final stack of the newly time-processed reflection profile, compared with the previously processed seismic. The new seismic had lateral resolution enhanced by ~30% in the near surface and vertical resolution enhanced by ~75%. In addition, the dipping nature of some events, which could not be established on the initial processing, was ascertained on the new seismic images. Likewise, some indistinguishable horizons lying immediately over the sub-Cretaceous unconformity were identified on the newly processed profile. This underscores the significance

of ensuring that primarily high-frequency signals are kept during the processing of near-surface reflection data.

Newly visible bright near-surface features, indicating the presence of gas, were imaged better in the new section. These features, which were not apparent in the old image, probably were muted during the previous processing sequence. Thus, we emphasize the importance of experimenting with different noise-suppression procedures before resorting to total muting of the noise cone. Aside from some washout zones in the data, the rock fabric and complex architecture of the channel and of the surrounding rock were imaged with better resolution in the newly processed stacked time section, and thus such processing is considered to produce a better result compared with that of the previous study.

Subsequently, we attempted a prestack depth migration on the noise-suppressed reflection-seismic data set, using the velocity field derived from the tomography. The quality of the result was poorer than we expected, with obvious reflection-continuity losses in some areas. We judged that problems in the tomographic model might be the reason for this. Refining the tomographic image iteratively as input into the PSDM algorithm might produce an image with unbroken reflections. The results, however, validate the importance of using a good velocity model for migration, and they underscore the challenge in obtaining the essential velocity accuracy from the shallow part of seismic data. In spite of the loss in reflection continuity, the depth to the sub-Cretaceous unconformity, as observed on the PSDM data, is consistent with the depth information obtained from two wellbores at distances of less than 3 km from the 2D seismic profile.

# Acknowledgments

We would like to thank Sam Kaplan and Todd Bown of the University of Alberta for assisting with the prestack-depth-migration processing and wellbore data gathering, respectively. We acknowledge also Colin Zelt and Uri ten Brink for providing copies of rayinvr and RayGUI software, respectively, for implementation of the seismic tomography inversion, and we thank GEDCO Ltd. for access to their VISTA® seismic data processing software via their university support program. The seismic-acquisition field crew included Jawwad Ahmad, Marek Welz, Len Tober, Gabrial Solano, Tiewei He, and Dean Rokosh (University of Alberta); John Pawlowicz, (Alberta Geological Survey, Edmonton); and Alain Plouffe (Geological Survey of Canada, Ottawa). Primary funding for the field programs was initiated by the Geological Survey of Canada and the

Alberta Geological Survey via the Targeted Geoscience Initiative–II programs. The work in this contribution was supported by the National Engineering and Research Council Discovery Grant and the Canada Research Chairs programs to D.R.S.

# References

Ahmad, J., 2006, High-resolution seismic and electrical resistivity tomography techniques applied to image and characterize a buried channel: M.S. thesis, University of Alberta.

Ahmad, J., D. R. Schmitt, C. D. Rokosh, and J. G. Pawlowicz, 2009, High resolution seismic and resistivity profiling of a buried Quaternary subglacial valley: Northern Alberta, Canada: Geological Society of America Bulletin, **121**, 1570–1583.

Andriashek, L. D., and N. Atkinson, 2007, Buried channels and glacial-drift aquifers in the Fort McMurray region, NE Alberta: Alberta Energy and Utilities Board, Alberta Geological Survey, EUB/AGS Earth Sciences Report 2007-01.

Auken, E., K. Sorensen, H. Lykke-Andersen, M. Bakker, A. Bosch, J. Gunnink, F. Binot, G. Gabriel, M. Grinat, H. M. Rumpel, A. Steuer, H. Wiederhold, T. Wonik, P. F. Christensen, R. Friborg, H. Guldager, S. Thomsen, B. Christensen, K. Hinsby, F. Jorgensen, I. M. Balling, P. Nyegaard, D. Seifert, T. Sormenborg, S. Christensen, R. Kirsch, W. Scheer, J. F. Christensen, R. Johnsen, R. J. Pedersen, J. Kroger, M. Zarth, H. J. Rehli, B. Rottger, B. Siemon, K. Petersen, M. Kjaerstrup, K. M. Mose, P. Erfurt, P. Sandersen, V. Jokumsen, and S. O. Nielsen, 2009, Buried Quaternary valleys — A geophysical approach: Zeitschrift der Deutschen Gesellschaft für Geowissenschaften, **160**, 237–247.

Baker, G. S., D. W. Steeples, and M. Drake, 1998, Muting the noise cone in near-surface reflection data: An example from southeastern Kansas: Geophysics, **63**, 1332–1338.

Belfer, I., I. Bruner, S. Keydar, A. Kravtsov, and E. Landa, 1998, Detection of shallow objects using refracted and diffracted seismic waves: Journal of Applied Geophysics, **38**, 155–168.

Benjumea, B., J. A. Hunter, J. M. Aylsworth, and S. E. Pullan, 2003, Application of high-resolution seismic techniques in the evaluation of earthquake site response, Ottawa Valley, Canada: Tectonophysics, **368**, 193–209.

Best, M. E., V. M. Levson, T. Ferbey, and D. McConnell, 2006, Airborne electromagnetic mapping for buried Quaternary sands and gravels in northeast British Columbia, Canada: Journal of Environmental and Engineering Geophysics, **11**, 17–26.

Biondi, B., 2003, Equivalence of source-receiver migration and shot-profile migration: Geophysics, **68**, 1340–1347.

Bradford, J. H., L. M. Liberty, M. W. Lyle, W. P. Clement, and S. Hess, 2006, Imaging complex structure in shallow seismic-reflection data using prestack depth migration: Geophysics, **71**, no. 6, B175–B181.

Bradford, J. H., and D. S. Sawyer, 2002, Depth characterization of shallow aquifers with seismic reflection, part II: Prestack depth migration and field examples: Geophysics, **67**, 98–109.

Büker, F., A. G. Green, and H. Horstmeyer, 1998, Shallow seismic reflection study of a glaciated valley: Geophysics, **63**, 1395–1407.

————, 2000, 3-D high-resolution reflection seismic imaging of unconsolidated glacial and glaciolacustrine sediments: Processing and interpretation: Geophysics, **65**, 1395–1407.

Canales, L. L., 1984, Random noise reduction: 54th Annual International Meeting, SEG, Expanded Abstracts, 525–527.

Chambers, J. E., O. Kuras, P. I. Meldrum, R. D. Ogilvy, and J. Hollands, 2006, Electrical resistivity tomography applied to geologic, hydrogeologic, and engineering investigations at a former waste-disposal site: Geophysics, **71**, no. 6, B231–B239.

Clague, J. J., J. L. Luternauer, S. E. Pullan, and J. A. Hunter, 1991, Postglacial deltaic sediments, southern Fraser River delta, British Columbia: Canadian Journal of Earth Sciences, **28**, 1386–1393.

Deen, T., and K. Gohl, 2002, 3-D tomographic seismic inversion of a paleochannel system in central New South Wales, Australia: Geophysics, **67**, no. 5, 1364–1371.

De Iaco, R., A. G. Green, H. R. Maurer, and H. Horstmeyer, 2003, A combined seismic reflection and refraction study of a landfill and its host sediments: Journal of Applied Geophysics, **52**, 139–156.

Fisher, T. G., H. M. Jol, and A. M. Boudreau, 2005, Saginaw Lobe tunnel channels (Laurentide Ice Sheet) and their significance in south-central Michigan, USA: Quaternary Science Reviews, **24**, 2375–2391.

Fradelizio, G. L., A. Levander, and C. A. Zelt, 2008, Three-dimensional seismic- reflection imaging of a shallow buried paleochannel: Geophysics, **73**, no. 5, B85–B98.

Francese, R. G., Z. Hajnal, and A. Prugger, 2002, High-resolution images of shallow aquifers — A challenge in near-surface seismology: Geophysics, **67**, 177–187.

Gabriel, G., R. Kirsch, B. Siemon, and H. Wiederhold, 2003, Geophysical investigation of buried Pleistocene subglacial valleys in northern Germany: Journal of Applied Geophysics, **53**, 159–180.

Greenhouse, J. P., and P. F. Karrow, 1994, Geological and geophysical studies of buried valleys and their fills near

Elora and Rockwood, Ontario: Canadian Journal of Earth Sciences, **31**, 1838–1848.

Henley, D. C., 1999, The radial trace transform: An effective domain for coherent noise attenuation and wave field separation: 69th Annual International Meeting, SEG, Expanded Abstracts, 1204–1207.

Hickin, A. S., B. Kerr, D. G. Turner, and T. E. Barchyn, 2008, Mapping Quaternary paleovalleys and drift thickness using petrophysical logs, northeast British Columbia, Fontas map sheet, NTS 94I: Canadian Journal of Earth Sciences, **45**, 577–591.

Hooke, R. L., and C. E. Jennings, 2006, On the formation of the tunnel valleys of the southern Laurentide ice sheet: Quaternary Science Reviews, **25**, 1364–1372.

Hunter, J. A., S. E. Pullan, R. A. Burns, R. M. Gagne, and R. L. Good, 1984, Shallow seismic reflection mapping of the overburden-bedrock interface with the engineering seismograph — Some simple techniques: Geophysics, **49**, 1381–1385.

Jørgensen, F., H. Lykke-Andersen, P. B. E. Sandersen, E. Auken, and E. Nørmark, 2003a, Geophysical investigations of buried Quaternary valleys in Denmark: An integrated application of transient electromagnetic soundings, reflection seismic surveys and exploratory drillings: Journal of Applied Geophysics, **53**, 215–228.

Jørgensen, F., and P. B. E. Sandersen, 2008, Mapping of buried tunnel valleys in Denmark: New perspectives for the interpretation of the Quaternary succession: Geological Survey of Denmark and Greenland Bulletin, **15**, 33–36.

Jørgensen, F., P. B. E. Sandersen, and E. Auken, 2003b, Imaging buried Quaternary valleys using the transient electromagnetic method: Journal of Applied Geophysics, **53**, 199–213.

Juhlin, C., H. Palm, C. Müllern, and B. Wållberg, 2002, Imaging of groundwater resources in glacial deposits using high-resolution reflection seismics, Sweden: Journal of Applied Geophysics, **51**, 107–120.

Kellett, R., 2007, A geophysical facies description of Quaternary channels in northern Alberta: CSEG Recorder, **32**, no. 10, 49–55.

Lennox, D. H., and V. Carlson, 1967, Geophysical exploration for buried valleys in an area north of Two Hills, Alberta: Geophysics, **32**, 331–362.

Levson, V., 2008, Geology of northeast British Columbia and northwest Alberta: Diamonds, shallow gas, gravel, and glaciers: Canadian Journal of Earth Sciences, **45**, 509–512.

Liner, C. L., 2004, Elements of 3D seismology: PennWell Corporation.

Miller, R. D., D. W. Steeples, and M. Brannan, 1989, Mapping a bedrock surface under dry alluvium with shallow seismic reflections: Geophysics, **54**, 1528–1534.

Montagne, R., and G. L. Vasconcelos, 2006, Extremum criteria for optimal suppression of coherent noise in seismic data using the Karhunen-Loève transform: Physica A, **371**, 122–125.

Morozov, I. B., and A. Levander, 2002, Depth image focusing in traveltime map-based wide-angle migration: Geophysics, **67**, 1903–1912.

Nitsche, F. O., A. G. Green, H. Horstmeyer, and F. Büker, 2002, Late Quaternary depositional history of the Reuss Delta, Switzerland: Constraints from high-resolution seismic reflection and georadar surveys: Journal of Quaternary Science, **17**, 131–143.

ÓCofaigh, C., 1996, Tunnel valley genesis: Progress in Physical Geography, **20,** 1–19.

Ogunsuyi, O., D. Schmitt, and J. Ahmad, 2009, Seismic traveltime inversion to complement reflection profile in imaging a glacially buried valley: 79th Annual International Meeting, SEG, Expanded Abstracts, 3675–3678.

Paulen, R. C., M. M. Fenton, J. A. Weiss, J. G. Pawlowicz, A. Plouffe, and I. R. Smith, 2005, Surficial Geology of the Hay Lake Area, Alberta (NTS 84L/NE): Alberta Energy and Utilities Board, EUB/AGS Map 316, scale 1:100000.

Pawlowicz, J. G., A. S. Hicken, T. J. Nicoll, M. M. Fenton, R. C. Paulen, A. Plouffe, and I. R. Smith, 2004, Shallow gas in drift: Northwestern Alberta: Alberta Energy and Utilities Board, EUB/AGS Information Series 130.

———, 2005a, Bedrock topography of the Zama Lake area, Alberta (NTS 84L): Alberta Energy and Utilities Board, EUB/AGS Map 328, scale 1:250000.

———, 2005b, Drift thickness of the Zama Lake area, Alberta (NTS 84L): Alberta Energy and Utilities Board, EUB/AGS Map 329, scale 1:250000.

Plouffe, A., I. R. Smith, R. C. Paulen, M. M. Fenton, and J. G. Pawlowicz, 2004, Surficial geology, Bassett Lake, Alberta (NTS 84L SE): Geological Survey of Canada, Open File 4637, scale 1:100000.

Pugin, A. J., T. H. Larson, S. L. Sargent, J. H. McBride, and C. E. Bexfield, 2004, Near-surface mapping using SH-wave and P-wave seismic land-streamer data acquisition in Illinois, U. S.: The Leading Edge, **23**, 677–682.

Pugin, A. J.-M., S. E. Pullan, J. A. Hunter, and G. A. Oldenborger, 2009, Hydrogeological prospecting using P- and S-wave landstreamer seismic reflection methods: Near Surface Geophysics, **7**, 315–327.

Pullan, S. E., J. A. Hunter, H. A. J. Russell, and D. R. Sharpe, 2004, Delineating buried-valley aquifers using shallow seismic reflection profiling and grid downhole geophysical logs — An example from southern Ontario, Canada, *in* C. Chen and J. H. Xia, eds., Progress in environmental and engineering geophysics: Proceedings of the International Conference on Environmental and Engineering Geophysics, 39–43.

Roberts, M. C., S. E. Pullan, and J. A. Hunter, 1992, Applications of land-based high resolution seismic reflection analysis to Quaternary and geomorphic research: Quaternary Science Reviews, **11**, 557–568.

Robertsson, J. O. A., K. Holliger, A. G. Green, A. Pugin, and R. De Iaco, 1996, Effects of near-surface waveguides on shallow high-resolution seismic refraction and reflection data: Geophysical Research Letters, **23**, 495–498.

Ronen, J., and J. F. Claerbout, 1985, Surface-consistent residual statics estimation by stack-power maximization: Geophysics, **50**, 2759–2767.

Sanderson, P. B. E., and F. Jørgensen, 2003, Buried Quaternary valleys in western Denmark — Occurrence and inferred implications of groundwater resources and vulnerability: Journal of Applied Geophysics, **53**, 229–248.

Schijns, H., S. Heinonen, D. R. Schmitt, P. Heikkinen, and I. T. Kukkonen, 2009, Seismic refraction traveltime inversion for static corrections in a glaciated shield rock environment: A case study: Geophysical Prospecting, **57**, 997–1008.

Sharpe, D. R., A. Pugin, S. E. Pullan, and G. Gorrell, 2003, Application of seismic stratigraphy and sedimentology to regional hydrogeological investigations: An example from Oak Ridges Moraine, southern Ontario, Canada: Canadian Geotechnical Journal, **40**, 711–730.

Sharpe, D., A. Pugin, S. Pullan, and J. Shaw, 2004, Regional unconformities and the sedimentary architecture of the Oak Ridges Moraine area, southern Ontario: Canadian Journal of Earth Sciences, **41**, 183–198.

Sheriff, R. E., 1980, Nomogram for Fresnel-zone calculation: Geophysics, **45**, 968–972.

Sloan, S. D., D. W. Steeples, and P. E. Malin, 2008, Acquisition and processing pitfall associated with clipping near-surface seismic reflection traces: Geophysics, **73**, no. 1, W1–W5.

Song, J. L., and U. ten Brink, 2004, RayGUI 2.0 — A graphical user interface for interactive forward and inversion ray-tracing: U. S. Geological Survey Open-File Report 2004–1426.

Spitzer, R., F. O. Nitsche, and A. G. Green, 2001, Reducing source-generated noise in shallow seismic data using linear and hyperbolic $\tau$-$p$ transformations: Geophysics, **66**, 1612–1621.

Spitzer, R., F. O. Nitsche, A. G. Green, and H. Horstmeyer, 2003, Efficient acquisition, processing, and interpretation strategy for shallow 3D seismic surveying: A case study: Geophysics, **68**, 1792–1806.

Steeples, D. W., A. G. Green, T. V. McEvilly, R. D. Miller, W. E. Doll, and J. W. Rector, 1997, A workshop examination of shallow seismic reflection surveying: The Leading Edge, **16**, 1641–1647.

Steeples, D. W., and R. D. Miller, 1990, Seismic reflection methods applied to engineering, environmental, and groundwater problems, *in* S. H. Ward, ed., Geotechnical and environmental geophysics, v.1: Review and tutorial: SEG Investigations in Geophysics Series No. 5, 1–30.

——, 1998, Avoiding pitfalls in shallow seismic reflection surveys: Geophysics, **63**, 1213–1224.

Steuer, A., B. Siemon, and E. Auken, 2009, A comparison of helicopter-borne electromagnetics in frequency- and time-domain at the Cuxhaven valley in northern Germany: Journal of Applied Geophysics, **67**, 194–205.

Stoffa, P. L., J. T. Fokkema, R. M. de Luna Freire, and W. P. Kessinger, 1990, Split-step Fourier migration: Geophysics, **55**, 410–421.

Widess, M. B., 1973, How thin is a thin bed?: Geophysics, **38**, 1176–1180.

Wiederhold, H., H. A. Buness, and K. Bram, 1998, Glacial structures in northern Germany revealed by a high-resolution reflection seismic survey: Geophysics, **63**, 1265–1272.

Wiederhold, H., H. M. Rumpel, E. Auken, B. Siemon, W. Scheer, and R. Kirsch, 2008, Geophysical methods for investigation and characterization of groundwater resources in buried valleys: Grundwasser, **13**, 68–77.

Yilmaz, Ö., 2001, Seismic data analysis: Processing, inversion, and interpretation of seismic data: SEG Investigations in Geophysics Series No. 10.

Zelt, C. A., A. Azaria, and A. Levander, 2006, 3D seismic refraction traveltime tomography at a groundwater contamination site: Geophysics, **71**, no. 5, H67–H78.

Zelt, C. A., and R. B. Smith, 1992, Seismic traveltime inversion for 2-D crustal velocity structure: Geophysical Journal International, **108**, 16–34.

Chapter 23

# Estimation of Chalk Heterogeneity from Stochastic Modeling Conditioned by Crosshole GPR Traveltimes and Log Data

Lars Nielsen[1], Majken C. Looms[1], Thomas M. Hansen[2], Knud S. Cordua[2], and Lars Stemmerik[1]

## Abstract

Rocks from the Chalk Group host important reservoirs of groundwater onshore Denmark and oil and gas in the North Sea. Fine-scale heterogeneity of rocks from the Chalk Group is investigated by stochastic modeling conditioned by first-arrival crosshole GPR traveltimes and lithology data from boreholes. The water-saturated carbonate-dominated rocks contain sharp contrasts among highly porous low-velocity carbonate, thin intercalations of flint, and hardened low-porosity carbonate of higher velocity. The stochastic simulation algorithm can model the contrasting lithologies effectively. It is found that bimodal distributions produce geologically plausible representations of the subsurface. Moreover, the variety of tested stochastic models that honors the specified data uncertainties and prior information provide a good overview of possible subsurface scenarios that the combined data set allows for. Results motivate future GPR-based tracer tests and time-lapse studies for generation of new knowledge of the dynamics of Chalk Group fluid flow.

## Introduction

The high resolution of subsurface heterogeneity provided by crosshole ground-penetrating-radar (GPR) measurements allows for detailed studies of characteristics and features critical for reservoir evaluation such as porosity variation, fluid content, and cavities (e.g., Binley et al., 2002; Becht et al., 2004; Tronicke et al., 2004; Day-Lewis et al., 2005). As a result, several time-lapse crosshole GPR studies that focused on hydrology applications have mapped patterns of fluid flow and estimated flow parameters (e.g., Binley et al., 2001; Binley et al., 2002; Looms et al., 2008).

Traditionally, first-arrival traveltimes from the crosshole GPR sections have been inverted to estimate minimum-variance radar-wave velocity structures using some type of ray-based, regularized least-squares inversion algorithm (cf. Menke, 1989; Binley et al., 2001). However, such methods are known to produce smooth representations of subsurface structures, which might not capture fine-scale heterogeneity that is important for proper reservoir characterization (Day-Lewis and Lane, 2004; Day-Lewis et al., 2005; Moysey et al., 2005). Recent advances in methodology have led to the development of full-waveform inversion algorithms that improve resolution of the interborehole region significantly (Ernst et al., 2007). Other approaches toward enhancing resolution of the target region rely on some type of stochastic inversion (e.g., Gloaguen et al., 2005; Hansen et al., 2006; Gloaguen et al., 2007a; Gloaguen et al., 2007b; Dafflon et al., 2009). A great advantage of stochastic schemes is that they allow for generation of multiple fine-scale models of subsurface heterogeneity that honor a specified data uncertainty and a priori information, which can be used for thorough model uncertainty assessment and multiple-scenario analysis.

Previous crosshole GPR studies conducted in limestone environments, for example, have aimed at resolving cavities (i.e., karsts) (Corin et al, 1997; Becht et al., 2004) and monitor the effects of steam injected into a fractured formation (Grégoire and Joesten, 2006). In this study, first-arrival traveltimes are picked from crosshole GPR sections collected in carbonate deposits from southeastern Denmark. Lithologic information and porosity data are extracted from cores from the boreholes. Characterization of the

[1]*University of Copenhagen, Department of Geography and Geology, Copenhagen, Denmark.*
[2]*Technical University of Denmark, Department of Informatics and Mathematical Modeling, Lyngby, Denmark.*

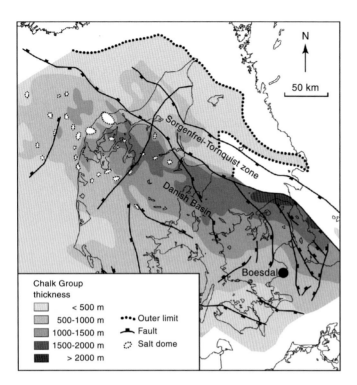

**Figure 1.** Map showing overall thickness variation of the Chalk Group in the Danish Basin area. The dot marks the study site in the Boesdal limestone quarry on the island of Sjælland. The outline and structure of the Chalk Group have been investigated in detail based on reflection-seismic data (Vejbæk et al., 2003). After Stemmerik et al., 2006. Used by permission.

heterogeneity of the interborehole region is obtained by jointly interpreting traveltime and borehole data. In particular, the effects of sharp lithologic boundaries with expected high contrasts in GPR wave velocity constrained by the borehole data are investigated. The data set is inverted by sequential simulation using the volume-average integration-simulation (VISIM) algorithm, which is described by Hansen and Mosegaard (2008) and is based on a theory by Hansen et al. (2006). This algorithm has been used previously in hydrogeophysical investigations of the unsaturated zone of sandy environments using only crosshole GPR data (Hansen et al., 2008; Cordua et al., 2009; M. C. Looms, personal communication, 2010).

Model simulations conditioned by traveltime data obtained from this procedure are used to estimate heterogeneity characteristics (i.e., variance and spatial correlation lengths of GPR wave-velocity fluctuations) at different depth levels of the carbonate rocks (cf. Hansen et al., 2008; M. C. Looms, personal communication, 2010). Next, stochastic images of the GPR wave-velocity fluctuations are generated for the interborehole region. These simula-

tions are conditioned either by the GPR traveltimes alone or by an integrated data set consisting of the crosshole GPR traveltimes and velocity contrasts inferred from lithology differences observed in the borehole core material. In addition, the stochastic images of the velocity structure are compared to reflection GPR data collected at the study site to test how fine-scale heterogeneity compares to overall layering of the subsurface. Finally, we discuss briefly how such high-resolution investigations can be used in the generation of improved reservoir models for hydrologic applications in similar settings and hydrocarbon production from the Chalk Group in the North Sea region.

## Study Site

The Upper Cretaceous–Danian Chalk Group forms an important reservoir for groundwater onshore eastern Denmark (cf. Nygaard, 1993) and for hydrocarbons in the Danish sector of the North Sea (Surlyk et al., 2003). Its regional thickness variation in the North Sea and Danish Basin area is well known from several reflection-seismic surveys (Vejbæk et al., 2003) (Figure 1). The upper part of the Chalk Group is well exposed along Stevns Klint in eastern Denmark, a coastal cliff that is more than 14 km long and as much as 40 m high. Stratigraphy and sedimentology of the exposed section are well known (Surlyk et al., 2006). The uppermost Maastrichtian (Upper Cretaceous) succession is subdivided into a mounded to horizontally bedded benthos-poor chalk of the Sigerslev Member and an upper mounded bryozoan chalk wackestone of the Højerup Member (Surlyk et al., 2006; Anderskouv et al., 2007). The boundary between the two members is at the top of a laterally extensive incipient hardground with hardening and cementation of the topmost 20 to 60 cm of the Sigerslev Member (Surlyk et al., 2006). The lowermost Danian strata (the Fiskeler and Cerithium Limestone Members) are overlain by a prominent lower Danian erosional hardground surface that is associated locally with hardening and cementation of the underlying strata of the upper part of the Maastrichtian Højerup Member. The lower Danian succession is composed of a succession of bryozoan mounds that are 9 m thick and 50 to 100 m long (Bjerager and Surlyk, 2007).

At the study site of Boesdal quarry (Figure 1), southern Stevns Klint, the Danian succession has been almost quarried away, and the quarry floor is close to the K-T boundary. The predominantly mounded nature of the upper part of the succession has been confirmed by reflection GPR data acquired using 100-MHz unshielded antennae provided by Sensors & Software (Figure 2). The mounds are detected down to a depth of approximately 5 to 6 m. Below this depth, weak subhorizontal, laterally continuous reflections are evident before the signal becomes strongly attenuated. The groundwater table was at depths of 2.14 to 2.74 m below the surface in the four boreholes at the study site. A constant velocity of 0.06 m/ns is used for migration and subsequent depth conversion of the reflection GPR data. This chosen velocity is slightly higher than the average velocity of the water-saturated part of the porous chalk succession (see next section). However, we selected the velocity of 0.06 m/ns for processing of the surface-reflection GPR data to accommodate the effects of the uppermost part of the studied section, which is above the groundwater table.

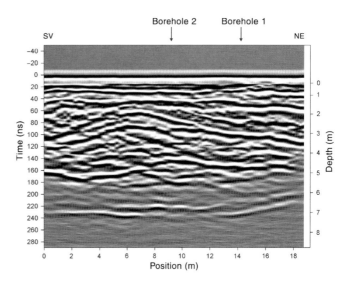

**Figure 2.** Reflection GPR data. The positions of boreholes 1 and 2 along the line are shown. The antenna frequency was 100 MHz, and the trace spacing was set to 0.25 m during data acquisition. For display purposes, the spatial sampling has been interpolated to a finer trace spacing. The reflection radar profile was dewowed using standard software provided by Sensors & Software, and a simple amplitude scaling was performed using an automatic-gain-control algorithm with a maximum scaling factor of 500. Migration and depth conversion were made using a constant velocity of 0.06 m/ns.

## Core Material and Porosity Data

The four boreholes used in our experiment were placed in a quadrangle with side lengths of approximately 5 m and were drilled and cored to 15-m depth. The studied section thus includes lower Danian bryozoan limestone of the Stevns Klint Formation, the mounded bryozoan chalk wackestones of the Højerup Member topped by a thin hardground, and the upper part of the horizontally bedded chalk of the Sigerslev Member. Core material was recovered from the boreholes for lithologic control, and subsequently, plugs with a nominal spacing of 1 m were extracted for measurement of porosity and grain density (Figure 3; Tables 1 through 4).

In borehole 1, the top of the incipient hardground at the top of the Sigerslev Member is at 5.20 m, and the low porosity of plug 106 (5.56 m) reflects syndepositional cementation of the chalk. Similarly, the prominent flint at 6.24 m is linked to this event of a stop in sedimentation (Tables 1 and 5). The low porosity values in plugs 101 and 102 at the top of the studied section (Table 1) are from the rather indurated lower Danian Stevns Klint Formation.

The upper 5 to 7 m of the cored section (corresponding to the lower Danian bryozoan limestone, the uppermost Maastrichtian mounded bryozoan chalk of the Højerup Member, and the top of the Sigerslev chalk) are characterized by abundant bands or nodules of flint and amorphous $SiO_2$. Below, flint nodules are less frequent and smaller, and the core material is composed almost entirely of unlithified, bioturbated chalk.

We have not found estimates of the dielectric properties of flint in the literature. In the following, it is assumed that the radar-wave velocity of pure flint is the same as the velocity of quartz (0.145 m/ns) (Reynolds, 1997). Inspection of the flint

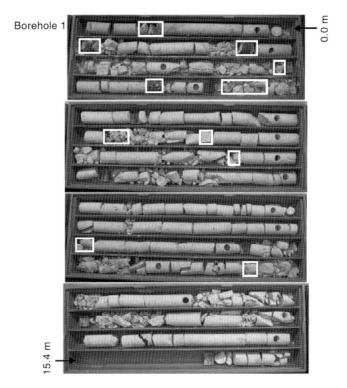

**Figure 3.** Core material from borehole 1. Open holes in the cored material show where plugs have been taken for porosity measurements of the carbonate. Flint intercalations are shown with white-outlined boxes.

**Table 1.** Petrophysical data and calculated radar-wave velocities for borehole 1.[3]

| Plug number | Depth (m) | Grain density (g/cm³) | Porosity (%) | Velocity (m/ns) |
|:---:|:---:|:---:|:---:|:---:|
| 101 | 0.21 | 2.685 | 32.71 | 0.0605* |
| 102 | 1.22 | 2.699 | 38.06 | 0.0568* |
| 103 | 2.75 | 2.695 | 42.01 | 0.0543 |
| 104 | 3.58 | 2.689 | 42.91 | 0.0538 |
| 105 | 4.39 | 2.688 | 44.32 | 0.0530 |
| 106 | 5.56 | 2.687 | 39.85 | 0.0556 |
| 107 | 6.51 | 2.691 | 45.61 | 0.0523 |
| 108 | 7.44 | 2.706 | 47.88 | 0.0511 |
| 109 | 8.53 | 2.684 | 48.63 | 0.0507 |
| 110 | 9.76 | 2.692 | 46.85 | 0.0516 |
| 111 | 11.27 | 2.688 | 47.94 | 0.0511 |
| 112 | 12.32 | 2.680 | 48.66 | 0.0507 |
| 113 | 13.02 | 2.680 | 46.60 | 0.0518 |
| 114 | 13.49 | 2.678 | 48.80 | 0.0506 |
| 115 | 14.32 | 2.692 | 48.73 | 0.0506 |
| 116 | 15.16 | 2.673 | 52.35 | 0.0489 |

[3]Velocities marked with an asterisk were calculated assuming a fully water-saturated carbonate formation even though those plugs were extracted from above the groundwater table.

**Table 2.** Petrophysical data and calculated radar-wave velocities for borehole 2.[4]

| Plug number | Depth (m) | Grain density (g/cm³) | Porosity (%) | Velocity (m/ns) |
|:---:|:---:|:---:|:---:|:---:|
| 201 | 0.21 | 2.691 | 37.26 | 0.0573* |
| 202 | 0.97 | 2.697 | 35.70 | 0.0583* |
| 203 | 1.90 | 2.700 | 43.35 | 0.0536* |
| 204 | 3.26 | 2.704 | 42.02 | 0.0543 |
| 205 | 4.52 | 2.707 | 44.08 | 0.0531 |
| 206 | 5.35 | 2.706 | 44.22 | 0.0531 |
| 207 | 6.76 | 2.703 | 42.09 | 0.0543 |
| 208 | 8.27 | 2.708 | 47.03 | 0.0515 |
| 209 | 8.87 | 2.700 | 47.96 | 0.0510 |
| 210 | 10.16 | 2.708 | 48.58 | 0.0507 |
| 211 | 12.71 | 2.699 | 47.35 | 0.0514 |
| 212 | 13.60 | 2.702 | 46.32 | 0.0519 |
| 213 | 14.53 | 2.706 | 45.71 | 0.0522 |

[4]Velocities marked with an asterisk were calculated assuming a fully water-saturated carbonate formation even though those plugs were extracted from above the groundwater table.

nodules indicates that they do not consist of pure flint but typically contain cavities filled with carbonate. Therefore, the velocity of flint intercalations of the formation is most likely smaller than the velocity of pure quartz. The depth to and the approximate thickness of the individual flint nodules found in boreholes 1 and 2 are summarized in Tables 5 and 6, respectively. However, the actual thickness of the flint nodules is uncertain because many flint nodules were fractured and broken during drilling.

Moreover, the depth to the center of the individual flint nodules is uncertain because material from the cores was lost during drilling. The flint depths have been interpolated according to fixpoints marked by the drilling team during operation, and we recognize that the average uncertainty of depth to individual nodules is probably on the order of 0.2 m. In addition, the core material recovered from borehole 2 suffered from significantly more loss of material than was the case for borehole 1. Thus, depths noted for borehole 2 are more uncertain than those listed for borehole 1.

Laboratory measurements show that porosity of the chalk varies from approximately 32% in the topmost, hardened part to more than 52% in the unlithified chalk at the base of the studied section (Tables 1 through 4). Matrix densities are close to but slightly below the density of pure carbonate of 2.71 g/cm³ (cf. Battey, 1988). These density values indicate that only minor amounts of noncarbonates are present in the plug samples.

Relative permittivities of the plugged material are calculated using the complex refraction index model (CRIM) (Lesmes and Friedman, 2005), assuming the formation is fully water saturated:

$$\sqrt{\kappa_f} - \phi\sqrt{\kappa_w} + (1 - \phi)\sqrt{\kappa_m}. \quad (1)$$

Here, $\kappa_w$ and $\kappa_m$ are the relative permittivities of water and the matrix material, respectively, and $\phi$ is the porosity. Relative permittivity $\kappa_w$ is set to 81 (cf. Reynolds, 1997), whereas $\kappa_m$ is set to 9, making the simplification that the matrix material consist of pure calcite (cf. Lesmes and Friedman, 2005). Radar-wave velocities are calculated for the plugs using the commonly used relation (cf. Annan, 2005a; Reynolds, 1997)

$$V = c/\sqrt{\kappa_f}, \quad (2)$$

where $c$ is the speed of light. The calculated radar-wave velocities for all cores are listed in Tables 1 through 4. None of the plugs was taken in flint nodules. The exact depth to the plugs is uncertain for the same reasons as discussed above for the depth to the flint nodules. Equation 2 holds only for low-loss materials, i.e., materials of low electrical conductivity (Annan, 2005b). In areas of high conductivity, it is important to account for the conductivity when describing the interrelation between GPR wave velocity and the relative permittivity (cf. Giroux and Chouteau, 2010). Further, the application of mixing laws such as equation 1 is subject to significant uncertainties depending on the assumptions made (Sambuelli, 2009a, 2009b; Giroux and Chouteau, 2010). Electrical resistivity measurements have not been made at the study site. However, measurements made in a deeper borehole approximately 300 m north of our study site show pore-water conductivities of 135 to 151 mS/m

**Table 3.** Petrophysical data and calculated radar-wave velocities for borehole 3.[5]

| Plug number | Depth (m) | Grain density (g/cm$^3$) | Porosity (%) | Velocity (m/ns) |
|---|---|---|---|---|
| 301 | 0.28 | 2.706 | 37.57 | 0.0571* |
| 302 | 2.20 | 2.703 | 40.70 | 0.0551* |
| 303 | 2.66 | 2.707 | 42.49 | 0.0541 |
| 304 | 3.72 | 2.707 | 42.51 | 0.0540 |
| 305 | 4.83 | 2.706 | 42.88 | 0.0538 |
| 306 | 8.21 | 2.711 | 47.69 | 0.0512 |
| 307 | 8.73 | 2.704 | 49.14 | 0.0504 |
| 308 | 9.58 | 2.707 | 48.40 | 0.0508 |
| 309 | 10.77 | 2.706 | 47.43 | 0.0513 |
| 310 | 12.07 | 2.718 | 47.13 | 0.0515 |

[5]Velocities marked with an asterisk were calculated assuming a fully water-saturated carbonate formation even though those plugs were extracted from above the groundwater table.

**Table 4.** Petrophysical data and calculated radar-wave velocities for borehole 4.[6]

| Plug number | Depth (m) | Grain density (g/cm$^3$) | Porosity (%) | Velocity (m/ns) |
|---|---|---|---|---|
| 401 | 0.13 | 2.690 | 37.32 | 0.0573* |
| 402 | 1.21 | 2.700 | 45.95 | 0.0521* |
| 403 | 1.89 | 2.704 | 45.10 | 0.0526* |
| 404 | 3.65 | 2.701 | 42.90 | 0.0538 |
| 405 | 5.42 | 2.703 | 43.02 | 0.0538 |
| 406 | 6.47 | 2.695 | 45.56 | 0.0523 |
| 407 | 8.01 | 2.695 | 44.71 | 0.0528 |
| 408 | 8.83 | 2.702 | 46.19 | 0.0520 |
| 409 | 10.26 | 2.693 | 46.71 | 0.0517 |
| 410 | 10.59 | 2.699 | 46.30 | 0.0519 |
| 411 | 11.67 | 2.691 | 46.85 | 0.0516 |
| 412 | 12.59 | 2.688 | 44.44 | 0.0529 |
| 413 | 13.38 | 2.699 | 46.15 | 0.0520 |

[6]Velocities marked with an asterisk were calculated assuming a fully water-saturated carbonate formation even though these plugs were extracted from above the groundwater table.

in the equivalent chalk section (S. L. Rasmussen, personal communication, 2010). Thus, we cannot rule out that the low-loss assumption might be violated to some degree. Thus, velocity values estimated using equation 2 could be biased as compared to the real subsurface velocity structure. In the following, we use only velocity values listed in Tables 1 through 4 for qualitative comparison to the overall 1D

**Table 5.** Borehole 1. Depths to the center of flint intercalations and approximate thicknesses of those intercalations.

| Depth to center of flint intercalations (m) | Thickness of flint intercalation (m) |
|---|---|
| 0.69 | 0.08 |
| 1.34 | 0.08 |
| 2.10 | 0.13 |
| 2.25 | 0.04 |
| 3.34 | 0.10 |
| 3.80 | 0.09 |
| 5.87 | 0.03 |
| 6.24 | (0.10) |
| 6.68 | 0.03 |
| 11.79 | 0.08 |
| 12.00 | 0.05 |

**Table 6.** Borehole 2. Depths to the center of flint intercalations and approximate thicknesses of those intercalations.

| Depth to center of flint intercalations (m) | Thickness of flint intercalation (m) |
|---|---|
| 0.68 | 0.08 |
| 1.71 | 0.14 |
| 5.06 | 0.09 |
| 8.09 | 0.07 |
| 8.42 | 0.07 |
| 12.15 | 0.06 |
| 13.38 | 0.25 |

velocity structure estimated from analysis of GPR travel-time data and as soft prior constraints on the background velocity during the subsequent 2D inversion of the GPR data set. Velocities in Tables 1 through 4 are not used as hard data during the inversions.

## Crosshole GPR Data

The crosshole radar data were collected using 100-MHz borehole antennae (the PulseEKKO® 100 system from Sensors & Software, Inc.). Distances between source and receiver positions in the boreholes were 1 m and 0.25 m, respectively. Data were collected using two commonly

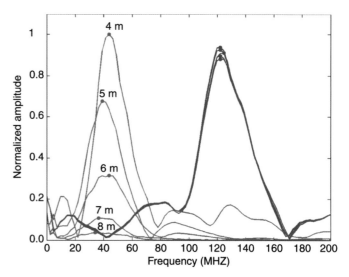

**Figure 4.** Normalized amplitude spectra for calibration data acquired above ground (blue) and data collected between boreholes 1 and 2 (red). For data acquired between boreholes, the transmitter antenna was held fixed at 4-m depth, and the receiver antenna was lowered gradually from 4- to 8-m depth, as indicated by the numbers above the red curve. Note how the signal amplitude decreases as the receiver is lowered and the distance between source and receiver increases. Red dots indicate the peak amplitude of the signals transmitted between boreholes.

used strategies (Annan, 2005a). Zero-offset profiling (ZOP), in which the source and receiver antennae were moved in parallel, was used to generate 1D velocity profiles between all borehole combinations. Multiple-offset gather (MOG) data with different acquisition angles were collected to estimate the 2D velocity variation between boreholes 1 and 2. The MOG data collected with acquisition angles outside the angle range of $\pm 45°$ were omitted from the recorded data set to minimize the undesired effects of wave guiding along the antennae (cf. Peterson, 2001; Irving and Knight, 2005). Using this procedure, 744 MOG traces were collected between boreholes 1 and 2.

Calibration data collected in free air above the boreholes show a well-defined signal peak at a frequency of about 120 MHz (exemplified for one calibration file shown with blue curves in Figure 4). Amplitude spectra calculated for the raw signals transmitted between boreholes below the groundwater table show a significantly lower peak frequency of approximately 40 MHz and an overall decrease in amplitude as the distance between the source and receiver antennae increases (exemplified for a selection of MOG data shown with red curves in Figure 4). Presumably, observed changes in frequency content are caused mainly by the combination of damping from the electrical conductivity of the formation and coupling between the borehole

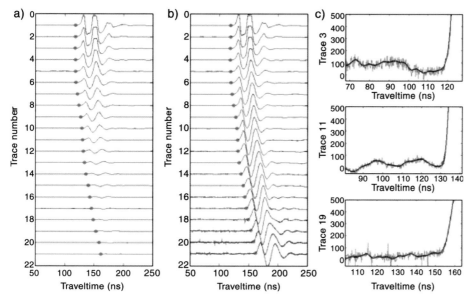

**Figure 5.** Raw (red) and frequency-filtered (black) MOG data. The source was at a constant depth of 4 m, and the receiver antenna was moved in steps from 4- to 9-m depth. (a) The complete source gather is scaled with a constant gain factor. (b) Trace-normalized data. (c) Detail of data (traces 3, 11, and 19) scaled with a constant gain factor showing how first-arrival traveltimes were picked (circles). The applied filter has a 5- to 100-MHz passband. Filtering was made using MATLAB.

antennae and the heterogeneous chalk formation, whereas the overall damping of the signal amplitudes also is influenced significantly by geometric spreading (cf. Reynolds, 1997; Lampe and Holliger, 2003).

Picking of first arrivals in the raw data set was a difficult task because of high-frequency noise leading to inconsistent picks. Therefore, a band-pass filter with a passband interval of 5 to 100 MHz was applied to suppress signals with frequencies above 100 MHz (cf. Oppenheim and Schafer, 1989). The filtering caused a significant improvement of the signal-to-noise ratio (S/N) (Figure 5). After frequency filtering, it was possible to pick first arrivals with a high degree of certainty despite the lowered peak frequency and the relatively strong damping of the signal amplitude caused by the studied formation. The standard deviation of the mismatch between 102 reciprocal traveltime measurements is less than 0.4 ns, which is comparable to results obtained for high-quality data collected in highly resistive, dry sandy material. (M. C. Looms, personal communication, 2010).

# Inversion of Data and Interpretation of Results

The crosshole GPR data are interpreted to assess the overall layering as well as fine-scale heterogeneity characteristics of the studied carbonate section. Only data collected with transmitter and receiver antennae positions below a depth of 3.5 m are included in the modeling to avoid the influence from the dry and faster formation above the groundwater table on first arrivals of refracted phases.

First, data collected in the ZOP configuration are interpreted to estimate the 1D velocity structure between the different borehole pairs and assess the overall layering characteristics at the study site. Second, spatial characteristics and variance of the velocity fluctuations are constrained by data-driven ergodic inference using the data acquired in the MOG configuration between boreholes 1 and 2 (M. C. Looms, personal communication, 2010). Third, we generate images of the velocity fluctuations of the subsurface by sequential simulation using the VISIM algorithm. Sequential simulations are constrained by first-arrival traveltime picks of the MOG data and information about the positions of high-velocity flint layers extracted from the boreholes.

## ZOP data

Distances between the boreholes is divided by the first arrivals picked in the ZOP data sets to generate 1D models of the velocity distribution between the different boreholes (Figure 6a). The 1D velocity analysis is supplemented with analysis of the maximum first-cycle amplitudes of the recorded waveforms (Figure 6b). In the depth range of 3.5 to 7.5 m, the 1D model velocity values obtained between the different pairs of boreholes fluctuate between 0.05 and 0.07 m/ns. Below 7.5-m depth, velocities of the 1D models fall in the range between 0.04 and 0.055 m/ns, and the individual 1D velocity profiles show less variation. The amplitude curves exhibit a relatively high degree of variation in the depth range of 3.5 to 6 m. Below 6-m depth, the amplitude curves show less variability, and they are centered at relatively small values.

**Figure 6.** (a) 1D velocity profiles and (b) amplitudes recorded in ZOP configuration between the four boreholes. Numbers in ZOP names indicate the boreholes used for the measurements. Note that the amplitudes measured for ZOP13 and ZOP24 are particularly small because boreholes 1 and 3 and boreholes 2 and 4 are placed diagonally with respect to one another in the quadrangle borehole geometry. Averages of velocities estimated from plug material (Tables 1 through 4) in the depth range between 3.5 to 7.5 m and below 7.5-m depth are shown with solid black lines.

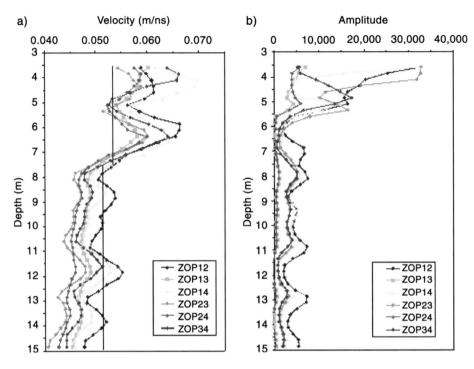

In combination, the 1D velocity model curves and amplitude values indicate a transition zone at a depth of 6 to 7.5 m. The bottom of this depth range coincides approximately with the penetration depth of the reflection GPR data and the transition from mounds with some flint nodules to carbonate with less flint intercalations and higher porosity values (Figures 2 and 3; Tables 1 through 4). Most likely, relatively high velocities above a depth of 7.5 m are caused by the flint and the hardened horizons present in this depth range. However, the 1D velocity models probably do not capture the full effect of the flint intercalations because the thicknesses of these flint bands are small compared to the dominant wavelength of the recorded wavefield (approximately 1.5 m).

In the depth interval of 3.5 to 7.5 m, average velocities estimated from the plug material (Tables 1 through 4) are low compared to the six 1D velocity profiles estimated from ZOP measurements (Figure 6a). This discrepancy between GPR measurements and calculated velocities based on the plug material is interpreted to be caused mainly by the high velocity of the flint, which is abundant in the upper part of the studied section but is not represented in the plugged material.

Below a depth of 7.5 m, where the attenuation of the GPR amplitudes is relatively strong, the 1D velocity profiles estimated from ZOP measurements (except the 1D profile calculated between borehole 1 and 2) are systematically low compared to the average velocity calculated for the plugs. At least to some extent, these misfits might be related to electrical conductivity that has a significant effect on the GPR wavefield; therefore, velocities calculated for plugs using low-loss criteria could be incorrect and biased (cf. Giroux and Chouteau, 2010).

## Linearized Gaussian inversion

The inverse problem of inferring radar-wave velocity distribution of the subsurface based on GPR traveltime and log data is solved using a probabilistic formulation of the inverse problem (Tarantola and Valette, 1982; Tarantola, 2005). This approach relies on an a priori assumption that the subsurface might be described by a Gaussian random model characterized by mean $m_0$ and covariance $C_m$. Here, an anisotropic a priori covariance model is used. This model is described by horizontal ($H_{max}$) and vertical ($H_{min}$) correlation lengths and a variance. Uncertainties of the data d are assumed to be Gaussian and are accounted for via the data-error covariance matrix $C_d$, as outlined by Tarantola and Valette (1982). The solution to the inverse problem is described fully by the a posteriori probability density function, which is characterized by its mean value $m_{est}$ and covariance $C_{m_{est}}$ (Tarantola, 2005):

$$\mathbf{m}_{est} = \mathbf{m_0} + (\mathbf{G}^T\mathbf{C_d}^{-1}\mathbf{G} - \mathbf{C_m}^{-1})^{-1}(\mathbf{G}^T\mathbf{C_d}^{-1})(\mathbf{d} - \mathbf{G}_{\mathbf{m_0}}). \quad (3)$$

The covariance of this solution also might be calculated (Tarantola, 2005):

$$\mathbf{C_{m_{est}}} = (\mathbf{G}^T\mathbf{C_d}^{-1}\mathbf{G} - \mathbf{C_m}^{-1})^{-1}. \quad (4)$$

Realizations of the a posteriori probability density function are made using the VISIM algorithm of Hansen and Mosegaard (2008). In our application, we regard average velocities along ray paths (observed GPR traveltimes divided by ray length between source and receiver) as data points. Matrix **G**, which expresses the link between model parameters (velocities) and data points, contains a weighting factor for each model cell hit by rays. These weighting factors are calculated from the ray length and traveltime delay in each cell. The sum of the individual rows of **G** equals 1. A linearized form of the solution described above is used. We linearize the inverse problem around a smooth representation of the velocity model.

The VISIM relies on a theoretical framework that combines inverse Gaussian theory and geostatistics (Hansen et al., 2006). This framework allows for non-Gaussian a priori distributions via direct sequential simulation (Hansen and Mosegaard, 2008). In this study, low-velocity carbonates with intercalations of flint are expected to yield a non-Gaussian a priori distribution. The forward modeling of traveltimes is based on solving the eikonal equation on a finite-difference grid using the algorithm described in Hole and Zelt (1995) and in Zelt and Barton (1998). Ray tracing is done by following the steepest gradient of the time field from a receiver back to the source (Vidale, 1988).

## Inference of variance and spatial correlation lengths

Values of the spatial correlation lengths and the variance of the velocity fluctuations of the formation between boreholes 1 and 2 are inferred from the MOG data using the data-driven ergodic inference methodology described by Hansen et al. (2008) and then expanded (M. C. Looms, personal communication, 2010). This methodology relies on estimating the likelihood of statistical parameters describing the subsurface based on a series of realizations drawn from the a priori distribution compared to realizations drawn from the a posteriori probability density function, conditioned by traveltime data. We refer to Hansen et al. (2008) and personal communication from M. C. Looms (2010) for an in-depth description of the applied methodology. A spherical covariance function is assumed to describe the velocity fluctuations during the data-driven ergodic inference. The spherical covariance function was chosen initially because it has the potential to capture a certain amount of roughness of the fluctuations, as might be expected based on observations made in the borehole cores.

Results from this analysis are shown in Figures 7 and 8. In case all GPR data between depths of 3.5 and 14.5 m are included, the variance of the velocity fluctuations is estimated to be $5.8 \times 10^{-5}$ $m^2/ns^2$ (peak likelihood in Figure 7a) and the vertical ($H_{min}$) and horizontal ($H_{max}$) correlation lengths are estimated to 2.6 m and 3.2 m, respectively (Figure 8a). However, the 1D velocity profiles and amplitude data extracted from the ZOP data indicate a transition at approximately 7.5-m depth. These observations indicate that the spatial characteristics of the formation above and below 7.5-m depth might be different. Inference of spatial correlation lengths and variance using only data collected above and below 7.5-m depth, respectively, also are shown in Figures 7 and 8. For the depth interval from 3.5 to 7.5 m, the variance is $3.2 \times 10^{-5}$ $m^2/ns^2$, and $H_{min}$ and $H_{max}$ are 1.1 m and 6.3 m, respectively. However, the exact value of $H_{max}$ is constrained poorly; the plot of Figure 7b indicates that this parameter in reality could take on any value between 3 and 9 m. This finding might be caused partly by the fact that the rays predominantly travel horizontally and therefore might better constrain $H_{min}$ than $H_{max}$ (M. C. Looms, personal communication, 2010). Analysis of the MOG data indicates that the variance is $2.8 \times 10^{-5}$ $m^2/ns^2$, whereas $H_{min}$ and $H_{max}$ are 2.7 m and 2.3 m, respectively, for the velocity fluctuation below 7.5-m depth.

Analysis of the whole data set results in a significantly higher variance of the velocity fluctuation as compared to the variances calculated for the zones above and below 7.5-m depth. This result combined with observations made based on the core material and ZOP data suggests that intervals above and below 7.5-m depth do have different variance and spatial correlation characteristics of fine-scale heterogeneity. Therefore, depth intervals above and below 7.5 m are treated as zones with different prior characteristics in subsequent 2D stochastic simulations of the fine-scale velocity distribution between boreholes 1 and 2.

## Stochastic inversion of MOG data

Initially, a series of unconditional realizations (i.e., not constrained by GPR data or by information from the boreholes) of the a priori probability density function is generated using the variance values and $H_{min}$ and $H_{max}$ values estimated from data-driven ergodic inference. Realizations generated for a spherical covariance function with Gaussian and bimodal a priori model-parameter value distributions, respectively, are shown in Figures 9 and 10. The bimodal a priori distribution of parameter values is defined in the following way: (1) Velocity distribution of

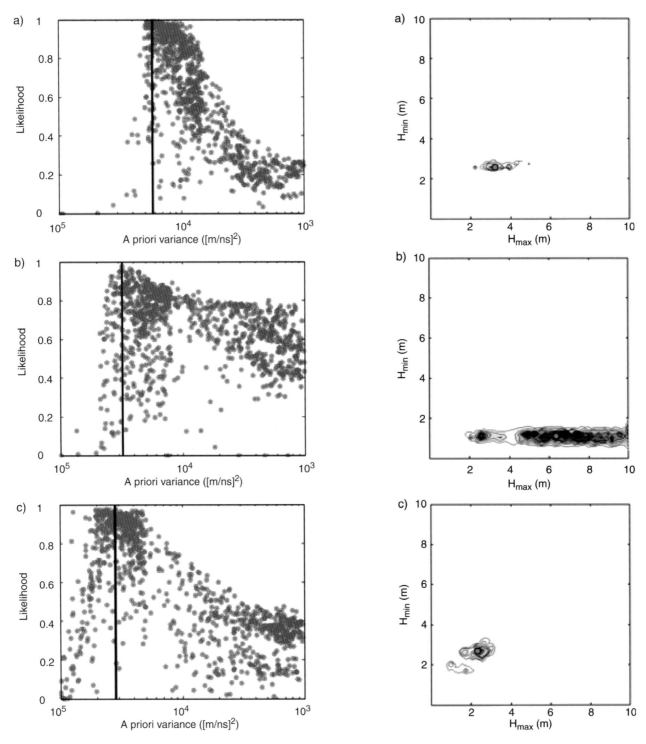

**Figure 7.** Variance of velocity fluctuations estimated from data-driven ergodic inference (M. C. Looms, personal communication, 2010). (a) All data (744 traveltime picks) were used (based on 1000 simulations). (b) Only data recorded above 8-m depth (170 traveltime picks) were used (based on 1000 simulations). (c) Only data recorded below 8-m depth (416 traveltime picks) were used (based on 1000 simulations).

**Figure 8.** Vertical and horizontal correlation length ($H_{min}$ and $H_{max}$) of velocity fluctuations estimated from data-driven ergodic inference (M. C. Looms, personal communication, 2010). (a) All data (744 traveltime picks) were used (based on 600 simulations). (b) Only data recorded above 8-m depth (170 traveltime picks) were used (based on 600 simulations). (c) Only data recorded below 8-m depth (416 traveltime picks) were used (based on 600 simulations).

the carbonate is modeled as a Gaussian distribution with a standard deviation of 0.0015 m/ns, a mean of 0.053 m/ns above a depth of 8 m, and a mean of 0.051 m/ns below 8 m. The mean velocities are estimated from the plug data. (2) For both zones, the distribution of the velocity of cells with flint is assumed to follow a uniform distribution that can take on values between 0.07 and 0.09 m/ns.

As described earlier, the GPR wave velocity of the flint nodules is not known from laboratory measurements or other prior knowledge but is assumed to be the same as the velocity of quartz (0.145 m/ns) (cf. Reynolds, 1997). The chosen distribution for the velocity of flint represents our best estimate for the present model parameterization with 0.25- × 0.25-m large cells describing the velocity field. The flint nodules typically have a thickness smaller

than the cell size (see Tables 5 and 6), and the chosen distribution for cells with flint is designed also to accommodate the effect of a significant fraction of low-velocity chalk. Further, the validity of the assumed velocity distribution for cells with flint will depend on the purity of the flint. The final bimodal velocity is obtained by weighing distributions for chalk and flint so that the variance of the bimodal distribution matches the variance found using ergodic inference above and below 8-m depth (Figure 7). The corresponding cumulative probability-distribution function above and below 8-m depth is shown in Figure 11. The corresponding a priori assumed mean velocity is then 0.055 m/ns above 8-m depth and 0.051 m/ns below 8-m depth. These a priori mean velocities correspond well to average velocities found from the 1D analysis of ZOP

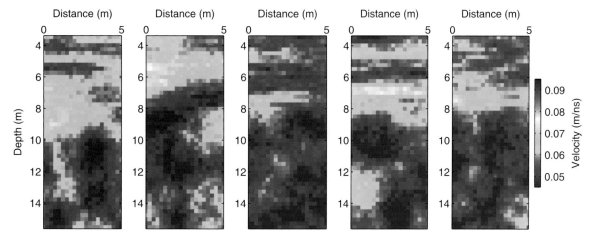

**Figure 9.** A selection of five unconditional simulations using a Gaussian a priori distribution. Upper medium (above 8-m depth): variance = $3.2 \times 10^{-5}$ m²/ns², $H_{min}$ (vertical direction) = 1.1 m, and $H_{max}$ (horizontal direction) = 6.3 m. Lower medium (below 8-m depth): variance = $2.8 \times 10^{-5}$ m²/ns², $H_{min}$ = 2.7 m, and $H_{max}$ = 2.3 m.

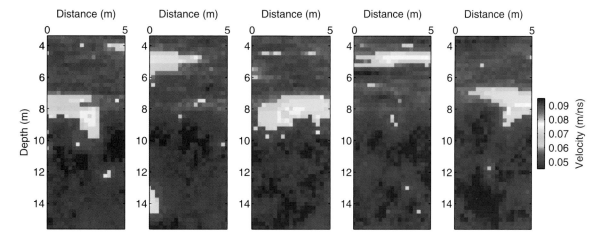

**Figure 10.** A selection of five unconditional simulations using a bimodal a priori distribution. Upper medium (above 8-m depth): variance = $3.2 \times 10^{-5}$ m²/ns², $H_{min}$ (vertical direction) = 1.1 m, and $H_{max}$ (horizontal direction) = 6.3 m. Lower medium (below 8-m depth): variance = $2.8 \times 10^{-5}$ m²/ns², $H_{min}$ = 2.7 m, and $H_{max}$ = 2.3 m.

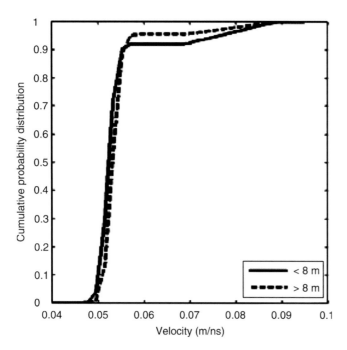

**Figure 11.** Bimodal target distribution. Note the difference in distribution of velocities above and below 8-m depth.

data, which indicate a mean velocity between boreholes 1 and 2 of about 0.051 m/ns below 8-m depth and 0.057 m/ns above 8-m depth (Figure 6a.)

Clearly, the media following a bimodal distribution have the potential to capture the heterogeneity observed in the borehole core material, i.e., alternating relatively thick layers of carbonate and thinner intercalations of flint and hardened horizons. The medium following a Gaussian distribution captures a lower degree of roughness, and it does not account for sharp boundaries to the same extent as the bimodal medium. Thus, the Gaussian medium does not show the degree of roughness that should be expected for the subsurface. Therefore, in the following, we base our conditional simulations on bimodal media. During the conditional simulations, we do not allow ray paths crossing the boundary between the zone above and below 7.5-m depth. Thus, the velocity structure between depths of 3.5 and 7.5 m will be conditioned to 170 travel-time picks, whereas the zone below 7.5-m depth will be conditioned to 416 GPR data points. Uncorrelated data uncertainties with a standard deviation of 2 ns are assumed during the conditional simulations. This uncertainty is higher than the picking uncertainty of approximately 0.4 ns and should accommodate the undesired effects of other error sources likely to affect the GPR data (Peterson, 2001), such as incorrect positioning of the antennae, systematic mispicking, and unknown borehole irregularities.

For sandy media, Cordua et al. (2008) and Cordua et al. (2009) find that irregularities at the borehole walls caused by cavities generated by loosely packed material that has fallen to the bottom of the borehole and unknown low-velocity clayey material are likely to result in strongly correlated data errors leading to artifacts in the tomographic analysis if not accounted for during inversion. In the present case, the chalk formation is not loosely packed, and no significant amount of material is believed to have fallen from the borehole walls. Further, we have knowledge of how the lithology changes over short vertical distances from the core material. Therefore, we do not see any reason to account for correlated data errors during our stochastic analysis, as outlined by Cordua et al. (2009).

The $H_{min}$ value of 1.1 m found for the medium above 8-m depth appears to generate anomalies that are too thick compared to the thin, relatively frequent flint intercalations found in this zone. The $H_{min}$ value of 1.1 m is similar to the wavelength of the dominant GPR signal and therefore might reflect the limited resolution of the GPR data rather than the actual $H_{min}$ of the velocity fluctuations of the subsurface. Layers of flint are often on the order of 0.1 m thick (cf. Tables 5 and 6). Therefore, an additional set of unconditional simulations with $H_{min}$ set to 0.1 m was generated (Figure 12). These simulations generate thinner layers with characteristics that are more consistent with the borehole observations as compared to the simulations made for an $H_{min}$ value of 1.1 m.

Realizations of the a posteriori distribution generated using sequential simulation conditioned by the picked first-arrival traveltimes of the MOG data are shown in Figure 13. We rely on the same ray-based forward-modeling algorithm of the wavefield as used in the data-driven ergodic inference study. For the stochastic inversions, we set the prior values of $H_{min}$, $H_{max}$, and variance of the velocity fluctuations to the values estimated from the data-driven ergodic inference studies with the exception that we assume $H_{min}$ to be 0.1 throughout the model section. This reduced value of $H_{min}$ is set so as to account for the observed thin layers of flint. The results show frequent horizontally elongated high-velocity bodies in the upper 6.5 m, whereas the bottom of the model is characterized by lower average velocities and only sporadic high-velocity bodies.

This result is consistent with the distribution of flint intercalations observed in the core material. However, the data-conditioned (conditional) simulations occasionally generate considerably thicker high-velocity anomalies than the corresponding unconditional simulations. This effect of the data is not surprising given the fact that the data-based inference resulted in an $H_{min}$ value of 1.1 m. As discussed above, this high value of $H_{min}$ might reflect

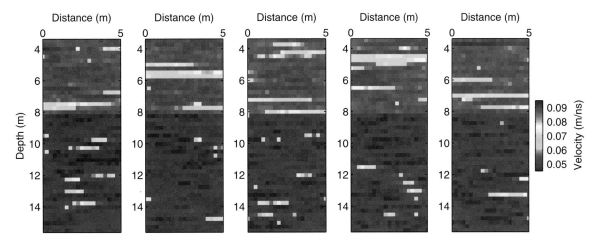

**Figure 12.** A selection of five unconditional simulations using a bimodal a priori distribution. Values of variance and spatial correlation lengths are the same as in Figures 9 and 10 except that $H_{min}$ is now 0.1 m throughout the model.

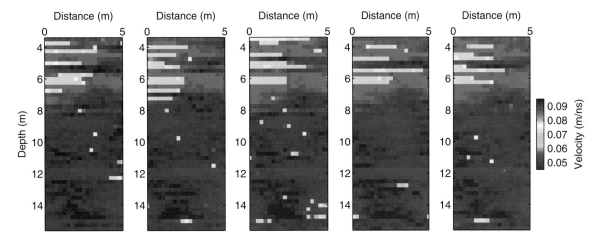

**Figure 13.** A selection of five conditioned simulations using a bimodal a priori distribution. Simulations are conditioned only to GPR data. Values of variance and spatial correlation lengths are the same as in Figure 12. Borehole 1 is to the right, and borehole 2 is to the left.

the inherent resolution limit of the GPR data but also could be a direct consequence of lack of resolution caused by the fact that we rely on a ray-based forward-calculation algorithm (c.f., Buursink et al., 2008).

To also constrain the position of the high-velocity flint anomalies along the model boundaries based on observations from the boreholes, simulations conditioned to both the traveltime data and a data set positioning the flint layers have been performed. For each observed location of the flint, we assume an a priori value of the flint of 0.09 m/ns with a standard deviation of 0.005 m/ns. In fact, this increased a priori velocity value is assigned to a region around the observed flint position corresponding to the assumed uncertainty of the depth reading.

Further, we have increased $H_{max}$ to 8 m to favor laterally coherent anomalies, which we would expect in the mounded interval based on the reflection GPR investigations and geologic observations made along cliff profiles in the vicinity of the study site (Surlyk et al., 2006). The results from this modeling strategy clearly indicate that the GPR data allow for laterally coherent anomalies of high velocities, which are consistent with observations made in the boreholes (Figure 14). Note that the individual simulations show the variability allowed for by the combined GPR traveltime and borehole data sets. Considering a Gaussian a priori model, the average result of a large number of simulations would be similar to the least-squares minimum variance estimate; however, as we consider a non-Gaussian velocity distribution, this is not the case here (cf. Hansen et al., 2008).

All unconditional heterogeneity simulations show a marked change in heterogeneity characteristics at 8-m

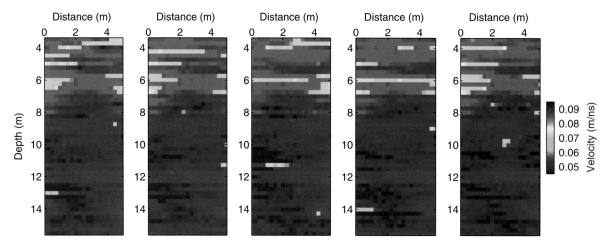

**Figure 14.** A selection of five conditioned simulations using a bimodal a priori distribution. Simulations are conditioned to GPR data and log data. Values of variance and spatial correlation lengths are the same as in Figure 12 except that $H_{max}$ is now 8 m throughout the model. Borehole 1 is on the right, and borehole 2 is on the left.

depth, consistent with the specified a priori information about the heterogeneity distribution. Models resulting from the conditional simulations fit the picked GPR first-arrival traveltimes equally well. The results obtained from simulations conditional only to the GPR data have an average misfit of $-0.73$ ns with respect to the picked traveltimes, whereas the average misfit is $-0.74$ ns for the model simulations conditioned to both GPR and anomaly positions in the boreholes. Thus, the conditional simulations fit the traveltime data within the prespecified uncertainty of 2 ns. A common feature of all conditional sequential simulations is that they show a marked change in heterogeneity characteristics at a depth of about 6 to 6.5 m. Above this depth, the conditional simulations show frequent high-velocity bodies, whereas high-velocity bodies are rare below 6-m depth. This result is consistent with the observations made based on lithologic log data, which show that flint layers and hardened horizons of low porosity occur relatively frequently in approximately the upper 6 m, whereas the bottom part of the section consists of purer chalk of higher porosity.

## Discussion

Numerous studies have demonstrated that, e.g., unsaturated sandy formations facilitate the collection of high-quality crosshole GPR data (e.g., Binley et al., 2001; Binley et al., 2002; Looms et al, 2008) because of the low electrical conductivity and resulting low damping of the GPR wavefield caused by such formations (cf. Reynolds, 1997). In the present data set from the limestone quarry at Boesdal, eastern Denmark, the high-frequency

part of the wavefield generated by the 100-MHz transmitter antenna is reduced significantly by the water-saturated carbonate formation such that the center frequency of the acquired signal is generally about 40 MHz. This effect clearly results in lowering of the resulting resolution of the collected data set. However, the GPR wave velocity of the carbonate-dominated section is low (approximately 0.05 to 0.06 m/ns). Therefore, the dominant wavelength of the acquired signals is about 1.5 m, which is comparable to what should be expected for 100-MHz antenna data collection in dry sands where the velocity is often significantly higher than 0.1 m/ns (e.g., Reynolds, 1997). The raw data collected in the studied carbonate sections appear to have a poorer S/N compared to what could be obtained for sandy formations (cf. Looms et al., 2008). However, good-quality first-arrival picking with reciprocal time inconsistencies with a standard deviation below 0.4 ns was feasible after frequency filtering designed to suppress high-frequency ($>100$-MHz) noise.

Two-dimensional waveform inversion techniques have the potential to increase the spatial resolution of the tomographic images significantly as compared to traditional ray-based regularized inversion approaches in which smooth representations of the subsurface variability are found (Ernst et al., 2007). In this study, data-driven ergodic inference (cf. Hansen et al., 2008; M. C. Looms, personal communication, 2010) is used to calculate statistical measures of the subsurface heterogeneity in terms of vertical and horizontal correlation lengths and variances of the GPR wave velocity. Application of unconditional and conditional stochastic simulation generates a span of models that the data and prior information allow for. This approach thus provides measures of uncertainty of the fine-grained heterogeneity of the studied carbonate-dominated section, which would not have been obtainable using present-day full-waveform inversion algorithms that do not provide uncertainty estimates.

In the present study, the data at hand indicated that the statistical parameters (spatial correlation lengths and variance of the velocity fluctuations) above and below 8-m depth, respectively, should be estimated separately to avoid inconsistencies in the resulting model. This was done by first performing conditional simulation for the model parameters above 8 m using one a priori model. Then the model parameters below 8 m were simulated using the other a priori model. The simulation of the lower medium was conditioned to the already simulated data above 8 m.

Further, the core material from the boreholes put strong constraints on the type of fluctuations that could be allowed for. The preferred model contains thin high-velocity bodies embedded in material of significantly lower velocity. Such models might result in apparent anisotropy, which could be accounted for quantitatively during tomographic inversion (Williamson et al., 1993). In a different example, Pratt and Chapman (1992) apply the method of Chapman and Pratt (1992) to estimate smooth anisotropic velocity models based on crosshole seismic data. Instead of inverting for additional anisotropy parameters, we have chosen an approach in which we generate models that could show a high degree of apparent anisotropy based on stochastic modeling in which model roughness is put in as a prior constraint. Moreover, the models presented here are characterized by correlation lengths $H_{min}$ and $H_{max}$ in the vertical and horizontal directions. The reflection GPR data collected between boreholes 1 and 2 show that dipping reflectivity dominates the depth interval of 2 to 3.5 m. Such a dipping structure would be accounted for better if prior information about velocity fluctuation dip were incorporated in the modeling procedure. Thus, the presented results are not optimal for imaging dipping layers.

The study site is less than 200 m from Stevns Klint, where several geologic studies have been made to unravel the development history of the carbonate succession (Surlyk et al., 2006; Anderskouv and Surlyk, 2007; Bjerager and Surlyk, 2007). Combined with interpretation of our lithologic log data, the studies made at Stevns Klint provide excellent geologic information about the studied formation. The geologic interpretation of our lithologic data predicts a geologic boundary at about 6-m depth at our study site. In the geophysical 2D models, we had put in an a priori boundary at 8-m depth separating two zones with different spatial characteristics and variance of the velocity fluctuations based on the results from 1D interpretation of the

ZOP data and the data-driven ergodic inference study. Therefore, all unconditional simulations of the velocity fluctuations show a boundary at about 8-m depth. However, the simulations conditioned to the crosshole GPR data alone as well as the simulations conditioned to the combined crosshole GPR and log-based data set show a boundary at a depth of 6 to 6.5 m. Thus, the acquired data are consistent with the geologic interpretation, and they pull the simulation results in a different direction than the a priori information regarding the separation of the two zones otherwise dictates.

The carbonates investigated here form part of the Chalk Group, which constitutes an important reservoir for hydrocarbons in the North Sea (Surlyk et al., 2003) and for groundwater onshore Denmark (Nygaard, 1993). The radar-wave velocity fluctuations of the subsurface between the boreholes appear to reflect variations in lithology as well as the effects of changes in porosity of the chalk. High-velocity anomalies are interpreted to represent flint occurrences and hardened horizons with low porosity, whereas low-velocity anomalies represent highly porous, water-saturated carbonate. Thus, the results presented here could have important implications for evaluating the reservoir potential of the studied rock type on a small scale. However, future studies, which should include tracer tests and time-lapse studies of fluid flow (cf. Looms et al., 2008), are important for forming a solid basis for generation of detailed small-scale reservoir models for modeling of fluid and gas flow in parts of the Chalk Group where the rock composition is similar to the one studied here. Furthermore, upscaling heterogeneity characteristics constrained by the combined GPR and log data to reservoir-scale heterogeneity characteristics that can be resolved by geophysical logging tools or seismic data is a challenging task (cf. Frykman and Deutsch, 2002).

# Conclusions

This study provides detailed images of the heterogeneity of a water-saturated section of the upper part of the Chalk Group in the eastern part of the Danish Basin based on joint stochastic modeling of crosshole GPR traveltimes and borehole data.

Initial 1D interpretation of ZOP traveltimes and amplitude data as well as data-driven ergodic inference studies of MOG data indicated that the 15-m-thick studied section should be divided into two zones with different characteristics of the radar-wave heterogeneity, one above and one below 7.5-m depth. The MOG-based inference studies relied on assuming a spherical covariance function to describe the subsurface heterogeneity and showed that the variance and the spatial correlation lengths of the media above and below 8-m depth in fact might be different.

Stochastic inversion of the crosshole GPR data alone as well as the joint stochastic inversion of crosshole GPR and core data shows that the heterogeneity characteristics change at a depth of 6 to 6.5 m, consistent with the observations made in the lithologic log data. Moreover, the different realizations illustrate the span of models that the combined data set and prior information allow for. Comparison to reflection GPR data collected at the study site and results from geologic mapping made by others at a prominent cliff profile near our study area show that the obtained results are geologically plausible for values of $H_{min}$ and $H_{max}$, set to 0.1 m and 6.2 to 8 m, respectively.

The results presented here are essential for designing future time-lapse tracer-test studies that should be made for estimating the fine-scale reservoir characteristics of the studied part of the Chalk Group.

# Acknowledgments

This study was supported financially by a faculty research grant from the Faculty of Science, University of Copenhagen, and a framework grant from the Danish Natural Science Research Council. Plugs extracted from core material were analyzed by the core laboratory of the Geological Survey of Denmark and Greenland (GEUS). We thank Finn Surlyk for comments on the geologic interpretation of lithologic data. Four anonymous reviewers provided constructive comments on an earlier version of this manuscript. The methodological approach referred to here as "M. C. Looms, personal communication, 2010," is described in an article by Looms et al. that is accepted for publication in GEOPHYSICS.

# References

Anderskouv, K., T. Damholt, and F. Surlyk, 2007, Late Maastrichtian chalk mounds, Stevns Klint, Denmark — Combined physical and biogenic structures: Sedimentary Geology, **200**, 57–72.

Annan, A. P., 2005a, GPR methods for hydrogeological studies, *in* Y. Rubin and S. S. Hubbard, eds., Hydrogeophysics: Springer, Water Science and Technology Library No. 50, 185–213.

———, 2005b, Ground-penetrating radar, *in* D. K. Butler, ed., Near-surface geophysics: SEG Investigations in Geophysics Series No. 13, 357–438.

Battey, M. H., 1988, Mineralogy for students, 2nd ed.: Longman Scientific & Technical.

Becht, A., J. Tronicke, E. Appel, and P. Dietrich, 2004, Inversion strategy in crosshole radar tomography using information of data subsets: Geophysics, **69**, 222–230.

Binley, A., G. Cassiani, R. Middleton, and P. Winship, 2002, Vadose zone flow model parameterisation using cross-borehole radar and resistivity imaging: Journal of Hydrology, **267**, 147–159.

Binley, A., P. Winship, R. Middleton, M. Pokar, and J. West, 2001, High-resolution characterization of vadose zone dynamics using cross-borehole radar: Water Resources Research, **37**, 2639–2652.

Bjerager, M., and F. Surlyk, 2007, Danian cool-water bryozoan mounds at Stevns Klint, Denmark — A new class of non-cemented skeletal mounds: Journal of Sedimentary Research, **77**, 634–660.

Buursink, M. L., T. C. Johnson, P. S. Routh, and M. D. Knoll, 2008, Crosshole radar velocity tomography with finite-frequency Fresnel volume sensitivities: Geophysical Journal International, **172**, 1–17.

Chapman, C. H., and R. G. Pratt, 1992, Traveltime tomography in anisotropic media, I. Theory: Geophysical Journal International, **109**, 1–19.

Cordua, K. S., M. C. Looms, and L. Nielsen, 2008, Accounting for correlated data errors during inversion of cross-borehole ground penetrating radar data: Vadose Zone Journal, **7**, 263–271.

Cordua, K. S., L. Nielsen, M. C. Looms, T. M. Hansen, and A. Binley, 2009, Quantifying the influence of static-like errors in least-squares-based inversion and sequential simulation of cross borehole ground penetrating radar data: Journal of Applied Geophysics, **68**, 71–84.

Corin, L., I. Crouchard, B. Dethy, L. Halleux, A. Monjoie, T. Richter, and J. P. Wauters, 1997, Radar tomography applied to foundation design in a karstic environment: Engineering Geology Special Publications, **12**, 167–173.

Dafflon, B., J. Irving, and K. Holliger, 2009, Simulated-annealing-based conditional simulation for the local-scale characterization of heterogeneous aquifers: Journal of Applied Geophysics, **68**, 60–70.

Day-Lewis, F. D., and J. W. Lane Jr., 2004, Assessing the resolution-dependent utility of tomograms for geostatistics: Geophysical Research Letters, **31**, doi: 10.1029/2004GL019617.

Day-Lewis, F. D., K. Singha, and A. M. Binley, 2005, Applying petrophysical models to radar traveltime and electrical resistivity tomograms: Resolution-dependent limitations: Journal of Geophysical Research, **110**, B08206, doi: 10.1029/2004JB003569.

Ernst, J. R., A. G. Green, H. Maurer, and K. Holliger, 2007, Application of a new 2D time-domain full-waveform inversion scheme to crosshole radar data: Geophysics, **72**, no. 5, J53–J64.

Frykman, P., and C. V. Deutsch, 2002, Practical application of geostatistical scaling laws for data integration: Petrophysics, **43**, 153–171.

Giroux, B., and M. Chouteau, 2010, Quantitative analysis of water content estimation errors using ground penetrating radar data and low-loss approximation: Geophysics, **75**, no. 4, WA241–WA249.

Gloaguen, E., B. Giroux, D. Marcotte, and R. Dimitrako-poulos, 2007a, Pseudo-full-waveform inversion of borehole GPR data using stochastic tomography: Geophysics, **72**, no. 5, J43–J51.

Gloaguen, E., D. Marcotte, and M. Chouteau, 2005, A non-linear GPR tomographic inversion algorithm based on iterated cokriging and conditional simulations, *in* O. Leuangthong and C. V. Deutsch, eds., Geostatistics Banff 2004: Springer, 409–418.

Gloaguen, E., D. Marcotte, R. Giroux, C. Dubreuille-Boisclair, M. Chouteau, and M. Aubertin, 2007b, Stochastic borehole radar velocity and attenuation using cokriging and cosimulation: Journal of Applied Geophysics, **62**, 141–157.

Grégoire, C., and P. K. Joesten, 2006, Use of borehole radar tomography to monitor steam injection in fractured limestone: Near Surface Geophysics, **6**, 355–365.

Hansen, T. M., A. G. Journel, A. Tarantola, and K. Mosegaard, 2006, Linear inverse Gaussian theory and geostatistics: Geophysics, **71**, no. 6, R101–R111.

Hansen, T. M., M. C. Looms, and L. Nielsen, 2008, Infer-ring a sub-surface structural covariance model using cross-borehole ground penetrating radar tomography: Vadose Zone Journal, **7**, 249–262.

Hansen, T. M., and K. Mosegaard, 2008, VISIM: Sequen-tial simulation for linear inverse problems: Computers and Geoscience, **34**, 53–76.

Hole, J. A., and B. C. Zelt, 1995, 3-D finite-difference reflection traveltimes: Geophysical Journal Interna-ional, **121**, 427–434.

Irving, J. D., and R. Knight, 2005, Effect of antennas on velocity estimates obtained from cross-borehole GPR data: Geophysics, **70**, no. 5, K39–K42.

Lampe, B., and K. Holliger, 2003, Effects of fractal fluctu-ations in topographic relief, permittivity and conduc-tivity on ground-penetrating radar antenna radiation: Geophysics, **68**, 1934–1944.

Lesmes, D. P., and S. P. Friedman, 2005, Relationships between the electrical and hydrogeological properties of rocks and soils, *in* Y. Rubin and S. S. Hubbard, eds., Hydrogeophysics: Springer, Water Science and Technology Library No. 50, 87–128.

Looms, M. C., K. H. Jensen, A. Binley, and L. Nielsen, 2008, Monitoring unsaturated flow and transport

using cross-hole geophysical methods: Vadose Zone Journal, **7**, 227–237.

Menke, W., 1989, Geophysical data analysis: Discrete inverse theory: Elsevier.

Moysey, S., K. Singha, and R. Knight, 2005, A framework for inferring field-scale rock physics relationships through numerical simulation: Geophysical Research Letters, **32**, doi: 10.1029/2004GL022152.

Nygaard, E., 1993, Denmark, *in* R. A. Downing, M. Price, and G. P. Jones, eds., The hydrogeology of the Chalk of north-west Europe: Clarendon Press, 186–207.

Oppenheim, A. V., and R. W. Schafer, 1989, Discrete-time signal processing, 2nd ed.: Prentice Signal Processing Series.

Peterson, J. E., 2001, Pre-inversion corrections and analysis of radar tomographic data: Journal of Environmental and Engineering Geophysics, **6**, 1–18.

Pratt, R. G., and C. Chapman, 1992, Traveltime tomogra-phy in anisotropic media — II. Application: Geophys-ical Journal International, **109**, 20–37.

Reynolds, 1997, An introduction to applied and environ-mental geophysics: John Wiley & Sons.

Sambuelli, L., 2009a, Uncertainty propagation using some common mixing rules for the modelling and inter-pretation of electromagnetic data: Near Surface Geo-physics, **7**, 285–296.

Sambuelli, L., 2009b, Corrigendum — Uncertainty propa-gation using some common mixing rules for the modelling and interpretation of electromagnetic data: Near Surface Geophysics, **8**, 95.

Stemmerik, L., F. Surlyk, K. Klitten, S. L. Rasmussen, and N. Schovsbo, 2006, Shallow core drilling of the Upper Cretaceous Chalk at Stevns Klint, Denmark: Geologi-cal Survey of Denmark and Greenland Bulletin, **10**, 13–16.

Surlyk, F., T. Damholt, and M. Bjerager, 2006, Stevns Klint, Denmark: Uppermost Maastrichtian chalk, Cretaceous-Tertiary boundary, and lower Danian bryo-zoan mound complex: Bulletin of the Geological Society of Denmark, **54**, 1–48.

Surlyk, F., T. Dons, C. K. Clausen, and J. Higham, 2003, Upper Cretaceous, *in* D. Evans, C. Graham, A. Armour, and P. Bathurst, eds., The millennium atlas: Petroleum geology of the central and northern North Sea: Millenium Atlas Company, Ltd., 213–233.

Tarantola, A., 2005, Inverse problem theory and methods for model parameter estimation: SIAM.

Tarantola, A., and B. Valette, 1982, Generalized nonlinear inverse problems solved using the least squares crite-rion: Reviews of Geophysics and Space Physics, **20**, 219–232.

Tronicke, J., K. Holliger, W. Barrash, and M. D. Knoll, 2004, Multivariate analysis of cross-hole georadar velocity and attenuation tomograms for aquifer zonation: Water Resources Research, **40**, W01519, doi: 10.1029/2003WR002031.

Vejbæk, O. V., T. Bidstrup, P. Britze, M. Erlström, E. S. Rasmussen, and U. Sivhed, 2003, Chalk structure maps of the central and eastern North Sea: Geological Survey of Denmark and Greenland (GEUS), Report 2003/106.

Vidale, J., 1988. Finite-difference calculation of traveltimes: Bulletin of the Seismological Society of America, **78**, 2062–2076.

Williamson, P. R., M. S. Sams, and M. H. Worthington, 1993, Crosshole imaging in anisotropic media: The Leading Edge, **12**, 19–23.

Zelt, C. A., and P. J. Barton, 1998, 3D seismic refraction tomography: A comparison of two methods applied to data from the Faraoe Basin: Journal of Geophysical Research, **103**, 7187–7210.

Chapter 24

# Clayey Landslide Investigations Using Active and Passive $V_S$ Measurements

F. Renalier[1], G. Bièvre[1, 2], D. Jongmans[1], M. Campillo[1], and P.-Y. Bard[1]

## Abstract

Clay slopes frequently are affected by gravitational movements. Such movements generate complex patterns of deformation that have slip surfaces located at different depths and are likely to modify geophysical parameters of the ground. Geophysical experiments performed on the large clayey Avignonet landslide (Western Alps, France) have shown that shear-wave velocity ($V_S$) is most sensitive to clay deconsolidation resulting from the slide. Values of $V_S$ at shallow depths exhibit an inverse correlation with the GPS-measured surface-displacement rates. Compared with measurements in stable zones, $V_S$ values in the most deformed areas of the slide can be reduced by a factor of two to three. Laboratory measurements on clay samples set in triaxial cells have shown that a strong decrease of $V_S$ values accompanies an increase in the void ratio, in a velocity range similar to that measured in situ. Although other factors (stress change, cementation, granularity) can modify $V_S$ values, these results justify the potential of $V_S$ imaging to map spatially the deformation induced by a landslide. Several active and passive techniques for measuring $V_S$ are tested and compared on the kilometer-size and 50-m-deep Avignonet landslide. The crosscorrelation technique, applied to seismic noise recorded by a large-aperture array and associated with shot records, turns out to be an effective tool for imaging the landslide in three dimensions. If permanent stations are installed, the same method also can be used to monitor the evolution of seismic velocity with time, as an indicator of landslide activity.

## Introduction

Shear-wave velocity ($V_S$) has emerged increasingly as a key geophysical parameter for characterizing soil layers in geotechnical engineering. Compared with compressional waves, shear waves offer the advantages of a shorter wavelength and its resulting better resolution and of little sensitivity to the fluid saturation (Dasios et al., 1999). Whereas P-wave velocity ($V_P$) contrasts are small in saturated soils (Mondol et al., 2007), shear waves exhibit a wide range of velocity values, thereby allowing better detection of changes in lithology and compactness at shallow depths. Numerous studies (e.g., Hegazy and Mayne, 1995; Andrus et al., 2004; Hasancebi and Ulusay, 2007) have suggested relationships between $V_S$ and the penetration resistance measured from the cone penetration test (CPT) and the standard penetration test (SPT), making $V_S$ a meaningful geotechnical parameter. As a result, $V_S$ has been used increasingly in a wide variety of applications, including delineation of geologic boundaries in the subsurface (e.g., Hunter et al., 2002; Ghose and Goudswaard, 2004), evaluation of ground densification (Kim and Park, 1999) and of landslide-related deconsolidation (Jongmans et al., 2009), and assessment of liquefaction potential (Finn, 2000). During earthquakes, ground motions may be amplified strongly at sites with soft layers that overlie bedrock. The main parameter controlling the dynamic-site response is shown to be the contrast in $V_S$ values (e.g., Bard and Riepl-Thomas, 1999; Sommerville and Graves, 2003), and $V_S$ in shallow layers is now considered to be the key

[1]*Laboratoire de Géophysique Interne et Tectonophysique (CNRS), Observatoire des Sciences de l'Univers, Université Joseph Fourier, Grenoble, France. Email: florence.renalier@obs.ujf-grenoble.fr; gregory.bievre@obs.ujf-grenoble.fr; denis.jongmans@ obs.ujf-grenoble.fr; michel.campillo@obs.ujf-grenoble.fr; pierre-yves.bard@obs.ujf-grenoble.fr.*
[2]*Centre d'Études Techniques de l'Equipement de Lyon, Laboratoire Régional d'Autun, Autun, France.*

parameter for site characterization in current building codes (Finn and Wightman, 2003).

Various in situ methods can be applied to derive shear-wave velocity: borehole tests, shear-wave refraction and reflection studies, and surface-wave techniques (Jongmans, 1992; Dasios et al., 1999; Hunter et al., 2002; Boore, 2006). Borehole tests have been used extensively in geotechnical engineering to a depth of a few tens of meters. Such tests provide accurate and well-resolved $V_S$ values with the following drawbacks: They are invasive, they are sensitive only to vertical variations of $V_S$, they are increasingly expensive with depth, and they offer only point estimates. Shear-wave velocities from noninvasive surface measurements can be obtained using transverse shear-wave (SH) refraction or reflection techniques (Dasios et al., 1999; Hunter et al., 2002; Ghose and Goudswaard, 2004). In a thick soil site in Canada, Hunter et al. (2002) applied high-resolution P and SH seismic reflection profiling to delineate the overburden-bedrock surface to a few hundreds of meters deep and obtained P- and S-wave velocities in the overburden layers from common midpoint (CMP) processing.

In recent years, surface-wave methods have been applied increasingly to measure $V_S$ vertical profiles, using the dispersion properties of these waves (for a review, see Socco and Jongmans, 2004). Surface-wave methods commonly are divided into two main categories, depending on whether active or passive sources are used. Active methods record vibrations generated by an artificial source, a limitation of which is the difficulty to generate low-frequency waves (Tokimatsu, 1997). Consequently, active sources usually offer a penetration that is limited to a few tens of meters (Park et al., 1999; Socco and Strobbia, 2004).

On the contrary, ambient vibrations are generated by sources of lower frequencies (e.g., Aki, 1957; Satoh et al., 2001; Okada, 2003), thereby making active and passive techniques complementary to each other for deriving the surface-wave dispersion curve (Wathelet et al., 2004; Park et al., 2007; Richwalski et al., 2007; Socco et al., 2008). Thus, $V_S$ profiles are estimated in two steps: (1) deriving the dispersion curve from the recorded seismograms (Lacoss et al., 1969) and (2) inverting the dispersion curve by using either direct search methods (Parolai et al., 2005; Dal Moro et al., 2007; Wathelet et al., 2008) or linearized inversion algorithms (Herrmann, 1987; Satoh et al., 2001).

Parallel to these developments, considerable interest has arisen during the past few years in the crosscorrelation of ambient seismic noise recorded at two distant receivers (Aki, 1957; Claerbout, 1968). Indeed, it has been demonstrated, both theoretically and experimentally, that such crosscorrelation converges (under certain conditions) toward the Green's function of the medium between these

two receivers (Shapiro and Campillo, 2004; Schuster et al., 2004; Sánchez-Sesma and Campillo, 2006; Gouédard et al., 2008; Wapenaar et al., 2008). Because ambient seismic noise is produced mainly by surface sources that generate predominantly surface waves, that property can be used to map the surface-wave group velocity at different frequencies and to derive a 3D $V_S$ image of a geologic structure, such as a 15- × 15-km-size volcano (Brenguier et al., 2007). Brenguier et al. (2007) also demonstrate that continuous ambient-noise records obtained over an 18-month period can detect very small seismic-velocity perturbations and can show decreases in seismic velocity before eruptions. At smaller scales, with interstation distances ranging between 50 and 500 m, recent studies have demonstrated the possibility of using ambient-noise crosscorrelation in an urban environment to retrieve the propagation functions (Nunziata et al., 2009) and to perform a 3D tomographic inversion for imaging shallow lateral heterogeneities (Picozzi et al., 2009).

The aim of the present paper is to show the benefit of using $V_S$ to characterize and image landslides that affect clay masses, with a focus on the large Avignonet landslide (Trièves area, France). Following a discussion on the landslide and the measurements performed, we investigate the relation between $V_S$ values and the damaging effect of this landslide at different scales, through laboratory and in situ tests. We then apply several active and passive seismic techniques at the Avignonet landslide to test the abilities of those methods to image a kilometer-size, 50-m-deep structure. Finally, seismic monitoring techniques are applied to crosscorrelated signals between two permanent stations, to detect changes in the medium that are related to the landslide activity during a three-year period of time.

# The Avignonet Landslide: Geologic Context and Seismic Investigation

## Geologic and geotechnical context

The large Avignonet slide ($40 \times 10^6$ m$^3$) is located in the Trièves area (French Alps) (Figure 1a). This 300-km$^2$ area is covered by a thick Quaternary clay layer (as thick as 200 m) that was deposited in a lake dammed by a glacier during the Würm period (Giraud et al., 1991). Those clayey deposits overlie compact old alluvial layers and marly limestone of Mesozoic age and are covered by thin till deposits. After the glacier melted, rivers cut deeply into the geologic formations, triggering numerous landslides (Giraud et al., 1991). Figure 1b shows the simplified geologic map of the study area, in the northern part of the

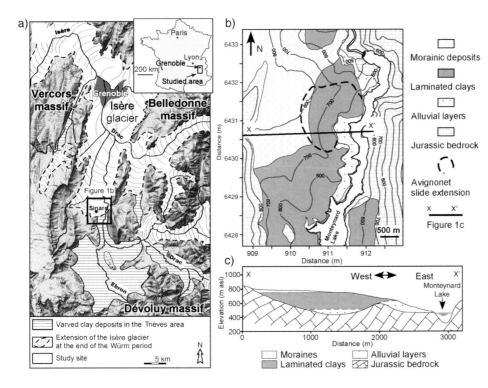

**Figure 1.** (a) Location map of the Trièves area and of the study site. (b) Geologic map of the study area, with the location of the Avignonet landslide (dashed line) and section XX'. (c) Geologic cross section XX'.

Trièves region, with the extension of the Avignonet landslide having occurred in the clay deposits. An east-west synthetic geologic section over the Avignonet landslide is presented in Figure 1c, showing the westward increase in thickness of the clay layer, from 0 m to more than 200 m.

The Avignonet landslide has been investigated by geologic, geotechnical, and geophysical tests since the beginning of the 1980s. Giraud et al. (1991) and Jongmans et al. (2009) provide a summary of the results. Five boreholes (labeled B0 to B4) were drilled in the southern part of the landslide (Figure 2), where a hamlet is settled. Table 1 synthesizes the geologic logs and the detected slip-surface depths. The contact between the clay and the underlying alluvial deposits was found at 14.5 m, 44.5 m, and 56 m in B2, B3, and B1, respectively.

On the contrary, B0 and B4 only encountered clay deposits, in agreement with the westward thickening of this clay formation. Inclinometer data and geologic logs have revealed at least three rupture surfaces: a shallow one at a few meters in depth, an intermediate one at 10–16 m, and a deep one at 42–47 m. Water-level measurements showed the presence of a very shallow water table (1 to 3 m below the ground level). The Avignonet slide has been monitored by biannual GPS measurements at 26 geodetic points since 1995. The velocity values averaged from the available GPS measurements were used to identify areas of increasing slide velocity (Figure 2). Slide velocities have varied from 0 to 2 cm/yr at the top of the slide to 7 to

15 cm/yr in the most active parts, around boreholes B2 and B4 (Jongmans et al., 2009). Most of the area is sliding southeastward, parallel to the general slope.

## Seismic investigation

The Avignonet landslide has been investigated recently via several geophysical campaigns. The first geophysical survey was performed in 2006–2007 (including seismic profiles P1 through P4, P6, and P7 in Figure 2) to test the sensitivity of three geophysical parameters (the electrical resistivity $\rho$, the seismic P-wave velocity $V_P$, and the S-wave velocity $V_S$) to the deformation resulting from the slide. It turned out that in such saturated clays, $\rho$ and $V_P$ are influenced strongly by the water level and are affected little by the landslide activity (Jongmans et al., 2009). On the contrary, $V_S$ showed significant variations both vertically and laterally that correlated with the landslide activity. For this reason, we have focused the present study on $V_S$ measurements.

Additional seismic profiles P5, P8, and P9 (Figure 2) were conducted in 2008–2009. Except for profiles P6 and P7, seismic profiles were performed using 24 vertical geophones (4.5 Hz) and 24 horizontal geophones (4.5 Hz for P5 and 14 Hz for all other profiles) spaced 5 m apart (2.5 m for profile P5), with a hammer striking a plate or a plank for the source, to record Rayleigh waves and SH-waves, respectively. Given the $V_S$ values in the medium, these geophone spacings did not alias the data spatially. Shots were located at every third geophone for the purpose of seismic tomography, and with offset at both ends for the surface-wave recordings. The two 470-m-long profiles P6 and P7 were conducted with 48 vertical geophones (resonance frequency 4.5 Hz) at a 10-m spacing. Signals were generated by nine and 14 explosive sources with an 80-m spacing and a 50-m spacing, respectively. Depending on the accessibility of the area, one or two offset shots were fired at each end.

In addition, three concentric arrays composed of six 5-s three-component seismometers were installed successively in 2007 close to profile P5 (see Figure 2 for location). Each array was composed of five Lennartz 5-s sensors distributed regularly around the same central sensor, with an increasing radius of 20 m, 40 m, and 60 m. Ambient vibrations were recorded during one hour for each array.

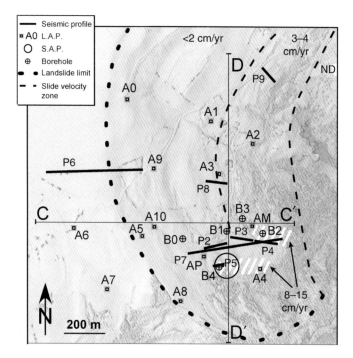

**Figure 2.** Locations of tests on the Avignonet landslide (delineated with a dotted line). Seismic profiles are labeled P2 through P9. Profile P1 is outside the represented area, about 300 m west of the top of the figure. Boreholes are labeled B0 through B4. Seismological stations of the large-aperture network (LAP) are A0 through A10. There are two permanent stations: AP and AM. The circle encompasses the small-aperture network (SAP). There are three zones, with different displacement rates (ND has no displacement, <2 cm/yr, and 3–4 cm/yr). They are indicated by dashed lines, whereas very active areas (7–15 cm/yr) are indicated with white hatching. Sections CC′ and DD′ refer to the two sections shown in Figure 9.

Finally, a large-aperture array (kilometer size) of eleven short-period seismological stations A0 to A10 (Figure 2) were installed during two weeks in 2007 on the southern part of the Avignonet landslide, which also is equipped with two permanent stations AP and AM (OMIV Observatory).

## Evolution of Clay Damage Inferred from $V_S$

A previous study on the $V_S$ value distribution along a section through the Avignonet landslide (Jongmans et al., 2009) showed an inverse relation between $V_S$ and displacement rate. Considering additional data acquired in 2008 and 2009 (profiles P5, P8, and P9), $V_S$ values measured at a depth of 10 m versus displacement rate are plotted in a semilogarithmic scale in Figure 3a. The point outside the landslide exhibits no measurable displacement and is represented as a bar with a maximum displacement rate of 0.01 cm/year. Data show a strong decrease of $V_S$ values with increasing displacement rates, from 630 m/s outside the slide to 225 m/s at the slide toe, where the ground surface is the most disturbed. These results obtained for meter-scale wavelengths suggest that the gravitational deformation strongly affects the shear-wave velocity in the clay.

The results are interpreted as the effect of an intensive cracking of the material at different scales, resulting from slip along the rupture surfaces at different depths. This cracking probably increases the void ratio and, consequently, increases the water content in the saturated material and also decreases the shear-wave velocity. Relations among $V_S$ (or shear modulus), void ratio, and mean effective stress have been derived from numerous laboratory studies for a wide variety of soil types (e.g., Bryan and Stoll, 1988). For a given depth (fixed confining pressure), they show that $V_S$ effectively decays with increasing void ratio. For increasing confining pressures, the laws predict an increase in shear-wave velocity, which still depends on the void ratio.

To investigate the relation among $V_S$, void ratio, and confining pressure in the clay of the Avignonet landslide, we performed laboratory measurements on three 0.4-m-long saturated soil specimens (two laminated, one nonlaminated) sampled during drilling at 8.3 m, 9.8 m, and 15.5 m in two boreholes. Clay specimens were submitted to isotropic stress conditions in a triaxial cell, with confining pressures varying between 0 and 400 kPa. With the low $V_S/V_P$ values (0.1 to 0.2) found in the landslide (Jongmans et al., 2009), Poisson's ratios are higher than 0.48 and horizontal effective stresses are close to the vertical effective stress.

**Table 1.** Geologic log and depth of the slip surfaces found in boreholes B0 through B4 (shown in Figure 2).[3]

| Borehole # | Depth (m) | Geologic log | Shallow slip surface (m) | Intermediate slip surface (m) | Deep slip surface (m) |
|---|---|---|---|---|---|
| B0 | 89 | 0–5 m: morainic colluvium<br>5–89 m: laminated clays | 5 | 10 | 47 |
| B1 | 59 | 0–5 m: morainic colluvium<br>5–56 m: laminated clays<br>56–59 m: alluvial deposits | — | 15 | 43 |
| B2 | 17 | 0–4 m: morainic colluvium<br>4–14.5 m: laminated clays<br>14.5–17 m: alluvial deposits | 1.5 and 4 | 12 | — |
| B3 | 59 | 0–4 m: morainic colluvium<br>4–44.5 m: laminated clays<br>44.5–59 m: alluvial deposits | — | 16.5 | — |
| B4 | 49 | 0–2.5 m: morainic colluvium<br>2.5–18.5 m: blocky clays<br>18.5–49 m: laminated clays | 5 | 10.3 and 14.5 | 42 |

[3]The inclinometer tube within B3 was sealed to a depth of 20 m. Dashes indicate that the presence of a slip surface at that depth was not shown during the time of the inclinometer measurements, which does not imply that a slip surface does not exist.

Confining pressures ranging between 0 and 400 kPa then approximately correspond to depths ranging between 0 and 40 m in saturated conditions.

Measurements of $V_S$ were made under saturated conditions using a GDS instrument associated with bender elements (Lee and Santamarina, 2005). Signals were generated at a frequency of 10 kHz and yielded a wavelength of about 0.7 to 3 cm. The volume, weight, and water content of the saturated samples were measured before introducing them into the cell, and the void ratio was derived from the water volume expelled during the experiment. Figure 3b shows the evolution of $V_S$ with the void ratio and the corresponding water content in the 0- to 400-kPa confining-pressure range. In the investigated void-ratio range, $V_S$ values exhibit a significant and regular decrease with increasing void ratio $e$ (and decreasing confining pressure $p$), from 300 m/s for $e = 0.52$ ($p = 400$ kPa) to 70 m/s for $e = 0.92$ ($p = 0$ kPa). These values will be compared later with the shear-wave velocities derived in situ in the depth range of 0 to 40 m.

## Active Methods

A 17-m-deep downhole test performed close to borehole B4 (Figure 2) in a highly disturbed area provided the $V_S$ vertical profile shown in Figure 4. The $V_S$ values in the first 3 m were obtained from Love-wave fundamental-mode velocities (see below). The $V_S$ increases from less than 150 m/s in the shallow layer to more than 300 m/s at the bottom of the profile, with two velocity breaks at 6 m and 15 m. These depths approximately fit two slip surfaces found in borehole B4 (Table 1). At 10 m deep, $V_S$ is approximately 250 m/s, which confirms the activity of the landslide at this site (Figure 3a).

Shear-wave velocity also can be measured from the surface by using active sources, with surface-wave methods or with transverse-shear-wave (SH) refraction or reflection techniques. The penetration depth of SH refraction tests usually is limited to 25–30 m because of the difficulty of generating SH-waves with a good signal-to-noise ratio over distances greater than 100 m (Jongmans et al., 2009). Figure 4a shows the 57.5-m-long SH refraction tomography profile P5, recently performed in the highly disturbed zone where borehole B4 was drilled and where the downhole test discussed in the previous section was conducted. The tomography (Dines and Lyttle, 1979) was performed on a grid that was 2.5 × 2.5 m, starting from a three-uniform-layer model derived from the classical refraction analysis. The seismic image shown in Figure 4a, with a least-squares misfit of <3% between measured and calculated traveltimes, was obtained after five iterations, after which the misfit remained stable.

The $V_S$ profile obtained from a downhole test is plotted in the same figure. Because signals were difficult to pick in the first 3 m, we constrained $V_S$ values close to the surface from Love-wave velocities along profile P5. Figure 4b and 4c shows the Rayleigh- and Love-wave dispersion curves, respectively, computed for a group of 15 geophones centered above B4. In the same figure, the normalized amplitudes of the smoothed Fourier spectra also are plotted. The Rayleigh waves (Figure 4b) are dominated by a higher mode in the 18- to 25-Hz frequency band, thereby preventing us from identifying the fundamental mode at these frequencies. On the contrary, the fundamental mode

dominates the Love waves (Figure 4c) over the whole frequency range, including values greater than 20 Hz, where its velocity corresponds to $V_S$ in the upper layer. Thus, the $V_S$ profile derived from borehole measurements was completed with this $V_S$ value in the first 3 m.

Downhole measurements and SH-wave refraction tomography show consistent results (Figure 4a), with a surficial low-velocity layer ($V_S$ lower than 200 m/s) over the shallow slip surface found at 5 m. In the depth range between the shallow and intermediate slip surfaces (5–15 m), $V_S$ is approximately 250–300 m/s and increases to 400 m/s below the intermediate slip surface. The SH-wave refrac-tion tomography provides the lateral variations in velocity, highlighting the eastward thickening of the shallow low-velocity unit (lower than 150 m/s, Figure 4a). On the other hand, this technique is unable to provide an

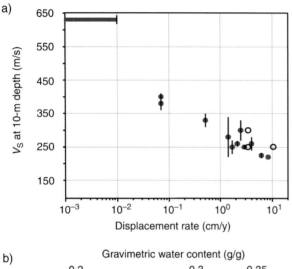

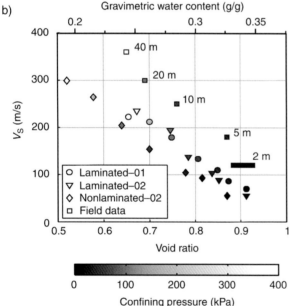

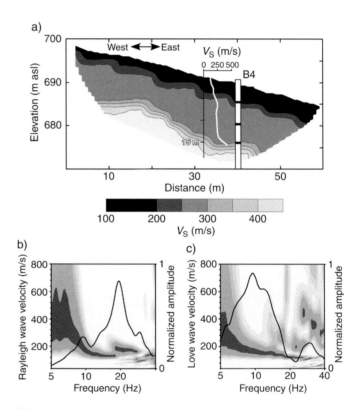

**Figure 3.** Measured $V_S$ variations. (a) Evolution of shear-wave velocity values (with error bars) as a function of displacement rates, at a depth of 10 m. Data are from surface-wave inversion and SH refraction studies for the profiles shown in Figure 2. The three empty circles correspond to the new measurements. The point outside the landslide ($V_S = 630$ m/s) exhibits no measurable displacement and is represented as a thick line with a maximum displacement rate of 0.01 cm/year. (b) Evolution of $V_S$ as a function of void ratio and water content for three saturated samples and field data. After Renalier et al. (2010). Used by permission.

**Figure 4.** (a) SH refraction tomography along profile P5 (located in Figure 2), performed with a three-uniform-layer initial model inferred from classical refraction analysis and a grid composed of cells that are 2.5 m on each side. The misfit is 2.4%. The $V_S$ profile derived from the combination of downhole test and inversion of Love waves is plotted in the same figure, as is borehole B4 with the shallow and two intermediate slip-surface positions. (b) Rayleigh-wave and (c) Love-wave dispersion maps with normalized Fourier spectra of signals recorded at B4. The Love-wave dispersion curve was used for constraining $V_S$ values of the downhole test in the first 3 m.

image down to the deepest slip surface of the landslide, which was found at 42 m in borehole B4.

The surface-wave inversion technique at a larger scale has been applied already to the slowly moving parts of the Avignonet landslide (Jongmans et al., 2009), which contributed to the results of Figure 3a. Here we investigate the possibility of inverting Rayleigh waves measured along the 470-m-long profile P7, which is oriented perpendicularly to the landslide motion and extends in the lower and more active part of the landslide. The first ten geophones are located in this highly sliding velocity zone (Figure 2 and group A in Figure 5a), which is characterized by low $V_S$ values and a rugged topography resulting from active surficial rupture surfaces. Figure 5a shows the seismograms recorded for a shot at 470 m (upper end of the profile) using 200 g of explosive. Surface waves are regular from 470 m to 80 m, but they exhibit attenuation and dis-

turbances in the active area from 80 m to 0 m, making any determination of dispersion curves difficult in this high-scattering medium.

Ground heterogeneity caused by the landslide activity also might affect the applicability of the surface-wave method along a linear array outside the disturbed zone. Several methods were proposed (Miller et al., 1999; Grandjean and Bitri, 2006; Lin and Lin, 2007; Socco et al., 2009) for obtaining pseudo-2D $V_S$ images from local 1D surface-wave inversion, using different source-receiver layouts. The main assumption is that locally the structure is layered horizontally (1D) around a receiver antenna (Grandjean and Bitri, 2006) or between one receiver and some shots (Lin and Lin, 2007). To test this hypothesis, we considered a group of 20 geophones located on P7 outside the disturbed area (from 200 to 400 m, group B in Figure 5a) and we applied the frequency-wavenumber

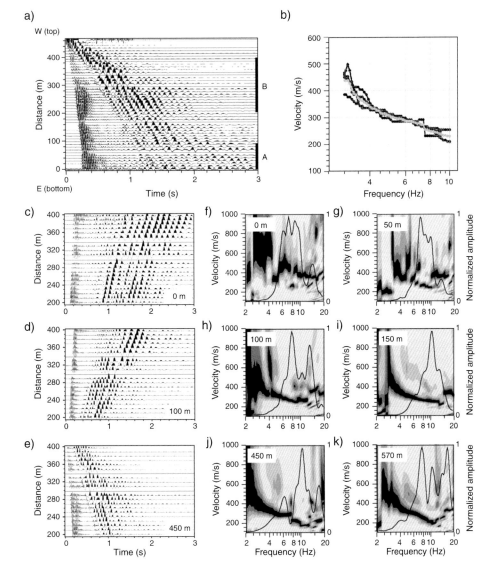

**Figure 5.** Active-source (explosive) experiment along profile P7. (a) Normalized signals recorded along profile P7 for the offset shot at 470 m. Vertical geophones are 10 m apart. The two groups of geophones located in the very disturbed and less disturbed areas are labeled A and B, respectively. (b) Dispersion curves extracted from the dispersion maps of Figure 5h through 5k, for shot positions greater than 100 m. Gray curve is the average. (c) through (e) Seismograms recorded by geophones of group B (200 m to 400 m) for shots at 0 m, 100 m, and 450 m, respectively. (f) through (k) Dispersion maps for shots at 0 m, 50 m, 100 m, 150 m, 450 m, and 570 m, respectively. Normalized Fourier spectrum of the signal recorded at 290 m is plotted on each dispersion map.

(*f-k*) analysis (Lacoss et al., 1969) to transform the data from the time-space domain into the frequency–phase velocity domain (dispersion map), from which dispersion curves can be extracted.

Figure 5 shows the seismograms for three shots, at 0 m, 100 m, and 450 m (Figure 5c, 5d, and 5e), along with the dispersion maps for these three shots and three additional ones (50 m, 150 m, and 570 m, Figure 5f through 5k). The Fourier spectrum of the signal in the center of the geophone spread also is shown in these graphs, revealing the filter effect of the geophones below 4 Hz. For shots located in the upper part of the profile (450 m and 570 m), Rayleigh waves propagate regularly through the geophone array and the dispersion curve of the fundamental mode is clearly defined between 3 Hz and about 10 Hz, with phase velocities decreasing from 400 m/s to 200 m/s. Higher modes seem to have been excited in some limited-frequency bands. The upper frequency limit of 10 Hz corresponds to the aliasing criterion ($\lambda_{min} = 2 \Delta x$, where $\Delta x$ is the 10-m geophone spacing).

On the other side of the spread, for the two shots fired in the lower active zone (0 and 50 m, Figure 5f and 5g), the first higher mode is excited at least as much as the fundamental mode, which cannot be identified clearly. Moreover, the signal duration is much longer than for sources located in a less disturbed area (compare seismograms in Figure 5c and 5d). For shot positions outside the very disturbed area (Figure 5h and 5i), the fundamental mode of Rayleigh waves again is well defined. Figure 5b shows the dispersion curves extracted for shot positions outside the very disturbed area (at $x > 100$ m), on both sides of the geophone array (which is between $x = 200$ m and 400 m). They are all similar in the 3- to 10-Hz frequency range, thereby highlighting the 1D geometry of the seismic ground structure below the spread. On the contrary, late arrivals on the seismograms of shots performed in the very disturbed area can be compared to coda waves, which result from scattering on seismic heterogeneities (Aki and Chouet, 1975; Aki and Richards, 1980).

Furthermore, higher modes may be excited in the case of inversely dispersive media for particular source-to-receiver distances (that can be determined theoretically; Tokimatsu et al., 1992), and in the case of deep sources or in the case of deep heterogeneities along the wave path (Herrmann, 1973; Keilis-Borok, 1986; Schlue and Hostettler, 1987; Uebayashi, 2003). In our case, excitation of the first higher mode only occurs for shots performed in the disturbed area and not on the other side of the geophone array. Moreover, borehole measurements indicated a normally dispersive medium. Both the long duration of signals and the excitation of the first higher mode on seismograms generated

in the disturbed area therefore probably are related to waves scattered on seismic heterogeneities linked to the landslide activity (scarps, open fissures, lateral velocity variations).

In the disturbed zone, the presence of scatterers in and out of the propagation plane turned out to make the use of the surface-wave method with linear arrays difficult. This conclusion also was reached during a seismic survey carried out on the large Séchilienne landslide affecting mica schists (Méric et al., 2005). These scatterers, which blur the dispersion curve here, can help to retrieve it when nonlinear arrays and ambient vibrations are used, however.

## Passive Methods

### Small-aperture array

Dispersion curves determined on long linear arrays with active sources (Figure 5b) are limited to 3 Hz in the low-frequency range, as a result of the low-cut filter effect from the 4.5-Hz geophones, combined with the source energy spectrum. The maximum penetration depth then is about 30 m, considering the rule of the thumb of one-third of the maximum wavelength. That penetration is not enough to reach the deepest slip surface of the landslide, which was found at a depth between 42 and 47 m in boreholes (Table 1). To obtain the dispersion curve at lower frequencies, we used surface-wave inversion of ambient vibrations recorded with an array of sensors (Asten and Henstridge, 1984; Satoh et al., 2001).

The main assumptions behind noise-array analysis are that ambient vibrations are composed mostly of surface waves and that the ground structure is stratified approximately horizontally (Tokimatsu, 1997). The geometry of the three concentric arrays (radii of 20, 40, and 60 m) is shown in Figure 6a. We applied the *f-k* method proposed by Lacoss et al. (1969), which assumes that plane waves travel across the array of sensors. The maximum output of the array is calculated by the summation, in the frequency domain, of signals shifted according to the time delays at the sensors. To estimate the uncertainty of the determination of apparent velocity at each frequency, signals are split into several short time windows (50 periods in this case), for each of which the output of the array is computed. Following Ohrnberger et al. (2004), a histogram of the velocities at the observed maxima is constructed for each frequency band, thereby allowing determination of the dispersion curve. The low- and high-frequency limits of the dispersion curve are controlled by several factors, including the maximum and minimum apertures of the array, filter effect of the medium, and excitation strength of the wavefield (Scherbaum et al., 2003). Resolution and aliasing limits (related to the geometry of the array) are derived from the array-response study (Wathelet et al., 2008). The resulting velocity histograms are shown in Figure 6b, 6c, and 6d for the three arrays A, B, and C, respectively, along with the resolution and aliasing limits.

The final dispersion curve, with the associated uncertainty, is presented in Figure 6f (black lines) and was completed at high frequencies with the Rayleigh and Love dispersion curves derived from profile P5 (Figure 4b). With the characteristics of the array, the Rayleigh dispersion curve obtained from ambient vibrations is reconstructed between 2.4 and 8 Hz and is complementary to the one derived from active seismics between 9 and 18 Hz. Although the frequency band is extended only slightly toward low frequencies compared with dispersion curves derived from large-scale active seismics (2.4 Hz for array C instead of 3 Hz for profile P7; see Figure 5), the ambient-noise analysis caught the increase of slope and reached a phase velocity of 680 m/s at 2.4 Hz, yielding a wavelength of about 280 m and a penetration greater than 90 m. Reaching as high as 25 Hz, the Love-wave dispersion curve measured on the small seismic profile P5 moreover gives information on the very shallow part of the profile. The parameterization chosen for the inversion was a model with a surficial layer of 3 m maximum and three other homogeneous layers, as suggested by previous investigation. Within the uncertainty of the data, all shear-wave velocity profiles explaining the data are given in Figure 6e.

In agreement with the results of the SH refraction tomography, the surficial layer has a very low velocity, between 120 and 140 m/s. Values of $V_S$ between 160 and 200 m/s and depth values ranging between 8 and 12 m characterize the second layer, which corresponds approximately to the highly disturbed horizon overlying the intermediate surface ruptures found between 10 and 14.5 m in borehole B4 and shown in the downhole test (Figure 4a). The third layer, with a velocity below 400 m/s, extends to a depth between 40 and 60 m and overlies a stiffer layer with $V_S$ values higher than 700 m/s. This interface probably corresponds to the deepest slip surface, which was found between 42 and 47 m in boreholes.

In situ measured $V_S$ values at 2, 5, 10, 20, and 40 m are plotted in Figure 3 against the void-ratio values determined in the borehole at the same depth. Compared with laboratory measurements, in situ $V_S$ values exhibit a similar linear decrease with increasing void ratio, for the same depth range. The covered velocity range is higher (from 120 to 360 m/s) than the laboratory measurements (70 to 300 m/s), which indicates that soil samples could have been remolded during extraction. Nevertheless, these re-

sults suggest that the clay material is probably as strongly damaged at the centimeter scale as it is at the meter scale, thus highlighting the penetrative characteristic of the cracking generated by the landslide activity. Comparison of Figure 3a and 3b also shows that the $V_S$ variation resulting from landslide damaging (at a given depth) is as strong as the one expected from the landslide surface to its bottom.

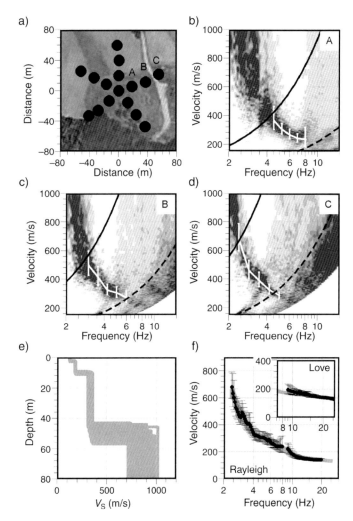

**Figure 6.** Surface-wave inversion. (a) Geometry of the three concentric arrays (A through C), composed of six 3-C seismometers with radius of 20 m, 40 m, and 60 m, plotted on top of an aerial photograph of the area. (b) through (d) Frequency-wavenumber analysis for arrays A through C. Continuous and dashed lines represent the resolution (continuous) and aliasing (dashed) limits of the dispersion curves. (e) Four-layer $V_S$ profiles with a misfit lower than 1. (f) Rayleigh (main plot) and Love (inset) dispersion curves. Black line: dispersion curve measured from passive (2.4–8 Hz) and active (9–18 Hz and 9–25 Hz for Rayleigh and Love waves, respectively) seismic experiments. Gray lines depict theoretical dispersion curves of profiles in (e).

Although $V_S$ variations probably also are affected by other factors, such as stress, cementation, or granularity, these results again highlight geoscientists' interest in using $V_S$ for imaging a clay landslide.

It has been shown that surface-wave inversion from ambient vibration measurements can provide locally valuable information on the velocity structure of a 50-m-deep landslide such as the Avignonet landslide. However, constraining the geometry of the entire landslide would require at least 100 arrays (one point every 100 m) for constructing a 3D model. The time necessary to cover the whole area by moving such arrays has been estimated to be about three months of work, with a team of three people, when considering the problems of forested and built areas. For that reason, we investigated the possibility of performing a surface-wave tomography on the landslide, by

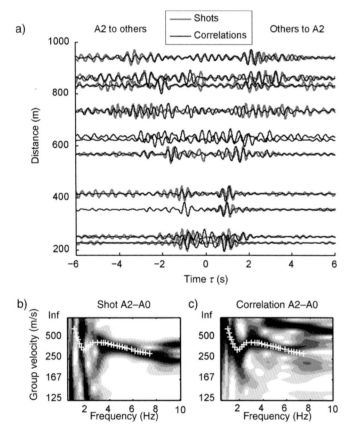

**Figure 7.** Ambient-noise crosscorrelation. (a) Comparison of the direct-shot signal (gray shading) and the noise-crosscorrelated signal (black shading) between A2 and the other stations in the 2- to 4-Hz frequency band. (b) Shot dispersion image computed for propagation from A2 to A0, and (c) crosscorrelation dispersion image computed for propagation from A2 to A0. The white crosses in (b) and (c) indicate the dispersion curve picked from the comparison of the two images.

retrieving wave-propagation times between stations using crosscorrelation of ambient noise. Our aim is to test the capacity of the technique to obtain a first-order 3D model of the landslide, using a limited number of sensors.

## Large-aperture array

We applied the crosscorrelation technique to the records of the large-aperture network (kilometer size) of 13 seismological stations installed on the southern part of the Avignonet landslide (A0 to A10, and AP and AM, in Figure 2). To evaluate the quality of the propagation functions computed from crosscorrelation of ambient noise, explosive shots were performed close to each station. The direct and crosscorrelated signals involving station A2 were band-pass filtered between 2 Hz and 4 Hz and the results are plotted in Figure 7a. The waveforms of the direct and correlated signals are not strictly equal because of differences in their frequency content. However, arrival times and waveforms of the most energetic wave (the direct Rayleigh wave) are comparable, showing the reliability of the crosscorrelation technique for reconstructing the Rayleigh-wave propagation between stations.

In the same figure are shown the Rayleigh-wave dispersion maps (group velocity) computed by using the S-transform (Stockwell et al., 1996) for the propagation from A2 to A0. On the shot image (Figure 7b), the fundamental mode of the Rayleigh wave can be picked between 3 Hz and 10 Hz, whereas it can be determined in the 1.5- to 5-Hz-frequency range on the crosscorrelation image (Figure 7c). The high-velocity mode observed between 4 and 10 Hz is visible only for this couple of stations and therefore cannot be used in the following part of the study.

On the whole network, noise crosscorrelations and shots gave complementary information on the Rayleigh fundamental mode. They were combined to obtain path-averaged (averaged over the path between two stations) group-velocity dispersion curves that cover the 1.5- to 7-Hz frequency range for the different pairs of stations. Interestingly, the frequency range for shots is similar in Figure 5 (a 470-m-long seismic profile with 4.5-Hz geophones) and 7 (a kilometer-size network with short-period seismological stations), showing that the low-frequency limit of the dispersion-curve frequency is controlled mainly by the frequency content of the source.

A tomographic inversion was performed on the arrival-time measurements given by the Rayleigh group-velocity dispersion curves, using the algorithm of Barmin et al. (2001). This simple algorithm considers straight rays on a regular grid. It includes a spatial smoothing, and it also includes a constraint on the amplitude perturbation that depends on local path density. The 15-cell × 15-cell tomographic grid was composed of cells that were 100 m long on each side. Because of the sparse ray coverage, we chose to give great importance to the spatial-smoothing parameter (Brenguier et al., 2007). Thus, a group-velocity map was obtained for frequencies between 1.5 and 7 Hz, using measurements at 0.1-Hz intervals. Figure 8a and 8b shows the maps for 1.7 and 4 Hz, together with all paths on which dispersion curves could be picked at these frequencies. These group-velocity maps highlight two features of lateral variations at high and low frequencies. The reliability of these results was evaluated using checkerboard tests with heterogeneities of various sizes (Figure 8c through 8f). The lateral variations are well reconstructed within the array, but their shapes are smoothed and their amplitudes are attenuated as a result of the smoothing parameter.

Subsequently, local group-velocity dispersion curves were reconstructed for each cell of the surface model. Each of those curves then was 1D inverted both for $V_S$ and thickness values with a neighborhood algorithm (Sambridge, 1999; Wathelet, 2008), using a parameter space that defined four layers over a half-space (Renalier, 2010). To reconstruct a 3D $V_S$ model of the area, we then computed for each cell the average of all resulting $V_S$ profiles that had a misfit smaller than 1.2 times the minimum misfit. In the following, we focus on the results and their comparison with those of other $V_S$ imaging techniques.

Figure 9 shows two perpendicular cross sections (CC′ and DD′) made through the 3D $V_S$ structure. The east-west cross section CC′ intersects boreholes B0 through B3, whereas the south-north section DD′ crosses boreholes B1 and B4. The headscarp and the slip surfaces found in boreholes also are located in Figure 9. The depth of the

deep slip surface (found between 42 and 47 m, Table 1) correlates very well with the 400-m/s velocity contour at which a vertical contrast is observed in both Figures 6 and 9. The landslide body appears to be characterized by a velocity $V_S$ lower than 400 m/s, and it thins significantly down to the west, in agreement with the headscarp location (Figure 9a). A regular thinning of the landslide body from 40 m to 20 m also is observed to the north (Figure 9b), but no geotechnical data are available for this area. Within the landslide mass, another vertical $V_S$ contrast can be seen at a depth of about 15 m, which correlates well with the intermediate slip surfaces detected in the boreholes (Table 1) and by the active seismic experiments (Figure 4). However, that depth is near the minimum penetration that is related to the maximum frequency of 7 Hz.

Finally, the faster velocities at depth below boreholes B1 through B3 hint at the shallow presence of compact alluvial layers and Jurassic bedrock. Further, the velocities of those alluvial layers and bedrock are underestimated because of the tomographic smoothing and the poor ray coverage at low frequencies in that area, which also explain the slight difference in $V_S$ values measured at depth with the small-aperture array.

This surface-wave inversion from a combination of ambient-noise crosscorrelation and explosive shot signals agrees well with independent geotechnical and geophysical results, which proves the potential of such a technique for imaging deep-seated landslides that affect clayey deposits, even in a rural environment. Further improvements could be obtained by using (1) more stations for a better ray coverage, (2) longer recordings for reconstructing the propagation functions at higher frequencies from noise correlation, or (3) specific processing techniques for recon-

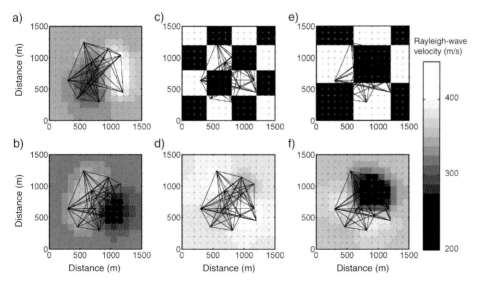

**Figure 8.** Group velocity maps resulting from the tomographic inversion: (a) at 1.7 Hz, and (b) at 4 Hz. Black lines indicate the available ray coverage at the corresponding frequency. (c) Checkerboard test with 400-m-side lateral heterogeneities and the ray coverage available at 4 Hz. (d) Group velocity map obtained from inversion of synthetic traveltimes computed from the model of (c). (e) Checkerboard test with 600-m-side lateral heterogeneities and the ray coverage available at 4 Hz. (f) Group velocity map obtained from inversion of synthetic traveltimes computed from the model of (e).

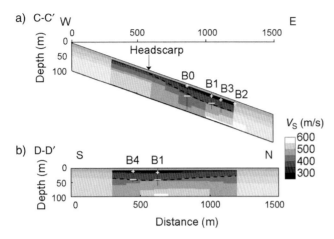

Figure 9. $V_S$ images resulting from inversion along sections CC′ and DD′ (see locations in Figure 2). (a) Parallel to the slope; the depth scale is exaggerated 2×. (b) Perpendicular to the slope. Deep and intermediate slip surfaces found in boreholes B0 through B4 are indicated by crosses and horizontal bars, respectively. Shaded areas on top and on both sides indicate low resolution.

structing those propagation functions, even in the case of directional noise (Roux, 2009).

## Landslide Monitoring

Because landslide characterization also includes a temporal dimension, we tested the ability of the crosscorrelation technique to monitor evolution of material degradation related to gravitational movement. For this purpose, we used the continuous seismic data available at the two permanent stations, AP and AM (Figure 2), from October 2006 to May 2009. Unfortunately, numerous gaps were present in the data, as a result of material robberies and of malfunctions.

After spectral whitening of the one-hour noise signals (sampled at 125 Hz), crosscorrelation functions were computed in three frequency ranges (1.3–3.5 Hz, 2–5 Hz, and 5–10 Hz) and summed to obtain one-day crosscorrelation functions (Figure 10, left column). In the same figure (right column) are continuous plots of 12 12-day

Figure 10. Crosscorrelation functions computed along time in different frequency ranges. (a) and (b) 1.3- to 3.5-Hz, (c) and (d) 2- to 5-Hz, (e) and (f) 5-to 10-Hz frequency bands. Left column: Temporal evolution of the one-day crosscorrelation functions. Right column: comparison of 12 12-day correlation functions.

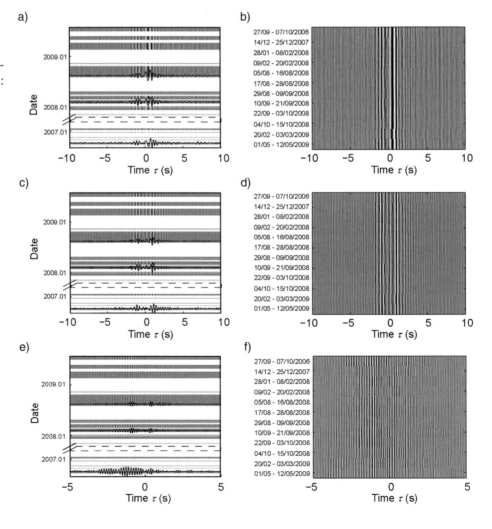

correlations whose dates are indicated on the left axis. In the two lowest-frequency ranges, 1.3–3.5 Hz and 2–5 Hz (Figure 10a through 10d), correlations are stable, with coherent arrivals as high as 10 s (especially at $\tau = -7$ s for both frequency bands). However, correlations at low frequencies (1.3–3.5 Hz) are asymmetrical, which indicates an anisotropic distribution of the noise (Stehly et al., 2006). In the high-frequency band (5–10 Hz), signals are asymmetrical and propagation functions are reconstructed only during the first 1 to 2 s, after which the different arrivals are less stable.

Next, the signals in the three frequency bands were processed, looking for temporal variations in the wave propagation. We used two methods for measuring the relative variations of the velocity $\Delta V/V$ during the period studied — the doublet technique (Poupinet et al., 1984; Brenguier et al., 2008) and the stretching technique (Sens-Schönfelder and Wegler, 2006; Hadziioannou et al., 2009). These two techniques rely on the hypothesis that observed waveform fluctuations result only from a homogeneous variation in seismic velocity in the medium. In such a case, the relative traveltime shift $\Delta \tau$ between the perturbed and reference correlations is proportional to the time-lapse $\tau$, and $\Delta V/V = -\Delta \tau/\tau$ and is constant along the whole seismogram (Poupinet et al., 1984). Because the two techniques yielded similar results (Renalier, 2010), only those of the doublet technique are presented here.

The crosscorrelation function obtained for the whole year 2008 was chosen as the reference correlation, and temporal evolution was evaluated by comparing this reference with current correlation functions computed by crosscorrelating the seismic noise in five-day and 12-day moving windows along the three-year time period. For each current correlation function, we assessed the relative time perturbations $\Delta \tau/\tau$ by measuring the slope of the traveltime shifts $\Delta \tau$ as a function of time $\tau$. Despite some little discrepancies relative to the hypothesis of homogeneous velocity variations ($\Delta \tau/\tau =$ constant), we could fit a linear trend on the curves $\Delta \tau = f(\tau)$ for time-lapse values $\tau$ between 1 and 7 s (see Poupinet et al., 1984; Brenguier et al., 2008 for details on the processing technique).

Figure 11 shows the relative arrival-time variations $\Delta \tau/\tau$ measured during 32 months between the two stations AP and AM, for five-day and 12-day windows. Results are similar for the two window lengths, and the time windows of 12 days are used in the following. Although measurements are discontinuous and are affected by short period variations (SP in Figure 11), the graphs for the three frequency bands show almost the same linear trend (highlighted with the dashed line in Figure 11), with an increase in arrival times of about 0.2% per year measured by linear regression for the 2- to 5-Hz frequency band.

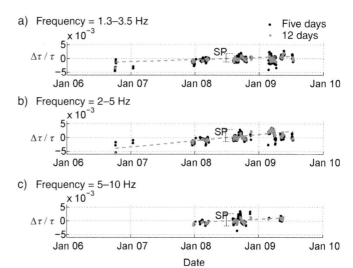

**Figure 11.** Evolution of relative arrival times $\Delta \tau/\tau$ between October 2006 and May 2009, measured with the doublet technique, for the three frequency bands: (a) 1.3–3.5 Hz, (b) 2–5 Hz, and (c) 5–10 Hz. When differences result only from homogeneous velocity variations, $\Delta \tau/\tau = -\Delta V/V$ (where $V$ is the seismic velocity in the medium). SP: short-period variations; gray dashed line: long-term tendency computed by linear regression on the 12-day moving windows.

This variation in the correlation function could result from distance variation resulting from differential motion of the two stations, which are installed on the landslide. However, the biannual GPS measurements (Jongmans et al., 2009), along with two aerial LIDAR campaigns in November 2006 and April 2009 (U. Kniess, personal communication, 2009), indicate that the distance of 250 m between the two stations has not varied by more than 15 cm during that period. This differential displacement corresponds to a traveltime increase of less than 0.06%, which is far lower than the 0.2% variation detected with seismic monitoring. This comparison indicates that the observed traveltime increase corresponds to a seismic ground-velocity decrease. That decrease probably is linked to the landslide continuously exerting a damaging effect, which has already reduced $V_S$ by a factor of 2 in the shallow active zone at the toe of the landslide (Figure 3). Although these results have to be confirmed by longer and continuous data, they highlight the potential of the crosscorrelation technique for monitoring landslide activity.

## Conclusions

Among the different geophysical parameters, shear-wave velocity $V_S$ has emerged as being of major importance for characterizing landslides because of its sensitivity to compaction or to the degree of fissuring in soils and rocks. In the case of the clayey Avignonet landslide, $V_S$ measurements have provided valuable information on the deformation state of the clay material, which resulted from the gravitational movement along slip surfaces at three depths (at about 5 m, 15 m, and 45 m).

Both lateral and vertical variations of $V_S$ have been related to the landslide activity and to the location of slip surfaces. Different techniques for in situ measurement of $V_S$ have been tested. Linear arrays of geophones with active sources offer the advantage of good resolution in the surficial layers and can detect the two shallower slip surfaces (at about 5 m and 15 m) highlighted by vertical $V_S$ contrasts. In addition, SH refraction tests provided 2D images that have relatively limited penetration (25–30 m) because of the weak energy of the source.

Surface-wave inversion with an active source was limited to a few tens of meters in penetration, too, because of the low energy of the source at low frequencies and/or the high-pass frequency filtering of the geophones used (4.5 Hz). In addition to the requirement that the local geometry be stratified horizontally, this study also pointed to the problem of secondary scattering sources (which are present in strongly perturbed zones) for linear arrays. Such scattering sources may generate higher modes and destroy the coherency of the surface-wave fundamental mode, even for a group of geophones located over a 1D medium.

A small-aperture 2D array of six short-period seismometers (maximum size of 120 m in this case), which measures ambient vibrations from all azimuths, was used successfully for surface-wave inversion in a horizontally stratified zone. This passive technique extracted lower-frequency surface waves from seismic records of one-hour duration and characterized the $V_S$ contrast at the bottom of the landslide. However, the technique was found to be excessively time- and effort-consuming for 3D mapping of the whole landslide. To overcome that limitation, we deployed a large-aperture array of 11 stations (1- × 1-km in size) covering the southern part of the landslide where geotechnical information was available.

Seismic noise was recorded during two weeks. Combining ambient noise crosscorrelation with records of explosive shots performed close to each station, a 3D $V_S$ image of the landslide was inverted from surface-wave group-dispersion curves. Although the lateral resolution is low (about 100 m) because of the limited number of stations, the $V_S$ image obtained is consistent with the depth of the landslide found in boreholes. Major advantages of this method are that manpower is required only during instrument installation and withdrawal, and that secondary scattering sources — which always exist in

active landslides — help to fulfill the diffuse wavefield assumption. Finally, the crosscorrelation technique was applied to the records from two permanent stations during a 30-month period. Associated with seismic monitoring techniques, the crosscorrelation technique allowed detection of a slight but significant decrease in surface-wave velocities. This result offers new perspectives for monitoring such landslides and for detecting rheological variations in clay when a slide evolves to a flow.

## Acknowledgments

The authors thank C. Pequegnat and E. Bourova for preprocessing the seismological data, R. Béthoux for maintaining instrumentation, and all the people who participated in the field investigation. This study was supported in part by the European project NERIES, by the French ANR projects ECOUPREF and SISCA, by the LCPC Sécheresse program, by the Isère Department through the Pôle Grenoblois des Risques Naturels, and by the ERC Advanced Grant Whisper. The French landslide observatory OMIV (www-lgit.obs.ujf-grenoble.fr/observations/omiv) provided data from the permanent seismological stations. We thank two anonymous reviewers for their valuable comments.

## References

Aki, K., 1957, Space and time spectra of stationary stochastic waves, with special reference to microtremors: Bulletin of Earthquake Research Institute, **35**, 415–456.

Aki, K., and B. Chouet, 1975, Origin of coda waves: Source, attenuation, and scattering effects: Journal of Geophysical Research — Solid Earth, **80**, 3322–3342.

Aki, K., and P. G. Richards, 1980, Quantitative seismology: Theory and methods: W. H. Freeman.

Andrus, R., P. Piratheepan, B. S. Ellis, J. Zhang, and C. H. Juang, 2004, Comparing liquefaction evaluation methods using penetration-$V_S$ relationships: Soil Dynamics and Earthquake Engineering, **24**, 713–721.

Asten, M. W., and J. D. Henstridge, 1984, Array estimators and use of microseisms for reconnaissance of sedimentary basins: Geophysics, **49**, 1828–1837.

Bard, P.-Y., and J. Riepl-Thomas, 1999, Wave propagation in complex geological structures and local effects on strong ground motion, in E. Kausel and G. Manolis, eds., Wave motion in earthquake engineering: International Series Advances in Earthquake Engineering, WIT Press, 37–95.

Barmin, M. P., M. H. Ritzwoller, and A. L. Levshin, 2001, A fast and reliable method for surface wave tomography: Pure and Applied Geophysics, **158**, 1351–1375.

Boore, D., 2006, Determining subsurface shear-wave velocities: a review: Paper presented at the Third International Symposium on the Effects of Surface Geology on Seismic Motion, Paper no. 103.

Brenguier, F., N. M. Shapiro, M. Campillo, V. Ferrazzini, Z. Duputel, O. Coutant, and A. Nercessian, 2008, Towards forecasting volcanic eruptions using seismic noise: Nature Geoscience, **1**, 126–130.

Brenguier, F., N. M. Shapiro, M. Campillo, A. Nercessian, and V. Ferrazzini, 2007, 3-D surface wave tomography of the Piton de la Fournaise volcano using seismic noise correlations: Geophysical Research Letters, **34**, L02305.

Bryan, G. M., and R. D Stoll, 1988, The dynamic shear modulus of marine sediments: Journal of the Acoustical Society of America, **83**, 2159–2164.

Claerbout, J. F., 1968, Synthesis of a layered medium from its acoustic transmission response: Geophysics, **33**, 264–269.

Dal Moro, G., M. Pipan, and P. Gabrielli, 2007, Rayleigh wave dispersion curve inversion via genetic algorithms and marginal posterior probability density estimation: Journal of Applied Geophysics, **61**, 39–55.

Dasios, A., C. McCann, T. Astin, D. McCann, and P. Fenning, 1999, Seismic imaging of the shallow subsurface: Shear wave case histories: Geophysical Prospecting, **47**, 565–591.

Dines, K., and J. Lyttle, 1979, Computerized geophysical tomography: Proceedings of the Institute of Electrical and Electronics Engineers, **67**, 106–1073.

Finn, W., 2000, State-of-the-art of geotechnical earthquake engineering practice: Soil Dynamics and Earthquake Engineering, **20**, 1–15.

Finn, W., and A. Wightman, 2003, Ground motion amplification factors for the proposed 2005 edition of the National Building Code of Canada: Canadian Journal of Civil Engineering, **30**, 272–278.

Ghose, R., and J. Goudswaard, 2004, Integrating S-wave seismic-reflection data and cone-penetration-test data using a multiangle multiscale approach: Geophysics, **69**, 440–459.

Giraud, A., P. Antoine, T. van Asch, and J. Nieyuwenhuis, 1991, Geotechnical problems caused by glaciolacustrine clays in the French Alps: Engineering Geology, **31**, 185–195.

Gouédard, P., L. Stehly, F. Brenguier, M. Campillo, Y. C. de Verdière, E. Larose, L. Margerin, P. Roux, F. J. Sánchez-Sesma, N. M. Shapiro, and R. L. Weaver, 2008, Cross-correlation of random fields: Mathematical approach and applications: Geophysical Prospecting, **56**, 375–393.

Grandjean, G., and A. Bitri, 2006, 2M-MASW: Multifold and multichannel seismic inversion of local dispersion of Rayleigh waves in laterally heterogeneous subsurfaces: Application to the Super-Sauze earthflow (France): Near Surface Geophysics, **4**, no. 6, 367–375.

Hadziioannou, C., E. Larose, O. Coutant, P. Roux, and M. Campillo, 2009, Stability of monitoring weak changes in multiply scattering media with ambient noise correlation: Laboratory experiments: Journal of the Acoustical Society of America, **125**, 3688–3695.

Hasancebi, N., and R. Ulusay, 2007, Empirical correlations between shear wave velocity and penetration resistance for ground-shaking assessments: Bulletin of Engineering Geology and the Environment, **66**, 203–213.

Hegazy, Y., and P. Mayne, 1995, Statistical correlations between $V_S$ and cone penetration data for different soil types: International Symposium on Cone Penetration Testing, 173–178.

Herrmann, R. B., 1973, Surface wave generation by the south central Illinois earthquake of November 9, 1968: Bulletin of the Seismological Society of America, **63**, no. 6, 2121–2134.

———, 1987, Computer programs in seismology, IV: Surface wave inversion: Saint Louis University.

Hunter, J., B. Benjumea, J. Harris, R. Miller, S. Pullan, and R. A. Burns, 2002, Surface and downhole shear wave seismic methods for thick soil site investigations: Soil Dynamics and Earthquake Engineering, **22**, 931–941.

Jongmans, D., 1992, The application of seismic methods for dynamic characterization of soils in earthquake engineering: Bulletin of the International Association of Engineering Geology, **46**, 63–69.

Jongmans, D., G. Bièvre, S. Schwartz, F. Renalier, N. Beaurez, and Y. Orengo, 2009, Geophysical investigation of a large landslide in glaciolacustrine clays in the Trièves area (French Alps): Engineering Geology, **109**, 45–56.

Keilis-Borok, V. I., 1986, Seismic surface waves in a laterally inhomogeneous earth: Kluwer Academic Publishers.

Kim, D., and H. Park, 1999, Evaluation of ground densification using SASW method and resonant column tests: Canadian Geotechnical Journal, **36**, 291–299.

Lacoss, R., E. J. Kelly, and M. N. Toksöz, 1969, Estimation of seismic noise using arrays: Geophysics, **34**, 21.

Lee, J., and J. Santamarina, 2005, Bender elements: Performance and signal interpretation: Journal of Geotechnical and Geoenvironmental Engineering, **131**, 1063–1070.

Lin, C.-P., and C.-H. Lin, 2007, Effect of lateral heterogeneity on surface wave testing: Numerical simulations and a countermeasure: Soil Dynamics and Earthquake Engineering, **27**, 541–552.

Méric, O., S. Garambois, D. Jongmans, J. Vengeon, M. Wathelet, and J. Châtelain, 2005, Application of geophysical methods for the investigation of the large gravitational mass movement of Sechilienne (France): Canadian Geotechnical Journal, **42**, 1105–1115.

Miller, R., J. Xia, C. Park, and J. Ivanov, 1999, Multichannel analysis of surface waves to map bedrock: The Leading Edge, **18**, 1392–1396.

Mondol, N. H., K. Bjørlykke, J. Jahren, and K. Høeg, 2007, Experimental mechanical compaction of clay mineral aggregates — Changes in physical properties of mudstones during burial: Marine and Petroleum Geology, **24**, 289–311.

Ohrnberger, M., E. Schissele, C. Cornou, S. Bonnefoy-Claudet, M. Wathelet, A. Savvaidis, F. Scherbaum, and D. Jongmans, 2004, Frequency wavenumber and spatial autocorrelation methods for dispersion curve determination from ambient vibration recordings: Paper presented at the 13th World Conference on Earthquake Engineering.

Okada, H., 2003, The microtremor survey method (K. Suto, trans.): SEG Geophysical Monograph Series No. 12.

Nunziata, C., G. De Nisco, and G. F. Panza, 2009, S-waves profiles from noise cross correlation at small scale: Engineering Geology, **105**, 161–170.

Park, C., R. D. Miller, and J. Xia, 1999, Multichannel analysis of surface waves: Geophysics, **64**, 800–808.

Park, C. B., R. D. Miller, J. Xia, and J. Ivanov, 2007, Multichannel analysis of surface waves (MASW) — Active and passive methods: The Leading Edge, **26**, no. 1, 60–64.

Parolai, S., M. Picozzi, S. M. Richwalski, and C. Milkereit, 2005, Joint inversion of phase velocity dispersion and $H/V$ ratio curves from seismic noise recordings using a genetic algorithm, considering higher modes: Geophysical Research Letters, **32**, L01303.

Picozzi, M., S. Parolai, D. Bindi, and A. Strollo, 2009, Characterization of shallow geology by high frequency seismic noise tomography: Geophysical Journal International, **176**, 164–174.

Poupinet, G., W. L. Ellsworth, and J. Fréchet, 1984, Monitoring velocity variations in the crust using earthquake doublets: An application to the Calaveras Fault, California: Journal of Geophysical Research, **89**, 5719–5731.

Renalier, F., 2010, Caractérisation sismique de sites hétérogènes à partir de méthodes actives et passives: Variations latérales et temporelles: Ph.D. thesis, Université de Grenoble, France.

Renalier, F., G. Bièvre, L. Valldosera, D. Jongmans, E. Flavigny, and P. Foray, 2010, Caractérisation de l'endommagement d'une argile par mesures de la vitesse des ondes de cisaillement: Journées Nationales de Géotechnique et de Géologie de l'Ingénieur, JNGG2010, 25–32.

Richwalski, S. M., M. Picozzi, S. Parolai, C. Milkereit, F. Baliva, D. Albarello, K. Roy-Chowdhury, H. van der Meer, and J. Zschau, 2007, Rayleigh wave dispersion curves from seismological and engineering-geotechnical methods: A comparison at the Bornheim test site (Germany): Journal of Geophysics and Engineering, **4**, no. 4, 349.

Roux, P., 2009, Passive seismic imaging with directive ambient noise: Application to surface waves and the San Andreas Fault in Parkfield, CA: Geophysical Journal International, **179**, 367–373.

Sambridge, M., 1999, Geophysical inversion with a neighbourhood algorithm, I: Searching a parameter space: Geophysical Journal International, **138**, 479–494.

Sánchez-Sesma, F. J., and M. Campillo, 2006, Retrieval of the Green function from cross correlation: the canonical elastic problem: Bulletin of the Seismological Society of America, **96**, 1182–1191.

Satoh, T., H. Kawase, and S. Matsushima, 2001, Estimation of S-wave velocity structures in and around the Sendai Basin, Japan, using array records of microtremors: Bulletin of the Seismological Society of America, **91**, 206–218.

Scherbaum, F., K.-G. Hinzen, and M. Ohrnberger, 2003, Determination of shallow shear wave velocity profiles in the Cologne/Germany area using ambient vibrations: Geophysical Journal International, **152**, 597–612.

Schlue, J. W., and K. K. Hostettler, 1987, Rayleigh wave phase velocities and amplitude values in the presence of lateral heterogeneities: Bulletin of the Seismological Society of America, **77**, 244–255.

Schuster, G. T., J. Yu, J. Sheng, and J. Rickett, 2004, Interferometric/daylight seismic imaging: Geophysical Journal International, **157**, 838–852.

Sens-Schönfelder, C., and U. Wegler, 2006, Passive image interferometry and seasonal variations of seismic velocities at Merapi Volcano, Indonesia: Geophysical Research Letters, **33**, L21302.

Shapiro, N., and M. Campillo, 2004, Emergence of broadband Rayleigh waves from correlations of the ambient seismic noise: Geophysical Research Letters, **31**, L07614.

Socco, L., and D. Jongmans, 2004, Special issue on seismic surface waves: Near Surface Geophysics, **2**, 163–165.

Socco, L. V., and C. Strobbia, 2004, Surface-wave method for near-surface characterization: A tutorial: Near Surface Geophysics, **2**, 165–185.

Socco, L., D. Boiero, C. Comina, S. Foti, and R. Wìsén, 2008, Seismic characterization of an Alpine site: Near Surface Geophysics, **6**, 255–267.

Socco, L. V., D. Boiero, S. Foti, and R. Wisén, 2009, Laterally constrained inversion of ground roll from seismic reflection records: Geophysics, **74**, no. 6, G35–G45.

Sommerville, P., and R. Graves, 2003, Characterization of earthquake strong ground motion: Pure and Applied Geophysics, **160**, 1811.

Stehly, L., M. Campillo, and N. M. Shapiro, 2006, A study of the seismic noise from its long range correlation properties: Journal of Geophysical Research, **111**, B10306.

Stockwell, R. G., L. Mansinha, and R. P. Lowe, 1996, Localization of the complex spectrum: the S transform: IEEE Transactions on Signal Processing, **44**, 998–1001.

Tokimatsu, K., 1997, Geotechnical site characterization using surface waves: First International Conference on Earthquake and Geotechnical Engineering, 1333–1368.

Tokimatsu, K., S. Tamura, and H. Kojima, 1992, Effects of multiple modes on Rayleigh wave dispersion charac-

teristics: Journal of Geotechnical Engineering, **118**, 1529–1543.

Uebayashi, H., 2003, Extrapolation of irregular subsurface structures using the horizontal-to-vertical ratio long-period microtremors: Bulletin of the Seismological Society of America, **93**, 570–582.

Wapenaar, K., D. Draganov, and J. O. A. Robertsson, 2008, Seismic interferometry: History and present status: SEG Geophysics Reprint Series No. 26.

Wathelet, M., 2008, An improved neighborhood algorithm: Parameter conditions and dynamic scaling: Geophysical Research Letters, **35**, L09301.

Wathelet, M., D. Jongmans, and M. Ohrnberger, 2004, Surface wave inversion using a direct search algorithm and its application to ambient vibration measurements: Near Surface Geophysics, **2**, 211–221.

Wathelet, M., D. Jongmans, M. Ohrnberger, and S. Bonnefoy-Claudet, 2008, Array performances for ambient vibrations on a shallow structure and consequences over $V_S$ inversion: Journal of Seismology, **12**, 1–19.

Chapter 25

# Detecting Perched Water Bodies Using Surface-seismic Time-lapse Traveltime Tomography

David Gaines[1], Gregory S. Baker[1], Susan S. Hubbard[2], David Watson[3], Scott Brooks[3], and Phil Jardine[4]

## Abstract

Applications of seismic time-lapse techniques generally are constrained to large-scale investigations associated with petroleum exploration and exploitation. There is growing interest in using geophysical methods to monitor near-surface phenomena, such as fluid flow in fractured or karstic bedrock, hydraulic infiltration, and anthropogenic manipulations during environmental remediation. Previous near-surface geophysical time-lapse studies have focused on electrical or electromagnetic (EM) techniques (including ground-penetrating radar) or borehole methods. To evaluate the utility of surface seismic time-lapse traveltime tomography, a site was monitored through time along a single 2D profile. The objective was to attribute increases in seismic P-wave velocity with the development of perched water bodies in the upper 4 m of the subsurface. The study was conducted in the Y-12 area of Oak Ridge National Laboratory in Tennessee, U.S.A., in conjunction with a broader multidisciplinary investigation on the fate and transport of contaminants. Because of previous anthropogenic alterations of the site associated with remediation efforts (e.g., replacing as much as 7 m of contaminated soil with poorly sorted limestone gravel fill during construction of a seepage basin cap), the near-surface hydrogeology was extremely heterogeneous and was hypothesized

to have a large influence on differential infiltration, contaminant distribution, and contaminant remobilization. The seismic data were processed using a wavepath eikonal traveltime (WET) tomography approach, and a modified trend-analysis technique was applied to remove the larger spatial component associated with geologic variability. The final "residual" velocity-anomaly images were compared with wellbore hydrologic data and error analyses and were used to interpret the presence and geometry of perched water in the shallow subsurface. The study suggests that velocity estimates obtained from surface-seismic traveltime tomography methods are effective for indicating the spatial and temporal distribution of perched water bodies at the Oak Ridge site in the upper 4 m of the subsurface.

## Introduction

We explore the utility of seismic time-lapse P-wave traveltime tomography (STLTT) for monitoring the presence or absence of perched water bodies in the upper 4 m of the subsurface at the S-3 ponds in the Y-12 site of the U. S. Department of Energy (DOE) Oak Ridge Reservation in Oak Ridge, Tennessee. In 2007, DOE established the Oak Ridge Integrated Field-Research Challenge (ORIFRC) to investigate the rates and mechanisms of in situ immobili-

[1]Earth and Planetary Sciences, University of Tennessee, Knoxville, Tennessee, U.S.A. E-mail: dgaines1@utk.edu; gbaker@utk.edu.

[2]Earth Sciences Division, Lawrence Berkeley National Laboratory, Berkeley, California, U.S.A. E-mail: sshubbard@lbl.gov.

[3]Environmental Sciences Division, Oak Ridge National Laboratory, Oak Ridge, Tennessee, U.S.A. E-mail: watsondb@ornl.gov; brookssc@ornl.gov.

[4]Biosystems Engineering and Soil Sciences, University of Tennessee, Knoxville, Tennessee, U.S.A. E-mail: pjardine@tennessee.edu.

zation of contaminants in the subsurface. A subset of this multidisciplinary research includes analyzing geophysical responses to hydrologic properties, because the flux during rainfall events is hypothesized to exert a large influence on contaminant remobilization and distribution. This study focuses on perched water bodies in the upper 4 m of the vadose zone at the ORIFRC, because they are believed to influence the rate and spatial location of hydrologic recharge that impacts deeper contaminated regions.

## Research site

The S-3 ponds (Figure 1) were constructed in 1951 for waste-disposal activities. They consist of four unlined ponds ~5 m deep and covering an area of ~120 × 120 m, with a total storage capacity of ~10,000,000 gal (37,854,000 L) for all four ponds (U. S. Department of Energy, 1997). A variety of liquid and sludge wastes, composed principally of nitrate, metals, and various radionuclides (e.g., uranium and technetium), is known to have been disposed of in the S-3 ponds (Watson et al., 2005). Total contaminant-mass estimates vary; however, from 1951 to 1983 the ponds received ~2,000,000 gal (7,571,000 L) of liquid waste per year that consisted of condensate mixed with nitric acid and aluminum nitrite in various concentrations (U. S. Department of Energy, 1997). Additional wastes, including sludge from clean-up activities, aerosol cans, and contaminated sediments, were disposed of in the S-3 ponds, but precise mass estimates were not available (U. S. Department of Energy, 1997). The ponds were neutralized and denitrified during 1983–1984, and they were filled and

capped in 1988 under the Resource and Conservation Recovery Act (U. S. Department of Energy, 1997). The ponds were unlined, so infiltration and density-driven flow has resulted in an extensive contaminated groundwater plume in the underlying geologic media. That contaminated groundwater has acted as a secondary contaminant source (Watson et al., 2005).

The research site described here is directly adjacent to the former S-3 ponds (in a southwesterly direction) and is oriented perpendicularly to the predominant direction of groundwater flow, parallel to geologic strike (Solomon et al., 1992). The underlying geology (Figure 2) is dominated by the presence of the Nolichucky Shale, a member of the Cambrian Conasauga Group, which dips ~45° to the southeast and strikes N55°E (Hatcher et al., 1992; Watson, et al., 2005). The competent Nolichucky Shale transitions toward the ground surface into a less-competent, weathered bedrock (i.e., saprolite) that retains the fracture and bedding attributes of its parent rock (Watson et al., 2005). The hydrology of the research site is extremely complex as a result of remnant anthropogenic alterations of the site, including as much as 7 m of poorly sorted limestone gravel fill, construction of a cap (causing highly variable water recharge during rainfall events), and construction of a gravel-lined drainage ditch adjacent to the ponds (causing unpredictable surface-water fluxes).

The current conceptual model is that large hydraulic conductivity contrasts between the highly permeable anthropogenic fill and the lower-permeability saprolite, in concert with the preferential recharge of the drainage ditch and impermeable cap, create a saturated perched

**Figure 1.** Aerial photo of the S-3 ponds at Y-12 Oak Ridge National Laboratory. The parking lot was constructed on top of the former ponds. Shot locations are indicated on the photo. Well locations and adjacent geophones are depicted in the inset.

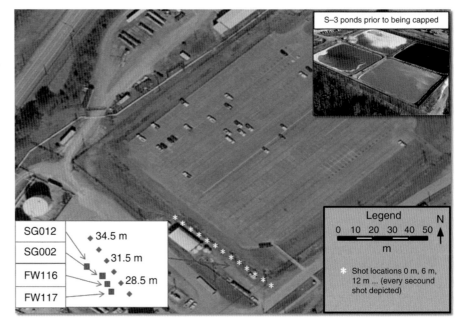

zone during rainfall events. Water-budget studies have indicated that much of the subsurface flow (~90%) at Oak Ridge takes place in the extreme near-surface (~0.8- to 1.2-m-deep) storm-flow zone (Solomon et al., 1992). At Oak Ridge, this storm-flow zone usually is associated with heavily vegetated areas. That contrasts significantly with the characteristics of the S-3 ponds, where water typically remains ponded in the drainage ditch adjacent to the S-3 ponds for several days after a large rainfall event. The perched water table at the ORIFRC varies but generally is 1–2 m below surface, and it ranges from ~305 m to 307 m above sea level, with ground level ranging from ~307 m to 308 m above sea level. The regional water table is less responsive to infiltration and is 3–5 m below surface (from ~303 to 304 m above sea level). The spatial locations of vertical recharge are affected by the distribution of the perched water bodies. The recharge is important at the ORIFRC because it provides a source of high dissolved-oxygen values and higher-pH water values relative to conditions at depth. In addition, the surface water readily mixes and reacts with contaminants.

## Previous characterization at ORIFRC

Multiple investigations have been performed at the ORIFRC using seismic techniques (Sheehan et al., 2005b; Watson et al., 2005; Chen et al., 2006). Much of the previous work has focused on characterizing low seismic-velocity zones and has demonstrated the coincidence between those zones and high-hydraulic-conductivity zones. For example, Chen et al. (2006) used joint stochastic inversion of crosswell seismic traveltimes and hydrologic data to estimate the distribution of high-hydraulic-conductivity zones. Beyond the benefits of local characterization of preferential flow paths, their study also demonstrated that a local high-permeability zone exists at depths between 10.5 m and 13.5 m. The existence of that zone has been hypothesized on the basis of surface seismic methods (Sheehan et al., 2005b).

Sheehan et al. (2005b) acquired seismic data in support of research at the Natural and Accelerated Bioremediation Research (NABIR) Field Research Center (FRC), which is the predecessor of the ORIFRC. The seismic data show the existence of a low-velocity zone at depth with very low ray coverage, presumed to indicate a high-hydraulic-permeability conduit, or pathway, in the underlying bedrock (Sheehan et al., 2005b). Although the cavities have not been drilled at the site, a separate anomaly located at the eastern end of Y-12 was located using seismic techniques and confirmed as a mud-filled cavity by driller's logs (Sheehan et al., 2005b). Watson et al. (2005) used coincident 2D

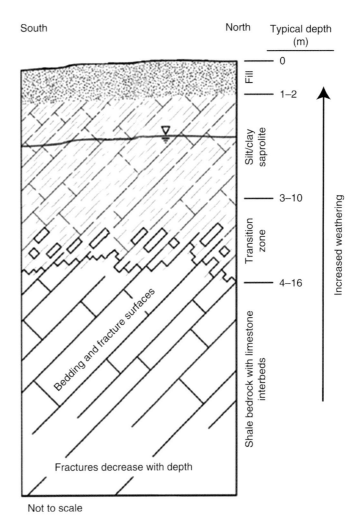

**Figure 2.** Conceptual model for underlying geology at the ORIFRC. After Watson et al., 2005. Used by permission of John Wiley & Sons, Inc.

surface-seismic and electrical data sets to select the location and depth of investigation for wellbores used in conjunction with a bioremediation study at the ORIFRC. Surface-seismic data were used in conjunction with borehole data to estimate the location of the transition zone between saprolite and competent bedrock as well as to locate probable areas of preferential flow (Watson et al., 2005). The surface electrical data were used to target areas of high electrical conductivity associated with elevated contaminant levels, and then they were integrated with the seismic data on the basis of spatial coincidence between seismically predicted preferential pathways and observed pathways using electrical methods (Watson et al., 2005). In addition, surface ground-penetrating radar was tested at the site to characterize near-surface anomalies, but that was not implemented because of the high electrical conductivity of the surface layers at the site (G. Baker, personal communication, 2007).

Although our work was influenced by Sheehan et al. (2005b), the depth of our investigation is significantly different (i.e., an ~1- to 6-m depth for our investigation). Also, our study focuses on perched water bodies that would appear as high-velocity zones because of presumed complete saturation, rather than on bodies that would appear as low-velocity zones (>10-m depth) because of variation in competency of bedrock or fracture density.

## Petrophysical model and error analysis

Conceptually, a change in saturation is expected to affect both density and effective bulk modulus, and consequently the P-wave velocity. The mathematical relationship between density, effective bulk modulus, and saturation has been documented previously (see, e.g., Domenico, 1974; Mavko et al., 1995; Bachrach and Nur, 1998), but it can be divided into a linear and a nonlinear domain (Figure 3). The linear domain is the region above the regional water table (i.e., it is the vadose zone) where a change in P-wave velocity is attributable to replacement of air with water in pore spaces. According to the Gassmann equation, the nonlinear domain is at 99% saturation and greater (i.e., typically at and below the regional water table), and a change in P-wave velocity is attributable to an increase of the effective bulk modulus as a result of fluid interactions. Descriptively, the linear domain is characterized by a gradual decrease in P-wave velocity with increasing saturation levels (i.e., as saturation increases, density increases and velocity decreases), whereas the nonlinear domain exhibits a rapid increase in P-wave velocity at 99–100% saturation. Recent laboratory measurements (George et al., 2009) have suggested that the change in seismic velocity is less abrupt than expected by these theoretical predictions and

that seismic P-wave velocity begins to increase before samples are completely saturated. In practice, this effect would introduce additional uncertainty in our measurements because it suggests that velocity increases may result from partial (rather than full) saturation only. However, George et al. (2009) indicate that seismic velocity in their heterogeneous soil mixture begins to increase gradually at <80% saturation, and not until ~90% saturation does seismic velocity begin to increase significantly. In our trend analysis (discussed below), we use a threshold P-wave velocity of 50 m/s, which partially accounts for the limited velocity increase at less than fully saturated levels. Thus, it is presumed that our trend-analysis images map out regions of saturation levels greater than 90% and that these regions indicate the target perched water bodies. Our hypothesis, therefore, is that the development of perched water bodies in the vadose zone will lead to changes in saturation (>90%) that will be detectable through P-wave velocity increases estimated using time-lapse surface-seismic data sets, thereby effectively highlighting the transition between the linear and nonlinear petrophysical model domains.

Sensitivity analyses of near-surface seismic tomography indicate the difficulty in accurately measuring subsurface velocities and provide a general estimate of the reliability of the tomography algorithm (Sheehan et al., 2005a; Hiltunen et al., 2007). The algorithm employed in this investigation is the wavepath eikonal traveltime (WET) algorithm, which is designed to account partially for wavepath effects by back-projecting traveltime residuals using a source weighting function (Schuster and Quintus-Bosz, 1993). The source weighting function in the WET algorithm is a gradient function that depends on the traveltime residuals, source and receiver locations, the slowness model, and geometric spreading terms. For a Ricker wavelet with a given peak frequency, the source weighting function is evaluated at all points in the traveltime grid, and the model is updated by distributing the residuals according to results of the source weighting function.

The generalized processing workflow includes picking first-arrival times for all sources and receivers, generating an initial slowness model and solving the eikonal equation (see Lecomte et al., 2000), then updating the model and continuing iteration until convergence. This is a computationally efficient way of partially accounting for "fat" rays, or the Fresnel volume associated with a propagating wave. Sheehan et al. (2005a) tested the WET algorithm using various synthetic models and found that the average rms error associated with a specific inversion relative to the true model was ~600 m/s. In most cases, the areas for which the rms error was calculated exhibited velocities

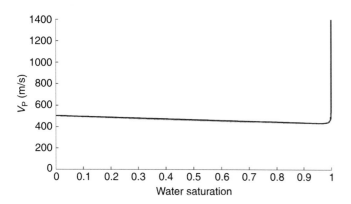

**Figure 3.** Petrophysical relationship between water saturation and P-wave seismic velocity according to the Gassmann equation.

ranging from 1000 m/s to 3000 m/s, although the area sampled and its associated rms error varied depending on initial synthetic model. As Sheehan et al. (2005b) stated, the rms error could be reduced significantly by altering smoothing parameters; however, this may result in the introduction of artifacts resulting from noise and inconsistent first-arrival picks. Although the tomographic images associated with this study are coupled via a shared initial model that is updated with minimal smoothing, they are wholly separate data sets that yield qualitatively similar images when processed independently with full smoothing. If successive STLTT data independently show an area of increased velocity above the regional water table, it is likely the anomaly is not an artifact caused by processing parameters.

If a 600-m/s rms error is accepted as the standard error associated with WET tomographic inversions, many shallow features, including near-surface seismic velocity features caused by variations in saturation, would be undetermined because of the expected error in velocity measurements. However, previous work has demonstrated the sensitivity of high-frequency seismic waves to small near-surface velocity variations. Additionally, shallow water-table reflections and refractions have been imaged at depths of <2 m (Baker et al., 1999a; Baker et al., 1999b; Baker et al., 2000a) that are sensitive to partial saturation (Birkelo et al., 1987; Bachrach and Nur, 1998; Baker et al., 2000b) and have been used successfully in mapping near-surface hydrostratigraphic reflectors (Lankston, 1989; Baker et al., 1999a; Baker et al., 2002; Garambois et al., 2002).

# Methodology and Results

The methodology for data acquisition is indistinguishable from many typical near-surface seismic-refraction surveys. However, the assumption that all first arrivals are to the result of refracted head waves may not be valid, so we avoid terminology involving "refraction" and instead consider "first arrivals" more appropriate. The seismic data presented here were collected during October and November 2007 and represent a selected time-slice during a period of increased rainfall. Hydrologic data, including precipitation, elevation of the regional water table, and elevation of the perched water table, were collected over the same time period and are nearly continuous at 15-minute intervals. However, occasional gaps in coverage exist where equipment malfunctioned, as well as artifacts whereby the measured groundwater table rapidly oscillates.

## Seismic data acquisition

Beginning in October 2007, we repeatedly collected coincident seismic P-wave traveltime 2D profiles and wellbore hydrologic data at the ORIFRC in order to monitor the development of perched water bodies that resulted from recharge. Each seismic profile is identical in acquisition geometry and spatial location and represents a time-dependent P-wave velocity image of the subsurface at the ORIFRC. The data used in the error analysis were collected during October 2007. The data for the time-lapse study of the perched water body were acquired during November 2007 and depict a period of increased rainfall. Seismic data acquisition was accomplished using two 24-channel Geometrics Geodes with 100-Hz vertical geophones located at 1.5-m intervals along the profile. The profile extends perpendicularly to geologic strike and parallel to the southwestern edge of the former S-3 ponds (see Figure 1). We stacked and recorded multiple seismic shots at each source location, using a conventional 2.2-kg (5-lb) sledgehammer and plate. The source-station interval was 3 m.

Geophones were removed and replaced for each acquisition period, and to maintain similarity between surveys, semipermanent plastic stakes were emplaced at 1.5-m intervals along the profile. During subsequent surveys, the semipermanent stakes were used to identify previous geophone locations, thereby ensuring minimal error caused by variation of geophone placements between surveys. Shot locations were directly adjacent to geophone locations and were similarly reproducible. The maximum error for relocating geophones was ± 5 cm; however, the positioning often was identical. To limit the effect of variable acquisition parameters or processing, data were not manipulated in the field (i.e., filters were not applied, and minimal pre-amp gain was applied). Because of anthropogenic noise outside the control of the experiment, stack numbers varied between acquisition periods. However, because we are not comparing amplitude but rather first-arrival times, we do not expect that variation to significantly affect first-arrival pick times or the subsequent tomographic inversions.

## Hydrologic data acquisition

The site is extremely well instrumented, and a variety of hydrologic information has been collected at the ORIFRC. We chose to use four wells in this study to provide ground truthing, on the basis of their proximity to the seismic profile and because they provide nearly continuous information for the period of November 2007. The hydrologic wells are offset perpendicularly from the profile by approximately 0.5 to 1 m and are located at positions that correspond to the 26- to 30-m points on the profiles. The wells are screened at different intervals, depending on their original purpose (i.e., FW116 and FW117 are screened at greater depth to measure the regional water table, whereas SG002 and SG012 are shallow-water-table wells designed to monitor the known perched water table). Water-table-elevation measurements were acquired at 15-minute intervals via semipermanent pressure transducers. Precipitation data also were acquired at 15-minute intervals via a rain gauge installed on site, and a secondary rain gauge located at the west end of the Y-12 plant, approximately 1 km away from the research site, is used during a gap in coverage. The water-table-elevation measurements provide information regarding the top of the perched water table and the regional water table, and although they are point measurements, they presumably provide estimates that can be extrapolated laterally to some degree. Additionally, it is important to clarify that the wells are not perfectly coincident with the seismic profiles, because there is some lateral offset and some variation may be expected.

## Seismic data

The tomographic inversion results associated with the October 2007 site-specific error analysis are presented in Figure 4, and the inversion results used to delineate the perched water body from November 2007 are presented in Figure 5. An example of typical data quality and associated first-arrival picks from 2 November 2007 is shown in Figure 6. The corresponding raypaths are presented in Figures 7 and 8, respectively. The inversion results in

Figure 4 depict three separate seismic surveys at the same location on the same day, and the corresponding raypaths are shown in Figure 7. The inversion results in Figure 5 are from data collected during November 2007, and the raypaths associated with these images are presented in Figure 8. In each case, the inversion results are ordered vertically, with the earliest survey located at the top. For both the error-analysis data and the November 2007 data, the raypath and inversion results are related by their component (i.e., the inversion results in Figure 4a correlate to the raypaths shown in Figure 7a). Both sets of data were processed using the WET algorithm (Schuster and Quintus-Bosz, 1993), employing commercially available software (Rayfract, Intelligent Resources Inc.). Data quality varied throughout each profile; however, in general, accurate first-arrival picks were made out to ~48-m shot-receiver offsets. Unfavorable noise conditions (e.g., traffic noise,

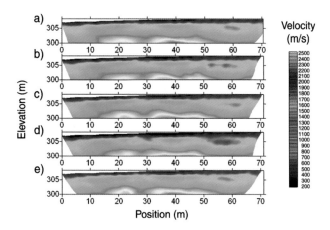

**Figure 5.** Seismic 2D profiles at the ORIFRC site, acquired on different days in November 2007. The lettering corresponds to profiles acquired on (a) 2 November 2007, (b) 7 November 2007, (c) 9 November 2007, (d) 16 November 2007, and (e) 21 November 2007.

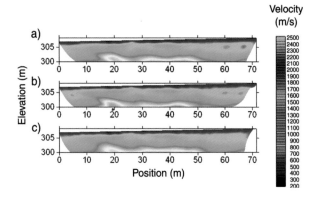

**Figure 4.** Repeated 2D profiles acquired on the same day in October 2007 with no precipitation between acquisitions. The lettering corresponds to (a) initial profile, (b) profile acquired two hours after initial profile, and (c) profile acquired four hours after initial profile.

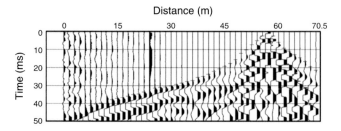

**Figure 6.** Seismic data acquired at the ORIFRC site on 2 November 2007. There are 48 channels spaced at 1.5-m increments, from 0 to 70.5 m. The source was located at 57 m. The seismic data are normalized and clipped to prevent adjacent traces from overlapping.

generators) occasionally resulted in degraded signal-to-noise at the outer offsets, but this principally affects only the resolved depth of the tomograms and not the resolution at the shallowest levels of the tomograms. The frequency range of the data sets is relatively high, with an average peak frequency of ~100 Hz for most shot gathers.

To couple each profile to a baseline, the output velocity tomogram of the inverted baseline data set is used as the initial velocity model for subsequent profiles (see Sarkar et al., 2003). The baseline data set for both the error analysis and the perched-water-body study is the first profile presented in Figures 4 and 5, respectively. Although it is possible to invert traveltime differences, previous studies suggest that assumptions regarding straight rays are required (Day-Lewis et al., 2002), which is not applicable for our experiment. In addition, traveltime differences likely are a function of changes in ray coverage and in velocity, implying that at the near-surface, differences between seismic surveys would not correlate spatially because of changes in raypaths. Thus, the alternative methodology of coupling profiles via an inverted baseline data set is adopted. Care was taken to begin both the time-lapse survey and the error-analysis survey during a lull in precipitation, to avoid abnormal starting conditions (i.e., no precipitation immediately preceding the profile).

## Hydrologic data

The water-level and precipitation data are displayed in Figure 9 and represent the automated acquisition during November 2007. There was a gap in coverage from 17 November through 21 November, and ancillary rain-gauge

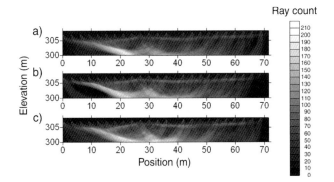

**Figure 7.** Ray tracing for repeated profiles acquired on the same day in October 2007, with no precipitation between acquisitions. The lettering corresponds to: (a) initial profile, (b) profile acquired two hours after the initial profile, and (c) profile acquired four hours after the initial profile.

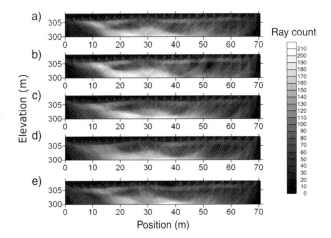

**Figure 8.** Ray tracing for profiles at the ORIFRC site, acquired on different days in November 2007. The lettering corresponds to profiles acquired on: (a) 2 November 2007, (b) 7 November 2007, (c) 9 November 2007, (d) 16 November 2007, and (e) 21 November 2007.

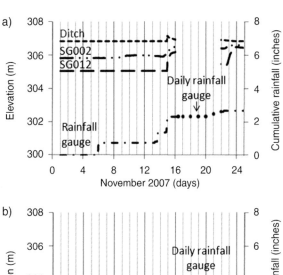

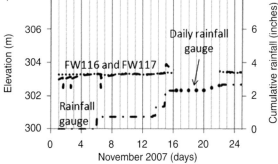

**Figure 9.** Hydrologic data acquired at the ORIFRC during November 2007. The left axis is a point measurement of the elevation of the water table in the wells, and the right axis is cumulative precipitation at the ORIFRC. (a) The shallow wells and ditch water levels, and (b) the regional water table. Precipitation includes both the rainfall gauge at the site and the daily measurements from a separate location during a period in which no data were collected at the S-3 ponds.

data from the aforementioned station (~1 km away) are used to fill this gap. The gap in hydrologic coverage means that the borehole measurements associated with the tomogram on 21 November actually were acquired on 22 November, and the lag between seismic measurement and hydrologic measurement is ~12 hours.

# Discussion

We first address the expected error and resolution of the tomographic inversions. A site-specific error analysis is performed prior to the collection of time-lapse profiles because of the expected difficulty in obtaining accurate velocity estimates for a near-surface, variably saturated perched water table. In addition to the site-specific error analysis, checkerboard resolution tests are performed for the baseline profiles of both the error analysis and the time-lapse study. Following discussion of the resolution and error analysis, we present the trend-analysis approach used to image the perched water table, and then we integrate the hydrologic data with the trend analysis.

## Inversion resolution

Inversion quality can be assessed by numerous different metrics, and we will present several metrics used on the October 2007 data. The typical mean unsigned error between modeled and picked times after processing was <1 ms (~0.3 ms). Maximum error between the observed and modeled traveltimes was ~2 ms, although it is expected that that error would be minimized in an iterative tomographic reconstruction. Ray coverage for the two data sets is variable (see Figures 7 and 8) but provides a general estimate of how well the model is constrained at specific grid points (see Zelt et al., 2006).

An additional technique for estimating tomographic-inversion sensitivity is to use a checkerboard resolution test (Humphreys and Clayton, 1988). A checkerboard test is performed in several steps, the first of which is inversion of the synthetic traveltimes of the tomogram to be tested (Figure 10a). A checkerboard pattern of velocity perturbations (10% of maximum at any grid node) is added to the inversion of the synthetic traveltimes (Figure 10b), and synthetic traveltimes are generated again for the updated model. The updated model is inverted again (Figure 10c), and the residuals between this final inversion and the initial model (the model in 10b) are calculated. These residuals are scaled from −100% to 100%, relative to the initial velocity perturbation, and are plotted for visual inspection (Figure 10d). The results of the checkerboard are a quantitative indicator of model resolution, and maintaining the shape and amplitude of the checkerboard pattern indicates the resolution of the tomogram (see Zhao et al., 1992; Zollo et al., 2002).

The checkerboard resolution test is performed for the initial profile used in the error analysis (October 2007 data; Figure 10) and for the time-lapse study (November 2007 data; Figure 11) to establish the resolution of the model. For the error analysis (Figure 10d), the checkerboard test indicates that the center of the profile is satisfactorily resolved and an expected loss of resolution occurs at either end of the profile. The checkerboard test for the time-lapse data (Figure 11d) shows greater model resolution, although some lateral and vertical smearing is visible in the center of the image. This smearing is presumed to represent the perched water table, because its location in the image is coincident with the presumed location of the perched water body.

## Velocity error analysis

To resolve the incongruity between high rms error and demonstrated seismic sensitivity to near-surface saturated zones, an intermediate interpretation was chosen so that

**Figure 10.** Checkerboard resolution test results for the baseline profile in the error analysis. The lettering corresponds to (a) results of inversion of synthetic data, (b) addition of sinusoidal velocity perturbation, (c) inversion of synthetic traveltimes of perturbed model, and (d) residual after subtracting profile A from profile C following normalization.

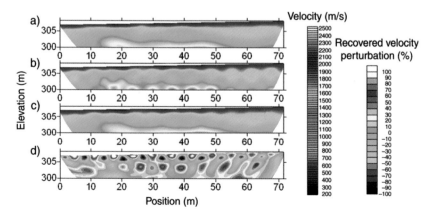

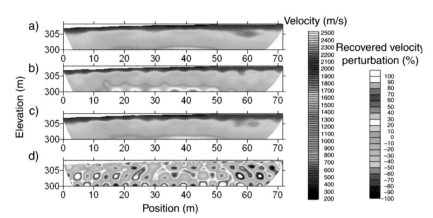

**Figure 11.** Checkerboard resolution test results for the baseline profile in the time-lapse data set. The lettering corresponds to (a) results of inversion of synthetic data, (b) addition of sinusoidal velocity perturbation, (c) inversion of synthetic traveltimes of perturbed model, and (d) residual after subtracting profile A from profile C following normalization.

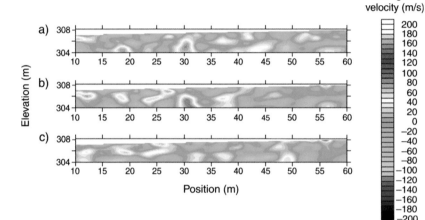
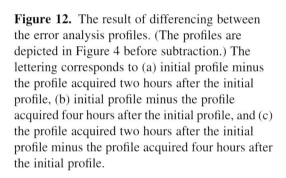

**Figure 12.** The result of differencing between the error analysis profiles. (The profiles are depicted in Figure 4 before subtraction.) The lettering corresponds to (a) initial profile minus the profile acquired two hours after the initial profile, (b) initial profile minus the profile acquired four hours after the initial profile, and (c) the profile acquired two hours after the initial profile minus the profile acquired four hours after the initial profile.

precise velocity estimates were not necessary. In other words, determining whether the velocity estimates were correct relative to the true model was considered unimportant, as long as they were represented consistently. The results of the checkerboard resolution test imply that the model is resolved satisfactorily in the region of interest (i.e., at the suspected location of the perched water bodies), indicating that the inverted model is reliable (see Zhao et al., 1992; Zollo et al., 2002).

To understand how velocity would change between surveys because of acquisition or inconsistent first-arrival time picks, a site-specific empirical analysis of error was performed at the ORIFRC by acquiring three surveys on the same day, with no precipitation, under the assumption that site conditions are static at that time scale. The inverted profiles (Figure 4) show very little variation of velocity in the vadose zone, although some variations are evident at depth. By taking the difference of the profiles, we can determine the quantitative change in measured velocity between surveys (Figure 12). Although profiles B and C are coupled to profile A in Figure 12, all three tomograms are presented

as being equally valid initially, and thus the differences between each are calculated. The differences in the images are dominated by specific anomalous regions with a maximum magnitude of ~83 m/s velocity for profile A (Figure 12a), ~89 m/s for profile B (Figure 12b), and ~108 m/s for profile C (Figure 12c). The rms error is 27 m/s for profile A, 29 m/s for profile B, and 28 m/s for profile C, although Figure 12 illustrates that most of the deviations are localized rather than averaged over the entire profile. Qualitatively, the profiles appear similar in the vadose zone and show some variation at depth. Thus, although there is clearly error in repeatability because of differences in first-arrival picks, acquisition, or processing artifacts, the error is qualitatively minimal in the target area within the vadose zone. It is possible that increased smoothing parameters would decrease the observed differences between profiles; however, the differences are negligible for a qualitative interpretation. As stated, much of the error is the result of specific anomalous regions and is not distributed broadly throughout the tomograms.

## Trend-analysis approach

To distinguish the perched water table visually, we explored the use of data differencing and trend analysis. Assuming an invariant background, the difference between two profiles would produce an image of time-variant changes (see Lumley, 2001), and the corresponding difference between two time-invariant profiles would be a homogeneous image. The observed differences (seen in

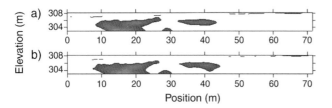

**Figure 13.** Trend-analysis images highlighting positive seismic-velocity zones greater than 50 m/s for the error-analysis profiles. The images are constructed by subtracting the regional velocity model from the profiles shown in Figure 4. (a) Image that corresponds to Figure 4b and is the profile acquired two hours after the initial profile. (b) Image corresponds to Figure 4c and is the profile acquired four hours after the initial profile.

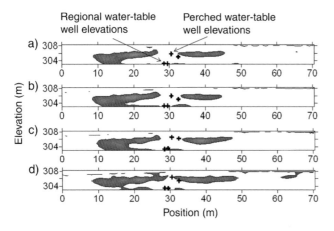

**Figure 14.** Trend-analysis images highlighting positive seismic-velocity zones greater than 50 m/s for the ORIFRC time-lapse profiles. The images are constructed by subtracting the regional velocity model from the profiles shown in Figure 5. Hydrologic data are overlain on the trend-analysis images using information from Figure 9. (a) Image that corresponds to Figure 5b and was acquired on 7 November 2007. (b) Image that corresponds to Figure 5c and was acquired on 9 November 2007. (c) Image that corresponds to Figure 5d and was acquired on 16 November 2007. (d) Image that corresponds to Figure 5e and was acquired on 21 November 2007.

Figure 12) are not homogeneous, thereby indicating the degree of difficulty in replication of surveys and processing. Although these differences are not significant for a qualitative interpretation, they complicated a quantitative, differencing approach.

As an alternative, trend surface analysis was investigated to delimit the spatial extent of the perched water table. Qualitatively, trend surface analysis is the process of separating data into a "regional" component and a "local" component (Davis, 2002). At the ORIFRC, the subsurface is heterogeneous, but the approximation at the near surface is that velocity increases with depth and that the regional component (tens of meters in scale) mirrors this trend. The local component (meter or submeter in scale) is the perched water table, which is an anomalous high-velocity zone above the regional water table. Trend surface analysis is similar to differencing, in that the regional component is subtracted from the image. However, the final product is read more intuitively as an image of the perched water table, compared with subtraction between profiles.

The typical methodology for separating the regional and local components in trend surface analysis is to fit polynomial functions to the data and to inspect the residuals (e.g., Davis, 2002; Evenick et al., 2008). This approach is similar functionally to previous work involving tunnel and near-surface velocity-anomaly detection (e.g., Belfer et al., 1998), but in this case, the background model is derived by using the horizontal average of the Delta-$t$-$V$ inversion (Gebrande and Miller, 1985). The resulting images (Figures 13 and 14) of the positive difference between the regional model and the local component indicate the presence of an anomaly above the regional water table. The trend-analysis images in Figure 13 are derived from the error-analysis data (Figure 4). The initial profile (Figure 4a) is not used here because it is a baseline profile, thus Figure 4b and 4c correlates to Figure 13a and 13b, respectively. The trend-analysis images in Figure 14 are derived from the data acquired in November 2007 (Figure 5). Similarly to the error-analysis data, the initial profile (Figure 5a) is not used for the trend analysis, and Figure 5b, 5c, 5d, and 5e correlates to Figure 14a, 14b, 14c, and 14d, respectively.

The maximum magnitude for the error-analysis value is ~100 m/s, and the expected change in seismic velocity from unsaturated to saturated loosely consolidated soils is significantly higher (i.e., ~700 m/s), on the basis of the petrophysical relationship shown in Figure 3. The difference images are generated by subtracting the regional component from the local component and are mapped showing seismic velocities greater than 50 m/s (Figures 13 and 14). The 50 m/s threshold is chosen to account for error in seismic-velocity measurements and the gradual increase in seismic velocity as a result of the high partial saturations discussed previously. As stated earlier, the maximum error magnitude in the error analysis was ~100 m/s; however, it was believed that this was confined to anomalous zones outside of the target area, and preference was given to maintaining the signal at the expense of possible noise sources.

## Integration of trend-analysis data with hydrologic data

The perched water bodies are interpreted as regions of higher velocity located above the regional water table and are identifiable visibly at elevations of ~305 m along the profile (see Figure 5), at approximately 1 m–2 m below the ground surface. Independent hydrologic point measurements (Figure 14) suggest that this interpretation of the perched water table is reasonable because the two shallow wells indicate an elevated water table at the same depth, ± 0.5 m. The hydrologic point measurements of the perched water table measure the elevation of the top of the water table, which is presumed to have some spatial extent around the well. Thus, the gap in the perched water table in the vicinity of the hydrologic wells, as seen on the tomograms, is not expected. It is possible that the perched water table is thin vertically in this region and that infiltrating water preferentially drains laterally in either direction from the wells. Alternatively, the installation or presence of the hydrologic wells may affect acquisition and perched-water drainage. From direct observation at the site, there is a change in fill characteristics in this region, possibly as a result of initial installation of the boreholes.

Although the perched water table and regional water table are separate phenomena, there are regions in which the tomograms are unable to differentiate between the two. Overlaying the hydrologic data on the trend-analysis images (Figure 14) indicates a separate issue, namely that the velocity anomalies appear to be slightly above the top of the perched water table in many of the images. This disagrees fundamentally with previous findings by Bachrach and Nur (1998), who observed a decrease in velocity as a result of partial saturation of beach sand and the theoretical predictions. A reasonable interpretation, assuming perfect accuracy of the hydrologic data, is that the seismic images are shifted or have a vertical error of approximately ± 1 m. Alternatively, because there are small-scale fluctuations in the measured elevations of the perched water table between the two shallow groundwater wells SG002 and SG012, and SG012 is located at a

greater distance from the profile and ditch (Figure 1), the tomograms could be representing the location of the perched water body accurately at the exact location of the profile. In this case, the wells indicate that the perched water table decreases in vertical extent, away from the S-3 ponds. This interpretation is favored, because SG002 is located directly adjacent to the profile whereas SG012 is offset by a larger distance, and SG002 measurements coincide with the top of the observed anomaly.

Regardless of the actual interpretation, it seems reasonable to conclude, on the basis of the measured elevations of the perched water body and the trend-analysis images, that the perched water body is imaged partially or wholly. Although the existence of the perched water table already was known, previously the spatial distribution of the perched zone was bounded only loosely. Additionally, although it was known that the perched water responded to rainfall events, the extent of that influence was unknown previously. Thus, as an initial interpretation, the trend-analysis images bound the vertical extent of the perched water body, assuming that the measured elevations at SG012 represent either a grading out or a change in elevation of the perched water body. The elevations at SG002 are representative of the perched water body at the profile itself, and SG012 represents the horizontal thinning out of the perched water body. The extent of the perched water body is defined only loosely toward the southern end of the profile (i.e., it extends approximately 0–10 m), because resolution tests indicate a lack of coverage.

The trend-analysis images (e.g., Figure 14) exhibit general similarity in the positions of the anomalies, although the dimensions are shown to change in time, particularly for the anomaly centered at the 40-m position. The current hypothesis is that infiltration and rainfall events cause the perched water table to expand in size laterally and vertically. Subsequent to rainfall events, the perched water table decreases in size and reequilibrates to its nominal configuration. The measured hydrologic data (see Figure 9) tentatively support this hypothesis, because water-table measurements in SG012 and SG002 show an increase in elevation after a rainfall event on 14–15 November and subsequently a gradual return to a lower elevation. The time between acquisition of a survey and the end of the previous rainfall event supports this hypothesis. For the initial two trend-analysis images (profiles B and C, seen in Figure 14a and 14b, respectively), the previous rainfall event (of ~0.74 in [1.88 cm] of precipitation) ended on 5 November. The lag between the times at which the surveys were collected was approximately 22 hours (~1 day) for profile B (Figure 14a), and approximately 70 hours (~3 days) for profile C (Figure 14b). Quantitatively, calculations show that the area of the mapped anomaly centered at position 40 m in Figure 14 decreased from $\sim$15 m$^2$ to 14 m$^2$ between profiles B and C (Figure 14a and 14b, respectively). The trend-analysis images show negligible qualitative differences between these two images, although the anomaly centered at the position of $\sim$40 m exhibits some lateral variation. The anomaly in profile C (Figure 14b) is shown to gradually shrink laterally in comparison with profile B (Figure 14a). Hydrologic information (see Figure 9) indicates no vertical increase in depth to the perched water table after the rainfall event. The third profile (profile D, Figure 14c) was acquired approximately 49 hours (2 days) after a prolonged rainfall event on 14–15 November ($\sim$1.4 in, or 3.56 cm, of precipitation). The perched water table in profile C (Figure 14b) is inflated vertically at the anomaly centered at the position of 40 m, although the anomaly located at 10–28 m exhibits less continuity. Calculations show the areal extent of the mapped anomaly at position 40 m to be $\sim$18 m$^2$. The vertical increase in groundwater elevation in the trend-analysis image is presumed to be real and mirrors the elevated water-table measurements at SG012. This indicates that the perched water table, which normally thins or grades out away from the S-3 ponds, is increasing vertically. Finally, the fourth profile (profile E, Figure 14d) was acquired on the same day as the rainfall event on 21 November. The well information for this profile is from 22 November because no well data were collected during this period. The approximate lag between seismic measurements and hydrologic measurements is 11 hours (one-half day). A separate rain gauge located at the Y-12 plant indicated precipitation during that time, with a precipitation amount listed as $\sim$0.3 in (0.76 cm) of rain. The top of the trend-analysis image correlates with the measured groundwater elevation at SG002, but SG012 again is measured below the anomaly. Quantitatively, the areal extent of the perched water body centered at the 40-m position is 25 m$^2$. In this case, the hydrologic data indicate that the perched water body thins, or grades out, away from the S-3 ponds. The lateral extent of the anomaly is significantly greater here relative to previous profiles and is presumed to indicate the relative timing between precipitation and acquisition.

## Conclusions

We have investigated the application of seismic P-wave time-lapse traveltime tomography (STLTT) for monitoring the presence or absence of perched water bodies in the vadose zone within 4 m of the subsurface. Comprehensive site-specific error analyses indicate that perched water would be within detection limits. The cutoff velocity of our experiment whereby perched water is considered "detected" is 50 m/s, on the basis of both the expected petrophysical relationship (see Figure 3) and subsequent error and resolution analysis specific to the site. On the basis of well measurements that are indicative of the perched water body thinning or pinching out away from the S-3 ponds, the trend-analysis images have submeter accuracy to the measured tops of the perched water bodies. The horizontal resolution is less clearly defined (because of the lack of available well coverage). However, because of the hydrologic complexity and lateral variability at the site, it is not unreasonable to assume that the trend-analysis images (Figure 14) are correct as a first approximation.

Our results have the direct impact that the existing comprehensive hydrologic model for the S-3 ponds region of Y-12 at Oak Ridge National Laboratory — used for evaluating remediation efforts and assessing long-term hazards — will be modified to include heterogeneous infiltration pathways. The hydrologic model for the S-3 ponds area previously represented the surface-water recharge in the vicinity of the secondary-contaminant source as a line source parallel to the edge of the ponds along the drainage ditch, in the region directly adjacent to our 2D seismic profiles. In future iterations of the model, the long line source will be modified and the dominant surface recharge will be characterized instead by several shorter line segments that represent regions of increased infiltration located between the identified perched water bodies. Those regions are based on the correlation lengths mapped out in our time-lapse experiment.

Our investigation indicates that seismic tomography can be used as a high-resolution geophysical tool for monitoring near-surface vadose-zone fluid transport. In this case, perched water bodies within 4 m of the subsurface are identified as high-velocity anomalies because of the expected petrophysical relationship at high levels of saturation ($>$90%). Future research will focus also on identifying time-varying changeability in partially saturated regions through corresponding low-velocity anomalies that are representative of regions having increased but $<$80% saturation, where density changes dominate.

## Acknowledgments

Funding for this study was provided by the U. S. Department of Energy, Biological and Environmental Research Program, as part of the Oak Ridge National Laboratory Integrated Field Research Center project (UT-B 4000059241), and by the Jones/Bibee Endowment at the University of Tennessee.

# References

Bachrach, R., and A. Nur, 1998, High-resolution shallow-seismic experiments in sand, Part I: Water table, fluid flow, and saturation: Geophysics, **63**, 1225–1233.

Baker, G. S., C. Schmeissner, D. W. Steeples, and R. G. Plumb, 1999a, Seismic reflections from depths of less than two meters: Geophysical Research Letters, **26**, 279–282.

Baker, G. S., D. W. Steeples, and C. Schmeissner, 1999b, In-situ, high-frequency P-wave velocity measurements within 1m of the earth's surface: Geophysics, **64**, 323–325.

———, 2002, The effect of seasonal soil-moisture conditions on near-surface seismic reflection data quality: First Break, **20**, no. 1, 35–41.

Baker, G. S., D. W. Steeples, C. Schmeissner, and K. T. Spikes, 2000a, Source-dependent frequency content of ultrashallow seismic reflection data: Bulletin of the Seismological Society of America, **90**, no. 2, 494–499.

———, 2000b, Ultrashallow seismic reflection monitoring of seasonal fluctuations in the water table: Journal of Environmental and Engineering Geoscience, **6**, no. 3, 271–277.

Belfer, I., I. Bruner, S. Keydar, A. Kravtsov, and E. Landa, 1998, Detection of shallow objects using refracted and diffracted seismic waves: Journal of Applied Geophysics, **38**, 155–168.

Birkelo, B. A., D. W. Steeples, R. D. Miller, and M. Sophocleous, 1987, Seismic reflection study of a shallow aquifer during a pumping test: Ground Water, **25**, 703–709.

Chen, J., S. Hubbard, J. Peterson, K. Williams, M. Fienen, P. Jardine, and D. Watson, 2006, Development of a joint hydrogeophysical inversion approach and application to a contaminated fractured aquifer: Water Resources Research, **42**, no. 6, W06425.

Davis, J. C., 2002, Statistics and data analysis in geology: John Wiley & Sons.

Day-Lewis, F. D., J. M. Harris, and S. M. Gorelick, 2002, Time-lapse inversion of crosswell radar data: Geophysics, **67**, 1740–1752.

Domenico, S. N., 1974, Effect of water saturation on seismic reflectivity of sand reservoirs encased in shale: Geophysics, **39**, 759–769.

Evenick, J. C., R. D. Hatcher, and G. S. Baker, 2008, Trend surface residual anomaly mapping and well data may be underutilized combo: Oil & Gas Journal, **106**, no. 4, 35–42.

Garambois, S., P. Senechal, and H. Perroud, 2002, On the use of combined geophysical methods to assess water content and water conductivity of near-surface formations: Journal of Hydrology, **259**, 32–48.

Gebrande, H., H. Miller, 1985, Refraction seismology, *in* F. Bender ed., Applied geosciences II: Ferdinand Enke Publishing House, 226–260.

George, L. A., M. M. Dewoolkar, and D. Znidarcic, 2009, Simultaneous laboratory measurement of acoustic and hydraulic properties of unsaturated soils: Vadose Zone Journal, **8**, no. 3, 633–642.

Hatcher, R. D., P. J. Lemiszki, R. B. Dreier, R. H. Ketelle, R. R. Lee, D. A. Leitzke, W. M. McMaster, J. L. Foreman, and S. Y. Lee, 1992, Status report on the geology of the Oak Ridge Reservation: Oak Ridge National Laboratory Report, ORNL/TM-12074.

Hiltunen, D. R., N. Hudyma, T. P. Quigley, and C. Samakur, 2007, Ground proving three seismic refraction tomography programs: Transportation Research Record, **2016**, 110–120.

Humphreys, E., and R. W. Clayton, 1988, Adaptation of back projection tomography to seismic travel time problems: Journal of Geophysical Research — Solid Earth and Planets, **93**, 1073–1085.

Lankston, R. W., 1989, The seismic refraction method: A viable tool for mapping shallow targets into the 1990s: Geophysics, **54**, 1535–1542.

Lecomte, I., H. Gjoystdal, A. Dahle, and O. C. Pedersen, 2000, Improving modelling and inversion in refraction seismics with a first-order Eikonal solver: Geophysical Prospecting, **48**, 437–454.

Lumley, D. E., 2001, Time-lapse seismic reservoir monitoring: Geophysics, **66**, 50–53.

Mavko, G., C. Chan, and T. Mukerji, 1995, Fluid substitution — Estimating changes in V-P without knowing V-S: Geophysics, **60**, 1750–1755.

Sarkar, S., W. P. Gouveia, and D. H. Johnston, 2003, On the inversion of time-lapse seismic data: 73rd Annual International Meeting, SEG, Expanded Abstracts, 1489–1492.

Schuster, G. T., and A. Quintus-Bosz, 1993, Wavepath eikonal traveltime inversion: Theory: Geophysics, **58**, 1314–1323.

Sheehan, J. R., W. E. Doll, and W. E. Mandell, 2005a, An evaluation of methods and available software for seismic refraction tomography analysis: Journal of Environmental and Engineering Geophysics, **10**, 21–24.

Sheehan, J., W. Doll, D. Watson, and W. Mandell, 2005b, Applications of seismic refraction tomography to karst cavities: US Geological Survey Karst Interest Group Proceedings, 29–38.

Solomon, D. K., G. K. Moore, L. E. Toran, R. B. Dreir, and W. M. McMaster, 1992, Status report: A hydrologic framework for the Oak Ridge Reservation: Oak Ridge National Laboratory Report, ORNL/TM-12026.

U. S. Department of Energy (US DOE), 1997, Report on the remedial investigation of Bear Creek Valley at the

Oak Ridge Y-12 Plant, Oak Ridge, Tennessee: DOE/OR/01-1455/V1&D2.

Watson, D. B., W. E. Doll, T. J. Gamey, J. R. Sheehan, and P. M. Jardine, 2005, Plume and lithologic profiling with surface resistivity and seismic tomography: Ground Water, **43**, 169–177.

Zelt, C. A., A. Azaria, and A. Levander, 2006, 3D seismic refraction traveltime tomography at a groundwater contamination site: Geophysics, **71**, no. 5, H67–H78.

Zhao, D., S. Horiuchi, and A. Hasegawa, 1992, Seismic velocity structure of the crust beneath the Japan Islands: Tectonophysics, **212**, 289–301.

Zollo, A., L. D'Auria, R. De Matteis, A. Herrero, J. Virieux, and P. Gasparini, 2002, Bayesian estimation of 2D P-velocity models from active seismic arrival time data: Imaging of the shallow structure of Mt. Vesuvius (Southern Italy): Geophysical Journal International, **151**, 566–582.

# Chapter 26

# Composite Moveout Correction to a Shallow Mixed Reflection/Refraction GPR Phase

J. F. Hermance[1], R. W. Jacob[2], and R. N. Bohidar[3]

## Abstract

Adapting elements that are common to exploration seismology and electromagnetic wave propagation is extremely useful for interpreting ground-penetrating-radar (GPR) data in a variety of forms, whether the application is environmental, geotechnical, agricultural, or archaeological. In many regions of the world, subsurface conditions are such that material nearest the surface has a low GPR velocity and is underlain by material that has a higher GPR velocity. This condition leads to a critically refracted GPR phase from the higher-velocity interface at depth. Such refracted phases might be particularly troublesome when one is routinely interpreting conventional GPR profiling data as reflections, if in fact the critical distance for refractions is less than the fixed transmitter-receiver offset used for the profile. Misinterpreting such a phase might lead to a subtle, even significant error in depth estimates if only the standard normal-moveout (NMO) correction is applied. However, a complementary wide-angle common-midpoint (CMP) or common-shotpoint (CSP) "calibration" sounding allows identification of such a phase and application of a simple composite moveout correction. The procedure can be illustrated with data from a stratified glacial-drift site in southeastern New England.

## Introduction

The intersection of ground-penetrating radar (GPR) with techniques of exploration seismology is no more ap-

parent than in the acquisition, processing, and interpretation of various modes of GPR refraction data, in which the toolboxes of seismologists have been plundered readily in the last decade, adapting the techniques of delay-time (or time-term) analysis, refraction profiling, and migration (Bohidar and Hermance, 2002; Hermance and Bohidar, 2002a). Historically, the cultural exchange between seismology and electromagnetics has been a two-way street. Now both disciplines routinely use signal-processing techniques with roots in aircraft and ship-ranging radar technology from World War II that led to the upsurge in data processing by the seismic industry of the 1950s. Although we have emphasized previously the importance of these intersecting technologies on GPR refraction methods (Bohidar and Hermance, 2002; Hermance and Bohidar, 2002a; Hermance et al., 2002), we describe in this report a particular type of problem that GPR refractions might create when one routinely interprets conventional GPR profile data.

To have a refracted phase from an interface in the subsurface, there needs to be an increase of velocity with depth, and such cases are well documented in the literature (cf. Annan and Davis, 1976; Arcone, 1984; Delaney et al., 1990; Fisher et al., 1992; Arcone et al., 1993; Arcone et al., 1998; Bohidar, 2001; Bohidar and Hermance, 2002). The water content of subsurface materials often plays a major role in such situations, and even a modest variation of soil water can lead to significant modulations of GPR velocity. One example of subsurface conditions that leads to GPR refractions is a fine-grained topsoil that has high soil water retention (exhibiting a low GPR velocity)

[1]*Brown University, Department of Geological Sciences, Environmental Geophysics/Hydrology Group, Providence, Rhode Island, U.S.A.*

[2]*Bucknell University, Lewisburg, Pennsylvania, U.S.A.*

[3]*Orissa Human Rights Commission, Orissa, India.*

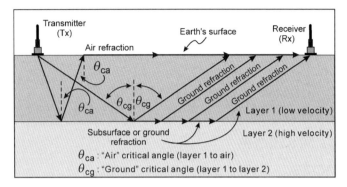

**Figure 1.** Typical raypaths for refractions from a high-velocity layer at depth in the earth and for refractions at the earth-air interface.

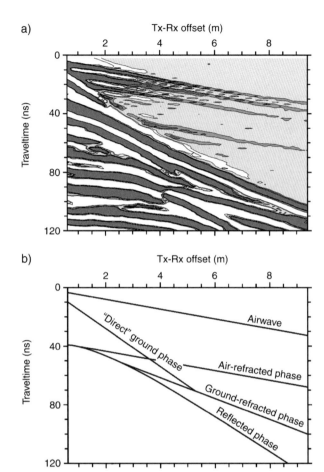

**Figure 2.** (a) Nonlinear variable gain-controlled data from a wide-angle GPR CMP sounding at a field site in south-central Rhode Island. Dark shades indicate positive phases. (b) Inferred traveltimes computed assuming a simple two-layer medium and parameters interpreted from the observed data in part (a).

underlain by a coarse-grained material (or bedrock) with relatively low soil water retention (exhibiting a higher GPR velocity). Another example would be at high latitudes (or in high-altitude mountainous terrains), where one might encounter a wet surface-thaw layer (low GPR velocity) underlain by permafrost or ice-cored material (higher GPR velocity). Even for a relatively uniform material, infiltration from a sudden precipitation event can result in a shallow wet zone underlain by a dryer host material. Thus, it is not uncommon to encounter a variety of stratigraphic settings around the globe where a low-velocity surface layer is underlain by higher-velocity material at depth.

In addition to these ground refractions (see Figure 1), Bohidar (2001) and Bohidar and Hermance (2002) describe the nature and application of air refractions (Figure 2), namely, that phase which is refracted at the earth-air interface because of the extremely high velocity of the atmosphere ($V_{air} = 0.2998$ m/ns) compared to typical values for the earth (generally $V_{earth} \leq 0.2$ m/ns). The interaction of wide-angle primary and multiple reflections with refractions is of increasing interest to those workers who are beginning to unravel the information in waveguide modes (van der Kruk et al., 2009).

## Example Data

The GPR data used here are from a site in south-central Rhode Island, in the northeastern United States, described by Bohidar and Hermance (2002) and by Hermance and Bohidar (2002a). To summarize those reports, the site is located on a kame terrace, a glacial deposit that (in this area) consists of a thick deposit of stratified sand and gravel underlying a fine-textured layer of topsoil. Nearby wells suggest that local bedrock is 25 to 30 m deep and is

beyond a depth that might influence our GPR data. The GPR velocity of the surface layer at our location is consistent with a partially saturated, fine-grained silt recovered in several shallow hand cores at the site, whereas the higher velocity at depth is consistent with a partially saturated, coarser-grained gravel seen in nearby excavations.

The particular problem we wish to address here is illustrated by the sequence of examples beginning with Figure 3, which shows a conventional, fixed transmitter-receiver (Tx-Rx) offset GPR reflection profile, plotted as usual in terms of traveltime versus distance (red denotes positive phases, and blue denotes negative phases). No gain or filtering has been applied to the data in Figure 3. The suppressed amplitude of the direct airwave is a consequence of the finite (2-m) Tx-Rx offset and the relatively low velocity of the surface layer, which tends to lead to a sharper roll-off of airwave amplitude with distance.

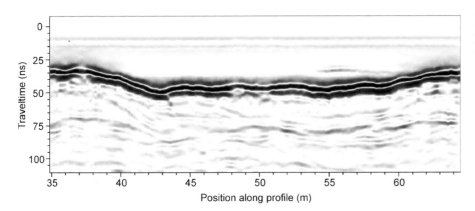

**Figure 3.** Conventional GPR profile plotted in terms of traveltime versus distance. The color code for this and other figures is that red denotes positive phases and blue denotes negative phases.

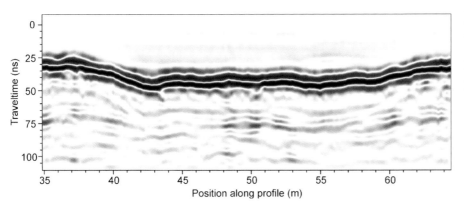

**Figure 4.** The data in Figure 3 after crosscorrelation with a native wavelet constructed according to the procedure of Hermance and Bohidar (2002b). The principal positive phase is the peak in the lagged correlation function associated with the arrival time of the original event.

These data were collected in a routine step-and-stop mode at an interval spacing of 0.2 m. For purposes of the present discussion, the data did not require corrections for instrument drift and timing effects as described by Jacob and Hermance (2004, 2005).

Both profile and common-midpoint (CMP) data were acquired as 32 field stacks per trace, using Systems & Software PulseEkko 200-MHz bistatic antennae. Our original CMP gather starts at a minimum Tx-Rx offset of 0.6 m and extends to a maximum Tx-Rx offset of 15 m, although it is clipped in Figure 2 to a smaller range. Whereas for the purposes of illustration the dynamic range of the CMP data in Figure 2 is enhanced significantly using autoscaling amplitude-dependent nonlinear gains, the profile data in Figure 3 are shown as recorded and with no gains applied.

Profile data in Figure 3 are relatively clean — interfering phases are minimal between zero time and the arrival of the reflector/refractor phase at 30 to 45 ns — and the interface we were attempting to delineate is indicated clearly in Figure 3. To better pick the arrival times of respective phases in the profile data, each trace in Figure 3 is crosscorrelated with an appropriate native wavelet (e.g., Hermance and Bohidar, 2002b), with the result shown in Figure 4. The principal positive phase in

Figure 4 is the peak in the lagged correlation function associated with the estimated arrival time of the principal event.

Complications developed during the next phase of our interpretation, shown in Figure 5, after we applied a conventional normal-moveout (NMO) correction to the crosscorrelation functions in Figure 4. The NMO correction uses the same data as Figure 4 and applies a simple NMO velocity of $V_1 = 0.076$ m/ns, allowing for the finite transmitter-receiver (Tx-Rx) offset used for the profiling (Tx-Rx = 2 m). Note the extreme waveform distortions at either end of the profile in Figure 5. This underscores the importance of problems associated with mixed reflection/refraction phases. For the NMO correction in Figure 5, we have assumed we are dealing with reflections only, but as we shall see, in fact, we also are dealing with refractions. Hence, a typical NMO correction (which assumes a *reflected* phase) is the wrong correction to apply to compensate for the finite Tx-Rx offset.

The object lesson here is that for many shallow investigations of interest to soil scientists, environmental engineers, and hydrologists, it is not infrequent for GPR Tx-Rx offsets to be on the order of (or greater than) the critical distance for the respective refraction. Thus, for Tx-Rx

**Figure 5.** Same data as Figure 4 after applying a simple NMO correction (NMO velocity = 0.076 m/ns). Note the extreme waveform distortions at either end of the profile.

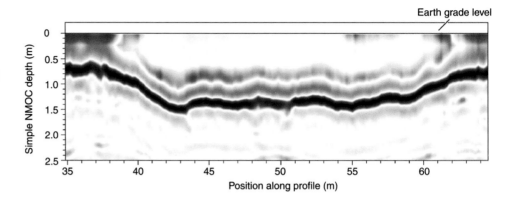

**Figure 6.** Comparison of automatic time picks of the uncorrected traveltimes for the maximum value of the crosscorrelated traces of phase X in Figure 4 to the traveltime of the wave transmitted directly to the receiver at a Tx-Rx offset of 2 m. For $V_1 = 0.076$ m/ns, T(direct) = 26.3 ns.

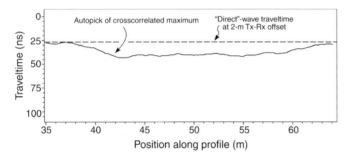

offsets beyond the critical distance, the principal and strongest phase in a conventional GPR profile might be a refraction, not a reflection. The reason for the aberration of the waveforms in Figure 5 is illustrated graphically in Figure 6, which compares the time picks of uncorrected traveltimes for the crosscorrelated traces of phase X in Figure 4 to the traveltime of the wave directly transmitted to the receiver; for $V_1 = 0.076$ m/ns, at a Tx-Rx offset of 2 m, $T_{direct} = 26.3$ ns. Figure 6 implies that the significant distortion at either end of the profile of the NMOC waveforms in Figure 5 occurs because the NMOC procedure geometrically confuses phase X with the direct mode. In fact, a conventional NMOC analysis at early traveltimes might lead to the paradox that what is inferred to be the "reflected" phase arrives *before* the direct ground phase, which is a physical impossibility for a uniform first layer. However, we should not be surprised at all to find that our phase X arrives prior to the direct Tx-Rx ground phase, but *not* having traveled as a simple reflected phase. It arrives before the direct phase because it has traveled for a portion of its travel path at the higher velocity of the refractor. In addition to the type of distortion of the waveforms shown in Figure 5, by misidentifying such a phase and applying the wrong

correction, subtle, even substantial errors might be introduced into the depth estimates that are not so readily apparent.

## Analysis of Phase X — A Mixed Reflection/Refraction Phase

From seismology (and physical optics, among other fields), we know that for a wave traveling from a low-velocity medium to a high-velocity medium, a critical refraction occurs when the angle of incidence of the signal is at the critical angle $\theta_c$, where $\theta_c = \sin^{-1}(V_1/V_2)$. For a refraction from an interface in the subsurface at depth $d$, the critically refracted ray will be returned to the surface at the critical distance $X_c = 2d \tan\theta_c$. We should note that for the purposes of the present discussion, all depths are relative to local grade or local elevation of the earth's surface.

The latter relations can be combined to determine the traveltime for the critically refracted phase to reach the critical distance

$$T_{X_c} = \frac{X_c V_2}{V_1^2}. \tag{1}$$

The essential point is that if we substitute the fixed transmitter-receiver offset ($X_{TxRx}$) used in conventional profiling for the critical distance $X_c$ in equation 1, we can ascertain the metric for determining which traveltime relation to use. For a particular fixed transmitter-receiver offset ($X_{TxRx}$), the critical traveltime for the transition of phase X from a reflected phase to a refracted phase is given by

$$T_X = \frac{X_{TxRx} V_2}{V_1^2}. \tag{2}$$

If the observed traveltime $T_{obs}$ is greater than $T_X$, then the phase is a reflection, and we should correct for moveout, assuming phase X is a reflected phase that has traveltime

$$T_{refl} = \frac{1}{V_1} \sqrt{X_{TxRx}{}^2 + (2d)^2}, \qquad (3)$$

leading to a NMOC depth estimate of

$$d_{NMO} = \frac{1}{2} \sqrt{(V_1 T_{refl})^2 - (X_{TxRx})^2}. \qquad (4)$$

Alternatively, if $T_{obs} < T_X$, the phase is a refraction, and we should correct for a linear moveout (LMO) in time following the traveltime relation

$$T_{refr} = \frac{X_{TxRx}}{V_2} + \frac{2d \cos \theta_c}{V_1}. \qquad (5)$$

In this case, equation 5 can be rearranged to provide a LMOC depth estimate of

$$d_{LMO} = \frac{V_1 T_{refr}}{2 \cos \theta_c} - \frac{V_1 X_{TxRx}}{2 V_2 \cos \theta_c}. \qquad (6)$$

The procedure that we refer to here as the composite moveout correction (MOC) is simply a lookup computer operation whereby, for a specific combination of $X_{TxRx}$, $V_1$, and $V_2$, the critical transition time $T_X$ is computed from equation 2. The observed sampling time $T_{obs}$ for each trace then is compared to $T_X$, and the appropriate composite MOC (either equation 4 or 6) is applied to the corresponding data value to infer the appropriate depth. Whereas the time base usually is sampled at uniform time increments, the corresponding depths from equation 4 or equation 6 will not be at uniform intervals. A simple interpolation to a uniform grid might be advisable when plotting MOC data in the vertical section as in Figure 7.

# Results of the Composite MOC

Figure 7 shows the resulting interpreted section based on a composite MOC assuming a surface velocity of $V_1 = 0.076$ m/ns and a refractor velocity of $V_2 = 0.135$ m/ns, determined from an independent wide-angle CMP. Obviously, selecting the appropriate MOC is possible only if the velocities of the surface layer and the refractor are known or can be estimated reasonably. In our case, velocities selected were provided by the velocity scans shown in Figure 8. Traces at various CMP offsets were segregated

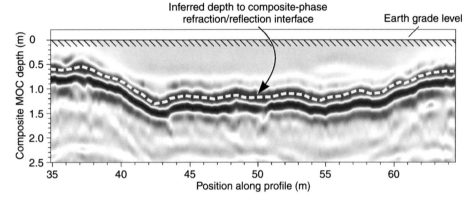

**Figure 7.** Interpreted section based on the composite moveout correction (MOC) described in the text. The picked depth to the principal interface (yellow dashes) is based on the local positive maximum of the respective MOC lagged-crosscorrelation function.

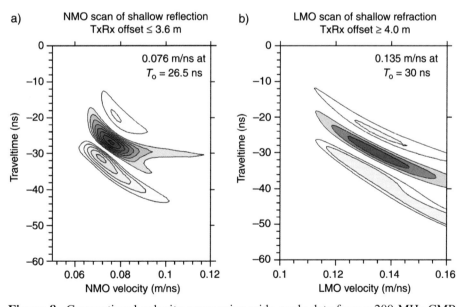

**Figure 8.** Conventional velocity scans using wide-angle data from a 200-MHz CMP station located at a position of 50 m along the profile. After a visual inspection, CMP traces were clipped and muted into two offset ranges such that (a) conventional hyperbolic NMO corrections were applied before stacking for Tx-Rx offsets ≤3.6 m, whereas (b) linear-moveout (LMO) corrections were applied to the refractions at Tx-Rx offsets ≥4 m.

into two groups, based on whether the dominant phase was a reflection (Tx-Rx $\leq$ 3.6 m, and an NMO velocity scan was performed) or a refraction (Tx-Rx $\geq$ 4.0 m, and an LMO velocity scan was performed). The picked depth to the principal interface in Figure 7 is based on a procedure for autopicking the local positive maximum of the cross-correlation function (Hermance and Bohidar, 2002b) — a standard seismic procedure.

## Conclusions

Although GPR profiling is a common tool used routinely by geophysical contractors, hydrologists, geotechnical engineers, and geophysical archaeologists, many of these profiles are interpreted without site-specific information on local field parameters, such as representative GPR velocities. Few serious and experienced practitioners would question the need for such supporting data, and this report underscores a particular set of issues arising if GPR ground refractions are present. Of course, ground refractions will be present only when there is a velocity increase with depth, and the particular complications from the mixed reflected/refracted phase discussed here is of concern only when the refractor is at a depth on the order of the Tx-Rx separation or less (or more precisely, when $T_{obs} < T_X$).

However, for the case where the refractor is shallow, we have shown that the phase one might typically infer to be a primary reflection should be qualified as a mixed reflection/refraction (referred to here as a phase X). This is because the finite Tx-Rx offset used could be beyond the local critical distance for shallow interfaces. Hence, the first arrival actually might be a refracted phase, not the conventional reflected phase usually inferred by casual interpreters. Caution is urged in such situations, and a strong case can be made for performing wide-angle CMP or CSP calibration runs at any site prior to (or following) conventional GPR profiling. If one identifies such a phase X on a profile, it is a straightforward matter to apply a composite MOC to obtain a more physically valid image of the subsurface.

## References

Annan, A. P., and J. L. Davis, 1976, Impulse radar sounding in permafrost: Radio Science, **11**, 383–394.

Arcone, S. A., 1984, Radar detection of ground water, *in* E. A. Dardeau Jr., ed., Proceedings of the ground-water detection workshop held at Vicksburg, Mississippi: NTIS No. AD A154 053, 68–76.

Arcone, S. A., D. E. Lawson, and A. J. Delaney, 1993, Radar reflection and refraction profiles of seasonal thaw over permafrost in Fairbanks, Alaska: Proceedings of the 2nd Government Workshop on GPR, 241–256.

Arcone, S. A., D. E. Lawson, A. J. Delaney, J. C. Strasser, and J. D. Strasser, 1998, Ground-penetrating radar reflection profiling of groundwater and bedrock in an area of discontinuous permafrost: Geophysics, **63**, 1573–1584.

Bohidar, R. N., 2001, Geophysical tools in hydrogeological studies of shallow unconfined aquifers: Gravity for reconnaissance survey and ground-penetrating radar for high-resolution applications: Ph.D. dissertation, Brown University.

Bohidar, R. N., and J. F. Hermance, 2002, The GPR refraction method: Geophysics, **67**, 1474–1485.

Delaney, A. J., S. A. Arcone, and E. F. Chacho Jr., 1990, Winter short-pulse radar studies on the Tanana River, Alaska: Arctic, **43**, 244–250.

Fisher, E., G. A. McMechan, and A. P. Annan, 1992, Acquisition and processing of wide-aperture ground-penetrating radar data: Geophysics, **57**, 495–504.

Hermance, J. F., and R. N. Bohidar, 2002a, Introduction to the GPR refraction method: 2002 Symposium on the Application of Geophysics to Environmental and Engineering Problems, EEGS, Proceedings.

Hermance, J. F., and R. N. Bohidar, 2002b, Better time picks = better traveltimes = better velocities: Progress in developing public domain software: 9th Symposium on Ground Penetrating Radar, UCSB, Expanded Abstracts, 262–267.

Hermance, J. F., R. N. Bohidar, and R. W. Jacob, 2002, Facilitating the interpretation of GPR refraction data for stratigraphic analysis: 72nd Annual International Meeting, SEG, Expanded Abstracts, 1424–1427.

Jacob, R. W., and J. F. Hermance, 2004, Assessing the precision of GPR velocity and vertical two-way travel time estimates: Journal of Environmental and Engineering Geophysics, **9**, 143–153.

———, 2005, Random and non-random uncertainties in precision GPR measurements: Identifying and compensating for instrument drift: Subsurface Sensing Technologies and Applications Journal, **6**, 59–71.

van der Kruk, J., H. Vereecken, and R. W. Jacob, 2009, Identifying dispersive GPR signals and inverting for surface wave-guide properties: The Leading Edge, **28**, 1234–1239.

Chapter 27

# Improving Fractured-rock Characterization Using Time-frequency Analysis of GPR Data Sets

Mehrez Elwaseif[1], Lee Slater[1], Mamdouh Soliman[2], and Hany Salah[2]

## Abstract

Resolution of the ground-penetrating-radar (GPR) method largely depends on the wavelength of the propagated electromagnetic waves, the depth, and the dimensions of the investigated targets. Locating small-scale targets such as fractures can be difficult, especially when the size of the fractures themselves and the spacing between them are smaller than the wavelength of the EM waves. In such cases, fractures will not be resolved individually in the time domain. Examining the frequency (spectral) content of GPR time series might provide additional information that can improve fracture location relative to the results from conventional examination of the time series alone. The S-transform, a powerful time-frequency analysis tool, can be applied to GPR time series extracted from synthetic data over 1D and 2D models of fractured limestone as well as to a data set collected at a fractured limestone site in Egypt. In synthetic scenarios, the value of the S-transform analysis can be investigated under two conditions: (1) discrete fractures spaced at distances greater than one-quarter wavelength and (2) closely spaced fractures separated by distances less than one-quarter wavelength. The synthetic studies include analysis of the dependence of S-transform results on the dielectric properties of a fracture, by simulating both air- and water-filled fractures. Time-frequency analysis can aid in determining the location of discrete, closely spaced fractures when fractures are separated by distances greater than one-quarter wavelength. However, the clarity of fracture expressions from the S-transform depends on the dielectric contrast between the fracture and the host material, so air-filled fractures are not always identified. In the case of fractures smaller than one-quarter wavelength, boundaries where fracture spacing (density) changes abruptly do not result in diagnostic shifts in the frequency content of the S-transform along the time series. However, diagnostic frequency shifts do occur when the dielectric properties of closely spaced fractures change between air-filled and water-filled fractures, with water-filled fractures displaying a higher-frequency content. This frequency-spectrum dependence on dielectric properties of fractures can be used to infer locations of water-filled fractures. In addition, this dependence can help to locate possible localized moisture transport between fracture zones, via capillary effects, at a field site.

## Introduction

Ground-penetrating radar (GPR) is employed routinely to obtain an image of subsurface structures and objects that have a detectable dielectric-permittivity ($\varepsilon_r$) contrast relative to the host medium. The success of the method in imaging targets, assuming the applicability of the low-loss assumption, is based on the frequency of the antennae used, physical properties of subsurface materials, and the

[1]*Department of Earth and Environmental Sciences, Rutgers, the State University of New Jersey, Newark, New Jersey, U. S. A. E-mail: Mehrez@andromeda.rutgers.edu; lslater@andromeda.rutgers.edu.*

[2]*National Research Institute of Astronomy and Geophysics, Cairo, Egypt. E-mail: amsoliman2001@yahoo.com; hmesbah2000 @hotmail.com.*

depth and dimensions of the investigated targets (e.g., Annan, 2006; Franseen et al., 2007). The minimum vertical resolution ($r_{min}$) of GPR commonly is approximated at one-quarter wavelength of the transmitted signal ($z_{min}$) and is a function of both antenna frequency and $\varepsilon_r$ of subsurface materials (e.g., Kallweit and Wood, 1982; Zeng, 2009), whereas the minimum horizontal resolution is a function of target characteristics depth and wavelength of the transmitted signal (e.g., Daniels, 1996). When the spacing between consecutive targets is below $z_{min}$, electromagnetic waves are scattered significantly. Furthermore, the reflected energy from individual fractures stacks together so that those targets will not be resolved individually as discrete reflected pulses in the time series.

Studying the spectral content (i.e., changes of amplitude, phase, and frequency content of a signal with time) of the GPR time series potentially can help in characterizing targets separated by distances smaller than $z_{min}$. Such an approach has been applied extensively to seismic data sets (e.g., McIvor, 1964; Dilay and Eastwood, 1995). Guha et al. (2005) first applied the fast-Fourier-transform (FFT) analysis to GPR traces collected over sedimentary deposits, in an effort to locate thin beds. They identified shifts toward higher frequencies in the peak amplitude of the frequency spectra (relative to the frequency content of the input signal) along the time series. They noted that this shift is the opposite of the typical shift toward lower frequencies that results from preferential attenuation of the higher-frequency components of the input signal with time. They suggested that this was the result of a transition to thinner layers (i.e., to more closely spaced dielectric boundaries) within the deposits. However, the FFT is applicable only to stationary signals, thereby providing a time-averaged amplitude response that does not capture changes of frequency content of a signal with time (Stockwell et al., 1996; Vatansevera and Ozdemir, 2008). To address those limitations, Guha et al. (2008) extended their previous study by applying the S-transform technique (Stockwell et al., 1996) to study variations of frequency content of a GPR time series with time for synthetic models of thin layers. They showed that the frequency content of a GPR time series can shift toward higher frequencies when layering in sedimentary deposits changes toward thinner (more closely spaced) layers. Guha et al. (2008) therefore concluded that the S-transform potentially is a powerful tool for characterizing sedimentary systems when layer thickness is less than $z_{min}$.

Fractured rock often is characterized by densely spaced fractures that are separated by distances shorter than $z_{min}$, which suggests that time-frequency analysis could improve the interpretation of GPR data sets collected over fractured rock. Fracture characterization often is required to assist the development of flow and transport models to support groundwater and oil-reservoir exploration. The GPR method therefore has been applied extensively to characterize fracture density (e.g., Seol et al., 2004; Liu, 2006), orientations (e.g., Seol et al., 2001; Mejia and Young, 2007) and fluid-filling properties (e.g., Olsson et al., 1992; Tsoflias and Becker, 2008). Seol et al. (2004) integrated the GPR results from surface, borehole-reflection, and tomography data to characterize fracture density within granite bedrock. Liu (2006) performed forward modeling of borehole radar data for a single-fracture model and multiple-fracture-network models to evaluate the possibility of identifying fracture locations, orientations, and pore-fluid content in field studies. They considered air-filled, freshwater-filled, and saline-water-filled fracture scenarios. They found that the reflection amplitude significantly increases when freshwater replaces air in fractures, thereby showing that the characterization of fractured rock using GPR depends on the dielectric properties of the fracture.

Motivated by the work of Guha et al. (2008) and its potential relevance to fracture characterization, we apply the S-transform to GPR time series extracted from synthetic data over 1D and 2D models of fractured limestone, as well as to a data set collected at a fractured limestone site in Egypt for which the problem is to distinguish water-filled from air-filled closely spaced fractures. In the synthetic scenarios, we investigate the value of the S-transform analysis under two conditions: (1) discrete fractures spaced at distances greater than one-quarter wavelength, and (2) closely spaced fractures spaced at distances less than one-quarter wavelength. By simulating both air- and water-filled fractures, the synthetic studies include an analysis of the dependence of S-transform results on the dielectric properties of a fracture.

## The S-transform Method

Spectral analysis is a powerful tool for enhancing interpretation of geophysical data sets (e.g., de Voogd and den Rooijen, 1983; Sinha et al., 2009). The Fourier transform has been applied extensively to geophysical data (e.g., Denker and Wenzel, 1983; Forsberg, 1985). The drawback of this technique is that it only provides time-averaged amplitude and does not provide information about changes in frequency content with time. That shortcoming often makes the method unsuitable for analyzing geophysical time series when we expect the spectral content of a signal to change as the signal travels through earth materials that have different physical properties (e.g., Stockwell et al., 1996; Tsoulis, 2003).

The S-transform is a powerful technique for predicting the spectral content of a signal in both time and frequency domains, where it uniquely combines a frequency-dependent resolution of the time-frequency space with absolutely referenced local phase information (Stockwell, 1999). Stockwell (1999) derives the S-transform from the short-time Fourier transform (STFT) as follows. The time-average spectrum $H(f)$ for a given time series $h(t)$ is given by

$$H(f) = \int_{-\infty}^{\infty} h(t)g(t)e^{-i2\pi ft}\, dt, \qquad (1)$$

where $f$ is the frequency of different periods within the signal, $t$ is time, and $g(t)$ is a Gaussian function used to window the time series $h(t)$.

The S-transform uses a Gaussian function with a width that depends on the local frequency under investigation, the dilation $\tau$, and translation of the wavelet. The Gaussian function $g(t)$ is given by

$$g(t) = \frac{f}{\sqrt{2\pi}} e^{\frac{(\tau-t)^2 f^2}{2}}. \qquad (2)$$

Following equations 1 and 2, the S-transform is defined as

$$S(\tau, f_1) = \int_{-\infty}^{\infty} h(t)\frac{f}{\sqrt{2\pi}} e^{\frac{(\tau-t)^2 f^2}{2}} e^{-i2\pi ft} dt \qquad (3)$$

As is shown in equation 3, $S(\tau, f_1)$ is a 1D vector that contains the changes of amplitude and phase over $h(t)$ for a particular frequency $f_1$. This equation is applied to all frequencies within the signal. The net result of the S-transform is a complex matrix; the rows of that matrix are different frequencies, and the columns are the time values.

We also perform a 2D S-transform analysis by simultaneously running the 1D S-transform analysis at a specific frequency on all individual traces of a radargram (to see how the amplitude for a specific frequency within the frequency spectrum changes with time on each individual trace) and by running the transform at the same arrival time in the horizontal direction across traces (to see how the amplitude at a specific frequency changes with distance). The 2D S-transform section is produced by averaging the observed results of two successive traces at each time step (Mansinha et al., 1997) and performing that across all traces. This 2D S-transform depicts the amplitude of the signal at a specific frequency across the radargram. A detailed description of the 2D S-transform analysis is given in Stockwell (1999).

# Synthetic Studies

## Numerical approach

We simulate 2D wave propagation for fractured limestone models using the finite-difference time-domain (FDTD) algorithm implemented in ReflexW (Sandmeier, 2007). We simulate limestone composed of air- and water-filled fractures to represent the basic field setting we describe later. In all models, the limestone is assigned an $\varepsilon_r$ value of 7, whereas fluid-filled fractures are assigned an $\varepsilon_r$ of 81 and air-filled fractures are assigned an $\varepsilon_r$ of 1. The model space is 6 m wide × 4 m thick. This model domain was considered to provide sufficient model space for modeling the complex fracture systems considered in the 2D scenario.

The central antenna frequency used for all the simulations was 200 MHz, resulting in a vertical resolution of 0.14 m (one-quarter of the dominant wavelength) in the limestone. We recognize that the received bandwidth actually will determine the true vertical resolution, but we make this simplification in the modeling effort. The total window length used in each simulation was varied to capture only the response from the area of interest. The trace increment and time increment, 0.01 m and 0.017 m, respectively, were the same for all models.

## Synthetic model A (1D fractures)

The first model (model A) is a 1D model of fractures, with differing fill material (air versus water), spaced at different distances: (1) nearly equal to $z_{min}$ and (2) below $z_{min}$. Thus the modeling is similar to that performed by Guha et al. (2008), except that here we simulate 2D wave propagation and vary the $\varepsilon_r$ contrasts as well as the spacing between dielectric boundaries. This model was selected to examine the potential of the S-transform approach for identifying changes in fracture spacing (i.e., fracture density) on the basis of shifts in the frequency spectrum with time, and for determining the dependence of such shifts on the $\varepsilon_r$ of the fractures.

Model A consists of four separate model scenarios (models A1 through A4), in which fill content and fracture spacing were varied as shown in Figure 1a. Each submodel consists of two sets of fractures: a set with vertical spacing of 0.19 m (slightly greater than $z_{min} = 0.14$ m) and a set of closely spaced fractures with a spacing of 0.04 m (much smaller than $z_{min}$). The length and width of each individual fracture is 6 m and 0.01 m, respectively. In the first three models the fill content of fractures within each set is identical, and the specific models examined are air-air (Figure 1a, model 1), water-water

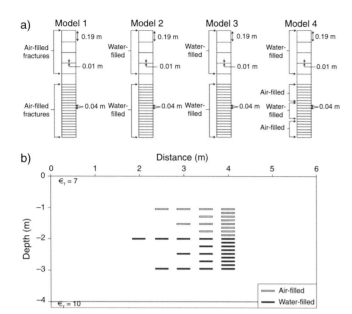

**Figure 1.** Subsurface fractured limestone models, with different fill material and spaced at different distances: (a) 1D fracture models, and (b) 2D fracture models.

(Figure 1a, model 2), and air-water (Figure 1a, model 3). The final model includes a transition from air to water to air within the closely spaced fracture set (Figure 1a, model 4).

## Synthetic model B (2D fractures)

The second model (model B) consists of a 2D fracture model, with fractures spaced at distances both considerably greater than $z_{min}$ (first four fracture sets) and less than $z_{min}$ (fifth set) (Figure 1b). This model was selected to simulate more complexity, as might be expected at real field sites, and to demonstrate how the S-transform analysis potentially can improve GPR interpretation with respect to fracture locations and fill content, even when fractures are spaced at distances greater than $z_{min}$ and therefore are directly resolvable in the time series.

The selected model consists of five horizontal fracture sets with variable spacings. All individual fractures have a length of 0.25 m and a width of 0.01 m. The horizontal spacing between the centers of these five fracture sets is 0.5 m. To minimize model boundary effects, fractures were modeled only toward the center of the model domain. To investigate the importance of fracture fill further, fractures below 2 m in depth were assumed to be filled with water whereas the shallower fractures were assumed to be filled with air. Fracture density and the vertical spacing between fractures varied as shown in Figure 1b.

## Processing of the time series

To make the synthetic time series easily interpretable and amenable to further S-transform processing without significantly modifying the spectral content of traces, we applied two preprocessing steps aimed at reducing noise and enhancing the frequency content of the signal. We first detrended the data to smooth high-frequency signals at the beginning of a time series. This was achieved by applying least-squares fitting of the time series to a parabola and subtracting that fit from the time series. A dewow filter was applied subsequently to remove low-frequency signals.

# Results of Time-frequency Analysis

## Time-series analysis of 1D fracture models

The resulting time series of all model A (1D) scenarios (Figures 2a and 2d and 3a and 3d) exhibit distinct reflections from the widely spaced fractures. Reflections from the individual closely spaced fractures are unresolved in the case of the air-filled fractures, but those from the water-filled fractures appear partly resolved. The better resolution of the water-filled fractures is expected, given the fact that the wavelength will be shortened by the presence of the low-velocity water in the fractures.

## S-transform analysis of 1D fracture models

The results of applying the S-transform on traces from the center of the four 1D fracture models are presented in Figures 2 and 3 (parts b and c and parts e and f). Figure 2b shows the results of S-transform analysis for model A1 (air-air-filled fractures). The frequency content of the radar signal throughout the time series shows little variation, including at the transition between the widely and closely spaced fracture sets at 16 ns. This is highlighted in Figure 2c, which plots the location (in frequency) of the maximum amplitude in the spectrum as a function of time. Anomalously high amplitudes are plotted as red (highest-amplitude) and green (high-amplitude) circles. Green anomalies occur at frequencies similar to those of the anomalies arising at the transition between the closely spaced fractures and unfractured limestone and likely result from the fractures. In contrast, red anomalies show shifts to lower-frequency content and likely represent artifacts in the processing — that is, from modeling errors, scattering losses, and so forth (Pinnegar, 2003).

Although we believe that individual high-amplitude anomalies (green circles in Figure 2c) result from individual fractures, these peaks in the spectrum do not occur at the exact times we would expect for the individual fractures on the basis of the model with known velocity. This is likely the result of contamination of the S-transform frequency spectrum by noise. Such a conclusion is supported by the work of Pinnegar (2003), who applied the S-transform on seismology time series in an effort to identify the first arrivals of P- and S-waves. Pinnegar observed a shift in the predicted locations from the S-transform and related

that shift to noise in the data. Although specific anomalies cannot be associated with specific fractures, we note that the number of amplitude anomalies associated with closely spaced fractures exceeds the number of anomalies associated with widely spaced fractures. Further research on relating such amplitude anomalies in the S-transform spectrum to distinct dielectric contrasts in models is needed but is beyond the scope of this paper.

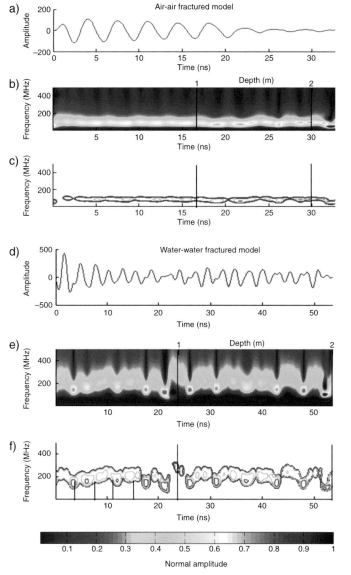

**Figure 2.** S-transform analyses for the air-air and water-water fracture models: (a) and (d) time-series data, (b) and (e) S-transform section, (c) and (f) variations of the frequency content at the maximum amplitudes, with time. The frequency content at the transition between the widely spaced and the closely spaced fracture sets does not change for the air-air model, whereas it is associated with a high-frequency peak for the water-water model (probably as a result of the significant $\varepsilon_r$ contrast between water and limestone).

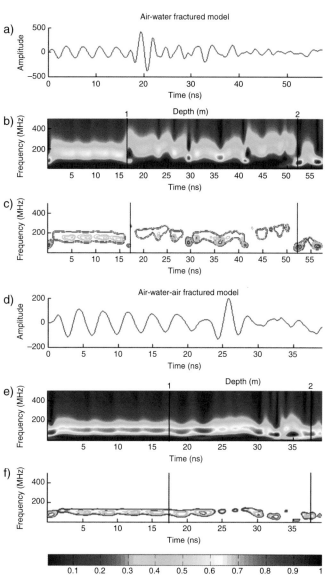

**Figure 3.** S-transform analyses for the air-water and air-water-air fracture models: (a) and (d) time-series data, (b) and (e) S-transform section, (c) and (f) variations of the frequency content at the maximum amplitudes, with time. The transition between air-filled and water-filled fractures shows a strong local increase in frequency content (note that there is no frequency shift at the transition between widely spaced and closely spaced air sets for the air-water-air model).

Figure 2e shows the S-transform analysis for the water-water model. The frequency content at the transition between the widely and closely spaced fracture sets, at approximately 23 ns, is associated with a high-frequency peak probably caused by the significant $\varepsilon_r$ contrast between water and limestone (Benedetto, 2010). Figure 2f is a plot of the variations of frequency content of the maximum-amplitude anomalies. The identified anomalies in the widely spaced fracture set are consistent with the actual location of fractures in this set.

Figure 3b shows the results of S-transform analysis for the air-water model. On the basis of the estimated 1D

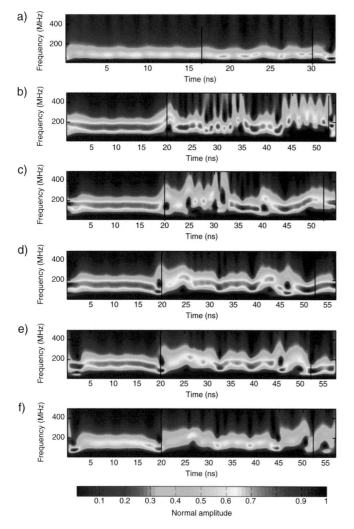

**Figure 4.** S-transform analyses for models of widely to closely spaced fractures, with a varying $\varepsilon_r$ of the closely spaced fractures (widely spaced fractures have a permittivity value of 1 for all models): (a) $\varepsilon_r$ of fractures = 1, (b) $\varepsilon_r$ of fractures = 10, (c) $\varepsilon_r$ of fractures = 20, (d) $\varepsilon_r$ of fractures = 40, (e) $\varepsilon_r$ of fractures = 60, and (f) $\varepsilon_r$ of fractures = 80. Note that the frequency shift increases as the $\varepsilon_r$ contrast increases.

velocity model for this scenario, the transition between the widely spaced and closely spaced fracture sets is at 16 ns, whereas the closely spaced fractures end at 52 ns. The frequency content does not change throughout the air-filled fracture set, but it shows significant variations throughout the water-filled fracture set, presumably caused by the significant Rayleigh scattering that occurred in this set compared with the air-filled fracture set. The transition between widely spaced and closely spaced fractures at 16 ns shows a strong local increase in frequency content, again probably related to the significant $\varepsilon_r$ contrast between water and limestone.

Figure 3c again shows the location (in frequency) of the maximum amplitude in the spectrum as a function of time. Localized high amplitudes (red anomalies) occur only in the water-filled fracture set and lead to wide frequency variations, whereas the frequency content for the signal through the air-filled fractures is uniform. The high amplitude at ~52 ns indicates the boundary between the closely spaced fractures and the massive limestone unit. Here too the green anomalies are assumed to indicate fractures, and the number of anomalies across the closely spaced water-filled fractures is significantly higher than the number across the widely spaced air-filled fractures. Although the number of anomalies does not accurately reflect the exact number of fractures, studying the changes of frequency content with time might give additional information about the spacing between fractures, as well as whether the fractures are air-filled or water-filled.

In the air-water-air model (Figure 3e), there is no frequency shift at the transition between widely spaced and closely spaced sets of air-filled-fractures. However, there is a significant frequency shift at the air-water transition (30 ns), even though the fracture spacing does not change. Again, the location (in frequency) of the maximum amplitude in the spectrum as a function of time (Figure 3f) indicates that the number of anomalies in the region of closely spaced fractures is higher than in the region of widely spaced fractures, although this could not be inferred from the time series. In addition, it shows that the red anomalies again are associated with a shift toward low frequencies and probably are related to noise.

To further demonstrate this dependence of the frequency shift on the $\varepsilon_r$ contrast, we ran the S-transform on a widely spaced versus closely spaced fracture model with different values for the $\varepsilon_r$ of the closely spaced fractures. Figure 4 shows the S-transform analysis for a model with wide spacing and close spacing, with a fixed $\varepsilon_r$ of 1 for the widely spaced fracture set, and with differing $\varepsilon_r$ values (1, 10, 20, 40, 60, and 80) for the closely spaced fractures. In all models, $\varepsilon_r$ for the limestone matrix is set at 7. It is obvious that as $\varepsilon_r$ values for the closely spaced fracture set increase, the frequency shift begins to emerge and strengthen across the set.

## Time-series analysis of 2D fracture models

The synthetic 2D radargram for model B (2D) displays multiple diffraction hyperbolae, preventing identification of distinct reflections arising from isolated fractures (Figure 5). Thus, finite-difference (FD) migration was applied using ReflexW (Sandmeier, 2007).

To perform the migration, seven 1D velocity models first were estimated over the massive limestone and over each of the six fracture sets along the model space, using the actual $\varepsilon_r$ values of limestone and fractures. Those models then were combined to construct a 2D velocity model, using the FD method (Figure 6). The velocity model displays a constant velocity of 0.11 m/ns for the limestone. However, it displays a wide velocity distribution within each fracture group because of the smoothing induced by interpolating the 1D velocities across two dimensions using the finite-difference method.

Using this velocity model, the raw data set was migrated using the FD approach (Figure 7). As expected, the first four sets of widely spaced fractures were resolved better than were the fifth and sixth closely spaced fracture groups. In contrast, resolution of individual fractures as distinct events in the time domain becomes increasingly challenging as the fracture density increases, as a result of the constructive interference between reflected wavelets from multiple reflection events. The air-filled fractures exhibit less scattering of EM waves than do the water-filled fractures, because they are shallower and less energy is reflected back from them (reflection coefficient = 0.45) than from water-filled fractures (reflection coefficient = −0.55).

## S-transform analysis of the 2D fracture model

A useful technique for displaying variations in the S-transform results across 2D space is to plot variations in amplitude for a specific frequency of interest identified in the frequency spectrum of the 1D S-transform. In our case, the 1D S-transform analysis revealed that water-filled fractures exhibit peak amplitudes at approximately 300 MHz. Because we are interested in locating water-filled fractures and distinguishing them from air-filled fractures, we plot the amplitude of the S-transform spectrum across the model space at a frequency of 300 MHz in Figure 8. Such analysis at a single frequency (1) aids in identification of specific targets (in this case, water-filled

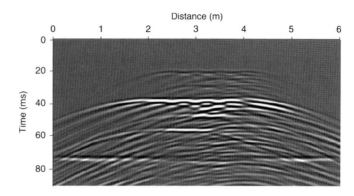

**Figure 5.** Radargram for the 2D synthetic model B. The model displays multiple diffraction hyperbolae, preventing identification of distinct reflections arising from isolated fractures.

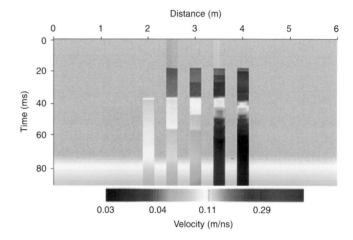

**Figure 6.** Constructed 2D velocity-distribution model for synthetic model B. The velocity model displays a wide velocity distribution within each fracture group and a constant velocity for the limestone matrix.

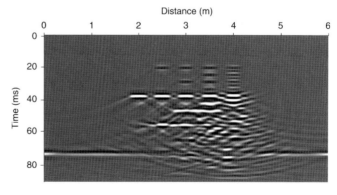

**Figure 7.** Migrated radargram for synthetic model B. The resolution of individual fractures as distinct events becomes increasingly challenging as the fracture density increases.

fractures), and (2) suppresses the effect of noise present at other frequencies. As expected, known water-filled fractures show higher amplitudes than do air-filled fractures. Because the first three widely spaced fracture sets are resolved perfectly in the radargram, they are identified as clear, isolated anomalies in the frequency spectrum.

However, when the number of fractures increases, interference between reflected energies increases and identification of water-filled fracture sets becomes more challenging. In addition, when the signal is contaminated with high noise levels, the real position of the identified anomalies shifts (Pinnegar and Mansinha, 2003). That shift is observed clearly in our results, especially for more closely spaced fracture sets (fourth and fifth columns), because the time series of those sets are contaminated with a significant amount of noise resulting from EM-wave scattering at the edges of individual fractures. Despite such limitations, the method still can be used to identify water-filled fractures

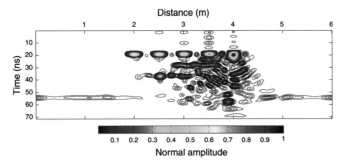

**Figure 8.** Results of applying the S-transform to the 2D fracture model for a frequency of 300 MHz. The water-filled fractures (showing high amplitudes) are distinguished from the air-filled ones.

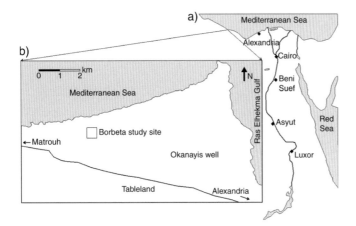

**Figure 9.** Field site map showing (a) the site location in Egypt, and (b) a close-up of the Borbeta study site.

from the characteristics of frequency shifts, and it can provide information on the approximate locations of fractures.

## Example application to field data: Characterizing fractures at Borbeta, Egypt

To illustrate the potential of such a signal-processing approach on field data, we applied the S-transform on measurements acquired at Borbeta, on the northwestern coast of Egypt (Figure 9). The site is 12 m above sea level and is 300 m south of the Mediterranean Sea. The study site is flat and the surface is covered with fine lagoonal and alluvial materials mixed with quartz sands and limestone crust. The surface layer extends approximately 6 m below surface and overlies a weathered, fractured limestone layer that is 4 m thick (Said, 1962). The shallow aquifer at this site is about 10 m deep and is composed of oolitic limestone (Said, 1962).

Geophysical surveys were conducted at this remote site in an effort to characterize subsurface fractures and to locate the depth to the shallow aquifer. These geophysical surveys include seven GPR lines, each 496 m long, with two common-midpoint (CMP) gathers acquired during a field effort in March 2002. Because of the desire to maximize the depth of investigation at this relatively conductive site, the parameters used in this field survey are different from those described in our synthetic study. Furthermore, because an accurate velocity model cannot be predicted from the acquired CMP gathers, the data were not migrated. Application of the S-transform technique as previously presented is still valid, but the fracture density at which fractures transition from being widely spaced to closely spaced differs. The GPR profiles were collected using the common-offset method. The GSSI Subsurface Interface Radar (SIR) System-20, equipped with 50-MHz unshielded antennae, provided an estimated penetration depth to 30 m. The number of traces, number of samples per trace, trace increment, and time increment were 13,693, 512, 0.035 m, and 0.976 ns, respectively.

The data were processed using the same steps employed for the 1D synthetic models. Figure 10 consists of a processed GPR profile (Figure 10a) and a photograph showing some near surface cracks, fractures, and moisture within a well near the study site (Figure 10b). As shown in Figure 10a, a strong GPR reflection appears at time 142 ns (at ~10 m in depth, on the basis of the average velocity of limestone of 0.13 m/ns). Also, on the basis of site geology, that strong reflection is believed to represent the upper surface of the shallow aquifer. Another strong GPR reflection at 360 ns (24 m in depth) is believed to indicate a fully saturated fracture zone. Several nearly vertical

other features in the earth that cause dielectric contrasts. However, our limited field study at a fractured-rock site has shown characteristic behavior that is consistent with what we observe in the synthetic studies. That encourages additional synthetic studies applying the S-transform technique under more complex conditions.

## Conclusions

We have shown that the S-transform method can be used to improve interpretation of GPR data collected in fractured-rock environments. The results of our study indicate that the S-transform can facilitate determination of fracture density and can help distinguish the locations of water-filled fractures from air-filled fractures. Time-frequency analysis therefore appears to contain important information about the characteristics of fracture-fill materials (i.e., air versus water). On the basis of this analysis, the frequency associated with specific targets (e.g., water-filled fractures) can be identified and subsequently mapped via the 2D S-transform approach by locating high amplitudes associated with those targets. We demonstrated this concept on field data sets collected at a site in Egypt, where a fractured limestone aquifer is suspected. We were able to isolate amplitudes at specific frequencies that are considered to be diagnostic of water-filled fractures and to predict likely changes in water saturation. Such a time-frequency analysis also can contain useful information about fracture density. However, the results appear to be affected significantly by noise, and further work is needed to suppress noise and better understand the predicted anomalies.

## Acknowledgments

We thank Evert Slob (Delft University) and two anonymous reviewers for their comments, which helped to improve the quality of this manuscript.

## References

Annan, A. P., 2006, Ground-penetrating radar, *in* D. K. Butler, ed., Near-surface geophysics, SEG Investigations in Geophysics Series No. 13, Chapter 11: 357–438.

Benedetto, A., 2010, Water content evaluation in unsaturated soil using GPR signal analysis in the frequency domain: Journal of Applied Geophysics, doi:10.1016/j.jappgeo.2010.03.001.

Daniels, D. J., 1996, Surface-penetrating radar: Institute of Electrical Engineers.

Denker, H., and H. G. Wenzel, 1987, Local Geoid determination and comparison with GPS results: Bulletin Géodésique, **61**, 349–366.

de Voogd, N., and H. den Rooijen, 1983, Thin-layer response and spectral bandwidth: Geophysics, **48**, 12–18.

Dilay, A., and J. Eastwood, 1995, Spectral analysis applied to seismic monitoring of thermal recovery: The Leading Edge, **14**, 1117–1122.

Forsberg, R., 1985, Gravity Field Terrain Effect Computations by FFT: Bulletin Géodésique, **59**, 342–360.

Franseen, E. K., A. P. Byrnes, J. Xia, and R. D. Miller, 2007, Improving resolution and understanding controls of GPR response in carbonate strata: Implications for attribute analysis: The Leading Edge, **26**, no. 8, 984–993.

Guha, S., S. Kruse, and P. Wang, 2008, Joint time-frequency analysis of GPR data over layered sequences: The Leading Edge, **27**, no. 11, 1454–1460.

Guha, S., S. Kruse, E. E. Wright, and U. E. Kruse, 2005, Spectral analysis of ground-penetrating radar response to thin sediments layers: Geophysical Research Letters, **32**, L23304.

Kallweit, R. S., and L. C. Wood, 1982, the limits of resolution of zero-phase wavelets: Geophysics, **47**, 1035–1046.

Liu, L., 2006, Fracture characterization using borehole radar: Water, air, and soil pollution: Focus, **6**, 17–34.

Mansinha, L., R. G. Stockwell, and R. P. Lowe, 1997, Pattern analysis with two-dimensional spectral localization: Applications of 2-dimensional S-transforms: Physica A, **239**, 286–295.

McIvor, I. K., 1964, Methods of spectral analysis of seismic data: Bulletin of the Seismological Society of America, **54**, 1213–1232.

Mejia, D. R., and R. A. Young, 2007, Fracture orientation determination in sedimentary rocks using multicomponent ground-penetrating radar measurements: The Leading Edge, **26**, no. 8, 1010–1016.

Olsson, O., L. Falk, O. Forslund, L. Lundmark, and E. Sandberg, 1992, Borehole radar applied to the characterization of hydraulically conductive fracture zones in crystalline rock: Geophysical Prospecting, **40**, 109–142.

Pinnegar, C. R., 2003, The S-transform with windows of arbitrary and varying shape: Geophysics, **68**, 381–385.

Pinnegar, C. R., and L. Mansinha, 2003, Time-local spectral analysis for non-stationary time series: The S-transform for noisy signals: Fluctuation and Noise Letters, **3**, L357–L364.

Said, R., 1962, The geology of Egypt: Elsevier.

Sandmeier, K. J., 2007, REFLEXW — Processing and interpretation software for GPR, seismic and borehole data: Sandmeier Scientific Software, version 4.5.

Seol, S. J., J. H. Kim, S. J. Cho, and S. H. Chung, 2004, A radar survey at a granite quarry to delineate fractures and estimate fracture density: Journal of Environmental and Engineering Geophysics, **9**, no. 2, 53–62.

Seol, S. J., J. Jung-Ho Kim, Y. Song, and S. Chung, 2001, Finding the strike direction of fractures using GPR: Geophysical Prospecting, **49**, no. 3, 300–308.

Sinha, S., P. Routh, and P. Anno, 2009, Instantaneous spectral attributes using scales in continuous-wavelet transform: Geophysics, **74**, 137–142.

Stockwell, R. G., 1999, S-Transform analysis of gravity wave activity from a small scale network of airglow imagers: Ph.D. thesis, University of Western Ontario.

Stockwell, R. G., L. Mansinha, and R. P. Lowe, 1996, Localization of the complex spectrum: The S-transform: IEEE Transactions on Signal Processing, **44**, no. 4, 998–1001.

Tsoflias, G. P., and M. Becker, 2008, Ground-penetrating radar response to fracture fluid salinity: Why lower frequencies are favorable for resolving salinity changes: Geophysics, **73**, no. 5, J25–J30.

Tsoulis, D., 2003, Terrain modeling in forward gravimetric problems: A case study on local terrain effects: Journal of Applied Geophysics, **54**, 145–160.

Vatansevera, F., and A. Ozdemir, 2008, A new approach for measuring RMS value and phase angle of fundamental harmonic based on Wavelet Packet Transform: Electric Power Systems Research, **78**, no. 1, 74–79.

Zeng, H., 2009, How thin is a thin bed? An alternative perspective: The Leading Edge, **28**, no. 10, 1192–1197.

Chapter 28

# Application of the Spatial-autocorrelation Microtremor-array Method for Characterizing S-wave Velocity in the Upper 300 m of Salt Lake Valley, Utah

William J. Stephenson[1] and Jack K. Odum[1]

## Abstract

Spatial-autocorrelation (SPAC) microtremor-array data acquired at 14 sites in Salt Lake Valley, Utah, characterize S-wave velocities to depths as great as 300 m. Three data sets acquired at each site were analyzed simultaneously using equilateral triangular arrays with sensors deployed at 33.3-m, 100-m, and 300-m separation. Of the 14 sites, eight were within 1.2 km of active-source (vibroseis) body- and surface-wave acquisition sites, and two were within 0.7 km of boreholes logged for S-wave velocity ($V_S$) to at least 50-m depth. A comparison to these existing active-source and borehole models indicates that these SPAC $V_S$ results typically differ by less than 10% on average to 100-m depth. At a majority of the investigation sites, SPAC modeling results can be interpreted confidently to more than 150-m depth. Linear ground-motion amplification spectra derived from these profiles of $V_S$ versus depth suggest amplification factors of more than three can occur at frequencies in the band of 0.5 to 4 Hz from the base of unconsolidated sediments in the upper 300 m.

## Introduction

Ambient noise techniques for estimating S-wave velocity ($V_S$) structure have expanded rapidly throughout the seismologic community in the past 10 years. Ambient noise tomography has revolutionized $V_S$ studies at crustal to mantle depths (Sabra et al., 2005; Shapiro et al., 2005; Bensen et al., 2007), and techniques such as passive multichannel analysis of surface waves (MASW) (Park et al.,

2007) and refraction microtremor (ReMi) (Louie, 2001) have gained widespread acceptance for investigations in the upper 30 m. There is only a limited number of published studies attempting a convergence of ambient-noise methodologies to investigate basin-scale $V_S$ structure, from 30 m to several kilometers in depth (Bettig et al., 2001; Delorey and Vidale, 2010).

Site-specific shallow $V_S$ has been recognized for years as a key element in predicting variable ground-motion amplification (Borcherdt, 1970). In addition to its traditional importance in earthquake-engineering design applications (IBC, 2006), seismic-hazard mapping investigations incorporating detailed $V_S$ through new methodologies is coming rapidly to the forefront of research (Cramer et al., 2004; Frankel et al., 2007; Frankel et al., 2009). Overall, the need for rapid, accurate, and inexpensive acquisition of shallow $V_S$ across large urban sedimentary basins is becoming critical for future earthquake-effects investigations, such as urban seismic-hazard mapping and community velocity-model development.

## Salt Lake Valley

In recent decades, the seismically active Intermountain West (IMW) region, from the Colorado Plateau to the Sierra Nevada, has experienced rapid urbanization and growth, exposing increasing populations and infrastructure to seismic hazards. Most IMW urban centers such as Salt Lake Valley lie atop fault-bounded sedimentary basins whose seismic velocity structures are characterized poorly. Consequently, detailed seismic hazards assessment using

[1]U. S. Geological Survey, Denver, Colorado, U.S.A.

**Figure 1.** Geologic map of Salt Lake Valley showing locations of spatial-autocorrelation microtremor sites. Map modified from Solomon et al. (2004) and Ashland and McDonald (2003). Acquisition sites in Salt Lake Valley discussed in this paper are shown by red diamonds. Inset: location of geologic map in Utah; GSL is Great Salt Lake. Quaternary units are annotated as Q01, Q02, Q03, Q04, and Q05. Dashed black lines are transects 1, 2, and 3, which are discussed in Figures 8, 9, and 10, respectively.

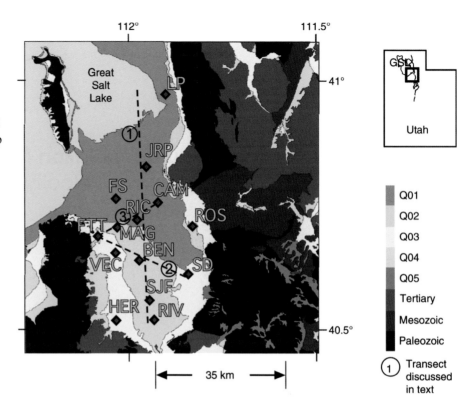

realistic velocity structures and earthquake sources is problematic. Because of logistical constraints and expense in conducting active-source seismic imaging to depths as great as 10 km, analysis of ambient seismic noise might be one of the few viable approaches for developing detailed 3D $V_S$ models beneath urbanized areas. The collection, processing, and modeling of ambient noise data have the potential to revolutionize our ability to characterize the 3D structure of these basins, which significantly improves our ability to address seismic hazards issues in these large economic centers of the western United States.

Within Salt Lake Valley (Figure 1), geologic deposits consist primarily of unconsolidated Quaternary and semiconsolidated Tertiary basin-fill deposits whose total thickness can exceed 1 km locally (Mabey, 1992). Deposits in the upper 200 m consist primarily of unconsolidated to semiconsolidated Quaternary sediments with Tertiary and older rock units near the surrounding mountains. Ashland and Mcdonald (2003) map three primary Quaternary surficial units, Q01, Q02, and Q03 (youngest to oldest) in the urbanized valley based in part on their engineering geologic properties including average shear-wave velocity to 30 m ($V_S30$), and they demonstrate that the $V_S30$ ranges of these units are statistically distinct throughout the valley. Results from this study and other sparse existing $V_S$ data acquired within Salt Lake Valley have been incorporated into the Wasatch community velocity model (WCVM; Magistrale et al., 2009).

The WCVM is a critical component for proposed realistic earthquake ground motion simulations to 1 Hz or greater; however, there is an essential need to obtain $V_S$ to depths greater than 30 m to make this evaluation. Within Salt Lake Valley, three main P-wave seismic impedance contrasts, defined as the R1, R2, and R3 boundaries, have been mapped by borehole and industry P-wave seismic reflection data (Hill et al., 1990). These reflectors within the basin are believed to be important seismic boundaries separating unconsolidated and semiconsolidated sediments (R1), semiconsolidated and consolidated sediments (R2), and consolidated sediments and basement rock (R3) and have been incorporated into the WCVM. To date, the $V_S$ contrasts across these boundaries only have been inferred from P-wave velocities, as described by Magistrale et al. (2009).

## Microtremor investigation in Salt Lake Valley

As part of an effort to characterize $V_S$ to depths as great as 300 km for improved earthquake ground-motion simulations, we acquired spatial-autocorrelation (SPAC) microtremor-array data at 14 sites across the urbanized basin underlying Salt Lake City, Utah (Figure 1). We analyzed the vertical-component SPAC microtremor data to estimate the subsurface structure of $V_S$ versus depth (henceforth, $V_S$ profile). Finally, we derive linear ground amplification and resonant frequencies based on these $V_S$ profiles at many of the sites as a step for further characterizing seismic hazards in Salt Lake Valley.

## The SPAC Array Method

Aki (1957) develops the theory for the SPAC method which, like many modern interferometric methods, relies on the coherency of the ambient noise wavefield acquired by an array of sensors to derive the subsurface $V_S$ structure. Traditionally, SPAC practitioners analyze the vertical wavefield acquired by sensors in a circular or semicircular array. Modern implementations have seen numerous advances (Okada, 2003), including nontraditional array designs (Cho et al., 2004; Asten and Boore, 2005; Chavez-García et al., 2006), three-component wavefield analysis (Köhler, et al., 2007), and other applications extending beyond Aki's original theoretical foundations (Bettig et al., 2001; Cho et al., 2004; Chavez-García et al., 2005; Tada et al., 2007; Ekstrom et al., 2009).

The fundamental equation governing SPAC can be written (Aki, 1957)

$$C(\text{r}, f) = J_0(2\pi f \text{r}/c(f)), \qquad (1)$$

where $C(\text{r}, f)$ is the spatially (and azimuthally) averaged coherency function at an interstation distance r and frequency $f$. By this derivation, the Bessel function $J_0$ (first kind, zero order) defines Rayleigh phase velocity $c(f)$ for a given frequency, or the dispersion relation, that can be modeled for $V_S$. Coherency in this study is defined as the normalized cross-power spectra and is calculated by

$$C_{ij}(f) = \{S_i(f) \cdot S_j^*(f)\}/\{S_i(f) \cdot S_i^*(f) \cdot S_j(f) \cdot S_j^*(f)\}^{1/2}, \qquad (2)$$

where $C_{ij}(f)$ is the complex spectral coherency and $S_i(f)$ and $S_j(f)$ are the complex Fourier spectra at stations i and j, respectively, and * denotes complex conjugate of $S$. Thus, cross-power spectra are calculated for each sensor pair at a given interstation distance r prior to averaging. The coherency function is ideally a real-valued function, but in practice, $C(f)$ often is complex-valued, primarily because of violations of theoretical assumptions such as insufficiently dense sensor spacing and nonisotropic wavefield (Asten, 2006). The real component of $C(f)$ is analyzed for $c(f)$ and $V_S$.

SPAC has been shown mathematically to be equivalent to modern crustal- to mantle-scale ambient-noise interferometry techniques, but to date, this equivalence has not been

used widely (Tsai and Moschetti, 2010). In practice, many crustal-scale microtremor techniques such as ambient-noise tomography rely on correlation analysis of data obtained with a single station pair to obtain empirical Green's functions (Bensen et al., 2007). Traditional SPAC techniques, although also relying on correlations between station pairs, additionally exploit acquisition on four to six sensor arrays for azimuthal averaging of the spectral coherency. As with all methodologies exploiting surface-wave dispersion for subsurface $V_S$ structure, the models derived from multimode spatial-autocorrelation (MMSPAC) analysis are inherently nonunique.

### Acquisition and processing

Although not considered the ideal geometry for SPAC acquisition (Asten, 2006), we deployed sensors in four-station equilateral triangular arrays because this configuration is well suited for using the existing roadway and sidewalk infrastructure commonly found in the urban environment (Asten and Boore, 2005; Stephenson et al., 2009). We acquired data at each site on three triangular arrays with dimensions of 33.3 m, 100 m, and 300 m (Figure 2a). Acquisition sites were selected based primarily on: (1) spatial distribution within the basin to sample the primary Quaternary geologic units (Q01, Q02, and Q03; Figure 1); (2) availability of sufficient space to deploy the array configurations; and (3) location of nearby preexisting $V_S$ data at depths greater than 50 m from methods such as S-wave reflection and refraction (Stephenson et al., 2007), with spectral analysis of surface waves (SASW) (Stokoe et al., 1994) or borehole logging (Williams et al., 1993) for comparison.

We acquired approximately 30 minutes of microtremor data on each array using four seismometers, with flat response to input velocity between 0.033 and 50 Hz, coupled to 24-bit analog-to-digital data loggers and GPS clocks. Data were sampled at 5 ms with a Nyquist frequency of 100 Hz. The sensors were placed directly on existing concrete or asphalt surfaces where possible. If the concrete or asphalt was of poor quality, the terrain sloped excessively, or the sensor location was directly on soil, the sensors were installed on a concrete paver (0.093 m²) leveled over a thin (less than 0.05 m) layer of sand.

All time-series data were preconditioned prior to SPAC analysis by detrending, demeaning, edge tapering, band-pass filtering from 0.1 to 20 Hz, and time-zero synchronization using SAC seismic-processing software (Goldstein and Snoke, 2005). The detrending step involved calculating a least-squares straight-line fit to the time-series ampli-

a)

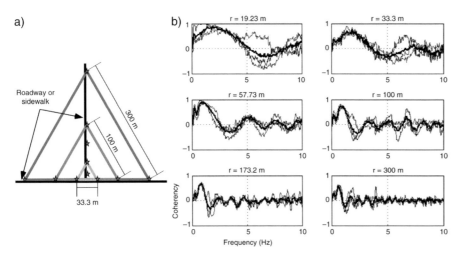

of the wavefield is forward-modeled by direct curve fitting in the spectral-coherency domain. The MMSPAC approach eliminates the requirement of picking dispersion points prior to modeling; thus the interpreter interactively adjusts layer $V_S$ and depths to optimize the fit. A key to many SPAC applications is mitigation of $V_S$ bias by azimuthal averaging in the spectral-coherency domain prior to modeling (Figure 2b). The multimode aspect of MMSPAC implies the ability to observe directly the predicted coherency of higher-mode Rayleigh phases, providing an important constraint for assessing frequency bands likely contaminated by higher-mode Rayleigh energy.

**Figure 2.** (a) Spatial-autocorrelation array geometries used at microtremor sites in Salt Lake Valley. Data were acquired on equilateral triangular arrays with dimensions of 33.3 m, 100 m, and 300 m at each site. Each array contained four sensors recording vertical wavefield components; the middle sensor for each array also recorded two horizontal components for calculation of horizontal-to-vertical spectral ratio (HVSR). The four sensor locations for each array are located at the vertices of equilateral triangles and in the triangle center, as shown by gray stars. (b) Representative frequency-coherency curves from site FS in Salt Lake Valley. Each plot shows three spectral-coherency curves (cross-power spectra) on each of three legs of a triangular array shown as thin black curves for a given interstation distance r. The heavy black line is the azimuthally averaged coherency curve used for SPAC analysis.

Each interpretation was conducted on data from a single time window. The time window was extracted from a given array data set to avoid amplitude spikes and other undesirable data phenomena. Each selected time series was edge-tapered with a Hanning window and then fast-Fourier-transformed for calculation of spectral coherency (equation 2). In general, all azimuthally averaged sensor correlation pairs were used in the modeling. However, in rare instances, one of the four sensors was omitted from analysis because of poor intra-array coherency. Poor coherency in these cases is believed to be from poor sensor-surface coupling and/or anomalous highly localized near-field effects.

tudes, then subtracting this linear trend from the data prior to mean removal. Edge tapering was conducted with 90-s Hanning windows. Band-pass filtering consisted of application of a four-pole Butterworth filter applied in two passes to maintain zero phase. Finally, the filtered waveforms were synchronized in time using the latest start time of all time series from a given array as time 0. All associated time series were truncated to begin at this time.

The nature of working with ambient seismic energy in a highly urbanized basin precludes understanding the exact sources of the analyzed surface waves; however, we anticipate that much of the energy in the frequency band of 0.1 to 20 Hz is from regional and local cultural sources within the basin such as automobiles, railroads, aircraft, and even pedestrians. The Salt Lake Basin might contribute to high-frequency ambient noise through mechanisms such as entrapment of local microtremor energy and generation of surface-wave phases from 3D geometric effects.

Average spectral-coherency curves from 0 to 20 Hz are shown for each site (Figure 3). The coherency curves at interstation distances as great as 100 m were interpretable at most sites. At interstation distances greater than 100 m, the coherency curves were more limited in their use for interpretation, particularly at sites HER and JRP. We believe that arrays with the 300-m dimension were at the limit of the acquisition parameters used to obtain useful information in the Salt Lake Valley geologic environment. Significantly longer recording times (greater than one hour) might improve the coherency of ambient data acquired on these larger arrays.

## Multimode SPAC modeling

We analyzed the microtremor data using the MMSPAC method of Asten (2006), in which the vertical component

During interpretation, the coherencies at each of six interstation distances r were used to constrain modeling. One can see the variability between the observed (solid line) and predicted (dashed line) coherency at given r distances for the models shown in Figure 3. Poor agreement

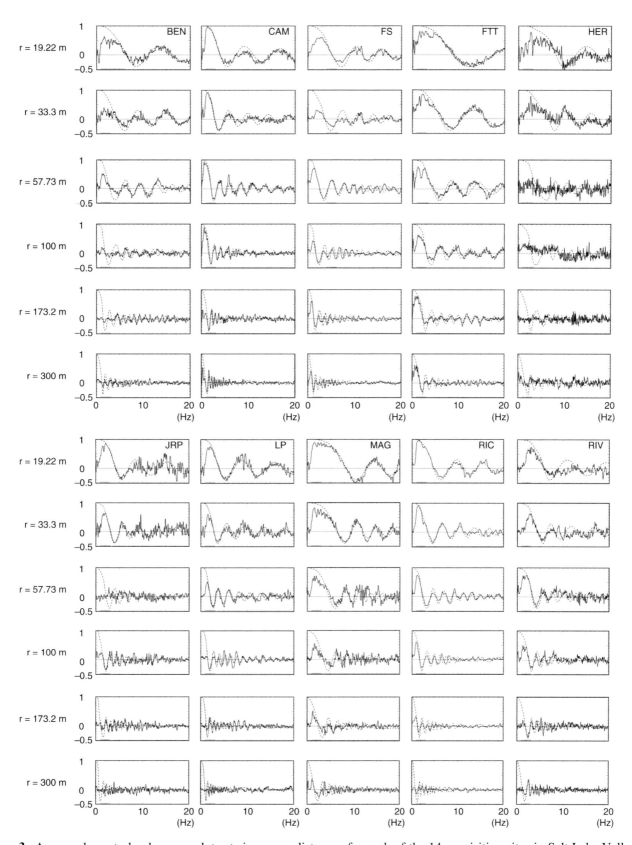

**Figure 3.** Averaged spectral-coherency plots at six sensor distances for each of the 14 acquisition sites in Salt Lake Valley. Horizontal axes are frequency from 0 to 20 Hz; vertical axes are spectral coherency from −0.5 to 1. The averaged spectral coherency at each interstation distance r is shown by a solid black line. The dashed black line is the spectral-coherency curve predicted from the final interpreted $V_S$ model shown in Figure 5 and discussed in the text. (Continued on p. 452.)

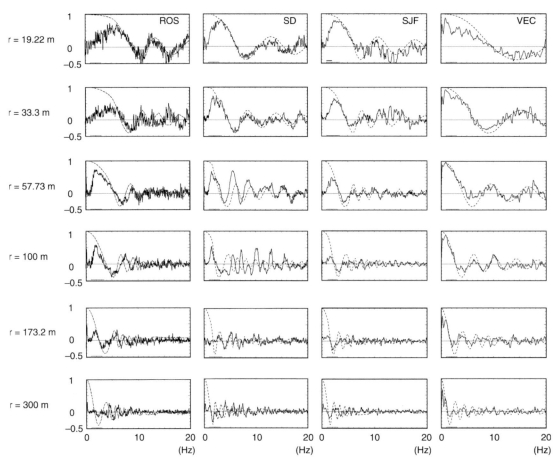

**Figure 3** (continued).

at some r distances probably is caused by departures from assumptions intrinsic in the SPAC method including nonplanar wavefields, 3D localized geologic effects (nonisotropic wavefield), and/or insufficient spatial sampling. Although array lengths differed by almost a factor of 10, we assume a single 1D $V_S$ profile can approximate the geologic structure.

In general, we assessed the goodness of fit between observed and predicted coherency by minimizing rms error over a frequency band of interest. In addition, the reliability of each spectral-coherency curve was evaluated and subjectively weighted during forward modeling. This weighting was based on knowledge of sensor site conditions, inferred geologic structure, and ambient-signal data quality. When conducting forward modeling, a given $V_S$ profile often would yield spectral fits at different interstation distances that required conflicting model changes to match the observed spectra better (e.g., predicted spectra at interstation distance r1 required a change of a given $V_S$ or layer depth that was opposite to that required at interstation distance r2 to better match observed

spectra). Sometimes this approach required splitting the difference to fit multiple curves partially if each curve was deemed equally reliable.

Using this approach, we estimated confidence bounds empirically on model $V_S$ and layer depths to be typically within 10%, which is consistent with the results of Stephenson et al. (2009) using an identical analysis process. A typical example of the sensitivity to changing the layer velocity for one layer of the preferred final model is presented in Figure 4. One can see spectral coherency is most sensitive to this layer $V_S$ around 5 Hz, and the $V_S$ is best resolved by coherency at an interstation distance of approximately 58 m. Deeper layer resolution was controlled by coherency modeling at interstation distances greater than those shown here. Resolution generally decreased with depth, with uncertainty for the velocity of the deepest interpreted layer as great as 25%, although depth of the deepest interpreted layer usually was constrained within 15%.

We present our results as $V_S$ versus depth because of its current nearly universal use in the geophysical and engineering communities. The final interpreted $V_S$ profile for each of the 14 sites is shown grouped by surface-geology map units Q01, Q02, and Q03 (Figure 5). In addition, we calculate average S-wave velocity to 100 m ($V_S100$), $V_S30$ and their standard deviations for each site within the unit groupings (Figure 5). The spectral coherency predicted for these models is displayed on the observed coherency as dashed curves in Figure 3.

In general, $V_S$ at sites on unit Q01 within the basin was consistently between 140 and 400 m/s to 100-m depth, and it never exceeded 1000 m/s to 300-m depth (Figure 5). Sites acquired on unit Q02 showed much more variability in $V_S$ in the upper 100 m, as indicated by the standard deviation of about 185 m/s for the mean $V_S100$ value of 527 m/s. The highest $V_S$ in the upper 20 m acquired on unit Q03 was on the west side of the valley at sites FTT and VEC. Although we have only four or five sites on each surface unit, Q01 looks more statistically unique compared to Q02 and Q03. In comparison with $V_S30$ means found by Ashland and McDonald (2003) of 198 m/s, 297 m/s, and 389 m/s for units Q01, Q02, and Q03, respectively, our mean values differ by 1.5%, 6%, and 2%, respectively. The sites located along the basin margins near the mountain front generally reveal shallower high-contrast soft-rock interfaces, and this is consistent with thinner unconsolidated and semiconsolidated geologic deposits along the valley periphery.

## Comparison to Existing Shallow $V_S$ Data

As a first step to verify our modeling results, we compare $V_S$ profiles derived from MMSPAC to preexisting models obtained from active-source surface and borehole methods to depths greater than 30 m. Although none of the sites was colocated precisely with the MMSPAC sites, there are several reasons why these comparisons are meaningful. First, because the arrays themselves are large (as much as 300 m for MMSPAC), in some cases, the active-source sensor array or borehole site was overlapped by the MMSPAC array. As such, the analyzed wavefields likely sample portions of the same shallow subsurface area. Second, the majority of the Salt Lake Valley shallow sediments are lacustrine, which tend to be laterally homogeneous because of the nature of their deposition. This, coupled with the demonstrated distinction in surface-geologic unit $V_S$ (Ashland and Mcdonald, 2003), as shown in Figure 1, suggests comparison over several hundred meters' distance is reasonable even though subtle changes in site geology over distances of less than 1 km clearly can contribute to differences in model results.

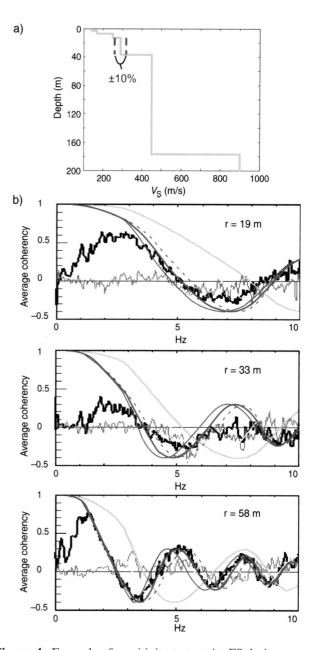

**Figure 4.** Example of sensitivity test at site FS during multimode spatial-autocorrelation modeling. (a) The fourth-layer $V_S$ was adjusted approximately ± 10% from the preferred value of 290 m/s. (b) The spectral-coherency curves for three interstation spacings r equal to 19, 33, and 58 m are presented, respectively, from top to bottom. The heavy red line is the fundamental-mode spectral-coherency curve for the preferred model, and yellow and green curves are predicted spectral-coherency curves for first and second higher-mode Rayleigh phases of the preferred model, respectively. The thin red solid line is the spectral-coherency curve with the fourth-layer $V_S$ at 260 m/s, and the thin red dashed line is with $V_S$ at 320 m/s. The orange line represents the imaginary component of the spectral-coherency curve, as discussed in Asten (2006). The model layer with $V_S$ at the preferred value of 290 m/s best fits the observed coherency spectra.

**Figure 5.** Interpreted $V_S$ profiles for Salt Lake Valley spatial-autocorrelation sites grouped by surface geologic units Q01, Q02, and Q03. Sites are coded by color underneath each plot. In general, models are interpreted to depths of 150 m greater, with some sites interpreted confidently to 300 m. The $V_S$ typically ranges from 140 m/s to more than 1000 m/s. Average S-wave velocities to 30 m and 100 m ($V_S 30$ and $V_S 100$) are shown below plots.

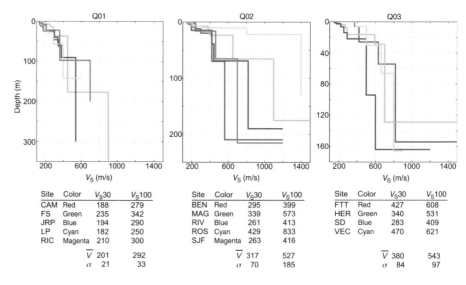

| Site | Color | $V_S 30$ | $V_S 100$ |
|------|-------|----------|-----------|
| CAM | Red | 188 | 279 |
| FS | Green | 235 | 342 |
| JRP | Blue | 194 | 290 |
| LP | Cyan | 182 | 250 |
| RIC | Magenta | 210 | 300 |
| $\overline{V}$ | | 201 | 292 |
| $\sigma$ | | 21 | 33 |

| Site | Color | $V_S 30$ | $V_S 100$ |
|------|-------|----------|-----------|
| BEN | Red | 295 | 399 |
| MAG | Green | 339 | 573 |
| RIV | Blue | 261 | 413 |
| ROS | Cyan | 429 | 833 |
| SJF | Magenta | 263 | 416 |
| $\overline{V}$ | | 317 | 527 |
| $\sigma$ | | 70 | 185 |

| Site | Color | $V_S 30$ | $V_S 100$ |
|------|-------|----------|-----------|
| FTT | Red | 427 | 608 |
| HER | Green | 340 | 531 |
| SD | Blue | 283 | 409 |
| VEC | Cyan | 470 | 621 |
| $\overline{V}$ | | 380 | 543 |
| $\sigma$ | | 84 | 97 |

**Figure 6.** Comparison of spatial-autocorrelation (SPAC) $V_S$ models to those from active-source spectral-analysis-of-surface-waves (SASW) acquisition (Wilder, 2007) and minivib refraction and reflection (Stephenson et al., 2007). The preferred SPAC $V_S$ models are shown as heavy gray lines, Models derived from SASW are shown as heavy black lines, and refraction and reflection are shown as thin black lines. Distances between compared array midpoints are in the upper right of each plot.

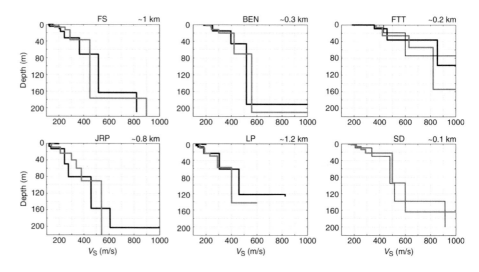

## Active-source models

SASW data were acquired at 10 sites in Salt Lake Valley using the NEESinc® Liquidator, a 31,000-kg vibroseis seismic-source vehicle (Wilder, 2007). Because of differing logistical constraints for the active-source SASW and the microtremor SPAC acquisition, sensor arrays could not be collocated perfectly. Instead, sensor array midpoints were between 0.3 and 1.2 km apart. Both methods are interpreted to depths of at least 100 m at sites BEN, FS, FTT, JRP, and LP (Figure 6). The deepest interpretation for each method was at site JRP, with depths interpreted to at least 200 m. From inspection, $V_S$ models for both methods are similar. Further, comparison of $V_S 30$ and $V_S 100$ shows they generally differ by less than 10%, with the greatest differences being 14% for $V_S 30$ at site FS and $V_S 100$ at site JRP (Table 1).

Although the MMSPAC models for sites BEN, FS, FTT, and LP generally resolve layers with $V_S$ greater than 600 m/s at depths slightly greater than models derived from SASW data, this might be coincidental because each of these SPAC sites is geographically basinward of the SASW sites and therefore this greater depth might be related to geologic layer differences between sites. For example, for SPAC site FS the layer boundary at about 180-m depth is approximately 20 m shallower and 1 km southward of the nearby SASW site. This implies a slope on the interpreted layer of about 1.1°, which is an acceptable dip for lacustrine deposits in the central part of the valley.

We next compare the results of two minivib S-wave refraction and reflection surveys (Stephenson et al., 2007) to our SPAC $V_S$ profiles at sites FTT and SD (Figure 6). SPAC site SD was near the midpoint of a 300-m array for refraction and reflection on the eastern side of Salt Lake Valley. Even though the array midpoints for the microtremor and refraction-reflection sites were separated by only 100 m, the agreement between the $V_S$ profiles is relatively poor at depths greater than 120 m. Site SD spectral-coherency curves at r distances greater than 33 m are anomalous with their highest amplitudes at frequencies above 4 Hz (Figure 3). We speculate this might be caused by a significant departure from an ideal triangular array for both the 100-m and 300-m array deployments, with r distances differing in length by almost 28%.

Site FTT was unique in having SPAC, refraction and reflection, and SASW sites in close proximity. The $V_S 30$ and $V_S 100$ averages are comparable (Table 1), although inspection of the $V_S$ models in Figure 6 suggests the body-wave active-source data apparently do not image the low $V_S$ layer (~360 m/s) inferred from the active-source surface-wave and SPAC methods. The three data sets were within 200 m of one another along a westward trend, with the SPAC, SASW, and refraction-reflection array midpoints aligned from east to west. However, the difference in depth of the interpreted base $V_S$ layer is consistent with a subsurface geologic contact shallowing westward to the range front (Figure 1; Magistrale et al., 2009).

## Borehole logs

We obtained SPAC microtremor-array data at sites near two boreholes logged to depths between 50 and 65 m within Salt Lake Valley (Williams et al., 1993). Site CAM was about 280 m from one borehole, although site MAG was about 700 m from another borehole. Significant urban development and SPAC field logistics prevented the SPAC sites from being closer to the boreholes, which were logged in the late 1980s. Comparisons of these borehole $V_S$ profiles to $V_S$ models from SPAC, shown in Figure 7 and Table 2, suggest a poorer fit than comparisons with active source models. The best correlation between a borehole and SPAC-derived $V_S$ profiles is at site CAM, where the locations of the individual methods are closest, at about 280 m.

**Table 1.** Spatial autocorrelation, spectral analysis of surface waves, (SASW) and refraction and reflection[2] $V_S 30$ and $V_S 100$ values for six sites in Salt Lake Valley. Percentages in parentheses are absolute difference.

| Site | $V_S 30$ (m/s) | $V_S 100$ (m/s) | Approximate distance between array centers (km) |
|---|---|---|---|
| FS | 235/202 (14%) | 342/312 (9%) | 1.0 |
| BEN | 295/300 (2%) | 399/409 (2%) | 0.3 |
| FTT | 427/414 (4%) | 608/623 (3%) | 0.2 |
| FTT (rr)* | 427/480 (12%) | 608/619 (2%) | 0.25 |
| JRP | 194/178 (8%) | 290/248 (14%) | 0.8 |
| LP | 182/188 (3%) | 250/287 (13%) | 1.2 |
| SD (rr)* | 283/280 (1%) | 409/397 (3%) | 0.1 |

[2]rr denotes refraction/reflection site comparison.

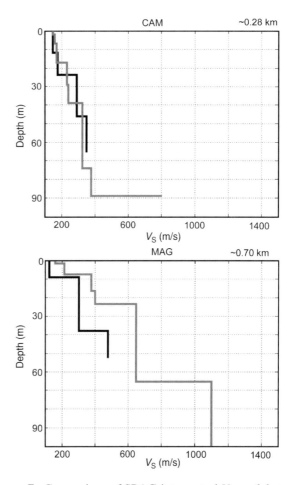

**Figure 7.** Comparison of SPAC-interpreted $V_S$ models to those from boreholes logged to depths greater than 30 m by Williams et al. (1993). Preferred SPAC $V_S$ models are shown by gray lines, and borehole models are shown as black lines. Distance between borehole and SPAC $V_S$ profile are in the upper right of each velocity-versus-depth plot.

## Ground-motion Prediction from SPAC $V_S$ Models

Having gained confidence in the SPAC $V_S$ profile interpretations from comparisons with other $V_S$ measurements, next we investigate variation in spectral amplification by calculating SH amplification at 11 of the 14 sites across Salt Lake Valley (transects 1, 2, and 3 are shown in Figure 1). The ground-motion spectral amplification caused by a 1D $V_S$ structure can be calculated by propagating body waves through that structure. We calculate theoretical 1D SH transfer functions for the interpreted MMSPAC models using the program RATTLE (C. Mueller, U. S. Geological Survey, personal communication, 1997). For

**Table 2.** Spatial autocorrelation (SPAC) and borehole $V_S30$ values for two sites in Salt Lake Valley.

| Site | $V_S30$ (m/s) | Approximate distance between SPAC array center and borehole (km) |
|------|---------------|------------------------------------------------------------------|
| CAM | 188/177 (6%) | 0.28 |
| MAG | 339/212 (37%) | 0.70 |

this modeling, each layer was assumed to have a constant attenuation Q value of 30.

First we interpret a layer boundary common to each site along transects 1, 2, and 3 as shown in Figures 8, 9, and 10, respectively. From analysis of the $V_S$ profiles in Figure 5, each interpreted profile has at least one layer with $V_S$ of 600 m/s or greater, $\pm 10\%$ uncertainty. After correcting the $V_S$ profiles to a common elevation datum of 1390 m, we interpret the common subsurface boundary at a depth where $V_S = 600$ m/s or greater (Figures 8 through 10). A $V_S$ of 600 m/s is consistent with a soft-rock shear-wave velocity profile for site characterization (IBC, 2006).

Over the southern extent of Salt Lake Valley, the R1 boundary as defined in the WCVM is typically in the upper 200 m (Magistrale et al., 2009); however, the R1 boundary is not as distinct as for the deeper R2 boundary and can be difficult to recognize strictly based on wave-speed contrast (e.g. Hill et al., 1990). In addition, because R1 is defined as the boundary between unconsolidated and semiconsolidated deposits, the lithologic characteristics of these deposits presumably can make the $V_S$ contrast quite variable. Comparison of the depth at which $V_S$ exceeds 600 m/s to the R1 boundary in the WCVM strongly suggests this interpreted interface and R1 are the same in Salt Lake Valley northward from site RIV to at least site RIC (Figures 8 and 10). At depths of less than 200 m, the interpreted subsurface boundary at sites JRP and LP on the more northern extent of transect 1 in Salt Lake Valley is too shallow to be the R1 boundary based on the WCVM and previous studies (Hill et al., 1990). The thickness of unconsolidated deposits interpreted from the SPAC profiles varies from about 30 m (sites FTT and VEC in Figure 9) to 125 m (site LP in Figure 8).

The initial reference $V_S$ used to calculate the relative amplification spectra was taken directly from the interpreted velocity model at each site. Using these base $V_S$, the peak amplifications vary from 1.5 to almost 3.5 between 0.5 and 4 Hz. Yet given the uncertainty in the magnitude of the reference $V_S$, empirically as high as 25%, next we calculate spectral amplifications relative to a common reference $V_S$ for comparison. The common reference $V_S$ we selected is the arithmetic mean of the interpreted reference $V_S$ from all sites. The

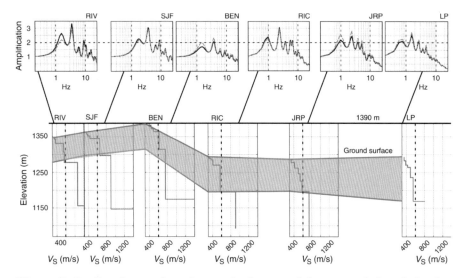

**Figure 8.** Predicted ground-motion results from spatial-autocorrelation-derived models projected onto transect line 1 (location shown in Figure 1). At bottom, transect 1 traverses Salt Lake Valley from south to north and includes sites RIV, SJF, BEN, RIC, JRP, and LP. The $V_S$ profiles are shown as gray lines at bottom; heavy dashed vertical line denotes 600 m/s. At top, black curves are amplification spectra relative to the interpreted site-specific reference velocity (SH wave amplitude at unity), and dashed gray curves are amplification spectra assuming the reference $V_S$ is the arithmetic mean for the interpreted subsurface layer. Log-log plots of amplitude-spectra scales are from 0.2 Hz to 25 Hz.

mean $V_S$ for the 11 sites used in this modeling is 660 m/s with a standard deviation of 86 m/s ($\pm$ 13%). The amplitude spectra using the mean reference $V_S$ give a generally consistent amplification pattern with the site-specific reference $V_S$ (Figures 8 to 10). The predicted peak amplifications are generally consistent with those derived from low-strain earthquake data by Pankow and Pechmann (2005).

## Discussion

The horizontal-to-vertical spectral-ratio (HVSR) method has proven to be a viable complementary technique to

SPAC for constraining $V_S$ interpretations (Scherbaum et al., 2003, Arai and Tokimatsu, 2005; Roberts and Asten, 2007). The HVSR gained prominence as a site characterization tool beginning with the work of Nakamura (1989), although research is still directed at resolving its theoretical basis (Bonnefoy-Claudet et al., 2008). We calculated HVSR using the vector magnitude of the horizontal-component spectra of the central sensor for a given array. The HVSR method was used to assess qualitatively the goodness of fit with the modeled theoretical Rayleigh ellipticity as part of MMSPAC analysis. This qualitative assessment can help constrain the depth to potentially resonant velocity boundaries. However, the HVSR method was not of great

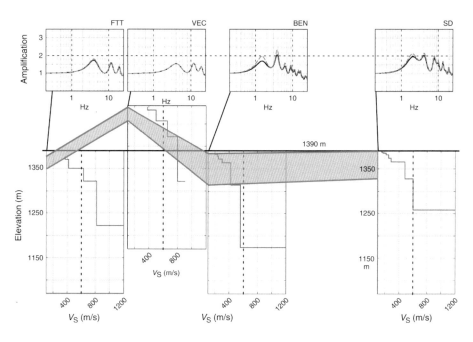

**Figure 9.** Predicted ground-motion results from spatial-autocorrelation-derived models projected onto transect line 2 (location shown in Figure 1). At bottom, transect 2 traverses Salt Lake Valley from northwest to southeast and includes sites FTT, VEC, BEN, and SD. The $V_S$ profiles are shown as gray lines at bottom; heavy dashed vertical line denotes 600 m/s. At top, black curves are amplification spectra relative to the interpreted site-specific reference velocity (SH-wave amplitude at unity), and dashed gray curves are amplification spectra assuming the reference $V_S$ is the arithmetic mean for the interpreted subsurface layer. Log-log plots of amplitude-spectra scales are from 0.2 Hz to 25 Hz.

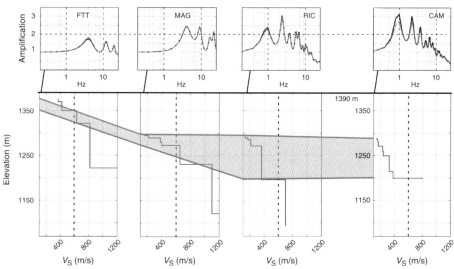

**Figure 10.** Predicted ground-motion results from spatial-autocorrelation-derived models projected onto transect line 3 (location shown in Figure 1). At bottom, transect 3 traverses Salt Lake Valley from southwest to northeast and includes sites FTT, MAG, RIC, and CAM. The $V_S$ profiles are shown as gray lines at bottom; the heavy dashed vertical line denotes 600 m/s. At top, black curves are amplification spectra relative to the interpreted site-specific reference velocity (SH-wave amplitude at unity), and dashed gray curves are amplification spectra assuming the reference $V_S$ is the arithmetic mean for the interpreted subsurface layer. Log-log plots of amplitude-spectra scales are from 0.2 to 25 Hz.

value in our analysis at many sites because the spectral peaks tended to be broad and of relatively low amplitude (SESAME, 2004).

The amplification spectra in Figures 8 to 10 demonstrate that moderate-contrast $V_S$ boundaries ($V_S$ ratio $<2.5$) in the upper 200 m can have a significant effect on ground-motion amplification in Salt Lake Valley in a frequency band often important to building-site response. The curves are simple representations of linear ground response and do not make approximations for nonlinearity that might induce a prominent effect during strong ground shaking. Each of the spectra displays a uniform spectral decay in amplitude with increasing frequency, which is a direct result of the assumed constant model Q. Lacking additional information on attenuation specific to Salt Lake Valley, $Q = 30$ should give a reasonable spectral decay rate for unconsolidated valley deposits.

Although $V_S30$ has been and will continue to be an important parameter for estimating seismic hazards (IBC 2006), $V_S$ information from depths greater than 30 m most likely will be equally critical for future site-characterization investigations. Urban-basin imaging is increasingly challenging and important in evaluation of seismic hazards for urban centers. Although active-source methods are important for resolving resonance boundaries and detailed lateral stratigraphic mapping, their use in heavily urbanized areas tends to be more logistically problematic than microtremor methods. The SPAC and other microtremor methods should continue to see technical advances that will further the capability to image to increasingly greater depths in urban areas.

## Conclusions

We have described an application of the SPAC microtremor-array method to obtain $V_S$ at depths greater than 30 m in the urbanized Salt Lake Valley. The SPAC microtremor-array data acquired at 14 sites in the valley characterize $V_S$ to depths consistently as deep as 150 m and to at least 300 m at sites FS and JRP. We analyzed three data sets acquired at all sites simultaneously using equilateral triangular sensor arrays and the MMSPAC modeling approach. A comparison of $V_S$ profiles from existing active-source and borehole models indicates these SPAC $V_S$ results often differ by less than 10% on average to 100-m depth. We believe that large differences in $V_S$ profiles between comparative sites occur in part because of 3D variability in the shallow geologic deposits beneath the active source/borehole and microtremor acquisition sites, as no sites were colocated precisely. Linear ground-

motion amplification spectra derived from SPAC $V_S$ profiles suggest amplification factors of over three within the frequency band of 0.5 to 4 Hz can occur from the base of unconsolidated sediments in the upper 300 m.

## Acknowledgments

The authors thank David Worley for immense technical assistance during data acquisition and Michael Asten for graciously supplying his MMSPAC modeling code for this research. The manuscript benefited greatly from reviews by Robert Williams, Morgan Moschetti, and three anonymous reviewers. Field logistical support from Utah Geological Survey, particularly from Gary Christensen and Greg Mcdonald, is also acknowledged gratefully. This work was supported by funding from the National Earthquake Hazards Reduction Program. Any use of trade, product, or firm names is for descriptive purposes only and does not imply endorsement by the U. S. government.

## References

Aki, K., 1957, Space and time spectra of stationary stochastic waves, with special reference to microtremors: Bulletin of the Earthquake Research Institute, **35**, 415–457.

Arai, H., and K. Tokimatsu, 2005, S-wave velocity profiling by joint inversion of microtremor dispersion curve and horizontal-to-vertical (H/V) spectrum: Bulletin of the Seismological Society of America, **95**, 1766–1778.

Ashland, F. X., and G. N. McDonald, 2003, Interim map showing shear-wave-velocity characteristics of engineering geologic units in the Salt Lake City, Utah metropolitan area: Utah Geological Survey, Open-file Report 424, 4–19.

Asten, M., 2006, On bias and noise in passive seismic data from finite circular array data processed using SPAC methods: Geophysics, **71**, no. 6, V153–V162.

Asten, M., and D. Boore, 2005, Microtremor methods applied to hazard site zonation in the Santa Clara Valley: Seismological Research Letters, **76**, 257.

Bensen, G., Ritzwoller, M. Barmin, A. Levshin, F. Lin, M. Moschetti, N. Shapiro, and Y. Yang, 2007, Processing seismic ambient noise data to obtain reliable broad-band surface wave dispersion measurements: Geophysical Journal International, **169**, 1239–1260.

Bettig, B., P.Y. Bard, F. Scherbaum, J. Riepl, F. Cotton, C. Cornou, and D. Hatzfeld, 2001, Analysis of dense array noise measurements using the modified spatial auto-correlation method (SPAC): Application to the

Grenoble area: Bolletino di Geofisica Teorica ed Applicata, **42**, 281–304.

Bonnefoy-Claudet, S., A. Köhler, C. Cornou, M. Wathelet, and P. Bard, 2008, Effects of Love waves on microtremor H/V ratio: Bulletin of the Seismological Society of America, **98**, 288–300.

Borcherdt, R. D., 1970, Effects of local geology on ground motion near San Francisco Bay: Bulletin of the Seismological Society of America, **60**, 29–61.

Chavez-García, F. J., M. Rodriguez, and W. R. Stephenson, 2005, An alternative approach to the SPAC analysis of microtremors exploiting stationarity of noise: Bulletin of the Seismological Society of America, **95**, 277–293.

——, 2006, Subsoil structure using SPAC measurements along a line: Bulletin of the Seismological Society of America, **96**, 729–736.

Cho, I., T. Tada, and Y. Shinozaki, 2004, A new method to determine phase velocities of Rayleigh waves from microseisms: Geophysics, **69**, 1535–1551.

Cramer, C. H., J. S. Gomberg, E. S. Schweig, B. A. Waldron, and K. Tucker, 2004, The Memphis, Shelby County, Tennessee, seismic hazard maps: U. S. Geological Survey, Open-file Report 2004-1294, 1–11.

Delorey, A. A., and J. E. Vidale, 2010, Seattle Basin shear-velocity model from noise correlation Rayleigh waves: Seismological Research Letters, **81**, 305.

Ekstrom G., G. A. Abers, and S. C. Webb, 2009, Determination of surface-wave phase velocities across USArray from noise and Aki's spectral formulation: Geophysical Research Letters, **36**, no. 18, L18301.

Frankel, A. D., W. J. Stephenson, D. L. Carver, R. A. Williams, J. K. Odum, and S. B. Rhea, 2007, Seismic hazard maps for Seattle, Washington incorporating 3D sedimentary basin effects, nonlinear site response, and rupture directivity: U. S. Geological Survey, Open-file Report 2007-1175, 1–77.

Frankel, A. D., W. J. Stephenson, and D. L. Carver, 2009, Sedimentary basin effects in Seattle, Washington: Ground-motion observations and 3D simulations: Bulletin of the Seismological Society of America, **99**, 1579–1611.

Goldstein, P., and A. Snoke, 2005, SAC availability for the IRIS community: DMS Electronic Newsletter, **7**, no. 1, http://www.iris.edu/news/newsletter/vol7no1/page1.htm, accessed 10 June 2010.

Hill, J., H. M. Benz, M. Murphy, and G. T. Schuster, 1990, Propagation and resonance of SH waves in the Salt Lake Valley, Utah: Bulletin of the Seismological Society of America, **80**, 23–42.

IBC, 2006, The 2006 international building code: International Code Council, 1–303.

Köhler, A., M. Ohrnberger, F. Scherbaum, M. Wathelet, and C. Cornou, 2007, Assessing the reliability of the modified three-component spatial autocorrelation technique: Geophysical Journal International, **168**, 779–796.

Louie, J. N., 2001, Faster, better: Shear-wave velocity to 100 meters depth from refraction microtremor arrays: Bulletin of the Seismological Society of America, **91**, 347–364.

Mabey, D. R., 1992, Subsurface geology along the Wasatch Front, in P. L. Gori and W. W. Hays, eds., Assessment of regional earthquake hazards and risk along the Wasatch Front, Utah: U. S. Geological Survey Professional Paper 1500-A-J, C1–C16.

Magistrale, H., K. Olsen, and J. Pechmann, 2009, Construction and verification of a Wasatch Front community velocity model: Collaborative research with San Diego State University and the University of Utah: U. S. Geological Survey Earthquake Hazards Program Final Technical Report, 1–14.

Nakamura, Y., 1989, A method for dynamic characteristics estimation of subsurface using microtremor on the ground surface: Quarterly Report of the Railway Technical Research Institute, **30**, 25–30.

Okada, H., 2003, The microtremor survey method (K. Suto, trans.): SEG Geophysical Monograph Series No. 12.

Pankow, K., and J. C. Pechmann, 2005, Determination of low-strain site-amplification factors in the Salt Lake Valley, Utah, using ANSS data: Basin and Range Province Seismic Hazards Summit II, Utah Geological Survey, Proceedings, Miscellaneous Publication 05-2, 1–15.

Park, C. B., R. D. Miller, J. Xia, and J. Ivanov, 2007, Multichannel analysis of surface waves (MASW) — Active and passive methods: The Leading Edge, **26**, 60–64.

Roberts, J., and M. Asten, 2007, Further investigation over Quaternary silts using the spatial autocorrelation (SPAC) and horizontal to vertical spectral ratio (HVSR) microtremor methods: Exploration Geophysics, **38**, 175–183.

Sabra, K. G., P. Gerstoft, P. Roux, W. A. Kuperman, and M. C. Fehler, 2005, Surface wave tomography from microseisms in Southern California: Geophysical Research Letters, **32**, L14311.

Scherbaum, F., K.-G. Hinzen, and M. Ohrnberger, 2003, Determination of shallow shear wave velocity profiles in the Cologne, Germany, area using ambient vibrations: Geophysical Journal International, **152**, 597–612.

SESAME, 2004, Guidelines for the implementation of the H/V spectral ratio technique on ambient vibrations: Measurements, processing and interpretation: WP12 European Commission — Research General Directorate Project No. EVG1-CT-2000-0026 SESAME, D23.12, 28–36.

Shapiro, N., M. Campillo, L. Stehly, and M. Ritzwoller, 2005, High-resolution surface wave tomography from ambient seismic noise: Science, **307**, 1615–1618.

Solomon, B. J., N. Storey, I. Wong, W. Silva, N. Gregor, D. Wright, and G. McDonald, 2004, Earthquake-hazards scenario for a M7 earthquake on the Salt Lake City segment of the Wasatch fault zone, Salt Lake City, Utah: Utah Geological Survey Special Study Report 111, 1–59.

Stephenson, W. J., S. Hartzell, A. D. Frankel, M. Asten, D. L. Carver, and W. Y. Kim, 2009, Site characterization for urban seismic hazards in lower Manhattan, New York City, from microtremor array analysis: Geophysical Research Letters, **36**, L03301.

Stephenson, W. J., R. A. Williams, J. K. Odum, and D. M. Worley, 2007, Miscellaneous high-resolution seismic imaging investigations in Salt Lake and Utah valleys for earthquake hazards: U. S. Geological Survey, Open-file Report 2007-1152, 1–29.

Stokoe, K. H. II, S. G. Wright, J. A. Bay, and J. A. Roësset, 1994, Characterization of geotechnical sites by SASW method, *in* R. D. Woods, ed., Geophysical characterization of sites, ISSMFE Technical Committee No. 10, Oxford & IBH Publishing, 15–25.

Tada, T., I. Cho, and Y. Shinozaki, 2007, Beyond the SPAC method: Exploiting the wealth of circular-array methods for microtremor exploration: Bulletin of the Seismological Society of America, **97**, 2080–2095.

Tsai, V. C., and M. P. Moschetti, 2010, An explicit relationship between time-domain noise correlation and spatial autocorrelation (SPAC) results: Geophysical Journal International, **182**, 1–7.

Wilder, B., 2007, SASW testing in Salt Lake Valley: M.S. thesis, University of Texas, Austin.

Williams, R. A., K. W. King, and J. C. Tinsley, 1993, Site response estimates in the Salt Lake Valley, Utah, from borehole seismic velocities: Bulletin of the Seismological Society of America, **83**, 862–889.

Chapter 29

# Integrated Approach for Surface-wave Analysis from Near Surface to Bedrock

Ali Ismet Kanlı[1]

## Abstract

A study of surface-wave analysis can be divided into two main parts. In the first stage, dispersive Rayleigh waves are extracted using multichannel analysis of surface waves (MASW) and then are inverted using a genetic-algorithm (GA) method to obtain shear-wave velocity profiles of the investigated site. A new interactive-based inversion algorithm that includes both GA and surface-wave inversion schemes was used in an MASW study. A special type of seismic source (SR-II, or Kangaroo) proved to be very effective in surface-wave studies. The standard-penetration-test data (SPT) and the shear-modulus distribution map derived from MASW data are compared with borehole results aimed for geotechnical applications. In the second stage, a microtremor survey carried out parallel to the MASW survey estimated lateral variations in sedimentary-basin depths up to bedrock. A shear-wave velocity profile of basin sediments is estimated from the GA inversion of the microtremor horizontal-to-vertical (H/V) spectrum based on surface waves from seismic noise at each site. Average shear-wave velocities estimated from the MASW survey are given as constraints in the microtremor inversion process. A new relationship between the resonance frequency $f_0$ and the thickness of the overlaying layer H is derived. A combination of active- and passive-source surface-wave analysis methods is proposed to obtain the optimum shear-wave velocity model from near surface to bedrock.

## Introduction

The effects of elastic properties of near-surface materials on seismic wave propagation are crucial in earthquake-engineering, civil-engineering, environmental, and earth-science studies.

One of the most important factors responsible for amplification of earthquake motion is the increase of amplitudes in soft sediments. Amplification is proportional to $1/\sqrt{V_S \cdot \rho}$, where $V_S$ is the shear-wave velocity and $\rho$ is the density of the investigated soil (Aki and Richards, 1980). Because the density is relatively constant with depth, the value of the shear-wave velocity can be used to represent site conditions.

In the generation of shear-wave-velocity ($V_S$) data, it is possible to carry out measurements in boreholes using crosshole or downhole methods. However, such measurements are not generally cost-effective because of the number of boreholes needed, and this approach tends to be problematic in urban areas. Surface seismic methods are the best alternative and are applicable for velocity determination. Both reflection and refraction data can be used for determination of compressional- and shear-wave velocity. However, refraction methods cannot handle velocity inversions and hidden layers effectively. In general, body-wave data are very sensitive to heavy traffic and culturally active environments. In addition, long arrays are necessary for deep investigations with refraction techniques, which makes it difficult to find suitable locations for measurements in populated urban areas. Similar problems exist for reflection data, which are time-consuming with dense data processing and higher cost of investigation.

In conventional body-wave surveys, ground roll is considered noise, but its dispersive properties can be used to obtain important information about near-surface elastic properties (Nazarian et al., 1983; Stokoe et al., 1994; Park et al., 1998; Kanlı et al, 2006). Surface waves have the

[1]Istanbul University, Department of Geophysical Engineering, Istanbul, Turkey. E-mail: kanli@istanbul.edu.tr.

highest-amplitude energy among all types of seismic waves and the highest signal-to-noise ratio (S/N) (Park et al., 2002), which makes the surface wave a powerful tool for near-surface characterization.

In this study, several aspects of surface-wave analysis will be discussed, based mainly on multichannel analysis of surface waves (MASW) and microtremor methods. New algorithm schemes are presented, including genetic-algorithm (GA) and interactive inversions, both in relation to the MASW study and microtremor inversion study. The importance of field data-acquisition procedures and the effects of source types also are discussed, including the introduction of the SR-II (Kangaroo), which is very effective in surface-wave surveys. To provide improved model-parameter constraints, P-wave refraction measurements also were performed parallel to surface-wave measurements.

The dispersion data of recorded Rayleigh waves from the MASW survey were inverted using a GA method to obtain shear-wave velocity profiles (Kanlı et al., 2006). Shear-wave velocity profiles from MASW processing also were used to correlate with the standard-penetration-test data (SPT) and then compared with borehole results. A shear-modulus distribution map was based on MASW survey results and was plotted for geotechnical purposes (Kanlı et al., 2004; Kanlı et al., 2008a).

The shear-wave velocity profile of basin sediments was estimated from inversion of the microtremor horizontal-to-vertical spectrum using a GA method calculated using surface waves recorded from seismic noise sources at each site. Average shear-wave velocities estimated from the MASW survey were used as constraints in the inversion, formulating a new algorithm scheme. Thickness of the sedimentary basin above the bedrock was mapped. Thereafter, a new relationship between the thickness of basin sediment and the main peak frequency in the horizontal-to-vertical spectral ratios (H/V) was derived (Kanlı et al., 2008b).

## History of Surface-wave Studies in Near-surface Characterization

Several types of surface-wave methods have been proposed for near-surface characterization using a great variety of testing configurations, processing techniques, and inversion algorithms. In general, two widely used techniques are preferred: spectral analysis of surface waves (SASW) and multichannel analysis of surface waves (MASW). Although there are still discussions concerning theoretical aspects, microtremor or passive-source-based techniques are seeing significant use. Their popularity is principally

because they provide more depth penetration without the need for an active source, in contrast to the SASW and MASW methods.

The original applications of surface waves in geotechnical studies were based on the SASW method (Nazarian and Stokoe, 1984; Stokoe et al., 1994). The basic idea is that signals recorded by two vertical receivers located a specific distance apart along a surface can be recorded and digitized using a dynamic signal analyzer. Using an FFT algorithm, each recorded time signal is transformed to the frequency domain with the phase difference between the two signals calculated for each frequency. Traveltime between the receivers can be obtained for each frequency by

$$T(f) = \Phi(f)/2\Pi f, \tag{1}$$

in which the phase difference $\Phi(f)$ and frequency $f$ are in radians and cycles per second, respectively. Because the distance between receivers (R2 to R1) is known, Rayleigh-wave velocity is calculated by

$$V_R = \frac{(R2 - R1)}{t(f)}. \tag{2}$$

Then, the corresponding surface wave wavelength can be calculated by the following equation:

$$\lambda = V_R/f. \tag{3}$$

Calculations from equations 1 through 3 are performed for each frequency with the results plotted in the form of a dispersion curve. Therefore, analysis is based on the dispersive nature of the surface waves in a layered or heterogeneous media.

Spectral analysis of surface waves is simple and is used widely for geotechnical applications, but it cannot be used indiscriminately. It is very sensitive to and is adversely affected by noise and receiver coupling. Interference of different wave types easily can lead to misinterpretation of phase velocities. Often, repeated measurements are needed in the field to build signal. Many disadvantages of SASW are solved by the MASW method (Park et al. 1999): the most obvious is the reduced measurement time in field studies. Shot records recorded for MASW applications allow separation of different wavefields in the frequency-wavenumber (f-k) domain and therefore the possibility to analyze fundamental and higher modes for more accurate S-wave velocities (Kanlı et al., 2006).

**Table 1.** Data-acquisition parameters used in MASW and refraction study.

| Parameters | Seismic refraction | Surface wave |
|---|---|---|
| Minimum number of channels | 24 | 24 |
| Geophone spacing | 4 m | 4 m |
| Array length | 92 | 92 |
| Sampling rate | 0.5 ms | 0.5 ms |
| Record length | 1024 ms | 2048 ms |
| Receiver type | 2.5-Hz vertical (spikes or plates) | 2.5-Hz vertical (spikes or plates) |
| Source type | 5-kg hammer (on metal plate) | SR-II (Kangaroo) |
| Minimum and optimum offset | 2 m | 30 m |

**Table 2.** Simple form of relationships among the parameters of frequency, wavelength, wavenumber, phase velocity, and penetration depth.

| Frequency (Hz) | Wavelength (m) | Wavenumber (1/m) | Phase velocity (m/s) | Penetration depth (m) |
|---|---|---|---|---|
| $f = c / \lambda$  **f**↘ | $\lambda = c / f$  $\lambda$↗ | $k = 1 / \lambda$  **k**↘ | $c = f / k$  **c**↗ | **z**↗ |

## Data Acquisition

Field configurations need to be optimized for survey requirements to obtain the best data quality. The principle task of our MASW survey was to estimate the shear-wave velocity profile for layers down to at least 30 m. This requirement necessitated that the frequency content of recorded surface waves be significantly below 15 Hz to allow calculation of phase velocities at wavelengths exceeding 60 m. The average shear-wave velocity for the top 30 m of soil is referred to as $V_S30$, and it is accepted by the National Earthquake Hazards Reduction Program (NEHRP) Uniform Building Code (UBC) and Eurocodes to classify sites according to the type of soil. Therefore, MASW is a suitable tool for accurate determination of both $V_S30$ values and shear-wave velocity profiles versus depths (Miller et al., 1999; Xia et al., 2002; Kanlı et al., 2006). In addition, the MASW technique makes it possible to plot soil-condition maps in accordance with NEHRP-UBC and Eurocode standards. These studies become more important in areas that are prone to seismic activity (Kanlı et al., 2004; Kanlı et al., 2006). Generally, the fundamental-mode Rayleigh wave dominates on the *f-k* spectrum, so the seismic array was optimized to obtain the fundamental mode in accordance with earlier experiences and the literature (Park et al., 2002; Kanlı et al., 2004; Kanlı et al., 2006).

During a field survey, it is strongly recommended that data-acquisition geometry, parameters, and near far-field effects consider the source type and site conditions. I use a quick preprocess and interpretation phase in the field or as soon as possible so the data-acquisition system including source type (or weight) can be adjusted according to soil conditions. There are several case studies and tests results that demonstrate the benefits of a tailored acquisition process and equipment in the literature (e.g., Xia et al., 2000; Park et al., 2001a; Park et al., 2001b; Park et al., 2002).

After evaluating several tests in the investigation area in context with previous experiences, not only in surface-wave surveys but also in seismic tomography studies (Kanlı et al., 2004; Kanlı et al., 2006; Kanlı 2008; Kanlı et al., 2008d), the most appropriate data-acquisition parameter scheme was developed (Table 1). One of the most important components in a seismic survey is the source. The source needs to be optimized for study objectives and field conditions. The specially designed SR-II (Kangaroo), developed by Eötvös Loránd Geophysical Institute of Hungary (ELGI), is a powerful source that has performed well in near-surface reflection studies and in the surface-wave surveys reported here.

Natural frequency of geophones and a powerful and broad-spectra source are very important for surface-wave data quality (Kanlı et al., 2004; Kanlı et al., 2006). For this study, 2.5-Hz (low) vertical geophones and an SR-II (Kangaroo) source were used to record surface-wave data. The P-wave refraction measurements also were acquired parallel to the surface-wave measurements. Low-frequency signals correspond to the longer wavelength of surface waves, which equates to greater potential depth of investigation. A generalized and simplified form of the relationship among frequency, wavelength, wavenumber, phase velocity, and penetration depth is summarized in Table 2.

The SR-II source was used for field measurements (Figure 1). It weighs 80 kg and is powered by an electrically fired blank 12-gauge shotgun cartridge. The weight of the source acting as a hold-down mass creates a more efficient coupling to the ground. Despite its heavy weight, it can elevate off the ground to a height of 2 m. The unit generates about 7850 J of energy, and its peak force is 25 t. Energy-transducer-type SR-IIs are operated by a hunting cartridge and were developed specifically for shallow exploration and geophysical-engineering applications. For our purposes, the source generates high-amplitude Rayleigh waves (Figure 2).

The amplitude spectrum of SR-II shot, as recorded by a broadband seismograph (CMG-40T) (Figure 3), is relatively flat in the frequency range of 0.3 to 30 Hz, making

**Figure 1.** The SR-II (Kangaroo) source used in this surface-wave survey is (a) fired by the electric starter device, (b) elevated from the ground, and (c) carried to the vehicle. (d). The SR-II source and the inside unit.

**Figure 2.** The seismic source (SR-II) generates several wave types such as refracted, guided, and Rayleigh waves. These are the main steps of the MASW process used in the study with one of the calculated velocity profiles. After Kanlı et al., 2006. Courtesy of John Wiley & Sons Ltd. Used by permission.

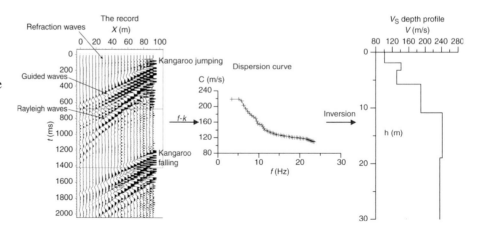

the SR-II a good source for surface-wave investigations. The decrease in amplitude near 30 Hz was the cutoff frequency used in the spectral analysis (Kanlı et al., 2006).

## Data Process and Signal Analysis of the MASW Method

Records were muted to reduce the random-noise effects and interference with other wave types of energy. After muting, only the surface-wave component of the shot gather was used for *f-k* transformation. To improve frequency and wavenumber resolution of the *f-k* transformation, zero padding was necessary in the space and time domains (Foti et al., 2003) (Figure 4). The dispersion curve was obtained from the maxima of the *f-k* spectrum. Only the fundamental mode of Rayleigh waves (from 3 to 5 Hz to 30 to 40 Hz) was investigated during this study. Figure 5 shows an example of the steps used to obtain the velocity profile, starting from the *f-k* spectra (Figure 4) to the velocity model, its *f-k* (frequency-wavenumber) plot, *c-f* (phase velocity–frequency) plot, and the λ-*f* (wavelength-frequency) plots. Extracted from the *f-k* spectra in the investigated area, the longest wavelengths observed are between 60 and 120 m, depending on the soil conditions at specific locations (Kanlı et al., 2006).

The inversion was performed in two steps. In the first stage, a starting model was constructed using an interactive inversion approach that takes into consideration only the $V_S$ and thickness parameters of the layered media. In the second stage, the GA inversion technique was used to calculate layer parameters. Both interactive and GA inversions are based on core computation of the dispersion curve, using well-known algorithms by Thomson (1950) and by Haskell (1953). The earth's structure is represented by horizontal layers of thickness $d$, body-wave velocities $V_S$ and $V_P$, and density $\rho$.

In this approach, $F(c, f)$ is a relatively complicated function constructed from the $4 \times 4$ layer matrices.

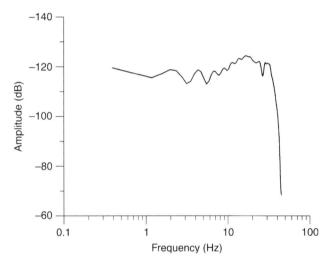

**Figure 3.** The amplitude spectrum of SR-II (Kangaroo) shot-recorded by broadband seismograph (type CMG-40T). From Kanlı et al., 2006. Courtesy of John Wiley & Sons Ltd. Used by permission.

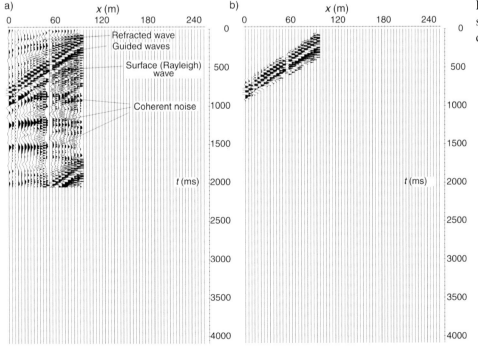

**Figure 4.** (a) Zero-padded record showing different waveforms. (b) The data retained for the analysis.

**Figure 5.** An example of the steps for obtaining the velocity model from MASW data. (a) The *f-k* spectra obtained from the record in Figure 4 and (b) the final velocity model. (c) Selected readings of the spectrum and (d) its *f-k* (frequency-wavenumber) plot. (e) The *c-f* (phase velocity–frequency) plot and (f) the $\lambda$-*f* (wavelength-frequency) plot.

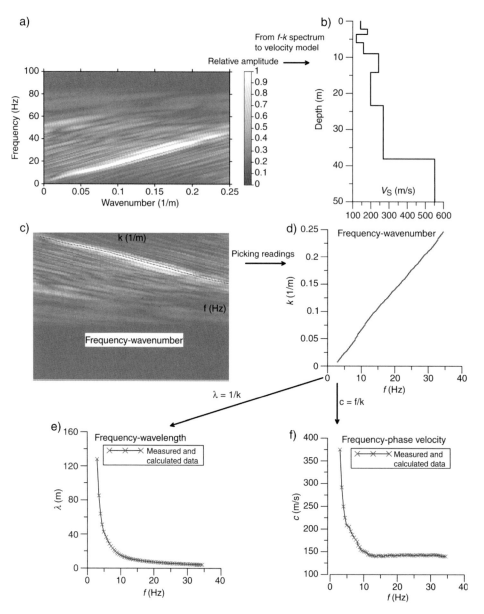

Phase-velocity curves can be determined by the $(c, f)$ root pairs of the equation

$$F(c, f) = 0. \tag{4}$$

The series of $c$ roots gives phase velocities of the fundamental and higher modes at fixed frequency. By repeating the process for all frequencies in the measured data, all $c$ values can be determined (Thomson, 1950; Haskell, 1953; Ben Menahem and Singh, 1981).

The most appropriate starting model was used to improve reliability and speed of the inversion in spite of the fact that GA is a global-optimization method. Earlier experiences have demonstrated that initial or starting-

model selection is a crucial aspect not only in the surface-wave inversion process but also in seismic refraction-tomography studies (Kanlı, 2009).

Rayleigh-wave dispersion curve $c(\lambda)$ is a scaled and smoothed form of the shear-wave velocity profile that can be approximated by the relation (Matthews et al., 1996)

$$V_S(z) = 1.1 \cdot c(\lambda = r \cdot z). \tag{5}$$

Here, $z$ represents depth, $V_S(z)$ is the shear-wave velocity at depth $z$, and $c(\lambda)$ is the phase velocity as a function of wavelength ($\lambda$), which is equal to the product of $z$ and $r$ ($r$ represents the depth-conversion factor, a value between two and four). Keeping the ratio of thickness to depth of a layer constant allows the number of layers to be estimated from the ratios $\lambda_{max}$ and $\lambda_{min}$ of the measured dispersion data. In this study, four to 12 layers were used with ratios between 0.2 and 0.4. In this interval, the $V_S$ of each layer can be determined as a geometric mean of the measured phase velocities for $c(\lambda)$ data,

$$V_s(i) = 1.1 \cdot \sqrt{\prod_j c(\lambda[j])}. \qquad (6)$$

Here, $V_S(i)$ is the velocity of the $i$th layer, $\Pi$ is the multiplication symbol, and $c(\lambda[j])$ is the phase velocity related to the $j$th wavelength

$$r \cdot H(i-1) < \lambda(j) < r \cdot H(i), \qquad (7)$$

where $H(i)$ is the depth to the $i$th layer and $r$ is the depth-conversion factor. The P-wave velocities of layers are calculated from the refraction data, and the ratio $\rho_i/\rho_1$ is estimated from $V_P(i)/V_P(1)$.

The $r$ value at each measured dispersion curve can be obtained by scaling down the depth data and searching for the best $r$ based on the overall fitness between the measured and calculated dispersion data. After determination of the scaling factor, an iterative feedback is performed to resolve the smoothness of $V_S$ data derived from $c$ values. The iterative feedback compares the measured and calculated dispersion curves on a layer-by-layer basis and then modifies the $V_S(i)$ value to improve the fit:

$$V_S(i)' = \sqrt{\prod_j c_m(\lambda[j])/c_c(\lambda[j])} \cdot V_S(i). \qquad (8)$$

Here, $c_m$ is the measured phase velocity and $c_c$ is the calculated phase velocity. It usually takes 16 to 32 iterations to converge on a solution (Kanlı et al., 2004; Kanlı et al., 2006).

## Genetic Algorithm

Genetic algorithms (GA) are known as global-optimization methods that use a blind search technique based on the natural process of evolution, in which nature constructs life-forms from genes (Sen and Stoffa, 1995). In geophysical applications, a straightforward scheme requires that the physical parameters control the observed or measured geophysical response. Each parameter in a model (e.g., thickness, P-S wave velocity, density) can be treated as genes that are used to find a solution that best agrees with observed

geophysical data. A good example of the use of GA for geophysical applications is the inverting of surface waves in pavement investigations (Al-Hunaidi, 1998).

Evolution of geophysical applications for GA principally involves the development of parameter controls through:

1) Coding: Represent the model parameters as chromosomes (binary string representation of each parameter, converting to $2^n$).
2) Reproduction and selection: A search is executed to optimize the fit between measured and calculated dispersion curves. Natural selection produces the best fit to the environment, allowing the optimal sets to reproduce in greater numbers. Therefore, following that line of reasoning, the next generation is better optimized than the previous one.
3) Crossing over: Genomes of mating partners are mixed to allow genetic information to be shared among the mated or paired models. This process results in the exchange of information among paired models through the generation of new pairs of models.
4) Mutation: Although mutation probability plays a smaller role and this more random process modifies chromosomes and creates distinct new individuals, mutation increases the variability of chromosomes. Mutation rapidly changes or alters certain parameters in a specific chromosome.

An initial population is selected at random, and over generations, the GA seeks to improve the fitness of the models. In our processing sequence, the starting model was included as the initial population. In general, only the $V_S$ parameters are calculated during the inversion process (Xia et al., 1999), thereby making the inversion more stable. In this study, although use of $V_P$ wave data is optional in the algorithm, P-wave velocity data can be included to increase reliability of the inversion process, especially when the groundwater table is within the depth of investigation.

In general, joint inversion of various data sets can be undertaken once layer parameters are constrained and their upper and lower limits are defined (e.g., Rayleigh dispersion and P-wave). For physical reasons, the $V_P/V_S$ ratio must be greater than $\sqrt{3}$.

In this study, the best-fit layer model had a population of 50 members and required 50 generations. The sum of the squares of the difference between measured and calculated values divided by frequency represents the misfit (error). Dispersion data were sampled uniformly in the frequency domain. Therefore, shallow depths (high frequencies) were overrepresented in the depth domain. To

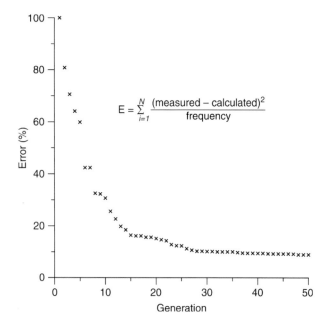

$$E = \sum_{i=1}^{N} \frac{(\text{measured} - \text{calculated})^2}{\text{frequency}}$$

**Figure 6.** Error versus generation (iterations) in the GA process. After Kanlı et al., 2006. Courtesy of John Wiley & Sons Ltd. Used by permission.

minimize the potential for the process to be dominated by shallow artifacts, misfit calculations for shallow depths were weighted by $1/f$ (Figure 6), (Kanlı et al., 2004; Kanlı et al., 2006).

## Use of MASW Data in Geotechnical Aspects

The soil classification in UBC (NEHRP) and Euro-codes is based principally on the ground type and lithology profile. Ground type is related directly to the lithologic profile of a site and is determined using one of three important parameters: $V_S30$ (m/s), standard-penetration-test $N_{SPT}$ (blow counts/30 cm), or undrained cohesion or undrained shear strength $c_u$ (kPa) values. To determine $N_{SPT}$ or $c_u$ values, the appropriate standard-penetration-test (SPT) or conic-penetration-test (CPT) special equipment is necessary. Compounding the feasibility of using SPT or CPT is the cost of drilling a borehole to depths in excess of 30 m at many sites, especially for large mapping projects. The fact that in situ measurements represent only material at the borehole location is also worth considering.

Profiling methods such as MASW can be used effectively in determining shear-wave velocity profiles and therefore the average shear-wave velocity in the upper

30 m of the subsurface in a cost-effective and more continuous manner. Results of an MASW survey allow the production of soil-condition maps in accord with the NEHRP-UBC and Eurocode standards (Figure 7) (Kanlı et al., 2006; Kanlı et al., 2008a).

The $N_{SPT}$ values within at least the upper 30 m can be calculated from MASW profiles, as discussed in the previous section. Correlations between SPT blow-count data from the borehole and the shear-wave profiles from MASW are fairly good (Figure 8).

An empirical equation between $V_S$ and the SPT blow counts derived for the same region (Dinar city) by Ansal et al. (2001) is given as

$$V_S = 51.5.\,N^{0.516}. \tag{9}$$

Here, $V_S$ is the shear-wave velocity and $N$ is the number of the blow counts in the standard penetration test. Using equation 9, an SPT blow-count distribution map was plotted and correlated with other studies conducted in the region (Figure 9). The SPT blow-count distributions calculated from our average shear-wave values match well with results from other researchers (Ansal et al., 2001; Bakır et al., 2002; Sucuoğlu et al., 2003).

Dynamic and elastic properties can be derived using seismic velocities and density. Therefore, it is possible to construct geotechnical maps from our results. In an elastic medium, the propagation velocity of S-waves is given as

$$V_S = \sqrt{\frac{\mu}{\rho}}, \tag{10}$$

where $\mu$ is the shear modulus and $\rho$ is the density (Dobrin and Savit, 1988). A shear-modulus distribution map of the region was produced using this relationship. These types of modulus maps are very important for geotechnical studies (Figure 10). As discussed, refraction measurements recorded parallel to the MASW study provide an advantage in stabilizing the inversion portion of the MASW method. These refraction data often can be instrumental many times at an investigated site for estimating additional properties necessary for geotechnical studies, such as Poisson's ratio and Young's modulus.

## Microtremor Study

The microtremor study was carried out in parallel with the MASW survey. For most engineering applications of microtremor studies, short-period seismometers with natural frequencies of about 1 Hz and a broadband response are suitable. However, the spectral content of seismograms recorded by short-period instruments usually lack the lower-frequency components which predominantly include longer periods for deeper sediments. Thus, a broadband seismometer was used to avoid this problem.

Because the penetration depth of the MASW method was limited by the lack of lower-frequency energy emitted by the SR-II source, examination of the deeper sediments

was problematic. Estimating the average shear-wave velocity of the sedimentary layers down to the depth of the sediment-bedrock interface exceeded the penetration depth of the MASW technique. Therefore, a microtremor-based study was performed in the second stage of the study to complement the results of the MASW method. The shear-wave velocity profile of the basin sediments was estimated by inverting the microtremor horizontal-to-vertical (H/V)

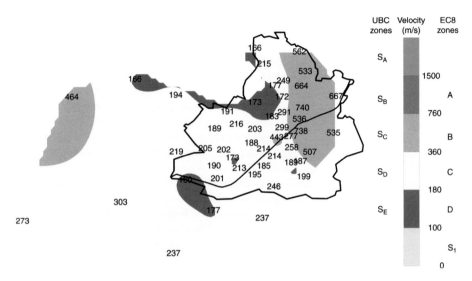

**Figure 7.** Soil-classification map overlaid by the calculated average shear-wave velocity distribution down to 30 m ($V_S30$), from MASW surveys according to UBC (NEHRP) and Eurocode-8 standards. After Kanlı et al., 2006. Courtesy of John Wiley & Sons Ltd. Used by permission.

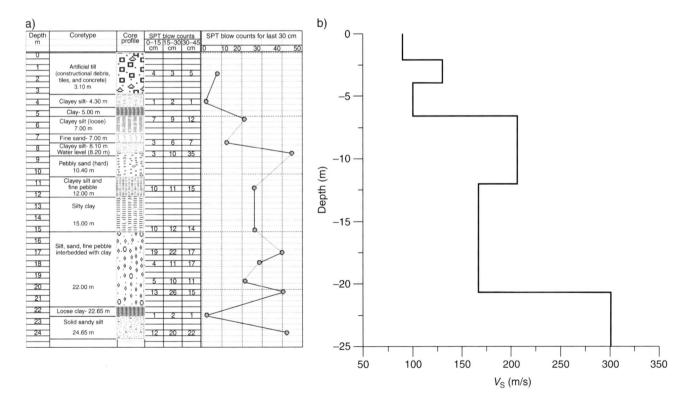

**Figure 8.** (a) Borehole data including SPT blow counts and geologic information of the core samples. (b) Shear-wave velocity-versus-depth profile. After Kanlı et al., 2008a. Courtesy of EAGE. Used by permission.

**Figure 9.** SPT blow-count distribution map ($N_{SPT}$). There are five contour intervals. After Kanlı et al., 2008a. Courtesy of EAGE. Used by permission.

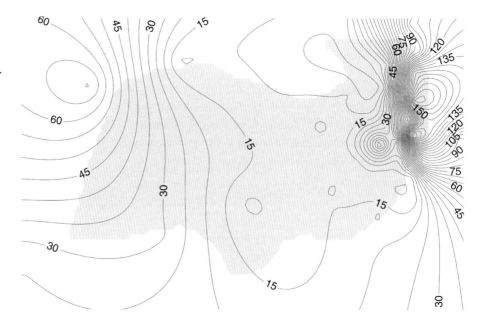

**Figure 10.** Shear-modulus distribution map. After Kanlı et al., 2008a. Courtesy of EAGE. Used by permission.

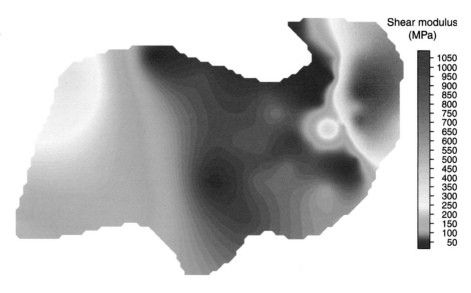

spectrum based on surface waves from seismic noise at each site using a GA. The average shear-wave velocities estimated from the MASW survey were used as constraints in the inversion scheme (Kanlı et al., 2008b).

## Microtremor Measurements

The study area was in an intermountain graben a few kilometers wide within an extensional regime in western Turkey filled with Pliocene-Quaternary unconsolidated sediments (pebbles, sand, silt, and clay). The mountainous region consists of relatively old limestone, flysch, and conglomerate bedrock (Kanlı et al., 2008b). A resistivity survey by Özpinar (1978) shows that the basin depth exceeds 200 m. Results from the MASW study show that the soil possesses relatively low shear-wave velocities, between 150 to 240 m/s in the basin zone (Figure 7).

Microtremor data were gathered at the MASW survey locations using a Guralp CMG-40T three-component broadband seismometer with a flat velocity response (Figure 3) between 0.03 and 50 Hz (Kanlı et al., 2008b). Figure 11 shows the study area and a few examples of H/V spectral ratios.

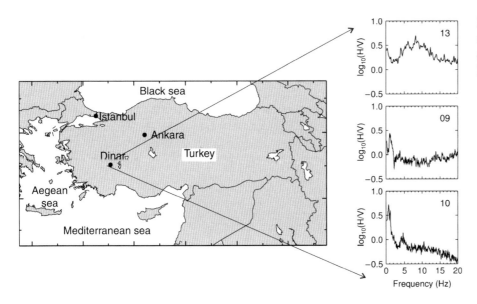

**Figure 11.** The survey area (Dinar City) and examples of H/V ratios. After Kanlı et al., 2008b. Courtesy of Taylor & Francis. Used by permission.

## Shear-wave Velocity Estimation and Mapping the Bedrock Interface from the Inversion of Microtremor Data

Microtremor energy has longer wavelengths and therefore penetrates deeper than the seismic energy of active source surveys. The microtremor H/V spectrum can be modeled successfully using the theoretical H/V surface-waves spectrum that depends on the shear-wave velocity profile above bedrock (Arai and Tokimatsu, 2004; Parolai et al., 2005). The theoretical development considers the effects of frequency changes in the fundamental and higher modes of both Rayleigh and Love waves. Therefore, shear-wave velocity structure at the investigated site can be estimated by inverting the microtremor H/V spectrum. The misfit between the measured and theoretical H/V spectrum for several stratified layers over a half-space is minimized to obtain the shear-wave velocity structure. However, the resulting profile requires a trade-off between layer thickness and average layer velocity during the inversion process (Scherbaum et al., 2003). In this study, the average shear-wave velocities ($V_S30$) estimated from the MASW survey were used (Figure 7) to constrain the inversion of the microtremor H/V spectrum.

To formulate the normal-mode solutions for surface waves, the propagator matrix method is used (Harkrider, 1964). The ratio of the horizontal power spectra of the Rayleigh and Love waves to the vertical spectrum of the Rayleigh waves describes the theoretical H/V spectrum of surface waves (Arai and Tokimatsu, 2004). According to the test results of Parolai et al. (2005), the ratio between the horizontal and vertical microtremor loading forces on the ground is fixed at unity. The forward calculation considers

the higher modes and fundamental mode of the surface waves for amplitude spectra representing the medium responses. The H/V ratios of Rayleigh waves on the free surface also are calculated in the corresponding modes. A GA scheme, which does not depend on the explicit initial model was performed in the inversion process (Carroll, 1996).

Several elastic, homogeneous, and isotropic layers with half-spaces were defined that are characterized by their thickness $h$, density $\rho$, and compressional-shear-wave velocities ($V_P$, $V_S$). The thickness and shear-wave velocity at each layer are the predominant parameters governing the theoretical H/V spectrum (e.g., Satoh et al., 2001; Arai and Tokimatsu, 2004). Thus, both the thickness and shear-wave velocity were defined as model parameters to be estimated in the inversion process. The compressional-wave ($V_P$) velocity of each layer was determined from the shear-wave ($V_S$) velocity using the empirical relations $V_P = 920.80 + 2.61\ V_S$ for $V_S \leq 400$ m/s and $V_P = 1677.7 + 0.72\ V_S$ for $V_S > 400$ m/s and mass density $\rho = 1570.32 + 0.31\ V_S$ (SI units are considered).

The dominant microtremor energy is from surface waves introduced by a random distribution of near-surface sources. Inverted velocity structures appear to be much more sensitive to S-wave velocity than to P-wave velocity. During the inversion, P-wave velocity information does not have a significant effect on the results. Therefore, the relationships were obtained simply from the linear regression of the results of the MASW surveys.

The shear-wave velocity and thickness of the horizontal layers composing the subsurface structure were assigned so that the theoretical microtremor H/V spectrum matches the observed one. The model was divided into six layers over a half-space. The thickness of each of the three top layers is freely variable within the range of 0.1 to 300 m and a range of 0.1 to 1000 m for the remaining three layers during the GA inversion process. The parameter space of the shear-wave velocity for each layer is within 2.0 km/s, and the half-space was assigned a maximum velocity for the whole model.

For the top layer, the results obtained from the MASW surveys were used as initial constraints for the inversion process. The resulting layer parameters that provided the best fit for the calculated H/V spectrum relative to the observed one were selected as optimal after 50 iterations. If the estimated parameters are at the boundary of the parameter space, where model parameters are allowed to vary, another 50 iterations are performed after modifications to the parameter space (Kanlı et al., 2008b). An example of the inversion results is given in Figure 12a and 12b. A fairly good fit was obtained between the calculated H/V spectra and the observed spectra. A flowchart is given to summarize the integration of MASW and the microtremor survey (Table 3).

Velocity contrasts between layers in the velocity model that possesses the best fits were investigated to determine sediment thickness overlying the bedrock. By taking into account the large velocity contrast between the sediment and the bedrock, the estimated thickness of the first or second layer could be considered the accumulated sediment overlying the bedrock.

Although it is a rather subjective conjecture, the validity of the estimated sediment depths was controlled by comparing them with the ones estimated from the resistivity survey of Özpinar (1978) conducted in the Dinar Basin. Bedrock depths estimated from the microtromer inversion results are very close to the resistivity survey. Therefore, the depth of the sedimentary layer beneath each observation site was used to map the lateral variation in the depths (Figure 12c).

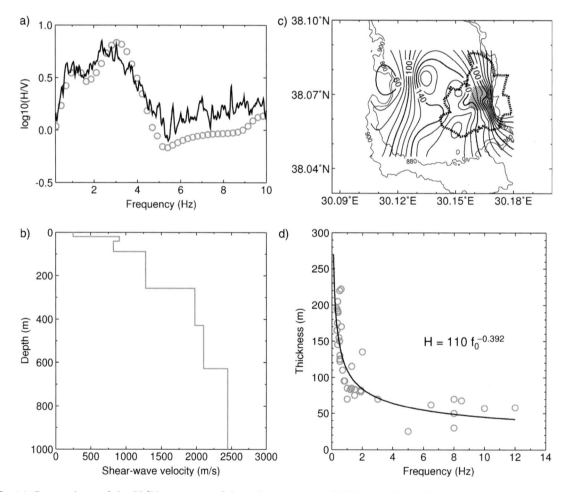

**Figure 12.** (a) Comparison of the H/V spectrum of the microtremor (solid line) with synthetic data (open circles) for (b) the inverted shear-wave velocity profile. (c) Lateral variation in the sediment thickness (contour intervals are 10 m). (d) The relationship between the resonance frequency and sediment thickness. After Kanlı et al., 2008b. Courtesy of Taylor & Francis. Used by permission.

**Table 3.** Flowchart of the integrated approach for MASW and the microtremor survey to obtain best-fit velocity model and H/V ratios.

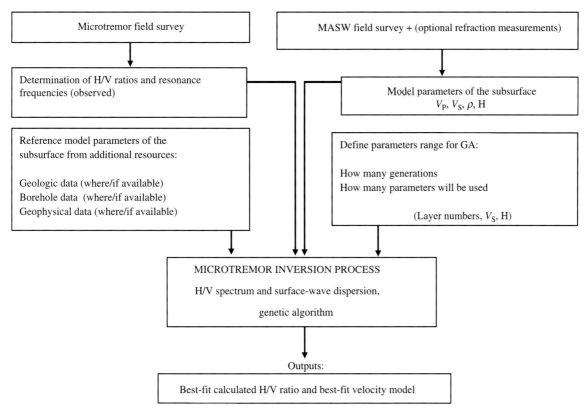

Integrated approach for estimation of the velocity model

A relationship between the resonance frequency $f_0$ and thickness H of the overlaying layer is derived as H $=$ $110f_0^{-0.392}$ (Figure 12d). If we know the resonance frequency of a site, then the thickness of the overlying layer can be estimated by using the equation derived for the region (Kanlı et al., 2008b; Kanlı et al., 2008c).

## Conclusions

Shear-wave velocity is an important parameter for evaluating the dynamic behavior of soil in the shallow subsurface. Determining near-surface shear-wave velocity values is a crucial task in most civil- and earthquake-engineering, environmental, and earth-sciences applications. Therefore, an integrated geophysical approach to improve accuracy and depth of sampling of shear-wave velocity profiles was introduced based on surface-wave analysis.

This integrated study tries to introduce and summarize a systematic approach for integrating active-source-type MASW and passive-source-type microtremor survey, based

on surface-wave data. A new interactive inversion algorithm is introduced that includes both GA and MASW surface-wave inversion schemes. A special active source was shown to be very effective in surface wave studies. The microtremor survey was carried out parallel to the MASW survey to determine the resonance frequencies and depths of the sedimentary layers. By using the average shear-wave velocities estimated from the MASW survey as constraints in the inversion process for the microtremor method, a new algorithm scheme has emerged.

It is advisable to use the MASW method with suitable source and data-acquisition parameters especially for near-surface characterization, geotechnical, and seismic hazard-assessment studies (e.g., shear wave-velocity mapping [$V_S$30], soil classification, and microzonation studies). Because of the MASW method's stability when used for near-surface characterization, shear-wave velocity estimation from MASW surveys is more robust than passive-source-based methods. On the other hand, at most sites where the MASW method experiences limited penetration, passive-source-based surface-wave methods are recommended to enhance the sampling depth. To increase the stability of microtremor-based studies, MASW data can be used to constrain the inversion process. Similarly, incorporating refraction results into the inversion stage of MASW improves stability and can be used to estimate several key geotechnical parameters.

During the inversion of microtremor data, MASW results were used to constrain the topmost layer of the shear-wave velocity profile. These additional constraints played an important role in stabilizing the inversion process as a priori information. It should be kept in mind that these additional measurements are representative of a different sampling aspect and/or property of the medium irrespective of active or passive source. This is the case because each technique captures or investigates a different size or orientation of the sampled earth volume. MASW surveys analyze information from a limited volume using a single source and associated receiver orientation, whereas the microtremor survey captures information from a wider range of azimuths that penetrate a volume from uncertain and multiple sources. Therefore, they can measure different velocity characteristics. The MASW results are used in the inversion of microtremor data to extend imaging depths of the MASW study. Generally, active-source models do not extend to depths as great as passive-source models do.

## Acknowledgments

I would like to express my sincere gratitude to P. Tildy, Z. Pronay, L. Hermann, B. Neducza, and E. Toros for their generous advice, friendship, and continuous guidance during our project term and my fellowship term. I also am indebted to them for their great contributions to our research. In addition, I am indebted to T. S. Kang for his valuable and great contributions to the research and his kind friendship. I would like to thank the Department of Engineering Geophysics of ELGI and Eötvös Loránd Geophysical Institute of Hungary for permitting me to use their resources. I would like to thank J. Xia for his constructive comments. I also would like to thank R. D. Miller, volume editor of this book, for his kind invitation and his kind editorial support.

## References

Al-Hunaidi, O., 1998, Evolution based genetic algorithm for the analysis of non-destructive surface wave test on pavements: NDT&E International, **31,** 273–280.

Aki, K., and P. G. Richards, 1980, Quantitative seismology: W. H. Freeman & Co.

Ansal, A. M., R. Iyisan, and H. Gullu, 2001, Mictrotremor measurements for the microzonation of Dinar: Pure and Applied Geophysics, **158,** 2525–2543.

Arai, H., and K. Tokimatsu, 2004, S-wave velocity profiling by inversion of microtremor H/V spectrum: Bulletin of the Seismological Society of America, **94,** 53–63.

Bakır, B. S., M. Y. Ozkan, and S. Ciliz, 2002, Effects of basin edge on the distribution of damage in 1995 Dinar, Turkey earthquake: Soil Dynamics and Earthquake Engineering, **22,** 335–345.

Ben-Menahem, A., and S. J., Singh, 1981, Seismic waves and sources: Springer-Verlag.

Carroll, D. L., 1996, Genetic algorithms and optimizing chemical oxygen-iodine lasers, *in* H. B. Wilson, R. C. Batra, C. W. Bert, A. M. J. Davis, R. A. Schapery, D. S. Stewart, and F. F. Swinson, eds., Developments in theoretical and applied mechanics, v. 18: School of Engineering, University of Alabama, 411–424.

Dobrin, M. B., and C. H. Savit, 1988, Introduction to geophysical prospecting, international ed.: McGraw-Hill.

Foti, S., L. Sambuelli, V. L. Socco, and C. Strobbia, 2003, Experiments of joint acquisition of seismic refraction and surface wave data: Near Surface Geophysics, **1,** 119–129.

Harkrider, D. G., 1964, Surface waves in multilayered elastic media, Part 1: Bulletin of the Seismological Society of America, **54,** 627–679.

Haskell, N. A., 1953, The dispersion of surface waves on multilayered media: Bulletin of the Seismological Society of America, **43,** 17–34.

Kanlı, A. I., 2008, Image reconstruction in seismic and medical tomography: Journal of Environmental and Engineering Geophysics, **13,** 85–97.

——, 2009, Initial velocity model construction of seismic tomography in near-surface applications: Journal of Applied Geophysics, **67,** 52–62.

Kanlı, A. I., L. Hermann, P. Tildy, Z. Pronay, and A. Pınar, 2008a, Multi-channel analysis of surface wave (MASW) technique in geotechnical studies: 14th European Meeting of Environmental and Engineering Geophysics, EAGE, Extended Abstracts, P13.

Kanlı, A. I., T. S. Kang, A. Pınar, P. Tildy, and Z. Pronay, 2008b, A systematic geophysical approach for site response of the Dinar region, south western Turkey: Journal of Earthquake Engineering, **12,** S2, 165–174.

——, 2008c, Combination of active (MASW) and passive (microtremor) source based geophysical approach: 14th European Meeting of Environmental and Engineering Geophysics, EAGE, A02.

Kanlı, A. I., Z. Pronay, and R. Miskolczi, 2008d, The importance of the spread system geometry on the image reconstruction of seismic tomography: Journal of Geophysics and Engineering, **5,** 771–785.

Kanlı, A. I., P. Tildy, Z. Pronay, and A. Pınar, 2004, Quantitative evaluation of seismic site effects: NATO

Science Program technical report, Grant No. EST. CLG.979847.

Kanlı, A. I., P. Tildy, Z. Prónay, A. Pınar, and L. Hermann, 2006, $V_S 30$ mapping and soil classification for seismic site effect evaluation in Dinar region, SW Turkey: Geophysical Journal International, **165**, 223–235.

Matthews, M., V. Hope, and C. Clayton, 1996, The use of surface waves in the determination of ground stiffness profiles: Proceedings of the ICE: Geotechnical Engineering, **119**, no. 2, 84–95.

Miller, R. D., J. Xia, C. B. Park, and J. Ivanov, 1999, Multichannel analysis of surfaces waves to map bedrock: The Leading Edge, **18**, 1392–1396.

Nazarian, S., and K. H. Stokoe II, 1984, In situ shear wave velocities from spectral analysis of surface waves: Proceedings of the 8th Conference on Earthquake Engineering, **3,** 31–38.

Nazarian, S., K. H. Stokoe II, and W. R. Hudson, 1983, Use of spectral analysis of surface waves method for determination of moduli and thicknesses of pavement systems: Transport Research Record, **930,** 38–45.

Özpınar, B., 1978, Geophysical resistivity investigation report on the Afyon-Dinar Plain: 18th District of the State Hydraulic Works, Isparta, Turkey (in Turkish).

Park, C. B., J. Ivanov, R. D. Miller, J. Xia, and N. Ryden, 2001a, Seismic investigation of pavements by MASW method — Geophone approach: Proceedings of the Symposium on the Application of Geophysics to Engineering and Environmental Problems (SAGEEP 2001), RBA-6.

Park, C. B., R. D. Miller, and H. Miura, 2002, Optimum field parameters of an MASW survey: 2nd International Symposium, SEG-J, Extended Abstracts.

Park, C. B., R. D. Miller, and J. Xia, 1999, Multichannel analysis of surface waves: Geophysics, **64,** 800–808.

——, 2001b, Offset and resolution of dispersion curve in multichannel analysis of surface waves (MASW): Proceedings of the Symposium on the Application of Geophysics to Engineering and Environmental Problems (SAGEEP 2001), SSM-4.

Park, C. B., J. Xia, and R. D. Miller, 1998, Ground roll as a tool to image near-surface anomaly: 68th Annual International Meeting, SEG, Expanded Abstracts, 874–877.

Parolai, S., M. Picozzi, S. M. Richwalski, and C. Milkereit, 2005, Joint inversion of phase velocity dispersion and H/V ratio curves from seismic noise recordings using a genetic algorithm, considering higher modes: Geophysical Research Letters, **32**, L01303, doi:10.1029/2004GL021115.

Satoh, T., H. Kawase, T. Iwata, S. Higashi, T. Sato, K. Irikura, and H. C. Huang, 2001, S-wave velocity structure of the Taichung Basin, Taiwan, estimated from array and single-station records of microtremors: Bulletin of the Seismological Society of America, **91**, 1267–1282.

Scherbaum, F., K. G. Hinzen, and M. Ohrnberger, 2003, Determination of shallow shear velocity profiles in the Cologne/Germany area using ambient vibrations: Geophysical Journal International, **152**, 597–612.

Sen, M. K., and P. L. Stoffa, 1995, Global optimization methods in geophysical inversion: Elsevier.

Stokoe, K. H. II, S. G. Wright, J. A. Bay, and J. M. Roësset, 1994, Characterization of geotechnical sites by SASW method, *in* R. W. Woods, ed., Geophysical characterization of sites: ISSMFE Technical Committee No. 10, Oxford & IBH Publishing, 15–25.

Sucuoğlu, H., J. G. Anderson, and Y. Zeng, 2003, Predicting intensity and damage distribution during the 1995 Dinar, Turkey, earthquake with generated strong motion accelerograms: Bulletin of the Seismological Society of America, **93**, 1267–1279.

Thomson, W. T., 1950, Transmission of elastic waves through a stratified solid medium: Journal of Applied Physics, **21**, 89–93.

Xia, J., R. D. Miller, and C. B. Park, 1999, Estimation of near-surface shear wave velocity by inversion of Rayleigh wave: Geophysics, **64**, 691–700.

——, 2000, Multichannel analysis of surface wave theory and applications: Kansas Geological Survey, Open-file Report 2000-25.

Xia, J., R. D. Miller, C. B. Park, J. A. Hunter, J. B. Harris, and J. Ivanov, 2002, Comparing shear-wave velocity profiles inverted from multichannel surface wave with borehole measurements: Soil Dynamics and Earthquake Engineering, **22**, 181–190.

# Index